29th VLSI Test Symposium

(VTS 2011)

Dana Point, California, USA
1 – 5 May 2011

IEEE Catalog Number:	CFP11029-PRT
ISBN:	978-1-61284-657-6

Copyright © 2011 by the Institute of Electrical and Electronic Engineers, Inc
All Rights Reserved

Copyright and Reprint Permissions: Abstracting is permitted with credit to the source. Libraries are permitted to photocopy beyond the limit of U.S. copyright law for private use of patrons those articles in this volume that carry a code at the bottom of the first page, provided the per-copy fee indicated in the code is paid through Copyright Clearance Center, 222 Rosewood Drive, Danvers, MA 01923.

For other copying, reprint or republication permission, write to IEEE Copyrights Manager, IEEE Service Center, 445 Hoes Lane, Piscataway, NJ 08854. All rights reserved.

***This publication is a representation of what appears in the IEEE Digital Libraries. Some format issues inherent in the e-media version may also appear in this print version.**

IEEE Catalog Number: CFP11029-PRT
ISBN 13: 978-1-61284-657-6
ISSN: 1093-0167

Additional Copies of This Publication Are Available From:

Curran Associates, Inc
57 Morehouse Lane
Red Hook, NY 12571 USA
Phone: (845) 758-0400
Fax: (845) 758-2633
E-mail: curran@proceedings.com
Web: www.proceedings.com

29th VLSI Test Symposium (VTS 2011)

Dana Point, California, USA
1-5 May 2011

IEEE Catalog Number: CFP11029-POD
ISBN: 978-1-61284-657-6

2011 29th IEEE VLSI Test Symposium

VTS 2011
Table of Contents

Foreword..ix

Organizing Committee..x

Program Committee...xiii

Steering Committee..xiv

Reviewers...xv

Acknowledgements..xvii

Test Technology Technical Council (TTTC)...xix

Test Technology Educational Program (TTEP) Tutorials......................xxii

Awards..xxv

Session 1

Session 1A: Post-Silicon Debug & Customer returns

Understanding Customer Returns From A Test Perspective...2
N. Sumikawa, D. Drmanac, L. Winemberg, Li-C. Wang and M. S. Abadir

A Distributed AXI-based Platform for Post-Silicon Validation..8
M. H Neishaburi and Z. Zilic

Efficient Trace Data Compression using Statically Selected Dictionary...............................14
K. Basu and P. Mishra

Session 1B: 3D ICS

A Built-In Self-Test Scheme for the Post-Bond Test of TSVs in 3D ICs................................20
Y.-J. Huang, J.-F. Li, J-J. Chen, D.-M Kwai, Y.-F Chou and C.-W Wu

Scan Chain and Power Delivery Network Synthesis for Pre-Bond Test of 3D ICs..................26
S. Panth and S. K. Lim

Exploiting Rotational Symmetries for Improved Stacked Yields in W2W 3D-SICs.................32
E. Singh

IP Session 1C: Test and Characterization of High-Speed Circuits..38
Organizer: S. Shaikh

Session 2

Session 2A: Power Issues in Test

Expedited Response Compaction for Scan Power Reduction...40
S. M. Saeed and O. Sinanoglu

Leakage Power Profiling and Leakage Power Reduction using DFT Hardware.......................................46
R. Sethuram, K. Arabi and M. Abu-Rahma

Levelized Low Cost Delay Test Compaction Considering IR-Drop Induced Power Supply Noise...........52
Z.Jiang, Z. Wang, J.Wang and D. M. H. Walker

Session 2B: Analog, Mixed-Signal & RF Test /Diagnosis

Automatic Test Stimulus Generation for Accurate Diagnosis of RF Systems Using Transient Response
Signatures..58
A. Banerjee, S. Sen, S. K. Devarakond and A. Chatterjee

Non-Linear Analog Circuit Test and Diagnosis under Process Variation using V-Transform Coefficients...............64
S. Sindia, V. D. Agrawal and V. Singh

A Diagnosis Testbench of Analog IP Cores Against On-Chip Environmental Disturbances....................70
T. Hashida, Y. Araga and M. Nagata

Session 3
Session 3A: Delay & Performance Test 1

Case Study: Efficient SDD Test Generation for Very Large Integrated Circuits.....................................78
K. Peng, F. Bao, G. Shofner, L. Winemberg and M. Tehranipoor

Static Test Compaction for Delay Fault Test Sets Consisting of Broadside and Skewed-Load
Tests...84
I. Pomeranz

Efficient and Product-Representative Timing Model Validation..90
E. J. Jang, A. Gattiker, S. Nassif and J. A. Abraham

Special Session 3B: Hot Topic: Multifaceted Approaches for Field Reliability..................96
Organizer: Y. Sato

IP Session 3C: Advanced Methods for Leveraging New Test Standards..........................97
Organizer: M. Laisne

Session 4

Special Session 4A: New Topics..99
Organizer: B. Kaminska

Parametric Yield and Reliability of 3D Integrated Circuits: New Challenges and Solutions
Siddharth Garg - Waterloo, Diana Marculescu

Session 4B: Panel: Security

Security-Aware SoC Test Access Mechanisms...100
K. Rosenfeld and R. Karri

Design and Analysis of Ring Oscillator based Design-for-Trust technique.............................105
J. Rajendran, V. Jyothi, O. Sinanoglu and R. Karri

IP Session 4C: The Buck Stops With Wafer Test: Dream Or Reality?......................111
Organizer: S. Natarajan and A. Sinha

Session 5

Special Session 5A: Apprentice, Season 4..113
Organizers: K. S. Kim, and R. Roy

Special Session 5B: Panel: How Much Toggle Activity Should We Be Testing With? 114
Organizer: Xiaoqing Wen

Session 6

Session 6A: Delay & Performance Test 2

A Novel Mechanism for Speed Characterization During Delay Test.................................116
A. Majumdar, A. Sinha, N. Patel, R. Setty, Y. Dong, S.-H. Chou

An Efficient Method to Screen Resistive Opens under Presence of Process Variation.......................122
S. Wang

On Clustering of Undetectable Transition Faults in Standard-Scan Circuits.........................128
I. Pomeranz

Session 6B: Memory Test and Repair

Designing a Fast and Adaptive Error Correction Scheme for Increasing the Lifetime of Phase Change
Memories..134
R. Datta and N. A. Touba

Programmable Extended SEC-DED Codes for Memory Errors...140
V. Gherman, S. Evain, F. Auzanneau and Y. Bonhomme

v

Training-Based Forming Process for RRAM Yield Improvement..146
H.-C. Shih, C.-Y. Chen, C.-W. Wu, C.-H. Lin and S.-S. Sheu

IP Session 6C: The Bang For The Buck With Resiliency: Yield Or Field? 152
Organizer: S. Natarajan and A. Sinha

Session 7

Session 7A: Low-Power IC Test

Modified Flip-flop Architecture to Reduce Hold Buffers and Peak Power during Scan Shift Operation................154
P. Narayanan, R. Mittal, S. Poddutur, V. Singhal and P. Sabbarwal

Power-Safe Test Application Using An Effective Gating Approach Considering Current Limits...........................160
W. Zhao, M. Tehranipoor and S. Chakravarty

Power-Aware Test Generation with Guaranteed Launch Safety for At-Speed Scan Testing.................................166
X. Wen, K. Enokimoto, K. Miyase, Y. Yamato, M. A. Kochte, S. Kajihara, P. Girard and M. Tehranipoor

Session 7B: On-line & System Testing

SLIDER: A Fast and Accurate Defect Simulation Framework...172
W. C. Tam and R. D. Blanton

An Industrial Case Study of Analog Fault Modeling..178
E. Yilmaz, A. Meixner, S. Ozev

A New Methodology for Realistic Open Defect Detection Probability Evaluation under Process
Variations..184
J. Moreno, V. Champac and M. Renovell

Session 8

Session 8A: Aging, Transients & Soft Errors

Impact of the Application Activity on Intermittent Faults in Embedded Systems.................................191
J. Guilhemsang, O. Héron, N. Ventroux, O. Goncalves and A. Giulieri

An Analytical Method for Estimating SET Propagation...197
S. Gangadhar and S. Tragoudas

Adaptive Error-Prediction Flip-flop for Performance Failure Prediction with Aging Sensors.....................203
C. V. Martins, J. Semião, J. C. Vazquez and V. Champac

Special Session 8B: New Topic: Solar Cells
Organizer: B. Kaminska and B. Courtois

Calibrated high-efficiency testing and modelling methodologies for concentrated multi-junction solar cells..........209
J. F. Wheeldon

Sessions 9

Special Session 9C: Panel: Coverage Closure in SoC Verification: Are We Chasing a Mirage? 211
Organizer: S. Vasudevan, University of Illinois at Urbana Champaign

Session 10

Session 10A: Design for Testability 1

A Scan Cell Architecture for Inter-Clock At-Speed Delay Testing....................213
K. Y. Cho and R. Srinivasan

Design and Implementation of A Time-Division Multiplexing Scan Architecture Using Serializer and Deserializer in GPU Chips....................219
A. Sanghani, B. Yang, K. Natarajan and C. Liu

Harmony Widget for X-Free Scan Testing....................225
D. K. Bhavsar

Session 10B: Error & Fault Tolerance 1

Localization of Damaged Resources in NoC Based Shared-Memory MP2SOC, using a Distributed Cooperative Configuration Infrastructure....................229
Z. Zhang, D. Refauvelet, A. Greiner, M. Benabdenbi and F. Pecheux

Exponent Monitoring for Low-Cost Concurrent Error Detection in FPU Control Logic....................235
M. Maniatakos, Y. Makris, P. Kudva and B. Fleischer

Enhancing Online Error Detection through Area-Efficient Multi-Site Implications....................241
N. Alves, Y. Shi, J. Dworak, R. I. Bahar, K. Nepal

Session 11

Session 11A: Design for Testability 2

Dynamic Scan Clock Control for Test Time Reduction Maintaining Peak Power Limit....................248
P. Shanmugasundaram and V. D. Agrawal

Structural Tests of Slave Clock Gating in Low-power Flip-flop 254
B. Wang, J. Rajaraman, K. Sobti, D. Losli and J. Rearick

Revival of Partial Scan: Test Cube Analysis Driven Conversion of Flip-Flops....................260
N. Alawadhi and O. Sinanoglu

Session 11B: Error & Fault Tolerance 2

Memory-Based Embedded Digital ATE....................266
D. Lee, S. P. Park, A. Goel and K. Roy

A Unified Test Architecture for On-Line and Off-Line Delay Fault Detections 272
S. Pei , H. Li, and X. Li

Design For Bit Error Rate Estimation of High Speed Serial Links..278
U. Guin, and C.-H. Chiang

Session 12

Session 12A: ATPG & Compression

An Efficient Test Data Reduction Technique Through Dynamic Pattern Mixing Across Multiple
Fault Models...285
S. Alampally, R. T. Venkatesh, P. Shanmugasundaram, R. A. Parekhji and V. D. Agrawal

Low Coverage Analysis using Dynamic Un-testability debug in ATPG..291
K. Chandrasekar, S. Bommu, and S. Sengupta

Prediction of Compression Bound and Optimization of Compression Architecturefor Linear
Decompression-based Schemes..297
J. Li, Y. Huang, and D. Xiang.

Session 12B: Reducing Test & Diagnosis Costs

Multi Domain Test: Novel Test Strategy to reduce the Cost of Test..303
Y. Takahashi, and A. Maeda

Low-Cost Diagnostic Pattern Generation and Evaluation Procedures for Noise-Related Failures.........309
J. Ma, N. Ahmed, and M. Tehranipoor

Sigma-Delta Modulation Based Wafer-Level Testing for TFT-LCD Source Driver ICs.......................315
W.-A. Lin, C.-C. Lee, and J.-L. Huang

Session 13

Special Session 13A: Practical Signal Processing at Mixed Signal Test Venues – Trend Removal, Noise Reduction, Wideband Signal Capturing –
....322
Organizer: H. Okawara

Session 13B: Hot Topic: Smart Silicon..323
Organizers: L. Winemberg, and M. Tehranipoor

Yin and Yang of Embedded Sensors for Post-Scaling-Era...324
A. Gattiker

Session 13C: Hot Topic: Design and Test of 3D and Emerging Memories..328
Organizer: C.-W. Wu

Foreword

Welcome to VTS 2011, the twenty-ninth in a series of annual symposia that focus on innovation in the field of testing of integrated circuits and systems.

The core of VTS 2011, the three day technical program, responds to the many trends and challenges in the semiconductor design and manufacturing industries with papers covering a diverse and seminal set of topics including, delay and performance test, design for testability, error & fault tolerance, ATPG and compression, power issues in test, analog, mixed-signal & RF test and diagnosis, low-power IC test, security issues, modeling and simulation of defects & faults, memory test & repair, post-silicon debug, 3D ICs, test & diagnosis costs, and reliability issues such as aging, transients and soft errors.

In addition to the three-day technical program, VTS 2011 features several special sessions including several panels addressing various hot topics, several new topic speakers, and two student activity sessions. VTS 2011 continue the tradition of featuring the Innovative Practices track. The sessions that make up this track highlight cutting-edge challenges faced by test practitioners, and innovative solutions employed to address them. This year, the Workshop on Design for Reliability and Variability (DRV 2011) will take place in conjunction with VTS.

Two tutorials are offered by the TTTC Tutorials & Education Group through the Test Technology Education Program (TTEP). This year the tutorials cover the exciting topics of: Advanced Topics and Recent Advances in Silicon Debug and Diagnosis; and Practices in Analog, Mixed-signal and RF Testing. Tutorials provide opportunities for design and test professionals to update their knowledge, and earn official IEEE TTTC accreditation.

The social program at VTS provides an opportunity for informal technical discussions among participants. Dana Point, in the beautiful Southern California coast, provides a very attractive backdrop for all VTS 2011 activities.

VTS is the result of the work of many dedicated volunteers: the reviewers, the best paper award judges, the Program Committee, the Organizing Committee, and the Steering Committee. We wholeheartedly thank them all. We also wish to thank all the authors who submitted their works to VTS 2011, and the program participants for their contribution at the symposium. We thank the IEEE Computer Society Test Technology Technical Council for its continued sponsorship and support. Finally, we thank the Corporate Supporters of VTS 2011.

We hope that you will find VTS 2011 enlightening, thought-provoking, rewarding, and enjoyable. We wish you all a fun-filled and productive week in the Dana Point area and hope that you will keep making VTS a success by actively participating in it, assisting in its organization, and letting us always know when we can do something better. Thank you all for coming.

General Chair
Cecilia Metra

Program Chair
Claude Thibeault

Conference Committees
Organizing Committee

General Chair
Cecilia Metra
University of Bologna

Program Chair
Claude Thibeault
E Tech Sup Montreal

Past General Chair
Magdy Abadir
Freescale

Vice-General Co-Chair
Peter Maxwell
Aptina

Vice-General Co-Chair
Michel Renovell
LIRMM

Vice-Program Co-Chair
Rajesh Galivanche
Intel Corporation

Vice-Program Co-Chair
Srivaths Ravi
Texas Instruments

Finance
Chen-Huan Chiang
Alcatel-Lucent

Special Sessions
Lorena Anghel
TIMA

Special Sessions
Kazumi Hatayama
STARC

Innovative Practices
Karim Arabi
Qualcomm

Innovative Practices
Subhasish Mitra
Stanford University

New Topics
Bernard Courtois
CMP

New Topics
Bozena Kaminska
Simon Fraser University

Audio/Visual
Rohit Kapur
Synopsys

Registration
Yiorgos Makris
Yale University

Publications
Dimitris Gizopoulos
University of Piraeus

Local Arrangements
Li-C. Wang
UC Santa Barbara

Corporate Support
Bruce Cory
Nvidia

Publicity
Giorgio Di Natale
LIRMM

Ex-Officio
Yervant Zorian
Synopsys

Publicity Committee Members

Stefano Di Carlo
Politecnico di Torino

Martin Omana
University of Bologna

Daniele Rossi
University of Bologna

Publications Committee Members

Ujjwal Guin
Temple University

International Liaisons

China & Taiwan
Cheng-Wen Wu
ITRI

Eastern Europe
Vladmir Hahanov
KHNURE Ukraine

India
C.P. Ravikumar
Texas Instrument

Japan
Yasuo Sato
Kyushu Institute of Technology

Latin America
Victor Champac
INAOEP Mexico

Midle East & Africa
Rafic Makki
UAE University

Western Europe
Erik Jan Marinissen
IMEC

Program Committee

J. Abraham - *University of Texas at Austin*
V. Agrawal - *Auburn University*
D. Appello - *STMicroelectronics*
B. Becker - *University of Freiburg*
J. Bhadra - *Freescale Semiconductor Inc.*
K. Chakrabarty - *Duke University*
A. Chaterjee - *Georgia Institute of Technology*
C.J. Clark - *Intellitech Corporation*
P. Girard - *LIRMM*
X. Gu - *Cisco*
S. Gupta - *University of Southern California*
I. Hartanto - *Xilinx*
S. Hellebrand – *University of Paderborn*
C.-T. Huang- *National Tsing Hua University*
A. Khoche - *Consultant*
H. Konuk - *Broadcom*
X. Li - *Chinese Academy of Sciences*
F. Lombardi - *Northeastern University*
M. Lubaszewski - *UFRGS*
A. Majumdar - *AMD*
S. Mourad - *Santa Clara University*
Z. Navabi – *Worcester Polytechnic*
A. Orailoglu - *UCSD*
J. Rajski - *Mentor Graphics Corp.*
S. Reddy - *University of Iowa*
K. Roy - *Purdue University*
J. Segura - *University of Illes Balears*
S. Shoukourian - *Synopsys*
M. Soma - *University of Washington*
P. Song - *IBM*
S. Sunter – *Mentor Graphics*
M. Tehranipoor - *University of Connecticut*
J. Tyszer - *Poznan University of Technology*
H. Wunderlich - *University of Stuttgart*

Steering Committee

Joan Figueras
*Universitat Politecnica
de Catalunya*

André Ivanov
*University of British
Columbia*

Michael Nicolaidis
TIMA

Paolo Prinetto
Politecnico di Torino

Adit Singh
Auburn University

Prab Varma
Blue Pearl Software

Yervant Zorian
Synopsys

Reviewers

Jacob ABRAHAM, *University of Texas at Austi*

Vishwani AGRAWAL, *Auburn Universit*

Lorena ANGHEL, *TIMA Laboratory*

Davide APPELLO, *STMicroelectronics*

Karim ARABI, *Qualcomm*

Bernd BECKER, *University of Freiburg*

Jay BHADRA, *Freescale Semiconductor*

Alberto BOSIO, *LIRMM*

Krishnendu CHAKRABARTY, *Duke University*

Abhijit CHATTERJEE, *Georgia Institute of Technology*

Chen-Huan CHIANG, *Alcatel-Lucent*

C.J. CLARK, *Intellitech Corporation*

Bruce CORY, *NVIDIA*

Stefano DI CARLO, *Politecnico di Torino*

Giorgio DI NATALE, *LIRMM*

Michele FABIANO, *Politecnico di Torino*

Joan FIGUERAS, *Universitat Politècnica de Catalunya*

Marie-Lise FLOTTES, *LIRMM*

Ghyslain GAGNON, *École de Technologie Supérieure*

Rajesh GALIVANCHE, *Intel Corporation*

Patrick GIRARD, *LIRMM*

Dimitris GIZOPOULOS, *University of Piraeus*

Xinli GU, *Cisco Systems, Inc.*

Sandeep GUPTA, *University of Southern California*

Ismed HARTANTO, *Xilinx*

Syed Rafay HASAN, *Ecole Polytechnique MTL*

Kazumi HATAYAMA, *Semiconductor Technology Academic Research Center*

Sybille HELLEBRAND, *University of Paderborn*

Christelle HOBEIKA, *École de Technologie Supérieure*

Chih-Tsun HUANG, *National Tsing Hua University*

André IVANOV, *University of British Columbia*

Palkesh JAIN, *Texas Instruments*

Niraj K. JHA, *Princeton University*

Bozena KAMINSKA, *Simon Fraser University*

Rohit KAPUR, *Synopsys Inc.*

David C. KEEZER, *Georgia Institute of Technology*

Ajay KHOCHE, *Independent*

Haluk KONUK, *Broadcom Corp.*

Nektarios KRANITIS, *University of Athens*

Xiaowei LI, *Chinese Academy of Sciences*

Fabrizio LOMBARDI, *Northeastern University*

Marcelo LUBASZEWSKI, *Federal University of Rio Grande do Sul (UFRGS)*

Amitava MAJUMDAR, *AMD Inc.*

Yiorgos MAKRIS, *Yale University*

Erik Jan MARINISSEN, *IMEC*

Peter MAXWELL, *Aptina Imaging*

Anne MEIXNER, *Intel Corporation*
Cecilia METRA, *Università di Bologna*
Subhasish MITRA, *Stanford University*
Sami MOURAD, *Santa Clara University*
Benoit NADEAU-DOSTIE, *Mentor Graphics Corporation*
Zainalabedin NAVABI, *Worcester Polytechnic Institute*
Nicola NICOLICI, *McMaster University*
Phil NIGH, *IBM*
Martin Eugenio OMANA, *University of Bologna*
Alex ORAILOGLU, *University of California at San Diego*
Antonis PASCHALIS, *University of Athens*
James PLUSQUELLIC, *University of New Mexico*
Ilia POLIAN, *Albert-Ludwigs-University of Freiburg*
Irith POMERANZ, *Purdue University*
Paolo PRINETTO, *Politecnico di Torino*
Mihalis PSARAKIS, *University of Piraeus*
MICHAEL M PURTELL, *Intersil*
Janusz RAJSKI, *Mentor Graphics Corporation*
Srivaths RAVI, *Texas Instruments*
C.P. RAVIKUMAR, *Texas Instruments*
Sudhakar REDDY, *University of Iowa*
Michel RENOVELL, *LIRMM*
Gordon ROBERTS, *McGill University*
Daniele ROSSI, *University of Bologna*
Kaushik ROY, *Purdue University*
Adoracion RUEDA, *Universidad de Sevilla/IMSE-CNM*
Yasuo SATO, *Kyusyu Institute of Technology*
Alberto SCIONTI, *Politecnico di Torino*
Jaume SEGURA, *Universidad des Illes Ballears*
Priyadharshini SHANMUGASUNDARAM, *Auburn University*
Samvel SHOUKOURIAN, *Virage Logic*
Suraj SINDIA, *Auburn University*
Adit SINGH, *Auburn University*
Mustapha SLAMANI, *IBM*
Peilin SONG, *IBM*
Vilas SRIDHARAN, *Northeastern University*
Stephen SUNTER, *Mentor Graphics Corporation*
Mehdi TAHOORI, *Northeastern University*
Mohammad TEHRANIPOOR, *University of Connecticut*
Nandu TENDOLKAR, *Freescale Semiconductor*
Claude THIBEAULT, *École de Technologie Supérieure*
Lionel TORRES, *LIRMM*
Nur TOUBA, *University of Texas at Austin*
Jerzy TYSZER, *Poznan University of Technology*
Devanathan VARDARAJAN, *Texas Instruments*
Prab VARMA, *Blue Pearl*
Rakshith VENKATESH, *Auburn University*

Duncan Moore Henry WALKER, *Texas A&M University*
Zhiyuan WANG, *Cisco Systems, Inc.*
Cheng-Wen WU, *Industrial Technology Research Institute*
Hans-Joachim WUNDERLICH, *Universität Stuttgart*
Hariri YASSINE LACIME, *École de Technologie Supérieure*
Lixing ZHAO, *Auburn university*

Acknowledgements

VTS, like any complex organization, is the result of the efforts of a large number of volunteers, who selflessly have volunteered their time and energy, with their only reward being the satisfaction of seeing a job well done, and the consciousness to have contributed to the dissemination of scientific knowledge through the continued success of a forum dedicated to the exchange of advances in both research and practice in VLSI Test. No words would compare to the magnitude of the efforts displayed by these volunteers. However, I would nonetheless like to register herein my personal note of thanks to the whole body of volunteers, who made it possible the organization of VTS 2011.

Among all VTS 2011 volunteers, I would like to thank all members of the VTS 2011 Technical Committee and Steering Committee. A special thanks goes to each member of the Organizing Committee, who excellently played a leading role in each aspect of VTS 2011 organization, with an enormous expenditure of energy and time: without their contribution VTS 2011 could have never taken place.

Then, I would like to express a very special thanks to those members of the VTS 2011 Organizing Committee, who contributed to the VTS 2011 organization quite beyond what expected according to their role, and who acted as continuous, trustworthy and active reference throughout the whole VTS 2011 organization. They are: Magdy Abadir, Chen Huan Cheng, Giorgio Di Natale and Yervant Zorian.

I would also like to thank Jacob Abraham and the University of Texas at Austin, for hosting the VTS Program Committee paper selection meeting.

Also, I would like to thank all of you, the VTS 2011 participants, the paper submitters, authors and speakers, reviewers, moderators, and special session organizers for making the VLSI Test Symposium a continued success and establishing it as the preeminent forum for the exchange of innovative ideas in all aspects of VLSI Test.

General Chair
Cecilia Metra

IEEE Computer Society

TTTC: Test Technology Technical Council

TTTC IN GENERAL

PURPOSE: The Test Technology Technical Council is a volunteer professional organization sponsored by the IEEE Computer Society. The goals of TTTC are to contribute to members' professional development and advancement and to help them solve engineering problems in electronic test, and help advance the state-of-the art. In particular, TTTC aims at facilitating the knowledge flow in an integrated manner, to ensure overall quality in terms of technical excellence, fairness, openness, and equal opportunities.

MEMBERSHIP: Membership is open to all individuals interested in test engineering at a professional level.

DUES: There are NO dues for TTTC membership and no parent-organization membership requirements.

BENEFITS: The TTTC members benefit from personal association with other test professionals. They may have the opportunity to be involved on a wide range of committees. They receive appropriate and updated information and announcements. There are substantial reductions in fees for TTTC-sponsored meetings and tutorials for members of IEEE and/or IEEE Computer Society.

TTTC ACTIVITIES

TECHNICAL MEETINGS: To spread technical knowledge and advance the state-of-the art, TTTC sponsors many well-known conferences and symposia and holds numerous regional and topical workshops worldwide.

STANDARDS: TTTC initiates, nurtures and encourages new test standards. TTTC-initiated Working Groups have produced numerous IEEE standards, including the 1149 series used throughout the industry.

TECHNICAL ACTIVITIES: TTTC sponsors a number of Technical Activity Committees (TACs) that address emerging test technology topics and guide a wide range of activities.

TUTORIALS and EDUCATION: TTTC sponsors a comprehensive *Test Technology Educational Program (TTEP)*. This program provides opportunities for design and test professionals to update and expand their knowledge base in test technology, and to earn official accreditation from IEEE TTTC, upon the completion of four full day tutorials proposed by TTEP.

TTTC CONTACT

TTTC On-Line: The TTTC Web Site at http://tab.computer.org/tttc offers samples of the TTTC Newsletter, information about technical activities, conferences, workshops and standards, and links to the Web pages of a number of TTTC-sponsored technical meetings.

Becoming a MEMBER: Becoming a TTTC member is extremely simple. You may either contact by phone or e-mail the TTTC office, or fill out and submit a TTTC application form, or visit the membership section of the TTTC web site.

TTTC OFFICE: 1 Marsh Elder Lane, Savannah, GA 31411, USA
Phone: +1-540-937-5066 Fax: +1-540-937-7848 E-mail:tttc@computer.org

TTTC Officers for 2011

Chair	**Adit D. SINGH** Auburn Univ. - USA	adsingh@eng.auburn.edu
1st Vice Chair	**Michael NICOLAIDIS** TIMA Laboratory - France	michael.nicolaidis@imag.fr
2nd Vice Chair	**Chen-Huan CHIANG** Alcatel-Lucent - USA	chen-huan.chiang@alcatel-lucent.com
President of Board	**Yervant ZORIAN** Virage Logic Corp. - USA	zorian@viragelogic.com
Past Chair	**André IVANOV** U. of British Columbia - Canada	ivanov@ece.ubc.ca
Senior Past Chair	**Paolo PRINETTO** Politecnico di Torino - Italy	paolo.prinetto@politolit
IEEE Design & Test EIC	**Krish Chakrabarty** Duke U - USA	krish@ee.duke.edu
ITC General Chair	**Ron PRESS** Mentor Graphics - USA	ron_press@mentor.com
Test Week Coordinator	**Yervant ZORIAN** Virage Logic Corp. - USA	zorian@viragelogic.com
Secretary	**Joan Figueras** UPC Barcelona Tech - Spain	figueras@eel.upc.edu
Vice Secretary	**Adam OSSEIRAN** Edith Cowan U. – Australia	a.osseiran@ecu.edu.au
Finance Chair	**Michael NICOLAIDIS** TIMA Laboratory - France	michael.nicolaidis@imag.fr
Finance Vice-Chair	**Don WHEATER** IBM Microelectronics - USA	dwheater@us.ibm.com

Group Chairs

Technical Meetings	**Chen-Huan CHIANG** Alcatel-Lucent- USA	chen-huan.chiang@alcatel-lucent.com
Technical Activities	**Patrick Girard** LIRMM – France	patrick.girard@lirmm.fr
Tutorials & Education	**Dimitris GIZOPOULOS** University of Piraeus - Greece	dgizop@unipi.gr
Standards	**Rohit KAPUR** Synopsys, Inc. - USA	rkapur@synopsys.com
Communications	**Cecilia METRA** U. of Bologna - Italy	cmetra@deis.unibo.it
Standing Committees	**André IVANOV** U. of British Columbia - Canada	ivanov@ece.ubc.ca
Industry Advisory Board	**Yervant ZORIAN** Virage Logic Corp. - USA	zorian@viragelogic.com
Electronic Media	**Alfredo BENSO** Politecnico di Torino - Italy	alfredo.benso@polito.it
Asia & Pacific	**Kazumi HATAYAMA** STARC - Japan	hatayama.kazumi@starc.or.jp
Europe	**Matteo SONZA REORDA** Politecnico di Torino – Italy	matteo.sonzareorda@polito.it
Latin America	**Victor Hugo CHAMPAC** Inst. Natl. de Astrofisica - Mexico	champac@inaoep.mx
North America	**André IVANOV** U. of British Columbia - Canada	ivanov@ece.ubc.ca
Middle East & Africa	**Ibrahim HAJJ** American U. of Beirut - Lebanon	ihajj@aub.edu.lb

Technical Activity Committees

Board Testing	**Bill EKLOW** Cisco Systems - USA	ben@dft.co.uk
Defect Tolerance	**Vincenzo PIURI** Politecnico di Milano - Italy	piuri@elet.polimi.it
Economics of Test	**Magdy S. ABADIR** Freescale, Inc. - USA	m.abadir@freescale.com
	Anthony P. AMBLER U. of Texas at Austin - USA	ambler@ece.utexas.edu
Embedded Core Test	**Yervant ZORIAN** Virage Logic Corp. - USA	zorian@viragelogic.com
FPGA Testing	**Michel RENOVELL** LIRMM - France	renovell@lirmm.fr
Freeware libraries	**Burnell WEST** NPTest - USA	west@ieee.org
IEEE 1149.1	**Christopher J. CLARK** Intellitech Corporation - USA	cjclark@intellitech.com
Infrastructure IP	**Yervant ZORIAN** Virage Logic Corp. - USA	zorian@viragelogic.com
Memory Testing	**Rochit RAJSUMAN** Advantest - USA	r.rajsuman@advantest.com
MEMs Testing	**Ronald D. BLANTON** Carnegie-Mellon U. - USA	blanton@ece.cmu.edu
	Bernard COURTOIS TIMA - France	bernard.courtois@imag.fr
Mixed-Signal Testing	**Bozena KAMINSKA** IMS Pultronics, Inc. - USA	bozena@pultronics.com
Nanometer Testing	**Jaume SEGURA** U. of the Balearic Islands - Spain	dfsjsf4@clust.uib.es
Nanotechnology Test	**Fabrizio LOMBARDI** Northeastern U. - USA	lombardi@ece.neu.edu
Network-On-Chip Test	**Erik Jan MARINISSEN** NXP – The Netherlands	erik.jan.marinissen@nxp.com
On-Line Testing	**Michael NICOLAIDIS** iRoC Technologies - France	michael.nicolaidis@iroctech.com
RF Testing	**Iboun Taimiya SYLLA** Texas Instruments - USA	isylla@ti.com
Silicon Debug and Diagnosis	**Michael RICHETTI** ATI Research, Inc. - USA	mike_ricchetti@ieee.org
System Test	**Ian HARRIS** UC Irvine - USA	harris@ics.uci.edu
3D chips & SiP Testing	**Yervant ZORIAN** Virage Logic Corp. - USA	zorian@viragelogic.com
Test Compression	**Rohit KAPUR** Synopsys, Inc. - USA	rkapur@synopsys.com
Test & Verification	**Magdy S. ABADIR** Freescale, Inc. - USA	m.abadir@freescale.com
Test Education	**Sule OZEV** Duke U. - USA	sule@ee.duke.edu
Test Synthesis	**Scott DAVIDSON** Sun Microsystems - USA	scott.davidson@eng.sun.com
Thermal Testing	**Bernard COURTOIS** TIMA - France	bernard.courtois@imag.fr

Standards Working Groups

IEEE 1149.4	**Bambang SUPARJO** Mentor Graphics - USA	bambang_suparjo@mentor.com
IEEE 1149.6	**Bill EKLOW** Cisco Systems, Inc. - USA	beklow@cisco.com
IEEE P1149.7	**Robert OSHANA** Texas Instruments – USA	roshana@ti.com
IEEE 1450-1999	**Gregory MASTON** Synopsys, Inc. - USA	gmaston@synopsys.com
IEEE 1450.1	**Tony TAYLOR**	t.taylor@ieee.org
IEEE 1450.2-2002	**Gregg WILDER** Texas Instruments - USA	gwilder@ti.com
IEEE P1450.3	**Tony TAYLOR**	t.taylor@ieee.org
IEEE P1450.4	**Doug SPRAGUE** IBM - USA	dsprague@us.ibm.com
	Jim O'REILLY Analog Devices - USA	jim_oreilly@ieee.org
IEEE P1450.6-1	**Bruce CORY** NVIDIA – USA	bcory@nvidia.com
IEEE P1450.6-2	**Saman ADHAM** LogicVision, Inc. - Canada	saman@logicvision.com
IEEE 1450.6-2005	**Rohit KAPUR** Synopsys, Inc. - USA	rkapur@synopsys.com
IEEE P1450.7	**Jean-Louis CARBONERO** STMicroelectronics - France	jean-louis.carbonero@st.com
IEEE 1500	**Yervant ZORIAN** Virage Logic Corp. - USA	zorian@viragelogic.com

IEEE 1532	**Neil JACOBSON** Xilinx Corp. - USA	neil.jacobson@xilinx.com
IEEE P1581	**Heiko EHRENBERG** GOEPEL Electronics - USA	h.ehrenberg@goepel.com
IEEE P1687	**Kenneth POSSE** AMD - USA	kepos@comcast.net
	Alfred CROUCH Asset InterTech - USA	al.crouch@asset-intertech.com

TTTC-Sponsored Technical Meetings in 2011

For the most current information, please visit the TTTC website (http://tab.computer.org/tttc)
or TTTC Events website (http://www.tttc-events.org)

1/17-1/19	Int'l Sym. on Electronic, Design, Test and Applications (DELTA), Queenstown, New Zealand	D. Bailey, S. Demidenko
3/14-3/18	Design, Automation and Test in Europe (DATE), Grenoble, France	B. Al-Hashimi
3/29-3/30	Workshop on Silicon Errors in Logic - System Effects (SELSE), U. of Illinois, IL, USA	A. Wood, R. Kumar
3/27-3/30	Latin American Test Workshop (LATW), Porto de Galinhas, Brazil	F. Vargas, Y. Zorian
4/13-4/15	Design & Diagnosis of Electronic Circuits & Systems Symposium (DDECS), Cottbus, Germany	H. T. Vierhaus
5/1-5/5	VLSI Test Symposium (VTS), Dana Point, CA, USA	C. Metra
5/4-5/5	Int'l Workshop on Design for Reliability and Variability (DRV), Dana Point, CA, USA	M. Nicolaidis, Y. Zorian
5/16-5/18	Int'l Mixed-Signals, Sensors, and Systems Test Workshop (IMS3TW), Santa Barbara, CA, USA	K.-T. Cheng
5/23-5/27	European Test Symposium (ETS), Trondheim, Norway	E. J. Aas
5/26-5/27	Int'l Workshop on Impact of Low Power Design on Test and Reliability (LPonTR)	A. Bystrow, P. Girard
5/26-5/27	Int'l Workshop on Processor Verification, Test and Debug (IWPVTD)	R. Aitken, V. Singh
6/5-6/6	Int'l Symposium on Hardware-Oriented Security and Trust (HOST), San Diego, CA, USA	K. Mai
6/6	Int'l Workshop on Design for Manufacturability & Yield (DfM&Y), San Diego, CA, USA	R. Aitken
7/13-7/15	International On-Line Testing Symposium (IOLTS), Athens, Greece	M. Nicolaidis, A. Paschalis
7/14	ATE: Vision 2020, San Francisco, CA, USA	E. Volkerink
9/09-9/12	East-West Design and Test Symposium (EWDTS), Sevastopol, Ukraine	V. Hahanov, Y. Zorian
9/18-9/23	International Test Conference (ITC), Anaheim, CA, USA	W. Eklow
9/22-9/23	Wkshop. on Testing Three-Dimensional Stacked Integrated Circuits (3D-Test), Anaheim, CA, USA	Y. Zorian
9/22-9/23	Int'l Defect and Adaptive Testing Workshop (DAT), Anaheim, CA, USA	TBD
10/5-10/7	Int'l Symp. on Defect & Fault Tolerance in VLSI and Nanotechnolohy Systems (DFT), Vancouver, Canada	G. Chapman, F. Salice
11/10-11/11	Int'l High Level Design Validation and Test Workshop (HLDVT), Napa, CA, USA	Z. Zilic
11/21-11/23	Asian Test Symposium (ATS), New Delhi, India	A. Chatterjee, Amit Patra
11/25-11/26	Workshop on RTL and High Level Testing (WRTLT), New Delhi, India	TBD
12/5-12/7	International Workshop on Microprocessor Test and Verification (MTV), Austin, TX, USA	M. Abadir
TBD	Board Test Workshop (BTW), Fort Collins, CO, USA	W. Eklow
TBD	Wkshop. on Test & Validation of High Speed Analog Circuits (TVHSAC), TBD	TBD
TBD	Int'l Workshop on Testing Embedded and Core-Based System-Chips (TECS), Online	Y. Zorian
TBD	International Workshop on Design & Test (IDT), TBD	Y. Zorian, TBD

TTTC Office

1 Marsh Elder Lane
Savannah, GA 31411
USA

Phone: +1-540-937-5066
Fax: +1-540-937-7848
E-mail: tttc@computer.org

http://tab.computer.org/tttc

TEST TECHNOLOGY EDUCATIONAL PROGRAM

TTEP 2011
OVERVIEW OF TUTORIALS OF VTS 2011

The Tutorials & Education Group of the IEEE Computer Society Test Technology Technical Council (TTTC) organizes in 2011 a comprehensive set of Test Technology Tutorials to be held in conjunction with TTTC sponsored technical meetings and included in the annual and expanding Test Technology Educational Program (TTEP). TTEP intends to serve the test and design professionals offering fundamental education and expert knowledge in state-of-the-art test technology topics.

Participation in TTEP-organized tutorials is credited by TTTC. Each full day tutorial corresponds to four TTEP units. Upon completion of each sixteen TTEP units official accreditation in the form of an "IEEE TTTC Test Technology Certificate" will be presented to the participants. In addition to the tutorials, certified university courses and industrial seminars related to test technology can also be included in TTEP and the participation in these credited similar to TTEP tutorials. For information on TTEP 2011 please visit the TTEP web site http://tab.computer.org/tttc/teg/ttep. The test technology tutorials of the VTS 2011 technical program are part of TTEP 2011.

TTTC Test Technology Educational Program (TTEP) 2011 – Committee

Chair   Dimitris Gizopoulos, *University of Piraeus*

Vice Chairs   Yiorgos Makris, *Yale University (Program)*
Anand Raghunathan, *Purdue University (Organization)*

Program Committee Members

M. Abadir, *Freescale*
T.Aikyo, *STARC*
R.Aitken, *ARM*
D.Appello, *ST Microelectronics*
U.Arz, *PTB*
S.Chakravarty, *LSI Logic*
V.Champac, *INAOE*
S.Davidson, *Oracle*
L. Fanucci, *University of Pisa*
A.Gattiker, *IBM*
P.Harrod, *ARM*
K.Hatayama, *STARC*
D.Josephson, *Intel*
T.M.Mak, *Intel*
H.Manhaeve, *Q-Star*
A.Osseiran, *Edith Cowan University*
A.Paschalis, *University of Athens*
M.Slamani, *IBM*
S.Sunter, *Mentor Graphics*

TUTORIAL 1

Advanced Topics and Recent Advances in Silicon Debug and Diagnosis

PRESENTERS

SRIKANTH VENKATARAMAN – *Intel*
ROBERT AITKEN – *ARM*
MIRON ABRAMOVICI – *Independent*

AUDIENCE

This tutorial is targeted towards IC designers, test, DFT, and product engineers, DA developers, validation, debug, and failure analysis engineers, researchers, managers, and anyone else determined to shorten the time-to-volume of a newly manufactured chip.

DESCRIPTION

The increasing design complexity along with the emergence of new failure mechanisms in the nanometer regime has significantly increased the complexity of verification, validation and manufacturing ramp of ICs. This tutorial covers the state of the art and the full spectrum of topics in silicon validation and debug and defect diagnosis ranging from the basic concepts to advanced applications and new DFD techniques. We will also describe successful debug and diagnosis methods used in real industrial products, industrial experiences, and case studies. Finally we will discuss future directions and challenges.

TUTORIAL 2

Practices in Analog, Mixed-signal and RF Testing

PRESENTERS

SALEM ABDENNADHER - *Intel*

SAGHIR A. SHAIKH - *Broadcom*

AUDIENCE

This tutorial is most suitable for design, test and DFT engineers involved in actual implementation of mixed-signal, analog, RF and wireless devices and systems. The architects and engineering managers would also greatly benefit from this tutorial.

DESCRIPTION

The objective of this course is to present existing industry ATE solutions and the alternative solutions to ATE testing for mixed-signal and RF SoCs. These techniques greatly rely upon DFT and BIST structures. Tutorial presents the basic concepts in analog and RF measurements (eye diagram, jitter, gain, power compression, harmonics, noise figure, phase noise, BER, EVM, etc.). Several industrial examples of production testing of mixed-signal and RF devices, such as, SERDES transceivers, PHYs, PMDs, and RF transceivers are also presented. The block-DFT solutions are presented for PLLs, delta-sigma converters, equalizers, filters, mixers, AGC, LNAs, DACs and ADCs. The testing of high speed IO interfaces, such as, PCI-Express, and XAUI, etc, and the new design trends in RF systems such as MIMO and SiP based systems and their testability are also presented in this tutorial.

VTS 2010
Best Paper
Award

Each year, VTS proudly presents the Best Paper Award to the author(s) of the most outstanding paper from those presented at the previous year's symposium. The candidates for this honor are initially selected based solely on the numerical ratings of the reviewers and symposium attendees, as recorded on the review forms and the session rating cards. The Best Paper Award Judges then carefully review the candidate papers as published in the proceedings. The judges provide numerical scores and comments for each candidate paper. The scores and comments are compiled to select the best paper.

The paper selected by VTS 2010 Best Paper Award Judges for the Best Paper Award is:

7B.1: Concurrent Autonomous Self-Test for Uncore Components in System-on-Chips

Y. Li, *Stanford U*
O. Mutlu, *Carnegie Mellon U*
D. Gardner, *Intel Labs*
S. Mitra, *Stanford U*

The VTS 2010 Best Paper Award selection committee is listed below. VTS extends special thanks to these individuals for reviewing the papers and offering invaluable comments.

Magdy Abadir, *Freescale (Chair)*
Sybille Hellebrand, *University of Paderborn* **Yiorgos Makris,** *Yale University*
Cecilia Metra, *University of Bologna* **Michel Renovell,** *LIRMM/CNRS*
Claude Thibeault, *E Tech Sup Montreal* **Li-C. Wang,** *U. of Cal., Santa Barbara*
Yervant Zorian, *Synopsys*

VTS 2010
Best Innovative
Practices Award

Each year, VTS recognizes the organizers and presenters of the Best Innovative Practices Session at the previous year's symposium. The selection is based entirely on audience feedback, as recorded on the attendee feedback forms.

For VTS 2010, the Best Innovative Practices Session Award goes to:

IP Session 3C: INDUSTRIAL PRACTICES OF TEST COST REDUCTION TECHNIQUES: IMPACT AND DESIGN TRADEOFFS

Organizer:

S. Tammali, *Texas Instruments Inc.*

Presenters:

S. Tammali, *Texas Instruments Inc.*
K. Arnold, *Pintail Technologies*

M. Hirech, *Synopsys*

VTS 2010
Best Special
Session Award

Each year, VTS recognizes the organizers and presenters of the Best Special Session at the previous year's symposium. The selection is based entirely on audience feedback, as recorded on the attendee feedback forms.

For VTS 2010, the Best Special Session Award goes to:

Special Session 11C: HOT TOPIC: DESIGN CONSIDERATION AND SILICON EVALUATION OF ON-CHIP MONITORS

Organizer:

S. Chakravarty, *LSI Corporation*

Presenters:

C. Kim, *U. of Minnesota*

S. Chakravarty, *LSI Corporation*

Session 1

Understanding Customer Returns From A Test Perspective

Nik Sumikawa, Dragoljub (Gagi) Drmanac,
Li-C. Wang
University of California, Santa Barbara

LeRoy Winemberg,
Magdy S. Abadir
Freescale Semiconductor, Inc

Abstract

Customer returns are defective parts that pass all functional and parametric tests, but fail in the field. To prevent customer returns, this paper analyzes wafer probe test data and tries to understand what it takes to screen them out during testing. Because these parts pass all tests, analyzing their signatures based on the original test perspective does not make sense. In this work, we search for a novel test perspective where the test signatures from parametric measurements can be used to separate the returned parts from the rest of population. Our study shows that in order to effectively screen customer returns during wafer test, a multivariate screening methodology is desired. This study is based on analyzing over 1000 parametric wafer probe tests and dies from seven lots, each lot containing one returned part. We demonstrate that analyzing customer returns from a multivariate test perspective leads to robust and conservative results.

1 Introduction

In a market that requires high quality products with near zero defective parts per million (DPPM), the effects of a test escape can be significant in terms of debug and diagnosis costs. More importantly, an excessive number of test escapes will damage a company's reputation and can lead to missed business opportunities. For this reason, quality is extremely important and often, it outweighs the costs of overkills. Examples are automotive products where a test escape can have significant consequences.

To reduce DPPM to near zero, the logical first step is to acquire a better understanding of customer returns. Traditionally, this is achieved by diagnosis to a root cause. This work does not follow a root cause analysis approach. Instead, our objective is to understand customer returns from a behavioral point of view where its behavior is reflected in the values of parametric measurements. We call this a test perspective. Instead of focusing on the root cause, we try to understand the behavior of customer returns in order to develop a new test strategy to screen them in the future.

It should be noted that there have been several proposed methods for improving the quality of parametric testing.

Generally, these methods employ test selection and/or some statistical learning techniques to detect potential failures. For example, the authors in [4] used test selection and outlier analysis to predict burn-in failures. First, a smaller set of tests was derived from known failures which was then used to compare dies residing within the same wafer residuals in order to predict other defective devices. This analysis was based on single test measurements to ensure simplicity and applicability in practice.

As another example, the work in [7] analyzed a dataset of functional and parametric results for screening RF devices. A subset of important tests was extracted and machine learning algorithms were used to identify defective devices. This screening strategy was effective and the work was a good comparison of various algorithms.

In this work, the problem context is different. First, the product is a SoC for the automotive market where a large portion of the design is memory and analog. Second, we analyzed seven customer returns that passed a comprehensive testing process where the quality requirements are very high. In other words, the DPPM was already close to zero, but not exactly zero. Finally, the seven customer returns belong to different lots and we had access to over 1000 parametric wafer probe test measurements for each die across the seven lots.

It is important to clarify the problem context for two reasons. First, a different context may imply a different problem. For example, the challenges in analyzing a situation with 250 DPPM with the goal of pushing to 100 DPPM could be very different from the challenges of pushing from 50 DPPM to 10 DPPM. Similarly, analyzing a SoC could be different from analyzing an analog device. Second, when applying the findings in this work, one should carefully consider the context for their respective application. For example, in a different DPPM range, a customer return can behave differently than the ones studied in this work. Hence, one should be cautious and check the assumptions in this work before applying our methods.

Instead of jumping into the development of a screening methodology, we first ask the key question: "How did we miss the seven customer returns?" After all, the testing was very comprehensive. This key question leads to the following three questions:

- Is it because we still lack the right test(s)?
- Is it because we did not set the test limits correctly? Were some of the limits set too conservatively?
- Is it because we did not look at the tests from the correct perspective? If so, what is the correct perspective?

Section 2 investigates these three questions. Our conclusion will be that we do not need more test(s), nor test limit adjustments, as long as the screening takes a multivariate perspective, i.e. making decisions based a collection of tests. Section 3 describes the process of selecting relevant tests, followed by section 4 that explains the role of test selection in multivariate outlier analysis. Section 5 discusses experimental results to show how such an outlier model works. Section 6 concludes.

2 The Three Questions

Questions 2 and 3 should be analyzed first because yes to either question means additional test(s) are not needed. In the following, we discuss these questions based on seven lots of data each consisting of ~12,000 dies.

Prior to analysis, the test data was cleaned by removing all dies that failed during wafer probe testing and those with missing measurements. The non-parametric tests were also removed. Each measurement value was normalized to zero mean and unit variance, which ensures compatibility between different test types. The resulting parametric test set contains various types of measurements including opens, shorts, leakages, Idd and memory tests.

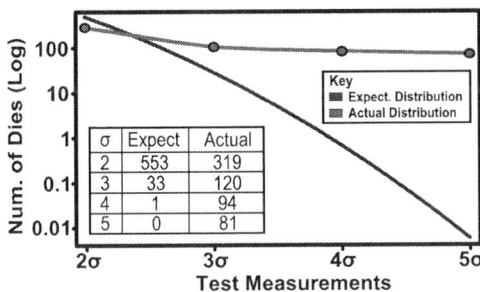

Figure 1. Gaussian is not a good assumption at the tail

2.1 How effective is using a kσ rule?

Due to a lack of knowledge, the parametric test limits are often set using the empirical $k\sigma$ rule, where an example may be k = 3, 6, etc. Figure 1 shows the behavior of a particular parametric measurement for one lot of dies. Assuming a Normal distribution and setting a 4σ limit, we would expect to see 1 die beyond this limit. However, we observed 94 dies. Keep in mind that all of these dies passed wafer testing (not necessarily after final testing), including the one

customer return. Hence, a 4σ test limit is too aggressive and it would result in the overkill of 94 good dies.

For this reason, test limits are usually set with a much larger σ to minimize overkill. Hence, the customer return may have been missed due to this conservatism.

Customer Return	applied to all tests	applied to only the specific test	specific test info	
			σ	Type
1	43.19%	0.71%	-3.02	Memory
2	42.41%	0.03%	3.45	Memory
3	13.02%	0.20%	8.08	Voltage
4	3.20%	0.20%	19.43	Leakage
5	5.33%	0.48%	4.80	Leakage
6	19.90%	0.23%	6.23	Leakage
7	27.26%	1.02%	4.01	Current

σ found with the best test to minimize overkill and screen out the return

Table 1. Overkill % based on applying the σ rule

Table 1 shows our hypothesis is not true. For each returned part, we found the σ value for each parametric test. For example, 5σ means the measured value using the test is at the 5σ point of the distribution whose mean and σ are calculated based on all dies in the lot. We then found the specific test whose σ value is the largest and used this value as a $k\sigma$ rule. For example, customer return #1 in Table 1 has a value of -3.02σ corresponding to a memory test.

Suppose we apply the -3.02σ rule to the specified test only in order to screen the dies (all dies with measured values whose absolute value $\geq 3.02\sigma$ are screened out), we would have an overkill of 0.71% in the lot, which is ~80 dies. This overkill is shown in the third column of Table 1.

This scenario assumes that we know which test is the best to use for the $k\sigma$ rule. Suppose we do not know this and we applied the same $k\sigma$ rule to all tests. In this case, the second column of the table show the overkill % for each lot containing the customer return.

Table 1 shows that with a $k\sigma$ rule, we cannot screen out a customer return without incurring significant overkill. Based on column 3, it is not desirable to capture customer returns by adjusting test limits. Using part #7 as an example; even in the best case we would have screened out more than 110 good dies.

2.2 Multivariate test perspective

In Table 1, we are examining the tests one at a time. What if we examine two or three tests collectively? Figure 2 shows such results based on customer return #4. Interestingly, the returned part looks more like an outlier when analyzing tests collectively, i.e. it's easier to separate from the rest of the dies.

Figure 3 shows that a hyperplane can separate customer return #5 from all good dies in three test dimensions. Based on these two figures, a customer return can be screened out with minimal overkill when examining the data using the right combination of tests. The big question is: From more than 1000 tests, which tests should we use?

978-1-61284-657-6/11 $26.00 © 2011 IEEE

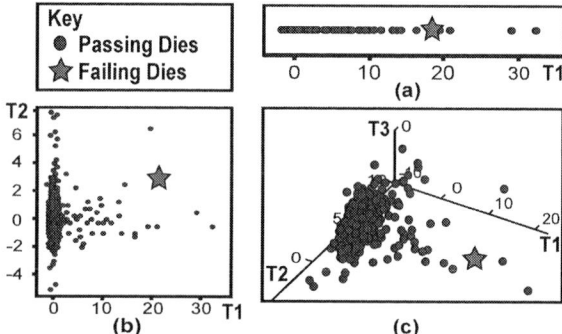

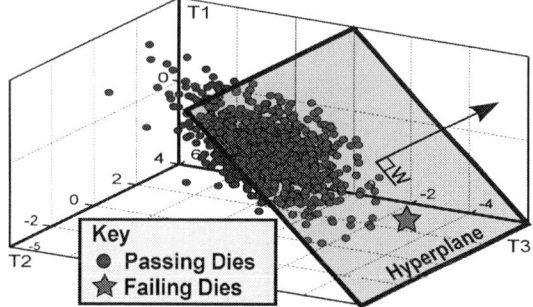

Figure 2. Moving to multivariate test perspectives

Figure 3. Hyperplane Separating the Return from Other Dies in 3 Test Dimensions

3 Learning the Relevant Tests

Based on the analysis above, we need to know the relevant tests in order to capture the customer return. In addition, it seems that the more tests we use, the more likely we are to be successful. For example, two tests are sufficient for return #5 while three tests are enough for return #4. Based on Table 1, we may conjecture that we will need even more tests for customer return #7.

The analysis also suggests a screening methodology to detect dies with similar behavior as the customer return. This method is based on the ability to learn a set of relevant tests from a returned device and show that the customer return can be identified as an outlier among all other dies in the lot. Using the same set of tests, outlier analysis can be performed on dies from future lots in order to identify potential customer returns that behave similar to the one used for learning.

To realize such a methodology, the first step is to have a learning method that can learn from the test data and select the relevant tests. The requirement for such a method depends on another question: Do we need to find a specific subset of tests relevant to the returned device for the outlier model to work effectively? In other words, how precise should the test selection process be? If we need a very precise set of tests; it can be challenging to develop such a learning method.

3.1 SVM and Chi Square

Given test data for the customer return and $\sim$12K dies, our goal is to rank the importance of the $\sim$1000 tests based on their ability to differentiate the returned part. This can also be thought as the following problem: Given two classes of samples and a set of features that describes the samples, rank the importance of the features in terms of their contribution to separate the two classes. This is commonly known as a feature ranking problem.

In the context of timing analysis, authors in [8] apply feature ranking to rank cells and nets by their contributions to path timing. This ranking is based on the linear Support Vector Machine (SVM), which is a binary classification approach described in [8]. Using the C-Support Vector Classification (C-SVC) algorithm with a dot-product kernel (i.e. a linear kernel) [5], we find an optimal hyperplane in n dimensions that best separates the two classes of samples. In our context, n is the number of tests examined and the samples are the dies. For example, Figure 3 shows a hyperplane in 3 dimensions. This hyperplane can be written as $f(T1, T2, T3) = w_1 T1 + w_2 T2 + w_3 T3 + b$ where b is the constant defining the location of the hyperplane, i.e. where it intercepts the $T2$-$T3$ plane.

For a linear model, the normal weight vector $\vec{W}$ of the hyperplane encodes the importance of each test [10]. Similarly, the components w_1, w_2, w_3 of the weight vector can be thought as the importance of tests $T1, T2, T3$, respectively. These components describe how much the hyperplane is tilted in the direction of the test in order to correctly classify the customer return. In Figure 3, tests $T1$ and $T3$ are more important than $T2$ because the weight vector is pointed toward test $T1$ and $T3$. Test $T2$ is irrelevant since the hyperplane is almost parallel to the $T2$ axis [10].

Another common method for ranking features is the Chi-Square method [2]. Chi-Square is an algorithm that does not build a binary classification model as does a linear SVM. Instead, it calculates the importance of each test as a chi-square statistic. This statistic tries to calculates the amount of separation between two classes of dies using a single test, i.e. measuring its separation power. Note that this separation power is measured based on each test individually. Hence, this method ignores correlation among tests.

3.2 Test Selection

SVM and Chi-Square can output the test importance for each customer return. Given such a ranking, one still needs to select a subset of tests. For example, Figures 2 and 3 show how two and three tests can separate customer returns. These tests were manually selected based on the outlying behavior of the return. When using SVM or Chi-Square to select tests automatically, we can think of various questions:

- Will the top two or three tests, ranked by either algorithm, match our manual selection?

- If not, how do we determine the size of the test set so the desired tests are included?
- Do we need to have specific subset of tests for outlier analysis to work? If not, how many "irrelevant tests" can we include before outlier analysis breaks down? i.e. how flexible must test selection be for outlier analysis to work?
- Do the SVM and Chi-Square rankings agree?

The following discussion is centered around these questions and will demonstrate several interesting points based on our data. First, results from SVM and Chi-Square usually do not agree. If we focus on the top two or three tests, their rankings do not agree with our manual ranking. At first, this was seen as a critical barrier but we found that for outlier analysis to work (i.e. to identify a customer return as a top outlier), it does not require the use of specific tests. In fact, there is a high degree of flexibility in test selection, which is enabled by the outlier analysis. This means that we do not have to worry about selecting an exact subset of tests. Instead, we can include many irrelevant tests. This is an important property to note because it enables the development of a practical methodology that can learn from returned parts and screen out similar dies in the future.

If there is a high degree of flexibility when selecting tests, why not include all tests? Later, we will show that this flexibility is bounded. Hence, if we select a test set that is too large, outlier analysis will lose its effectiveness.

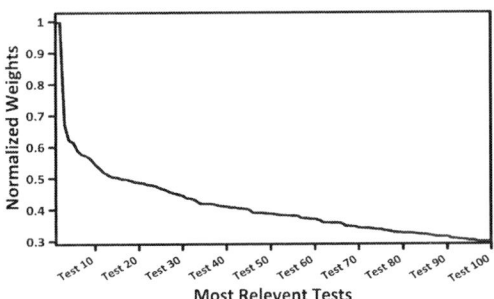

Figure 4. Test Importance for the Top 100 Tests

To demonstrate the output of the test ranking algorithm, Figure 4 shows the normalized importance of the top 100 tests using SVM ranking for customer return #1. From the weight curve in Figure 4, we can see that the importance starts to level out after the first 10 tests.

The behavior of customer return #1 is shown in Figure 5, where the tests are shown on x-axis. The tests are ordered by their importance, which is determined by the Chi-Square ranking this time. The measured values for each test are normalized by the variance. The behavior of the customer return is shown (red). The average measured value (mean) of each test, across all dies in the same lot, is shown (dark blue). One standard deviation (one σ) on either side of the mean is shown (light blue) for each test. A clear trend can be observed where the customer return deviates further from

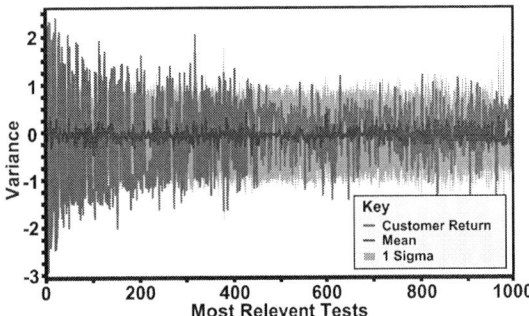

Figure 5. Effects of Diminishing Test Importance

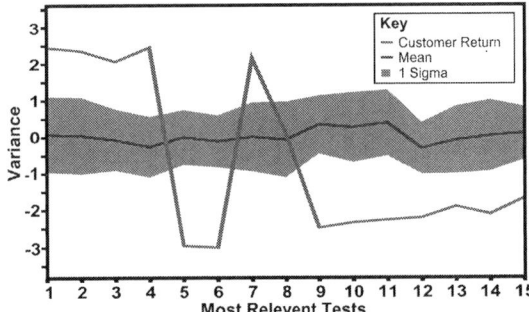

Figure 6. Return #1 as an Outlier on top 15 Tests

the mean in the higher ranked tests (on the left). Also, the measurements for the customer return reside within 3.02σ limit as was also shown in Table 1.

Figure 6 zooms in on the top 15 tests. The customer return curve (red) is clearly different from the expected trend, which is shown as the mean$\pm 1\sigma$ band (blue). From this figure, the outlying behavior of the returned part is easily seen. Since the customer return's measurements are within $\pm 3\sigma$, this outlying behavior is only seen when examining the 15 tests collectively. Hence, this is a multivariate outlier.

4 Multivariate Outlier Analysis

It is important to note the following properties. If we present Figure 6 by removing the tests 11-15, it does not alter our ability to declare the customer return as an outlier. On the other hand, if we consider all tests in Figure 5 together, it is not clear if the customer return is an outlier. These two figures hint at an interesting property. There is a certain degree of flexibility in test selection that allows a customer return to be identified as an outlier. In the following section, we will illustrate this property further.

In this work, multivariate outlier analysis is performed using the one-class SVM algorithm on a subset of relevant tests. We use the one-class ν-SVM algorithm [6] with a Gaussian kernel [9] and a modified version of the open source LibSVM software package [1]. All experiments were performed under the software framework RapidMiner [3]. Here, we do not intend to describe the details of the outlier analysis algorithm. Instead, we are interested in studying the effects of test selection on outlier analysis.

To study this impact, we perform the following experiments. For each customer return, we apply a ranking method to rank the tests. From this ranking, we select the top k tests and create a dataset D_k. Each test in D_k contains the measurements for all the dies within the lot. Outlier analysis is performed on D_k using the one-class SVM algorithm. The results of outlier analysis is a ranking of the dies. Using these results, we identify the rank R_k of the customer return. In the best case we would have $R_k = 1$ and in the worst case we would have $R_k \approx 12,000$. In general, a $R_k \leq 20$ is considered very good.

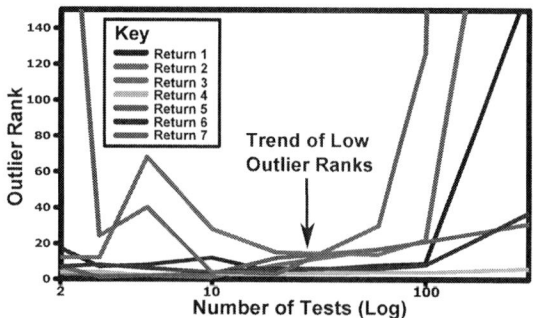

Figure 7. Using Chi-Squared Test Selection

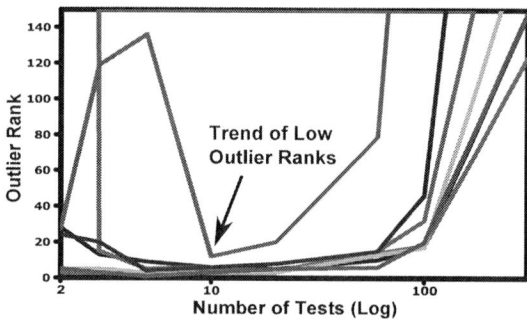

Figure 8. Using SVM Test Selection

Figure 7 shows the R_k (y-axis) for $k = 2 \ldots 300$ tests (x-axis) where the x-axis is in log scale. In this image, the results are based on Chi-Square test ranking and each customer return corresponds to a curve. In general, the trend says that if we use too few or too many tests, the results are not good. On the other hand, there is not much difference between using 10 to 50 tests.

For returns #1, 3, 4, 5 and 6, using a few tests is fine and using more (up to 50) does not hurt. This is not true for return #7, which requires using more than 10 tests. From Figure 7 we see that return #7 is a high-dimensional outlier (≥ 10 test dimensions). Figure 8 shows similar results based on SVM test ranking. It can be clearly seen that SVM results do not exactly agree with Chi Square results, but a similar trend exists.

Customer Return	1	2	3	4	5	6	7
SVM; $k, R_k =$	20,5	20,4	3,3	10,3	5,1	5,4	10,12
Chi-Square; $k, R_k =$	20,4	3,12	10,1	5,2	3,1	10,4	20,3

Table 2: k that gives the smallest R_k (k, R_k) in Figures 7 and 8

Customer Return	1	2	3	4	5	6	7
SVM; $R_k =$	6	6	4	3	3	6	12
Chi-Square; $R_k =$	12	28	1	3	3	4	4

Table 3: R_k Based on top 20 tests in Figures 7 and 8

Table 2 shows the k value that gives the best R_k in Figures 7 and 8. For example, using SVM ranking on customer return #1, with $k = 20$, the returned part is identified as the 5th outlier ($R_k = 5$) according to the outlier analysis ("20,5"). To achieve the best result for different returns, we would need different numbers of tests. The best R_k is usually small except for return #7 using SVM and return #2 using Chi Square, where both are ranked $R_k = 12$.

Table 3 shows results (R_k) when selecting the top 20 tests (fix $k = 20$) for all returns. These results are similar to the best R_k presented in Table 2. This method further demonstrates the flexibility of the test selection because a small variation on the size of the test set does not have a significant impact on outlier analysis.

5 Screening Potential Returns

Suppose we learn the 20 most relevant tests for a customer return and verify that outlier analysis can rank the returned part as a top outlier. Suppose we take the 20 tests and perform outlier analysis on another lot. Can we identify dies that behave similar to the customer return?

This question can be studied from two perspectives: How similar is a die's behavior to the customer return and how much variability exists from one lot to another. In the following experiments, we report various results using customer return #1 and SVM test ranking.

In the first experiment, we take all of the test measurements from the customer return and inject $i\sigma$ noise on each value in order to make a "simulated customer return." This simulated return is put back into the dataset and we perform outlier analysis on the new dataset to see how the simulated return is ranked. Table 4 shows results for $i = 0, 0.5, 1.0$ and 1.5. For $i = 0$, the simulated return is the original customer return. The experiment is iterated 100 times thus simulating 100 returns.

In Table 4, "Rank" is the average R_k across 100 simulated returns. "E" is the number of simulated returns whose ranks exceed 50, i.e. assuming top 50 outliers are screened out, these dies would become test escapes. It is interesting to see (the "Rank" column) that using 2 tests does not tolerate the noise injected on the simulated returns as the average R_k is large. Using 2 tests can capture some simulated returns, but many (23-35) have a $R_k > 50$ and thus escapes detection. This shows that using 2 tests is not robust. This is still more effective than using 300 tests where outlier analysis is neither robust nor effective in capturing simulated returns.

978-1-61284-657-6/11 $26.00 © 2011 IEEE

Noise	2 Tests		10 Tests		100 Tests		300 Tests	
	Rank	E	Rank	E	Rank	E	Rank	E
$0.0\,\sigma$	13.0	0	5.0	0	18.0	0	147	100
$0.5\,\sigma$	294.8	23	6.0	0	16.7	0	415	100
$1.0\,\sigma$	2392.3	35	5.6	0	14.7	0	347	100
$1.5\,\sigma$	2545.6	32	4.7	0	10.9	0	205	88

Rank: Average R_k over 100 simulated returns, E: Number of Escapes

Table 4: Simulating Failures via Noise Injection

If we use 10 to 100 tests, the average R_k of the simulated returns does not change much with respect to different amounts of injected noise. Based on the results in Table 4, if we want a robust solution that detects potential returns whose behavior is similar to the known customer return, we cannot use too few or too many tests.

In the second experiment, we randomly select one good die and change its test measurements for only the top 2 tests based on SVM ranking. In particular, we replace the measured values of this good die with values similar to the customer returns. This die is added back to the dataset and outlier analysis is performed using various numbers of tests. Since the altered die is mostly good and behaves nominally for all but 2 tests, we expect it to have a poor rank and it should not be screened. The experiment was iterated 100 times and the results are shown in Table 5. According to this table, when using 2 or 3 tests, the altered good die would be classified as a top outlier and it would be screened. When more tests are used, the rank increases and fewer altered good dies are screened (20-33). Hence, using more tests ensures conservatism by capturing dies that are most similar to the customer return, which is a desired property to avoid overkill.

2 Tests		3 Tests		5 Tests		10 Tests	
Rank	S	Rank	S	Rank	S	Rank	S
1	100	1	100	570	33	630	20

Rank: Avg. Rank of Altered Good Die, S: # of Altered Good Dies Screened

Table 5: Conservatism Study in Outlier Analysis

Table 6 repeats the experiments shown in Table 4 with additional noise injected on the good dies. We altered the good dies by injecting 2% random noise on all of the test measurements. As it can be seen, the results in Table 6 are similar to the results in Table 4. This shows that small random variability across lots does not impact the effectiveness of outlier analysis as long as we do not use too few tests.

Noise	2 Tests		10 Tests		100 Tests		300 Tests	
	Rank	E	Rank	E	Rank	E	Rank	E
$0.5\,\sigma$	429.6	25	5.1	0	17.7	0	407	100
$1.0\,\sigma$	1794.2	35	5.1	0	17.1	0	352	100
$1.5\,\sigma$	2594.6	37	3.2	0	15.3	0	211	88

Rank: Average R_k over 100 simulated returns, E: Number of Escapes

Table 6: Noise Injection on All Good Dies

In the last experiment, we take the top 50 tests learned from one lot and perform outlier analysis on another lot. Table 7 shows the results. If we learn from the returned part in lot 1, outlier analysis on lot 5 is able to classify its customer return as the 51st outlier. If we can tolerate 0.4% overkill, we can capture this customer return. In another

case, we learn from the return in lot 2 and we can screen out the customer return in lot 4 if we are willing to tolerate 0.7% overkill. It is interesting to note that among the top outliers in lots 5 and 4, there were 4 and 10 dies actually failed final test, respectively.

Train Lot	Predict Lot	Rank	Overlap of top 50 tests
1	5	51	12
2	4	100	11

Table 7: Cross-lot Fortuitous Prediction

If we examine the top 50 tests learned from these four lots individually, lot 1 and lot 5 share 12 tests and lot 2 and lot 4 share 11 tests. This sharing may be used to explain the cross-lot fortuitous customer return detection.

6 Conclusion

In this paper, we study how to learn from existing customer returns and how to develop a methodology to screen other potential customer returns. Findings, based on studying seven customer returns from seven lots of data and more than 1000 parametric tests, are: (1) Customer returns can be effectively screened by multivariate outlier analysis. (2) To perform such an analysis on a lot of dies, we must first determine which relevant tests to use. (3) Relevant tests can be learned from existing customer returns. (4) The selection of relevant tests is not strict and has a degree of flexibility. In fact, using too few tests is not robust. When using at least 10 tests, outlier analysis becomes much more robust and using up to 100 tests can still be effective. (5) Using more tests, i.e. performing a high-dimensional outlier analysis, ensures both robustness and conservatism in capturing potentially defective parts whose behavior is similar to the returns we have learned from.

References

[1] Chih-Chung Chang and Chih-Jen Lin. *LIBSVM: a library for support vector machines*, 2001. Software available at http://www.csie.ntu.edu.tw/ cjlin/libsvm.

[2] G. Forman. An Extensive Empirical Study of Feature Selection Metrics for Text Classification. In *Journal of Machine Learning*, 2003.

[3] I. Mierswa, M. Wurst, R. Klinkenberg, M. Scholz, and T. Euler. YALE: Rapid Prototyping for Complex Data Mining Tasks. In *SIGKDD*, 2006.

[4] A. Nahar, K. Butler, J. Carulli Jr., C. Weinberger. Quality Improvement and Cost Reduction Using Statistical Outlier Methods. In *IC-CAD*, 2009.

[5] B. Scholkopf, A. Smola, R. Williamson, and P. L. Bartlett. New support vector algorithms. *Neural Computation*, 2000.

[6] B. Scholkopf, et al. Estimating the Support of a High-Dimensional Distribution. In *Neural Computation*, 2001.

[7] H. Stratigopoulos, P. Drineas, M.Slamani, and Y. Makris. Non-RF To RF Test Correlation Using Learning Machines: A Case Study. In *VTS*, 2007.

[8] L. Wang, P. Bastani, and M. Abadir. Design-Silicon Timing Correlation - A Data Mining Perspective. section 4.2, Figure 8, *DAC*, 2007, pp. 384-389.

[9] S. Wu, D. Drmanac, and L. Wang. A Study of Outlier Analysis Techniques for Delay Testing. In *ITC*, 2008.

[10] D. Drmanac, N. Sumikawa, L. Winemberg, L. Wang, and M. Abadir. Parametric Test Set Optimization of Wafer Probe Data for Predicting in Field Failures and Setting Tighter Test Limit. In *DATE*, 2011.

A Distributed AXI-based Platform for Post-Silicon Validation

M. H Neishaburi, Zeljko Zilic

McGill University, 3480 University Street, Montreal, Quebec Canada H3A-2A7

Mh.neishabouri@mail.mcgill.ca, zeljko.zilic@mcgill.ca

Abstract: *With a significant increase in the design complexity of cores and associated communication among them, post-silicon validation has become a demanding task in System on Chips (SoCs) design. To ensure that final products are fault-free and ready for market, the post-silicon validation goal is to catch bugs and pinpoint the root causes of errors that could escape from pre-silicon verification tools. Post-silicon validation involves running a hardware prototype in an environment that is similar to its final platform with its expected workload. As new SoCs tend to have many cores, the interactions among these cores are becoming so complex that post-silicon debug techniques should address not only validation of the functional aspects of a design but such techniques have to "bulletproof" the communication and synchronization among cores inside an SoC. In this paper, we propose an AXI based environment for post-silicon validation. The proposed environment involves Local Debugging Unit (LDU) and Shared Debugging Unit (SDU). LDU monitors trace of transactions issued by the hardware prototype and detect undesired conditions on bus. SDU combines debug traces from different LDUs. We embed the proposed SDU inside an AXI configurable interconnect. Major benefits of using our proposed debug platform over traditional techniques for silicon validation are as follows: 1) it detects and bypasses real time severe faulty conditions such as deadlocks resulting from design errors or electrical faults 2) there is no need for internal trace memory because SDU can communicate to the external memory through slave ports 3) it enables online monitoring of the trace buffer.*

I. INTRODUCTION

With the advances in technology and CAD tools, and driven by reckless demands of consumers for new functionality, new SoCs require to have many cores. To carry out their demanding functions, SoCs typically require substantial number of communication links among its embedded cores.

Verification tools have to ensure that such a complex system is error-free and the final products meet the strict time-to-market deadline. Despite new enhancements in pre-silicon verification techniques, bugs slip to the first hardware prototype in significant percentage. In general, once a first-silicon is placed in its intended target environment and the actual workload (software) is exercised, errors arise in a hardware prototype. *Design errors* as well as the *electrical errors* are two major source of failure in first-silicon.

A design error results from designer mistakes in implementing the high level specifications. To manifest these errors, we need to exercise corner cases of a design. Such errors may cause failures in intercommunications among cores; consequently, monitoring transactions among cores (maters/slaves) in a bus-based system contributes to their detection.

An electrical error emanates from transient errors inside storage elements of a system. Such increases in rate of electrical errors are due to crosstalk, low voltage levels, high frequency and small noise margins. As we will illustrate later on monitoring bus transaction with our proposed unit, we can not only detect and bypass design errors but also mitigate the severe consequences of electrical errors.

Conventional debug methods and tools tend to focus more on the computational part of a system, e. g. the processor and its interaction with main memory. However, many SOCs contain processing cores, e.g., Digital Signal Processors (DSPs) and interfaces and a considerable complexity resides in the interactions between these processors and other system components. Hence, a platform for post-silicon validation should enable not only functional validation of a prototype but it also has to guarantee that the prototype can realize the required compatibility with other blocks. Fig. 1 illustrates some of the components that are usually part of a post silicon validation test board [2]. The prototype under validation has to plug into this platform. Other peripherals expected to have interconnection with such prototype should also get connected through the proper bus. For instance, in the realm of microprocessors, a platform for post-silicon validation must enable the processor to start executing an application that resides in memory. Moreover, to emulate a real working condition, commercially available peripherals must be able to interrupt a processor to communicate with it. As Fig. 1 shows, a programmable interrupt generator can play the role of a peripheral by interrupting a processor in a suitable manner.

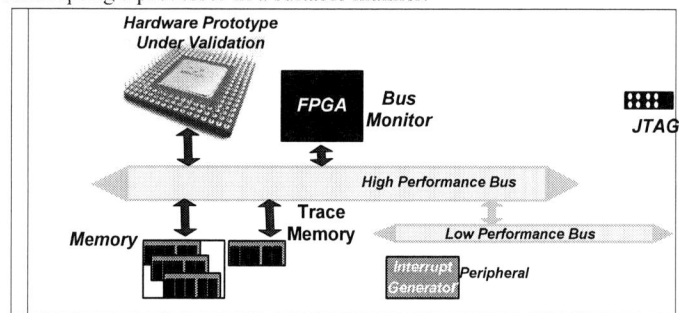

Fig. 1 Post-silicon validation environment

In this paper, we propose a distributed Advanced Extensible Interface (AXI) based platform for post silicon validation. A Local Debugging Unit (LDU) and Shared Debugging Unit (SDU) are the main components of our platform. An LDU is placed inside the AXI Master Interface that connects the hardware prototype and other masters to the AXI Bus.

LDUs are distributed among AXI Master Interfaces that connect master devices to the bus. Armed with two levels of hardware checkers that are generated automatically using MBAC checker generator [14], [15] in a way described as in [16], LDUs trace the transactions and signals issued by the hardware prototype and other modules connected to the bus, and detect the undesired conditions. The concept of assertions is taken from Assertion-Based Verification (ABV), the well-accepted pre-silicon verification techniques. LDUs are connected to the central SDU and transmit their debug traces to that module. SDU combines debug traces from different LDUs, and it schedules properly the extracted debug data from several masters to an external trace memory. We implant the proposed SDU inside an AXI configurable interconnect. The proposed post-silicon validation provides mechanisms to detect and bypass severe faulty conditions such as a deadlock resulting from design error or electrical errors.

Moreover, such a platform enables other master devices to access trace memory in an online manner; therefore, online debugging is achievable. To the best of our knowledge this works is the first study that provides the comprehensive distributed debug environment based on the AXI protocol.

II. RELATED WORK

The main goal of Design for Debug (DfD) methods is to increase real-time observability and controllability of internal signals during post-silicon validation [1]. We can categorize broadly previous works on the SoC debug into two groups: 1) techniques that enhance observability of signals inside a core and 2) DfDs that enable monitoring inter cores communication.

The previous studies belonging to the first group mainly strive to employ conventional test techniques and resources available, such as scan chains and Test Access Mechanism (TAM) to achieve the required real-time observability [13], [19]. Once the specific trigger or hardware checker fires, internal state elements of a system are captured in parallel using available scan chains; subsequently, the captured data is offloaded serially utilizing scan-out operation. Finally, post processing algorithms analyze offloaded data. Such consecutive stops and resumptions are no longer practical for debugging of a complex system [1]. Another DfD technique dedicated to improving real time observability of signals inside the core is Embedded Logic Analyzer (ELA); ELA utilizes on-chip trace buffers and trigger units to capture signals in real-time [19]. In [6] a use of Assertion checkers in the context of wireless systems has been considered; however, authors provide no mechanism for the placement of the checkers inside a system. In [4][5] authors provide a hybrid HW/SW mechanism to reduce the effect of electrical errors in the SoCs that use Real-Time Operating Systems (RTOS). Further, synthesis of assertions that provides debug functionality was considered in [7]. Such methods mainly focus on debugging the functionality of a SoC, without any emphasis on communication. However, as more cores with various communication protocols constitute modern SoCs it is required to apply debug beyond the functionality check of single core and create a new debug mechanism.

Regarding debugging of inter-core communication, we have to consider bus protocols, handshaking mechanism, blocking and no blocking transaction etc. In [8] authors review the benefits of debugging and diagnosing at transaction level; Authors propose a debug system for network-on-chip (NoC) communications in [9]. A distributed performance analysis mechanism in an AXI-based system is introduced in [20]; however, their proposed environment cannot be utilized for debug and validation. In [15], we investigate a method for clustering assertion checkers inside the bus; however, there was no mechanism to trace transaction inside bus and detect and bypass undesired condition during the debug.

In [4] authors propose a method to raise the debug abstraction level of communication to transaction level in post-silicon validation. They utilize an offline method to find the erroneous sequences of transactions. However, considering the fact that as soon as the first severe failure such as a deadlock occurs, the system cannot proceed further. Therefore, there is a need for an online monitoring unit that extracts and stores erroneous transactions, thereafter takes proper measures concerning the failures. As we will see later on, such failures resulting from design errors and electrical errors are typical in new complex buses. In [10] authors explore the effect of 1-bit transient error on AXI-based interconnect. Here, we consider effect of both design errors and transient errors. Our proposed post-silicon validation provides efficient mechanisms to detect, store and bypass severe faulty conditions such as deadlock and live lock result from design error or electrical errors. Moreover, we relax the requirements of having transaction level assertions by utilizing SystemVerilog Assertions (SVA) language and the MBAC checker generator [15].

III. PRELIMINARIES

In this section, we briefly review a few concepts that will be used during the paper.

1.1 Assertions and Checkers

Assertion is a statement that indicates how a given circuit should behave under different circumstances. Assertions represent a complex range of behaviors. Assertion-Based Verification (ABV) is one of the most practical and efficient RTL verification techniques for pre-silicon verification. System designers are able to define both expected and unexpected behaviors of a design using the temporal logic and the extended regular expressions using assertion languages such as PSL (Property Specification Language, IEEE 1850 standard) and the SVA. For example, the PSL assertion below specifies that once the request signal goes high, the arbiter is expected to grant the bus to the client within three clock cycles. The client must also keep its request signal active until it receives the 'grant' signal. This assertion will fire if any of these conditions not happen.

- assert always ({$rose(req)} |=>{req[*0:3] ; req&grant});

The |=> operator is a temporal implication, with pre and post conditions appearing as left- and right-hand arguments, respectively. $rose(b) as a precondition evaluates to true in the case of any changes in "b" from false to true. In this example, the post-condition is a regular expression consisting of a temporal concatenation ";" of two sub-expressions, the left of which contains a repetition range and the right expression is a Boolean expression. Assertions once converted to RTL module usually called checker.

Here, we use MBAC tool to generate checkers [15]. First, MBAC generates an automaton with respect to each assertion statement; thereby, a set of automata for properties and sequences are generated. A generated automaton is a directed graph, where vertices are states, and edges among states shows the conditions for transitions among them. Fig. 2 shows the generated automata from the previous mentioned PSL assertion. It has been shown in [15] that the properties in PSL and SVA can be converted in a recursive manner to an equivalent finite automaton. Assertion violation is activated whenever an automaton representing an assertion reaches its final (failure) state.

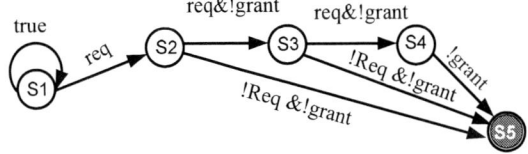

Fig. 2 Generated automaton from the PSL statement

In this paper, we model a transaction level assertion with two assertion statements. The first assertion represents the sequences of events that maintain non-periodic trigger signals for activation of the second assertion. The second assertion follows the sequences of transactions that cause erroneous status such as deadlock or livelock.

1.2 AXI Protocol

The AXI bus protocol is an enhanced bus protocol of the existing Advanced High-performance Bus (AHB). AXI is targeted at high-performance, high-frequency system designs. AXI protocol has five independent unidirectional channels that carry the address/control and data. Each channel uses a two-way valid and ready handshake mechanism. The five independents channels are the Address Read (AR) channel, Address Write (AW) channel, Read Data (RD) channel, Write Data (WD) channel, and write response channel (B). AW and AR channel convey the address and control of write and read transactions. The control signals of such channels describe the nature of the read and write transactions.

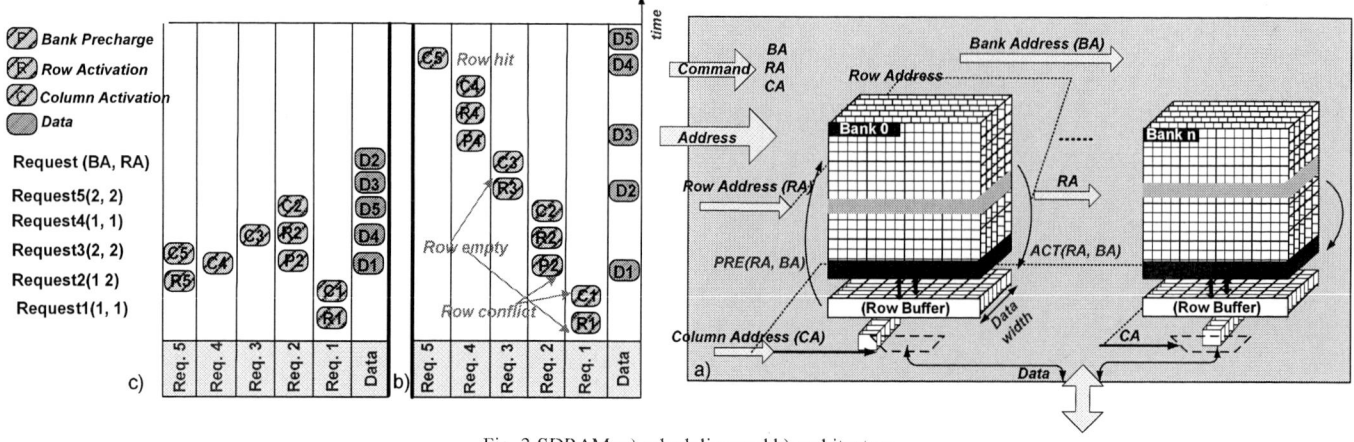

Fig. 3 SDRAM: a) scheduling and b) architecture

A transaction can be a burst of a different length, or it can be atomic. A burst is composed of a number of transfers whose length is defined. The data is transferred between master and slave, and vice versa using WD and RD channels respectively. Write response channel (B) allows a salve to signal completion of the write transaction or an error. One of the features of AXI is a burst transaction with only the start address issued. The split transaction AXI protocol enables out-of-order transaction completion; it provides a "transaction ID" field assigned to each transaction. Transactions from the same master IP core, but with different IDs have no ordering restriction while transactions with the same ID must be completed in order. Out-of-order transaction transactions completion improves system performance in that it allows a complex slave like memory return data out-of order. For instance, a data item of an earlier access might be available from an internal buffer sooner that of later access (temporal locality).

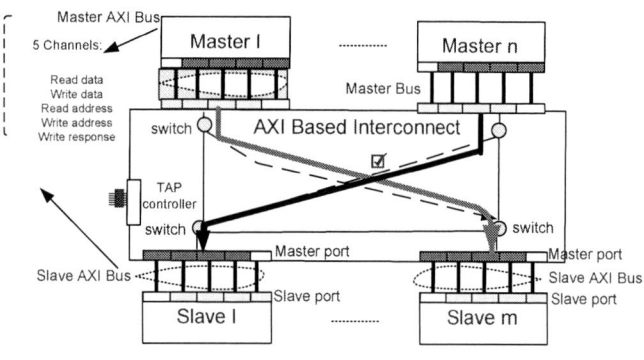

Fig. 4 An AXI-based system

Out of order execution and interleaving are the two main features of AXI bus that provide high throughput, but also increase the susceptibility of a system to bugs.

1.3 SDRAM

Design errors and bugs in a hardware prototype may lead to deadlock, livelock or other undesired conditions inside a bus. To understand how these failures take place, we explain the rescheduling mechanisms of the SDRAM in the sequel. SDRAM as one of the common complex slaves provides high bandwidth by executing memory requests in parallel. Memory requests are served by memory controller. It issues the required commands corresponding to each request and schedules them on SDRAM buses. By providing request reordering, the scheduler can provide higher bandwidth throughout.

SDRAM has a 3-D structure that involves banks, rows, and columns. Having multiple independent banks in such a 3-D structure enables memory scheduler to service serial requests in parallel; moreover, commands to different banks can be pipelined. Address

bus inside SDRAM is divided into three parts: Bank Address (BA), Row Address (RA) and Column Address (CA)[12][3]. BA specifies one of the banks inside an SDRAM, while RA and CA points to the particular row and column on that bank. SDRAM controller works with three commands: Activate (ACT), Read/Write(R/W) and Precharge (PRE). Taking the RA and BA, the ACT command activates the particular row (RA) inside the bank (BA) and places that row on the row buffer of that bank after tRCD. The row buffer serves as a cache to reduce the latency of subsequent accesses to that row. PRE command gets the BA and after tPR copy the content of row buffer to its related row in the bank, then it makes that bank idle. The R/W commands can be executed only after a bank is activated and the row buffer contains the particular row that they want to access. After either read latency called column access strobe (CAS) latency (CL) or write latency (WL), successive data go from or to SDRAM. In this paper, we assume timing constraint of DDR2-512MB which is 2-2-2 (tRP-tRCD-tCL) [22]. Latency of a memory request depends on whether the requested row is in the row buffer of the bank or not, a memory request could be a row hit, row conflict or row empty with different latencies. Memory performance suffers from bank conflict and data contention. To improve its performance and decrease bank conflict, memory scheduler should reschedule the requests.

As Fig. 4 (b) and (c) illustrates scheduling affects the performance. We consider five memory requests. As it was shown in Fig. 4 (b), Request 1 and 3 are row empties, and request 1 and 2 are row conflicts, and request 5 and 4 are row hit.

If the memory controller schedules these memory requests in order, it will take 24 memory cycles to complete them Fig. 4 (b). However, in Fig. 4(c) the same five requests are scheduled out of order. As can be seen, request 5 is scheduled before request 2, 3 and 4. In addition, request 4 is pipelined after request 5, called bank interleaving, since it has the different bank address from the bank address of request 4. As a result, only 14 memory cycles are needed to complete the four requests. Therefore, out of order memory scheduling provides better memory utilization 5/14= 35% over 5(data)/24(cycle) = 20%. However, as we will see out of order memory scheduling leads to increase in susceptibility of the memory controller to design errors and bug.

IV.   FAILURES IN AXI BUS

In this section, we introduce common errors that may happen during the validation of the hardware prototype. These problems that target communications among cores are mainly result of a design or electrical errors in bus interfaces (masters/slaves), memory controller and microprocessors. It is responsibility of LDU and SDU to detect these errors and apply a proper measure to relax their destructive effects and expedite root cause analysis of errors. Such errors fall into four categories: race condition, deadlock, livelock and data inconsistency.

a. DEADLOCK

Both design errors and electrical errors might lead to a deadlock in AXI. It is obvious that a debug unit should detect the deadlock, dismiss it, and let a system proceed with its operation; otherwise, the system under debug should get restarted. As it is illustrated in Fig. 6, supposing that at time 4, master1 issues transaction A1_2, which is the burst write. Meanwhile, due to a temporary error, the value of "LEN", which indicates a number of transfers in write transaction, gets changed. As a consequence of such an error, once slave2 starts receiving data, it expects either more or less than the number of actual data transfer. As soon as slave2 receives the expected number of data ("LEN"), it sends an acknowledgement on the B channel.

However, in the first scenario illustrated in Fig. 6, where a slave expects more data transfer than the actual one, not only slave2 gets block at time 9, while waiting to receive more data, but also master1 which is waiting to receive an acknowledgement on its B channel can no longer proceed.

The second scenario which slave2 expects less data transfers than the actual one causes Data-inconsistency which we will explain in section D. Deadlocks not only emanate from intermittent errors but it might also results from design errors. For instance, as it was shown in Fig.6, master 2 has issued following in order transactions: B2, C2 and D2 at time 2, 3 and 4. Assuming there is a design error inside the reordering unit of Mem 1. Consequently, Mem1 at time 7 provide data related to read transaction D2 sooner than that of C2. However, Master2 while expecting to receive C2 stops data D2 at time 7 that has the same transaction ID as C2. Thereafter data C2 cannot advance due to the blocked D2, which is the case of deadlock.

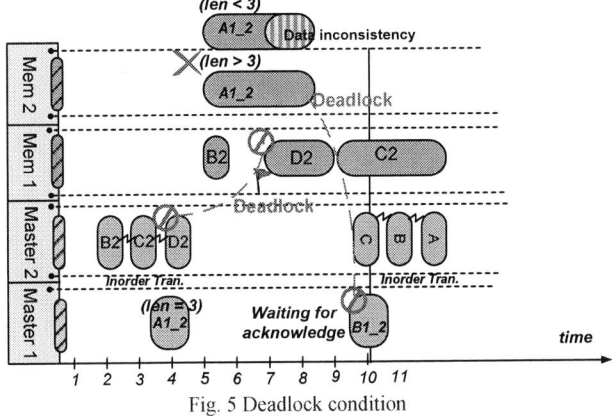

Fig. 5 Deadlock condition

b. RACE CONDITION

Race condition usually occurs when multiple masters are writing to the same place with overlapped or consecutive transactions, or a master is to process two in-order write transactions (not necessarily consecutive) that has the same address. Any undesirable changes in execution order of such transactions may cause inconsistency in the content of memory (memory corruption). For example temporarily errors on the transaction ID field of one of the sequence of in-order write transactions makes memory controller consider that transaction out-of order; subsequently, random execution orders of these transaction defines the data consistency. As it was shown in Fig.7, due to the temporarily error in transaction ID of C2 or D2, memory controller at time 11might schedule them differently. Their execution order may change the result of the read transaction D2 at time 11 in that it can get the newer value of D2 provided that write transaction D2 on Mem1 execute at time 12. Design errors might also create race condition. For example, if master side interface without considering the dependency mistakenly assigns transaction ID to the read or write transactions the arbitrary scheduling of memory scheduler determine whether read or write transactions executed properly.

c. LIVELOCK

Livelock is a situation that a system performs continuously the same sequence of operations without any changes in the status of that system. Livelock is different from deadlock in that once deadlock occurs in a system, such system no longer can proceed. For example, in Fig. 7 Master1 issues the A1_2 write transaction at time 3 and starts sending data. Once it finalizes its data transmission at time 7, it waits to receive an acknowledgement from its Channel B.

However, because of a design error in the memory controller, Mem2 is unable to detect completion of data transfer; consequently, it sends an error signal instead of an acknowledgement. Such a circumstance causes Mester1 initiates again its previous A1_2 write request at time 8 and the same scenario occurs continually.

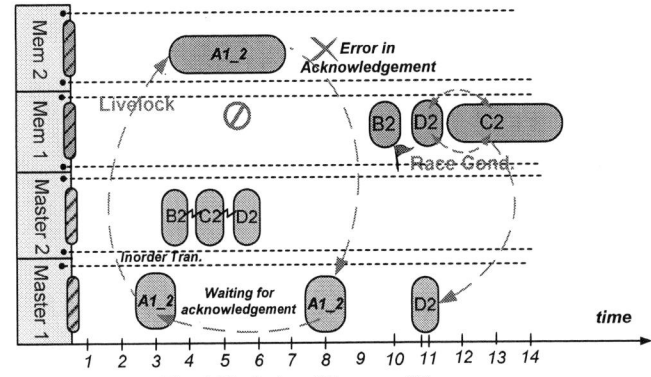

Fig. 6 Livelock and Race condition

d. DATA INCONSISTNCY

Data inconsistency could cause failure later on, if the data is used at some point. Considering the previous example, which deals with the temporarily error on the value of "LEN", As fig.6 illustrates, If Mem2 samples value of "LEN" lower than the actual value at time 5, the memory controller of Meme2 will have problem generating memory address to store the data coming from Master 2. The data either will be restored in the same place or they will be restored in other parts of the memory. Design errors might also cause data inconsistency. For example, design errors that lead to race condition in Fig.7 might cause data inconsistency for the read transaction D2 at time 11.

V. PROPOSED DEBUG PLATFORM

Our proposed AXI based debug environment consists of two main units: Local Debug Unit (LDU) and Shared Debug Unit (SDU). In the sequel, we will explain the architecture of LDU an SDU.

a. LOCAL DEBUGING UNIT (LDU)

Architecture of the proposed AXI master interface is illustrated in Fig. 7. A master that initiates several in order requests expect to receive requested data concerning to a read transaction or an acknowledgment to a write transaction in order.

However, such in-order transactions might get served out of order by slaves e.g. memories. For instance, as we explained before, SDRAM controller as a typical slave module performs transactions reordering to gain high performance and lower latency.

An AXI master relies on the AXI-based memory controller to receive its in-order requests properly. However, in case of bugs inside the memory controllers or other slave interfaces, a master no longer can communicate with its memory or other slaves due to the incident of the one of the failures, which is mentioned in previous section.

Not only validation process might get stuck because of bug in slave side, but bugs inside the master modules also hamper the validation process.

Our proposed debug mechanism (LDU and SDU) takes advantages of the observability of transactions inside the AXI

Interface. LDU monitors follow of transactions inside the wrapper. By placing LDU inside the AXI Interface, we maintain resource efficiency in that many components are shared between the units.

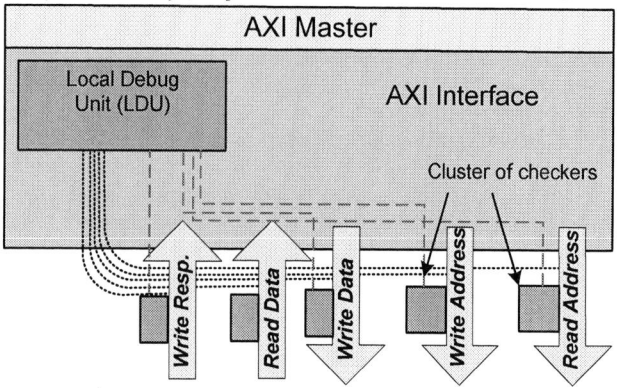

Fig. 7 AXI Interface Master side and LDU

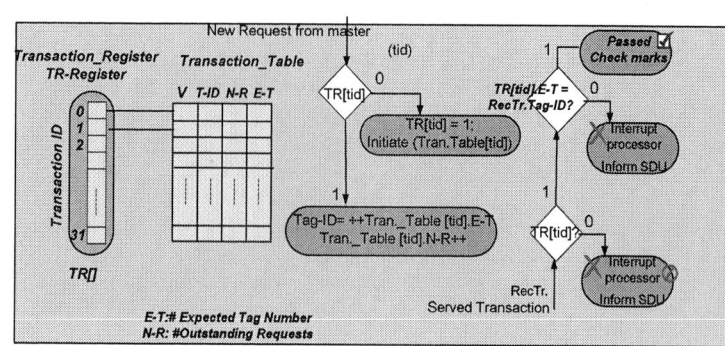

Fig. 8 LDU transaction control

The proposed LDU consists of three main units. 1) Transaction Register (TR) 2) Transaction Table 3) Cluster of checkers. TR keeps track of all the in-order transactions. The master shows its intention for having several in-order requests (transactions) by assigning the same transaction ID to them. It is slave responsibility to ensure such transactions which may complete out of order are issued in order to the master. Any deviation from such rules must be detected by LDU. To carry out its task, LDU assigns different Tag-IDs to these in-order transactions. TR is a 32-bit register, where each bit corresponds to one of the AXI transaction IDs. To record states of the outstanding transactions, LDU adopts a table called Transaction-Table [16]. Each row in this table corresponds to a request with the same transaction ID. Every row involves valid tag (V), Transaction ID (T-ID), Number of outstanding transactions (N-T) and Expecting Tag number (E-T). As Fig. 8 shows a bit at the specific location inside the TR register has a direct correspondence with a request that master has already issued.

Once a master initiates a request on AXI read address, LDU compares T-ID of this new request with the concerning bit at TR register. If that bit was set before, LDU realizes that either one or more than one request with the same transaction ID have been issued before, otherwise LDU updates TR register by assigning 1 to that particular bit. In the first case, a new Tag_ID is produced by LDU and once request wants to get buffered at the next level, Tag_ID will be attached to it. Tag_ID indicates the order of the request within the transactions with the same transaction ID.

On the other hand, once a master receives either data concerning its read transaction or acknowledgment with respect to its write transaction, LDU again explores and updates the TR register and Transaction table. In fact it is a critical moment for LDU that can detect deadlock or other erroneous condition.

As Fig. 8 illustrates LDU first investigates whether the new served transaction has been issued before or not by looking at the TR register. Then if such a transaction has been initiated before it will be considered whether its Tag-ID matches with the Expected-tag-ID from the Transaction Table. If there is no such match, LDU will detect the deadlock, interrupt the processor and will send trace of data contain information of transaction-table as well as TR-register.

As Fig. 8 illustrates, once a newly served transaction passes all the checks without any inconsistencies, SDU updates Expected-Tag-ID field of the entry in the transaction-table that is associate with the Transaction-ID of the served request. Meanwhile, if the Expected-Tag-Id becomes zero, that means that all the in-order transactions with that particular Transaction-ID were served properly and the corresponding entry in the transaction table will be freed.

Not only an LDU monitors the connected hardware prototype at transaction level, but it also continuously checks actions of the attached hardware prototype such as handshaking and X-propagations by means of hardware checkers. We have used the method in [16] to cluster these checkers; however, here we placed each checker cluster as close as possible to the source of signals that need to get monitor. These assertions are selected among 83 assertion statements available in [21]. We have used MBAC [15] to convert these assertion to hardware checkers and then we synthesize them using Xilinx ISE 9. These cluster of checkers are connected to LDU, as soon as one of the checker inside each cluster get fired, LDU interrupts the processor and send the debug trace which involves both cluster-ID and checker-ID to SDU.

b. SHARED DEBUGGING UNIT (SDU)

Architecture of general SDU is illustrated in Fig. 9. SDU combines trace information from LDUs and schedules them to the trace memory. SDU traces failure patterns that might lead to erroneous conditions on bus. Since SDU is connected directly to the trace memory, in Fig. 9, we pictured both units. In fact SDU can be programmed to perform DMA based operation. In our future work, we plan to extend SDU with new functionalities that will enhance the proposed post-silicon validation environment.

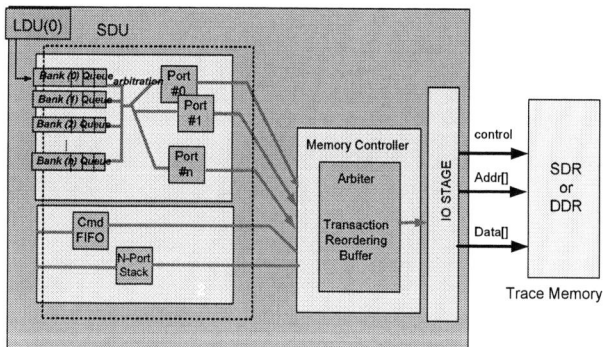

Fig. 9 SDU architecture

VI. EXPERIMENTAL RESULTS

To evaluate our proposed post-silicon validation environment, we have implemented an environment illustrated in Fig. 10. The platform consists of two instances of 16-bit SAYEH processors [23] with their cache controllers. These two instances of processor are connected to our proposed post-silicon validation using an AXI bus. We utilized a memory controller module from Gaisler ip-cores[24].

The interrupt generator unit is periodically asserted during the test execution. Interrupt Service Routine that resides in memory will be executed once the interrupt pin of a processor gets activated by Interrupt generator unit. Once an interrupt pin is activated, the time stamp and its status is stored to the previously assigned memory location. LDUs that are placed inside the AXI interface all monitor the local transactions. We intentionally inserted design errors inside the processors by changing the functionality of some parts of each core. We injected these faults inside SAYEH processor, memory controller and AXI interface. We assumed that LDU and SDU are fault-free. We consider design error at the RTL level. In other words, we injected errors in RTL specification of the processors and the memory controller. The errors are tuned to represent the corner cases. For example, inside the memory controller, we target memory scheduler and in the processor, we injected faults inside the forwarding unit, cache controller and interrupt controller such that the expected functionalities of these units get changed.

During the debug phase, the processors start running the program residing in the memory. Table 1 shows the result of synthesizing one SAYEH processor on Virtex IV family of Xilinx FPGAs.

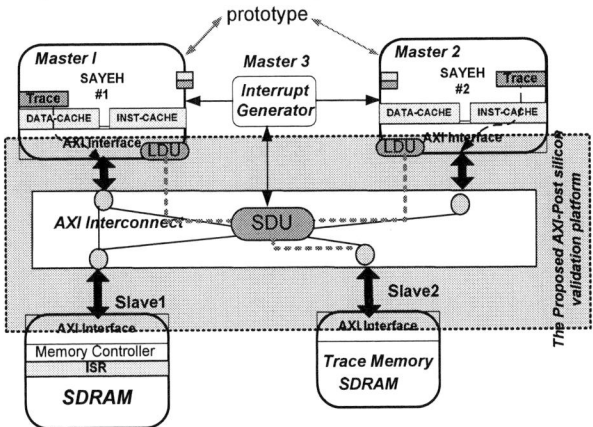

Fig. 10 our experimental environment

Table 1

32 bits Architecture	CLB SLICE	Utilization
SDU	290	4.70%
LDU (45 checkers)	929	7.57%
AXI Interface	271	4.43%
SAYEH with Cache and Interrupt	1162	18.94%

During the validation of faulty hardware prototype, deadlock never happened in our proposed AXI–based validation platform. It turned out that 94% of the errors inside the memory controller have been detected using the LDU; moreover, 87% of injected errors the AXI interface inside have been detected. Fig. 11 illustrates the patterns of failures inside the proposed validation platform system. As it was illustrated in this figure 40% and 35% of design errors lead to data inconsistency and deadlock respectively.

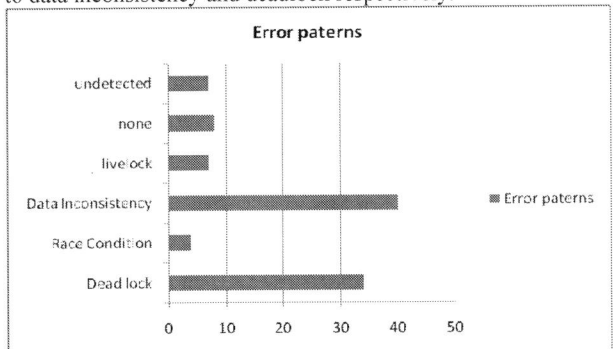

Fig. 11 error patterns

VII. CONCLUSIONS

We presented in this paper an AXI-based post-silicon validation platform. The proposed post-silicon validation provides mechanisms to detect and bypass severe faulty conditions such as deadlocks resulting from design error or electrical errors. Plus, our platform enabled online debugging by letting other master devices access trace memory in an online manner.

References:

[1] M. Abramovici, P. Bradley, K. Dwarakanath, P. Levin, G. Memmi, and D. Miller, "A reconfigurable design-for-debug infrastructure for SoC," *Proc. 43rd Design Automation Conference* (43rd DAC), pp. 7-12, 2006.

[2] V. Bertacco, "Post-silicon debugging for multi-core designs," in *Proceedings of 15th Asia and South Pacific Design Automation Conference* (ASP-DAC), pp. 255 – 258, 2010.

[3] M. Daneshtalab, M. Ebrahimi, P. Liljeberg, J. Plosila, H. Tenhunen, "A Low-Latency and Memory-Efficient On-chip Network," *Proc. Intl. Symposium on Networks-on-Chip (NOCS)*, pp. 99 – 106, 2010.

[4] M. H. Neishaburi, M. R. Kakoee, M. Daneshtalab and S. Safari, "HW/SW architecture for soft-error cancellation in real-time operating system," *IEICE Electron. Express*, Vol. 4, No. 23, pp.755-761, (2007)

[5] M. H. Neishaburi, M.R. Kakoee, M. Daneshtalab, S. Safari, Z. Navabi, "A HW/SW Architecture to Reduce the Effects of Soft-Errors in Real-Time Operating System Services," *Proceedings of DDECS 2007*, pp. 247-250, 2007.

[6] V. Kallankara, M. H. Neishaburi, Z. Zilic and K. Radecka, "Using Assertions for Wireless System Monitoring and Debugging," *Proc. of IEEE International NEWCAS Conference*, pp. 401-404, 2010.

[7] M. Boulé, J-S. Chenard and Z. Zilic, "Adding Debug Enhancements to Assertion Checkers for Hardware Emulation and Silicon Debug", *Proceedings of IEEE International Conference on Computer Design*, ICCD 06, pp. 294-299, Oct. 2006.

[8] A.M. Gharehbaghi, M. Fujita, "Transaction-based debugging of system-on-chips with patterns." *Proceedings of IEEE International Conference on Computer Design* (ICCD), pp. 186-192, 2009.

[9] K. Goossens, B. Vermeulen, R. van Steeden and M. Bennebroek, "Transaction-Based Communication-Centric Debug," *Proceedings of Intl. Symposium on Networks on Chip*, NOCS'07, pp. 95-106, 2007.

[10] W. Kwon, S.Yoo, J. Um; S.Jeong, "In-network reorder buffer to improve overall NoC performance while resolving the in-order requirement problem," *Proceedings of Design, Automation & Test in Europe Conference* DATE '09, pp. 1058 - 1063, 2009.

[11] D. Graham, P. Strid, S. Roy, F. Rodriguez, "A low-tech solution to avoid the severe impact of transient errors on the IP interconnect," in *proceeding of IEEE/IFIP International Conference on Dependable Systems & Networks* (DSN), pp 478-483, 2009.

[12] W. Jang, D. Z Pan, "An SDRAM-Aware Router for Networks-on-Chip," *IEEE Transactions on Computer-Aided Design of Integrated Circuits and Systems* (TCAD), Volume: 29 , Issue: 10, pp. 1572 - 1585, 2010

[13] B. Vermeulen and S. K. Goel, "Design for debug: catching design errors in digital chips," *IEEE Design & Test of Computers*, vol. 19, pp. 35-43, 2002.

[14] M. Boulé and Z. Zilic, "Efficient Automata-Based Assertion-Checker Synthesis of SEREs for Hardware Emulation", *Proc. Asia South Pacific Design Automation Conference*, ASP-DAC2007, pp. 324-329, 2007.

[15] M. Boule, and Z. Zilic, *Generating Hardware Assertion Checkers: for Hardware Verification, Emulation, Post-Fabrication Debugging and On-Line Monitoring.* Springer, 2008.

[16] M. H Neishaburi, Z. Zilic, "Reliability aware NoC router architecture using input channel buffer sharing," *Proceedings of Great Lake Symposium on VLSI* (GLSVLSI), pp 511-516, 2009.

[17] S. Rixner, W. J. Dally, U. J. Kapasi, P. Mattson, and J. D. Owens, "Memory access scheduling," *Proc. of ISCA'00*, pp. 128-138, US, 2000.

[18] M. H. Neishaburi and Z. Zilic, "Enabling efficient post-silicon debug by clustering of hardware-assertions," *Proceedings of IEEE Design, Automation & Test in Europe (DATE)*, pp. 985 – 988, 2010.

[19] H. F. Ko, A. B. Kinsman and N. Nicolici, "Distributed Embedded Logic Analysis for Post-Silicon Validation of SoC," *Proceedings of IEEE International Test Conference (ITC)*, 10 pages , 2008.

[20] H.-m. Kyung et al. Design and Implementation of Performance Analysis Unit (PAU) for AXI-based multi-core system on a chip (SoC), *Microprocess. Microsyst.* (2010), doi:10.1017/j.micpro.2010.03.001.

[21] ARM AMBA 3 specification and assertions. http://www.arm.com/products/solutions/axi_spec.html.

[22] Micron Technology, *Micron 512Mb: x4, x8, x16 DDR2 SDRAM Datasheet*, 2006.

[23] Z. Navabi, *"Digital Design and Implementation with Field Programmable Devices,"* Springer, May 2004.

[24] *Gaisler IP Cores, http://www.gaisler.com/products/grlib/, 2009.*

Efficient Trace Data Compression using Statically Selected Dictionary

Kanad Basu and Prabhat Mishra
Computer and Information Science and Engineering
University of Florida, Gainesville FL 32611-6120, USA
email: {kbasu, prabhat}@cise.ufl.edu.

Abstract

Post-silicon validation and debug have gained importance in recent years to track down errors that have escaped the pre-silicon phase. Limited observability of internal signals during post-silicon debug necessitates the storage of signal states in real time. Trace buffers are used to store these states. To increase the debug observation window, it is essential to compress these trace signals, so that trace data over larger number of cycles can be stored in the trace buffer while keeping its size constant. In this paper, we propose several dictionary based compression techniques for trace data compression that takes account of the fact that the difference between golden and erroneous trace data is small. Therefore, the static dictionary selected based on golden trace data can provide notably better compression performance than the dynamic dictionaries selected in the current approaches. This will also significantly reduce the hardware overhead by reducing the dictionary size. Our experimental results demonstrate that our approach can provide up to 60% better compression compared to existing approaches, while reducing the architecture overhead by 84%.

I. Introduction

Increase in System-on-Chip (SoC) design complexity has made design verification an essential step before a chip is released in the market. However, with the time-to-market constraints, it is not possible to spend a lot of time in this process. The verification must be foolproof and should be accomplished as fast as possible. Pre-silicon debug has been a popular method for identifying the design errors before the chip is actually fabricated. Formal verification and extensive simulation are the two main approaches of pre-silicon verification. However, many physical parameters cannot be modeled correctly in early stages; as a result, pre-silicon validation and debug fail to actually detect and fix all the errors.

Manufacturing tests are used to detect manufacturing defects, like shorts and opens. However, they are not designed to capture any functional bugs that might have escaped the pre-silicon phase. Post-silicon validation techniques are used to capture these design errors that are present even after the chip is fabricated.

Post-silicon debug operates on a fabricated chip. Therefore, it is not possible to record the values of each and every internal signal. Various techniques have been proposed to observe some selected internal signals that will aid in the debug process [1], [2], [11]. These internal signal states are stored in a trace buffer during execution. During debug, the trace buffer content is analyzed, where it is matched against a set of ideal values to check for possible errors.

The trace buffer has two parameters, width and depth. Width refers to the total amount of debug data that can be stored per cycle,

while depth refers to the total number of cycles over which debug data is to be stored. In order to keep the trace buffer size constant, while increasing the amount of trace data that can be stored, either the depth or the width has to be compressed. An efficient lossless trace data compression technique is necessary which can provide both fast compression and high compression efficiency, with minor impact on the architecture overhead. Different techniques of trace data compression, either by depth [3], [4] or by width [5] have been proposed. Depth compression approaches deal with selecting the cycles where the data are erroneous, and store the data for only those cycles. The problem with depth compression algorithms [3], [4] is that they assume rerunning the same set of tests on the same system produce the same output, that is, repeatable; which may not always be true. It would be better if the optimized trace data can be generated by running the tests only once. Width compression [5] utilizes this observation and run the tests just once in order to generate the trace data and compress them.

We have proposed a lossless dictionary based width compression scheme that operates on real-time to compress the trace data. Unlike [5], our method chooses the dictionary offline, which provides a better compression performance as well as huge reduction in compression architecture overhead. Three different compression algorithms have been proposed to trade-off between compression performance and architecture overhead.

We have used *Compression Ratio*, defined in Equation 1, as a metric to measure the efficiency of a compression algorithm. A higher compression ratio implies a better compression.

$$Compression\ Ratio = \frac{Uncompresssed\ Data\ Size}{Compressed\ Data\ Size} \quad (1)$$

The rest of the paper is organized as follows. Section II describes the related works in post-silicon debug and compression field. Section III describes our trace data compression techniques. Section IV describes the experimental results. Section V concludes the paper.

II. Related Work

The primary problem concerning post-silicon debug is the limited observability of the internal signals. Once the values of signals are known, they are analyzed using some algorithms like failure propagation tracing [6] to identify the errors in the circuit.

In order to obtain real-time observability of the internal chip signals, the obvious choice would be to use chip pins to observe them [7]. However, this method is difficult to implement in the presence of high frequency internal clocks and limitation of the chip pins. As a result, trace buffer based debugging techniques have been proposed [8] to counter this problem. In these techniques, the Embedded Logic Analyzer (ELA) acquires the signal sample and stores them in an on-chip trace buffer. The data from the trace buffer

---

* This work was supported in part by NSF grants CAREER-0746261 and CNS-0915376.

is then loaded to a processor via low bandwidth interface, which analyzes them to find out the error in the circuit. Since the trace buffer is used only for debugging, it is better to keep its size as small as possible to reduce the overall cost, area and energy requirements. Thus, to increase the amount of data that can be stored in a trace buffer, trace compression techniques have been proposed [3], [4], [5], which compress the trace data before storing them into trace buffer.

In order to compress the trace data, a compression technique is required which can provide a good compression with very low architecture overhead. In general, complex statistical algorithms produce best possible compression but introduce huge area overhead. On the other hand, simple dictionary-based techniques introduce acceptable area overhead but sacrifices compression efficiency. The dictionary based compression scheme was improved to remember mismatches with the help of bitmasks by Seong et al. [9] for code compression in embedded systems. Bitmask based compressions were also used in [10] for test compression. In this paper, we have explored the standard dictionary based compression, bitmask-based compression, as well as a variation of the BSTW compression scheme to provide much better performance in terms of compression and area overhead compared to existing approaches.

III. Trace Data Compression

The existing compression techniques compress the trace data by selecting a dictionary dynamically during execution. This not only results in inferior compression performance (due to non-optimal dictionary selection), but also increases the architecture overhead. This section describes our trace data compression techniques. The overview is shown in Figure 1.

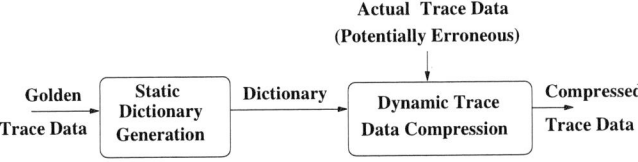

Fig. 1. Overview of our trace compression procedure

Our approach is based on an important observation. In any post-silicon debug environment, after the trace data is collected from the chip, it is validated by checking with a set of ideal trace data, that is obtained from a golden model. Since very few (2-5%) bugs actually remain to be tracked during the post-silicon debug phase, there are a few cycles which produce erroneous values [3], [4], that are different from the ideal ones. We utilize this information to design our approach. Since the difference between the ideal and the actual trace data is very small, the same dictionary applicable for ideal trace data compression can be reused for compression of the actual trace data. This takes care of the two problems by providing a better compression performance, and reducing the architecture overhead as well[1]. These compressed data are then read out through a channel to a debugger, where they are checked against the ideal trace data. Any discrepancy in the trace data is reported as error. As can be seen from our analysis in Section III-C, introduction of 2-5% error in trace data results in 2-6% penalty in compression performance, which is acceptable. It can be seen from the discussions in Section IV, even with the introduction of errors, our technique provides less compression penalty compared to the methods described by [5]. The remainder of this section describes our dictionary selection algorithms and also performs a theoretical analysis of the maximum

[1]no need to implement a dynamic dictionary selection algorithm

penalty possible when the dictionary from the ideal trace data is used to compress the actual (potentially erroneous) trace data.

A. Dictionary Selection Algorithms

We have explored three compression algorithms for compression of the trace data, namely Dictionary based compression (DC), Bitmask based compression (BMC) and fixed Dictionary MBSTW (fMBSTW) based compression. All these three techniques use a dictionary for compression. The dictionary selection is extremely vital since it would be reused to compress the actual trace data. We will now describe how the dictionaries are selected in order to achieve the maximum compression performance.

1) Dictionary based compression (DC): Algorithm 1 outlines the dictionary selection method. In a dictionary based compression, the main aim is to include in the dictionary all the unique entries which have maximum repetitions in the dataset. Therefore, the first step determines all the unique entries in the dataset. We then find the number of repetitions for each entry. The unique entries are sorted in a descending order of the number of repetitions. The entries with the highest number of repetitions are included in the dictionary.

Algorithm 1 Dictionary selection algorithm for DC

$M = Number\ of\ unique\ entries$
$N = Number\ of\ Dictionary\ Entries$
$DIC = Dictionary$
for *each entry in* M **do**
    Calculate the number of repetitions in the entire dataset
end for
Sort the M entries in decreasing order of repetition count
Include the first N entries in DIC

---

2) Bitmask Based Compression (BMC): The dictionary selection for bitmask based compression follows the same trend as the dictionary based compression, that is, select dictionary entries giving the maximum savings. However, there is a minor difference between the two. While savings for DC corresponds to just the repetitions, for BMC it includes those due to bitmask based matchings as well. Hence, the savings for each unique entry should be calculated based on the direct as well as bitmask based matches. The entries are then sorted in order of savings and included in the dictionary. The dictionary selection algorithm is shown in Algorithm 2.

Algorithm 2 Dictionary selection algorithm for BMC

$M = Number\ of\ unique\ entries$
$N = Number\ of\ Dictionary\ Entries$
$DIC = Dictionary$
for *each entry in* M **do**
    Calculate the savings due to repetition and bitmask based matching in the entire dataset
end for
Sort the M entries in decreasing order of total savings
Include the first N entries in DIC

---

3) Fixed Dictionary MBSTW compression (fMBSTW): The compression technique for fMBSTW algorithm follows the same technique as MBSTW compression [5]. The difference from MBSTW is that the dictionary is selected statically and the number of dictionary entries is limited. We would now explain the dictionary selection steps for fMBSTW in Algorithm 3. This algorithm is shown for a 2-fMBSTW (2-strings are encoded together, similar to 2-MBSTW). This can be further extended to 3-fMBSTW, where 3 strings are encoded together.

Algorithm 3 Dictionary selection algorithm for fMBSTW

> $M = No.\ of\ unique\ entries$
> $N = No.\ of\ Dictionary\ Entries$
> $DIC = Set\ of\ Dictionaries$
> $first\_entry = last\_entry = NULL$
> Create a 2-tuple for each pair of entries in M
> **for** each 2-tuple **do**
> > Calculate the savings across the entire dataset assuming only this tuple is in the dictionary
>
> **end for**
> Find the 2-tuple with the highest savings and add it to DIC
> $first\_entry$ = first entry of 2-tuple
> $last\_entry$ = last entry of 2-tuple
> $N = 2$
> **while** Size of DIC less than N **do**
> > find the 2-tuple that starts with $last\_entry$ and produces maximum savings
> > **if** such a 2-tuple exists **then**
> > > Include the 2-tuple in DIC, $N = N + 1$
> >
> > **else**
> > > Find any 2-tuple (not containing an entry already in DIC) which has the highest savings and include it in DIC
> > > $last\_entry$ = last entry of the 2-tuple, $N = N + 2$
> >
> > **end if**
>
> **end while**

Figure 2 shows an illustrative example for dictionary selection using Algorithm 3. In this example, the strings in the trace data are represented using p, q, r, s, t. The amount of savings for each 2-tuple is shown in Figure 2. We want to have a dictionary of size 4. As can be seen, the highest savings is obtained from the 2-tuple $< r, s >$. Both of these are now included in the dictionary. The $last\_entry$ is s here. Now, we proceed to see which 2-tuple with the first entry s has the maximum savings. $< s, p >$ is selected as the 2-tuple and included in the dictionary. When searching for the next 2-tuple, it is seen that $< p, r >$ gives the highest savings. However, r is already present in the dictionary. Hence, $< p, r >$ is avoided. The 2-tuple having the next highest savings is $< p, t >$. Therefore, t is selected for the dictionary. In this way, the dictionary is built up.

Tuple	\<p,q\>	\<p,r\>	\<p,s\>	\<p,t\>	\<q,p\>
Savings	12	25	4	17	11

Tuple	\<q,r\>	\<q,s\>	\<q,t\>	\<r,p\>	\<r,q\>
Savings	12	12	7	19	14

Tuple	\<r,s\>	\<r,t\>	\<s,p\>	\<s,q\>	\<s,r\>
Savings	28	21	15	9	12

Tuple	\<s,t\>	\<s,p\>	\<s,q\>	\<s,r\>	\<s,t\>
Savings	12	14	7	9	11

Final Dictionary:
r
s
p
t

Savings for each tuple   Final Dictionary

Fig. 2. Example of dictionary selection in fMBSTW

B. Dynamic Trace Data Compression

Our final goal is to debug the DUT, for which we need the trace data from it. Application of a set of tests produces the trace data from the DUT which are compressed to reduce the size of the trace buffer. The overview of the compression architecture is shown in

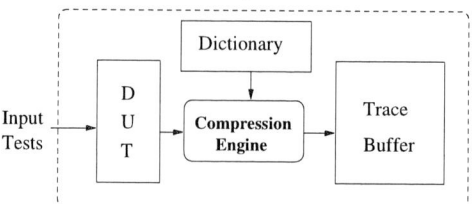

Fig. 3. Actual Trace Data Compression

Figure 3. As can be seen, the compression architecture consists of two parts, the dictionary and the actual compression engine. Depending on the design and associated constraints, a specific compression algorithm and its respective dictionary is used. For example, when BMC is most suitable for a design, the compression engine will have BMC in it and the dictionary will be the one selected for BMC. It should be noted, that the dictionary size is fixed here and not variable as in the case of dynamic dictionary selection [5]. Actually, [5] tried to include every single unique string in the dictionary. This increases the dictionary size, thereby introducing significant architecture overhead and also degrades the compression performance (since the number of bits used to index the dictionary increases with an increase in dictionary size). Our approach eliminates these disadvantages by keeping a limited number of profitable entries in the dictionary.

C. Performance Analysis with Erroneous Trace Data

Our approach is promising due to use of statically selected dictionary. However, this dictionary will be used to compress actual (potentially erroneous) trace data. This section analyzes our procedure and determines the performance degradation that may occur when the dictionary obtained from ideal trace data is used to compress the actual trace data. We have kept the trace data length constant at 32 bits. We introduce a term *compression penalty*, which is the ratio of the number of extra bits needed for compression when error is introduced, compared to the original trace data length. Obviously, a lower compression penalty signifies less number of bits needed to accommodate the error, and hence, a better compression performance.

$$Compression\ Penalty\ (CP) = \frac{Number\ of\ extra\ bits\ needed}{Size\ of\ original\ trace\ data}$$

We first analyze the compression penalty for DC and BMC. Next, similar analysis is performed for fMBSTW.

1) Compression Penalty for DC and BMC: We try to obtain the compression penalties for the two methods DC and BMC. In this section, we make two important observations.

Theorem 1: When statically selected dictionary (based on golden trace data) is used, the compression penalty is bounded by the percentage of error introduced in the actual trace data.

Proof: Let there be x strings in the original trace data. Let the percentage of error in case of actual trace data be l, expressed as a fraction ($l < 1$). The introduction of error changes $l \times x$ strings. In the worst case, all these $l \times x$ strings will be among the strings originally compressed, and these will now be uncompressed due to contamination. Let the number of bits required to compress the rest (that is $(1 - l) \times x$ strings) in the dataset be M^2. It should be noted that these strings are not affected due to error injection and hence, the value of M remains constant in both cases. Let the number of dictionary entries be 2^d, so that d bits are needed to represent the

---

[2]Some of the strings may be compressed, while the rest uncompressed

978-1-61284-657-6/11 $26.00 © 2011 IEEE

dictionary. The $l \times x$ strings were compressed in the ideal case using $(1 + d)$ bits each. If y_{ideal} be the number of bits after compression for the ideal trace data, it can be rewritten as,

$$y_{ideal} = M + l \times x \times (1 + d) \qquad (1)$$

Now, let's analyze the actual trace data. In the worst case, all of the $l \times x$ strings remain uncompressed. Each of these strings will require 33 bits[3] to be represented. The M bits required to represent the $(1 - l) \times x$ strings will remain the same. If y_{faulty} is the number of bits needed to represent the strings now, it can be represented as

$$y_{faulty} = M + l \times x \times (33) \qquad (2)$$

which implies,

$$y_{faulty} = y_{ideal} + l \times x \times (32 - d) \qquad (3)$$

Therefore, number of extra bits needed, represented as y_{extra}, is

$$y_{extra} = y_{faulty} - y_{ideal} = l \times x \times (32 - d) \qquad (4)$$

If CP_{DC} is the compression penalty for DC, then from the definition,

$$CP_{DC} = \frac{l \times (32 - d)}{32} \qquad (5)$$

As can be seen CP_{DC} is always less than l, and hence is bounded by it. ∎

For example, with 8 dictionary entries, we have $d = 3$, and assuming the error rate is 5%, (which is the maximum error rate in these scenarios [3], [4]), we get

$$CP_{DC} = 4\% \qquad (6)$$

Thus, we see that a very slight compression penalty is introduced in DC even in the worst case. It can be seen from Equation (5) that increase in dictionary size can lessen this degradation.

Theorem 2: Compared to the ideal case (if dictionary was selected using erroneous trace data), the compression penalty using statically selected dictionary (using golden trace data) will be bounded by the twice the percentage of error.

Proof: We would like to see if the actual trace data were compressed without the help of ideal dictionary, how much compression would be obtained. In this case, the dictionary entries might differ from the ideal dictionary. If n is the extra number of strings that can be compressed in the actual case and m is the number of strings that were compressed in the ideal case, then the total number of strings compressed are $m + n - l \times x$. It is obvious that the maximum value of n can be $l \times x$, otherwise, these new strings would have been compressed in case of ideal trace compression, that is, these new strings would have been represented in the ideal dictionary. Therefore, the maximum number of strings compressed is m, which is the same case as in golden trace data.

As an example, consider a hypothetical trace data set of 20 entries. Suppose we choose the best 2 entries in the dictionary, each of which can compress a total of 5 entries. Therefore the total number of compressed entries will be 10. Corresponding to the symbols described above, $x = 20$, $m = 10$ and $d = 2$. Let the error rate be 10%, that is $l = 0.1$. When error is introduced, the number of strings contaminated is $l \times x$, that is, 2. In the worst case, both these strings were part of m and are now left uncompressed due to errors. The number of compressed strings now are $m - l \times x$, which is equal to 8. Now, if we try to compress these erroneous data with a different set of dictionary, let the number of extra strings being compressed be n. It is obvious that if n is greater than 2, the new

[3]32 bits (original size), plus one bit to indicate not compressed

dictionary would have been selected in the first place, so that the value of m would be different. So, the maximum value of n is bounded by $l \times x$.

However, in the best case, these contaminated strings can be all compressed using some other entry, which is not part of the dictionary now. Let us reiterate our previous example to explain this. For example, all of the $l \times x$ contaminated entries can be compressed using some other entry. Now, if that entry has high enough frequency, it will be included in the dictionary. In this example, the maximum frequency (original, without contamination) that an entry can have is 5; otherwise, it would have been included in the original dictionary. Therefore, the maximum number of strings that can be compressed with the new dictionary is $m + l \times x$, that is, 12 in this case. Hence, the maximum number of strings that can be compressed extra using the dynamically selected dictionary is $(m + l \times x) - (m - l \times x)$, that is, $2 \times l \times x$, which means the difference in compression ratio should be $2 \times l$. Therefore, the difference in compression efficiency between the dictionary based on golden data and dictionary based on actual data, will be bounded by twice the error rate in the data. ∎

It can be noted that the analysis for BMC will be similar to DC. This is because, even for BMC, the worst case comes when some strings which were completely compressed (not using bitmasks) change to uncompressed due to error introduction.

2) Compression Penalty for fMBSTW: To find the compression penalty, we analyze the worst case condition for 2-fMBSTW here. The worst case scenario can be divided in two parts. The first part is similar to that of DC and BMC, that is, the worst part comes when some completely compressed strings become uncompressed. The second part of the condition is explained as follows. Suppose, two consecutive strings correspond to two consecutive dictionary entries a, b. Therefore, all the two strings will be compressed using the 11 prefix, followed by the dictionary entry corresponding to a. However, if either a or b gets contaminated by error, in the worst case, one of them is uncompressed and the other one gets compressed separately, which requires more bits to compress the trace data. We now investigate the compression penalty in this approach.

Let the error rate and the number of strings be l and x as before. Let d be the number of bits to represent the dictionary index. Therefore, the number of such strings changed is $l \times x$. Each of these string corresponds to a $< a, b >$ tuple which is broken due to perturbation. Before the introduction of error, the number of bits required to compress these is given as y_{ideal} in Equation (7)

$$y_{ideal} = l \times x \times (2 + d) \qquad (7)$$

Here, 2 bits are needed to represent the prefix 11 and d bits for the dictionary index of a. After perturbation (of b), in the worst case, a is independently compressed as single bits using the prefix 01[4]. Therefore, the number of bits needed to represent are $(36 + d)$[5]. There will be $l \times x$ such occurances. Therefore, the total number of bits needed to represent the erroneous tuples is given by y_{faulty} as

$$y_{faulty} = l \times x \times (36 + d) \qquad (8)$$

As before, let M be the number of bits required to compress the other $(1 - l) \times x$ strings. Since M is unchanged in either case, the number of extra bits needed, is given by

$$y_{extra} = y_{faulty} - y_{ideal} = l \times x \times 34 \qquad (9)$$

[4]as discussed in Section III-A.3

[5]$(2 + d) + (2 + 32)$, where $2 + d$ bits are needed to compress a and $2 + 32$ bits are needed to represent the uncompressed string b

Therefore, the compression penalty is given by,

$$CP_{fMBSTW} = \frac{l \times 34}{32} \quad (10)$$

With a 5% error rate we can see that,

$$CP_{fMBSTW} = 5.31\% \quad (11)$$

Thus, even with an introduction of 5% error, the compression penalty is small. These analysis will be later verified with experimental results in Section IV-C.

IV. Experiments

We have compared the compression performance of our approach with the algorithms proposed by Anis et al. [5] (MBSTW and WDLZW). We have also investigated our compression performance when the number of dictionary entries are varied. We have shown that our methods require much less compression architecture overhead compared to those in [5]. Finally, in Section IV-C, we have also analyzed the effect of introduction of errors on compression ratio and validated the equations developed in Section III-C. We have applied all the algorithms on the 5 largest ISCAS 89 benchmarks.

A. Compression Performance

First, we compare the compression performance of our algorithms with the algorithms in [5] using the traces obtained from ISCAS 89 benchmark circuits. The traces were obtained by following the approach outlined in [11]. The results are reported in Figure 4. We have fixed the dictionary entry to be 8 in each of the two compression algorithms, DC and BMC. For MBSTW, we have used the 2-MBSTW algorithm[6]. For the fMBSTW algorithm, the number of dictionary entries is floored to the nearest integer which is a power of 2. It can be seen that the fMBSTW approach works best in all cases except s38584. This is because the traces of s38584 has very less number of unique entries. As a result, even with 8 dictionary entries, a large portion of the circuit can be compressed using DC. DC works better than MBSTW in most cases and worse only in some cases (s9234 and s35932). The reason for this is the large number of unique entries in those trace data, which are effectively captured by MBSTW, but not by the 8-entry dictionary used in DC. If the number of bits needed to represent the compressed data is analyzed, it can be seen that fMBSTW provides up to 60% reduction in compressed data size compared to MBSTW and 70% compared to WDLZW. WDLZW provides worst performance for almost all the benchmarks. The high redundancy in the trace dataset is responsible for its somewhat good performance in s38584 and s38417.

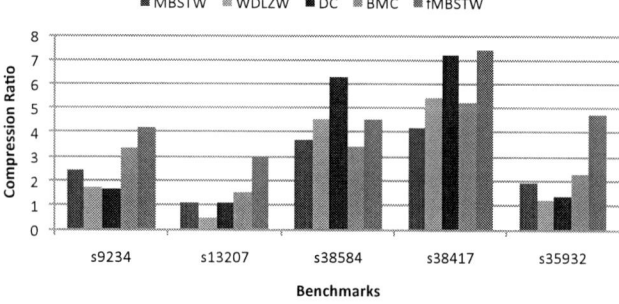

Fig. 4. Comparison of compression performance

Next, we vary the dictionary size of DC to see the effect on compression ratio. The results are shown in Figure 5. We have varied the number of dictionary entries from 8, 16, 32 and 64. As can be seen from Figure 5, the variation is not uniform for all the benchmarks. For s9234, s13207 and s35932, the compression ratio increases with increase in dictionary entries. On the other hand, for s38584 and s38417, increase in number of dictionary entries worsens the compression ratio and the optimal compression is achieved at 8 dictionary entries. Once we reach an optimal compression ratio, any increase in the number of dictionary entries will add to the total compressed data size both due to the increased number of entries in the dictionaries and increase in the number of bits representing the dictionary index.

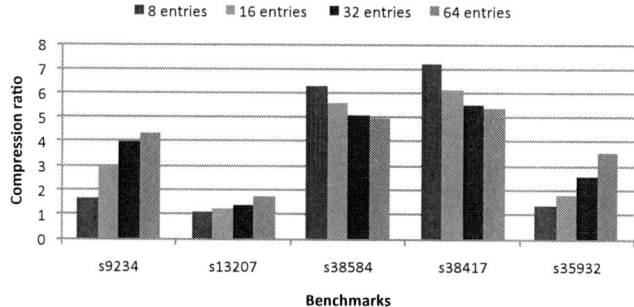

Fig. 5. Compression performance with dictionary entries

B. BRAM Requirement (Hardware Overhead)

The dictionary for compression has to be stored in on-chip 32-bit BRAMs. We have computed the total size of BRAMs needed for compression using each of these algorithms. Figure 6 compares the requirements for each of these approaches. It can be seen that since the dictionary size is always fixed (8 entries) for DC and BMC, the number of BRAMs required in these two algorithms is significantly less than any other approaches. WDLZW has the highest number of BRAM requirements since it captures all the double symbol repetitions (which worsens the compression performance). For fMBSTW, the number of BRAMs is kept floored to the nearest higher power of 2 for the number of unique entries in the stream[7]. From Figure 6, it can be seen that our two methods (DC and BMC) provides almost 96% less compression architecture overhead compared to MBSTW and almost 99% less than WDLZW.

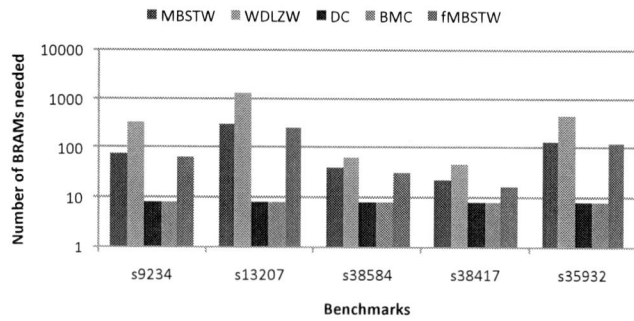

Fig. 6. BRAM requirements

It can be seen that there is a tradeoff between better compression ratio and lower architecture overhead. As can be seen from Figure 4 and Figure 6, either of the two techniques BMC or fMBSTW can be applied based on priority - BMC can be used for least area overhead (up to 96% reduction) with reasonable compression improvement (10%) compared to MBSTW, whereas fMBSTW should be used for best possible compression (up to 60%) while providing reasonable (up to 84%) reduction in BRAM requirement.

---

[6]Provides better performance than the 3-MBSTW algorithm

[7]Results in Figure 4 are also reported using this configuration

C. Compression Performance with Erroneous Trace Data

We now like to validate the analysis done in Section III-C. Errors have been inserted randomly at a rate of 2% to 10% in steps of 2% in the trace data, and the same is compressed using DC, BMC and fMBSTW. Figure 7 shows the comparison of compression penalty in DC with varying percentage of error. It can be seen that the change in compression penalty complies with Equation (5) in Section III-C. For example, putting a value of $l = 2\%$ in Equation (5) will result in a compression penalty of less than 2%, which matches in the figure for all the benchmarks. We have conducted similar experiments for

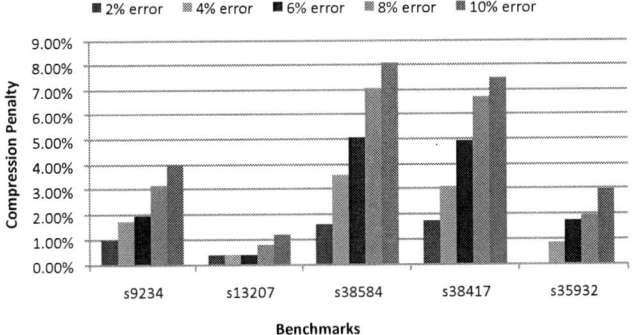

Fig. 7. Comparison of compression penalty for DC

BMC based compression technique as well. The results in Figure 8 shows that the compression penalty also follows Equation (5).

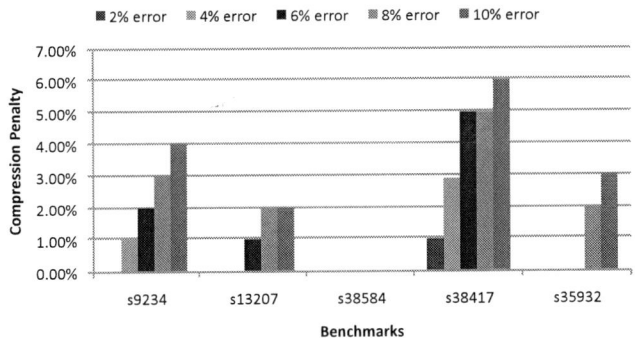

Fig. 8. Comparison of compression penalty for BMC

Now, we would like to verify the last part of the discussion in Section III-C, that is, the change in compression penalty with error rate for fMBSTW. We have conducted similar experiments and the results are shown in Figure 9.

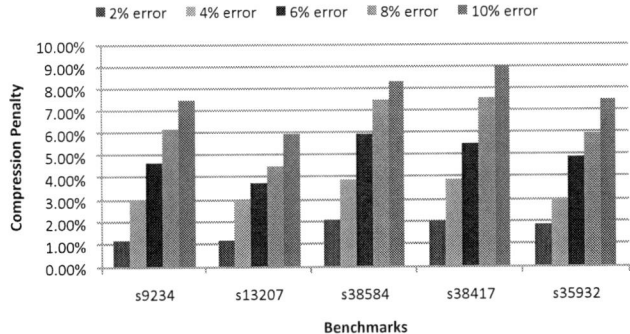

Fig. 9. Comparison of compression penalty for fMBSTW

An important observation here is that the change in penalty is sharper than the case of Figure 7 or Figure 8. This is quite obvious as per the discussion in Section III-C, since in fMBSTW, 2 strings

are affected when an error is introduced, whereas in DC or BMC, only 1 string is affected.

Finally, we compare how the introduction of errors affect the compression performance in cases of MBSTW and fMBSTW. The results are shown in Figure 10. We have introduced 2% error for every benchmark's trace data. It can be seen that the compression penalty obtained using fMBSTW is always less than MBSTW, the maximum difference being 4% for s38417. The reason for higher penalty in MBSTW is that if error gets introduced early, MBSTW cannot benefit from a profitable sequence. In summary, our approach (fMBSTW) will perform significantly better irrespective of the percentage of errors in the dataset.

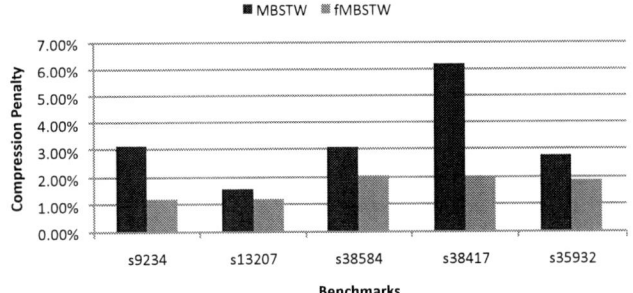

Fig. 10. Comparison of compression penalty

V. Conclusions

Post-silicon validation is extremely complex and time consuming in overall design methodology. To aid in debug, trace data obtained from the chip are stored in the trace buffer. However, the trace buffer size is limited due to area/cost constraints. Trace data compression schemes have been popular which deals with dynamic dictionary based compression that enables to store larger traces. We have proposed a trace data compression technique, which employs a statically computed dictionary. We have used three compression algorithms for compressing the trace data. Our approaches can produce up to 60% better compression performance, and reduce the compression architecture overhead up to 84% compared to best-known existing approaches.

References

[1] H. Ko and N. Nicolici, "Algorithms for state restoration and trace-signal selection for data acquisition in silicon debug," *IEEE TCAD*, vol. 28, no. 2, pp. 285–297, Feb. 2009.

[2] X. Liu and Q. Xu, "Trace signal selection for visibility enhancement in post-silicon validation," in *DATE*, 2009.

[3] J. Yang and N. Touba, "Expanding trace buffer observation window for in-system silicon debug through selective capture," in *VTS*, 2008.

[4] E. Anis and N. Nicolici, "Low cost debug architecture using lossy compression for silicon debug," in *DATE*, 2007, pp. 225–230.

[5] ——, "On using lossless compression of debug data in embedded logic analysis," in *ITC*, 2007, pp. 1–10.

[6] O. Caty, P. Dahlgren, and I. Bayraktaroglu, "Microprocessor silicon debug based on failure propagation tracing," in *ITC*, 2005, pp. 10–293.

[7] B. Vermeulen and S. Goel, "Design for debug: Catching design errors in digital chips," *IEEE Des. Test*, vol. 19(3), pp. 37–45, 2002.

[8] R. Leatherman and N. Stollon, "An embedded debugging architecture for socs," *IEEE Potentials*, vol. 24, no. 1, pp. 12–16, February 2005.

[9] S. Seong and P. Mishra, "Bitmask-based code compression for embedded systems," *IEEE TCAD*, vol. 27(4), pp. 673–685, April 2008.

[10] K. Basu and P. Mishra, "Test Data Compression Using Efficient Bitmask and Dictionary Selection Methods," *IEEE TVLSI*, vol. 18, no. 9, pp. 1277–1286, 2010.

[11] K. Basu and P. Mishra, "Efficient Trace Signal Selection for Post Silicon Validation and Debug," in *International Conference on VLSI Design*, 2011.

A Built-In Self-Test Scheme for the Post-Bond Test of TSVs in 3D ICs

Yu-Jen Huang, Jin-Fu Li
Department of Electrical Engineering
National Central University
Jhongli, Taiwan 320

Ji-Jan Chen*, Ding-Ming Kwai*, Yung-Fa Chou**, and Cheng-Wen Wu**
*Information and Communication Research Lab. (ICL)
Industrial Technology Research Institute (ITRI)
*Department of Electrical Engineering
National Tsing-Hua University
Hsinchu, Taiwan 310

Abstract—**Three-dimensional (3D) integration using through silicon via (TSV) has been widely acknowledged as one future integrated-circuit (IC) technology. A 3D IC including multiple dies connected with TSVs offers many benefits over current 2D ICs. However, the testing of 3D ICs is much more difficult than that of 2D ICs. In this paper, we propose a cost-effective built-in self-test circuit (BIST) to test TSVs of a 3D IC. The BIST scheme, arranging the TSVs into arrays similar to memory, has the features of low test/diagnosis time and low silicon area cost. Simulation results show that the area overhead of the BIST circuit implemented with $0.18\mu m$ CMOS technology for a 16×32 TSV array in which each TSV cell size is $45\mu m^2$ is 2.24%. Also, the BIST needs only 130 clock cycles to test the TSV array with stuck-at faults. In comparison with the IEEE 1500-based test approach, the BIST scheme can achieve 85.2% area cost and 93.6% test time reduction.**

I. INTRODUCTION

Three-dimensional (3D) integration using through silicon via (TSV) has been proposed as a very good alternative to cope with the challenges faced by the current 2D technology [1]. A 3D IC using TSV is implemented by stacking multiple dies which are vertically connected by TSVs. This enables that the global interconnects in the 3D chip can be shortened such that high performance improvement can be achieved. Furthermore, high bandwidth can be achieved due to the significant increase of IO interconnection density provided by the TSVs. In addition, the 3D integration technology provides many advantages over 2D integration technology, such as high functionality, low power, small form factor, and so on [2]. However, some challenges, e.g., in technology, yield improvement, thermal management, infrastructure construction, and so on, should be overcome before volume production of 3D ICs using TSV becomes possible [2].

Among these challenges, testing of 3D IC is a key issue [3]–[5]. Recently, several research works of 3D IC testing have been reported [6]–[16]. In [6]–[10], the authors proposed test optimization approaches to minimize the test cost of 3D ICs. In [11], testability of 3D random access memories and content addressable memories was explored. In [12], two effective test methods were proposed to test the TSVs of 3D ICs before bonding. The two test methods can be used to screen out defective TSVs in pre-bond phase to reduce the stacking yield loss. A structured and scalable test access

architecture for TSV-based 3D ICs was proposed in [13]. Two different test interfaces handle the test controlling of the 3D IC. An IEEE standard 1149.1-based and an IEEE standard 1500-based test interfaces are used for the bottom die and non-bottom dies, respectively. In [16], Chou *et al.* proposed a test integration method for handling the test operations of 3D ICs. An IEEE 1149.1-based and a modified IEEE 1149.1-based test interfaces control the design-for-testability circuits of the 3D IC. In [17], four analog self-test circuits were proposed to test the TSVs with pinhole defects. Leakage current of each self-test circuit is utilized to measure the TSV-to-substrate resistance, such that the short defect caused by a pinhole can be detected.

Clearly, the testing of TSVs is essential for 3D ICs. One straightforward test approach to cover the defects of TSVs between dies is to wrap the terminals of each die with wrapper cells (e.g., IEEE 1149.1 boundary scan register or IEEE 1500 wrapper boundary register). But, this approach results in high area overhead since the number of TSVs in a 3D IC is usually very huge. A cost-effective test approach thus is imperative. In this paper, we present a BIST scheme to test the TSVs of a 3D IC. The scheme can test and diagnose the defective TSVs with very short test time and very low test cost.

II. PRELIMINARY AND MOTIVATION

A. 3D IC Testing and TSV Defects

The test flow for the 3D IC typically consists of three testing phases [4]: pre-bond test, known-good stack (KGS) test, and final test (i.e., post-bond test). The pre-bond test is performed after the wafer fabrication. The quality of pre-bond test has a heavy impact on the final yield of 3D ICs. The KGS test is performed once a die is stacked. This can check weather the dies are damaged during the stacking process or not. If the faulty stacked dies are identified, then the remaining dies will not be stacked. This can avoid wasting good dies. Finally, the final test is performed when the stacking process is completed.

Several previous works have listed some possible defects in TSVs [12], [13]. TSV might be broken and have an open defect or might have voids or pinholes along the TSV sidewall during the fabrication or metal filling process and have an open defect or a short defect. Before wafers are stacked, wafer thinning is executed. TSVs might crack due to the stress during wafer stacking and bonding. Furthermore, an interconnection between two dies of a 3D IC may consist of the TSV, the microbump, and the redistribution layer (RDL). The post-bond test of TSVs should cover not only the defects of the TSVs, but also the defects of the microbumps and the RDL lines.

B. Motivation

Existing works deal with the post-bond test of TSVs by IEEE 1149.1 or IEEE 1500 test standards. Each TSV thus requires at least one register. Typically, the number of TSVs in a 3D IC is very huge. The IEEE 1149.1 and 1500-based test approaches thus result in high area cost. Furthermore, in the case of 3D stacking of random access memory, a RAM die is seldom wrapped by an IEEE 1149.1 or an IEEE 1500 test wrapper. Fig. 1 shows an exemplary 3D IC in which multiple memory dies are stacked on top of a processor die. Each RAM die encompasses a memory BIST (MBIST) circuit to perform the pre-bond and post-bond test of the memory dies. For the 3D RAMs, a cost-effective approach for testing TSVs is imperative. In this paper, we propose a BIST scheme to test and diagnose the TSVs of 3D ICs.

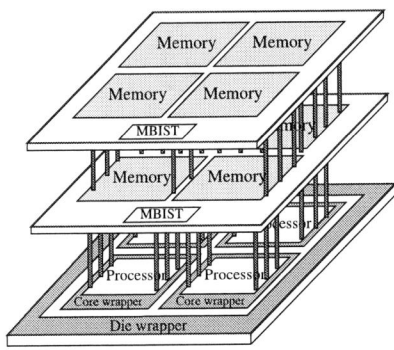

Fig. 1.   3D processor and memory stacking

III. PROPOSED TESTING SCHEME FOR TSV ARRAYS

A. Proposed BIST Architecture

For reducing the area cost of design-for-testability circuits, we design the proposed BIST circuit by considering that the TSVs are arranged in regular arrays in a 3D IC. As shown in Fig. 1, the TSVs between the RAM dies or the TSVs between the RAM and the processor dies are placed regularly. In case that the TSVs may not be placed regularly, for example, between logic dies, we can still connect them logically as in arrays. Fig 2 gives an example where the TSVs are randomly distributed on a die as in Fig. 2(a) and we group them into two logical arrays as in Fig. 2(b).

Fig. 3 shows a 4×4 TSV array in which 8 TSV signals are from Die 1 to Die 2 and 8 TSV signals are from Die 2 to Die 1. We define a TSV transmitting a signal to the other die as an outbound TSV and a TSV receiving a signal from the other die as an inbound TSV. A bidirectional TSVs in the functional mode can be regarded either as an inbound or an outbound TSV in the test mode. In other words, Die 1 has 8 outbound TSVs and 8 inbounds TSV, so does Die 2. We

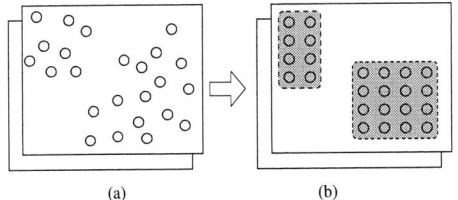

Fig. 2.   (a) physical TSVs in a 3D-IC (b) logical TSV arrays

configure 4 TSVs with same type in a row. Therefore, row 0 and row 1 both have 4 TSVs from Die 1 to Die 2; row 2 and row 3 both have 4 TSVs from Die 2 to Die 1.

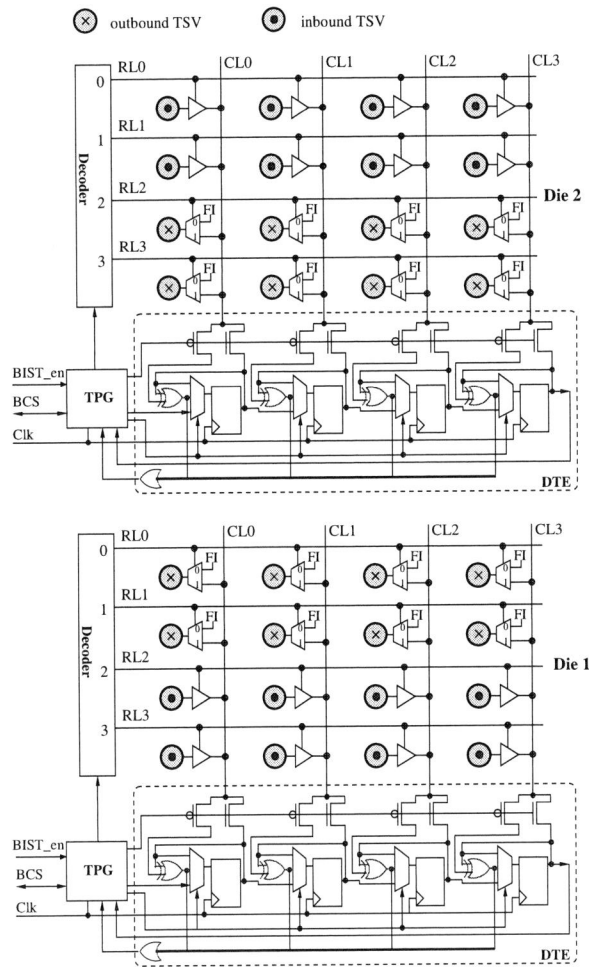

Fig. 3.   Proposed test architecture for a 4×4 TSV array.

As Fig. 3 shows, the proposed BIST scheme for TSVs consists of a decoder, a test pattern generator (TPG), and a test data transportation and evaluation (DTE) circuit. The BIST control signals (BCS) and enable signal (BIST_en) handle the test operations of the BIST circuit. The decoder selects a row for test pattern application or test response evaluation. The TPG generates control signals and test patterns for the decoder and the DTE. The DTE consists of shift registers and a comparator such that it can perform the test pattern/response

978-1-61284-657-6/11 $26.00 © 2011 IEEE   21

transportation using the shift register and the test response evaluation using the comparator. TSVs in the same column are connected to a column line (CL) through a multiplexer or a tri-state buffer. For an outbound TSV, a multiplexer is used to select a signal from the functional circuit (FI) or the CL. For an inbound TSV, a tri-state buffer is used to pass or block the signal to the test circuit. Only one row is selected at a time through the row line (RL), which is activated by the decoder. The DTE is connected to the CLs of the TSV array. Each CL is connected to each shift register through a switch. If a row with outbound TSVs is under test, the test pattern is applied to the TSVs under test through the DTE. If a row with inbound TSVs is under test, the test response from TSVs is captured to the DTE for evaluation.

B. Test Operation Flow

The proposed BIST scheme can support the test and diagnosis of the TSV arrays. The BIST circuit only exports the result of pass/fail in the test mode. But it exports the syndrome of each row in the diagnosis mode. Fig. 4 shows the test operation flow of the BIST circuit. The test operation flow consists of 6 states, namely, MON, GEN_PAT, SHIFT_PAT, CAPTURE_PAT, SHIFT_SIG, and NEXT_ADDR. In the beginning, the BIST circuit is at MON state and waits until the test or diagnosis begins. As the test or diagnosis mode is asserted, the BIST circuit goes to the GEN_PAT state. The BIST circuit generates the test pattern in this state. Once the test pattern is generated, it is shifted to the shift register of DTE in the SHIFT_PAT state. So far, the BIST circuits in both dies execute the same operations; the shift registers in both DTEs have the same test patterns. After the pattern shifting is done, the decoder enables the row under test according to the address. For example, if row 0 in Fig. 3 is under test, only RL0s in both dies are activated. RL1, RL2, and RL3 are all disabled.

In the CAPTURE_PAT state, if the BIST is in the test mode, the test pattern is delivered through TSVs in RL0 of Die 1 to Die 2. The BIST circuit in Die 2 compares the data of TSVs with the expected in the shift registers of Die 2. The OR-gate tree evaluates the outputs of the XOR gates and reports a pass or fail outcome through the Pass/Fail signal. If any TSV fails, the test procedure is terminated. Otherwise, the BIST circuit changes the address of row under test in the NEXT_ADDR state and repeats the CAPTURE_PAT and NEXT_ADDR operations till every TSV is tested. The early abort shortens the test time during the final production test, if no repair or tolerance strategy is applicable.

If the BIST circuit executes the diagnosis function, the test syndromes need to be exported. After the SHIFT_PAT state, in the CAPTURE_PAT state, the test pattern is applied to another die. But different from the test mode, the TSV response is evaluated and captured to the shift registers. Refer again to Fig. 3. The shift registers of Die 2 store the syndrome of TSVs in this state, if row 0 is under test. On the other hand, if row 2 or 3 is under test, the syndrome is stored in the shift registers of Die 1. In the following SHIFT_SIG state, the data in each shift

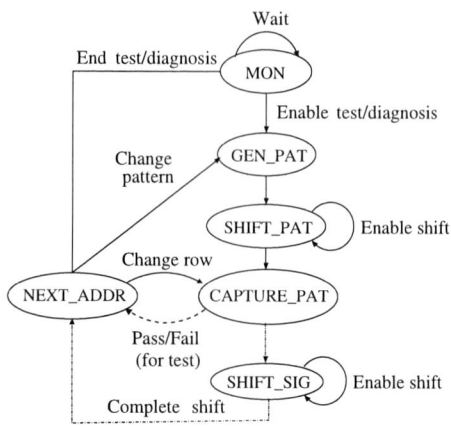

Fig. 4.   The test operation flow of the proposed BIST circuit

register of the DTE are shifted out to check whether the tested TSVs are fault-free or faulty. During the response exportation, the next test data are imported to both dies simultaneously. After the response is exported, the BIST proceeds to the NEXT_ADDR state to check whether the last row is tested. If no, the BIST circuit returns to the CAPTURE_PAT state. The iteration among the NEXT_ADDR, the CAPTURE_PAT, and SHIFT_SIG states is repeated until the test application of a test pattern for the TSV array under test is completed. Subsequently, if another test pattern is applied, then the BIST circuit returns to the GEN_PAT state and repeats the same test operations described above. For the stuck-at fault test, for example, all-0 and all-1 patterns can be applied. If all the TSVs are diagnosed or tested by applying one test pattern, e.g., the all-0 pattern, one can change to all-1 test pattern or other test patterns and repeat the diagnosis or test operation flow again. If the test/diagnosis is completed, the BIST circuit returns to the MON state from the NEXT_ADDR state.

The proposed BIST scheme can support the testing of stuck-at and stuck-open faults. Many existing works have reported the test algorithms, and test pattern generations for these faults. One can apply the required test patterns through the proposed BIST scheme. The test controlling of the BIST circuits in different dies can be performed by the following two possible approaches. First, one can design a control circuit in the bottom die of a 3D IC to control the BIST circuit in the other dies. Second, one can use the test access architecture of a 3D IC, e.g., the test access architecture reported in [16], to control all the BIST circuits of the 3D IC.

C. BIST Programmability

The number of TSVs of different layers may be different. A BIST circuit in a die may have to support two different test configurations. For example, as Fig. 5(a) shows, when two dies are stacked, 8 TSVs are formed between Die 1 and Die 2. Thus, Die 1 and Die 2 both have a 4×2-TSV array. When the third die, Die 3, is stacked as shown in Fig. 5, four TSVs run through Die 1 to Die 3, and four TSVs are formed only between Die 2 and Die 3. Therefore, the BIST circuit for the Die 2 should have two configurations for the testing of TSVs between Die 1 and Die 2, and between Die 2 and Die 3, respectively.

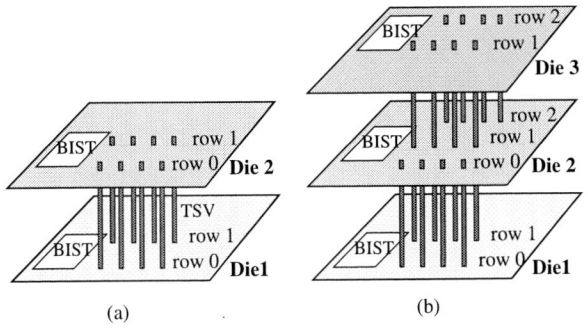

Fig. 5. (a) Two dies are stacked (b) Third die is stacked

One can separate the 12 TSVs in Die 2 into two arrays in which each TSV array has an individual BIST circuit. This results in undesirably large area overhead. To cope with this problem, the proposed BIST circuit is designed to be programmable. As Fig. 5(a) shows, the BIST circuits in Die 1 and Die 2 only need to test 2 TSV rows when the testing of the TSVs between Die 1 and Die 2 is performed. If Die 3 is stacked, however, the BIST circuits in each die need to test 3 TSV rows, as shown in Fig. 5(b). Although Die 1 has only 2 TSV rows, the test circuit in Die 1 performs redundant operation when the TSVs between Die 2 and Die 3 are tested. Similarly, the BIST circuit in Die 3 performs redundant operation when row 0 is tested. The logical position of each TSV needs to be predefined. For example, the TSVs between Die 1 and Die 2 can be defined as row 0 and row 1. The TSVs between Die 2 and Die 3 can be defined as row 2. Each BIST circuit needs to generate address 0 and 1 for the testing of TSVs between Die and Die 2, and address 0 to 2 for the testing of the TSVs between Die 2 and Die 3. Also, if each row does not have the same amount of TSVs, the number of columns is designed to be the widest.

IV. SIMULATION RESULTS AND ANALYSIS

A. Hardware Implementation

Fig. 6 shows the block diagram of the proposed BIST scheme in detail. The signals of the BIST circuit include BIST_en, Test/Diag, Cmd, Pass/Fail, Cmd_done, and Syn_out. The BIST_en is used to enable the BIST function. The Test/Diag is used to set the BIST circuit in the test or diagnosis mode. The FSM in the BIST circuit controls the test/diagnosis flow, as shown in Fig. 4. The BIST circuit also has an address counter and a command register. The address counter generates the address of the row under test. The Capture and Shift_en signals are used to control the operation of the shift registers. The test patterns are shifted to the shift registers through Pattern_in signal. The test syndrome is exported through Syn_out signal. The command register is used to program the address counter for the testing of TSVs between different layers by shifting in command from Cmd signal. The OR-gate

tree compresses the test syndrome during the test mode and exports one bit Error signal to the BIST. If the BIST operates in the test mode, the Pass/Fail output signal will be activated according to the Error signal. If the BIST operates in the diagnosis mode, the Pass/Fail signal is always disabled. Once the test/diagnosis operation is finished, then the Cmd_done signal is pulled up.

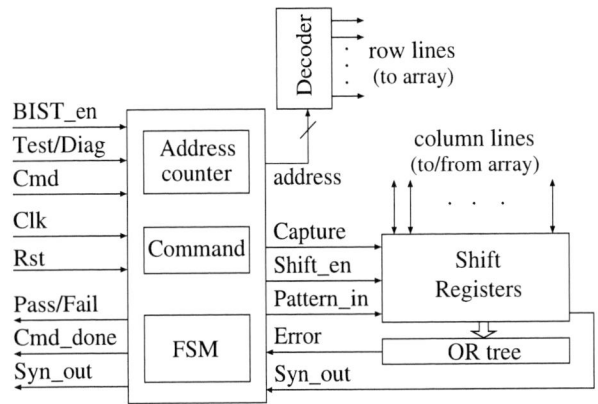

Fig. 6. The block diagram of the proposed BIST circuit

B. Simulation Results

We have implemented the proposed BIST scheme using TSMC $0.18\mu m$ standard cell library. Here we implemented two BIST circuits for two dies. Note that only the BIST circuits are implemented and the TSVs are replaced by interconnections. Also, each TSV is wrapped with a multiplexer or a tri-state buffer. The delay incurred by the tri-state buffer or multiplexer is less than 0.1 or 0.15ns in $0.18\mu m$ technology. The configuration of a TSV array has the impact on the test time and the area overhead. Different configurations of TSV arrays thus are simulated. For a 16×8 TSV array, the area cost of the BIST circuits is $10890 \mu m^2$, where 16 and 8 represent the numbers of rows and columns, respectively. In the array, a half of TSVs are outbound ones and the others are inbound ones. If the diameter of TSV is $45\mu m$, for example, then the area of a TSV is $45 \times 45 \mu m^2$. Then the area overhead of the proposed BIST scheme is 4.2% for the 16×8 TSV array, a total of 128 TSVs. Clearly, the area overhead is reduced if the size of the TSV array under test grows. For example, the area overhead of the BIST circuit for a 16×32 TSV array becomes 2.24%. Fig. 7 shows the area overhead of the proposed BIST scheme with respect to different array configurations. The solid line and dash-dotted line represent the area overhead of the proposed BIST scheme for the TSV arrays with constant number of rows and columns, respectively.

As the figure shows, the area overhead decreases as the size of TSV array grows in both cases. However, for an array with the same number of TSVs but different configurations, the area overhead might vary. For example, if the array size is 32×16, a total of 512 TSVs, the area overhead of the proposed BIST is 2.48%. If the array size is 16×32, the area overhead of the proposed BIST becomes 2.24%. The area overhead of TSV

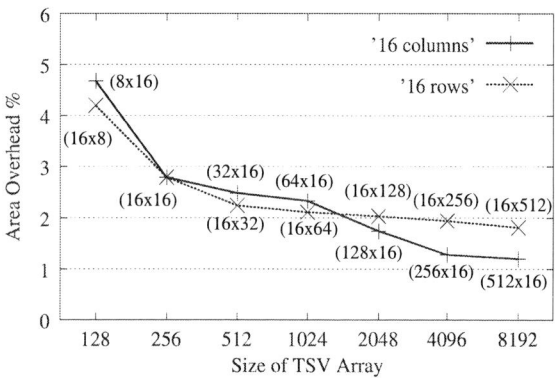

Fig. 7. Area overhead of the proposed BIST scheme with respect to different TSV array configurations

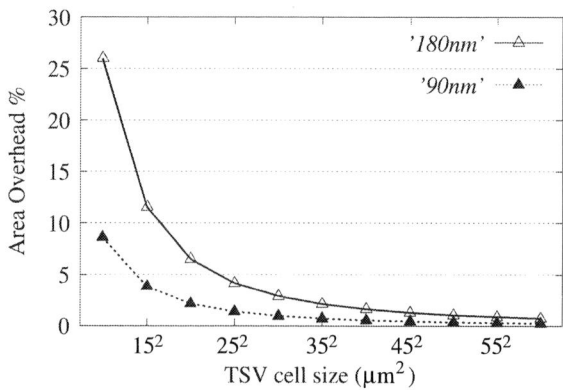

Fig. 8. Area overhead of the proposed BIST circuit for a 256×16 TSV array with respect to different CMOS and TSV technologies

arrays with a constant number of rows is less than that with a constant number of columns when the number of TSVs of the array is not larger than 1024. After that, the area overhead saturated. This is because if the number of columns becomes large, the shift registers dominate the overall area of the BIST circuits. Although the area cost of the BIST circuits for the TSV array is increased with the increasing of the number of columns, the test/diagnosis time is decreased. Subsequently, the test/diagnosis time with respect to different configurations of the TSV arrays is estimated. Assume that a TSV array has C columns and R rows. The test and diagnosis time for a TSV array by one test pattern can be estimated as follows:

$$\text{Test:} \quad (1 + C + 2R) \text{ cycles} \quad (1)$$

$$\text{Diagnosis:} \quad (1 + C + (C + 2)R) \text{ cycles} \quad (2)$$

For a 512×16-TSV array, for example, the diagnosis time is 9233 cycles. For the 16×512-TSV array, the diagnosis time is 8737 cycles. Whether to arrange a wider TSV array (the number of columns is larger than that of rows) or a longer TSV array (the number of rows is larger than that of columns) depends on cost of the test time and silicon area. As the number of TSVs in the array increases, the area overhead of the proposed BIST scheme is less significant.

With different TSV manufacturing processes, the size of a TSV cell may range from few micrometers to a hundred micrometers. Thus, the area overhead of the proposed BIST scheme is also different for different TSV technologies. Furthermore, the area cost of the BIST circuit is related to the CMOS technology. Fig. 8 shows the area overhead of the BIST circuit for a 256×16 TSV array with respect to different sizes of TSVs and CMOS technologies. We take the implemented BIST scheme for example. The area of the BIST circuit implemented using $0.18\mu m$ CMOS technology is $106478\mu m^2$. Therefore, the area overhead ranges from 26% to 0.72% corresponding to different sizes of TSVs. However, if the BIST circuit is implemented using $90nm$ CMOS technology, its area is $35271\mu m^2$. Then, the area overhead of the BIST circuit ranges from 8.61% to 0.24% corresponding to different sizes of TSVs.

C. Comparison Results

As aforementioned, one can use the IEEE 1149.1 or 1500 test standard to support the post-bond test of TSVs. However, this comes with the cost of silicon area and test time. Using IEEE 1149.1 or 1500 test architecture, each TSV is wrapped with a wrapper cell, either with one or two storage cells. Thus, the area overhead for testing TSVs is at least a storage cell for each TSV. We have implemented a wrapper cell with one and two storage cells including the decoding circuit for WIR signals. The area is about $206\mu m^2$ for one-storage wrapper cell and $306\mu m^2$ for two-storage wrapper cell using $0.18 \ \mu m$ technology. We compare the area overhead of the proposed BIST scheme and the test approach using IEEE 1500 test wrapper. The configurations of TSV arrays in which the number of columns in each array is 16, as the solid line of Fig 7 shows, are considered. For the IEEE 1500 architecture, only the area overhead of the wrapper cells are calculated.

Fig. 9 shows the comparison result of area overhead of the BIST circuit and the IEEE 1500 test wrapper with respect to different configurations of TSV arrays. Here we assume that the TSV cell size is $45\mu m^2$. For the IEEE 1500 architecture, each TSV is equipped with a wrapper cell, so the area overhead is one wrapper cell to one TSV cell. For one storage wrapper cell, the area overhead is 10.17%. For a two-storage wrapper cell, the area overhead is 15.11%. However, the area overhead of the proposed scheme is only from 4.6% to 1.7%. Compared with the 1500 two-storage wrapper cell, the proposed BIST scheme can save 70% to 88.7% area cost. Clearly, the area overhead of the proposed BIST scheme is much lower than that of the IEEE 1500 test wrapper.

For the test approach of IEEE 1500 test wrapper, the serial test interface layer or the parallel test interface layer can be used to perform the testing of TSVs. Although the test pattern and response also can be transported in parallel by the test access mechanism (TAM), multiple TSVs must be implemented for the TAM. The required test TSVs also results in a portion of area overhead of the test circuit. In the aspect of test/diagnosis time, therefore, the IEEE 1500 test wrapper apply test patterns by importing the test pattern and exporting

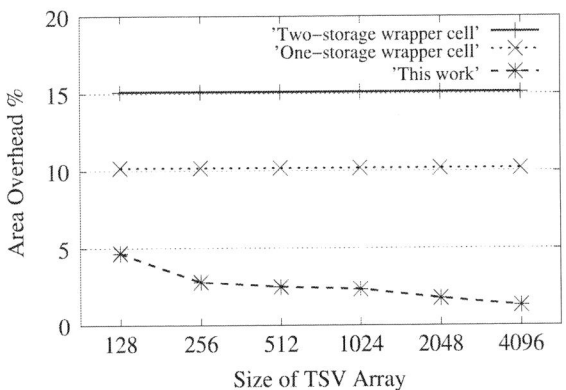

Fig. 9. Area overhead of the BIST circuit and the IEEE 1500 test wrapper with respect to different TSV array configurations

the test response through the wrapper cells in serial. This requires $(2 \times C \times R)$ test clock cycles for one test pattern.

Fig. 10 shows the test time of the proposed BIST scheme and IEEE 1500 scheme for different numbers of TSVs. 1500-1bit and 1500-2bit are the test time of the IEEE 1500 test wrapper using one test input port and two test input ports, respectively. Diagnosis/Test represents the operation mode of the proposed BIST. The row or col represents the arrays with a constant number of rows or columns, since the configuration of an array affects the test time. The constant number is 16. As Fig. 10 shows, the test time of the proposed BIST is much lower than that of the test approach of IEEE 1500 test wrapper. For example, the 16×32 TSV array requires 65 test clock cycles for one test pattern, while the 1500 test wrapper using one test input port requires 1024 test clock cycles. The proposed BIST scheme can save 93.6% test cycles in comparison with the 1500 approach. Although the IEEE 1500 test wrapper can reduce the test time by adding more test inputs, this also leads to higher area overhead.

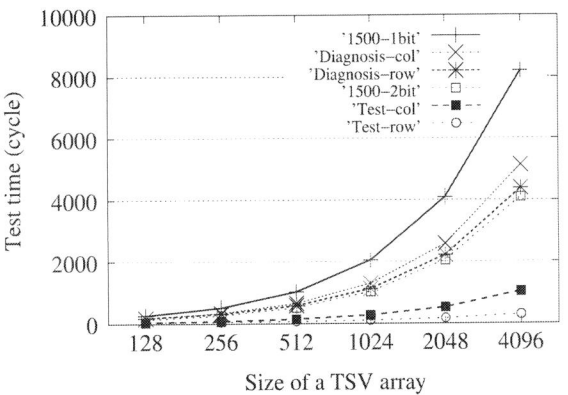

Fig. 10. Test time of the proposed BIST scheme and IEEE 1500 test scheme with respect to different sizes of TSV arrays

V. CONCLUSIONS

A cost-effective BIST scheme for testing TSVs of 3D ICs has been presented in this paper. The proposed BIST scheme offers the features of low test/diagnosis test time and area overhead. The area overhead and test time with respect to different array configurations, CMOS technologies, and TSV diameters are discussed. The area overhead and the required test/diagnosis time of the proposed BIST is lower than that of the test approach using IEEE 1500 test architecture.

ACKNOWLEDGMENT

This work was supported in part by the National Science Council, R.O.C., under Contract NSC 97-2221-E-008-094-MY3 and NSC 97-2221-E-008-095-MY3.

REFERENCES

[1] V. F. Pavlidis and E. G. Friedman, "Interconnect-based design methodologies for three-dimensional integrated circuits," *Proceedings of the IEEE*, vol. 97, no. 1, pp. 123–140, Jan. 2009.

[2] J.-Q. Lu, "3-D hyperintegration and packaging technologies for micronano systems," *Proceedings of the IEEE*, vol. 97, no. 1, pp. 18–30, Jan. 2009.

[3] H.-H. S. Lee and K. Chakrabarty, "Test challenges for 3D integrated circuits," *IEEE Design & Test of Computers*, vol. 26, no. 5, pp. 26–35, Sept.-Oct. 2009.

[4] E. J. Marinissen and Y. Zorian, "Testing 3D chips containing through-silicon vias," in *Proc. Int'l Test Conf. (ITC)*, Austin, Nov. 2009, ET1.1, pp. 1–11.

[5] J.-F. Li and C.-W. Wu, "Is 3D integration an opportunity or just a hype?" in *Proc. Asia and South Pacific Design Automation Conf. (ASP-DAC)*, Taipei, Jan. 2010, pp. 541–543.

[6] X. Wu, P. Falkenstern, and Y. Xie, "Scan chain design for three-dimensional integrated circuits 3D ICs," in *Proc. IEEE Int'l Computer Design Conference (ICCD)*, Oct. 2007, pp. 208–214.

[7] X. Wu, Y. Chen, K. Chakrabarty, and Y. Xie, "Test-access mechanism optimization for core-based three-dimensional SOCs," in *Proc. IEEE Int'l Computer Design Conference (ICCD)*, Oct. 2008, pp. 212–218.

[8] B. Noia, K. Chakrabarty, and Y. Xie, "Test-wrapper optimization for embedded cores in TSV-based three-dimensional SOCs," in *Proc. IEEE Int'l Computer Design Conference (ICCD)*, Oct. 2009, pp. 70–77.

[9] X. Wu, P. Falkenstern, K. Chakrabarty, and Y. Xie, "Scan chain design and optimization for three-dimensional integrated circuits," *ACM Journal on Emerging Technologies in Computing Systems*, vol. 5, no. 2, pp. 9:1–9:26, July 2009.

[10] L. Jiang, Q. Xu, K. Chakrabarty, and T. M. Mak, "Layout-driven test architecture design and optimization for 3D SoCs under pre-bond test-pin-count constraint," in *Proc. IEEE/ACM Int'l Conf. on Computer-Aided Design (ICCAD)*, Nov. 2009, pp. 191–196.

[11] Y.-J. Huang and J.-F. Li, "Testability exploration of 3-D RAMs and CAMs," in *IEEE Asian Test Symp. (ATS)*, Taichun, Nov. 2009, pp. 397–402.

[12] P.-Y. Chen, C.-W. Wu, and D.-M. Kwai, "On-chip testing of blind and open-sleeve TSVs for 3D IC before bonding," in *Proc. IEEE VLSI Test Symp. (VTS)*, May 2010, pp. 263–268.

[13] E. J. Marinissen, J. Verbree, and M. Konijnenburg, "A structured and scalable test access architecture for TSV-based 3D stacked ICs," in *Proc. IEEE VLSI Test Symp. (VTS)*, May 2010, pp. 269–274.

[14] E. J. Marinissen, C.-C. Chi, J. Verbree, and M. Konijnenburg, "3D DfT architecture for pre-bond and post-bond testing," in *IEEE Int'l 3D Systems Integration Conference (3DIC)*, Munich, Nov. 2010, pp. 1–8.

[15] C.-Y. Lo, Y.-T. Hsing, L.-M. Denq, and C.-W. Wu, "SOC test architecture and method for 3-D ICs," *IEEE Trans. on Computer-Aided Design of Integrated Circuits and Systems*, vol. 29, no. 10, pp. 1645–1649, Oct. 2010.

[16] C.-W. Chou, J.-F. Li, J.-J. Chen, D.-M. Kwai, Y.-F. Chou, and C.-W. Wu, "A test integration methodology for 3D integrated circuits," in *IEEE Asian Test Symp. (ATS)*, Shanghai, Dec. 2010, pp. 377–382.

[17] M. Tsai, A. Klooz, A. Leonard, J. Appel, and P. Franzon, "Through silicon via (TSV) defect/pinhole self test circuit for 3D-IC," in *Proc. IEEE Int. Conf. 3D System Integration*, Sept. 2009, pp. 1–8.

Scan Chain and Power Delivery Network Synthesis for Pre-Bond Test of 3D ICs

Shreepad Panth and Sung Kyu Lim
School of Electrical and Computer Engineering
Georgia Institute of Technology
Email: {spanth, limsk}@ece.gatech.edu

Abstract—Pre-bond testing of 3D ICs improves yield by preventing bad dies and/or wafers from being used in the final 3D stack. However, pre-bond testing is challenging because it requires special scan chains and power delivery mechanism. Any 3D scan chains that traverse multiple dies will be fragmentized in each individual die during pre-bond testing. In this paper we study the scan chain and power delivery network synthesis for pre-bond testing of 3D ICs. The testing of individual dies is facilitated by the addition of dedicated probe pads for power delivery and scan IO as a form of design-for-testing. We investigate the impact of scan-chain Through-Silicon-Vias (TSVs) on power consumption and voltage drop. We also study the requirements of power probe pads for power delivery during pre-bond structural test.

Index Terms—3D ICs, pre-bond test, structural test, probe test, power delivery network

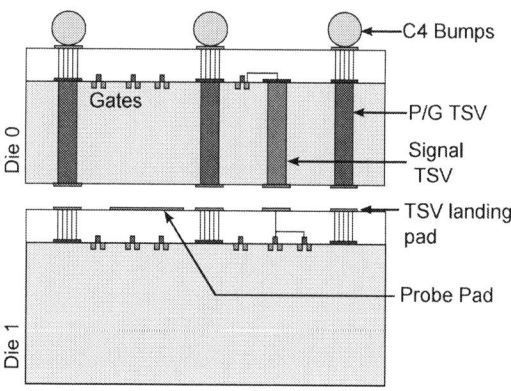

Fig. 1.   The Structure of a 3D Integrated Circuit

I. INTRODUCTION

Three Dimensional (3D) Integrated Circuits have emerged as a solution to the interconnect scaling problem. A Through-Silicon Via (TSV) is used to connect multiple dies in the vertical dimension, and this extra dimension offers us the possibility of reducing the length of the longest wires, the total wirelength, increasing the system clock frequency, and reducing power consumption. However, these through silicon vias have non-negligible area, which affects the placement results of other surrounding logic or memory [1]. Thus, they cannot be treated simply as an interconnect, but their impact on the layout has to be considered as well.

Testing of 3D ICs remains one of the major EDA challenges facing their widespread adoption [2]. Testing of the entire 3D stack is known as post-bond test, and is not overly challenging as we have a completely functional circuit, as well as solder bumps for providing test access. However, pre-bond test, which is the testing for each die prior to stacking, is challenging because each die may not have complete logic, and there are no solder bumps to connect to. A solution to the first problem is performing structural test, using scan chains, where the high-level functionality of the circuit is irrelevant, and a solution to the second problem is the addition of probe pads so that probe needles can touchdown during wafer test.

We address the following issues related to the pre-bond testing for 3D ICs. First, we study the power delivery re-

This material is based upon work supported by the National Science Foundation under Grant No. CCF-0917000, the FCRP Interconnect Focus Center (IFC), and Intel Corporation.

quirements and methods for the structural post-bond testing of a 3D Stacked IC. Second, we investigate the impact of scan-chain TSV count on the scan chain length as well as the system parameters such as total circuit wirelength. Third, we investigate the impact of probe pad usage on the power delivery during pre-bond testing.

II. PRELIMINARIES AND MOTIVATION

A. Prior Work in 3D IC Testing

Although the testing of 3D IC's is an important challenge facing their adoption, only limited work has been done in the field. Wu et al [3] compare several scan chain schemes, and provide a genetic algorithm and ILP based algorithms for routing them for post bond test. Zhao et al [4] provided a scheme for clock tree synthesis to facilitate pre-bond test. At the architecture level, Lewis and Lee [5] proposed a scan island based methodology to test incomplete circuits during pre-bond test. The 3D test flow, TSV defects, and possible changes to the test architecture were presented in [6]. The authors of [7], [8] provide test architecture design for 3D SoCs.

B. Motivation

Although some work has been done in the area of pre-bond test, none have yet considered the need for power supply during pre-bond test. The structure of a typical 3D IC is shown in Figure 1. Usually, only one die has C4 bumps, and all other dies receive power through P/G TSVs. During pre-bond test of these dies, no wire bond pads for power or ground exist, and probe pads need to be added to facilitate their testing as

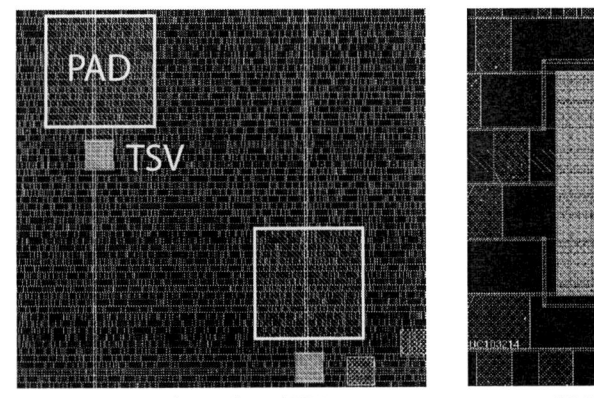
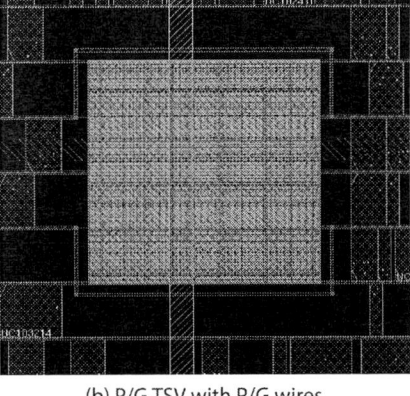
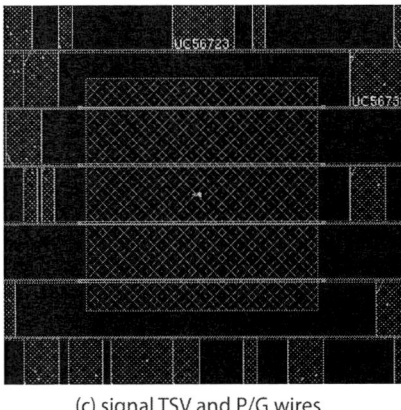

| (a) probe pads and TSVs | (b) P/G TSV with P/G wires | (c) signal TSV and P/G wires |

Fig. 2. Cadence Encounter image shots. (a) probe pads and TSVs, (b) P/G TSVs and P/G wire detours, (c) signal TSVs and P/G wires. P/G wires can be routed over signal TSVs.

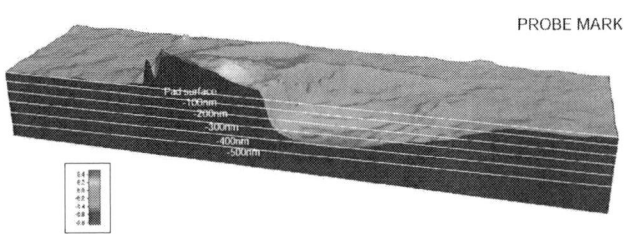

Fig. 3. Damage caused to the probe pad after a single probe touchdown [11]

shown. In addition, any scan chains that traverse multiple dies will be broken, which may leave several unbalanced smaller chains which also need to be provided with test access in the form of probe pads.

Fine grained touchdown probe needles are unlikely to be available at least for another decade [9], and today's probe pads are limited by available technology [10] to a minimum pitch of $35 - 40\mu m$ for cantilever probing, and $100\mu m$ for vertical probing with a minimum pad size of around $25\mu m$. As seen from Figure 1, not only do these probes occupy significant area on the die in which it is placed, any TSVs in the previous die cannot be placed in the same location as the probe pad in order to avoid overlap with its landing pad. Therefore, several layout implications exist to the addition of probe pads, and their locations have to be chosen carefully.

III. LOCATION OF PROBE PADS

Ideally, we would like to place a probe pad over the landing pad of the TSV to which we wish to connect . This would minimize the area overhead, as well as provide a low resistance connection in the case of power delivery. However, when the probe makes contact with the probe pad, it creates a scrub mark, which significantly affects its planarity, as shown in Figure 3 [11]. It is still unclear how this scrub mark will affect the TSV bonding process, and for the sake of reliability, a certain distance has to be maintained between the probe pad and the TSV landing pad. Figure 2(a) shows such an arrangement, with Figure 2(b) showing a close up shot.

- **P/G Probe Pads:** In this case, we wish to have as low a resistance path as possible. Thus, the probe pad has to be placed close to the P/G landing pad and connected to it using a thick strip of the top metal layer. We must be careful to not place it too close, because the P/G TSV landing pad is connected to the power grid through an array of local vias. Such an array present near the probe needle at touchdown is a major source of dielectric cracking [12].
- **Other Probe Pads:** These probe pads include probe pads added for the scan chain, as well as those for a die level wrapper. We add a IEEE 1500 compliant [13] die level wrapper, which gives us test access to the TSVs during pre-bond test. Since the resistance is not paramount for these probe pads, it can be left up to the router to connect the probe pad to the TSV landing pad. The location of these probe pads can be chosen based on ATE constrains, or to minimize the change in the layout of other dies.

IV. 3D SCAN CHAINS

Constructing a scan chain in a 3D fashion has several advantages over a 2D approach. Wu et al [3] have shown that around a 40% reduction in the scan wirelength can be achieved with 3D scan chains compared with 2D chains for post-bond testing of 3D ICs. This can significantly reduce the test time of the circuit. A 3D scan chain relies on the use of TSVs, and since TSVs occupy silicon area, the total number of scan TSVs that we can use is limited.

In this section, we present a greedy heuristic to stitch a 3D scan chain, in order to minimize its wirelength. The constraints are the number of scan TSVs that can be used, and a fixed scan-in and scan-out pin. This is shown in Figure 4. Here, C represents the TSV constraint for each die, and there are k dies. Assuming that we use Face-to-Back (F2B) bonding, TSVs are absent on the last die, and we have $k-1$ constraints. X represents the set of all scan cells, we have m scan cells. x_0 represents the scan-in pin, and x_{m+1} represents the scan out pin. Next, the cost function between two cells are initialized. This cost function is given by Equation (1), where z represents

978-1-61284-657-6/11 $26.00 © 2011 IEEE

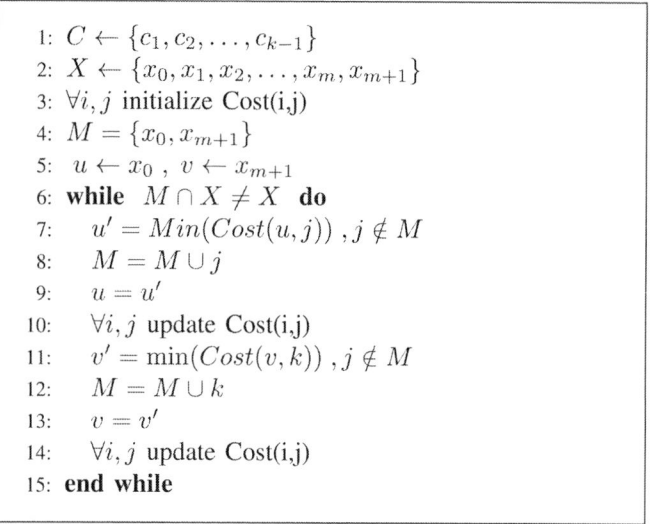

```
 1: C ← {c₁, c₂, ..., c_{k-1}}
 2: X ← {x₀, x₁, x₂, ..., x_m, x_{m+1}}
 3: ∀i, j initialize Cost(i,j)
 4: M = {x₀, x_{m+1}}
 5: u ← x₀ , v ← x_{m+1}
 6: while M ∩ X ≠ X do
 7:    u' = Min(Cost(u, j)) , j ∉ M
 8:    M = M ∪ j
 9:    u = u'
10:    ∀i, j update Cost(i,j)
11:    v' = min(Cost(v, k)) , j ∉ M
12:    M = M ∪ k
13:    v = v'
14:    ∀i, j update Cost(i,j)
15: end while
```

Fig. 4.   Greedy algorithm to construct a 3D scan chain

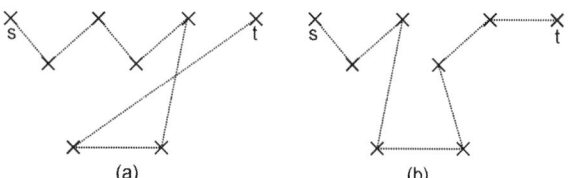

Fig. 5.   (a) Scan Chain grown from one direction, and (b) Scan Chain grown from two directions

all dies between x_i and x_j, and R_z represents the remaining number of TSVs that we are allowed to use for that die.

$$Cost(i, j) = \begin{cases} d_{ij} & \text{i,j in same die} \\ \dfrac{d_{ij}}{\min R_z / C_z} & \text{otherwise} \end{cases} \quad (1)$$

Next, set M represents the set of marked cells, and the scan-in and scan-out pin are initially marked. Next, we stitch the scan chain from two sides, both from the scan-in and the scan-out pins. We choose the cell with minimum cost in each iteration, and this process continues until all cells are marked. The cost function is dynamically updated, and as more TSVs are used in a particular die, it becomes less attractive to use them, until the cost eventually becomes infinity when all the TSVs are used. It is important to note that when this happens, it may not be possible to stitch all the scan cells without using more TSVs due to the presence of isolated chains. In this case, extra TSVs may be used , which will not exceed two TSVs per die. Each constraint may be subtracted by two to compensate for this effect. Although it is possible to grow the scan chain from one direction only, we grow it from both directions in order to ensure smaller wirelength, as shown in Figure 5.

A. Re-Use of Signal TSVs

So far, we have assumed that when a scan chain goes from one die to another, we require a dedicated scan TSV. In a scan chain, the output of a flip-flop is connected to the scan input of

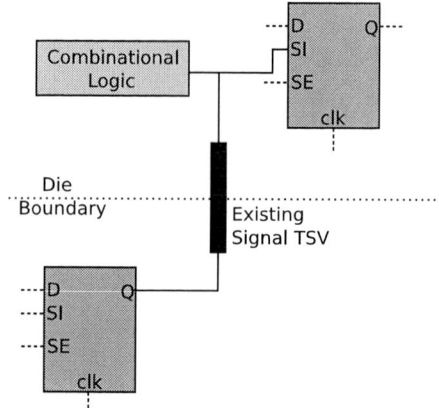

Fig. 6.   Re-use of existing signal TSVs for scan chain

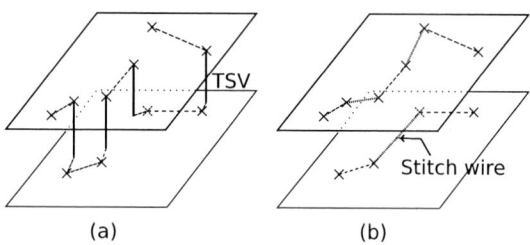

Fig. 7.   (a) A 3D scan chain, and (b) multiple fragments connected together

the next flip flop, as well as to some combinational logic that is of no consequence during the test mode. It might be possible as shown in Figure 6, that a flip flop drives some combinational logic on another die through an existing signal TSV. In this case, we do not need to insert an additional TSV just for the purpose of the scan chain, but we can re-use the existing signal TSV as shown. A careful choice of scan ordering can make use of several existing signal TSVs, thereby reducing the overall scan chain wirelength, without suffering the penalty of inserting a large TSV into the layout.

B. Broken Scan Chains

Once we have inserted a 3D scan chain into the design, it is used during post-bond test, and its scan-in and scan-out pins are accessed through solder bumps. However, if we wish to perform pre-bond test on each die, we are left with several scan chain fragments, whose number depend on the number of scan TSVs used. It is not feasible to probe all these fragments due to the large size of the required probe pads. Thus, we have to stitch together different fragments as shown in Figure 7 so that the test-pin count is reduced. We use tri-state buffers to stitch together the broken fragments, and they are enabled by a pre-bond test signal.

V. DESIGN AND ANALYSIS FLOW

A. Design Flow

The addition of probe pads complicates the design flow process. This is because the probe pads added in a die not only affects components in its own die, but also affects the placement results of the adjacent die. Initially scan cells are

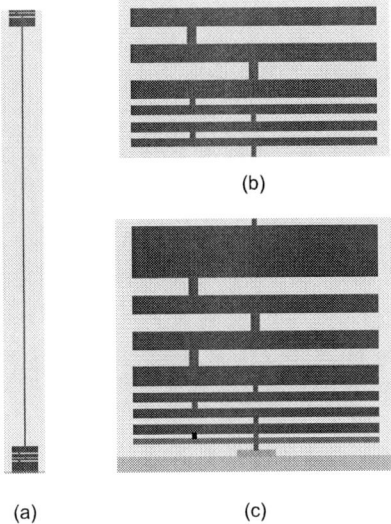

Fig. 8. (a) Scale representation of ICT along y-axis, (b) top die, (c) bottom die

inserted into the 2D netlist, either during or after synthesis. Next, the original netlist is partitioned into as many dies as required, and individual netlists are obtained for each die. In the next step, we perform power planning where we decide the configuration and pitch of the number of the power and ground TSVs, as well as the location and number of power and ground probe pads to be used. Once the power plan is completed, we perform an initial placement of each die individually using Cadence Encounter to get the location of scan flip-flops. We then stitch the scan chains together using the greedy algorithm discussed in Section IV. This process introduces additional scan TSVs into the design, which change the placement results. In the next step, we decide the location of the additional probe pads as scan-in and scan out pads for both the scan chain and die-level wrapper, a clock pad, and the scan enable pad. Once the locations of the pads are determined, placement is again carried out to accommodate the new TSVs, as well as to avoid placing any TSVs at the location of probe pads. Finally, the design is routed on a die-by-die basis.

B. 3D Power Analysis

In order to analyze 3D power, we create a testbench for the two dimensional netlist using pseudo-random test vectors, and then conduct logic simulation using Synopsys VCS to obtain the 2D VCD (value change dump) file. With the information from the die netlists, this 2D VCD file is converted into several VCD files, one for each die. Once we have the VCD file for a given die, the TSV pins on that die are annotated with the TSV capacitance, and power simulations for each die are carried out using Cadence VoltageStorm.

C. 3D Voltage Drop Analysis

Once the 3D IC design is completed, we perform voltage drop analysis for the entire 3D IC to obtain IR-drop noise values. Our 3D IR-drop analysis tool is based on Cadence

VoltageStorm, which is designed for 2D ICs. We take the following steps to handle 3D designs using VoltageStorm:

1) We modify the interconnect technology file (ICT), which contains information on all layers (device layer, dielectric layer, metal layer, vias, and TSVs) and their relative position and resistance values, to model our two die–stacked configuration as shown in Figure 8.

2) We create a 3D technology file (TCH), which contains resistive and capacitive information for all and between layers, using Cadence Techgen.

3) We generate a 3D library exchange format (LEF) file so that layers and gates in different dies can be distinguished by the tools. For example, M1 in top-most die and M1 in bottom-most die should be differentiated so that 2D tools distinguish these M1 layers.

4) Power consumption data and layer mapping files, which maps design to appropriate LEF and GDS layers, are modified as well to be used for different dies accordingly.

5) Finally, we create a 3D design exchange format (DEF) file from the final layout of each die to form a single 3D design. After all these preparations are ready, we can run 3D IR-drop analysis on this combined design using Cadence VoltageStorm.

Using our IR-drop estimation method, we were able to match both 2D and the two die-stacked 3D IR-drop results from the VoltageStorm within 7% error. Our computed resistance value based on the ICT file for each P/G wire segment overestimated by 6% compared to VoltageStorm. Since this is deterministic error, resistivity value is tuned to match the results. Due to the tools limitation on number of layers it can process, we validated our algorithm up to two die-stacked 3D ICs.

VI. EXPERIMENTAL RESULTS

The greedy heuristic for scan chain insertion was implemented in C++, and we choose a FFT circuit from [14] for our analysis. Synthesis was carried out in Synopsys Design Compiler using NCSU $45\mu m$ technology. The design was placed in two dies, and the number of TSVs is chosen to occupy around 20% of the entire die area. Statistics about the design used are as follows: the number of gates is $400,213$, signal TSV is 2953, Flip-Flops is $75,723$, power TSV is 81, and ground TSV is 64. We use TSVs with $6\mu m$ diameter and $10\mu m$ landing pad for both P/G and signal. Its height is assumed to be $50\mu m$, and its resistance is $50m\Omega$. Power and Ground TSVs are inserted at a pitch of $150\mu m$. The probe pad width is assumed to be $40\mu m$, and is assumed to have a minimum pitch of $100\mu m$. The inserted wrapper scan elements occupy 1.96% of the total die area, and have a total stitched wirelength of $75054\mu m$. This corresponds to 0.48% of the baseline wirelength reported in Figure 9(a).

A. Impact of Scan Chain TSV on Wirelength

In order to study the impact of the number of scan TSV on the wirelength of the design, we construct three different scan chains as shown in Table I. Since we cannot construct a 3D

978-1-61284-657-6/11 $26.00 © 2011 IEEE

TABLE I
SCAN CHAIN CONFIGURATIONS

Name	No. TSVs	#TSV reused	Stitch WL (μm)
scan0	2	0	4.75
scan100	100	2	26595
scan200	200	4	34296

scan chain with no TSVs, we insert 2 TSVs in "scan0", which is the minimum number required. Column 3 shows that even without using any specific algorithm to re-use existing signal TSVs, it is possible to re-use around 2% of the TSVs required for the scan chain. The number of scan-chain fragments formed per die is exactly half of the number of TSVs, and Column 4 gives the amount of additional wirelength that is required to stitch all the fragments together into a single scan chain. We observe that with an increase in the number of fragments, the wirelength required to stitch them together also increases.

Next, we study the impact of the scan chain TSV count on the scan wirelength and the total wirelength of the 3D design. This is plotted in Figure 9(a). First, we observe that an increase in the number of scan TSVs always helps reduce the scan wirelength. But in the case of signal and total wirelength, adding more scan TSVs helps reduce them initially. But beyond a certain point they start to worsen. The initial improvement is achieved due to the lower scan wirelength reducing the routing blockage. With a further increase in the number of TSVs, either the die area or standard cell density increases. If the die area increases, the average distance between gates increases, increasing the overall wirelength. An increase in the cell density also increases routing congestion.

B. Power and IR-drop Analysis during Post-bond Testing

We vary the number of scan chain TSVs and study the impact on the power consumption and voltage drop of the 3D die stack during normal operation, and post-bond test. During the normal operation, we generate vectors at the primary inputs of the design. During post-bond structural test, we generate patterns at the scan-in terminal and shift them through the scan chain. Figure 9(b) shows the results of power analysis. As expected, we observe that structural test always has higher power consumption, when compared with the normal mode of operation. During test mode, we observe an initial drop in the power consumption of the circuit when we increase the number of scan TSVs. As we increase this further, no more gains are achieved. This can be attributed to the initial decrease in the scan wirelength, but further gains are offset by the increase in the total wirelength of the 3D FFT design.

Next, the same study is carried out for voltage drop and shown in Figure 9(c). We again see worse IR drop during test. When we vary the number of scan TSVs, the same trends are observed here for both normal operation and test mode: an initial increase in voltage drop, followed by a decline as the scan TSV count increases. These results seem counterintuitive as they are exactly the opposite of the power analysis. However, two factors are noteworthy. First, the magnitude of

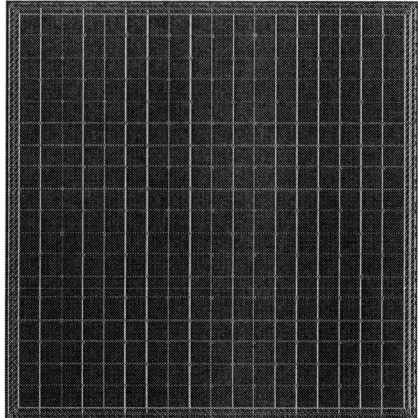

(a) power distribution network

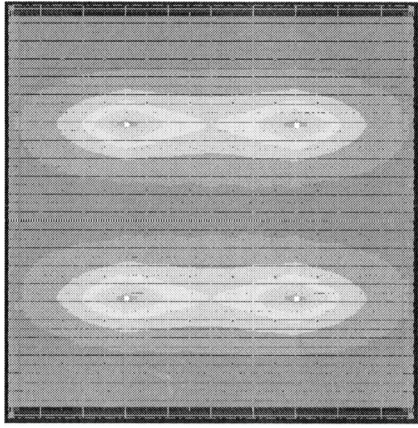

(b) IR drop map

Fig. 10. (a) GDSII image of our PDN for pre-bond testing, (b) IR drop map under P/G probe pad pitch of 600um using Cadence VoltageStorm.

the change in voltage drop is quite small, with the maximum change being $6mV$. Secondly, the voltage drop depends not only on the power results, but also on the resistance of the power network. This resistance is affected by the number of TSVs and their location. Inserting more TSVs into the design increases the resistance of the power network due to M1 power strips having to take a detour around them. This accounts for the initial increase in the voltage drop. However, the maximum voltage drop depends on the worst case resistance path to areas of highest power consumption. It is unlikely that all TSV's inserted will fall into this path. It is also possible that the TSVs inserted will avoid this path completely. This accounts for the seemingly unpredictable nature of the voltage drop.

C. IR-drop Analysis during Pre-bond Testing

In this section, we study the number of probe pads that are required to provide power during pre-bond test. The bottom die has 7 metal layers, and the top die has 6. Metal 7 on the bottom die is used exclusively for TSV landing pads and probe pads. Testing of the top die, the die which has solder bumps is trivial. Power can simply be provided to all the power pins. The voltage drop for Die0 is obtained as follows: scan0 =

978-1-61284-657-6/11 $26.00 © 2011 IEEE

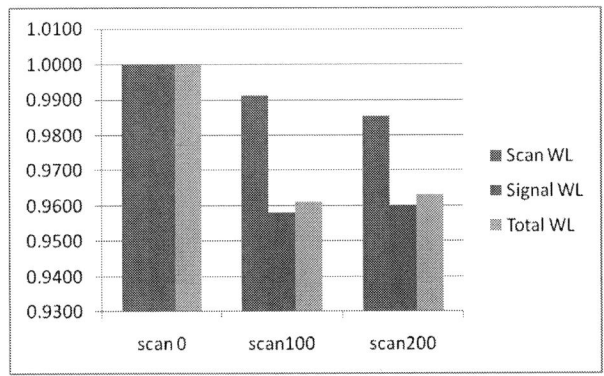

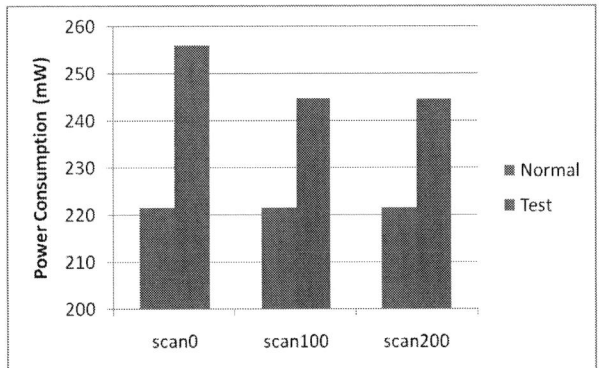

(a) Impact of Scan TSV count on wirelength

(b) Impact of scan TSV count on power

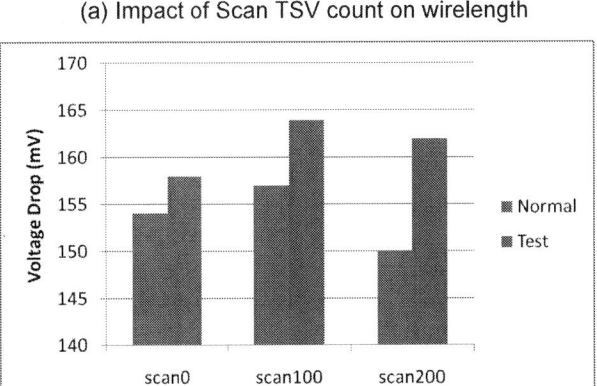

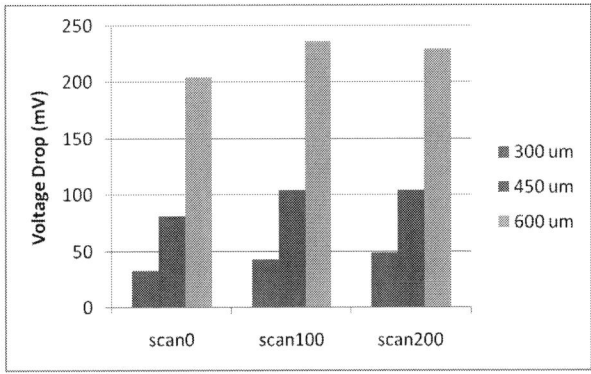

(c) 3D IR-drop (post-bond-testing)

(d) 2D IR-drop (pre-bond testing)

Fig. 9. Various results

23mV, scan100 = 18mV, scan200 = 18mV. As expected, the voltage drop is well within noise margins, and power delivery for the top die is not an issue.

In the case of the bottom die, however, probe pads have to be added at select locations to provide power. Additionally, it is not possible to smoothly control their pitch as they have to align with the power and ground TSVs. We add probe pads at two, three and four times the P/G TSV pitch and show the results in Figure 9(d). We observe that with an increase in pitch, there is a sharp increase in the voltage drop of die1. Thus, the P/G probe pads need to be inserted carefully to control voltage drop. Assuming a 150mV constraint, we see that a pitch of 600um violates the noise constraint.

VII. CONCLUSION

In this work, we explored the impact of scan chain TSVs on wirelength, power and voltage drop of 3D ICs. We also explored the design options and requirements for power delivery during pre-bond testing. Experimental results show that increasing the number of scan TSVs upto a certain point helps reduce both wirelength and power consumption. We also studied the impact of P/G probe pads used for pre-bond testing on voltage drop.

REFERENCES

[1] D. Kim, K. Athikulwongse, and S. Lim, "A study of Through-Silicon-Via Impact on 3D Stacked ICs," in *Proc. IEEE Int. Conf. on Computer-Aided Design*, 2009.

[2] "T.Vucurevich. The Long Road to 3D Integration: Are we there yet?" *Key note speech at the 3D Architecture Conference*, 2007.

[3] X. Wu, P. Falkenstern, K. Chakrabarty, and Y. Xie, "Scan Chain Design and Optimization for Three-Dimensional Integrated Circuits," *ACM Journal on Emerging Technologies in Computing Systems*, 2009.

[4] X. Zhao, D. Lewis, H.-H. S. Lee, and S. K. Lim, "Pre-bond Testable Low-Power Clock Tree Design for 3D Stacked ICs." in *IEEE International Conference on Computer-Aided Design*, 2009.

[5] D. Lewis and H. H. S. Lee, "A Scan Island Based Design Enabling Pre-Bond Testability in Die Stacked Microprocessors," in *IEEE International Test Conference*, 2007.

[6] E. J. Marinissen and Y. Zorian, "Testing 3D Chips Containing Through Silicon Vias," in *IEEE International Test Conference*, 2009.

[7] L. Jiang, L. Huang, and Q. Xu, "Test Architecture Design and Optimization of Three-Dimensional SoCs," in *Design, Automation and Test in Europe*, 2009.

[8] L. Jiang, Q. Xu, K. Chakrabarty, and T. Mak, "Layout-Driven test-Architecture design and Optimization for 3D SoCs under Pre-Bond Test-Pin-Count Constraint," in *IEEE International Conference on Computer Aided Design*, 2009.

[9] "International Technology Roadmap for Semiconductors 2009," *http://www.itrs.net/* , 2009.

[10] W. R. Mann, F. L. Taber, P. W. Seitzer, and J. J. Broz, "The leading edge of Production Wafer Probe Test Technology," in *IEEE International Test Conference*.

[11] K. Karklin, J. Broz, and B. Mann, "Bond Pad Damage Tutorial," in *IEEE Semiconductor Wafer Test Workshop*, 2008.

[12] T. Hauck, I. Schmadlak, C. Argento, and W. H. Muller, "Damage Risk Assessment of Under-Pad structures in Vertical Wafer Probe Technology," in *IEEE European Microelectronics and Packaging Conference*, 2009.

[13] E. Marinissen, J. Verbree, and M. Konijnenburg, "A structured and scalable test access architecture for TSV-based 3D stacked ICs," in *IEEE VLSI Test Symposium*, 2010.

[14] http://www.opencores.org/.

978-1-61284-657-6/11 $26.00 © 2011 IEEE

Exploiting Rotational Symmetries for Improved Stacked Yields in W2W 3D-SICs

Eshan Singh
Electrical Engineering
Stanford University
esingh@stanford.edu

Abstract – **Three-dimensional Stacked Integrated Circuit packages interconnected using high speed Through-Silicon Via technology can be efficiently manufactured using a wafer-to-wafer stacking process. Efforts to mitigate degradation in the composite yield of the stacked die are primarily focused on matching defect maps while assigning the pre-tested wafers from the available wafer repository to individual wafer stacks. In this paper we show how rotational symmetry can be exploited to increase the number of available wafer defect maps by a factor of four, thereby significantly improving matching possibilities and hence yield. We further apply our approach to processor-memory stacks, with relatively high memory die yield from redundancy and repair, for which we also present new heuristic matching algorithms. Altogether, our new approach shows viable absolute yields, with yield improvement of better than 25% over stacking without any matching. This is a significant advance over earlier results.**

I. INTRODUCTION

Three-dimensional Stacked Integrated Circuit packages (3D-SICs) comprising multiple stacked die interconnected using high speed Through-Silicon Via (TSV) technology are currently the focus of considerable development effort by the industry[1-4]. Such packages promise high integration density along with exceptional performance and low power because of the small (micrometer) dimensions and minimal electrical loading of the TSVs as compared to traditional off chip interconnections [5, 6]. Importantly, they also allow heterogeneous die, each implemented in its own individually optimized fabrication technology, to be compactly integrated together at efficiencies approaching monolithic fabrication [7, 8].

Three basic die stacking approaches can be used in fabricating 3D-SICs from traditionally processed wafers: Wafer-to-Wafer (W2W), Die-to-Wafer (D2W), and Die-to-Die (D2D) [1, 2]. Of the three, the W2W approach appears most attractive since considerable additional processing (including careful alignment) is required while fabricating the TSV connections during the die stacking and bonding steps associated with assembling the 3D-SICs. Handling, stacking and processing wafers in a W2W process rather than individual die required by the other approaches is generally much easier [9].

Unfortunately, W2W stacking suffers from the problem of yield degradation from compounding. For example, if the yield of good die on a wafer is 50%, the yield of a two such wafers stacked together can only be expected to be 25%, i.e. both stacked die at only 25% of the sites on the stacked wafer can be expected to be defect free. For a four wafer stack, the compound yield drops to an unacceptable 6.25%. Indeed, unless the 3D-SIC requires die from a heterogeneous mix of technologies, there is little benefit, from a yield perspective, in going to a 3-D technology over a single traditional die with a larger area. W2D or D2D stacked technologies, on the other hand, can allow individual pretested good die to be stacked, achieving much higher yields and silicon use efficiency. However, working with individual bare die during fabrication is extremely challenging. Consequently, techniques to mitigate the yield degradation from compounding in W2W technologies are being extensively researched [10-14].

The main approach being considered to improve the yield of stacked is wafer matching [11, 14]. This assumes the availability of a repository of pre tested wafers of each type to be stacked. The size of this repository is generally taken to be the one or two standard lots of 25 wafers of each type to be stacked. These can be reasonably expected to be available at the wafer stacking step during 3D-SIC fabrication. Wafers are then carefully matched in constructing the stacks so as to maximize the yield of defect free stacked 3D-SICs based on the known test results for each die location in the pretested wafers. However, in a fair evaluation of the effectiveness of such a wafer matching approach towards enhancing the 3D-SIC yield over random wafer stacking, it must be assumed that the matching procedure utilizes all the wafers in each lot, i.e. none of the wafers are scrapped or wasted.

Research so far has suggested that the benefits of wafer matching in terms of yield improvement are quite modest. Early researchers even questioned the usefulness of such a strategy [10]; subsequent work has, however, shown yield improvements up to about 10% in some applications. The reason for this limited yield improvement is the large number of possible combinations in which defective die can be distributed over a wafer. Even optimum wafer matching, using relatively small repositories of 25 or 50 wafers, typically results in stacked arrangements where relatively few additional defective die locations on the wafers are aligned, when compared to stacking random wafers. The availability of a larger pool of wafers of each type to perform the matching can obviously help improve yield further. *A key contribution of this paper is to show that, with die appropriately orientated on the wafer during wafer processing, rotational symmetry can be exploited to increase the number of wafer defect maps available to perform the wafer matching by a factor of four, without any increase in the size of the wafer repository.* Our simulations presented here show that this can lead to a significant improvement in the 3D-SIC yield at virtually no additional cost.

While earlier studies have mostly assumed similar wafer die yields for the different wafers to be stacked, in this paper we particularly focus on the important processor-memory 3D-SIC applications that offer powerful computing systems in a single package [15]. Such 3D packages typically comprise of one or two low yielding processor die integrated with a larger number of DRAM die that can achieve much higher die yields through redundancy and repair. Multiple identical memory die in the 3D-SIC also allow for the possibility of selecting the wafers for stacking multiple memory tiers from a single consolidated memory wafer repository. For example, 50 wafer stacks of a 6 tier 3D-SIC comprising a processor die and five identical memory die can be assembled from 50 (2 lots) processor wafers and 250 (10 lots) memory wafers. Consolidating the memory wafers in a single repository in this manner offers additional selection possibilities while matching wafer defect maps and can further enhance the yield. Given that optimum wafer matching is known to be a NP hard problem which becomes computationally intractable in practice for as few as four stacked wafers[11], *we investigate new heuristic wafer matching algorithms for this application where multiple tiers of identical wafers are to be selected from a single consolidated repository.* The results presented here show here that realistic processor memory 3D-SIC designs can achieve viable yields, with a yield improvement of up to 25% over unmatched wafer stacking.

The rest of this paper is organized as follows. Section II reviews prior work on wafer matching. We present our proposed new approach of using rotational symmetry to enhance the repository of available wafer defect maps in Section III, along with simulation results evaluating the resulting improvement in stacked die yield. We apply this methodology to processor-memory 3D-SICs in Section IV, where we also introduce new heuristic wafer matching algorithms for this key application. Section V concludes the paper.

II. PRIOR WORK ON WAFER MATCHING

Using wafer matching to increase the compound yield of stacked wafers was first suggested in [10]. However, for the relatively small die size and 25 wafer (one standard lot) repository considered, the yield improvement for a simple 2-tier stack from matching was quite limited, less than 2%, leading to questions regarding the usefulness of such an approach. A more comprehensive study of wafer matching was presented in [11], which also considered multi-tier stacks. In [11], optimum wafer matching was shown to be a NP hard problem for stacks with more than two tiers, which can quickly become computationally intractable for many practical applications. Computationally feasible heuristic wafer matching approaches are presented in [11] and also in [14]. These are yield greedy iterative methods that first match two layers of the wafer stack to obtain an initial collection of partially stacked wafers, and then iteratively build up the rest of the stack by adding an additional wafer to the stack during each subsequent matching step. The algorithm for stacking two wafers based on bipartite graph matching presented in [11] is optimal in enhancing yield and has only polynomial time complexity $O(N^3)$. For stacking additional wafers the authors suggest a greedy heuristic which selects the next wafer lot to be matched to a partially formed stack by considering all the remaining lots and picking the lot that maximizes the stack die yield at each step of the iteration. (It is important to note that this selection of the wafer stacking order is virtual, employed only for the purpose of obtaining a good matching for yield optimization among the wafers to be stacked. Once a wafer matching is obtained, actual physical stacking of wafers during fabrication is free to follow a different order as determined by the by the adjacency requirements of the die in the design.)

Unfortunately, the yield benefits from wafer matching in simulation experiments presented in [11] and [14] using repositories of one to two wafer lots remain modest. For example, in the best case presented in the simulation plots in [14], that of a six-tier stack of 50mm die with 81.66% die yield, the stacked die yield increases from 29.6% to 32.5% with wafer matching, an improvement of less than 10%. While somewhat larger percentage improvements are possible for multi-tier stacks of large low-yielding die, the absolute stacked yield in such cases becomes unacceptably low. The effectiveness of wafer matching can obviously be improved by increasing the size of the wafer repository beyond the 25 or 50 wafers (one or two standard wafer lots) used in the above experiments. However, this can present logistical challenges and manufacturing inefficiencies in managing inventories. Our proposed approach of exploiting wafer rotational symmetry described in the next Section allows a four-fold increase in the wafer defect-maps available for matching without any increase in the physical size of the repository.

Earlier research on wafer matching [11, 14] has also implicitly assumed that die yields in each of the wafer lots to be stacked to be the same. Given that W2W stacking requires that the die being stacked are all the same size, this appears to be a reasonable assumption if the defect densities for each wafer lot are the same. However, in practice, the yields for the different die in the stack can be quite different for reasons such as differences in fabrication technology, process maturity, and die area utilization (feature density), defect clustering etc. Perhaps the largest contributor to a difference in die yields is memory redundancy and repair, which can raise memory die yield to 90% or more, while complex processors with similar die area may only achieve yields in the region of 50%. Since 3D-SICs comprising of processor memory stacks represent a particularly important application area that have attracted much research interest [3, 15], in this work we also focus on wafer matching algorithms and simulation results for stacking such wafers with significantly differing yields.

III. MULTIPLE DEFECT MAPS FOR WAFER ROTATIONAL SYMETRY

The basic idea behind obtaining four different defect maps from a single wafer for optimizing matching is straightforward, although it does require that the die be appropriately (and differently) oriented in each quadrant of the wafers to be stacked. Figure 1 illustrates the concept. Notice that if all wafers are fabricated in this manner, then the die orientation at each location on a wafer will look identical for each 90 degree rotation of the wafer. Thus each wafer can be mapped and stacked in any one of four different rotational positions, each offering a different defect map. This results in a four-fold increase in the size of the available "virtual" repository for any size of the physical wafer repository. In fact, even the last remaining wafer to be stacked in any layer now offers four possibilities to increase yield by overlapping defects; without the possibility of rotation, there are no choices at this final mapping stage.

To directly compare and evaluate the benefit of our new approach that exploits wafer symmetry, we repeated the simulation experiments in [14]. These experiments assume 300mm wafers, a defect density of 0.5 defects/cm² and a

clustering parameter α of 0.5. (Because the clustering parameter is highly nonlinear, this value indicates very strong defect clustering which can lead to optimistic estimates of yield. The SIA Roadmap [16] suggests using an α of 2.0. However, we have used the 0.5 value to ensure that our work is consistent with the experiments in [14] for purposes of comparison) Different die sizes, from 25mm² to 125mm², are considered; this die area dictates the number of die on the wafer, and the die yield. Note that while the defect clustering parameter is used in determining individual die yield based on Stapper's yield formula[17], the clustering of defective die on the wafer is not considered in both our simulations and earlier work, i.e. it is assumed that defective die are uniformly distributed over the wafer. (It can be argued that with wafer defect map matching, the projected die stack yield under this uniform defect distribution assumption is pessimistic, much in the same way as the simpler Poisson yield model is known to underestimate die yield. However, this remains to be verified.) To smooth out the impact of random variations in the simulated wafer defect maps, each experiment reported here was run 1000 times and the results averaged. As in prior work, in these first experiments we use an iterative approach for mapping wafers in a multi-tier stack, using the greedy algorithm described in [14] to match wafers in the two lots at each iterative wafer matching step. This algorithm first considers all pairs of wafers in the two repositories to be matched, and matches the two wafers that give the highest yield. The process is repeated (with the now smaller repositories) until all wafers are matched.

Figure 2 shows simulation results for stacked die yield versus the size of the wafer repository for a 50mm² die size, and a corresponding die yield of 81.65%. The plots show 2, 4, and 6 tier stacks, both with and without the benefit of using wafer rotation to increase the number of available defect maps.

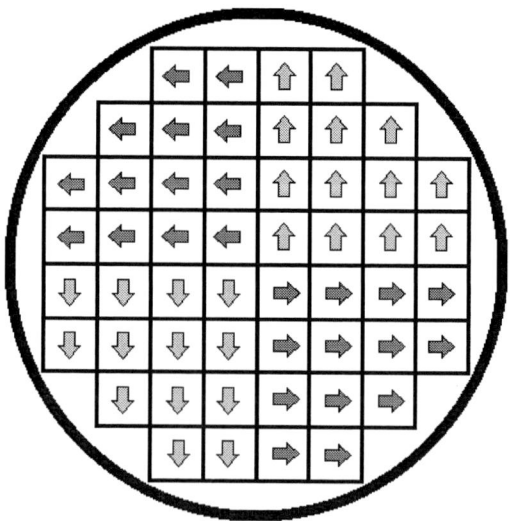

Figure 1: Orientation of Die on the Wafer

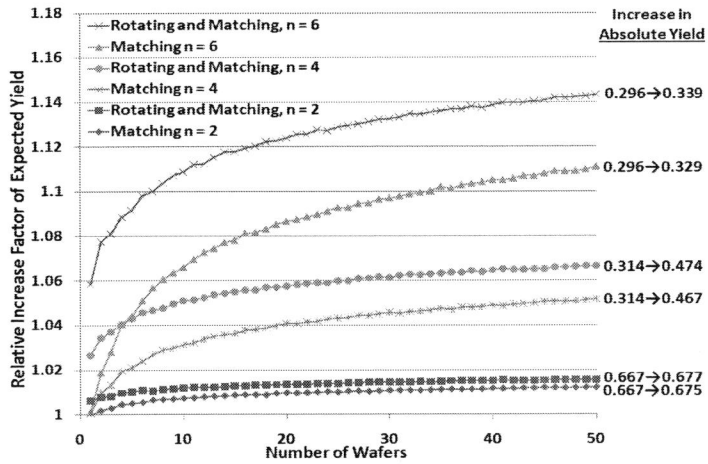

Figure 2: Increase in Yield from Rotating and Matching Compared to Just Matching for a Stack Height of (n) Wafers. Die Yield = 81.67%, 1296 Die per Wafer

The Figure clearly shows the yield improvement from rotation. This benefit is most pronounced for a repository size (number of wafers) of 1 where without rotation there are no options for defect mapping. As the number of wafers in the repository grows, the relative benefits of rotation gradually decrease, but still remain significant.

Figure 3 shows results for a two-tier stack and varying die area and yield. As mentioned earlier, varying the die area also varies the wafer yield as modeled by Stapper's yield formula [17], and the number of die that can be accommodated on the 300mm wafer. In this simulation the wafer yields were 89.44% for the 25 mm^2 die, 75.59% for the 75 mm^2 die and 66.67% for the 125 mm^2 die. The Figure shows that rotation continues to provide noticeable improvement in the total yield as the area is varied. Once again the improvement is most apparent in the lower yielding stacks, which are made up of the largest area die with the lowest yield.

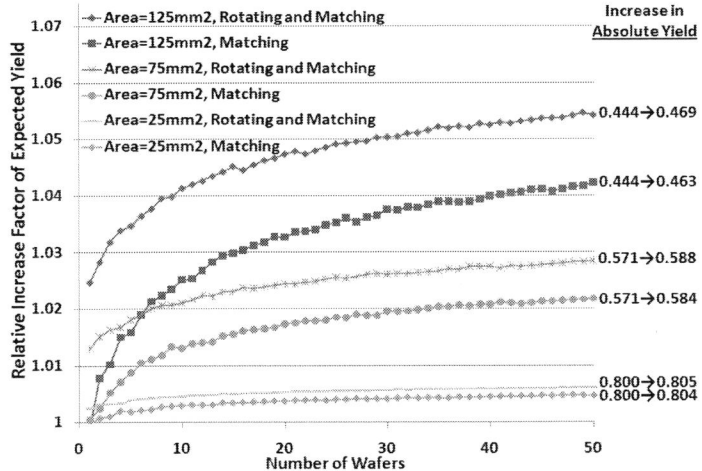

Figure 3: Increase in Yield for Different Die Areas for Two-Tier Stacks

IV. PROCESSOR MEMORY 3D-SICs

Despite the significant improvement in yield from exploiting wafer rotational symmetries, the results in the previous section indicate that multi-tier W2W stacking still faces serious cost challenges. If the die size is very small, there appears to be little benefit from moving to a high cost 3D-SIC implementation over just increasing die area. (Although, the latter approach may not practical if the functionality in different die require incompatible technologies for their implementation.) For large die, 3D multi-tier stacking offers significant advantages in size and form factor, but such die are typically low yielding and will suffer from prohibitively low compounded yield for multi-tier stacks, even with optimum wafer matching. Observe from Figure 3 that a 2-tier stack of a large die with 66.67% yield at best achieves 46.9% yield with rotation and mapping. The yield for a 4-tier stack will be less than 25%, which already appears to be on the margin of commercial viability. Taller stacks of die, of this size or larger, are clearly impractical.

However, there may be one important application involving large die that offers an important exception. Processor-memory 3D-SICs hold out the potential of powerful computing systems in a single package and are being aggressively pursued [3, 15]. Such 3D packages typically comprise one or two large low yielding processor die integrated with a larger number of DRAM die. The DRAMs, despite their large area, can achieve much higher die yields through redundancy and repair. Consequently, stacking a number of high yielding DRAM die, on a lower yield processor dies, can still result in acceptable 3D-SIC yield. We now show how our improved mapping approach with wafer rotation can be effective in improving the stack yield for this important application.

The primary focus of our simulations in this Section are processor memory stacks typically comprising one or two complex processor die and a larger number of high yielding (repairable) memory die. A further assumption is that the die sizes are large: 125-200 sq mm. A 300 mm wafer can accommodate approximately 400 die of 150 sq mm area each. For simplicity we assume a square 20x20 die array on each wafer, although it is easy to see that the exact shape of the die array in each wafer quadrant (assuming that the four quadrants are identical as shown in Figure 1) does not affect the simulation results. In general, our simulations again follow the experimental approach from [14] described earlier in the paper.

Figure 4 shows yield results for 4 and 6 tier stacks comprising of a single processor die with 50% yield and the rest memory

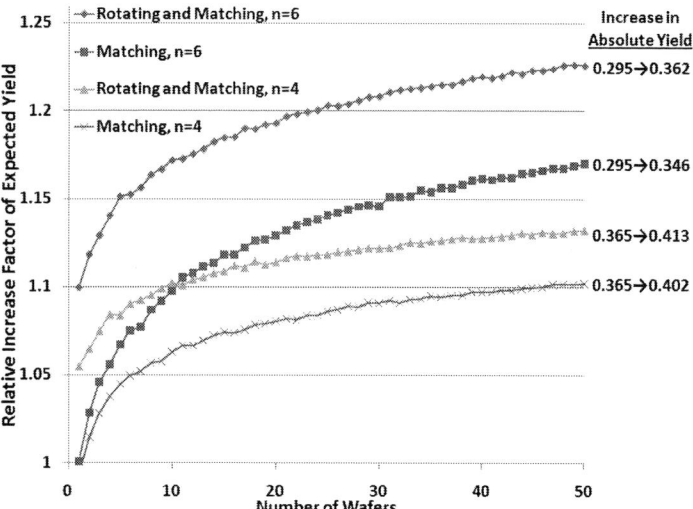

Figure 4: One 50% Yield Processor Stacked with 90% Yield Memories for a Total Stack Height of (n) Wafers

die with 90% yield. In this simulation, we have used the same matching algorithm as before, with each tier iteratively added to the stack. Observe that the yield improvement from mapping with defect map rotation is 24%. More importantly, the stacked die yield for even a 6-tier stack is a very viable 36.2%, not too far below the 50% yield of the processor die alone.

Figure 5 shows the same results where the processor memory stack comprises of two low yielding wafers with 65% yield and the remaining memory wafers with 90% repaired yield. Here the yield improvement from rotation and matching is even better and approaches 30% for a 50 wafer repository; the composite yield is 35.4%.

A. Mapping from a Consolidated Memory Wafer Repository

Stacking multiple identical memory die on one (or two) processor die additionally opens up the additional possibility of treating all the identical memory wafers that go into the different layers of the stack as a single pool for wafer matching. Thus, for example, when working with a 50 wafer repository to fabricate 6-tier 3D-SICs comprising a processor wafer and 5 memory die, the 250 memory wafers can all be treated as a single repository.

We next evaluate two different heuristic algorithms for wafer matching in such an environment. We assume here that if there are multiple processor die in the 3D-SIC, these are matched first. Our focus is on mapping memory wafers from a single consolidated repository with processor wafers or partially completed stacks of processor wafers. *Algorithm 1* builds up the wafer stacks layer by layer using memory wafers from the consolidated memory wafer repository. Memory wafers are assigned to all wafer stacks in a given layer before moving on to assign wafers for the next layer. At each assignment step at a given stack level, all currently unassigned stacks are compared with all available memory wafers to find the highest yielding match; this match is committed and the process is then repeated. *Algorithm 2* relaxes the requirement that all wafer stacks in a given layer be assigned before moving on to assign wafers for the next layer. Here at each assignment step all partially completed stacks are compared with all available memory wafers to find and commit the highest yielding match. When a stack reaches the required number of memory wafers, it is removed from further consideration. The process continues until all stacks are fully assigned.

Figure 6 shows the simulation results for 6 tier stacks comprising of one low yielding wafer with 50% yield and five

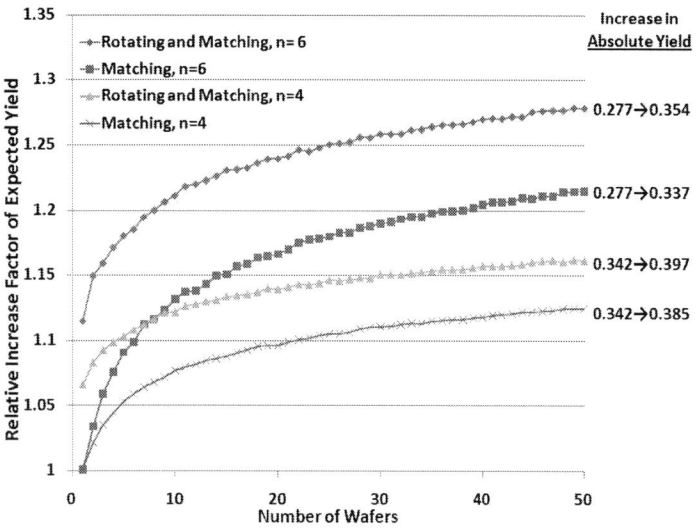

Figure 5: Two 65% Yield Processors Stacked with 90% Yield Memories for a Total Stack Height of (n) Wafers

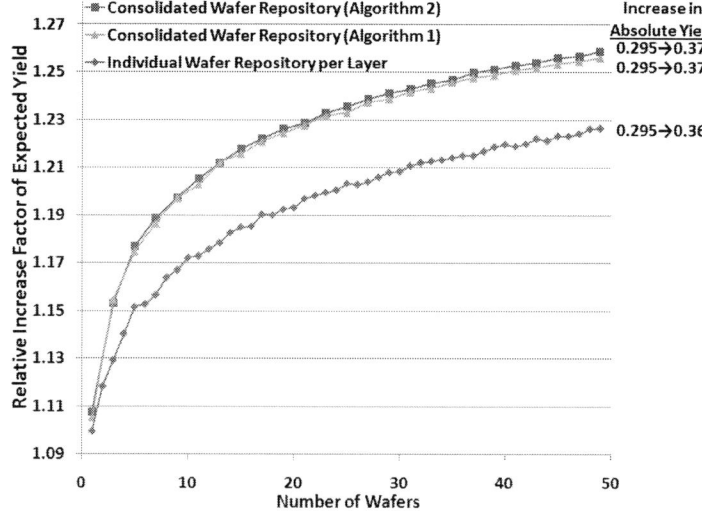

Figure 6: Yield Improvement from Consolidated and Individual Memory Repositories

memory wafers with 90% repaired yield as in Figure 4, stacked with rotation and matching. The benefits from having a common memory wafer repository over a separate repository for each memory layer can be clearly observed. The yields from the two algorithms for matching wafers from the consolidated repository are very close. *Algorithm 2* marginally improves over *Algorithm 1* by allowing for the flexibility in matching wafers to any partially completed stack.

V. CONCLUSION

The W2W stacking approach appears most attractive for fabricating 3D-SICs from individually processed wafers since considerable additional processing with careful die alignment is required while forming the TSV connections and other specialized inter die structures. Handling bare die is difficult. However, W2W stacking suffers from serious yield degradation due to die yield compounding. The main approach being considered to mitigate this problem is wafer matching. It is reasonably assumed that a repository of one or two lots of pre tested wafers of each type to be stacked is available at assembly. Wafer defect maps are then carefully matched in forming the wafer stacks so as to align the good die locations on the different wafers and maximize stacked die yield. Unfortunately, the best results previously reported for stacked die yield improvement from such a mapping strategy were limited to about 10%. Furthermore, for multi-tier stacks of low yielding wafers of complex die, the absolute stacked die yield was often unacceptably low.

In this paper we have shown that with proper wafer preparation, rotational symmetries on the wafer can be exploited to increase the number of available wafer defect maps by a factor of four. This can increase the effective size of available wafer repository by the same factor, thereby allowing for significantly more mapping defect possibilities and much improved yield. We have also applied our new wafer matching approach to important processor-memory 3D-SIC applications that offer powerful computing systems in a single package. Such 3D packages typically comprise one or two low yielding processor die integrated with a larger number of DRAM die that can achieve much higher die yields through redundancy and repair. Multiple identical memory die in the 3D-SIC also allow for the possibility of selecting the wafers for stacking multiple memory tiers from a single consolidated memory wafer repository. For this situation we have developed and evaluated new heuristic wafer matching algorithms. Altogether, our new approach shows quite viable absolute yields for this important application, with yield improvement of better than 25% when compared to stacking without any matching. This is a significant advance over earlier results.

REFERENCES

[1] K. Banerjee, S. J. Souri, P. Kaput, and K. C. Saraswat, "3-D ICs: A novel chip design for deep-submicrometer interconnect performance and systems-on-chip integration," in *Proc. IEEE*, vol. 89, no. 5, pp. 602–633, May 2001.

[2] J. A. Burns, B. F. Aull, C. Chen, C.-L. Chen, C. L. Keast, J. Knecht, V. Suntharalingam, K. Warner, P. Wyatt, and D.-R. Yost, "A wafer-scale 3-D circuit integration technology," in *IEEE Trans. Electron Devices*, vol.53, no. 10, pp. 2507–2516, Oct. 2006.

[3] R. S. Patti, "Three-dimensional integrated circuits and the future of systems-on-chip designs," in *Proc. IEEE*, vol. 94, no. 6, pp. 1214–1224, Jun. 2006.

[4] E. Beyne, B. Swinnen, "3D System Integration Technologies," in *Proceedings of IEEE International Conference on Integrated Circuit Design and Technology (ICICDT)*, pp. 1-3, June 2007.

[5] Jan Van Olmen et al., "3D Stacked IC Demonstration using a Through Silicon Via First Approach," in *Proceedings IEEE International Electron Devices Meeting (IEDM)*, pp. 1–4, December 2008.

[6] K. Banerjee et al, "3-DICs: A Novel Chip Design for Improving Deep-Submicrometer Interconnect Performance and Systems-on-Chip Integration," in *Proceedings of the IEEE*, vol. 89, no. 5, pp. 602–633, May 2001.

[7] G. H. Loh, Y. Xie, B Black, "Processor Design in 3D Die-Stacking Technologies," in *IEEE Micro*, vol. 27, no. 3, pp. 31–48, May/June 2007.

[8] R. Weerasekera et al, "Extending Systems-on-Chip to the Third Dimension: Performance, Cost and Technological Tradeoffs," in *Proceedings International Conference on Computer-Aided Design (ICCAD)*, pp. 212–219, November 2007.

[9] X. Dong and Y. Xie, "System-Level Cost Analysis and Design Exploration for Three-Dimensional Integrated Circuits (3DICs)," in *Proceedings IEEE Asia South Pacific Design Automation Conference (ASP-DAC)*, pp. 234-241, January 2009.

[10] L. Smith, G. Smith, S. Hosali, and S. Arkalgud, "3-D: It all comes down to cost," presented at the *3-D Architectures for Semiconductor Integration and Packaging*, 2007.

[11] S. Reda, G. Smith, and L. Smith, "Maximizing the Functional Yield of Wafer-to-Wafer 3-D Integration," *IEEE Transactions on VLSI Systems*, vol. 17, pp. 1357–1362, September 2009.

[12] C. Ferri, S. Reda, and R. I. Bahar, "Strategies for improving the parametric yield and profits of 3-D ICs," in *Proc. Int. Conf. Computer Aided Design*, 2007, pp. 220–226.

[13] L. Smith, G. Smith, S. Hosali, and S. Arkalgud, "Yield considerations in the choice of 3D technology," in *Proc. IEEE Int. Symp. Semiconductor Manufacturing*, 2007, pp. 535–537.

[14] J. Verbree, E. J. Marinissen, P. Roussel, D.Velenis, "On the Cost-Effectiveness of Matching Repositories of Pre-Tested Wafers for Wafer-to-Wafer 3D Chip Stacking," in *Proceedings 2010 European Test Symposium*, May 2010.

[15] D. S. Kung, Y. Xie, "Guest Editors' Introduction: Opportunity and Challenges of 3D Integration," in *Design and Test of Computers IEEE*, vol. 26, no. 5, pp. 4-5, Sept.-Oct. 2009.

[16] "International Technology Roadmap for Semiconductors," SIA, 2009.

[17] C. H. Stapper "On a Composite Model to the IC Yield Problem," in *IEEE Journal of Solid-State Circuits*, vol. 10, no. 6, pp. 537-539, Dec 1975.

Test and Characterization of High-Speed Circuits

Organizer: Saghir Shaikh, Broadcom

Abstract: Test, validation and characterization of high-speed circuits is a becoming a complex issue due to increase in circuit marginality, higher fallout, and more complex test solutions. The issue is compounded by the complex interactions between packaged components; interconnect design and customer board designs. This session will address the challenges in characterizing high-speed circuits including PLLs and SERDES. It describes test methodologies to overcome those challenges using industrial test cases.

"Characterization Of The Digital PLLs On An 8-Core Microprocessor Using Electrical And Optical Techniques"
By: Kevin Stawiasz, Keith Jenkins, Peilin Song*, Franco Stellari, Jose Tierno, Alexander Rylakov and Daniel Friedman (IBM)

"Challenges in High Volume Manufacturing Test and System Correlation for High Speed IO"
By: Anne Meixner* (Intel Corporation)

"High-Speed SerDes Characterization"
By: Dongwoo Hong* and Matthew Isaacs (Broadcom)

Session 2

978-1-61284-657-6/11 $26.00 © 2011 IEEE

Expedited Response Compaction for Scan Power Reduction

Samah Mohamed Saeed
Computer Science Department
New York University - Polytechnic Institute

Ozgur Sinanoglu*
Computer Engineering Department
New York University - Abu Dhabi

Abstract

Transitions embedded in between consecutive stimulus/response bits toggle scan cells during shift operations. The consequent switching activity in the scan chains further propagate into the combinational logic, resulting in elevated power dissipation levels, and thus, endangering the reliability of the chip being tested. Based on the observation that the content of scan chains during shift operations is irrelevant and unimportant, we propose an expedited response compaction technique in order to reduce power dissipation during scan operations. Parallelized (and expedited) compaction operations help compress the entire capture response onto a single reference chain during the first portion of shift cycles, enabling a simultaneous constant-0 feed to all the remaining chains, in which no scan-out power is dissipated during the subsequent shift cycles. This DfT-based approach is non-intrusive for design flow, requires a very minor investment in area, and in turn delivers significant savings in test power. The proposed solution reduces test power without resorting to x-filling, enabling orthogonal x-filling techniques to be applied in conjunction, while retaining the observed responses intact. Experimental results justify the efficacy of the proposed technique in attaining test power reductions.

1 Introduction

Serial shift operations during the scan-in of stimulus and the scan-out of response bits result in a switching activity in the scan chains, which propagate into the combinational logic, dissipating further dynamic power unnecessarily. The end-result of this problem, unless treated properly, could be an unexpected behavior of the design, thereby resulting in a yield loss, or reliability problems. Elevated levels of *peak power*, which is the maximum instantaneous power throughout the entire test process, is the cause of the former problem, while the underlying reason for the latter problem is rather *average power* that is the total power dissipation averaged over the duration of the test application process [1]. As the test application process is dominated by shift operations, average power mostly depends on scan power, and thus, the impact of capture power on average power is negligible. Capture power is more of a concern when peak power is the targeted issue.

Numerous scan power reduction methodologies have been proposed; most of these methodologies are outlined in [1]. The utilization of externally controlled gates or modified scan

---

*Email: ozgursin@nyu.edu, Phone: +1 (347) 309 8079

---

cell designs [2, 3, 4, 5] at the expense of functional performance degradation, appropriate primary input assignments during shift cycles [6, 7], test vector ordering and scan-latch clustering/ordering techniques [8, 9], modification of test cube compaction [10], test generation and don't care bit specification [11, 12, 13, 14, 15, 16, 17, 18, 19], scan chain design [20, 21, 22, 23], shift clock spreading [24], test pattern scrubbing [25], and scan chain segmentation via clock gating [26, 27, 28, 29, 30, 31, 32, 33] have all been proposed to deliver savings in test power.

A variety of low-power test solutions targeted for compression-based scan architectures have been proposed recently. X-filling solutions for addressing capture power [34, 35, 36], or both shift and capture power [37], have attained reductions at the expense of an increase in pattern count. A similar end-result have been observed also with DfT-based solutions in the form of filling some of the chains with constant-0's and disabling capture in scan chains [38], or by disabling the clocks of scan chains [39, 40].

In an effort towards identifying a test power reduction solution that retains compression level intact and that does **not** interfere with the design flow via intrusive techniques such as clock gating, we note the following simple observation. The content of the scan chains during scan operations is irrelevant and unimportant, enabling reduction of transitions in the scan chains during shift cycles. As long as the intended stimulus is delivered prior to the capture cycle and responses are compacted the same way, the quality/application of test remains intact. Very recently, a DfT-based approach [41], for reducing scan-in power in Illinois scan architecture, has been proposed to reduce test power based on this observation. In this approach, which we shall refer to as the *Deferred-Broadcast (DB)*, the broadcast stimulus is distributed from only one reference chain into the other chains during the final small fragment of the shift process, thus allowing all-but-one chains to receive constant-0's for the majority of shift cycles. As a result, scan-in power is reduced, while the intended stimulus is delivered intact; and, this is achieved without clock gating.

The shortcoming of the DB architecture [41] is that it only targets scan-in power reduction and overlooks scan-out power. While each stimulus and response transition equally contributes to switching activity during test, scan-out power typically dominates test power; stimulus don't care bits (x's) that remain post-compression can be filled properly (0-fill or repeat-fill) to leash the scan-in power, while such a direct control over response transitions, with the exception of probabilistic and inexact simulations, does not exist. Thus, although the DB architecture [41] may attain significant savings in scan-in power, these savings may correspond to only a small fraction of the overall scan power.

In this work, we propose a complementary solution, *Expedited-Compact (EC)*, that targets scan-out power reduction. The expedited-compact feature in the proposed architecture enables the collection of the compacted responses in one chain by utilizing that chain as a buffer. Overwriting of the captured response (upon its expedited compaction) in all the other scan chains with shifted constant-0 values in turn delivers reductions in scan-out power. For industrial cases that employ 0-fill so as to eliminate transitions in stimuli, the proposed technique is **5 to 66 times** more effective than DB [41] in reducing average test power. The proposed features can be implemented at a very minor area cost, yielding significant power savings cost-effectively. Furthermore, as the proposed EC and the previous DB approach are complementary and orthogonal, they can be utilized in conjunction to reap both scan-in and scan-out power reduction.

Most importantly, the proposed EC approach can be utilized with *any* compression-based scan architecture. While such an extension has not been mentioned in [41] for the DB approach, it is also extendable, with further area investment, for utilization with compression types other than fanout based compression. Given the dominance of scan-out power, however, area cost to be expended for integrating DB into other compression environments may not justify the return in overall power reduction, rendering area investments in EC more favorable.

The proposed technique has the following features:

- Expedited-compact operations do not require design-flow intrusive hardware such as clock gating logic, retaining the clock tree intact, **which differentiates the proposed solution from the traditional scan chain segmentation techniques** [26, 27, 28, 29, 30, 31, 32, 33]. If clock tree modifications can be afforded, the proposed scheme can be used in conjunction with clock gating to disable the clock of the non-reference chains for most of the shift cycles, capable of delivering commensurate power reductions in the clock tree as well.

- Expedited-compact operations impose no modifications on the test development process, enabling the application of a given set of test data as is.

- Expedited-compact operations have no impact on the observed responses (identical compacted responses are observed with or without EC), rendering the test quality and test application process (time) intact.

- Expedited-compact operations enable the filling of most of the scan chains with all 0's after a small fraction of the shift operations, delivering scan-out power reductions.

- Expedited-compact operations deliver **average scan power reductions between 70-85%** for the industrial test cases that we have experimented with.

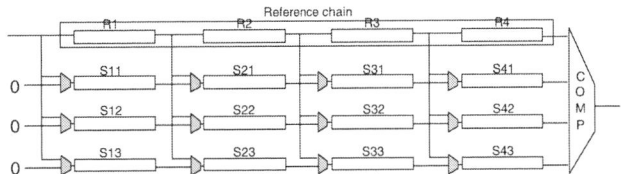

Figure 1. Deferred-Broadcast (DB) - 4 blocks

- Expedited-compact operations reduce power without resorting to x-filling, presenting a solution that is orthogonal to x-filling [11, 12, 13, 14, 15], and hence, can be applied in conjunction to either further boost the average power reduction levels, or more importantly, pursue co-optimizations with peak shift and/or peak capture power.

- Expedited-compact operations can be implemented at a projected area cost of less than 0.1% for large-sized industrial circuits. Furthermore, the inserted hardware has no impact on the functionality of the circuit timing-wise.

2 DB and EC architectures

In this section, we present the DB [41] and the proposed EC architectures. We also elaborate on how these architectures deliver power savings. For simplicity of presentation, we illustrate DB with fanout-based compression and EC with XOR-based compaction.

The DB architecture [41] for a single scan-in channel fanning out to four scan chains is provided in Figure 1. The topmost chain in the architecture is referred to as the *reference chain* (R), while the other (three) chains are referred to as the *shadow chains* (S). Also, in this example, the DB architecture decomposes every scan chain into four *blocks*.

As the chains are decomposed into four blocks, so are the shift cycles. In the first three quadrants of the shift cycles, the broadcast stimulus is inserted only into the reference chain, filling in the first three blocks of the reference chain, while the shadow chains receive constant-0 as the stimulus. In the last (fourth) quadrant of the shift cycles, the deferred broadcast operation is performed; the stimulus in R_{i-1} is broadcast into R_i and S_{ij} blocks, while the scan-in channel broadcasts stimulus into R_1 and S_{1j} blocks. By the end of the last quadrant of shift cycles, the intended broadcast stimulus will have been delivered into all the chains. In the DB architecture, a small and trivial counter-based controller is required in order to control the select lines of the multiplexers, imposing no changes in the test interface (and number of tester channels).

Power reduction in the DB architecture stems solely from the constant-0 stimuli pumped into the shadow chains, delivering only scan-in power reductions. As the responses are shifted out intact, however, scan-out power remains the same in the DB architecture. It can be shown that DB attains a reduction factor of $\frac{b \cdot c}{(b+c-1)}$ in scan-in power where b and c denote the number of blocks and chains, respectively.[1]

---

[1]We follow a simplified power model wherein the number of transitions in scan cells defines the power value as the two strongly correlate [11]; more

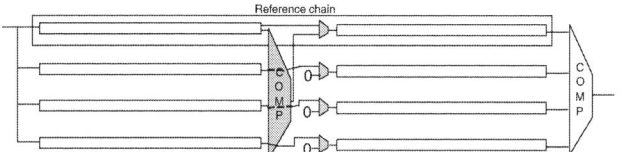

Figure 2. Expedited-Compact (EC) - 2 regions

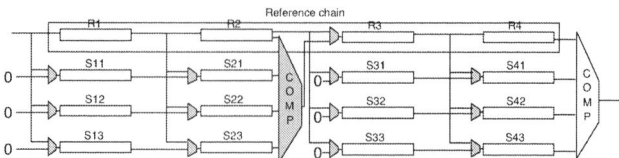

Figure 3. DB + EC (4 blocks, 2 regions)

The proposed EC architecture is provided in Figure 2, for a single scan-out channel and four scan chains, this time each decomposed into two *regions*. The expedited compaction operation is performed by the additional compactor (shaded color) introduced in between the regions, which feeds the reference chain of the right region with the compressed response of the chains of the left region, while the original compactor propagates the compressed response of the rightmost region to the scan-out channel. Also in the first half of shift operations, constant-0 stimulus feeds the shadow chains of the right region. By the end of the first half of shift cycles, the chains in the left region consist of broadcast stimulus, the reference chain in the right region consists of compacted response, and the shadow chains of the right region consist of 0's.

In the second half of shift cycles, constant-0 feed into the shadow chains in the right region continues, while the compacted responses in the reference chain of right region are passed to the scan-out channel. Simultaneously, the broadcast stimuli is inserted into all the chains. In the EC architecture, the select lines of the multiplexers can be controlled in a similar fashion as in the DB architecture.

While the proposed EC architecture is illustrated for only two regions in Figure 2, a larger number of regions can be employed to increase the scan-out power savings. EC with r regions enables the filling of all the shadow chains, except for those in the leftmost region, to be filled with 0's subsequent to one r^{th} of the shift cycles, collecting all the compacted responses in the reference chain at this time. Thus, during the remainder of shift cycles (the last $\frac{r-1}{r}$ portions), the scan-out power dissipation occurs only in the reference chain. It can be shown that EC attains a reduction factor of $\frac{r \cdot c}{(r+c-1)}$ in scan-out power. As can be expected intuitively, the formulation hints that a larger number of regions and/or chains yield higher savings in scan-out power.

The DB and EC tehcniques can be utilized in conjunction, as illustrated in Figure 3, to gain savings in both scan-in and scan-out power. We also note that both DB and EC architectures can accommodate uneven block sizes.

As a multiplexer driven by a constant-0 can be simplified down to an AND gate, the cost of DB per scan chain is approximately 1 AND gate and $b-1$ multiplexers. The cost of

EC per chain, assuming a simple XOR tree as the compactor for instance, is approximately $r-1$ XOR gates and $r-1$ AND gates. Based on the area constraints and targeted power reduction levels, b and r can be appropriately adjusted, enabling a cost-effective trade-off between area and power; larger values for b and r deliver larger savings in scan-in and scan-out power, respectively, yet at the expense of higher area cost.

Response compactors that consist of multiple XOR trees [42] feeding multiple scan-out channels are quite commonly used, in order to cope with response x's. Such compactors necessitate a slight modification in the proposed EC architecture; as many reference chains as the number of scan-out channels needs to be utilized. While replication of these compactors is more costly compared to replicating a simple XOR tree, significant power reductions can still be attained cost-effectively by properly adjusting the number of regions.

3 Experimental Results

We present the experimental results in this section. Three sets of experiments, which are on randomly generated test data, on test data of ISCAS89 circuits, and on industrial test data provided by Cadence, have been performed.

3.1 Setup

The underlying scan architecture (base case) assumed in our experiments consists of a single scan-in channel fanning out to a number of scan chains, which drive a single scan-out channel through an XOR tree. The DB architecture, always with 12 blocks, and the proposed EC architecture with varying number of regions are evaluated; although DB targets only scan-in power reductions and EC targets only scan-out power reductions, percentage reductions in *overall* scan power are reported for both approaches. The utilization of both DB and EC together helps attain a power reduction that equals the sum of the reductions attained by these techniques individually.

In the first set of experiments, test stimuli are generated randomly based on a given don't care bit probability, and responses on a given 0-bit probability (fully specified response vectors are generated).[2] The probability of generating a 0-bit and a 1-bit in the stimulus are presumed identical. Once a partially specified test stimulus is generated, it is checked for encodability. If the stimulus is encodable, some of its x's are specified as dictated by the encodability requirements. The remaining x's are specified based on a given x-fill option, which can be repeat-fill, 0-fill or random-fill. Test power computations are conducted for 100 encodable patterns.

---

elaborate and accurate models can also be used to improve the accuracy of the analysis. In traditional testing, all chain fragments in the ith block from the left receive stimulus in the ith portion of the shift cycles, and stimulus is scanned in and out of the chain fragments in this block thereafter. In the DB scheme, the non-reference chains dissipate scan-in power only during the last portion of the shift cycles. While this discussion sheds light to the derivation of the scan-in power reduction formula, details are omitted due to space constraints.

---

[2]In these experiments, test data is generated with probability values reported in industry.

chains	depth	x-fill	P_x (stim)	P_0 (resp)	DB (12 blocks) [41]	EC (2 regions)	EC (3 regions)	EC (4 regions)	EC (6 regions)	EC (12 regions)
4	480	repeat	0.98	0.6	4.3	34.0	45.3	51.0	56.6	62.4
8	480	repeat	0.98	0.6	9.8	37.8	50.4	56.6	63.0	69.3
4	480	repeat	0.98	0.7	5.1	31.8	42.2	47.6	52.8	58.1
8	480	repeat	0.98	0.7	11.2	36.0	48.1	54.0	60.1	66.1
4	480	repeat	0.97	0.7	7.6	30.3	40.4	45.5	50.5	55.6
8	480	repeat	0.97	0.7	15.3	33.9	45.2	50.8	56.5	62.1
4	480	repeat	0.95	0.5	9.2	32.1	42.8	48.2	53.6	59.0
8	480	repeat	0.95	0.5	17.4	34.1	45.6	51.1	56.9	62.5
4	960	repeat	0.98	0.6	4.7	34.0	45.3	51.0	56.6	62.3
8	960	repeat	0.98	0.6	9.8	37.8	50.5	56.7	63.1	69.4
4	480	0-fill	0.98	0.6	8.7	31.8	42.4	47.7	53.0	58.3
8	480	0-fill	0.98	0.6	17.0	34.1	45.4	51.0	56.8	62.5
4	480	random	0.98	0.6	35.3	17.7	23.7	26.6	29.5	32.5
8	480	random	0.98	0.6	41.6	20.7	27.7	31.1	34.6	38.0

Table 1. Average scan power reductions (%) with randomly generated responses.

In the second set of experiments, ATALANTA is utilized to generate a test cube for a given ISCAS89 benchmark circuit. Encodability check, x-specification for ensuring encodability, and filling of the remaining x's based on a given x-fill option are all conducted similarly as in the first set of experiments. In the second set of experiments, however, response of the benchmark circuit is obtained via the execution of a fault simulator, HOPE. Test power computations are conducted for all the encodable test cubes.

In the third set of experiments, industrial test data provided from Cadence is utilized. The data consists of 100 fully specified (0-filled) broadcast stimulus patterns and their responses for three industrial designs. Test power computations are conducted for all 100 patterns for each of the three designs.

3.2 Randomly Generated Test Data

Table 1 provides the average power reduction results obtained in the case of randomly generated test data. Columns 1 through 5 denote the number of chains, scan depth, x-fill option, probability of a don't care bit in the randomly generated stimulus, and probability of a 0-bit in the randomly generated response, respectively. Column 6 reports the power reductions delivered by the DB architecture, while the remaining columns denote average power reductions attained by the proposed EC architecture, where the number of regions is increased from left to right.

We can state the following observations:

- With more aggressive compression ratios (higher degree of fan-out, and thus, more chains), higher reductions in average power can be delivered by all instances of the proposed EC architecture, consistent with the power reduction ratio $\frac{r \cdot c}{(r+c-1)}$.

- With higher don't care bit probabilities and/or with 0-bit probabilities in responses closer to 0.5, more x's remain in stimuli after compression and more transitions will exist in responses. As the repeat-fill or 0-fill options can reduce transitions in stimuli, scan-out power becomes the dominant factor in overall scan power. The DB architecture, even with 12 blocks, fails to deliver reasonable power reductions, while the proposed EC architectures with a larger number of regions can deliver very high and consistent reductions. The variation in effectiveness

among EC versions with different numbers of regions becomes higher.

- Scan depth has a very minor impact on power reductions.

- More efficient x-fill options (0-fill and repeat-fill) lead to fewer transitions in stimuli, and thus to lower scan-in power. The end-result is again more dominant scan-out power, rendering DB ineffective and variation in effectiveness for EC versions higher.

- In the case of random-fill, DB with 12 blocks and EC with 12 regions deliver similar savings in overall scan power reduction.

- With the exception of the random-fill option, which is typically not employed in any case, DB delivers less than 20% scan power reductions, EC with only three regions delivers 40-50% scan power reductions, and EC with maximal number of regions delivers 55-70% scan power reductions.

3.3 Results on ISCAS89 Benchmark Circuits

Table 2 provides the average power reduction results in the case of ISCAS89 benchmark circuits. Columns 1 through 3 denote the circuit name, number of chains, and the x-fill option for stimuli, respectively. Columns 4 through 9 denote average power reductions attained by the DB and the proposed EC architectures, where the number of regions for EC is varied. We can state the following observations:

- With more aggressive compression ratios (larger number of chains), higher reductions in average power can be delivered by all instances of the EC architecture.

- More efficient x-fill options lead to fewer transitions in stimuli, and thus to lower scan-in power. The end-result is more dominant scan-out power, rendering DB ineffective and variation in effectiveness for EC higher for different numbers of regions.

- With the exception of the random-fill option, on average, DB delivers a 10% reduction at around 0.1% area cost, EC with only three regions delivers a 40% reduction at around 0.1% area cost, and EC with maximal number of regions delivers a 55% reduction at 0.6% area cost.

978-1-61284-657-6/11 $26.00 © 2011 IEEE

Circuit	chains	x-fill	DB (12 blocks) [41]	EC (2 regions)	EC (3 regions)	EC (4 regions)	EC (6 regions)	EC (12 regions)
s35932	4	repeat	1.5	36.7	50.2	54.2	61.4	66.7
s35932	8	repeat	0.4	38.1	54.2	59.6	65.3	71.6
s35932	4	0-fill	5.5	35.7	48.4	52.2	59.0	64.1
s35932	8	0-fill	10.0	34.4	48.8	53.5	58.9	64.3
s35932	4	random	35.9	18.6	24.4	27.6	30.7	33.7
s35932	8	random	41.4	21.1	28.7	32.0	35.5	39.1
s38417	4	repeat	7.6	19.6	27.9	30.8	35.0	38.2
s38417	8	repeat	-0.6	34.2	48.1	53.3	59.5	65.5
s38417	4	0-fill	19.9	24.7	33.1	36.0	39.6	43.1
s38417	8	0-fill	9.1	36.7	51.0	56.1	61.7	67.1
s38417	4	random	36.2	16.1	21.3	24.7	27.2	30.1
s38417	8	random	39.7	20.8	27.7	31.0	35.1	38.8
s38584	4	repeat	10.7	22.3	28.7	32.2	36.0	39.3
s38584	8	repeat	15.0	26.7	38.6	41.9	46.9	51.9
s38584	4	0-fill	18.3	17.7	23.8	25.0	29.6	31.7
s38584	8	0-fill	24.8	22.1	32.5	35.1	39.5	43.7
s38584	4	random	34.8	18.8	25.1	28.2	31.3	34.4
s38584	8	random	40.3	21.3	28.3	31.9	35.4	38.9

Table 2. Average scan power reductions (%) for ISCAS89 circuits.

Circuit	#bits	chains	DB (12 blocks) [41]	EC (2 regions)	EC (3 regions)	EC (4 regions)	EC (6 regions)	EC (12 regions)
A	15,669	10	14.1	36.5	48.3	54.7	60.7	66.9
B	22,213	20	15.7	38.7	51.5	57.9	64.4	70.8
C	61,298	30	1.3	47.1	62.9	70.6	78.4	86.2

Table 3. Average scan power reductions (%) for industrial circuits.

3.4 Results on Industrial Test Data

Table 3 provides the average power reduction results with the proposed EC technique applied on test data of three industrial designs. Columns 1 through 3 denote the circuit name, number of scan cells, and number of chains, respectively. Columns 4 through 9 denote average power reductions where the number of regions is varied.

The results hint that the variation in reductions attained by different EC versions (different number of regions) increases as the size of the circuit increases. For the largest circuit C, for instance, DB delivers almost no reduction, while the full-capacity 12-region EC delivers a reduction close to 90%. On the other extremal point, the proposed EC delivers 35%-50% reductions in scan power for these designs with only a single replication of the compactor (2 regions) cost-effectively. In between these two extremal points, the cost-effective 3-region EC delivers 45-65% reductions; for design C that has 61K registers, 3-region EC with an area cost of 2 XOR gates + 2 AND gates per scan chain[3], with each chain having more than 2K scan cells, delivers a power reduction of 63%.

4 Conclusions

In this paper, we propose a DfT-based solution that can reduce average test power significantly in a cost-effective manner without resorting to any x-filling techniques. The proposed solution is simple, scalable, and retains test data and quality intact, as observed responses are the same with or without EC. Furthermore, EC is non-intrusive for design flow, as it does **not** require clock gating for power savings. The proposed EC architecture advances the response compaction operations, ensuring that only the reference chain holds the compacted response during the majority of shift cycles, thus enabling a constant-0 feed into all the other chains. The proposed EC architecture also offers a power-area co-optimization for designs with a very tight area budget. Significant reductions in test power can still be delivered at reduced area costs. The experimental results also confirm the efficacy and the cost-effectiveness of the proposed architecture. For industrial test cases we have experimented with, we observe 70-85% reductions in test power, boding well for even larger-sized circuits.

References

[1] P. Girard, N. Nicolici and X. Wen, *Power-Aware Testing and Test Strategies for Low Power Devices*, Springer, 2010.

[2] S. Gerstendörfer and H.-J. Wunderlich, "Minimized power consumption for scan-based BIST," in *International Test Conference*, 1999, pp. 77–84.

[3] R. Sankaralingam and N. A. Touba, "Inserting test points to control peak power during scan testing," in *International Symposium on Defect and Fault-Tolerance in VLSI Systems*, 2002, pp. 138–146.

[4] S. Bhunia, H. Mahmoodi-Meimand, D. Ghosh, S. Mukhopadhyay, and K. Roy, "Low-power scan design using first-level supply gating," *IEEE Transactions on Very Large Scale Integration (VLSI) Systems*, vol. 13, no. 3, pp. 384–395, 2005.

[5] M.-H. Chiu and J. C.-M. Li, "Jump scan: A DFT technique for low power testing," in *VLSI Test Symposium*, 2005, pp. 277–282.

[6] T.-C. Huang and K.-J. Lee, "An input control technique for power reduction in scan circuits during test application," in *Asian Test Symposium*, 1999, pp. 315–320.

[7] N. Nicolici, B. M. Al-Hashimi, and A. C. Williams, "Minimisation of power dissipation during test application in full scan sequential circuits using primary input freezing," in *IET Computers and Digital Techniques*, 2000, pp. 313–322.

[8] V. Dabholkar, S. Chakravarty, I. Pomeranz, and S. M. Reddy, "Techniques for minimizing power dissipation in scan and

---

[3]We cannot provide exact percentage area costs, as only test data has been made available to us.

combinational circuits during test application," *IEEE Transactions on Computer-Aided Design of Integrated Circuits and Systems*, vol. 17, no. 12, pp. 1325–1333, 1998.

[9] Y. Bonhomme, P. Girard, L. Guiller, C. Landrault, S. Pravossoudovitch, and A. Virazel, "Design of routing-constrained low power scan chains," in *Design, Automation and Test in Europe Conference*, 2004, pp. 62–67.

[10] R. Sankaralingam and N. A. Touba, "Controlling peak power during scan testing," in *VLSI Test Symposium*, 2002, pp. 153–159.

[11] R. Sankaralingam, N. A. Touba, and B. Pouya, "Reducing power dissipation during test using scan chain disable," in *VLSI Test Symposium*, 2001, pp. 319–324.

[12] S. Remersaro, X. Lin, Z. Zhang, S. M. Reddy, I. Pomeranz, and J. Rajski, "Preferred fill: A scalable method to reduce capture power for scan based designs," in *International Test Conference*, 2006, pp. 32.2.1–32.2.10.

[13] S. Kajihara, K. Ishida, and K. Miyase, "Test vector modification for power reduction during scan testing," in *VLSI Test Symposium*, 2002, pp. 160–165.

[14] J. Saxena, K. M. Butler, V. B. Jayaram, S. Kundu, N. V. Arvind, P. Sreeprakash, and M. Hachinger, "A case study of IR-drop in structured at-speed testing," in *International Test Conference*, 2003, pp. 1098–1104.

[15] A. Chandra and R. Kapur, "Bounded adjacent fill for low capture power scan testing," in *VLSI Test Symposium*, 2008, pp. 131–138.

[16] X. Wen, Y. Yamashita, S. Kajihara, L.-T. Wang, K. K. Saluja, and K. Kinoshita, "On low-capture-power test generation for scan testing," in *VLSI Test Symposium*, 2005, pp. 265–270.

[17] X. Wen, Y. Yamashita, S. Morishima, S. Kajihara, L.-T. Wang, K. W. Saluja, and K. Kinoshita, "Low-capture-power test generation for scan-based at-speed testing," in *International Test Conference*, 2005, pp. 1019–1028.

[18] X. Wen, S. Kajihara, K. Miyase, T. Suzuki, K. W. Saluja, L.-T. Wang, K. S. Abdel-Hafez, and K. Kinoshita, "A new ATPG method for efficient capture power reduction during scan testing," in *VLSI Test Symposium*, 2006, pp. 58–65.

[19] H.-T. Lin and J. C.-M. Li, "Simultaneous capture and shift power reduction test pattern generator for scan testing," *IET Computers and Digital Techniques*, vol. 2, no. 2, pp. 132–141, 2008.

[20] O. Sinanoglu, I. Bayraktaroglu, and A. Orailoglu, "Test power reduction through minimization of scan chain transitions," in *VLSI Test Symposium*, 2002, pp. 166–171.

[21] O. Sinanoglu, I. Bayraktaroglu, and A. Orailoglu, "Scan power reduction through test data transition frequency analysis," in *International Test Conference*, 2002, pp. 844–850.

[22] O. Sinanoglu and A. Orailoglu, "Test power reductions through computationally efficient, decoupled scan chain modifications," *IEEE Transactions on Reliability*, vol. 54, no. 2, pp. 215–223, 2005.

[23] O. Sinanoglu and A. Orailoglu, "Modeling scan chain modifications for scan-in test power minimization," *International Test Conference*, pp. 602–611, 2003.

[24] K. Joshi and E. MacDonald, "Reduction of instantaneous power by ripple scan clocking," in *VLSI Test Symposium*, 2005, pp. 271–276.

[25] K. M. Butler, J. Saxena, T. Fryars, and G. Hetherington, "Minimizing power consumption in scan testing: Pattern generation and DFT techniques," in *International Test Conference*, 2004, pp. 355–364.

[26] T. Yoshida and M. Watari, "A new approach for low power scan testing," in *International Test Conference*, 2003, pp. 480–487.

[27] L. Whetsel, "Adapting scan architectures for low power operation," in *International Test Conference*, 2000, pp. 863–872.

[28] P. Girard, L. Guiller, C. Landrault, S. Pravossoudovitch, and H. J. Wunderlich, "A modified clock scheme for a low power BIST test pattern generator," in *VLSI Test Symposium*, 2001, pp. 306–311.

[29] Y. Bonhomme, P. Girard, L. Guiller, C. Landrault, and S. Pravossoudovitch, "A gated clock scheme for low power scan testing of logic ICs or embedded cores," in *Asian Test Symposium*, 2001, pp. 253–258.

[30] P. M. Rosinger, B. M. Al-Hashimi, and N. Nicolici, "Scan architecture with mutually exclusive scan segment activation for shift- and capture-power reduction," *IEEE Transactions on Computer-Aided Design of Integrated Circuits and Systems*, vol. 23, no. 7, pp. 1142–1153, 2004.

[31] K.-J. Lee, T.-C. Haung, and J.-J. Chen, "Peak-power reduction for multiple-scan circuits during test application," in *Asian Test Symposium*, 2000, pp. 453–458.

[32] P. Girard, L. Guiller, C. Landrault, and S. Pravossoudovitch, "Circuit partitioning for low power BIST design with minimized peak power consumption," in *Asian Test Symposium*, 1999, pp. 89–94.

[33] S. Almukhaizim and O. Sinanoglu, "Dynamic scan chain partitioning for reducing peak shift power during test," *IEEE Transactions on Computer-Aided Design of Integrated Circuits and Systems*, vol. 28, no. 2, pp. 298–302, 2009.

[34] J. Li, X. Liu, Y. Zhang, Y. Hu, X. Li, and Q. Xu, "On capture power-aware test data compression for scan-based testing," in *International Conference on Computer-Aided Design*, 2008, pp. 67–72.

[35] X. Liu and Q. Xu, "A generic framework for scan capture power reduction in test compression environment," in *International Test Conference*, 2008, poster 20.

[36] M.-F. Wu, J.-L. Huang, X. Wen, and K. Miyase, "Power supply noise reduction for at-speed scan testing in linear-decompression environment," *IEEE Transactions on Computer-Aided Design of Integrated Circuits*, vol. 28, no. 11, pp. 1767–1776, 2009.

[37] X. Liu and Q. Xu, "On simultaneous shift- and capture-power reduction in linear decompressor-based test compression environment," *International Test Conference*, p. 9.3, 2009.

[38] D. Czysz, M. Kassab, X. Lin, G. Mrugalski, J. Rajski, and J. Tyszer, "Low-power scan operation in test compression environment," *IEEE Transactions on Computer-Aided Design of Integrated Circuits*, vol. 28, no. 11, pp. 1742–1755, 2009.

[39] C.-W. Tzeng and S.-Y. Huang, "QC-Fill: Quick- and cool x-filling for multicasting-based scan test," *IEEE Transactions on Computer-Aided Design of Integrated Circuits*, vol. 28, no. 11, pp. 1756–1763, 2009.

[40] C. G. Zoellin, and H. J. Wunderlich, "Low-Power Test Planning for Arbitrary At-Speed Delay-Test Clock Schemes," in *VLSI Test Symposium*, 2010, pp. 93–98.

[41] A. Chandra, F. Ng, and R. Kapur, "Low power Illinois scan architecture for simultaneous power and test data volume reduction," in *Design, Automation and Test in Europe Conference*, 2008, pp. 462–467.

[42] S. Mitra and K. S. Kim, "X-Compact: An Efficient Response Compaction Technique for Test Cost Reduction", in *IEEE International Test Conference*, pp. 311–320, 2002.

Leakage Power Profiling and Leakage Power Reduction using DFT Hardware

Rajamani Sethuram, Karim Arabi, Mohamed Abu-Rahma

Qualcomm CDMA Technologies, San Diego - 92121

Email: {rsethura, karabi, marahma}@qualcomm.com

Abstract—**In a CMOS logic circuit, the leakage power dissipated depends on the state of the design. In this paper we propose a novel technique to use the Q-gating logic that are added to reduce power during shift to also reduce leakage power during functional standby mode of the circuit. First, we propose** *leakage-aware* **test (λ-test) vector generation that can be used to profile leakage power consumed by the circuit. This is used to identify blocks that drains excessive standby leakage power. We also propose a new partial Q-gating technique that uses the λ-test to determine the subset of flops that should be gated-off to achieve maximum simultaneous reduction in shift mode dynamic power and standby mode leakage power. A fast, test relaxation and test cube merging algorithm is used for this purpose. Experiments conducted on ISCAS and ITC benchmarks show up to 43.6% reduction in leakage power. For the partial gated design, we obtained up to 15.3% leakage power reduction and up to 6.1$\times$ reduction in shift power.**

I. Introduction

Ever-increasing device integration and the need for advanced mobile applications is driving the need to drastically reduce power consumption, particularly for mobile battery operated devices. Mobile devices are often in the standby mode for long periods of time, during which the leakage currents drain the battery charge. Leakage current has drastically increased [11], [12] due aggressive scaling of CMOS geometry, thereby, becoming a significant portion of the overall power consumed. This is because leakage current has exponential relationship with both the threshold voltage, channel length, and the gate oxide thickness that reduces with shrinking feature size. The leakage power of a CMOS gate vary by more than an order of magnitude over the set of input states as shown in Fig. 1. A 2-input NAND gate along with its leakage power for different input combination is shown. The leakage value of the input combination 00 is small because both the transistors $T1$ and $T2$ are OFF. This is called *transistor*

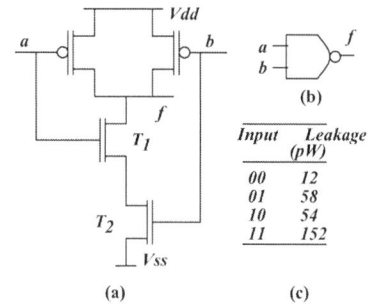

Fig. 1. Example Illustrating Transistor Stacking Effect

stacking effect. One method to reduce leakage during standby

mode is to apply an input vector such that the entire circuit is parked in its least leakage state. This technique is referred to as input vector control (IVC) [6]. In this paper, the Q-gating logic is used to park the circuit in its least leakage state during the functional standby mode.

From product test standpoint, power delivery to *device-under-test* (DUT) and power issues in test have posed several problems. The number of pins that deliver power to the DUT in ATE is lesser than the total number of power pins used during functional mode; it is getting even worse with multi-site testing where multiple dies are tested simultaneously sharing the tester's power pin resources. Mismatch in the power delivery strength of ATE and on-board regulators/power management ICs have caused test escapes. Another main issue that has recieved a lot of attention is switching activity during scan based test is several fold higher than during functional mode causing large IR drop and $L\,di/dt$ voltage drop (L=inductance of power pins and di/dt=instantaneous change in current) in the power/ground mesh [5].

To reduce the dynamic power consumption during test, several solutions have been proposed such as novel power-aware scan architecture [14] and scan ordering, core-based DFT methodology [10], power-aware ATPG, vector re-ordering [4], efficient X-filling and *Q-gating* [7], [13]. [1] In gated-Q designs, q and nq pins of scan flip-flops (SFFs) are clamped by the $scan\_en$ pin such that during the scan shift, the outputs of these flops do not toggle to reduce the switching activity in their fanout cones. To reduce the area overhead and/or the timing impact due to the extra gating hardware inserted partial gating technique is used where only a subset of the SFFs are chosen for gating purpose. Efficient algorithms are proposed to choose the SFF set and to determine if they should be clamped low or high. Lin *et al.* [9] also proposed a technique to re-use Q-gating logic to also reduce power consumption during capture. In this paper, we show that the Q-gating logic can also be used to reduce leakage power during functional mode to improve the battery life of the device. To the best of our knowledge, our work presents the first unified approach that reduces both shift power and leakage power using the Q-gating logic for different circuit operating modes.

There are two main contributions in this paper: a) Leakage-aware test vector generation. We generate several scan vectors using a commercial ATPG tool that can park the circuit from

---

[1] Please note that due to space limitation we are unable to cite all of the prior work. Interested readers may refer to the recent book edited by Girard *et al.* [5] for detailed treatment. We only described the widely used Q-gating technique here because it is more relevant to this paper.

978-1-61284-657-6/11 $26.00 © 2011 IEEE

its least leakage state to its maximum leakage state. These vectors are then simulated to identify blocks that dissipate excessive leakage power. b) We also propose a novel Q-gating technique that also takes into account leakage current. Furthermore, we provide control mechanism by which the Q-gating logic of a block is activated during the functional standby mode, thereby, reducing leakage power during the functional mode.

The rest of this paper is organized as follows. We describe algorithms to generate λ-test vectors in Section II. A new leakage fault model is introduced in Section II-A which is then used to generate the leakage-aware test set. Section III explains hardware modification required to reduce leakage power using the Q-gating hardware. Section IV gives the experimental results that demonstrate the scalability of the proposed method and its effectiveness in reducing leakage current in different operating modes. Section V concludes the paper.

II. Leakage-aware Test Vectors

The leakage current of a circuit depends on the state of the circuit. Understanding the leakage variation for different blocks of the design based on different states can give designers insight on where to put their effort to optimize the design for overall leakage power reduction. The leakage-aware tests (λ-tests) are simulated to compute the range of leakage power consumed by different blocks in the circuit. Note that generating single *minimum leakage vector* (MLV) is an NP-hard problem and large computational times are reported for MLV generation for ISCAS combinational benchmarks [3]. The problem solved in this paper is even more complex where we generate several leakage-aware vectors that can park the circuit from its least leakage state to maximum leakage state. Hence, accurate techniques [3] that consumes enormous run time are not applicable for our problem.

One solution for this problem is a modified *podem* algorithm or SAT-based approach that additionally considers the leakage values of standard cells. However, creating (yet) another ATPG tool just for the purpose of generating leakage-aware test is less desirable. Moreover, we want to build a tool that is scalable for large circuits. Hence, we propose to generate λ-tests using an existing commercial ATPG tool that does not understand leakage power! Our solution comprises creating a new fault model called *leakage fault* (λ-fault). We then use an existing ATPG tool to generate vectors for all λ-faults in the circuit. The resulting vector represents the λ-test set. Before describing these procedures, we first describe few terms that will be used extensively in the rest of the paper.

Minimum/Maximum Leakage State: of a gate represents the Boolean input combination that should be applied such that the gate drains its minimum/maximum current. Also, the notation $S^{min}(g_i)$ and $S^{max}(g_i)$ will be used to denote these states and $\kappa^{min}(g_i)$ and $\kappa^{max}(g_i)$ denotes the corresponding leakage currents. Here, g_i represents any gate. $\Delta_\kappa(g_i)$ is used to denote their difference.

$$\Delta_\kappa(g_i) = \kappa^{max}(g_i) - \kappa^{min}(g_i) \qquad (1)$$

Fig. 2 shows and example circuit and the S^{min}, S^{max}, and Δ_κ values for all the gates.

Implied Neighbor ν: Signal lines l_i and l_j are said to be implied neighbors, $\nu(l_i, l_j)$ if implying a Boolean 0 or 1 at l_i generates a known Boolean value at l_j or vice-versa.

ν-**Region:** A ν-region, ν^ρ, is a set of lines such that every line $l_i \in \nu^\rho$ has at least one implied neighbor $l_j \in \nu^\rho$.

Leakage Sensitive ν-Region: is a region, ν^L, such that a) every gate $g \in \nu^L$ can be independently parked in their S^{min} (S^{max}) state without generating a conflict, b) $\sum_{g \in \nu^L} \Delta_\kappa(g)$ is above certain threshold value K, and c) Cardinality, $|\nu^L|$, should be below certain threshold. These thresholds are described later.

A connected component: of an undirected graph, G, is a subgraph, C_i, in which any two vertices are connected to each other by paths, and to which no more vertices or edges from G can be added while preserving its connectivity.

A. Leakage Faults

λ-Fault is a static fault which is a tuple $\lambda(\nu^L, s_i)$ ($i=1,2, \ldots$) where ν^L is a leakage sensitive ν-region in the circuit and s_i is $S^{min}(g_i)$ or $S^{max}(g_i)$ $\forall$ $g_i \in \nu^L$ state that should be justified at the gate inputs for λ-fault detection. Note that λ-faults are merely an artifact created for λ-test generation; they are not abstractions of any physical defect. The leakage sensitive regions, ν^L, of the circuit is first identified and λ-faults are injected $\forall$ gates $g \in \nu^L$. An example of λ-fault is shown in Fig. 2 where gates labeled 1 and 3 produces more leakage variation compared to other gates in the circuit. Hence, gates 1 and 3 constitutes a leakage sensitive region ν_1^L. The fault corresponding to this region is $\lambda_1^{\nu_1}(g_1=00 \; g_3=01)$ that generates κ^{min} and $\lambda_2^{\nu_1}(g_1=11 \; g_3=11)$ that generates κ^{max} for ν_1^L. Gate 6 and 7 of Fig. 2 cannot be grouped together to form a leakage sensitive region because of the their conflicting S^{min} condition requirements. Hence, gates 6 and 7 are two separate single-gate leakage sensitive regions.

Algorithm 1 describes the method to create all λ-faults in the circuit. The algorithm considers the leakage sensitive gates of the circuit. It creates a *conflict graph* G_X which is then used to create a *compatibility graph* G_E based on Boolean implication of S^{min} states. Details of conflict graph and compatibility graph is given later in this section. Identifying connected components in G_E determines the leakage sensitive ν-regions in the circuit, which is used to create the λ-faults. We will now describe Algorithm 1 in detail. Line 1 of the algorithm greedily chooses the top $K\%$ of the gates with maximal $\Delta_\kappa(g_i)$. Next, for each node (or gate) ig in the circuit, we create two gate queues called *comp_queue0* and *comp_queue1* that represents the 0-compatibility and 1-compatibility information. More precisely, a gate $g \in ig \to comp\_queue0$ (*comp_queue1*) means applying $S^{min}(g)$ state at the inputs of g implies Boolean 0(1) at the output of ig. Lines 3-13 shows how these queues are constructed that are then used to create the conflict graph G_X. A conflict graph is an un-directed graph whose vertices V_G represents the leakage sensitive gates (selected in Line 1) and edges $e(g_i, g_j) \in E_X$ (where $g_i, g_j \in V_G$) means that setting S^{min} states for gates g_i and g_j, simultaneously, will result in a conflict. Lines 15-21 constructs the edges of G_X.

Algorithm 1 Create λ-fault set (K)

1: Obtain, G which is set of $K\%$ of gates g_i with maximum $\Delta_\kappa(g_i)$

2:

3: **for each** gate $g_i \in G$ **do**

4:     Set the $S^{min}(g_i)$ state at g_i and obtain implied neighbor set IG

5:     **for each** gate $ig \in IG$ **do**

6:         Let v_{ig} denote the Boolean value implied at node ig

7:         **if** $v_{ig} == 0$ **then**

8:             Add g_i into $ig \rightarrow$ comp_queue0

9:         **else**

10:             Add g_i into $ig \rightarrow$ comp_queue1

11:         **end if**

12:     **end for**

13: **end for**

14:

15: **for each** node ig in the ckt. **do**

16:     **for each** gate g^0 in $ig \rightarrow$ comp_queue0 **do**

17:         **for each** gate g^1 in $ig \rightarrow$ comp_queue1 **do**

18:             Add undirected edge $e\ (g^0, g^1)$ in G_X

19:         **end for**

20:     **end for**

21: **end for**

22:

23: **for each** node ig in the ckt. **do**

24:     **for each** gate pair g_i,g_j in $ig \rightarrow$ comp_queue0 **do**

25:         **if** Edge $e(g_i, g_j) \notin E_X$ **then**

26:             Add undirected edge $e\ (g_i, g_j)$ in G_E

27:         **end if**

28:     **end for**

29:

30:     **for each** gate pair g_i,g_j in $ig \rightarrow$ comp_queue1 **do**

31:         **if** Edge $e(g_i, g_j) \notin E_X$ **then**

32:             Add undirected edge $e\ (g_i, g_j)$ in G_E

33:         **end if**

34:     **end for**

35: **end for**

36: **for each** connected component $\nu_{cc} \in G_E$ **do**

37:     Create fault $\lambda(\nu_{cc}, s_i)$ where s_i represents the $S_i^{min}(g_i)$ $\forall$ gate $g_i \in n u_{cc}$

38: **end for**

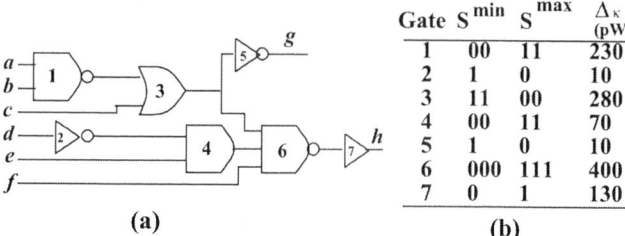

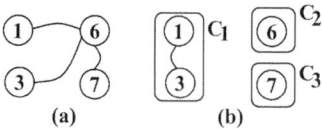

Gate	S^{min}	S^{max}	Δ_κ (pW)
1	00	11	230
2	1	0	10
3	11	00	280
4	00	11	70
5	1	0	10
6	000	111	400
7	0	1	130

Fig. 2. (a) Example Circuit and (b) Min./Max. Leakage State

Fig. 3. (a) Conflict Graph, and (b) Compatibility Graph with Connected Components for the Example Circuit

Lines 23-35 describes the construction of compatibility graph G_E. G_E is an undirected graph where vertices, V_G, of G_E is similar to G_X but edges $e(g_i, g_j)$ means applying S^{min} conditions at g_i and g_j will: a) not cause conflict at any internal node, and b) will imply same Boolean value at least at one internal node to make sure that g_i and g_j are structurally located close to each other. We then identify connected components C_i of G_E. Note that, by construction, gates $g_i \in C_i$ represents a leakage sensitive ν-region and hence a λ-fault can be created using the S^{min} conditions for all gates $g_i \in C_i$. From our experiments, we found that for certain circuits the number of vertices in C_i can be very large and ATPG may not be able to detect the λ-fault corresponding

to large C_is. Hence, we used a threshold T (set to 3 in our experiments) so if $|C_i| > T$ we divided C_i into smaller sets before creating a λ-fault. Note that we have only considered S^{min} state in our algorithm. All of these steps will have to be repeated with S^{max} conditions that will result in more λ-faults. This is not shown in Algorithm 1.

To clarify the different steps of Algorithm 1 we will illustrate it using an example circuit (see Fig. 2). Fig. 2(b) shows the minimum leakage state and leakage power reduction achieved by applying S^{min} for all gates in the example circuit. We assume that only gates 1, 3, 6, and 7 are selected for leakage power reduction (Line 1). The conflict graph, compatibility graph and the connected components after running Algorithm 1 are shown in Fig. 3. The connected components, C_1, C_2, and C_3, identifies the region where λ-faults are injected. The leakage faults corresponding to C_1 are $\lambda^{C_1^{min}}$ ($a=0$ $b=0$ $c=1$) and $\lambda^{C_1^{max}}$ ($a=1$ $b=1$ $c=0$). Leakage faults generated for C_2 are $\lambda^{C_2^{min}}$ ($g3=0$ $g4=0$ $f=0$) and $\lambda^{C_2^{max}}$ ($g3=1$ $g4=1$ $f=1$), and for C_3 the two faults are $\lambda^{C_3^{min}}$ ($g6=0$) and $\lambda^{C_3^{min}}$ ($g6=1$).

B. λ-test generation

So far, we have described how to create the λ-faults. In this section, we will describe two ways of using an existing ATPG tool to generate tests to detect λ-faults. First method is the circuit transformation method where an additional gate, g, is added such that λ-faults can be mapped into an equivalent stuck-at fault at g. For example, consider the fault $\lambda^{C_1^{min}}$ ($a=0$ $b=0$ $c=1$) from our previous example. We can insert a 3-input OR gate g and an inverter ig. The inputs of g will come from a, b and inverted c; its output is a new primary output of the circuit. Now, sa-1 fault at the output of g is equivalent to $\lambda^{C_1^{min}}$ ($a=0$ $b=0$ $c=1$). This concept, which can be easily extended for λ-faults of any size, is however not new. [8] modeled two-line bridge faults by inserting a four-gate circuit into the netlist. Another approach to generate λ-test is using fault tuples (also referred as universal faults) proposed by Blanton *et al.* [1]. Cadence's Encounter Test (ET) ATPG tool [2] supports universal fault vector generation. In ET, this feature is called *pattern faults* and is widely used for diagnostic ATPG and defect-aware fault modeling. In this

```
Entity = demo_circuit
Static {
    Required
    {
        pin demo_circuit . a 0
        pin demo_circuit . b 0
        pin demo_circuit . c 1
    }
}
```

Fig. 4.   Pattern Fault Representing $\lambda^{C_1^{min}}$ (a=0 b=0 c=1)

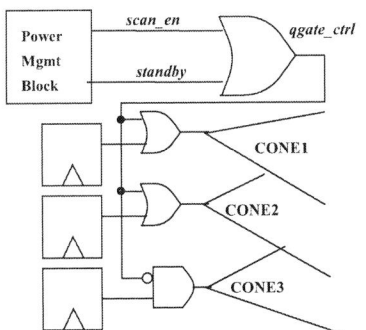

Fig. 5.   Reducing Leakage Power in Gated-Q Design

paper, we modeled λ-faults using pattern faults. Fig. 4 gives an example pattern fault representing $\lambda^{C_1^{min}}$ (a=0 b=0 c=1). Here, *demo_circuit* is the name of the top module. Our tool simply generates pattern fault description for all of the λ-faults. All λ-faults are grouped into bins comprising λ^{min}-fault bin and λ^{max}-faults bin. ATPG vectors are generated for faults in different bins. For all ATPG runs, we set *compactioneffort* = high, so that the tool generate vector that detects as many λ-faults as possible.

III. Hardware Modifications

The λ-test generated from the previous section is simulated against the design to identify the leakage variation at the block level and the system level. If the leakage variation based on Boolean states is large, then leakage power can be reduced using the technique described in this section. We propose to use the Q-gating hardware to park the circuit in its least leakage state. This is illustrated in Fig. 5 example circuit with three scan flip-flops (SFF). Also, we assume that 110 is the minimal leakage state of this circuit. Hence, two OR gates and an AND gate with an invertor at one of its input leg is used for the purpose of Q-gating. Also, shown in the figure is a 3-input OR gate labeled *qgate_control* with inputs *scan_en*, *standby*, and *iddq_dont_test*. During the normal functional operation, all the inputs of *qgate_control* are 0 and hence, the q-gates are transparent. During scan shift, *scan_en* goes high and hence, these gates behave as regular Q-gates.

During functional mode, if the block is idle then *standby* signal is asserted. This will cause the Q-gating logic to park the entire circuit/block in its least leakage state so stand-by leakage power for this block is minimized. Compared to using the scan chain hardware where the minimum leakage state vector is stored in memory, which is then scanned into the flops to park the circuit in its least leakage state [3], the proposed approach has the following advantages: a) lesser latency for going into or waking up from the standby mode, and b) no additional dynamic power consumed while scanning-in minimal leakage state vector.

A. Partial Gating and Leakage Power Reduction

In the previous section we showed how full-gating can be done using the minimal leakage λ-test vector. However, adding a Q-gating cell for all SFFs in the design incurs unacceptable area overhead. Hence, we propose a partial-gating approach by solving the following problem:
Identify a subset of SFF and their values to be gated-off that will drastically reduce shift power during scan mode while

also reducing leakage power during the functional mode.
Note that, this problem is not same as minimizing the sum of dynamic power, P_D, and leakage power, P_L, because: a) we are reducing power in different circuit operation mode and b) dynamic power during shift is very large compared to the functional leakage power; so when we used a greedy approach to minimize P ($=P_D + P_L$) we saw that the algorithm prioritized reducing P_D over P_L. Hence, in this paper we propose a 2-phased iterative solution. In phase 1, we use an iterative greedy algorithm similar to the one used by Jayaraman *et al.* [7] to reduce the dynamic power during scan shift. In phase 2, we use a combination of *test relaxation* and *test cube merging* on a subset of λ-tests (see Section II) to determine the care bits to reduce the leakage power to an acceptable limit without impacting the achievable shift power reduction.

Phase I. Dynamic Shift Power Reduction: We borrowed algorithms from [7] to greedily select the best 5% flops for gating to reduce shift power. In this paper, we only choose the SFF for gating even though Jayaraman *et al.* gated-off both SFFs and internal circuit lines. The algorithm involves computing gain function for each SFF that quantifies the benefit of gating a particular SFF and greedily choose the best available SFF for gating. An *update* algorithm then updates the gain function for all of the remaining SFF to reflect the reduction in the gain of neighbors of the selected flop.

These steps of selection followed by updation is performed till 5% of flops are chosen. Thus, at the end of *Phase I* we obtain a partially specified (5%) bit vector, denoted by v_s, that tells how the selected SFFs should be gated-off. For details on gain and update function, please refer [7].

Phase II. Leakage Power Reduction: As mentioned earlier, we use a combination of test relaxation and test cube merging algorithms to identify care bits for reducing leakage power. These algorithms are commonly used in static and dynamic compaction step of vector generation. However, in this paper our criteria for using them is different. We start with the λ-test that generates minimum leakage power. We then flip each bit, b_i, of a λ-test v_l to its opposite value, one at time and imply the value into internal nodes. We associate a cost function $c(b_i)$, which quantifies the increase in leakage power due to the flipping of b_i to its opposite value. We then greedily choose bits such that flipping them results in little or no impact in the leakage power. We perform this step with the best M% λ-tests to obtain vector set V_L that has L% specified bit and

Algorithm 2 Overall Algorithm for Partial Gating (L, M)

1: Generate λ-fault set (see Algorithm 1)
2: Use any ATPG tool to generate λ-test V_λ and obtain the best $M\%$ of the vectors, V_M, with minimum leakage power
3: Obtain v_s, a partially specified vector representing flops to be gated-off to reduce shift power
4: **for each** vector $v_m \in V_M$ **do**
5:     Build cost function $c(b_i)$ that represents the leakage power increase due to flipping of bit b_i
6:     Relax $L\%$ of the bits with minimum cost $c(b_i)$ to obtain the relaxed vector v_l. Add it to set V_L
7: **end for**
8: **for each** vector $v_l \in V_L$ **do**
9:     Merge test cubes of v_l with that of v_s to create v_f and add it to the set V_F. In case of care-bit collision, give higher priority for care bits of v_s.
10: **end for**
11: The vector $v_f \in V_F$ with least leakage power is used for partial gating

---

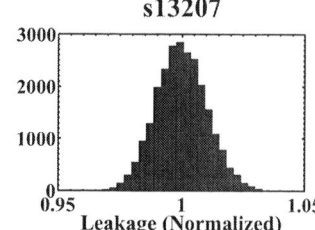 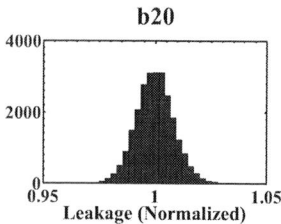

Fig. 6. Leakage Current Distribution for $s13207$ and $b20$

has minimum leakage power. Finally, we perform test cube merging of vector v_s obtained from *Phase I* with each vector $v_l \in V_L$. During the test cube merging step if care bit collision occurs, we give more priority for the care bit of v_s than v_l. This is because the number of care bits used for the purpose of shift power reduction is much less than L (L was set to 50% in this paper). The final vector set, V_F, obtained after merging is simulated and the best vector $v_f \in V_F$ is selected to determine the SFFs to be gated. This is outlined in Algorithm 2.

IV. Experimental Results

We used ISCAS'89 and ITC'99 benchmark circuits and conducted several experiments to validate the proposed approach. These circuits were synthesized and mapped on to a commercial $65nm$ standard cell library. The leakage power values (both sub-threshold and gate oxide leakage) for all standard cell gates for all input combinations are obtained from accurate SPICE simulation and represented in the *Synopsys liberty file*. While synthesizing, all scan cells in the design are stitched into a single scan chain driven by an external shift clock tck. Note that for this work, presence of multiple scan chains and/or on-chip test compressor in the design will have little impact to the quality of our results. The proposed approach is implemented in Cadence's Encounter Test (ET) Test Bench Extension (TBX) language platform [2]. Its built-in TBX APIs are used to parse the netlist and its implication engine was used to implement the algorithms described in Sections II and III. The leakage profiling tool was also built in TBX that has a liberty file parser to obtain the leakage numbers and simulates (built using TBX simulation APIs) the λ-test to compute leakage for the overall circuit.

A. Leakage profiling

First, we generated large number of random vectors to understand the distribution of leakage currents for different benchmarks. We see Gaussian distribution for the leakage

variation for all benchmarks (see Fig. 6). Next, we generated λ-tests and compared against the work of Halter and Najm [6]. In [6], they presented a statistical confidence statement for an observed leakage current. Their results can be summarized as follows: By simulating sufficiently large number of random and independent vectors, n,

$$n \geq \frac{ln(1 - \alpha)}{ln(1 - \epsilon)} \qquad (2)$$

it can be stated with α confidence that less than ϵ fraction of the vector population has leakage current which is lesser (greater) than the least (maximum) observed leakage from n trials. In their paper, they presented the results for the case of α=0.99 and ϵ=0.01. We compare our results with the case of (α=0.99, ϵ=0.01, n=458) and (α=0.99, ϵ=0.001, n=5295). Our experiments with large n values ($n = 25$K) also produced similar results. The columns 2 and 3 of Table I show this result. Note that the percentage values represents the difference of maximum and the minimum leakage power compared to the minimum leakage power. Column 4 represents the percentage

TABLE I
COMPARISON OF λ-TEST WITH [6]

	[6] α=0.99		Proposed λ-test		Vector-less
Ckt.	ϵ=0.01 (%)	ϵ=0.001 (%)	%	CPU (s)	($\times$)
s1196	37.11	44.39	43.66	5.80	7.29
s1423	19.93	25.59	29.48	6.07	3.52
s1488	18.18	19.63	29.60	6.06	3.98
s5378	10.10	12.14	14.85	11.79	2.56
s13207	5.78	8.62	9.95	6.21	2.72
s15850	18.20	19.33	18.69	27.55	3.39
s35932	9.67	9.67	12.55	272.34	2.19
s38417	3.86	4.77	5.06	238.58	2.99
s38584	6.46	6.70	6.47	99.70	2.82
b14	8.29	11.91	12.81	61.13	4.22
b18	4.29	4.49	4.96	51.62	4.53
b19	3.34	3.52	13.29	411.46	4.83
b20	7.99	8.07	8.57	61.42	3.70

difference in the maximum and minimum leakage achieved by the proposed λ-tests compared to the minimum leakage value and shows that we perform better than [6](col. 2). *Thus, statistically, we claim with 99% (α=0.99) confidence that the proposed technique can achieve better results compared to the 99% (ϵ=0.01) of the vector population and comparable results for 99.9% (ϵ=0.001) of the overall vector population.* In our experiments, we set K (see Algorithm 1) to pick the gates contributing top 40% of leakage power and set T=3 (see Section II-A). Note that larger size of T results in more

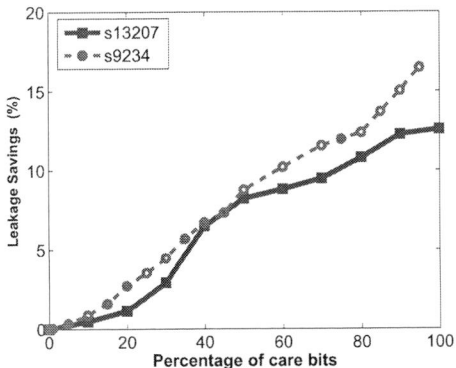

Fig. 7. Reducing Leakage Power with Partial Q-gating

difficult to test λ-faults. From our experiments, we see that increasing T also results in more untestable λ-faults.

The next column shows the CPU time for for the proposed approach which includes the CPU time for generating the λ-faults and perform the λ-test generation. Note, these numbers are an order of magnitude less compared to [3]. Unlike prior work that uses accurate algorithms, the main advantage of our approach is that it is a heuristic approach that uses a commercial ATPG tool for generating the vectors making it a scalable solution.

The last column shows the results for vector-less approach in which we added the minimum and maximum leakage state power of all gates in the design without considering the correlation among any of the internal gates. The number represents the $max.$ to $min.$ leakage ratio. This is a large number, which suggests that by breaking the correlation among internal signal lines in the circuit, we can further reduce leakage power.

B. Leakage Power Reduction with Partial Gating

To understand the care bit requirement and their relationship to achievable leakage power reduction, we conducted several experiments. We used test relaxation technique on λ-tests to obtain a partially specified vector that produces maximum reduction in leakage power compared with the average power

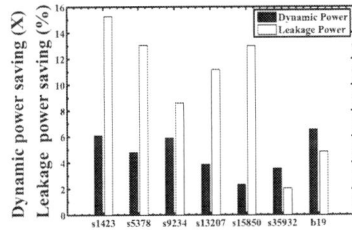

Fig. 8. Shift Power and Leakage Power Reduction

consumption. This is shown in Fig. 7 for the circuits $s9234$ and $s13207$. Here, x-axis represents the % specified care bits in the vector and y-axis represents the maximum achievable percentage reduction in leakage power. Note, that during our simulation we modeled the unspecified bits in the vector as an X. When these X's propagate to internal gates in the design, we estimated its leakage using the cell average leakage

value. Average leakage value or default leakage values can be obtained from the liberty file.

We then generated vectors with $L=50\%$ (see Algorithm 2) to reduce both dynamic power during scan shift and standby leakage power during the functional mode (see Section III-A). Results are shown in Fig. 8. For the circuit $s1423$, we see 15.3% leakage power reduction and 6.1 $\times$ reduction in shift power. To the best of our knowledge, this is the first work that reduce both test mode scan shift power and functional mode leakage power using the Q-gating logic.

V. Conclusion

In this paper, we presented a scalable technique to generate leakage aware test (λ-test) vectors that can park a design in its least leakage state to maximum leakage state. Experiments on ISCAS and ITC benchmarks show that the proposed heuristic technique can achieve up to 43.6% leakage power reduction while performing better than prior work. We then showed a technique to re-use the Q-gating logic to also reduce leakage power in different circuit modes such as functional standby and I_{ddq} core test mode. We also presented a partial Q-gating technique that uses the proposed λ-tests to pick a subset of scan flip-flops for gating. Experiments show that up to 6.1$\times$ reduction in shift power can be obtained and leakage power can be reduced by up to 15.3%.

References

[1] R. D. Blanton, K. N. Dwarakanath, and R. Desineni. Defect Modeling Using Fault Tuples. *IEEE Trans. on Computer Aided Design*, 25(11):2450–2464, Nov. 2006.

[2] Cadence. Encounter dft architect. http://www.cadence.com/products/ld/test_architect/pages/default.aspx.

[3] K. Chopra and S. B. K. Vrudhula. Implicit Pseudo Boolean Enumeration Algorithms for Input Vector Control. In *Proc. of the IEEE/ACM Design Automation Conf.*, pages 767–772, 2004.

[4] V. Dabholkar, S. Chakravarty, I. Pomeranz, and S. Reddy. Techniques for Minimizing Power Dissipation in Scan and Combinational Circuits During Test Application. *IEEE Trans. on Computer Aided Design*, 17(12):1325–1333, Dec 1998.

[5] P. Girard, N. Nicolici, and X. Wen. *Power-Aware Testing and Test Strategies for Low Power Devices*. Springer, New York, 2010.

[6] J. P. Halter and F. N. Najm. A Gate-level Leakage Power Reduction Method for Ultra-low-power CMOS Circuits. In *Proc. of the Custom Integrated Circuits Conf.*, pages 475–478, May 1997.

[7] D. Jayaraman, R. Sethuram, and S. Tragoudas. Scan shift power reduction by gating internal nodes. *J. on Low Power Electronics*, 6(2):311–319, Aug 2010.

[8] J. F. Kaposi and A. A. Kaposi. Testing switching networks for short-circuit faults. *Electron. Lett.*, 8(24):586–587, Nov. 1972.

[9] X. Lin and J. Rajski. Test Power Reduction by Blocking Scan Cell Outputs. In *Proc. of the Asian Test Symp.*, pages 329–336, Nov. 2008.

[10] E. J. Marinissen, R. G. J. Arendsen, G. Bos, H. Dingemanse, M. Lousberg, and C. Wouters. A structured and scalable mechanism for test access to embedded reusable cores. *Proc. of Int'l. Test Conf.*, pages 284–293, 1998.

[11] S. Mukhopadhyay and K. Roy. Modeling and estimation of total leakage current in nano-scaled-cmos devices considering the effect of parameter variation. pages 172 – 175, aug. 2003.

[12] K. Roy, S. Mukhopadhyay, and H. M Meimand. Leakage current mechanisms and leakage reduction techniques in deep-submicrometer cmos circuits. *Proceedings of the IEEE*, 91(2):305–327, 2003.

[13] S. Vishwanath, M. A. Shukoor, S. K. Vooka, and S. Ravi. Fan-out and statistical power estimation based scan cell gating for combinational shift power reduction. *Proceedings in SNUG India,*, 2009.

[14] L. Whetsel. Adapting scan architectures for low power operation. In *Proc. of the Int'l. Test Conf. (ITC)*, page 863, 2000.

978-1-61284-657-6/11 $26.00 © 2011 IEEE

Levelized Low Cost Delay Test Compaction Considering IR-Drop Induced Power Supply Noise

Zhongwei Jiang[1], Zheng Wang[1], Jing Wang[2], D. M. H. Walker[1]

[1]Department of Computer Science and Engineering
Texas A&M University
College Station TX 77843-3112
Tel: (979) 862-4387
E-mail: {jzweiwei, wangz, walker}@cse.tamu.edu

[2]Advanced Micro Devices
Austin, TX
Tel: (512) 602-0425
E-mail: jing.y.wang@amd.com

Abstract – **Power supply noise is very important in delay testing. Excessive noise can cause circuit delay increases that lead to test overkill. Test patterns that are too quiet can lead to test escapes. In this work, we introduce a realistic low cost delay test compaction flow that guardbands circuit delay during test using a sequence of estimation metrics. Significant reductions in CPU time are demonstrated over prior work.**

Keywords – **delay test, power supply noise, static compaction, dynamic compaction, IR drop**

1. Introduction

Delay testing of integrated circuits is essential for achieving product quality, due to reduced timing margins and increased clock rates. Small delay defects can be tested using the path delay fault model [1]. However, as semiconductor technology has been scaled, designs have become increasingly sensitive to noise [2], such as leakage, crosstalk and power supply noise. Too much supply noise can result in excessive noise-induced circuit delay increase, leading to overkill during delay test.

Several techniques have been proposed for estimating power supply noise during timing analysis [3][4]. These methods focused on supply network and circuit models to achieve reasonable accuracy. Most prior work in testing while considering power supply noise adopts a vectorless strategy due to the high simulation cost of the power supply noise model on large circuits. Tirumurti et al. [5] added power noise to a generalized fault model [6]. Pant et al. [7] used a vectorless approach and Krstic et al. [8] a vector-based approach to compute the maximum delay under power supply noise. Lee et al. [9][10] proposed novel techniques to maximize supply noise during automatic test pattern generation (ATPG). Maximum noise can be considerably greater than the mission-mode worst-case noise. Moreover, maximizing noise may be in competition with other goals, such as maximizing crosstalk, that may have greater impact on path delay.

In this work, we focus on power supply noise modeling and estimation during delay test pattern compaction. Our prior work [11] introduced a simplified power region model and circuit switching model. Delay estimation results were verified by circuit simulation and measurement on industrial circuits during static test compaction. The major drawback of this approach was the large number of logic simulations required. The number of logic simulations was reduced by skipping simulation of test patterns with low care bit density. This approach has also been used by other researchers [12][13], but our recent experiments have shown that some patterns with low care bit density can be noisy. We recently developed a new dynamic compaction procedure [14] for path delay test that reduces pattern count by as much as 4x over static compaction, but at the cost of producing some very high noise patterns that could result in test overkill.

In this work, we first introduce a levelized low cost static compaction flow for delay test by reusing the noise and delay model in [11], and then apply it to dynamic compaction. Experimental results on ISCAS89 circuits show that our approach is up to 5x faster than [11]. Simulation results also verify that the low cost model correctly guardbands the extra noise-induced delay on every path. Section 2 summarizes our delay and circuit switching models. Then we analyze the relationship between delay and voltage drop, and between effective weighted switching activity (WSA) and voltage drop. Based on these correlations, we introduce the low cost delay test pattern static compaction framework in Section 3. In Section 4 this framework is applied to dynamic compaction. Parameter setting is discussed in Section 5. Experimental results and discussion are given in Section 6, and conclusions in Section 7.

2. Delay Modeling and Analysis

In this research we only consider power supply noise caused by IR (resistive) voltage drop in the on-chip power grid. This permits modeling the power grid as an RC network. To accurately model and analyze Ldi/dt (inductive) drop, a RLC network is necessary, which is computationally too expensive. We use a modified version of our prior power region model [11]. Tirumurti [5] created a table of peak power and ground currents for different values of gate output load and input slope by circuit simulation. We adopt a similar strategy where the waveform of each cell library is approximated as triangular if the load is small, otherwise as a trapezoid, in order to compute the total charge consumed by each signal transition [11].

2.1 Delay vs. Supply Voltage Drop

Our model of rising transition delay increase is as follows:

$$\Delta delay / delay = \delta \Delta V / V_{DD}$$

where *delay* is the nominal delay, ΔV is the estimated voltage drop at the cell, and V_{DD} is the ideal supply voltage. A table of coefficients δ under different output loads and input slopes is obtained by simulation for each cell type. The accuracy of these models was verified with circuit simulation on circuit s1488 [11] and from measurement on an industrial design [15].

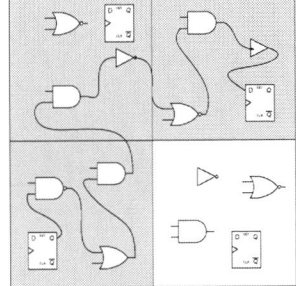

Fig. 1. Effective regions associated with a path.

We conducted experiments using these models to determine the correlation between voltage drop in the effective regions and delay increase. Here the *effective regions* are the power regions that the circuit path under test traverses. The three gray regions in Fig. 1 shows a chip divided into four power regions (shown as rectangular for illustration). The regions colored gray are the effective regions for the path shown. By the definition of region construction, only the voltage drop in these three regions can affect the delay of the target path.

Fig. 2 shows the correlation of voltage drop in effective regions to delay increase for ISCAS89 circuit s38417 for more than 14k paths generated by the *CodGen* ATPG [1], with minimum transition fill of the don't care bits. The correlation is 0.97, so voltage drop is a good estimate of extra delay.

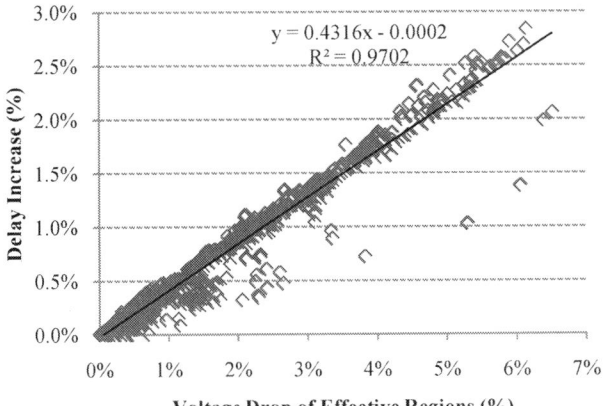

Fig. 2. Voltage drop vs. delay increase for s38417.

2.2 Supply Voltage Drop vs. Effective WSA

WSA can be used to estimate test power [16] and voltage drop. We conducted experiments to find the correlation between regional voltage drop and effective WSA. Here effective WSA means the WSA in those regions traversed by the target path. Fig. 3 shows near perfect correlation between voltage drop and effective WSA for s38417.

2.3 Delay Distribution Analysis

Our prior work [11] did not distinguish the path length during compaction, so much time was spent unnecessarily checking short paths, and rejecting compaction attempts that did not increase circuit delay. Fig. 4 shows the delay distribution of the paths in Fig. 2. The cell-to-cell standard

delay format (SDF) delay was generated using Synopsys *PrimeTime* with 180 nm technology. We can see that many paths are short enough that noise-induced delay will not cause them to exceed the delay of the critical path, and so they can be ignored during compaction.

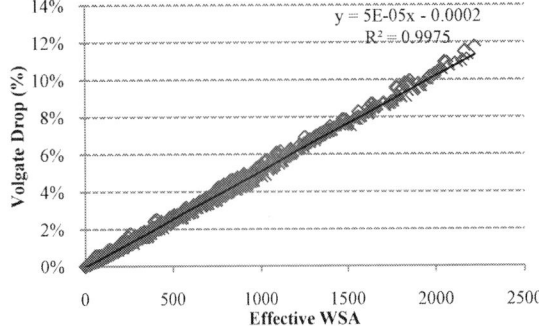

Fig. 3. Voltage drop vs. effective WSA for s38417.

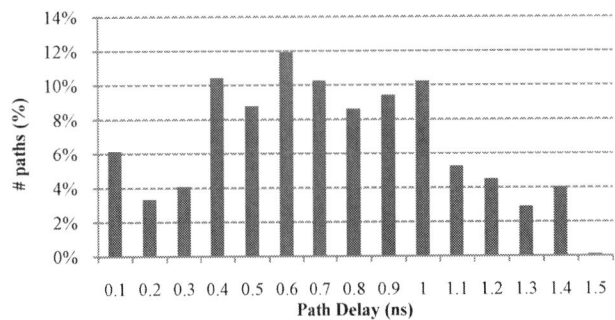

Fig. 4. Path delay distribution for s38417.

As patterns are compacted, one test pattern can contain tests for many paths. As explained above, we will only focus on the longer paths tested in that pattern. The 'long' paths are those paths that are longer than a threshold, which can be a fraction of the critical path length. During static compaction, since we know all the paths and test patterns, we can set the threshold before compaction. But since we are compacting paths as they are generated in dynamic compaction, we must first find the global longest path, using either ATPG or the structurally-longest path.

3. Supply Noise-Aware Static Compaction Framework

We improved on the high cost delay test static compaction algorithm in [11] by exploiting the correlations discussed in Section 2. Fig. 5 shows our proposed delay test compaction framework that consists of two major steps, with each step having a four-level estimation flow embedded.

Step 1: Uncompacted vectors are loaded and their supply noise is checked in a levelized fashion, with each level having higher accuracy at higher computational cost. In Level 1, we check if the SDF delay of the vector (or path) m is >*threshold1*, if not, we go to Step 2; else start Level 2, where if the WSA is <*threshold3*, go to Step 2, else go to Level 3, which is similar to [11]. Accurate WSA requires logic simulation, which is expensive compared to Level 1. WSA can be estimated without using simulation with the help of ATPG, as discussed in Section 4. In Level 3, the voltage drop is computed after logic simulation, when cell switching activity is known. If the voltage drop is less than

978-1-61284-657-6/11 $26.00 © 2011 IEEE

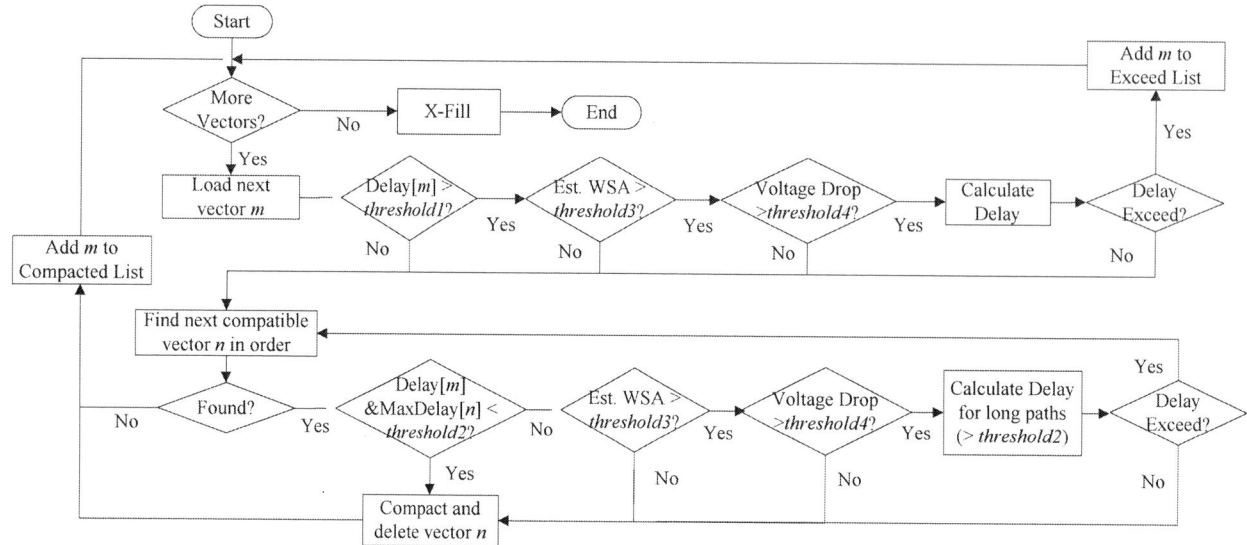

Fig. 5. Levelized low cost static compaction flow for delay test considering power supply noise.

threshold4, we go to Step 2, otherwise we go to Level 4. Level 4 computes the path delay. If the path delay exceeds the delay constraint, this vector is too noisy all by itself, so we put it on an *exceed list*. The high noise of these vectors is due to ATPG, not compaction. Such vectors should be rare.

Step 2: We try to find a compatible vector *n* for vector *m* from Step 1. If the SDF delay of the longest target paths for vectors *m* and *n* are both <*threshold2*, from our previous knowledge, they can safely be compacted. The reason is that two very short paths being compacted will not generate extra delay sufficient to slow the circuit. Here we require that *threshold2* < *threshold1*, since during compaction, the care bit density and gate switching increases. If the delay is >*threshold2*, we will follow an approach similar to Step 1. In Level 4 of this step, we compute the delay of the long paths using look-up tables.

If the supply noise for vectors *n* and *m* together is within limits, compaction is performed and the new vector is added to the set of compacted vectors. If the compaction is rejected, the next compatible vector is considered.

4. Power Supply Noise-Aware Dynamic Compaction

Dynamic compaction [14] has been used in K Longest Path Per Gate (KLPG) delay test ATPG [1]. We modified the supply noise framework described in Section 3 and embedded it into the dynamic compaction algorithm [14]. The supply noise aware dynamic compaction flow is shown in Fig 6. The major difference from the compaction flow in [14] is that 'Initial Too Noisy' and 'Pass Supply Noise Check' steps (marked in gray) have been added to the flow.

The basic idea of dynamic compaction is that for each path, we retain the set of necessary assignments (NAs), rather than primary input justification values, since the NAs are unique to each path. When checking two paths for compatibility, the NAs are first checked, and if they are compatible, a justification is done to verify compatibility. If they are compatible, the new pattern is placed into a Path Pool [14], with each pattern retaining knowledge of the set of paths it contains. After we check pattern compatibility, we perform the noise check before we accept this compaction.

Since dynamic compaction is performed during ATPG, we know the necessary assignments (NAs) of all the internal gates along the new path being considered for compaction. We performed experiments to find the correlation between the WSA of the NAs and the entire circuit. Fig. 7 shows the correlation for s38417. The correlation is high enough that it can be used to estimate WSA without doing simulation.

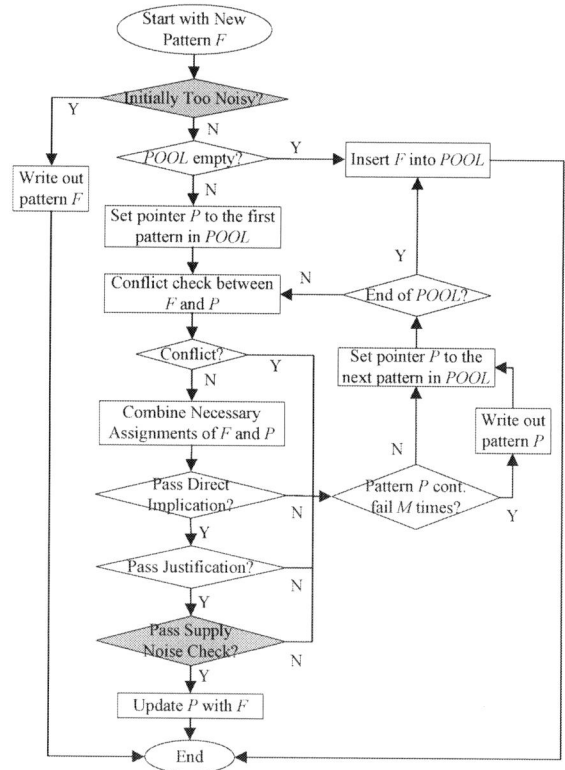

Fig. 6. Power supply noise-aware delay test dynamic compaction flow.

5. Parameter Setting

As discussed in previous sections, there are four parameters used in the compaction flow: *threshold1*, *threshold2*, *threshold3* and *threshold4*. The following rules are proposed on how to set those parameters.

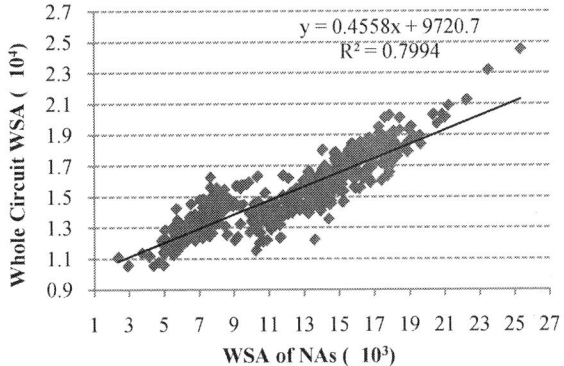

Fig. 7. Correlation between WSA of whole circuit and necessary assignments for s38417.

Rule 1: *threshold1* is normally set to 75% of the delay of the longest testable path or the delay of system clock period. This is based on the experimental results shown in Fig. 8. The delay increase was caused by compaction and for all the paths generated for s38417, we can see that the delay increase for most paths is 4% to 8% of the maximum delay. Similar results are found in other circuits. Setting *threshold1* to 75% provides enough guardband to prevent estimation escape since the maximum delay increase observed is less than 20%. A smaller *threshold1* can be used to be conservative.

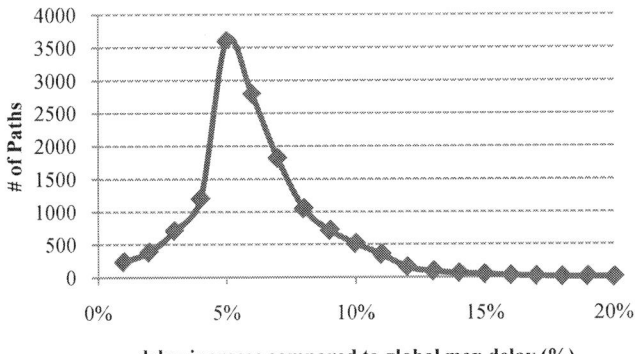

Fig. 8. Delay increase distribution for paths in s38417.

Rule 2: *threshold2* is set to 50% of the delay of the longest testable path or the delay of system clock period. The reason that *threshold2* is smaller than *threshold1* is that during compaction, one pattern can test multiple paths, which increases the supply noise of all paths in the pattern. To be conservative, a smaller value for *threshold2* can be used. Rule 3: *threshold4* can be estimated by doing a cell delay library simulation prior to compaction. As with the correlation between voltage drop and delay (Fig. 3), we can do a pre-simulation for our delay model by using a sample set of test patterns. For most libraries, we expect to see a correlation similar to Fig. 3. For example, suppose we have a relationship between voltage drop (x) and delay (y) of $x = 2 \cdot y$ with good correlation (>90%). The formula to set *threshold4* will then be *threshold4* = $2 \cdot delay\_constraint$. However, if the correlation is not high, a guardband would be added to the formula. In general this analysis must be done only once per cell library.

Rule 4: *threshold3* is set by using the correlation between WSA and voltage drop, as shown in Fig. 4. This correlation can come from simulation of a sample of test patterns. Suppose the WSA (x) and voltage drop (y) has a relationship of $x = 2000 \cdot y$ with good correlation (>90%) and *threshold4* is 10%. The formula to set *threshold3* will be *threshold3* = $2000 \cdot threshold4$. Then we can set *threshold3* to be $2000 \cdot 10\%$ = 200. We can add a guardband if the correlation is lower.

6. Experimental Results

The low cost compaction framework was implemented in C++ and rim on a PC with 2.66 GHz Core 2 Duo and 2 GB memory. The circuit layouts were created using Cadence *SOC Encounter* in 180 nm technology. As discussed in Section 5, we set *threshold1* to be 0.75 and *threshold2* to be 0.5 of the maximum delay. We set *threshold3* to 0.5 of the maximum WSA and *threshold4* to be the same as the circuit delay constraint, since we want to be conservative in using the correlation in Fig 3. The higher the thresholds we set, the faster the compaction flow, but the greater the risk of creating test patterns that exceed the desired noise levels.

Table I shows the static compaction results for our new low-cost framework and the prior high-cost framework. Due to limited space, we only show the results for the largest ISCAS89 circuit s38417. Similar results were seen for other ISCAS89 circuits, such as s1488, s15850 and s15932. The 14,405 path delay patterns were generated with the *CodGen* KLPG ATPG [1] with K set to 1 and launch-on-capture clocking. Initially, we used a greedy forward-order procedure to compact all the patterns into 977 patterns without considering supply noise. We term this procedure "force compaction." We term the noise-aware compaction approach "veto compaction" because we will veto any compaction that violates our delay constraint. Column 'Total Time' is the total compaction time while "Delay Estimate Time" is only the time used in the delay estimation including logic simulation and table lookup.

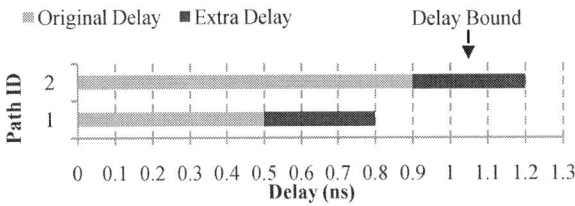

Fig. 9. Delay constraint effect on different paths.

Since the high cost model in [11] considers all paths without looking at their nominal delay, it will reject many compactions that should be approved. For example, as shown in Fig. 9, suppose the delay bound of a circuit is 1 *ns* and the maximum path delay is 0.9 *ns*. Then for safe compaction, we should set the delay constraint to be 11% of the maximum delay, or 0.1 *ns*. However, a compaction is still safe if it adds 0.3 *ns* extra delay to short path #1 with nominal delay of 0.5 *ns*. But if the extra 0.3 *ns* is added to the 0.9 *ns* path #2, it would violate the delay constraint, so this compaction should be vetoed. During static compaction, we know the maximum nominal delay of all paths, so we know how much extra delay we can tolerate. As a result, the low cost model can reduce unnecessary simulations and accept some compactions that were rejected by [11].

For all the delay increase constraints considered in Table I, the low cost model has smaller compacted pattern count than the high cost model. For a 5% delay increase constraint, the low cost model is 5x faster than the high cost model in delay estimation. For a 3% delay constraint, the low cost model has a pattern count half the size of the high cost model. The number of simulations also has been greatly reduced. For a 3% delay constraint, more than 210k simulations are needed in the high cost model, but only 45k simulations in the low cost model. Also, the high cost model needs the delay constraint to be relaxed to 15% to generate the 977 patterns of force compaction, while the low cost model achieves this pattern count while meeting a 7% delay constraint.

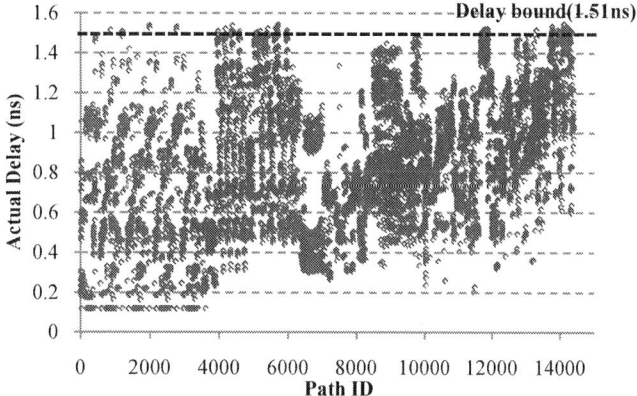

Fig. 10. Actual path delay after compaction for s38417.

We simulated the delay of all paths after static compaction. Before compaction, the max path delay is 1.44 *ns*, and the 5% delay constraint is set to 0.07 *ns*, so the delay bound is 1.51 *ns*. Fig. 10 shows that our low cost model can keep the max path delay of almost all paths within the delay bound except for those 30 patterns that are originally too noisy. Those noisy patterns come from ATPG, not from test

compaction, because testing these paths requires excessive switching in their effective regions, which reduces the supply voltage and introduces extra delay that exceeds the delay constraint. However, these are real paths that must still be tested, even if the estimation models are predicting they will fail the test. The low cost framework did not have any estimation escapes that produced test patterns with excessive delay, due to excessive compaction. Fig. 11 shows that the extra delay induced by compaction. We can see that for short paths, we can tolerate them having extra delay larger than 0.07 ns while for long paths, they must strictly obey the constraint.

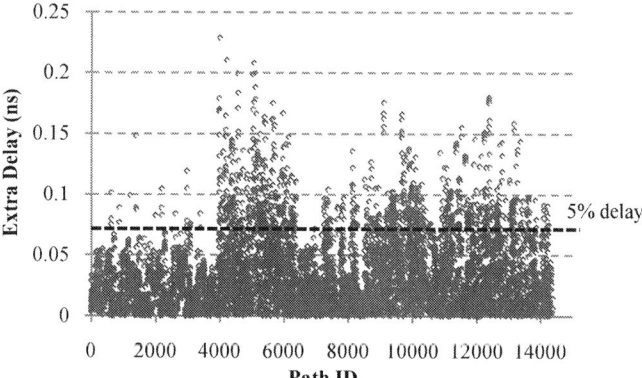

Fig. 11. Extra path delay after compaction for s38417.

Dynamic compaction (DC) results are shown in Table II. As with static compaction (SC), the high cost framework is slower due to more simulations. Before applying the frameworks, we conducted force compaction, which will generate 389 patterns in 22.5 minutes. For the low cost framework, at a 7% constraint, only 4 additional test patterns are generated. For a 3% delay constraint, 349 extra patterns are generated, using considerably more CPU time. The large CPU time increase is due to the many patterns near their delay increase thresholds, requiring many more simulations. Also, each time a compaction is rejected due to the noise, we

TABLE I. LOW COST vs. HIGH COST DELAY ESTIMATION FRAMEWORK DURING STATIC COMPACTION FOR s38417.

Delay constraint	Low Cost Framework				High Cost Framework			
	Total Time(m:s)	Delay Estimate Time(m:s)	# patterns after compaction	# simulations	Total Time(m:s)	Delay Estimate Time(m:s)	# patterns after compaction	# simulations
3%	3:32	2:39	1093	19338	30:24	29:31	1941	210646
5%	3:18	2:25	996	16679	14:54	14:01	1265	100713
7%	3:08	2:15	977	14845	7:36	6:43	1103	49280
10%	3:07	2:14	977	14826	6:15	5:22	999	38957
15%	3:07	2:14	977	14826	4:24	3:31	977	24974
No constraint	0:53	0:00	977	0	0:53	0:00	977	0

TABLE II. LOW COST vs. HIGH COST DELAY ESTIMATION FRAMEWORK DURING DYNAMIC COMPACTION FOR s38417.

Delay constraint	Low Cost Framework				High Cost Framework			
	Total Time(m:s)	Extra Time(m:s)	# patterns after compaction	# simulations	Total Time(m:s)	Extra Time(m:s)	# patterns after compaction	# simulations
3%	40:34	18:03	742	36983	64:46	32:15	1427	80969
5%	35:05	12:34	469	20146	46:42	24:11	698	48982
7%	33:04	10:33	393	15979	39:08	16:37	527	31480
10%	32:22	9:51	389	15749	34:44	12:13	413	28721
15%	32:22	9:51	389	15749	34:28	11:57	389	25536
No constraint	22:31	0:00	389	0	22:31	0:00	389	0

must find another compatible pattern and pass both the direct implication and final justification phases, which requires significant CPU time. Column 'Extra Time' includes all these efforts and delay estimation. For the high cost framework, about half an hour was spent in delay estimation for 3% delay constraint which is 2x slower than the low cost model and the pattern count is around twice that of the low cost model. For delay constraints 5% and 7%, delay estimation of the low cost model is still 2x faster than the high cost model, together with a large reduction in pattern count. By comparing Tables I and II, we can see that DC has higher execution time, but has smaller pattern count than SC. Also in both cases the low cost framework works well until the delay constraint becomes so stringent that many patterns are close to the constraint, and so require detailed analysis. The reason that DC has fewer simulations and higher compaction than SC in the 3% constraint is the following: if two short paths are compatible during DC, they do not need noise estimation and are compacted. If the short paths are not compatible in SC, but one or both are compatible with a long path, then we need to estimate supply noise in SC for the compacted patterns.

TABLE III
DELAY ESTIMATION DURING STATIC COMPACTION FOR s38417 WITH DIFFERENT THRESHOLD1 AND THRESHOLD2. (DELAY CONSTRAINT=3%, THRESHOLD4=3%)

Threshold1	Threshold2	Delay Estimate Time (m:s)	# Patterns After Compaction	# Exceed Paths	# Simulations
0.55	0.5	3:04	1093	157	22780
0.65	0.5	2:49	1093	157	20631
0.75	0.5	2:39	1093	157	19338
0.75	0.6	2:30	1093	157	18467
0.75	0.7	2:01	1093	157	14350
0.8	0.7	1:54	1093	157	13882
0.9	0.7	8:14	1091	134	62061

We also analyzed the effect of different thresholds on compaction. Table III shows that by changing *threshold1* and *threshold2*, the number of simulations rises as the thresholds fall. When we fixed *threshold1* to 0.75 but changed *threshold2*, we can see that the higher *threshold2*, the lower the number of simulations because we are reducing the number of patterns requiring delay estimation by filtering out even more short paths. Similar results can be seen when we fixed *threshold2* to 0.5 and altered *threshold1*. Reducing *threshold2* increases the number of simulations, because we increased the number of patterns requiring delay estimation, including shorter paths. Note that from the data in Table III, we can see that our default setting of [*threshold1, threshold2*] to [0.75, 0.5] is pessimistic because [0.8, 0.7] achieves the same pattern count with 15% less delay estimation time.

A threshold pair of [0.9, 0.7] is too optimistic. The '# Exceed Paths' for [0.9, 0.7] is 23 fewer than the threshold pair [0.8, 0.7]. This means that some paths that exceeded the delay constraint were skipped at the delay check and were added to the final pattern set. In real designs, the percentage of short paths might not be as significant as in the ISCAS89 benchmark circuits, but the benefit of filtering short paths can still be seen.

7. Conclusions and Future Work

In this work, we have introduced a realistic low cost delay test compaction flow that guardbands the circuit delay using a sequence of estimation metrics. Significant improvements using both static compaction and dynamic compaction are demonstrated over prior work using benchmark circuits. Current work targets larger designs. The veto compaction process can be also applied as a guardband for other constraints, such as test power [17]. In the future we want to also consider off-chip Ldi/dt effects during delay test ATPG and compaction because of its impact of voltage fluctuation during capture cycle [18].

Acknowledgements

This research was funded in part by the Semiconductor Research Corporation under contract 2007-TJ-1618 and by the National Science Foundation under grant CCF-0702669.

References

[1] W. Qiu, et al, "K Longest Paths Per Gate (KLPG) Test Generation for Scan-Based Sequential Circuits," *IEEE Int'l Test Conf.*, Charlotte, NC, Oct. 2004, pp. 223-231.

[2] K. L. Shepard and V. Narayanan, "Noise in Deep Submicron Digital Design," *IEEE Int'l Conf. CAD*, San Jose, CA, Nov. 1996, pp. 524-531.

[3] Y.-S. Chang, S. K. Gupta and M. A. Breuer, "Analysis of Ground Bounce in Deep Sub-Micron Circuits," *IEEE VLSI Test Symp.*, Monterey, CA, Apr. 1997, pp. 110-116.

[4] H. H. Chen and D. D. Ling, "Power Supply Noise Analysis Methodology for Deep Submicron VLSI Chip Design," *ACM/IEEE Design Auto. Conf.*, Anaheim, CA, June 1997, pp. 638-643.

[5] C. Tirumurti, S. Kundu, S. Sur-Kolay and Y.-S. Chang, "A Modeling Approach for Addressing Power Supply Switching Noise Related Failures of Integrated Circuits," *Design, Automation and Test in Europe*, Paris, France, Feb. 2004, pp. 1078-1083.

[6] S. T. Zachariah, Y.-S. Chang, S. Kundu and C. Tirumurti, "On Modeling Cross-talk Faults," *Design, Automation and Test in Europe*, Munich, Germany, Mar. 2003, pp. 490-495.

[7] S. Pant, D. Blaauw, V. Zolotov, S. Sundareswaran and R. Panda, "Vectorless Analysis of Supply Noise Induced Delay Variation," *IEEE/ACM Int'l Conf. CAD*, San Jose, CA, Nov. 2003, pp. 184-191.

[8] A. Krstic, et al, "Pattern Generation for Delay Testing and Dynamic Timing Analysis Considering Power Supply Noise Effects," *IEEE Trans. CAD*, vol. 20, no. 3, Mar. 2003, pp. 416-425.

[9] J. Ma, J. Lee, and M. Tehranipoor, "Layout-Aware Pattern Generation for Maximizing Supply Noise Effects on Critical Paths," *IEEE VLSI Test Symp.*, May. 2009, pp. 221-226.

[10] J. Lee and M. Tehranipoor, "A Novel Test Pattern Generation Framework for Inducing Maximum Crosstalk Effects on Delay-Sensitive Paths," *IEEE Int'l Test Conf.*, Oct. 2008, pp.1-10.

[11] J. Wang, et al, "A Vector-based Approach for Power Supply Noise Analysis in Test Compaction," *IEEE Int'l Test Conf.*, Austin, TX, Nov.2005, pp. 517-526.

[12] A. Kokrady and C. P. Ravikumar, "Fast, Layout-Aware Validation of Test-Vectors for Nanometer-Related Timing Failures," *Int'l Conf. on VLSI Design*, Bombay, India, Jan. 2004, pp. 597-602.

[13] A. Kokrady and C. P. Ravikumar, Static Verification of Test Vectors for IR Drop Failure, *ICCAD*, Nov. 2003, pp. 760-764.

[14] Z. Wang and D. M. H. Walker, "Dynamic Compaction for High Quality Delay Test", *IEEE VLSI Test Symp.*, Apr. 2008, pp. 243-248.

[15] J. Wang, et al, "Power Supply Noise in Delay Testing", *IEEE Int'l Test Conf.*, Santa Clara, CA, Oct. 2006, pp. 1-10.

[16] Z. Jiang and D. M. H. Walker, "An Efficient Algorithm to Achieve Constant Test Power", *IEEE Workshop on Defect and Data Driven Testing*, Santa Clara, CA, Oct. 2008.

[17] Z. Jiang and D. M. H. Walker, "Enhancement Approaches for Constant Test Power Algorithm", *International Test Synthesis Workshop*, Austin, TX, Mar. 2009.

[18] P. Pant and J. Zelman, "Understanding Power Supply Droop During At-Speed Scan Testing", *VLSI Test Symp.*, May 2009, pp. 227-232.

Automatic Test Stimulus Generation for Accurate Diagnosis of RF Systems Using Transient Response Signatures

Aritra Banerjee, Shreyas Sen, Shyam Kumar Devarakond and Abhijit Chatterjee

School of Electrical and Computer Engineering, Georgia Institute of Technology, Atlanta, Georgia 30332, USA

Email: aritra.banerjee@ece.gatech.edu, abhijit.chatterjee@ece.gatech.edu

Abstract—**Low cost diagnosis of RF systems has become an important problem due to increased process variability effects on the performance of RF devices and the need to ramp-up RF IC yield rapidly. In the recent past, there has been work on diagnosing RF device model parameters from random "frequency-rich" test stimulus. In this paper, we develop a novel test stimulus generation approach which produces a compact, *deterministic* test stimulus in such a way that the RF DUT model parameters can be *computed directly from the DUT response* (called the DUT *signature*). This is achieved through use of a non-linear solver that adjusts the DUT model parameters iteratively until the model response to the applied test matches the observed DUT test response signature. It is shown that a small set of optimized tones in the frequency domain or an optimized transient waveform in the time domain can be used as test stimulus. It is shown how the use of embedded sensors in the RF design can expedite model parameter diagnosis. The practicality and accuracy of the proposed diagnosis approach is shown through simulations and hardware measurements.**

I. INTRODUCTION

Diagnosis of RF systems for yield ramp-up and post manufacture tuning is becoming a critical problem due to high frequencies of operation (1-60 GHz and beyond) and the use of scaled CMOS technologies that are increasingly susceptible to manufacturing process variations. Measuring all the test specifications of RF circuits is difficult due to the different test setups and test instrumentation needed for measuring different RF specifications and incurs significant test time as well as test cost. To alleviate this problem, the concept of "alternate testing" has been proposed in the past, which allows *multiple* specifications of the device under test (DUT) to be predicted from a *single* data acquisition [1, 2]. This has the benefit of incorporating the inaccuracies of the DUT simulation model as well as those of the test measurement system into the test data analysis algorithms (these inaccuracies are difficult to model). However, the technique involves the development of regression models [3] and the "training" procedure used to develop the regression models requires the use of standard testing procedures for calibration purposes. A key benefit of the alternative testing approach when applied to RF circuits is that test response analysis *does not require explicit signal demodulation* (except for training). Rather, the RF DUT specifications can be predicted from the raw downconverted data using a *single data acquisition* (resulting in a very fast test procedure). Also, it is possible to determine specification values from the alternate testing procedure such as bias current, for which behavioral simulation models are *difficult to derive.*

In contrast to the alternative testing approach, iterative techniques have been proposed in the past for determining the specifications of mixed-signal/RF systems. These methods assume that (accurate) mathematical (MATLAB/Simulink) DUT models exist from which the relevant specification values can be extracted easily. The DUT model parameters and the relevant specifications are determined by stimulating the DUT in such a way that the DUT model parameters can be estimated accurately from the observed test response data. Algorithms (iterative) for determining the parameters of predistortion filters to compensate for amplitude and phase nonlinearities of RF power amplifiers are well known and used extensively in wireless handsets [4, 5]. In [6], Senguttuvan has shown how the model parameters of a transmitter and thereby the inverse distortion characteristics of a predistortion filter can be derived from transient response analysis of the transmitter output through use of an *iterative least mean squares (LMS) solver*. In [7] and [8], Cherubal and Chatterjee have proposed a device parameter computation based diagnosis methodology for analog integrated circuits. A *nonlinear solver* is used to determine the critical Spice-level model parameters of devices such as amplifiers directly from the transient response of the device to an optimized test stimulus. In [9] Park and Abraham show how the parameters of a Volterra series model of nonlinear mixed-signal devices can be estimated from the results of pseudorandom tests.

The problem of loopback testing of RF transceiver systems was investigated in [10] by Halder and it was shown for the first time how the Gain and IIP3 values of the transmitter and receiver could be decoupled from the results of loopback tests. This was further applied to *decouple the nonlinearities of cascaded RF modules* (e.g. mixer, PA) from envelope detector data by Han, Bhattacharya and Chatterjee in [11]. Recently, in [12], Erdogan and Ozev describe a method for characterizing the RF transceiver parameters from analysis of demodulated loopback data. In addition to the fundamental nonlinearity parameters diagnosed in [10, 11], they are also able to diagnose other parameters such as I-Q mismatch and DC offset values of the transmitter and receiver from the looped back test data. While the methods of [10, 11] use regression based techniques to decouple embedded RF module parameters, the method of [12] uses a nonlinear solver to do the same (as in [6-8]). In [10, 11], *optimized multitones* are used as test stimulus but only the Gain and IIP3 specs are targeted (very short test time: DC offset and I-Q mismatch can be determined for the transmitter and receiver in loopback using other tests). In [12], a larger set of specs is targeted, but *OFDM frames containing random data are used as test stimulus*, resulting in longer test times. In [13] a multitone test generation approach for behavioral model parameter computation is shown. In each of the above methods [6-13] the device parameter values are extracted from observed test data using accurate simulation models of the DUT that incorporate the parameters being diagnosed.

In contrast to prior research, the key contributions of this paper are as follows:

• We show that a *few optimized "deterministic" multitones* (as opposed to random OFDM data which incurs longer test times) are sufficient to diagnose complex transmitter and receiver parameters (Gain, 3rd and 5th order nonlinearity, DC offset, I-Q imbalance, AM-PM distortion) from loopback test data. Prior loopback test methods have not handled AM-PM distortion in the presence of the other stated nonidealities. Note that the selected tones *need not be necessarily orthogonal*. Also, minimizing the number of tones minimizes the "intermodulation noise" due to nonlinearities in the data converters used to drive the test. This can be a problem if on-chip converters are used for driving the test signals.

• We show how optimized time-domain stimulus can also be used to achieve all the test objectives as discussed above.

• We show that with the use of an envelope detector sensor attached to the output of the transmitter, it is possible to accurately diagnose transmitter (AM-AM and AM-PM distortion) and receiver specifications along with DC offset and I-Q imbalance values.

Note that as in [6-8, 12-13], a nonlinear solver is used to determine the transmitter and receiver behavioral model parameters *directly from the observed time-domain test response*. The time taken to converge to a solution is optimized via intelligent assessment of the likely solution corresponding to an observed signature. This time is also significantly reduced through the use of an embedded envelope detector at the output of the transmitter.

The rest of the paper is organized as follows. In Section 2, the key principles of signature test are discussed. This is followed in Section 3 by mathematical analysis of unique signature of transceiver system and relevant concepts. In Section 4, signature test stimulus generation is presented followed by description of model in Section 5. Test setup and model parameter computation technique are explained in Section 6 and 7 respectively. Section 8 presents the simulation results while hardware measurements are shown in Section 9. Finally the paper is concluded in Section 10.

II. SIGNATURE TEST: KEY PRINCIPLES

The term 'Signature Test' signifies the generation of a unique time or frequency domain signature of the component under test in response to an optimized signature test stimulus. This is derived from the concept of alternate test [1, 2] except that a nonlinear solver is used to solve for the DUT model parameters and thereby its specifications, as opposed to the use of regression models for directly predicting the DUT specifications [1, 2]. Clearly, the signature must be constructed in such a way that all the DUT model parameters can be uniquely determined from the observed signature. This is possible if "close" ranges of model parameter values in which the solution lies are known a-priori and "local" search within the selected ranges is performed to find the exact solution using a nonlinear solver (the solver tries to match the simulated model response with the observed DUT response). Such determination of range values is done through preliminary analysis of the DUT signature using regression models (this does not require training on hardware data, rather extensive model simulation). This also determines if a unique solution is at all feasible much in the spirit of the "defect filter" of [14]. However, a detailed discussion of this is not possible in this paper and we focus primarily on test stimulus generation issues. The following steps are involved.

System Modeling: We construct MATLAB models for RF simulation that include the following nonidealities of WLAN OFDM systems: I/Q gain and phase mismatch, AM-AM and AM-PM effects of power amplifiers, DC offset and third and fifth order nonlinearity of the receive/transmit chains.

Signature Test Optimization: A genetic algorithm based test stimulus generation technique is used to optimize the test stimulus for signature based test. This algorithm has two versions: (a) one in which the test signal is a multitone waveform (b) one in which the test waveform is a transient waveform sampled at specified time points.

Computation of Model Parameters: The model parameters are computed by minimizing the least squares error between the model-generated and observed DUT test response signature using a nonlinear optimizer. In loopback test, the I and Q output signatures of the receiver are used for analysis.

III. UNIQUE SIGNATURE OF TRANSCEIVER SYSTEM

As discussed in [10, 11, 12], the presence of nonlinearity in an RF system allows the computation of the individual specifications of two cascaded RF modules from end-to-end measurements on the cascaded chain. Let us assume that two RF blocks are connected in cascade and their behaviors can be expressed by (1) and (2) respectively:

$$r(t) = m_1.x(t) + m_2.[x(t)]^2 + m_3.[x(t)]^3 \qquad (1)$$

$$s(t) = n_1.r(t) + n_2.[r(t)]^2 + n_3.[r(t)]^3 \qquad (2)$$

where $x(t)$ is the input to the first block, $r(t)$ is the output of the first block and $s(t)$ is the final output coming out of the second block. If (1) and (2) are combined then the behavior of the complete system can be expressed as:

$$s(t) = \sum_{i=1}^{9} c_i.[x(t)]^i \qquad (3)$$

$$c_i = f_i(m_1, m_2, m_3, n_1, n_2, n_3) \qquad (4)$$

Equations (3) and (4) show that third order nonlinear behavior of the two blocks effectively results in a 9th order nonlinear transfer function of the total system where each coefficient of $[x(t)]^i$ is a function of the m and n coefficients of individual blocks. If using the model fitting technique all the c_i coefficients are accurately calculated by analyzing the input and output waveforms, then a set of 9 equations are obtained in terms of the parameters of Equation (4). These 9 equations containing 6 unknown variables (m and n coefficients) define a system of equations from which the approximate values of the unknown variables can be calculated using standard least square techniques.

Three important facts have to be considered while decoupling parameters of individual components from system response. First let us consider the case when the second block is completely linear i.e. $n2 = 0$ and $n3 = 0$. The combined equation becomes:

$$s(t) = n_1.m_1.x(t) + n_1.m_2.[x(t)]^2 + n_1.m_3.[x(t)]^3 \qquad (5)$$

Here we have 3 equations with 4 unknown variables which mean multiple solutions are possible. If the applied signal amplitude level is very small and one or both of the blocks operate in the linear region (if $x(t)$ is very small then $r(t) \approx m_1.x(t)$ and $s(t) \approx n_1.r(t)$) then decoupling will not be possible. By optimizing the test stimulus we ensure that the nonlinear behavior of both of the blocks are excited.

The second issue in diagnosis is accuracy. The polynomial coefficients (c_i) of (3) are calculated in such a way that for different time points of the input and output waveform, the computed coefficients represent the most accurate approximation of system behavior. The optimized test stimulus must traverse all regions of the output dynamic range of the system so that the unique combination of coefficients can be found during analysis.

Third, even if the system under consideration shows 9th order nonlinear behavior as shown in Equation (3), the test must be designed in such a way that information upto the 9th order harmonic and relevant intermodulation terms are preserved in the observed output signature. In the presence of filters inherent in RF systems, the effects of higher order nonlinear behavior should be captured in the intermodulation frequency components within the passband of the system.

IV. TEST STIMULUS GENERATION

A genetic algorithm based test stimulus optimization algorithm is used in this work and is shown in Figure 1. Genetic Algorithms (GA) are a class of evolutionary algorithms that are extensively used for solving optimization problems [15]. Crossover and Mutation techniques are applied

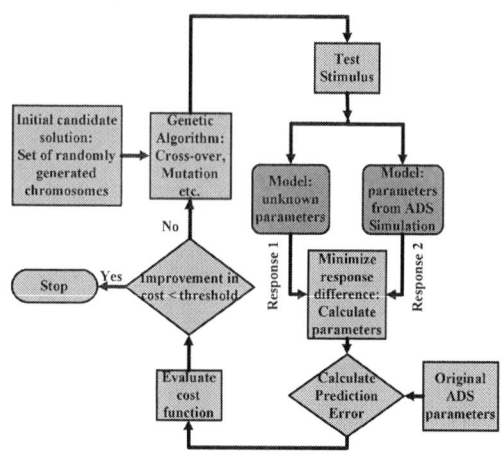

Fig. 1. Test generation approach

on a set of selected chromosomes (candidate solutions) and in each iteration of the algorithm, a new generation of chromosomes are created. The GA determines the best solution corresponding to a fitness value (cost function) upon convergence of the algorithm. Figure 1 shows the steps involved in the test generation process. For multi-tone test stimulus generation, the amplitudes and phases of the tones are optimized by GA, whereas for time domain transient test stimulus generation the amplitude values of the signal at different time points are optimized. Monte-Carlo simulation is performed using the Advanced Design System (ADS) simulation tool and process variation effects are introduced to create M transceiver instances at critical performance (process) corners ($M = 50$ in our experiments). Using a nonlinear equation solver [16, 17, 18], the parameter values for these instances are calculated for each candidate stimulus. From this data, a weighted mean square error across all the instances and all the DUT specifications is determined. The GA finds the "best" stimulus that minimizes this least square error. Three types of test signals are optimized and used for model parameter computation. The GA is first tailored to find the amplitude and phase of least number of orthogonal and non-orthogonal tones that meet a specific accuracy criterion. Finally, the GA is tailored to find a transient waveform across 128 time points or less (to make the test time comparable to or less than that of an OFDM frame with 64 sub-carriers).

V. CONSTRUCTION OF TRANSCEIVER MODEL

The RF MATLAB models for transceiver nonidealities used in this research are described below.

A. Nonlinearity of LNA and Mixer

The nonlinear characteristics of Low Noise Amplifier and up and down-conversion mixers are modeled as third order polynomials:

$$v(t) = \sum_{p=0}^{3} \alpha_p.[u(t)]^p \qquad (6)$$

where $u(t)$ and $v(t)$ are the input and output signal of the nonlinear block respectively.

B. I/Q Imbalance in transmitter

In I-Q modulation scheme the carrier used for modulating I and Q signals should have same amplitude and one should be phase shifted from the other by 90°. But due to quadrature mismatch [19, 20] the relative amplitude and phase difference of the carriers on two paths change from their ideal values. The upconverted signal with I/Q imbalance is modeled as:

$$Y(t) = I(t).cos(2\pi f_c t) \\ - (1 + \varepsilon_{TX}).Q(t).sin(2\pi f_c t + \phi_{TX}) \tag{7}$$

where ε_{TX} is the magnitude mismatch and ϕ_{TX} is the phase mismatch in transmitter and f_c is the carrier frequency.

C. AM-AM and AM-PM of PA

A complex envelope driven memory-less behavioral modeling of the power amplifier is done in this paper. This model reflects the instantaneous variation in the output envelope of the PA due to change in the input envelope [21]. The input signal to the PA can be written as:

$$s(t) = r(t).cos(\omega_c t + \psi(t)) = Re\{r(t).e^{j(\omega_c t + \psi(t))}\} \tag{8}$$

where the complex envelope $r(t).e^{j\psi(t)}$ modulates the RF carrier of frequency ω_c. If the envelope of the input signal is expressed as $r'(t)$ then the output of the power amplifier is modeled as:

$$s'(t) = f(r'(t))\{cos(\omega_c t + \psi(t) + g(r'(t)))\} \tag{9}$$

where $f(r'(t))$ implements the AM-AM and $g(r'(t))$ models the AM-PM behavior of the PA. These functions can be represented as fifth order and third order polynomials respectively:

$$f(r'(t)) = \sum_{i=0}^{5} \alpha_{i(PA)}.[r'(t)]^i \tag{10}$$

$$g(r'(t)) = \sum_{k=0}^{3} \beta_{k(PA)}.[r'(t)]^k \tag{11}$$

where $\alpha_{1(PA)}$ is the gain of the power amplifier. Now the input to the power amplifier is given by (7) which is to be expressed in the form of (8) to introduce AM-AM and AM-PM effects. (7) can be written as:

$$Y(t) = cos(\omega_c t).[I(t) - (1 + \varepsilon_{TX}).Q(t).sin(\phi_{TX})] \\ - sin(\omega_c t).[(1 + \varepsilon_{TX}).Q(t).cos(\phi_{TX})] \tag{12}$$

Now, let us define,

$$V_1(t) = I(t) - (1 + \varepsilon_{TX}).Q(t).sin(\phi_{TX}) \tag{13}$$

$$V_2(t) = (1 + \varepsilon_{TX}).Q(t).cos(\phi_{TX}) \tag{14}$$

We can define two other variables in terms of $V_1(t)$ and $V_2(t)$ as:

$$A(t) = \sqrt{V_1(t)^2 + V_2(t)^2} \tag{15}$$

$$\gamma(t) = tan^{-1}(V_2(t)/V_1(t)) \tag{16}$$

Now from (12) and using (13), (14), (15) and (16) we can write:

$$Y(t) = A(t).cos[\omega_c t + \gamma(t)] \tag{17}$$

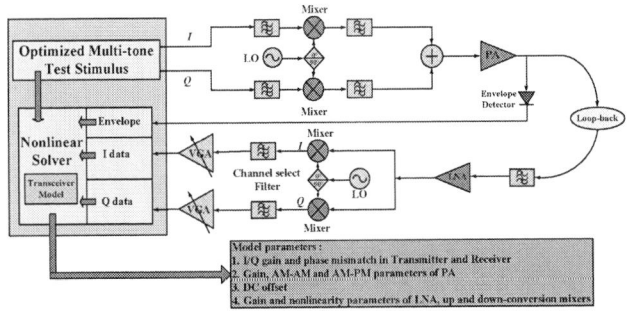

Fig. 2. Test setup

Equation (17) is exactly in the form of (8) and AM-AM and AM-PM effects are included in (17) by finding the envelope of $Y(t)$ and using (10) and (11) to calculate amplitude and phase components.

D. I/Q Imbalance in receiver

The I/Q magnitude and phase mismatch in the receiver is modeled in the same way as done for the transmitter. The Local Oscillator signal with the imbalance can be written as:

$$s_{LO}(t) = cos(\omega_{LO}t) - j.(1 + \varepsilon_{RX}).sin(\omega_{LO}t + \phi_{RX}) \tag{18}$$

where ε_{RX} is the magnitude mismatch and ϕ_{RX} is the phase mismatch of the receiver LO signal.

E. DC Offset in transmitter and receiver

Due to capacitive feed through from the Local Oscillator (LO leakage) and self-mixing, a dc signal component appears in the signal path [20]. This phenomenon is modeled by adding a dc component with the signal which depends on the capacitive coupling.

F. Computation of specifications from model parameters

The gain and input IP3 specifications of nonlinear modules can be calculated from model parameters using the following equations [20]:

$$Gain = 20.log_{10}[\alpha_1] \text{ dB} \tag{19}$$

$$A_{IP3} = \sqrt{\frac{4}{3} \times \frac{|\alpha_1|}{|\alpha_3|}} \text{ volts} \tag{20}$$

VI. TEST SETUP

The test setup is shown in Figure 2. The OFDM transmitter and receiver are connected in loopback mode. In addition to the use of loopback, the usefulness of an envelope detector connected to the output of the transmitter for test purposes is also studied. This reduces the complexity of the problem: the transmitter parameters are calculated from the envelope detector output and using those values as known parameters, the receiver parameters are computed from the down converted I and Q data. Without the envelope detector, the coefficients of transmitter and receiver nonlinearity can be decoupled from the downconverted I and Q outputs of loopback setup. When the envelope detector is placed, nonlinearity parameters of upconversion mixer and PA can

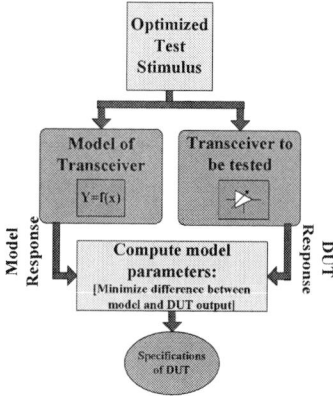

Fig. 3. Model parameter computation from signature

be calculated from the captured envelope of the transmitted signal. And using these known parameters, the nonlinearity coefficients of downconversion mixer and LNA are decoupled from I and Q outputs.

VII. COMPUTATION OF MODEL PARAMETERS

The optimized test stimulus is applied to the DUT and its model. The DUT test response signature is captured and a MATLAB based nonlinear data fitting problem solver [16] is used to find the model parameters corresponding to the observed DUT signature (see Figure 3). The non-linear solver adjusts the DUT model parameters iteratively until the simulated model response to the applied test matches the observed DUT test response. The solver uses "trust-region-reflective" nonlinear equation solving algorithm to compute the parameters of the transceiver model. The algorithm is based on trust-region method [17] and interior-reflective Newton method [18]. In trust-region method a model function is built which is an approximation of the original objective function around the current point. The neighborhood of the current point where the approximated function reflects the behavior of the original function is called the "trust-region". Then the approximated model function is solved in the trust-region. If the candidate solution does not improve the original objective function then the trust region is shrunk and the algorithm searches for a new solution.

VIII. SIMULATION RESULTS

Simulation results for different steps of signature testing are described in this section.

A. Measurement Noise

To introduce measurement noise in simulation, the down-converted I and Q outputs of the receiver and envelope detector response (while diagnosing nonlinearity of different components of transmitter and receiver separately) are digitized with a 10 bit analog-to-digital converter.

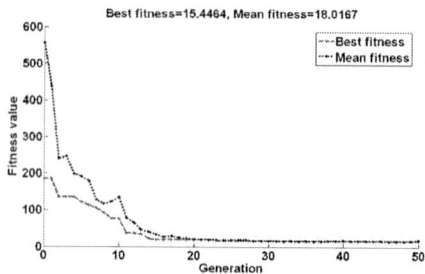

Fig. 4. Convergence of genetic algorithm

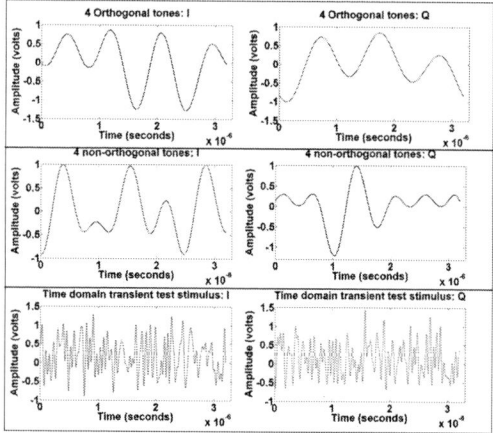

Fig. 5. Different optimized test inputs

B. Test Stimulus Optimization

Figure 4 shows the optimization of the cost metric vs. the number of GA iterations. The optimization is repeated for 3 types of test stimuli for the I and Q inputs of the transmitter (result of optimization shown in Figure 5): orthogonal tones (top, Figure 5), non-orthogonal tones (middle, Figure 5) and time domain transient waveform (bottom, Figure 5). The left and right waveforms in each case represent the optimized I and Q inputs, respectively. Test time for all of the optimized stimuli is $3.6\mu s$. It is seen that for both cases of multitone stimulus, the use of 4 tones results in significant accuracy in parameter computation.

C. Model Parameter Computation

MATLAB based nonlinear equation solver is used for computation of model parameters using all 3 types of test stimuli shown in Figure 5. The computation is performed in two ways: (a) with envelope detector connected to the output of the transmitter and (b) loopback without the envelope detector. In the first case, the nonlinear solver is able to decouple the nonlinearity parameters of the mixer and PA from envelope detector output. Using the computed transmitter parameters as known variables, mixer and LNA nonlinearity parameters of the receiver are decoupled from downconverted I and Q output. In the second case, effective nonlinearity parameters of transmitter and receiver are computed from I and Q outputs. The parameters that are computed for case (b) are: I/Q gain and phase mismatch of transmitter and receiver,

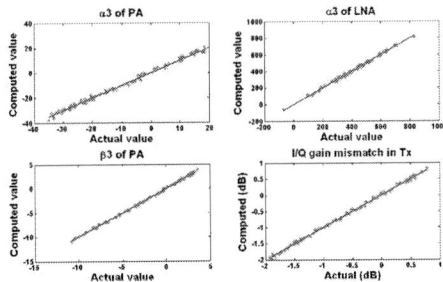

Fig. 6. Accuracy of parameter computation

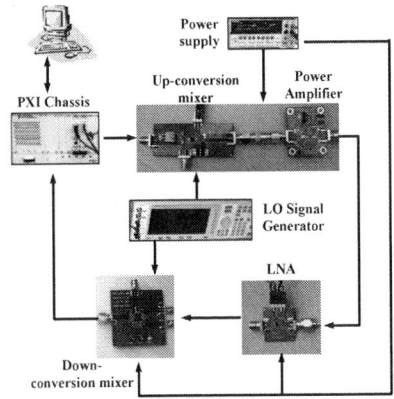

Fig. 7. Hardware setup

all coefficients of 5th order nonlinearity of the transmitter, AM-PM distortion of the PA, DC offset in transmitter and receiver and coefficients of 3rd order nonlinearity of the receiver. For case (a), in addition to the parameters above, the 5th order nonlinearity parameters of PA and 3rd order nonlinearity parameters of the up-conversion mixer, down conversion mixer and LNA are also determined. For both cases and with all 3 test stimuli, the average error in model parameter computation ranges from 0.5% to upto 10% (see Figure 6 for some RF parameters diagnosed from loopback using envelope detector).

IX. HARDWARE MEASUREMENTS

The validity of the proposed approach is shown by conducting experiment on a hardware prototype. The hardware setup is shown in Figure 7. A data acquisition module by National Instruments (NI PXIe-1073) is used to interface the RF front-end with PC. The optimized 4 tone (orthogonal) test stimulus is up-converted by a mixer RF2638 by RFMD. The upconverted signal passes through a power amplifier (MAX2247) followed by a low noise amplifier (RF2370) connected in loopback. The LNA output is downconverted by another mixer MAX2039. Results of the hardware measurements are shown in Table 1.

TABLE I
HARDWARE MEASUREMENT: ERROR IN SPECIFICATION COMPUTATION

Specifications	PA Gain	PA IIP3	LNA Gain	LNA IIP3
Average error	5.85%	11.72%	4.61%	9.88%

X. CONCLUSION

A test stimulus optimization technique is proposed for model parameter computation of individual modules of RF systems from its response ('signature'). A few optimized multitone (orthogonal and non-orthogonal) signal and time domain optimized test waveform are generated using genetic algorithm and transmitter and receiver parameters (AM-AM and AM-PM of PA, I/Q mismatch, DC offset, nonlinearity of LNA and mixers) are diagnosed from the response of the transceiver using a nonlinear equation solver. Simulation results show the accuracy in computation of the individual specifications of RF modules from end-to-end measurements on the cascaded chain. Experiments performed on hardware prototype also validates the proposed concept.

ACKNOWLEDGEMENT

This work was funded in part by GSRC/FCRP 2009-DT-2049 and NSF CCR-0635016.

REFERENCES

[1] P. N. Variyam et al., "Prediction of analog performance parameters using fast transient testing," *IEEE TCAD*, Vol. 21, Mar. 2002, pp. 349 -361.

[2] R. Voorakaranam et al., "Signature testing of analog and RF circuits: Algorithms and methodology," *IEEE TCAS I*, Vol 54, Issue 5, May 2007, pp. 1018-1031.

[3] J. H. Friedman, "Multivariate Adaptive Regression Splines", *The Annals of Statistics*, vol. 19, 1991, pp. 1-141.

[4] K. Mekechuk et al., "Linearizing Power Amplifiers Using Digital Predistortion, EDA Tools and Test Hardware", *High Frequency Electronics*, Vol. 3, No. 4, pp. 18-30, April 2004.

[5] Thomas H. Lee, "The design of CMOS radio-frequency integrated circuits", *Cambridge University Press*, 2004.

[6] R. Senguttuvan, "Low-cost test, diagnosis and tuning for adaptive radio frequency systems", Ph.D. dissertation, ECE, Georgia Tech, 2008.

[7] S. Cherubal and A. Chatterjee, "Parametric fault diagnosis for analog systems using functional mapping", *DATE*, 1999, pp. 195 - 200.

[8] S. Cherubal and A. Chatterjee, "Test generation based diagnosis of device parameters for analog circuits", *DATE*, 2001, pp. 596 - 602.

[9] J. Park and J. A. Abraham, "Pseudorandom test for nonlinear circuits based on a simplified Volterra series model", *ISQED 2007*, pp. 495-500.

[10] A. Halder, "Efficient Alternate Test Generation for RF Transceiver Architectures", Ph.D. dissertation, ECE, Georgia Tech, 2006.

[11] D. Han et al., "Low-Cost Parametric Test and Diagnosis of RF Systems Using Multi-Tone Response Envelope Detection", *IET Proceedings on Computers & Digital Techniques*, Vol. 1, No. 3, May 2007, pp. 170-179.

[12] E. S. Erdogan and S. Ozev, "Detailed Characterization of Transceiver Parameters Through Loop-Back-Based BiST", *IEEE Transactions on VLSI Systems*, Vol. 18, No. 6, 2010, pp. 901 - 911.

[13] A. Banerjee et al., "Optimized Multitone Test Stimulus Driven Diagnosis of RF Transceivers Using Model Parameter Estimation", *IEEE 24th International Conference on VLSI Design*, January 2011, pp. 274 - 279.

[14] H. Stratigopoulos, S. Mir, E. Acar, S. Ozev, "Defect Filter for Alternate RF Test", *IEEE ETS 2010*, pp. 265-270.

[15] David E. Goldberg, "Genetic Algorithms in Search, Optimization, and Machine Learning", *Addison-Wesley Publishing Company*, 1989.

[16] http://www.mathworks.com/help/toolbox/optim/ug/bqnk0r0.html.

[17] J. Nocedal S. J. Wright, "Numerical Optimization", *Springer*, 1999.

[18] T. F. Coleman and Y. Li, "On the Convergence of Reflective Newton Methods for Large-Scale Nonlinear Minimization Subject to Bounds" *Mathematical Programming*, Vol. 67, No. 2, pp. 189-224, 1994.

[19] Chia-Liang Liu, "Impacts of I/Q imbalance on QPSK-OFDM-QAM detection", *IEEE Transactions on Consumer Electronics*, Vol. 44, Issue 3, 1998, pp. 984 - 989.

[20] Behzad Razavi, "RF Microelectronics", *Prentice Hall*, 1998.

[21] J. C. Pedro and S. A. Maas, "A comparative overview of microwave and wireless power-amplifier behavioral modeling approaches", *IEEE T-MTT*, Vol. 53, No. 4, Part 1, 2005, pp. 1150 - 1163.

Non-Linear Analog Circuit Test and Diagnosis under Process Variation using V-Transform Coefficients

Suraj Sindia*, Vishwani D. Agrawal†
Department of Electrical and Computer Engineering
Auburn University, Alabama, AL, USA
*Email: szs0063@auburn.edu
†Email: vagrawal@eng.auburn.edu

Virendra Singh
Supercomputer Education
and Research Centre
Indian Institute of Science, Bangalore, India
Email: viren@serc.iisc.ernet.in

Abstract—**Parametric fault testing of non-linear analog circuits based on a new mathematical transform is presented. The V-Transform acts on the polynomial expansion of the circuit's function. Its main properties are: 1) to make the polynomial coefficients monotonic, 2) to reduce masking of parametric faults due to process variation, and 3) to increase the sensitivity of polynomial coefficients to the circuit parameter variation, thus enhancing diagnostic resolution. We show that the sensitivity of V-Transform Coefficients (VTC) with respect to circuit parameter variation is up to 3 to 5 times greater than the sensitivity of polynomial coefficients. Fault diagnosis of parametric faults under process variation using VTC is then presented. We also propose a scheme to distinguish between circuit specifications failures due to process variation versus manufacturing defects which manifest as parametric faults. To validate our approach, we apply the test and diagnosis procedures to a benchmark fifth order elliptic filter. We use SPICE program for fault injection, with about 50,000 Monte Carlo simulation runs to demonstrate fault detection-diagnosis under process variation. The test scheme uncovers 95% of all injected single parametric faults whose sizes deviate 5% from the nominal values of circuit components corrected for process variation, while the procedure successfully diagnosed all component faults under $\pm 3\sigma$ process variation with 88% confidence level.**

I. INTRODUCTION

Non-linear circuit testing has been well studied and different methods have been proposed for finding parametric faults [1], [2], [3], [4], [5], [6], [7], [8]. Prominent among them in the industry is the I_{DDQ} based testing where current from the supply rail is monitored and sizable deviation from its quiescent value is reported. However this requires augmentation of the CUT. For example, in the simplest case a regulator supplying power to any sizable circuit has to be augmented with a current sensing resistor and an ADC (for digital output) and then there is subsequent analysis to be performed on sensed current. Further I_{DDQ} is suitable only for catastrophic faults as the current drawn from the supply is distinguishable only when there is some "big enough" fault so as to change the current drawn from the supply from its quiescent value to a region where it is distinguishable. For example with resistor R_2 being open in Figure 1, the current drawn from supply can change by 50% of its quiescent value. Such faults can typically be

found by monitoring I_{DDQ} using a current sensor. However parametric deviations say lesser than 10% from its nominal value cannot be observed using this scheme, specially so in the deep submicron era where the leakage currents can be comparable with defect induced current [9]. The other approach for testing parametric faults that can be found in literature [10], [11], [12], [13], [14] is based on the use of neural networks. Neural network based approaches propose the use of circuit observer blocks to track the output for a set of input signals which is used for training the neurons. The trained set of neurons is then used to estimate variations in the output for a standard input stimulus. This method, however, suffers from large amounts of training required and the consequent increase in test application time that the scheme is prohibitive for even medium sized analog circuits at production. More recently, the use of Volterra series coefficients was proposed to estimate non-linear characteristics of the system. These coefficients are then used for testing the circuit with a pseudo random input stimulus [15], [16]. This method however suffers from the high computational requirement of estimation of Volterra series coefficients for every circuit at production which can increase the test cost significantly. It is therefore interesting to develop a method to detect parametric faults with little circuit augmentation while keeping the test access mechanism simple and the test application time to a minimum.

To address the issue of parametric deviation, we would typically need more observables to have an idea about the parametric drift in circuit parameters. This would mean an increase in complexity of the sensing circuit. However, we would also want only little augmentation to tap any of the internal circuit nodes or currents. To overcome these seemingly contrasting requirements the method intended should have some way of "seeing through" the circuit with only the outputs and inputs at its disposal. References [17], [18] have accomplished this sort of a strategy for linear circuits in a different context as described next.

Savir and Guo describe a method [17] based on transfer function of a circuit under test (CUT). The transfer function,

$H(s)$, of the CUT is expressed as:

$$H(s) = \frac{\sum_{i=0}^{M} a_i s^i}{\sum_{i=0}^{N} b_i s^i} \quad (M < N) \quad (1)$$

Here, a_i and b_i are referred to as transfer function coefficients (TFCs). The CUT is subjected to frequency rich input signals and the output at these frequencies is observed. With these input-output pairs they estimate the TFCs of CUT. These coefficients are now compared with the ideal circuit TFCs, which are known a priori. The CUT is classified faulty if any of the estimated coefficients are beyond the tolerable range. This method necessarily needs the CUT to be linear, as transfer functions are possible only for LTI systems.

To extend the above idea to more general non-linear circuits we adopted a strategy in [19] where we expand the function of the circuit as a polynomial by the Taylor's series expansion about the input voltage mangitude v_{in} at a given frequency, as follows:

$$v_{out} = f(v_{in}) = f(0) + \frac{f'(0)}{1!}v_{in} + \frac{f''(0)}{2!}v_{in}^2 + \frac{f'''(0)}{3!}v_{in}^3 + \cdots + \frac{f^{(n)}(0)}{n!}v_{in}^n + \cdots \quad (2)$$

where $f(v_{in})$ is a real function of v_{in}. Ignoring the higher order terms in (2), we can expand v_{out} up to the nth power of v_{in}, which gives us the approximation in (3):

$$v_{out} = a_0 + a_1 v_{in} + a_2 v_{in}^2 + \cdots + a_n v_{in}^n \quad (3)$$

where $a_0, a_1, a_2, \ldots, a_n$ are all real-valued functions of circuit parameters p_k, $\forall k$. Further assume that normal parameter variations (normal drift) in a good circuit are within a fraction α of their nominal value, where $\alpha << 1$. This means that every parameter p_i is allowed to vary within the range $p_{k,nom}(1 - \alpha) < p_k < p_{k,nom}(1 + \alpha)$, $\forall k$, where $p_{k,nom}$ is the nominal value of parameter p_k. Whenever one or more of the coefficient values slip outside its individual hypercube we get a different set of coefficients that reflects a detectable fault. Therefore, equation (4) describes a hypercube for all parameters that correspond to either good machine values or undetectable parameter faults [2], [8], [17]:

$$a_{i,\min} < a_i < a_{i,\max} \quad \forall a_i, \ 0 \le i \le n \quad (4)$$

In the latter portion of this paper we address an important problem that has kept analog circuit test cost high [20], namely, distinguishing between faults induced due to process variation. For example, we would like to distinguish random drifts in t_{ox}, W, L, and doping densities of devices in an integrated circuits from those resulting from manufacturing defect induced (parametric) faults (e.g., Lithographic errors, etching errors, etc.) that lead to a substantial deviation of a circuit from its nominal behavior but are not large enough to render the circuit dysfunctional. We quantify an error distance measure between faults induced due to process variation with those induced due to manufacturing defects. We can then estimate the probabilities of a detected fault being caused by

process variation or by a manufacturing defect. We make this estimation based on maximizing a posteriori probabilities of the two kinds of errors conditioned on the event that a fault is detected.

This paper is organized as follows. Section II we state previously published results [21] on the polynomial expansion of function $f(v_{in})$ and notions of detectable fault sizes. Section III outlines V-Transform and the resulting sensitivity improvement. In Section IV we describe the problem at hand and discuss the proposed solution with an example. In Section V we generalize the test solution to an arbitrarily large circuit. Section VI establishes, 1) a method to distinguish between process variation induced faults and those induced due to manufacturing defects and 2) identify the fault site if it is of the latter kind. Section VII presents the simulation results for a standard elliptic filter. We conclude in Section VIII.

II. BACKGROUND

The coefficients a_i, $\forall 0 \le i \le n$, are in general non-linear functions of circuit parameters p_k, $\forall k$. The rationale in using these coefficients as metrics in classifying CUT as faulty or fault free is based on the premise of dependence of coefficients on circuit parameters.

Theorem 1. If coefficient a_i is a monotonic function of all parameters, then a_i takes its limit (maximum and minimum) values when at least one or more of the parameters are at the boundaries of their individual hypercube.

Lemma 1. If coefficient a_i is a non-monotonic function of one or more circuit parameters p_i, then a_i can take its limit values anywhere inside the hypercube enclosing the parameters.

By Theorem 1 and Lemma 1 it is clear that by exhaustively searching the space in the hypercube of each parameter we can get the maximum and minimum values of the polynomial coefficient. Typically this can be formulated as a non-linear optimization problem to find the maximum and minimum values of coefficient with constraints on parameters allowing only a normal drift.

Theorem 2. In polynomial expansion of Non-Linear Analog circuit there exists at least one coefficient that is a monotonic function of all the circuit parameters.

In conclusion, from Lemma 1 and Theorem 2, circuit parameter deviations have a bearing on coefficients and the monotonically varying coefficients can be used to detect parametric faults of the circuit parameters [19].

Definition: A minimum size detectable fault (MSDF), ρ, for a parameter is defined as the minimum fractional deviation of the circuit parameter from its nominal value for it to be detectable with all other parameters held at their nominal values. The fractional deviation can be positive or negative and is named upside-MSDF (UMSDF) or downside-MSDF (DMSDF) accordingly.

If ψ is the set of all coefficient values spanned by the parameters while varying within their normal drifts, i.e.,

$$\psi = \{v_0, v_1, \cdots, v_n \,|\, v_0 \in A_0, v_1 \in A_1, \cdots, v_n \in A_n\}$$
$$\forall_k \quad p_{k,nom}(1 - \alpha) < p_k < p_{k,nom}(1 + \alpha)$$

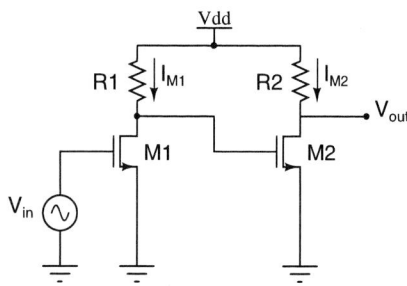

Fig. 1. Cascaded amplifier.

then by definitions of MSDF, ψ includes all possible values of coefficients that are not detectable. Any parametric fault inducing coefficient value outside the set ψ will result in a detectable fault.

III. THE V-TRANSFORM

We define V-Transform coefficients as follows: if $C_1, C_2 \cdots C_n$ are polynomial coefficients of CUT then their V-Transform coefficients, $V_{C_1}, V_{C_2} \cdots V_{C_n}$, are

$$V_{C_i} = e^{\gamma C_i'} \ \forall \ 0 \le i \le n \tag{5}$$

where C_i' are the modified polynomial coefficients defined indirectly as follows

$$\frac{dC_i'}{dp_j} = \left| \frac{dC_i}{dp_j} \right| \ \forall \ 0 \le i \le n \tag{6}$$

The modification C_i' according to (6) ensures that the modified polynomial coefficients are monotonic with the polynomial coefficients. Further, the V-Transform coefficients (VTC) are exponential functions of the modified polynomial coefficients and γ is a sensitivity parameter chosen according to the desired sensitivity. The gain in sensitivity of V-Transform coefficients to circuit parameters over the sensitivity of ordinary polynomial coefficients is given by

$$\frac{S_{p_i}^{V_{C_i}}}{S_{p_i}^{C_i}} = \frac{\left| \frac{dC_i}{dp_i} \right| \gamma e^{\gamma C_i'} \bullet \frac{p_i}{e^{\gamma C_i'}}}{\frac{dC_i}{dp_i} \bullet \frac{p_i}{C_i}} = \gamma C_i \tag{7}$$

Choices of $\gamma = 3$, for instance, results in a 3 times more sensitive coefficient to circuit parameters.

IV. PROBLEM AND APPROACH

We shall first illustrate with an example the calculation of limits of the polynomial coefficients for a simple circuit using MOS transistors. We shall follow this up with MSDF values for the circuit parameters.

Example 1. Two stage amplifier: Consider the cascaded amplifier shown in Figure 1. The output voltage V_{out} in terms of input voltage results in a fourth degree polynomial:

$$V_{out} = a_0 + a_1 v_{in} + a_2 v_{in}^2 + a_3 v_{in}^3 + a_4 v_{in}^4 \tag{8}$$

where constants a_0, a_1, a_2, a_3 are defined symbolically in (9) for transistors M1 and M2 operating in the saturation region.

Nominal values of $V_{DD} = 1.2V$, $V_T = 400mV$, $\left(\frac{W}{L}\right)_1 = \frac{1}{2}\left(\frac{W}{L}\right)_2 = 20$, and $K = 100\mu A/V^2$ are used for this example.

$$a_0 = V_{DD} - R_2 K \left(\tfrac{W}{L}\right)_2 \left\{ \begin{array}{l} (V_{DD} - V_T)^2 + \\ R_1^2 K^2 \left(\tfrac{W}{L}\right)_1^2 V_T^4 - \\ 2(V_{DD} - V_T) R_1 \left(\tfrac{W}{L}\right)_1 V_T^2 \end{array} \right\}$$

$$a_1 = R_2 K \left(\tfrac{W}{L}\right)_2 \left\{ \begin{array}{l} 4R_1^2 K^2 \left(\tfrac{W}{L}\right)_1^2 V_T^3 \\ +2(V_{DD} - V_T) R_1 K \left(\tfrac{W}{L}\right)_1 V_T \end{array} \right\}$$

$$a_2 = R_2 K \left(\tfrac{W}{L}\right)_2 \left\{ \begin{array}{l} 2(V_{DD} - V_T) R_1 K \left(\tfrac{W}{L}\right)_1 \\ -6R_1^2 K^2 \left(\tfrac{W}{L}\right)_1^2 V_T^2 \end{array} \right\}$$

$$a_3 = 4V_T K^3 \left(\tfrac{W}{L}\right)_1^2 \left(\tfrac{W}{L}\right)_2^2 R_1^2 R_2$$

$$a_4 = -K^3 \left(\tfrac{W}{L}\right)_1^2 \left(\tfrac{W}{L}\right)_2^2 R_1^2 R_2 \tag{9}$$

To find the limit values of the coefficient a_0 we assume that parameters R_1 and R_2 deviate by fractions x and y from their nominal values, respectively. To maximize a_0 we have the objective function (10) subject to constraints (11) through (15). Note that here we have set out to find MSDF of R_1. Similar approach can be used to find the MSDF of R_2.

$$1.2 - R_{2,nom}(1+y) \left\{ \begin{array}{l} 2.56 \times 10^{-3} + \\ 1.024 \times 10^{-7} R_{1,nom}^2 (1+x)^2 \\ -5.12 \times 10^{-4} R_{1,nom}(1+x) \end{array} \right\} \tag{10}$$

$$\begin{array}{l} 4.096 \times 10^{-9} R_{1,nom}^2 (1+x)^2 R_{2,nom} \ (1+y) \\ +5.12 \times 10^{-6} R_{1,nom}(1+x) R_{2,nom}(1+y) \\ \quad = 4.096 \times 10^{-9} R_{1,nom}^2 (1+\rho)^2 R_{2,nom} \\ \quad + 5.12 \quad \times \quad 10^{-6} R_{1,nom}(1+\rho) R_{2,nom} \end{array} \tag{11}$$

$$\begin{array}{l} 1.28 \quad \times \quad 10^{-5} R_{1,nom}(1+x) R_{2,nom}(1+y) \\ -1.536 \times 10^{-8} R_{1,nom}^2 (1+x)^2 R_{2,nom} \ (1+y) \\ \quad = 1.28 \times 10^{-5} R_{1,nom}(1+\rho) R_{2,nom} \\ \quad - 1.536 \times 10^{-8} R_{1,nom}^2 (1+\rho)^2 R_{2,nom} \end{array} \tag{12}$$

$$\begin{array}{l} 2.56 \times 10^{-8} R_{1,nom}^2 (1+x)^2 R_{2,nom}(1+y) \\ \quad = 2.56 \times 10^{-8} R_{1,nom}^2 (1+\rho)^2 R_{2,nom} \end{array} \tag{13}$$

$$\begin{array}{l} 1.6 \times 10^{-8} R_{1,nom}^2 (1+x)^2 R_{2,nom}(1+y) \\ \quad = 1.6 \times 10^{-8} R_{1,nom}^2 (1+\rho)^2 R_{2,nom} \end{array} \tag{14}$$

$$-\alpha \le x, y \le \alpha \tag{15}$$

The extreme values for x and y are obained by solving the set of equations (10-15). We get $x = -\alpha$ and $y = -\alpha$ and this gives the MSDF for R_1, as

$$\rho = (1-\alpha)^{1.5} - 1 \approx 1.5\alpha - 0.375\alpha^2 \tag{16}$$

Table I gives the MSDF for R_1 and R_2 based on the above calculation.

978-1-61284-657-6/11 $26.00 © 2011 IEEE

TABLE I
MSDF FOR CASCADED AMPLIFIER OF FIGURE 1 WITH $\alpha = 0.05$.

Circuit parameter	%upside MSDF	%downside MSDF
Resistor R_1	10.3	7.4
Resistor R_2	12.3	8.5

V. GENERALIZATION

The computation of the previous section is too complex for arbitrarily large circuits. Such circuits are handled by first obtaining a nominal numeric polynomial expansion for them. This is done by sweeping the input voltage across all possible values and noting the corresponding output voltages. The output voltage is plotted against the input voltage. A polynomial is fitted to this curve and the coefficients of this polynomial are taken to be the nominal coefficients for the desired polynomial. A V-Transform curve is now obtained based on the polynomial curve using the transformation in equation (5). The circuit is simulated for different drifts in the parameter values at equally spaced points from inside the hypercube enclosing each circuit parameter, spaced ϵ apart. Polynomial coefficients and hence V-Transform coefficients are obtained for each of these simulations. The maximum and minimum values of coefficient in this search are taken as the limiting values for that coefficient. Once the limiting values for all coefficients have been determined the CUT is subjected to a DC sweep at the input and the output response is curve-fitted using a polynomial of the same order as that used for the fault free circuit. The V-Transform coefficients for CUT are now obtained. If there are any coefficients that lay outside the limiting values of the corresponding coefficients of the fault free circuit, we conclude that CUT is faulty. The converse need not be true as there could be other specifications, the circuit needs to meet, which are not captured by polynomial based test. Flowchart I in Figure 2 summarizes the process of numerically finding the V-transform coefficients and their bounds. Flowchart II in Figure 2 outlines a procedure to test CUT using the V-Transform coefficients. The bounds on coefficients of fault free circuit are found a priori as shown in Flowchart I of Figure 2.

VI. FAULT DIAGNOSIS

Fault diagnosis involves the location of likely fault sites in a CUT given that the CUT has failed an applied test giving a particular response. We use V-transform coefficients (VTC) to characterize the response of the circuit at different frequencies (about 4 or 5 frequencies are sufficient for most circuits with less than 100 circuit elements), by obtaining its input-output response over the entire input range. Process variation induces a fault-free variation of say σ, about the mean value of every VTC. Any value beyond σ from the mean μ of VTC indicates a circuit failure. Assuming a normal distribution for circuit parameter variation, we can find the probability distribution of the coefficients by Monte-Carlo simulation for process variation of all circuit parameters. Once the Monte-Carlo distributions for the coefficients of fault free circuit are

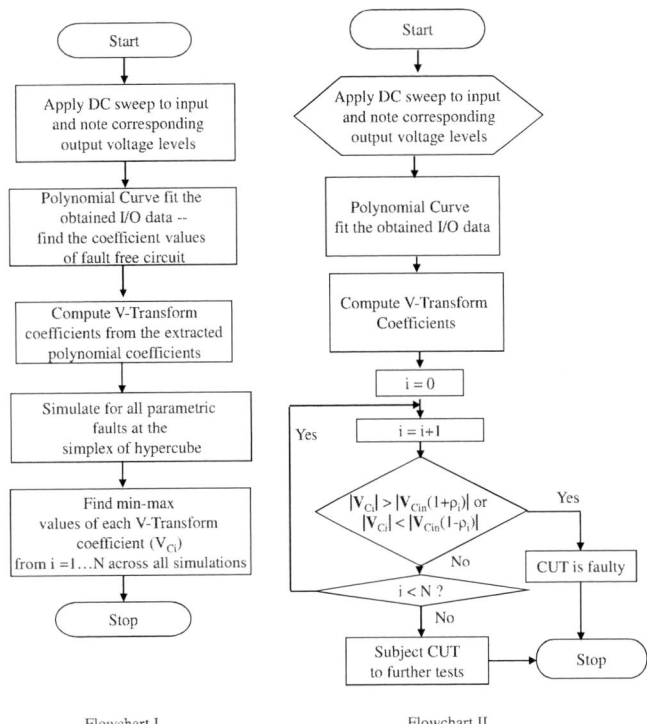

Fig. 2. Fault simulation process and bounding of coefficients (Flowchart I), and complete test procedure (Flowchart II).

obtained we can inject desired sizes of parametric faults (those that are induced due to manufacturing defects) and obtain the new probability distribution of faulty circuit under process variation.

As an illustration, Figure 3 shows the probability density distributions obtained with (broken line) and without (solid line) parametric fault. There are three distinct regions in the probability space of any coefficient C_k. Region R is the fault-free space because coefficients at all frequencies are within the desired limits. Region 1 where dominant mechanism of faults are due to PV of circuit parameters and Region 2 where dominant mechanism of faults is due to manufacturing defects (also called parametric fault). The cross-over point of these two distributions gives the equiprobable region of faults, where we can have faults due to either of the mechanism with the same likelihood. We denote this point on the coefficient axis as C_{th}. Measuring the value of coefficient C of CUT, we can now determine the likelihood of the nature of fault mechanism. That is, $C \in [\mu, C_{th}] \implies$ failures due to PV are more in number and $C \in [C_{th}, \mu'] \implies$ failures due to parametric faults are more in number. The confidence of this distinction is given by the relative magnitudes of the two probability density function G_1 and G_2 at the point C on coefficient axis. Once we know, that the fault mechanism is due to a manufacturing defect, we can predict the fault site based on knowledge of the sensitivity of the coefficient to various circuit parameters at different frequencies [22], [23]. A fault dictionary is maintained for faults against circuit parameters at different frequencies. On measuring a parametric fault, the most likely fault site is deduced by intersection of fault sites that can contribute to

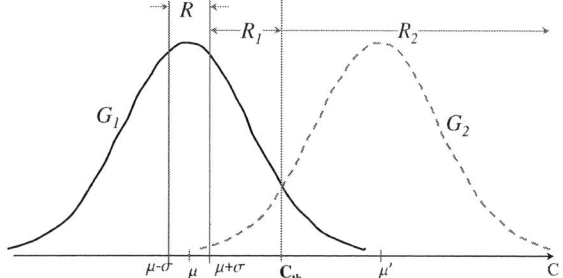

Fig. 3. Probability distribution of polynomial coefficient C under a parametric fault (broken line) as opposed to that with only process variation (solid line).

this fault at most of frequencies. The confidence level (P) of this deduction is given by:

$$P = 1 - \prod_{i=1}^{i=N} (1 - P_i) \qquad (17)$$

where N is the number of frequencies (including DC) which the circuit is diagnosed and P_i is the confidence of fault diagnosis at i^{th} frequency.

VII. SIMULATION RESULTS

We simulated an elliptic filter shown in Figure 4 for V-Transform coefficient based test. The circuit parameter values are as in the benchmark circuit maintained by Stroud et al. [24]. Our Monte-Carlo simulation included 50,000 circuit instances, with process variations sampled as zero mean and standard deviation = ±10% of nominal circuit component value. This was repeated for different injected parametric faults to obtain distribution of the coefficients under both parametric faults and process variation (PV) of circuit components. We used parametric faults of sizes $\alpha = 5\%$ from their nominal value to find min-max values of coefficients. Figure 5 shows the computed response and the estimated polynomial obtained by curve fitting:

$$\begin{aligned} v_{out} = {} & 4.5341 - 3.498v_{in} - 2.5487v_{in}^2 \\ & + 2.1309v_{in}^3 - 0.50514v_{in}^4 + 0.039463v_{in}^5 \end{aligned} \qquad (18)$$

The combinations of parameter values leading to limits on the coefficients are as shown in Tables II and III. Some of the circuit parameters are not shown in the table because they do not appear in any of the coefficients and are kept at their

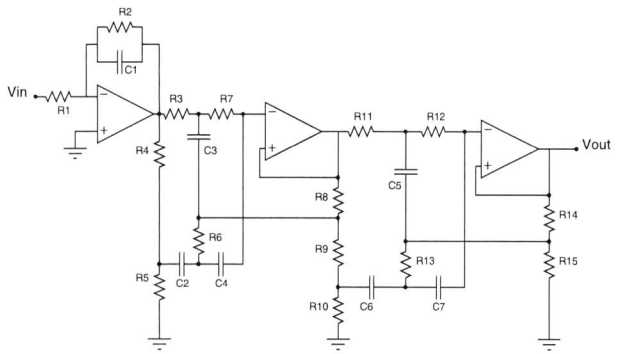

Fig. 4. Elliptic filter.

TABLE II
PARAMETER COMBINATIONS LEADING TO MAXIMUM VALUES OF V-TRANSFORM COEFFICIENTS WITH $\alpha = 0.05$.

Circuit Parameter (Ω)	V_{c_0}	V_{c_1}	V_{c_2}	V_{c_3}	V_{c_4}	V_{c_5}
R_1 = 19.6k	18.6k	20.5k	20.5k	20.5k	18.6k	18.6k
R_2 = 196k	186k	205k	186k	186k	186k	205k
R_3 = 147k	139k	154k	154k	154k	139k	154k
R_4 = 1k	950	1010	1010	1010	1010	1010
R_5 = 71.5	70	80	80	70	80	70
R_6 = 37.4k	37.4k	37.4k	37.4k	37.4k	37.4k	37.4k
R_7 = 154k	161k	161k	146k	161k	146k	146k
R_{11} = 110k	115k	115k	104k	115k	104k	104k
R_{12} = 110k	104k	115k	104k	104k	104k	104k

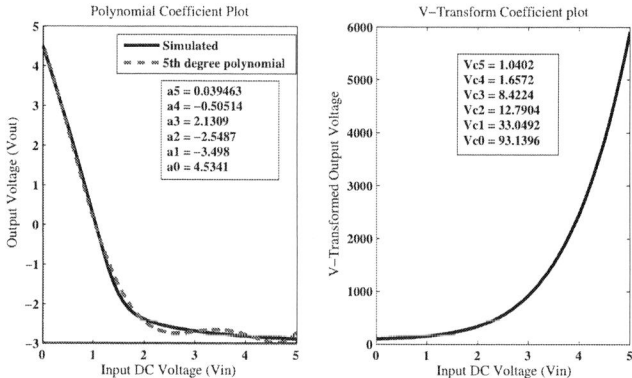

Fig. 5. DC response of elliptic filter with curve fitting polynomial and V-Transform plot.

nominal values. Further, results on pass/fail detectability of few injected faults are tabulated in Table IV. In the cases where coefficient deviation lies in the region R_1 for a coefficient C_k, the fault is attributed to PV as opposed to parametric fault. The same procedure is repeated for VTC and the number of cases in which the fault is diagnosed to be in the region R_1 and incorrectly attributed to PV is reduced. This is due to he enhanced sensitivity of V-transform coefficients to circuit parameters. Table V shows the diagnosed results of a few injected faults using sensitivity of V-transform coefficients to circuit parameters as described in Section VI.

VIII. CONCLUSION

A new approach for test and diagnosis of non-linear circuits based on a transformation of polynomial expansion of the circuit is demonstrated. The V-Transform renders the polynomial coefficients monotonicity and enhances their sensitivity. The

TABLE III
PARAMETER COMBINATIONS LEADING TO MINIMUM VALUES OF V-TRANSFORM COEFFICIENTS WITH $\alpha = 0.05$.

Circuit Parameter (Ω)	V_{c_0}	V_{c_1}	V_{c_2}	V_{c_3}	V_{c_4}	V_{c_5}
R_1 = 19.6k	20.5k	18.6k	18.6k	20.5k	20.5k	20.5k
R_2 = 196k	205k	186k	205k	205k	205k	186k
R_3 = 147k	150k	139k	139k	146k	154k	139k
R_4 = 1k	1010	950	950	950	950	950
R_5 = 71.5	80	70	70	80	70	80
R_6 = 37.4k	39.2k	39.2k	39.2k	39.2k	35.5k	39.2k
R_7 = 154k	146k	146k	161k	146k	161k	161k
R_{11} = 110k	104k	104k	115k	104k	115k	115k
R_{12} = 110k	115k	104k	115k	115k	115k	115k

TABLE IV
RESULTS FOR SOME INJECTED FAULTS.

Circuit Parameter	Out of bound polynomial coefficient	Fault detected?	Out of bound V-Transform coefficient	Fault detected?
R_1 down 5%	a_3, a_4	Yes	$V_{c_0} - V_{c_4}$	Yes
R_2 down 10%	a_2	Yes	V_{c_2}, V_{c_5}	Yes
R_3 up 5%	a_3	Yes	V_{c_1}, V_{c_2}, V_{c_3}	Yes
R_4 down 10%	a_0	Yes	$V_{c_0} - V_{c_4}$	Yes
R_5 up 10%	a_4	Yes	V_{c_0}, V_{c_4}	Yes
R_7 up 5%	None	PV	V_{c_1}, V_{c_2}	Yes
R_{11} up 5%	None	PV	V_{c_4}, V_{c_5}	Yes
R_{12} down 5%	None	PV	V_{c_4}, V_{c_5}	Yes

TABLE V
PARAMETRIC FAULT DIAGNOSIS WITH CONFIDENCE LEVELS OF $\approx 88\%$

Injected fault	Diagnosed fault sites					Deduced fault
	DC	100Hz	900Hz	1000Hz	1100Hz	
R_1 dn 15%	R_1 R_4	R_1	R_1, R_2	R_1, R_2 C_1	R_1 C_1	R_1
R_2 dn 10%	R_2	R_2 C_1	R_2, R_3 C_1	R_2, R_3	R_2 C_1	R_2
R_3 up 5%	R_1 R_3	R_3 C_3	R_3, R_4 C_3	R_3	R_3, C_3 C_3	R_3
R_4 dn 20%	R_1 R_4	R_1 R_4	R_2, R_4 C_1	R_1, R_2 R_4	R_1, R_2 R_4	R_4
R_5 up 15%	R_5	R_5 C_2	R_4, R_5	R_4, R_5 C_2	R_5, R_6 C_3	R_5
R_7 dn 10%	R_3 R_7	R_7 C_3	R_3, R_7	R_3, R_6 R_7	R_3, R_7 C_3	R_7

minimum sizes of detectable faults in some of the circuit parameters are as low as 5% which implies that impressive fault coverage can be achieved with VTC. The use of VTC shows a reduction in masking of parametric faults due to process variation. The method is then extended to sensitivity based fault diagnosis by evaluating VTC at different frequencies. Our future work will attempt to reduce the "false alarms" in fault detection when there is no manufacturing defect but just process variation.

ACKNOWLEDGMENT

The authors would like to thank Professor Vittala Rao, formerly of Mathematics Department at the Indian Institute of Science for his comments and help in monotonicity arguments of V-Transform coefficients.

REFERENCES

[1] A. Abderrahman, E. Cerny, and B. Kaminska, "Optimization Based Multifrequency Test Generation for Analog Circuits," *Journal of Electronic Testing: Theory and Applications*, vol. 9, no. 1–2, pp. 59–73, Mar 1996.

[2] S. Chakravarty and P. J. Thadikaran, *Introduction to IDDQ Testing*. Kluwer Academic Publishers, 1997.

[3] S. Cherubal and A. Chatterjee, "Test Generation Based Diagnosis of Device Parameters for Analog Circuits," in *Proc. Design, Automation and Test in Europe Conf.*, 2001, pp. 596–602.

[4] G. Devarayanadurg and M. Soma, "Analytical Fault Modeling and Static Test Generation for Analog ICs," in *Proc. Int. Conf. on Computer-Aided Design*, Nov. 1994, pp. 44–47.

[5] S. L. Farchy, E. D. Gadzheva, L. H. Raykovska, and T. G. Kouyoumdjiev, "Nullator-Norator Approach to Analogue Circuit Diagnosis Using General-Purpose Analysis Programmes," *Int. Journal of Circuit Theory and Applications*, vol. 23, no. 6, pp. 571–585, Dec. 1995.

[6] R. K. Gulati and C. F. Hawkins, *IDDQ Testing of VLSI Circuits*. Springer, 1993.

[7] W. L. Lindermeir, H. E. Graeb, and K. J. Antreich, "Analog Testing by Characteristic Observation Inference," *IEEE Trans. Comp. Aided Design*, vol. 23, no. 6, pp. 1353–1368, Jun. 1999.

[8] R. Rajsuman, *IDDQ Testing for CMOS VLSI*. Artech House, 1995.

[9] J. Figueras, "Possibilities and Limitations of IDDQ Testing in Submicron CMOS," in *Proc. Innovative Systems in Silicon Conf.*, Oct. 1997, pp. 174–185.

[10] P. Kabisatpathy, A. Barua, and S. Sinha, "A Pseudo-Random Testing Scheme for Analog Integrated Circuits using Artificial Neural Network Model-Based Observers," in *Proc. 45th Midwest Symp. Circuits and Systems*, vol. 2, 2002.

[11] J. Kaderka, V. Musil, J. Povazanec, and P. Simek, "Neural Network Based System for Testing and Diagnostics of Analogue Integrated Circuits," in *Proc. 3rd IEEE Int. Electronics, Circuits, and Systems Conf.*, vol. 2, 1996, pp. 1198–1201.

[12] V. Stopjakova, D. Micusik, L. Benuskova, and M. Margala, "Neural Networks-Based Parametric Testing of Analog IC," in *Proc. 17th IEEE Int. Symp. Defect and Fault Tolerance in VLSI Systems*, 2002, pp. 408–416.

[13] V. Stopjakova, P. Malosek, M. Matej, V. Nagy, and M. Margala, "Defect Detection in Analog and Mixed Circuits by Neural Networks using Wavelet Analysis," *IEEE Transactions on Reliability*, vol. 54, no. 3, pp. 441–448, 2005.

[14] H.-G. Stratigopoulos and Y. Makris, "Error Moderation in Low-Cost Machine-Learning-Based Analog/RF Testing," *IEEE Transactions on Computer-Aided Design of Integrated Circuits and Systems*, vol. 27, no. 2, pp. 339–351, 2008.

[15] J. Park, J. Chung, and J. A. Abraham, "LFSR-Based Performance Characterization of Nonlinear Analog and Mixed-Signal Circuits," in *Proc. IEEE Asian Test Symp.*, Nov. 2009, pp. 373–378.

[16] J. Park, H. Shin, and J. A. Abraham, "Pseudorandom Test for Nonlinear Circuits Based on a Simplified Volterra Series Model," in *Proc. International Symposium on Quality Electronic Design*, Mar. 2007, pp. 495–500.

[17] Z. Guo and J. Savir, "Analog Circuit Test Using Transfer Function Coefficient Estimates," in *Proc. Int. Test Conf.*, Oct. 2003, pp. 1155–1163.

[18] V. Panic, D. Milovanovic, P. Petkovic, and V. Litovski, "Fault Location in Passive Analog RC Circuits by Measuring Impulse Response," in *Proc. 20th International Conf. on Microelectronics*, Sep. 1995, pp. 12–14.

[19] S. Sindia, V. Singh, and V. D. Agrawal, "Polynomial Coefficient Based DC Testing of Non-Linear Analog Circuits," in *Proc. 19th ACM Great Lakes Symp. on VLSI*, May 2009.

[20] S. Mir, H.-G. Stratigopoulos, and A. Bounceur, "Density Estimation for Analog/RF Test Problem Solving," in *Proc. 28th VLSI Test Symp.*, Apr. 2010, pp. 41–41.

[21] S. Sindia, V. Singh, and V. D. Agrawal, "Multi-tone Testing of Linear and Nonlinear Analog Circuits Using Polynomial Coefficients," in *Proc. Asian Test Symp.*, Nov. 2009, pp. 63–68.

[22] M. Slamani and B. Kaminska, "Analog Circuit Fault Diagnosis Based on Sensitivity Computation and Functional Testing," *IEEE Design & Test of Computers*, vol. 19, no. 1, pp. 30–39, 1992.

[23] S. Sindia, V. Singh, and V. D. Agrawal, "Parametric Fault Diagnosis of Nonlinear Analog Circuits Using Polynomial Coefficients," in *Proc. 23rd Int. Conf. VLSI Design*, Jan. 2010, pp. 288–293.

[24] R. Kondagunturi, E. Bradley, K. Maggard, and C. Stroud, "Benchmark Circuits for Analog and Mixed-Signal Testing," in *Proc. 20th Int. Conf. on Microelectronics*, Mar. 1999, pp. 217–220.

978-1-61284-657-6/11 $26.00 © 2011 IEEE

A Diagnosis Testbench of Analog IP Cores Against On-Chip Environmental Disturbances

Takushi Hashida, Yuuki Araga and Makoto Nagata
Graduate School of System Informatics, Kobe University
{hashida, araga, nagata}@cs26.scitec.kobe-u.ac.jp

Abstract

Analog IP cores exhibit a multivariate response to dynamic variations of an operation environment, that are typically represented by power and substrate voltage changes. A testbench provides a silicon area to embed and diagnose custom IP cores with power delivery and substrate networks, where the area is surrounded by on-chip precision waveform capturing and configurable power and substrate noise generation circuits. The coefficients of noise propagation and noise coupling are quantitatively derived for fabless IP cores processed in a target technology, that will be further linked with EDA tooling for the successful adoption of such IP cores in SoC integration.

I. Introduction

System integration using VLSI technology needs to adopt a broad range of analog and mixed-signal functionality, such as analog-to-digital/digital-to-analog conversion, digital data linkage, and wireless connectivity. Because of the high complexity of analog circuits, chip designers consider to reuse existing designs as well as to purchase designs from external suppliers. These reusable designs are often called IP cores and facilitate integration of systems in a single chip.

There is always a potential risk of design failures due to undesired power and substrate noise coupling among such analog IP cores and digital elements, as has been continuously discussed in the area of substrate noise aware design technologies [1][2]. Furthermore, since the fluctuation of circuit performance due to the noise coupling may cause instability of testing, the design of complex SoCs is required to be high noise tolerant even from the viewpoints of testability and yield managements.

Suites of power and substrate noise analysis are provided by electronic design automation (EDA) software vendors, with the capability of capturing generation, propagation, coupling, and interaction of noises in the whole chip [3][4]. This will help to reduce the impact of noises in SoC integration, as long as noise numbers are properly predicted. In order to successfully accomplish such noise analysis, the response and tolerance of IP cores need to be pre-characterized against environmental voltage variations in a target silicon technology.

This paper is dedicated to the design of a testbench to diagnose the response of analog IP cores against power and substrate noises as well as to derive noise parameters. Section II details the structure of a testbench and designs of constituent elements. Section III discusses physical design issues and silicon results in a 90 nm CMOS technology. Section IV proposes strategies to diagnose analog IP cores. A brief summary of this paper will be given in Section V.

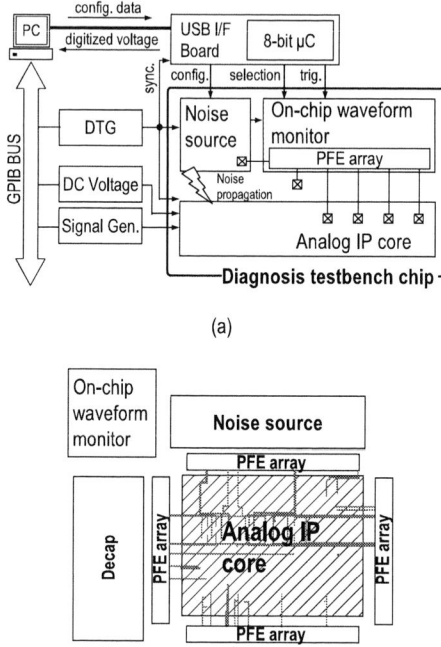

Fig. 1. System overview of diagnosis testbench. (a) Block diagram and (b) physical floor plan.

II. Diagnosis testbench

A) System overview

A diagnosis testbench provides a silicon area to embed IP cores to diagnose, where the area is surrounded by on-chip waveform monitors (OCM) and configurable noise sources (NS), as depicted in Fig. 1. IP cores located in the area include physical couplings to a silicon substrate and experience the transfer response of power delivery networks.

The noises from a configurable NS is injected into substrate as well as power delivery networks, and responded by IP cores when they couple to internal circuits. The propagation of noises is monitored by probing channels of OCM at a variety of locations in the testbench. The response of IP cores against the noises are monitored by OCM as well. Physical connections of probing channels to circuit nodes will be discussed more in detail in Sect. III.

The diagnosis system is entirely governed by an external personal computer (PC). The configuration variables of NS and OCM are automatically set for an intended set of measurements. The captured waveforms are post-processed to extract noise parameters, namely, the coefficients of noise propagation

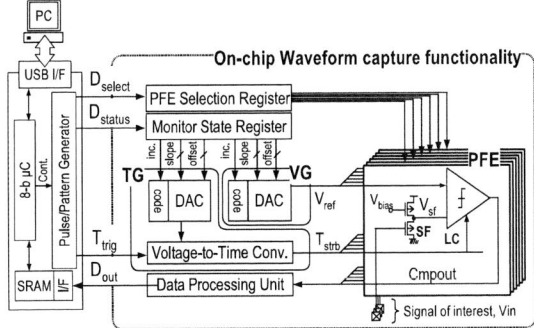

Fig. 2. On-chip waveform capture functionality.

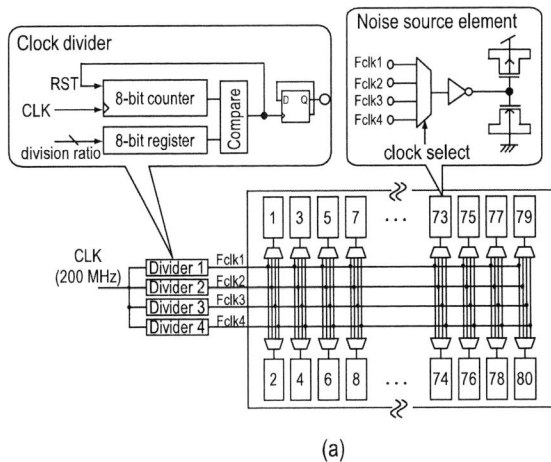

(a)

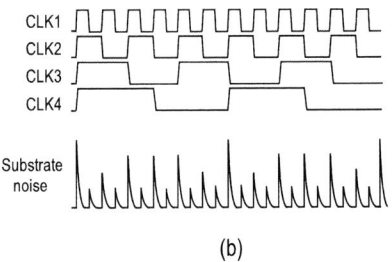

(b)

Fig. 3. Configurable noise generator. (a) Circuit diagram and (b) multi-tone noise generation.

in power delivery and substrate networks as well as those of noise coupling to circuit nodes.

B) Precision waveform capturer

On-chip waveform capture [5] is embodied, as given in Fig. 2. A probing front-end circuit (PFE) senses and digitizes voltage at the probing point of interest, Vin. A set of PFE is provided to capture voltage variations at substrate taps (V_{sub}), ground (V_{ss}) and supply voltage (V_{dd}) of circuits, as well as signals (V_{sig}) at the circuit nodes of interest.

A voltage generator, VG, converts a digital code to the corresponding voltage step, V_{step}. A timing generator, TG, places precision sample timing, T_{smp}, by using a function of voltage-to-time conversion where the voltage corresponding to a given digital code is linearly translated into the time difference. Both VG and TG have the resolution of 10 bits. In capturing a waveform, the nearest voltage of V_{in} at T_{smp} is algorithmically searched across the space of V_{step}, under iterative circuit operation. The sample timing of T_{smp} is then moved to the next position once the nearest V_{step} is determined. The details of measurement algorithm and sequence are discussed in [6].

The monitor processor (8-bit μC) selects a single PFE at the point of interest and executes time-domain waveform acquisition through resister level transactions with VG and TG. The processor can be placed off-chip and implemented in FPGA, saving silicon areas of the testbench. The processor also communicates with system PC.

The voltage and timing resolution of VG and TG, respectively, are appropriately chosen. Long-term waveforms use coarse resolution for overviewing trends of noise. Zoom-in waveforms finely capture noise coupling processes and precisely derive noise parameters. The voltage offsets are also adjustable against the voltage domains of interest, V_{sub}, V_{ss}, V_{dd}, V_{sig}. The high reconfigurability of OCM is advantageous over the past implementation of on-chip oscilloscope functions [7][8][9] and essential to the realization of a diagnosis testbench.

C) Configurable noise generator

A configurable noise generator (NS) is newly developed to create multi-tone noise waves. The circuit diagram is

given in Fig. 3. Noise source channels in parallel share a common digital clock signal input, while individually having a programmable divider and an array of noise source elements. The division ratio to the clock signal at Fclk and the number of noise source elements to activate are configurable in each of the channels.

A digital circuit changes internal activity accordingly to the contents of signal processing, while constantly being clocked. This creates power and substrate noises including the tones at the clock frequency and its harmonics, as well as low frequency tones corresponding to the envelope of activity variation. The configurable NS follows to this consideration and captures power and substrate noise generation in digital circuits such as a sequential data path and a scan chain. This is in contrast to an arbitrary digital noise emulator [10] that was motivated to capture comprehensive operation of general digital circuits such as a micro processor. The power noise on a ground (V_{ss}) wirings is naturally injected into a p-type silicon substrate through distributed substrate contacts and thus appears as a substrate noise.

III. Physical design consideration

A) Prototype design and evaluation

A prototype of the diagnosis testbench has been developed with a 90-nm CMOS technology. A sample layout is given in Fig. 4. A set of a configurable NS and an OCM adjoining each other is located on the top of a silicon area that is reserved for

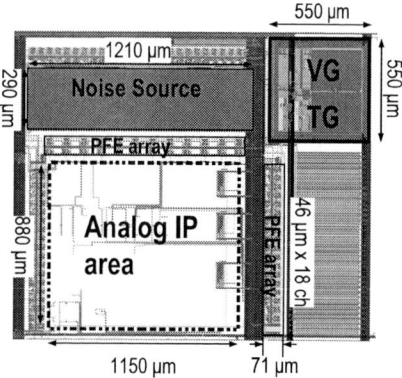

Fig. 4. Prototype of diagnosis testbench with vacant silicon areas.

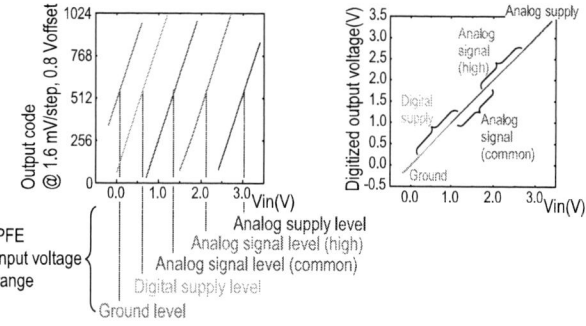

Fig. 5. DC transfer of PFE channels with digitized outputs.

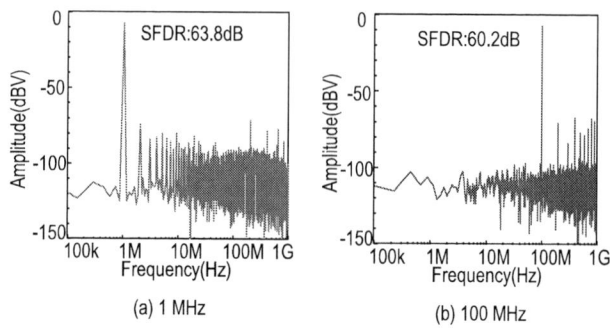

Fig. 6. AC transfer of waveform capture with sinusoids at (a) 1 MHz and (b) 100 MHz.

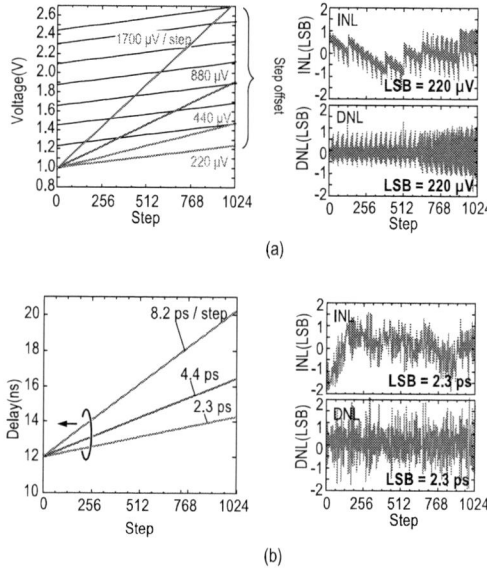

Fig. 7. Voltage and timing generation of waveform capture, (a) VG and (b) TG responses.

IP cores to diagnose. Probing channels (PFEs) of 56 in total form arrays surrounding the area, with probe wirings drawing to the positions of interest.

DC input-output transfer of PFE channels of different voltage domains are summarized in Fig. 5, where the output is on-chip digitized through the nearest search of VG voltages. The purity of waveform capturing is better than 60.2 dB in spurious-free dynamic range (SFDR), as shown in Fig. 6, corresponding to the effective number of bits (ENOB) of 9.7. This comes from the 10-bit linearity of on-chip voltage and timing generation. The resolutions of VG and TG range from 220 μV to 1700 μV and from 2.3 ps to 8.2 ps, respectively, as shown in Fig. 7. As for VG, INL of ± 1.3 LSB and DNL of ± 1.5 LSB were measured among 1024 steps with LSB of 220 μV. Similarly, INL of ± 1.8 LSB and DNL of ± 2.0 LSB were achieved by TG among 1024 steps with LSB of 2.3 ps.

The whole of waveform capturing circuits uses high-voltage I/O MOS transistors in a deep N well option and operates under the supply voltage of 3.3 V, for the better immunity against background noise coupling.

A configurable NS with four signal channels is prototyped. The fastest operating frequency at 200 MHz is achieved by the configurable NS. Voltage peaks remain mostly constant in the substrate noise waveform given in Fig. 8, where noise source channels share an identical configuration. On the other hand, the modulated noise peaks with Fclk of 200 MHz exhibits the lower frequency tone at 50 MHz, given in Fig. 9, in addition to the Fclk and higher harmonics.

The substrate noise was measured at the probing point in the area of diagnosis, with 150 μm distance from NS. The relation of noise peaks to the number of active noise source elements is plotted in Fig. 10, achieving a wide noise dynamic range of 10.6 dB useful for diagnosis.

B) Background coupling

The physical placements of noise source and monitor circuits impact the usability of a diagnosis testbench. The array of PFE is located between the configurable NS and the area of diagnosis for a wide coverage of probing in the area. While the generated substrate noise is expected to reach IP cores, PFE circuits needs to be isolated from background power line and substrate coupling. Since the entire OCM is independently supplied on a chip as well as on an evaluation board, power-line noise coupling is considered negligible. On the one hand, each PFE circuit is individually covered by a deep N well for the suppression of substrate noise coupling.

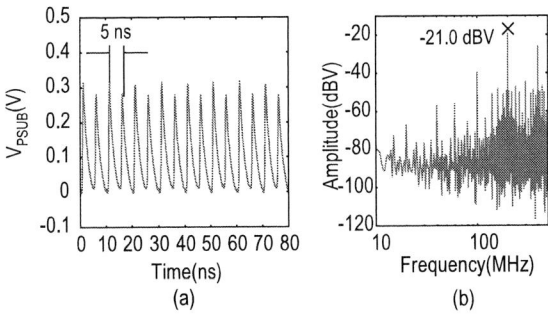

(a)   (b)

Fig. 8. Noise generation at Fclk of 200 MHz with constant logic activity. (a) Time domain noise waveforms and (b) frequency ingredients of substrate noise.

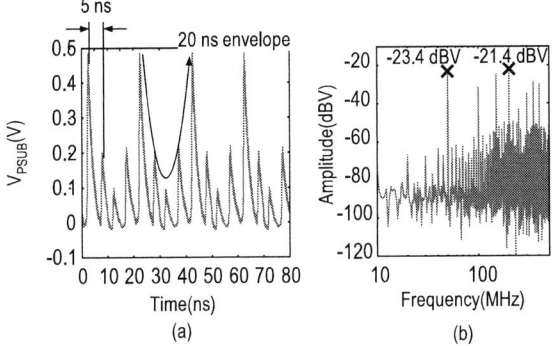

(a)   (b)

Fig. 9. Noise generation at Fclk of 200 MHz with modulated logic activity. (a) Time domain noise waveforms and (b) frequency ingredients of substrate noise.

PFE rows with different array shapes were experimentally compared for the balance of substrate noise isolation and propagation, shown in Fig. 11. The high suppression is expected when PFE cells are densely placed in a line and guarded by deep N well bands for blockage [11] in the 1st case. In contrast, PFE cells are intermittently placed in the 2nd case, allowing substrate noises from NS to reach the area of diagnosis through non-blocked tracks in the lattice.

Measured attenuation of 200 MHz substrate noise components is summarized in Fig. 12. The peak noise number of -21.0 dBV is probed on the ground wirings of the configurable NS. This noise is not significantly reduced at the nearest substrate taps with 10 μm distance from NS. On the other hand, the full attenuation of 15 dB in the case 1 is measured at the distance of 45 μm, that is almost equivalent to the distance of 150 μm in the case 2 with the partial attenuation.

The background coupling of the whole waveform capture system to a configurable NS is evaluated in the case 2, as additionally plotted in Fig. 12. The coupled tone at 200 MHz is decomposed from the captured waveforms when PFE circuit is given a pure DC voltage at its input. The overall attenuation reaches -25 dB for 200 MHz tone, proving the effective isolation even in the compact placements of Fig. 4.

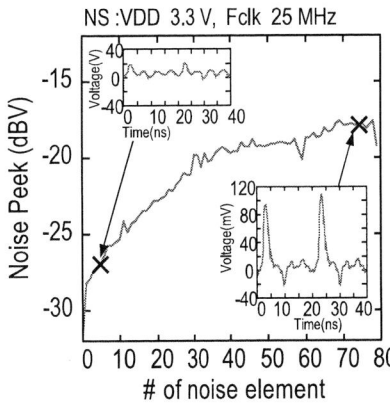

Fig. 10. Noise peaks versus number of active noise source elements.

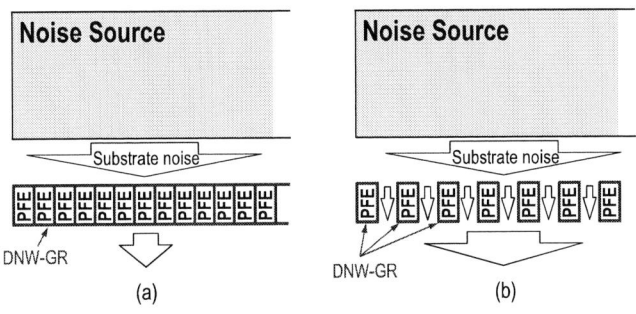

(a)   (b)

Fig. 11. Placements of PFE in an arrays; with (a) high and (b) partial attenuation of substrate noise propagation.

C) Routing probe wirings

The probing points of interest are in the vicinity as well as at the inside node of IP cores to diagnose. The metal wirings from PFE circuits need to be carefully drawn over as well as through the IP cores, not to interfere with circuit operation. Figure 13 shows an example of such wiring. It is routed along p+ guard bands within an analog IP core and stitched between bundled wirings of the guard bands. This also prevents from background coupling to probe wirings, thanks to very stable voltage of the guard bands.

IV. Diagnosing analog IP cores

A) Noise propagation

Noise parameters of a fabless IP core are initially unknown in a given silicon technology. The propagation and coupling coefficients are not only a function of impurity profile of the technology but also strongly impacted by the physical layout of IP cores as well as by the structures of assembly. Therefore, it is desirable to evaluate power and substrate noise propagation and coupling during the silicon validation of IP cores, where the proposed diagnosis testbench is quite applicable.

Two prototypes of the diagnosis testbench were prepared. In the 1st prototype, a product-level analog IP is experimentally placed and fully routed in the area of diagnosis. The other prototype removes the core while leaving all the probe wirings.

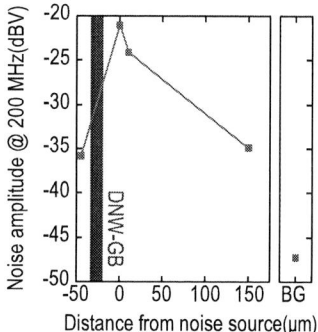

Fig. 12. Attenuation of 200 MHz components of substrate noise.

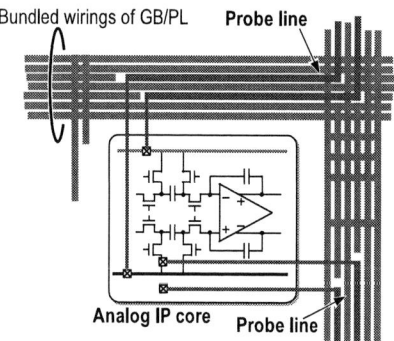

Fig. 13. Drawing probe wirings to point of interest in IP core.

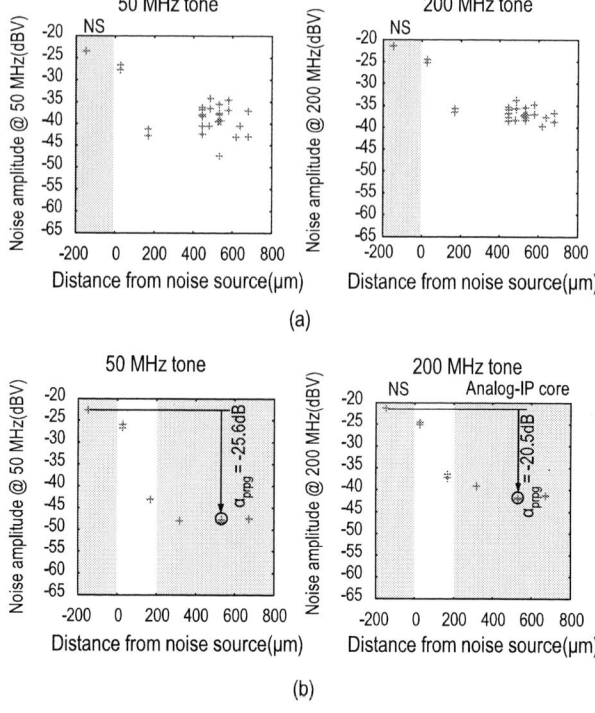

(a)

(b)

Fig. 14. Substrate noise maps in the area of diagnosis for 200 MHz and 50 MHz tones. (a) Before and (b) after placements of analog IP core.

Since the power supply and ground wirings of IP core and related I/O rings are isolated from the configurable NS, the propagation and coupling of substrate noise is focused in the later discussions.

The propagation of substrate noise in the area of diagnosis is compared among the two prototypes, with the presence of an analog IP core and the vacancy of the same area, in Fig. 14. While the configurable NS generates mixed 50 and 200 MHz tones, their voltage magnitudes are measured from substrate noise waveforms probed at the various locations of p+ taps within the IP core of interest. The waveforms in the vacant area were captured at the exactly same locations as in the analog IP core.

It is obviously shown that substrate coupling is significantly suppressed as well as shaped in a flat distribution, with the presence of the IP core. The attenuation reaches -25.6 dB and -20.5 dB for 50 MHz and 200 MHz noise ingredients, respectively, in reference to the noise magnitude at the vicinity of the configurable NS (distance = 0 μm). The attenuation by distance is further enhanced by the placements of IP core. This suggests the effectiveness of densely placed guard bands inside the physical layout. The difference in attenuation among frequency components is due mainly to AC impedance of power delivery and substrate networks.

The coefficient of substrate noise propagation at the frequency of interest, $\alpha_{\mathrm{prpg}}(f)$ is defined as the attenuation number of noise magnitudes from the source of noise to the particular substrate tap of interest, as indicated in Fig. 14(b).

This tap was intentionally chosen, in the immediate proximity of a sample capacitor in the product IP core.

B) Noise coupling

The impact of environmental noises to circuit performance is externally evaluated from macroscopic quantities such as effective number of bits (ENOB), spurious free dynamic range (SFDR) [12], and jitters. On the other hand, the proposed diagnosis testbench provides the observation of microscopic response to noises at the particular internal nodes of interest in a circuit. Here, such nodes are selected not to disturb circuits, such as (i) power lines and substrate taps for indirect sensing and (ii) low-impedance nodes driven by buffers for direct evaluation. A sample-and-hold (S/H) circuit is chosen for demonstration of diagnosis in the particular analog IP core. It is generally considered to be sensitive to substrate voltage variation since capacitors have large capacitive coupling at their bottom plates and transistors in switches and amplifiers experience body bias modulation as well. The indirect nodes of substrate, power, and ground lines of a S/H circuit are internally probed in the analog IP core. In addition, the bottom plates of a sample capacitor are also monitored for direct observation. Figure 15 plots acquired waveforms, overviewing the operation of S/H circuit and zooming finely to evaluate coupling processes among internal nodes.

In reference to the substrate noise voltage seen at the nearest p+ taps to the S/H circuit, the coupling coefficients at the frequency of interest, $\beta_{\mathrm{cplg}}(f)$, is defined for the internal nodes of interest, such as V_{dd}, V_{ss}, and the bottom plate of a

978-1-61284-657-6/11 $26.00 © 2011 IEEE

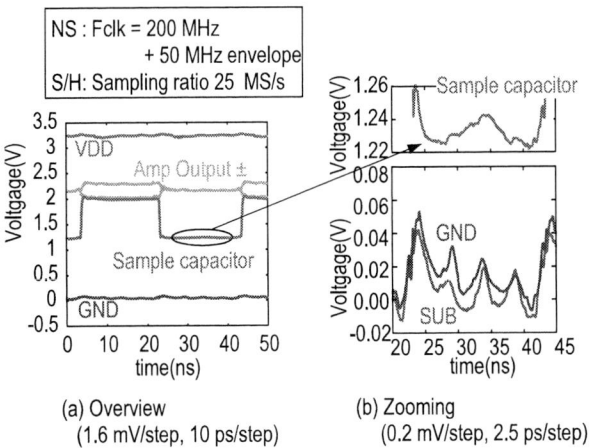

(a) Overview
(1.6 mV/step, 10 ps/step)

(b) Zooming
(0.2 mV/step, 2.5 ps/step)

Fig. 15. Substrate noise coupling at a particular node in analog IP core. (a) Overview and (b) zoom in waveforms.

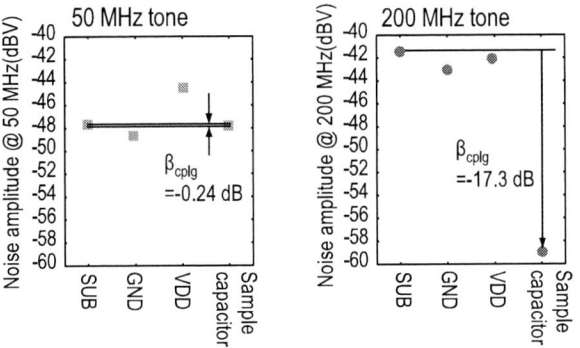

Fig. 16. Coupling coefficients of internal nodes to substrate noise derived from waveforms.

sample capacitor, as shown in Fig. 16. The obvious difference of $\beta_{\mathrm{cplg}}(f)$ among noise components at 50 MHz and 200 MHz at the signal node suggests the multiple modes of noise coupling, needing further investigation with the help of power and substrate noise analysis.

C) Noise analysis

The previous subsections demonstrate noise diagnosis of a fabless IP core, in regard to (i) AC propagation of noises through power delivery and substrate networks in a given technology, and (ii) transient response of circuit nodes to voltage variations around the circuit. The noise parameters derived from diagnosis, $\alpha_{\mathrm{prpg}}(f)$ and $\beta_{\mathrm{cplg}}(f)$, are related with the noise magnitudes that are measured at the source of noise and at the internal node of interest, $\mathrm{Vn}_{\mathrm{source}}(f)$ and $\mathrm{Vn}_{\mathrm{node}}(f)$, respectively, according to (1).

$$\mathrm{Vn}_{\mathrm{node}}(f) = \alpha_{\mathrm{prpg}}(f) \times \beta_{\mathrm{cplg}}(f) \times \mathrm{Vn}_{\mathrm{source}}(f) \quad (1)$$

The linkage of on-chip diagnosis and full-chip noise analysis technologies essentially supports noise managements in IP-based system level integration. The relation of (1) can be directly applied for the validation and calibration of full-chip noise analysis technologies, where simulation elements of noise generation, propagation, and coupling, are individually tuned.

The translation of microscopic noise coupling to macroscopic impacts on circuit performance can be quantitatively traced by the calibrated noise analysis. Furthermore, the interference of system operation to IP cores will be accurately predicted by the analysis, which accelerates the design for high robustness of system level integration.

V. Summary

An on-chip diagnosis testbench provides opportunities of quantitative evaluation of power and substrate noise coupling in a fabless IP cores when they are embodied in a given silicon technology. Various observations on propagation and interaction of noises with circuit operation were demonstrated by using a 90-nm CMOS prototype. The linkage of on-chip diagnosis and full-chip noise analysis technologies accomplishes quality noise managements in system level integration.

Acknowledgements

This work was partly supported by Ministry of Internal Affairs and Communications (MIC) in the development program of technical examination services concerning frequency crowding.

REFERENCES

[1] D. K. Su, M. J. Loinaz, S. Masui, B. A. Wooley, "Experimental results and modeling techniques for substrate noise in mixed-signal integrated circuits," IEEE J. Solid-State Circuits, Vol. 28, pp. 420-430, Apr. 1993.

[2] A. Afzali-Kusha, M. Nagata, N. K. Verghese, D. J. Allstot, "Substrate Noise Coupling in SoC Design: Modeling, Avoidance, and Validation," Proceedings of the IEEE, Vol. 94, No. 12, pp. 2109-2138, Dec. 2006.

[3] W. K. Chu, N. Verghese, H. J. Cho, K. Shimazaki, H. Tsujikawa, S. Hirano, S. Dosho, M. Nagata, A. Iwata, and T. Ohmoto, "A Substrate Noise Analysis Methodology for Large-Scale Mixed-Signal ICs," in Proc. IEEE 2003 Custom Integrated Circuits Conference, pp. 369-372, Sept. 2003.

[4] D. Kosaka, Y. Bando, G. Yokomizo, K. Tsuboi, Y. S. Li, S. Lin, M. Nagata. "A Full Chip Integrated Power and Substrate Noise Analysis Framework for Mixed-Signal SoC Design, " in Proc. IEEE 2009 Custom Integrated Circuits Conference, pp. 219-222, Sept. 2009.

[5] T. Hashida, M. Nagata, "On-Chip Waveform Capture and Diagnosis of Power Delivery in SoC Integration." in 2010 Symp. on VLSI Circuits, Dig. of Tech. Papers, pp. 121-122, June 2010.

[6] Y. Araga, T. Hashida M. Nagata, "An On-Chip Waveform Capturing Technique Pursuing Minimum Cost of Integration," in Proc. IEEE Intl. Symp. on Circuits and Systems, pp. 3557-3560, May 2010.

[7] M. M. Hafed, N. Abaskharoun, G. W. Roberts,"A 4-GHz Effective Sample Rate Integrated Test Core for Analog and Mixed-Signal Circuits," IEEE J. Solid-State Circuits, vol. 37, no. 4, pp. 499-514, Apr. 2002.

[8] Y. Zheng, K. Shepard, "On-Chip Oscilloscopes for Noninvasive Time-Domain Measurement of Waveforms in Digital Integrated Circuits," IEEE Trans. VLSI Systems, vol. 11, no. 3, pp. 336-344, June 2003.

[9] K. Noguchi, M. Nagata, "An On-Chip Multichannel Waveform Monitor for Diagnosis of Systems-on-a-Chip Integration," IEEE Trans. VLSI Systems, vol. 15, no. 10, pp. 1101-1110, Oct. 2007.

[10] T. Matsuno, D. Fujimoto, D. Kosaka, N. Hamanishi, K. Tanabe, M. Shiochi, M. Nagata, "A 6-bit Arbitrary Digital Noise Emulator in 65nm CMOS Technology," in Proc. 2009 Custom IC Conference, pp. 187-190, Sept. 2009.

[11] D. Kosaka, M. Nagata, Y. Hiraoka, I. Imanishi, M. Maeda, Y. Murasaka, and A. Iwata, "Isolation Strategy against Substrate Coupling in CMOS Mixed-Signal/RF Circuits", in 2005 Symp. VLSI Circuits, Dig. of Tech. Papers, pp. 276-279, June 2005.

[12] T. Blalack, B. A. Wooley, "The Effects of Switching Noise on an Oversampling A/D Converter," in IEEE 1995 Intl. Solid-State Circuits Conf., Dig. Tech. Papers, pp. 200-201, Feb. 1995.

978-1-61284-657-6/11 $26.00 © 2011 IEEE

978-1-61284-657-6/11 $26.00 © 2011 IEEE

Session 3

978-1-61284-657-6/11 $26.00 © 2011 IEEE

Case Study: Efficient SDD Test Generation for Very Large Integrated Circuits

Ke Peng[1], Fang Bao[1], Geoff Shofner[2], LeRoy Winemberg[2], Mohammad Tehranipoor[1]

[1]ECE Department, University of Connecticut

[2]Freescale Semiconductor

Abstract—**Semiconductor industry has come to the era to rely heavily on detecting small-delay defects (SDDs) for high defect coverage of manufactured digital circuits and low defective parts per million (DPPM). Traditional timing-unaware transition-delay fault (TDF) ATPGs are proven to be inefficient in detecting SDDs. The commercial timing-aware ATPGs have been developed for screening SDDs, but they suffer from large pattern count and CPU runtime. The previously proposed methodologies are either inefficient or too complex in terms of memory and runtime to be applied to large industry designs (>few million gates). In this paper, we present a new SDD-based pattern grading and selection procedure to meet the SDD test challenges in practice. We propose techniques to reduce the runtime and memory complexity and make the procedure applicable and scalable to large industry designs. Experimental results on both academic and industry circuits demonstrate the efficiency of our procedure; it detects a greater number of SDDs with a much lower pattern count and CPU runtime.**

I. INTRODUCTION

Testing for delay defects, also referred to as timing-related defects, has become extremely vital for ensuring product quality and in-field reliability in the very deep-submicron (VDSM) regime. Such defects can fail the chip by introducing extra signal-propagation delay to produce an invalid response when the chip operates at its operating frequency [1] [2]. Small-delay defect (SDD) is one kind of such timing defects. SDDs were not being seriously considered when testing designs at higher technology nodes since they only introduced a small amount of extra delay to the design, which seldom results in failure of the design with comparable lower frequency and larger slack margins. However, when technology scales to 45nm and below, resulting in an increase in design density and operating frequency, SDDs requires a serious consideration [3]. Since a SDD only introduces a small amount of extra delay to the design, it is commonly recommended to detect them via the long paths, or least-slack paths running through the fault sites.

The transition-delay fault (TDF) model has been widely used in industry for testing delay defects. Experiments have demonstrated that TDF test patterns can achieve a defect coverage level that stuck-at patterns alone cannot [1]. Unfortunately, the TDF pattern set has a limited ability for detecting SDDs in the device and meeting the high SDD test coverage requirements. Per the demands from industry, timing-aware ATPG tools [4] [5] have been developed for SDD detection. However, its significantly increased CPU runtime and pattern count has limited its usage in real applications. n-detect TDF ATPG can be considered as an alternate solution for screening SDDs [6] [7]. However, its extremely large pattern count makes it impractical for large industry designs.

Figure 1 presents the normalized pattern count, number of detected SDDs, and CPU runtime of 1-detect, n-detect ($n = 5, 10, 20$) and timing-aware (ta) pattern sets for the IWLS benchmark ethernet (138,012 gates and 11,617 flip-flops) [8]. The detected SDD is defined as a detected TDF with slack equal or smaller than $0.3T$, where T is the clock period of the design. It can be seen that n-detect and timing-aware pattern sets can detect more SDDs than traditional 1-detect timing-unaware pattern set (1.5-1.9X). However, the penalty is large pattern count (3.6-12.2X) and CPU runtime (3.0-12.0X). It is obvious that as n increases, the increase in pattern count and CPU runtime of n-detect ATPG are approximately linear. For this design, the timing-aware ATPG results in a pattern count comparable to 5-detect ATPG. But its CPU runtime is even larger than 20-detect ATPG.

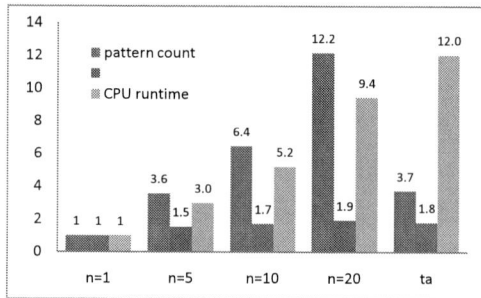

Fig. 1. Normalized pattern count, detected SDDs and CPU runtime of different pattern sets for ethernet. The reuslts are normalized with respect to 1-detect pattern set.

A. Related Prior Work

Several techniques have been presented for screening SDDs. A method to generate K longest paths per gate for testing transition faults was proposed in [9]. In [10], a static-timing-analysis based method was proposed to generate and select patterns that sensitize long paths by masking the observation points of short paths and intermediate paths. The authors in [11] proposed a delay fault coverage metric to detect the longest sensitized path affecting a TDF fault site. It is based on the robust path delay test and attempts to find the longest sensitizable path passing through the target fault site. The authors in [12] proposed path-based and cone-based metrics for estimating the path delay under test, which can be used for path length analysis. All these long path-based methods are suffering from high complexity, extended CPU runtime, and limited by the shortcomings of path delay fault model.

Two hybrid methods were proposed in [13] using 1-detect and timing-aware ATPGs to detect SDDs based on fault classification. Another circuit topology-based fault classification method was proposed in [14]. It differentiates faults to be critical and non-critical for timing-aware and timing-unaware ATPGs, respectively. The efficiency of these fault classification-

---

*The work was supported in part by NSF under Grants no. ECCS-0823992 and CCF-0811632. The work of Ke Peng for implementation on industry circuits was done at Freescale Semiconductor.

based methods is questioned since it still results in a pattern count much larger than 1-detect TDF ATPG.

The output-deviation based method was proposed in [15] [16]. This method uses the delay defect probability matrix (DDPM) of each gate to calculate the output deviations, which are used for pattern selection. However, in case of a large number of gates along the paths, the output deviation metric can saturate and make the procedure inaccurate.

In [17], a SDF-based hybrid method was proposed to generate patterns with minimized pattern count and large long path sensitization. This method was enhanced in [18], by taking the impacts of power supply noise and crosstalk into consideration when calculating the path delay. A similar pattern grading and selection procedure was presented in [19], which is based on statistical timing analysis and takes process variations and crosstalk effects into consideration. All these methods use sensitized long paths as a criteria to grade and select patterns, which makes them difficult to scale for large industry designs with millions of gates.

B. Contributions and Paper Organization

In this paper, we proposed a new pattern grading and selection procedure for screening SDDs, which is based on detected SDDs and their actual slack, rather than sensitized long paths like [17], [18], [19]. This significantly saves the CPU runtime of path tracing and hardware for storing sensitized paths. Several techniques are proposed in this paper to enable the procedure scalable to very large industry circuits. Before generating the original pattern repository, static timing analysis (STA)-based timing-critical fault selection is performed to save ATPG runtime and hardware resources. After pattern generation, fault simulation is performed on each individual test pattern for the pattern evaluation and selection. Parallel fault simulation is used to further reduce the CPU runtime. A new fault merging technique is also proposed to reduce the hardware resources and data processing runtime in the pattern selection procedure.

The pattern selection procedure will minimize the overlap of detected SDDs between patterns. This can ensure that only the most effective patterns with minimum overlap between detected SDDs can be selected and can further reduce the selected pattern count. 1-detect top-off ATPG is performed after pattern selection to ensure that our final pattern set can detect all the testable TDFs in the design. Several new metrics are used to evaluate the efficiency of our final pattern set for screening SDDs on large designs.

The remainder of this paper is organized as follows. Section II presents our proposed techniques used for reducing CPU runtime and memory. Our pattern grading and selection procedure is presented in Section III. The proposed procedure is applied to several academic and industrial circuits and the experimental results are presented in Section IV. Finally, we conclude our work in Section V.

II. TECHNIQUES FOR REDUCING RUNTIME AND MEMORY

A. Critical Faults Identification

We use n-detect pattern set as the original pattern repository since it has demonstrated its efficiency for screening SDDs. However, the CPU runtime of n-detect ATPG may still be significant when n is large. Furthermore, the n-detect ATPG on

TABLE I
COMPARISON BETWEEN TF- AND CF-BASED METHODS FOR 10-DETECT ATPG ON TWO ACADEMIC CIRCUITS AND ONE INDUSTRY CIRCUIT.

Circuit		TF-based	CF-based	Percent
wb_ conmax	# faults	347,300	28,981	8.34%
	# patterns	3,302	771	23.35%
	CPU	2m11s	38s	29.01%
ether- net	# faults	868,248	28,209	3.25%
	# patterns	17,582	5,906	33.59%
	CPU	8m37s	2m26s	28.24%
Circuit_ A	# faults	3,396,938	377,857	11.12%
	# patterns	>500K	56,784	<11.36%
	CPU	>5days	17h30m03s	<14.58%

the entire fault list results in a significantly large pattern count, which requires large hardware resources and CPU runtime for the following fault simulation step. In fact, there is a large portion of faults in a design that may never be timing-critical, and it is not necessary to run n-detect ATPG on them. Therefore, we identify and select the timing critical faults before running n-detect ATPG to avoid unproductive consumption on CPU runtime and hardware resources. In practice, the proposed procedure can be applied to any kind of pattern set.

We use the static timing analysis (STA) tool for critical fault selection. Note that the STA tool reports the minimum fault slack by calculating the length of the longest path running through the fault site. In reality, the actual fault slack after pattern generation may not necessarily be equal to this minimum value since (i) the longest path running through the fault may not be testable, and (ii) the ATPG tool does not generate patterns to detect the target fault via the longest path. A critical fault selection slack threshold is needed for the timing critical fault selection. All the TDFs with minimum slack equal or smaller than the pre-defined slack threshold will be selected as timing-critical faults.

Table I shows the efficiency of the proposed critical fault selection method on two academic circuits (ethernet and wb_conmax) and one industry circuit (shown as "Circuit_A"). The slack threshold for critical fault selection is $0.3T$, where T is the clock period. 10-detect ATPG is performed on both total fault (TF) list and selected critical fault (CF) list of the circuits. It is clearly seen that the number of faults for 10-detect ATPG is significantly reduced after critical fault selection for all these three circuits. It can also be seen that the CPU runtime and pattern count is significantly reduced after critical fault selection. This is because the selected critical fault list is significantly shrunk. Note that the fault coverage between the TF-based and CF-based pattern sets is different. Since top-off ATPG will be performed after our pattern selection procedure for compensation, it does not impact the fault coverage of our final pattern set.

B. Parallel Fault Simulation

After pattern generation, fault simulation is performed on each of the test patterns in the pattern repository. The individual fault simulation is based on the entire fault list of the design. With the fault simulation, the detected fault list of each pattern and the actual slack of each detected fault can be obtained, based on which we grade and select patterns. The individual fault simulation may consume a large CPU runtime, especially for the modern industry circuits with a large pattern count. For instance, fault simulation on a single test pattern of "Circuit_A" may take 50-65 seconds. Therefore, for the 10-detect pattern set with 56,784 patterns, it may need 39 days for the fault simulation if running one by one. To reduce the CPU runtime, parallel fault simulation is proposed in this paper.

Since the individual fault simulation is independent of each other, we divide the pattern repository into groups and run fault simulation on each group simultaneously as shown in Figure 2. In this example, i ($i \in [0, pat.count)$) represents the pattern ID. Therefore, each group runs fault simulation on k test patterns according to the pattern IDs, except for the last fault simulation group, which runs fault simulation on the remaining test patterns in the pattern repository. The fault simulation on all these groups can be run simultaneously. In this way, if the pattern repository is equally divided into $k+1$ different groups, the fault simulation can be sped up approximately $(k+1)$X. For example, if we divide the above 10-detect pattern set of "Circuit_A" into 100 fault simulation groups and run them simultaneously, it can be finished in less than 10 hours.

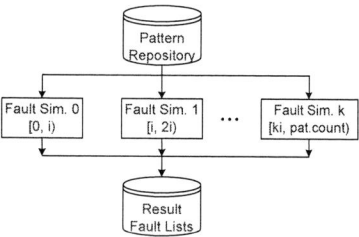

Fig. 2.   Parallel fault simulation.

C. Fault Merging

After fault simulation, each test pattern will result in an fault list. The fault lists of all the test patterns may require a large hardware resources, especially for large industry circuits. Again, take the above "Circuit_A" as an example, the size of the fault list from a single test pattern may reach to 2M bytes on average, even after file compression. Therefore, for its 10-detect pattern set with 56,784 patterns, over 100G bytes hardware resources is needed to store the fault lists. We propose a fault merging technique in the paper to meet this challenge and save hardware resources.

Assuming that there are n test patterns in a pattern repository, and they detect m TDF faults in total. Then we can merge the fault lists of these n test patterns to be a pattern-fault matrix like Figure 3. In the matrix, the n patterns are named as $P_{1,2\cdots n}$ according to their pattern IDs and the m faults are named as $F_{1,2\cdots m}$ according to their assigned IDs. If fault F_i is detected by pattern P_j, then the element a_{ij} represents the actual slack of fault F_i when detected by pattern P_j. Otherwise, if it is not detected by pattern P_j, a_{ij} goes to 0 or some other value that the fault slack may never reach to so that it can be differentiated from the actual slack of the fault. This pattern-fault matrix can be compressed by matrix compression methods [20]. Note that a fault may be detected by one or several test patterns, therefore the pattern-fault matrix may come to a sparse matrix, which can be compressed with a considerable ratio.

$$
\begin{array}{c}
\quad P_1 \quad P_2 \cdots P_j \cdots P_n \\
\begin{array}{c} F_1 \\ F_2 \\ \vdots \\ F_i \\ \vdots \\ F_m \end{array}
\left[
\begin{array}{ccccc}
a_{11} & a_{12} & \cdots & a_{1j} & \cdots & a_{1n} \\
a_{21} & a_{22} & \cdots & a_{2j} & \cdots & a_{2n} \\
\vdots & \vdots & & \vdots & & \vdots \\
a_{i1} & a_{i2} & \cdots & a_{ij} & \cdots & a_{in} \\
\vdots & \vdots & & \vdots & & \vdots \\
a_{m1} & a_{m2} & \cdots & a_{mj} & \cdots & a_{mn}
\end{array}
\right]
\end{array}
$$

Fig. 3.   Pattern-fault matrix used for fault merging and compression.

Since only detected SDDs are considered for pattern grading and selection, we can only include SDDs in the pattern-fault matrix to further reduce its size. With the pattern-fault matrix and sparse matrix compression technique, we can merge the fault list and reach a 60-100X reduction in the data volume for both academic and industry circuits. This fault merging and compression technique can be easily integrated into current commercial EDA tools.

III. PATTERN EVALUATION AND SELECTION

A. Pattern Evaluation

After pattern generation and fault simulation, we can evaluate each test pattern for our pattern selection procedure. Since only SDDs are used for pattern evaluation, a slack threshold (SL_{thr}) is needed to differentiate the detected TDFs and SDDs. If the actual slack of a detected TDF is equal or smaller than SL_{thr}, it will be considered as an SDD and will be counted for pattern evaluation. Otherwise, it will be considered as an gross TDF and will be ignored in the pattern evaluation procedure. In this paper, we use $0.3T$ as the slack threshold to differentiate the gross TDFs and SDDs, where T is the clock period. In practice, some other slack thresholds can be used as well. It is obvious that the greater number of detected SDDs, the more effective the pattern is considered. Our pattern evaluation and selection procedure is based on this observation.

During pattern evaluation, we assign a weight to each detected SDD based on its slack, and then calculate the weight of the target test pattern using Equation (1).

$$W_{Pi} = \sum_{i=1}^{N} W_{sddi} \tag{1}$$

Where W_{Pi} is the weight of pattern Pi, N is the total number of detected SDDs of pattern Pi, and W_{sddi} represents the weight of i-th detected SDD. In this paper, we assign a weight of 1.0 for each detected SDD. The gross TDFs are ignored in the pattern evaluation procedure. Therefore, in this work, the weight of a pattern represents the number of its detected SDDs. Similar to the long path-based method in [17], the definition of pattern weight in this paper is also open ended. For example, given the available slack for each SDD, a metric with actual slack representing W_{sddi} can also be developed.

B. Pattern Selection

The pattern selection procedure is shown in Figure 4. Before pattern selection, the inefficient patterns without SDD detection are removed since they will never be selected. This can save the following pattern selection runtime. During pattern selection, the pattern with the largest weight will be selected first. Then the newly detected SDDs by the selected pattern will be removed from the remaining patterns for re-evaluation. After that, the pattern with the largest weight in the remaining pattern set is secondly selected. This procedure is repeated to select patterns from the original pattern repository until the pattern weight threshold (shown as "W_{thr}") is met (i.e., $WP_{max} < W_{thr}$). Therefore, only patterns with weight equal or larger than this threshold can be selected. The pattern selection procedure will check the overlap of detected SDDs between selected patterns and ensure that only the unique detected SDDs of a pattern are

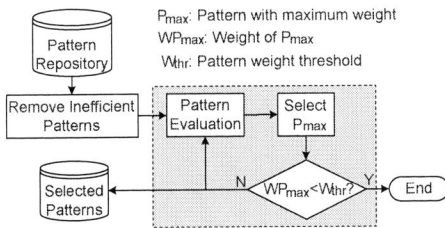

Fig. 4. Selection and sorting procedure for TDF patterns based on their unique weight.

used for the pattern evaluation and selection, which can further reduce the selected pattern count.

Assume that there are N patterns in the original pattern repository, and a maximum of M SDDs are detected by a pattern. The worst-case time complexity of the pattern selection algorithm is $O(N^2M)$ where $N >> M$ for large designs. In fact, this is the worst-case scenario which may never be met in real applications since several new techniques are added to the procedure to speed it up: (i) The inefficient patterns are removed before performing pattern selection; (ii) Once a pattern is selected, all its detected SDDs will be removed from the detected SDD lists of the remaining patterns. This will reduce the SDD list of each pattern significantly after several patterns are selected; (iii) After re-evaluation, the new inefficient patterns in the remaining pattern set will be removed since they will never be selected; and (iv) the pattern selection will be terminated if $WP_{max} < W_{thr}$, no matter how many patterns are left in the repository. According to our experiment on the 10-detect pattern set of "Circuit_A" (56,784 patterns), the CPU runtime for pattern selection is just *1 hour and 6 minutes*.

IV. EXPERIMENTAL RESULTS

In this section, we will present experimental results on both academic and industry circuits. The characteristics of these circuits are shown in Table II in a form of total number of cells (gates+FFs). For the academic circuits, 180nm Cadence Generic Standard Cell Library was used for physical design. 90nm technology library was used for the industry circuits. All patterns are generated using TDF launch-off-capture (LOC) method. After pattern selection, top-off ATPG is run on the undetected faults to ensure that all the testable faults in the design are detected by our final pattern set. The entire flow for our pattern selection procedure is shown in Figure 5.

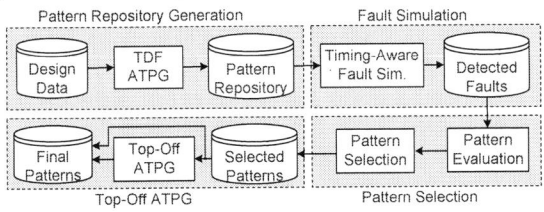

Fig. 5. The entire SDD-based pattern selection flow.

TABLE II
CHARACTERISTICS OF THE EXPERIMENTAL CIRCUITS.

Circuit	wb_conmax	ethernet	Circuit_A	Circuit_B
# total cells	47,757	149,638	1.7 million	3.7 million

A. Pattern Set Analysis

In this subsection, we compare our pattern set, which is selected patterns plus top-off patterns, with 1-detect and timing-

TABLE III
COMPARISON OF SELECED PATTERNS FROM SDD-BASED AND LP-BASED METHODS IN [17] ON WB_CONMAX AND ETHERNET.

Pat. Set		# ori. Pat.	SDD-based		LP-based [17]	
			# Sel. Pat.	CPU	# Sel. Pat.	CPU
wb_conmax	n=5	433	274	3s	404	1m21s
	n=10	771	351	6s	674	2m22s
	n=20	1,446	423	11s	1,122	4m29s
ethernet	n=5	3,048	350	4s	1,094	8m50s
	n=10	5,906	428	7s	1,429	14m16s
	n=20	11,558	532	15s	1,739	26m21s

aware pattern sets generated using a commercial ATPG tool. In this experiment, 1-detect and timing-aware (ta) pattern sets are generated based on the entire fault list. After applying our pattern selection procedure to 5, 10, 20-detect pattern sets of ethernet, 1-detect top-off ATPG is run to ensure that all these pattern sets have the same fault coverage. The pattern count and number of detected SDDs of each pattern set is shown in Figure 6.

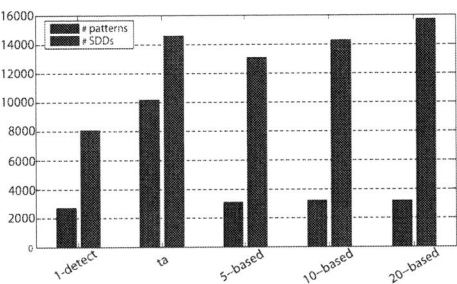

Fig. 6. Pattern set comparison in terms of pattern count and detected SDDs for ethernet. Here "5-based" represents selected patterns from 5-detect pattern set plus its corresponding top-off patterns.

It can be seen from Figure 6 that timing-aware ATPG can detect a lot more SDDs compared with 1-detect ATPG. However, the penalty is that it also results in a much larger pattern count. In contrast, the pattern set generated with our proposed method can detect SDDs comparable to timing-aware pattern set but with a much lower pattern count, which is very close to 1-detect pattern set. In fact, our pattern set generated from 20-detect patterns (shown as "20-based") can even detect more SDDs than timing-aware pattern set. This is because 20-detect pattern set generated on critical faults can detect more SDDs and our pattern set generated from it maintains its SDD detection efficiency. Since only high-quality patterns are selected, our pattern count is significantly reduced.

B. Comparison with LP-Based Method

In this subsection, we compare the proposed SDD-based method with the previous long path-based (LP-based) method [17] in terms of pattern selection efficiency and CPU runtime. To make a fair comparison, we apply both methods to the same n-detect pattern set, and ensure that (i) the slack threshold SL_{thr} in the SDD-based pattern selection is equivalent to the long path threshold LP_{thr} in the LP-based method, i.e., $SL_{thr} + LP_{thr} = T$, where T is the clock period; (ii) the SDD-based method selects patterns that can detect all the SDDs detected by the original n-detect pattern set; and (iii) the LP-based method selects patterns that can sensitize all the long paths sensitized by the original n-detect pattern set, which ensures the selected patterns to detect all the SDDs detected by the original pattern set. Therefore, the selected patterns from both methods have the same SDD coverage.

Table III presents the pattern selection results from both SDD-based and LP-based methods. In this experiment, both methods are applied to 5-detect, 10-detect, and 20-detect pattern sets of IWLS benchmarks ethernet and wb_conmax, respectively. It is clearly seen that the selected pattern set (shown as "# Sel. Pat.") from SDD-based method is much smaller than LP-based method. This is due to the fact that sensitizing an extra unique long path may not necessarily contribute to detect more new SDDs, if these SDDs have been detected by some other already-selected patterns. The LP-based method did not take this issue into consideration and consequently results in a comparable large pattern count. When comes to CPU runtime, the SDD-based method is significantly better than the LP-based method. This is because (i) the pattern selection procedure is much faster after introducing several new techniques as discussed in Section III-B; and (ii) the SDD-based method saves the runtime for path tracing, which is dominant in the LP-based method.

C. Multiple Detection Analysis for Critical Faults

In presence of uncertainties such as process variations, on-chip temperature, power supply and crosstalk noises, the length of a path can vary to a large extent. As a result, it is desirable to detect each critical fault via various long paths. If a pattern set can detect the critical faults via a large number of different long paths, it is superior to keep the high test quality in presence of uncertainties mentioned above. Figure 7 presents a comparison between our n-detect-based pattern sets ($n = 5, 10, 20$) and timing-aware pattern set of ethernet benchmark. In this figure, we compare the average number of sensitized long paths running through each detected critical fault (shown as "NLP_{DCF}"), which is calculated using Equation (2), and the average number of sensitized long paths running through each critical fault (shown as "NLP_{TCF}"), which is calculated using Equation (3).

$$NLP_{DCF} = \frac{\sum_i NLP_{CFi}}{N_{DCF}} \quad (2)$$

$$NLP_{TCF} = \frac{\sum_i NLP_{CFi}}{N_{TCF}} \quad (3)$$

where NLP_{CFi} represents the number of sensitized long paths running through detected critical fault i, N_{DCF} represents the number of detected critical faults, N_{TCF} represents the total number of critical faults in the design.

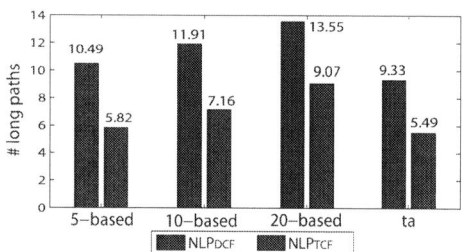

Fig. 7. Comparison of the average number of long paths sensitized through each critical fault for ethernet.

It can be seen from the figure that the number of unique sensitized long paths running through each detected critical fault or each critical fault of our pattern set increases as n increase from 5 to 20. In any case, our pattern set is superior to

TABLE IV
ATPG AND PATTERN SELECTION RESULTS ON INDUSTRY CIRCUITS.

	Circuit	1-detect	10-detect	EMD_3-20	EMD_10-20	ta
Circuit_A	# ori. Pat.	5,456	56,784	14,125	49,137	16,940
	CPU(ATPG)	2.4h	17.5h	8.0h	26.2h	26.3h
	# Sel. Pat.	631	1,542	1,050	1,348	1,099
	CPU(Sel.)	4m19s	1h5m41s	15m33s	58m53s	20m43s
	# SDDs	63,097	80,950	74,755	80,110	73,650
Circuit_B	# ori. Pat.	2,113	15,249	4,456	11,749	5,370
	CPU(ATPG)	1.1h	5.6h	3.6h	6.5h	7.0h
	# Sel. Pat.	593	1,013	737	1,066	1,066
	CPU(Sel.)	22m58s	34m51s	51m1s	28m1s	4m51s
	# SDDs	57,278	67059	61,783	69,125	76,211

timing-aware pattern set in detecting critical faults via various long paths, although they detect comparable number of SDDs as shown in Figure 6. Therefore, our pattern set can ensure a more reliable SDD detection capability.

D. Experiments on Industry Circuits

In this subsection, we present the experimental results on industry circuits. 1-detect, 10-detect, timing-aware, and embedded multi-detection (EMD) pattern sets are generated as the original pattern repositories for our pattern grading and selection procedure. The EMD technique tries to increase the number of detection of each testable fault without pattern count increase. But it does increase the ATPG runtime [21]. To save CPU runtime, the original pattern sets are generated based on selected timing-critical faults, which were selected with $0.3T$ slack threshold. Table IV shows the pattern selection results on "Circuit_A" and "Circuit_B". The "EMD_3-20" and "EMD_10-20" pattern sets are generated using EMD technique. For example, the "EMD_3-20" is generated to guarantee each testable fault to be detected at least 3 times, but desired to be detected 20 times or more. The pattern selection procedure ensures that the selected patterns can detect all the SDDs detected by the original pattern sets. For instance, both 1-detect pattern set of Circuit_A (5,456 patterns) and its selected patterns (631 patterns) detect the same 63,097 SDDs. It can be seen from the table that, our pattern selection procedure can reduce the pattern count significantly while keeping the original SDD detection efficiency. For example, only 11.6% patterns (631 out of 5,456) selected from the 1-detect pattern set of Circuit_A can detect all the SDDs detected by the original pattern set. For its 10-detect pattern set, only 2.7% patterns (1,542 out of 56,784) are selected. This is because with larger pattern repository, 10-detect has a good chance to include patterns with higher SDD test quality. It can also be seen that 10-detect and EMD pattern sets detect more SDDs than 1-detect pattern set, and sometimes even more than timing-aware pattern set (Circuit_A). For Circuit_B, timing-aware pattern set detects more SDDs than both 10-detect and EMD pattern sets, and consequently performed better than our selected patterns. This is because the SDD detection capability of our selected patterns is bounded by its original pattern repository. However, when comparing the pattern count between our selected patterns and the original timing-aware patterns ("# ori. Pat." in "ta" column), the benefit of our procedure is obvious. When applying our pattern selection procedure to timing-aware pattern set, it can also reduce the pattern count significantly for both circuits. Also, the CPU runtime of our pattern selection procedure is much smaller compared with the ATPG runtime.

After pattern selection, 1-detect top-off ATPG is run on the entire undetected faults in the target clock domain to ensure

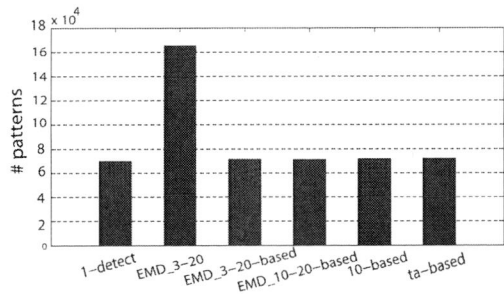

Fig. 8. Pattern count comparison between different pattern sets for Circuit_A. Here "EMD_3-20-based" represents selected patterns from CF-based EMD_3-20 pattern set plus its corresponding top-off patterns.

all the testable faults in this domain can be detected by our final pattern set. Figure 8 shows the pattern count comparison between our final pattern sets (selected + top-off patterns) and 1-detect and EMD_3-20 pattern sets (based on the entire fault list) of Circuit_A. All the pattern sets shown in this figure have the same test coverage (i.e., detect all the testable TDFs). Due to the complexity of runtime and memory, 10-detect, EMD_10-20 and timing-aware ATPG on the entire fault list cannot be finished in a reasonable time. But we would expect them to provide a much larger pattern count than EMD_3-20 pattern set. It can be seen from the figure that our final pattern count, regardless of the pattern repository, is very close to 1-detect pattern set, which is much smaller than EMD_3-20 pattern set. However, with different original pattern repository, the SDD detection efficiency of our final pattern set is varied, as can be seen from Table IV. Please note that our final pattern set may detect a slightly more SDDs than the corresponding selected patterns shown in Table IV, since the top-off patterns may detect some extra SDDs fortuitously.

We also run our procedure on faults crossing clock domains, and the results are shown in Table V. In this experiment, we run EMD_3-20 and timing-aware ATPG on faults crossing domain CLK_1 and CLK_2 of Circuit_B, and apply our pattern selection procedure on both pattern sets. Similar to the experiments in Table IV, the pattern selection procedure ensures the selected patterns can detect all the SDDs detected by the original pattern repository. It can be seen that for this case, EMD_3-20 detects more SDDs than timing-aware pattern set (53,133 vs. 52,824) with slightly smaller pattern count (10,339 vs. 10,479). This is because there are fewer paths running through the crossing-domain faults, and therefore, EMD ATPG has a good chance to detect the faults via long paths runing through them. Note that the performance of both EMD and timing-aware ATPG are all design-dependent. However, in any case, our pattern selection can reduce the pattern count significantly while keeping the SDD detection efficiency of the original pattern repository. The proposed procedure can also be used to evaluate the effectiveness of a pattern set for screening SDDs.

TABLE V
EXPERIMENTAL RESULTS ON CLK_1 TO CLK_2 FOR CIRCUIT_B.

Pat. Set	# Pat.	CPU (ATPG)	# Sel. Pat.	CPU (Sel.)	# SDDs
EMD_3-20	10,339	15.7h	1,248	48m0s	53,133
ta	10,479	56.3h	1,196	45m0s	52,824

V. CONCLUSIONS

In this paper, we have presented an efficient SDD-based pattern grading and selection procedure for screening SDDs.

n-detect pattern sets were used as the original pattern repository to take its advantage for SDD detection. Techniques were proposed to reduce the runtime and memory complexity significantly and therefore make the procedure applicable to very large industry circuits. Compared with previous LP-based procedure, this SDD-based method also reduces the number of selected patterns significantly. Experimental results on both academic and industry circuits demonstrate the efficiency of the proposed procedure.

VI. ACKNOWLEDGMENT

The authors would like to thank Dr. Mahmut Yilmaz from Advanced Micro Devices, Inc. for his assistance in developing the tool for path tracing, and Prof. Krishnendu Chakrabarty from Duke University for his valuable suggestions and discussions on the pattern selection procedure.

REFERENCES

[1] J. Savir and S. Patil, "On Broad-Side Delay Test," in *Proc. VLSI 2 Symp. (VTS'94)*, pp. 284-290, 1994.
[2] M. Bushnell and V. Agrawal, *"Essentials of Electronics Testing"*, Kluwer Publishers, 2000.
[3] R. Mattiuzzo, D. Appello C. Allsup, "Small Delay Defect Testing," *http://www.tmworld.com/article/CA6660051.html* Test & Measurement World, 2009.
[4] Synopsys Inc., "TetraMAX ATPG, Automatic Test Pattern Generation," Synopsys Datasheet, 2010.
[5] Mentor Graphics, "Tessent FastScan, Advanced Automatic Test Pattern Generation," Silicon Test and Yield Analysis Datasheet, 2009.
[6] M. E. Amyeen, S. Venkataraman, A. Ojha, S. Lee, "Evaluation of the Quality of N-Detect Scan ATPG Patterns on a Processor", *IEEE International Test Conference (ITC'04)*, pp. 669-678, 2004.
[7] Y. Huang, "On N-Detect Pattern Set Optimization," in *Proc. IEEE the 7th Int. Symp. on Quality Electronic Design (ISQED'06)*, 2006.
[8] IWLS 2005 Benchmarks, "http://iwls.org/iwls2005/benchmarks.html".
[9] W. Qiu, J. Wang, D. Walker, D. Reddy, L. Xiang, L. Zhou, W. Shi, and H. Balachandran, "K Longest Paths Per Gate (KLPG) Test Generation for Scan Scan-Based Sequential Circuits," in *Proc. IEEE ITC*, 2004.
[10] N. Ahmed, M. Tehranipoor and V. Jayaram, "Timing-Based Delay Test for Screening Small Delay Defects," *IEEE Design Automation Conf.*, pp. 320-325, 2006.
[11] A. K. Majhi, V. D. Agrawal, L. Jacob, L. M. Patnaik, "Line coverage of path delay faults," *IEEE Trans. on Very Large Scale Integration (VLSI) Systems*, vol. 8, no. 5, pp. 610-614, 2000.
[12] H. Lee, S. Natarajan, S. Patil, I. Pomeranz, "Selecting High-Quality Delay Tests for Manufacturing Test and Debug," in *Proc. IEEE Int. Symp. on Defect and Fault-Tolerance in VLSI Systems (DFT'06)*, 2006.
[13] S. Goel, N. Devta-Prasanna and R. Turakhia, "Effective and Efficient Test pattern Generation for Small Delay Defects," *IEEE VLSI Test Symposium (VTS'09)*, 2009.
[14] S. Goel, K. Chakrabarty, M. Yilmaz, K. Peng, M. Tehranipoor, "Circuit Topology-Based Test Pattern Generation for Small-Delay Defects," to appear in *IEEE Asian Test Symposium (ATS'10)*, 2010.
[15] M. Yilmaz, K. Chakrabarty, and M. Tehranipoor, "Test-Pattern Grading and Pattern Selection for Small-Delay Defects," in *Proc. IEEE VLSI Test Symposium (VTS'08)*, 2008.
[16] M. Yilmaz, K. Chakrabarty, and M. Tehranipoor, "Interconnect-Aware and Layout-Oriented Test-Pattern Selection for Small-Delay Defects," in *Proc. Int. Test Conference (ITC'08)*, 2008.
[17] K. Peng, J. Thibodeau, M. Yilmaz, K. Chakrabarty, and M. Tehranipoor, "A Novel Hybrid Method for SDD Pattern Grading and Selection," in *Proc. IEEE VLSI Test Symposium (VTS'10)*, 2010.
[18] K. Peng, M. Yilmaz, K. Chakrabarty, and M. Tehranipoor, "A Noise-Aware Hybrid Method for SDD Pattern Grading and Selection," to appear in *Proc. IEEE Asian Test Symposium (ATS'10)*, 2010.
[19] K. Peng, M. Yilmaz, M. Tehranipoor, and K. Chakrabarty, "High-Quality Pattern Selection for Screening Small-Delay Defects Considering Process Variations and Crosstalk," in *Proc. IEEE Design, Automation & Test in Europe Conference & Exhibition (DATE'10)*, 2010.
[20] David Salomon, *"Data Compression, The Complete Reference, Forth Edition"*, Springer Publishers, 2007.
[21] J. Geuzebroek, E.J. Marinissen, A. Majhi, A. Glowatz, F. Hapke, *"Embedded multi-detect ATPG and Its Effect on the Detection of Unmodeled Defects"*, in *Proc. Int. Test Conference (ITC'07)*, pp. 1-10, 2007.

Static Test Compaction for Delay Fault Test Sets Consisting of Broadside and Skewed-Load Tests

Irith Pomeranz
School of Electrical & Computer Eng.
Purdue University
W. Lafayette, IN 47907, U.S.A.
pomeranz@ecn.purdue.edu

Abstract - **Test sets that consist of both broadside and skewed-load tests provide improved delay fault coverage for standard-scan circuits. This paper describes a static test compaction procedure for such mixed test sets. The unique feature of the procedure is that it can modify the type of a test (from broadside to skewed-load or from skewed-load to broadside) if this contributes to test compaction. Experimental results demonstrate that the procedure is able to reduce the sizes of available mixed test sets significantly. Moreover, it modifies the types of significant numbers of tests before including them in the compacted test set.**

Keywords - Broadside tests, scan circuits, skewed-load tests, static test compaction, transition faults.

I. INTRODUCTION

Skewed-load tests [1] and broadside tests [2] are two types of two-pattern tests that can be used for detecting delay faults in standard-scan circuits. Both types of tests can be represented as $<s_1v_1, s_2v_2>$, where s_1v_1 is the first pattern of the test, and s_2v_2 is the second pattern of the test. For $i = 1$ and 2, s_i is a state and v_i is a primary input vector. The primary input vectors can be empty if all the inputs of the combinational logic are state variables.

The first pattern of a two-pattern test is applied under a slow clock, and the second pattern is applied in functional mode under a fast clock. Skewed-load and broadside tests differ in the way s_2 is obtained. In a broadside test, s_2 is the next-state obtained for present-state s_1 and primary input vector v_1. In a skewed-load test, s_2 is obtained by a single shift of s_1.

Skewed-load tests require a scan enable signal that can change at the circuit speed. This is due to the fact that the first pattern is applied in scan mode, and the second pattern is applied in functional mode under a fast clock. Methods to generate fast scan enable signals locally from the global scan enable signal can be used for removing this shortcoming of skewed-load tests [3]-[4].

Skewed-load tests typically achieve a higher fault coverage than broadside tests. However, there are faults that only be detected using broadside tests. Moreover, some of these faults are irredundant, and they can affect

the circuit during functional operation. Therefore, in order to maximize the fault coverage, both types of tests should be applied [2]. The advantages of combining the two types of tests were also noted in [5]-[7].

A test set that consists of both broadside and skewed-load tests can be generated by first generating one type of tests, and then generating the other type of tests for faults that remain undetected. Test generation procedures such as the ones described in [8]-[13] can be used for this purpose. The test sets considered in this work were generated by a test generation procedure that attempts to predict for every target fault whether a broadside or skewed-load test is more likely to exist. It then attempts to generate a test of this type. The resulting test set consists of a mix of broadside and skewed-load tests depending on the types of tests found to be most suitable for the target faults.

Regardless of the way the test set is generated, static test compaction can be an important contributor to the generation of compact test sets. The goal of static test compaction is to reduce the number of tests in a test set without reducing its fault coverage. It can complement or replace dynamic test compaction where the test generation process is modified to produce fewer tests [14]-[19]. Different types of static test compaction procedures were described in [20]-[24]. Reverse order fault simulation is the simplest static test compaction procedure that requires a single fault simulation pass over the test set in reverse order. Procedures that merge test cubes (incompletely specified tests) in order to reduce the number of tests were described in [20] and [24]. A procedure that performs limited test generation during the static test compaction process was described in [21].

Considering test sets for standard-scan circuits that consist of broadside and skewed-load tests, a dimension of the static test compaction process that has not been explored earlier is the ability to replace a test of one type with a test of another type. Thus, a skewed-load test may be replaced with a broadside test, and a broadside test may be replaced with a skewed-load test, if the test with a different type detects more faults and is, therefore, likely to contribute to test compaction. The static test compaction procedure described in this paper can replace broadside tests with skewed-load tests, and skewed-load tests with broadside tests, when this is effective in reducing the number of tests.

The procedure is based on the concept of test vector improvement introduced in [25]. This concept as well as the proposed procedure can be applied to fully-specified tests or to test cubes used for test data compression [26]-[27]. In this work the proposed procedure is applied to fully-specified tests.

The paper is organized as follows. Section II reviews the concept of test vector improvement. Section III describes a static test compaction procedure based on test vector improvement that is able to modify the type of a test. Section IV presents experimental results of static test compaction. The results demonstrate that the procedure changes the types of a significant number of tests in order to obtain the final compacted test set.

II. TEST VECTOR IMPROVEMENT

This section describes a test vector improvement procedure that accepts a test t of a given type (skewed-load or broadside) and increases the number of faults it detects. This is done in this section without changing the test type. Changing the test type will be considered in the next section. The procedure also accepts a set of target faults F. Some of these faults may be detected by t while others may not.

In general, test vector improvement is achieved by complementing bits of t one at a time, undoing the complementation of bits that decrease the number of detected faults out of F, and otherwise accepting the complementation. A more detailed description requires the following notation.

The number of primary inputs is denoted by n, and the number of state variables is denoted by k. Considering a primary input vector v_i, the value of primary input j is denoted by $v_i(j)$, for $0 \leq j < n$. Considering a state s_i, the value of state variable j is denoted by $s_i(j)$, for $0 \leq j < k$.

We assume that all the flip-flops are included in a single scan chain and that the scan chain is shifted to the right. This implies that, in a skewed-load test $t = <s_1 v_1, s_2 v_2>$, the value of $s_2(0)$ is determined by the scan-in value, and that $s_2(j) = s_1(j-1)$ for $0 < j < k$.

Broadside and skewed-load tests differ in the bits that can be complemented during test vector improvement. For a broadside test $t = <s_1 v_1, s_2 v_2>$, the procedure attempts to complement the bits of s_1, v_1 and v_2. The state s_2 is always computed as the next-state of s_1 and v_1. For a skewed-load test $t = <s_1 v_1, s_2 v_2>$, the procedure attempts to complement the bits of s_1, v_1 and v_2, as well as $s_2(0)$. The state s_2 is always computed as a shifted version of s_1, with $s_2(0)$ keeping its value. For ease of reference we denote by $B(t)$ the set of bits that can be comple-

mented in t, depending on its type.

The test vector improvement procedure starts by simulating t under F in order to find the number of detected faults, denoted by d_{best}. If $d_{best} = 0$, consideration of t ends without modifying it. This is motivated by the following observation. A test set T under consideration is expected to contain enough tests to detect all or most of the detectable target faults. To reduce the size of T, it is necessary to ensure that each test will detect a large number of faults. Starting from a test that does not detect any faults, test vector improvement may not be able to produce a test that detects a large number of faults. In this case, the procedure does not consider t further. If $d_{best} > 0$, consideration of t proceeds as follows.

In an iteration of test vector improvement, the procedure considers the bits included in $B(t)$ one at a time. When bit b is considered, its value is complemented and the state s_2 is updated according to the test type. The procedure simulates the modified test t under F to find the number of detected faults, denoted by d. If $d < d_{best}$, the value of b is complemented again, and s_2 is recomputed if necessary. This is done in order to restore the previous test. Otherwise, the complementation of b is accepted. If $d > d_{best}$, d_{best} is updated by setting $d_{best} = d$.

The procedure performs a constant number of iterations, denoted by N. After N iterations it stops with an improved test t.

III. STATIC TEST COMPACTION PROCEDURE

The static test compaction procedure accepts a test set T that consists of broadside and skewed-load tests, and a set of target faults F.

The unique feature of the procedure is that it changes the type of a test $t \in T$ if this contributes to test compaction. Next, we describe how the change of type is accomplished.

Let $t = <s_1 v_1, s_2 v_2>$. If t is a skewed-load test, in order to change it into a broadside test, s_2 is recomputed as the next-state of s_1 and v_1. For example, ISCAS-89 benchmark $s298$ has a skewed-load test

    <11010101111010 101, 11101010111101 111>.

The broadside test obtained from it is

    <11010101111010 101, 00000000000000 111>,

where 00000000000000 is the next-state obtained for present-state 11010101111010 and primary input vector 101.

If t is a broadside test, in order to change it into a skewed-load test, s_2 needs to be a shifted version of s_1. We satisfy this condition by setting $s_2(j) = s_1(j-1)$ for $0 < j < k$, and keeping the same value for $s_2(0)$. For example, ISCAS-89 benchmark $s298$ has a broadside test

    <11110001110110 011, 00000010010001 111>.

The skewed-load test obtained from it is

    <11110001110110 011, 01111000111011 111>.

978-1-61284-657-6/11 $26.00 © 2011 IEEE

In itself, changing the type of a test may reduce the number of faults detected by the test. Combined with test vector improvement, changing the test type can result in a test that detects more faults than the original test. For example, the broadside test <11110001110110 011, 00000010010001 111> of $s298$ detects 46 transition faults. The skewed-load test <11110001110110 011, 01111000111011 111> obtained from it detects 45 transition faults. Starting from the broadside test, test vector improvement with $N = 4$ yields the broadside test <11111011100100 010, 00001011100101 001> that detects 83 transition faults. Starting from the skewed-load test, test vector improvement yields the skewed-load test <10110101010101 101, 01011010101010 010> that detects 121 transition faults.

Next, we describe the static test compaction procedure.

The static test compaction procedure starts by applying forward-looking reverse order fault simulation to T in order to reduce its size without reducing its fault coverage [22]. At the end of the static test compaction procedure, forward-looking reverse order fault simulation is applied again to obtain the final test set.

After the first application of forward-looking reverse order fault simulation to T, the procedure copies T into another test set W and sets $T = \phi$. The procedure adds tests to T one at a time based on the tests in W in order to obtain a compacted version of T that has the same or improved fault coverage. For this purpose, the procedure considers all the tests in W repeatedly. The need for repetition will be clarified later. In an iteration of this process, test vector improvement is applied to every test $w \in W$ as follows.

Regardless of whether w is a broadside or skewed-load test, w is copied into a test t_{brd} and turned into a broadside test. Separately, w is copied into a test t_{skw} and turned into a skewed-load test.

Test vector improvement is applied to t_{brd} as a broadside test and to t_{skw} as a skewed-load test using the set of target faults F. The numbers of faults detected by t_{brd} and t_{skw} after test vector improvement are denoted by d_{brd} and d_{skw}, respectively. The procedure applies the following rules to decide whether one of t_{brd} and t_{skw} will be added to T.
(1) If $d_{brd} = d_{skw} = 0$, neither t_{brd} nor t_{skw} is added to T.
(2) Else, if $d_{brd} \geq d_{skw}$, t_{brd} is added to T.
(3) Else, t_{skw} is added to T.

If either t_{brd} or t_{skw} is added to T, the test is simulated under F, and detected faults are removed from F. In this way, the set of target faults F decreases with every additional test included in T.

After all the tests in W are considered once, it is possible that T will not detect all the faults detected by W. This can happen since test vector improvement targets only the number of detected faults. It does not require that all the faults detected by a test $w \in W$ will be detected by the test t added to T based on w.

To compensate for this the procedure considers all the tests in W repeatedly until no new test is added to T based on any one of the tests in W. When a test $w \in W$ is considered again, F contains fewer faults, and test vector improvement yields different tests t_{brd} and t_{skw} based on w.

At the end of the static test compaction process it is guaranteed that T will detect all the faults detected by W (and by the initial test set T). This can be seen as follows. Suppose that W detects a fault f that is not detected by T. This implies that there is a test $w \in W$ that detects f. As long as w detects a fault that is not detected by T, a test based on it will be added to T. It is possible that faults other than f will be detected by tests based on w. However, every time w is considered, at least one such fault will be detected. Eventually, only f will remain that is detected by w but not by T, and the test added to T based on w will detect f.

For illustration we describe the application of the static test compaction process to ISCAS-89 benchmark $s298$. The test set T after forward-looking reverse order fault simulation contains 44 tests, and it detects 566 transition faults. After setting $W = T$, we have that $W = \{w_0, w_1, \cdots, w_{43}\}$. We use $N = 4$ for test vector improvement.

The test w_0 is the broadside test considered above. The broadside test based on w_0 detects 46 faults. Test vector improvement increases this number to 83. The skewed-load test based on w_0 detects 45 faults. Test vector improvement increases this number to 121. The skewed-load test is included in T that now detects 121 faults.

The test $w_1 \in W$ is also a broadside test. The broadside test based on w_1 detects 44 faults. Test vector improvement increases this number to 57. The skewed-load test based on w_1 detects 43 faults. Test vector improvement increases this number to 99. The skewed-load test is included in T that now detects 220 faults.

The test $w_2 \in W$ is a skewed-load test. The broadside test based on w_2 detects 9 faults. Test vector improvement increases this number to 47. The skewed-load test based on w_2 detects 9 faults. Test vector improvement increases this number to 36. The broadside test is included in T that now detects 267 faults.

After w_6 is considered, T detects 412 faults. The test $w_7 \in W$ is a broadside test. The broadside test based on w_7 does not detect any faults. Test vector improvement is not applied in this case. The skewed-load test based on w_7 detects 6 faults. Test vector improvement increases this number to 18. The skewed-load test is included in T that now detects 430 faults.

After w_{23} is considered, T detects 540 faults. The test $w_{24} \in W$ is a broadside test. The broadside test based on w_{24} does not detect any faults. Test vector improvement is not applied in this case. The skewed-load test based on w_{24} does not detect any faults either. Test vector improvement is not applied to this test. Consequently, no test is added to T based on w_{24}.

After $w_{43} \in W$ is considered, T contains 32 tests and it detects 556 faults. A second iteration over the tests in W adds tests based on w_1, w_3, w_{11}, w_{13} and w_{29}. The size of T reaches 37, and the number of detected faults reaches 564. A third iteration over the tests in W adds a test based on w_{11}. The size of T reaches 38, and the number of detected faults reaches 566. A fourth iteration does not increase the number of detected faults further, and the process ends. Forward-looking reverse order fault simulation reduces the number of tests to 37.

Different values of N (different numbers of iterations of test vector improvement) may result in different compacted test sets. If the initial test set does not detect all the detectable faults, the compacted test sets may detect different numbers of additional faults. However, in all the cases, the compacted test sets will detect at least all the faults detected by the initial test set.

IV. EXPERIMENTAL RESULTS

This section reports the results of the application of the static test compaction procedure to available test sets consisting of both broadside and skewed-load tests for transition faults in benchmark circuits.

We first consider the value of N, or the number of iterations of test vector improvement applied to every test. A higher value of N allows the number of detected faults to be increased more significantly for every test. However, it also increases the computational effort. In addition, beyond a certain number of iterations, the improvements in the numbers of detected faults are low, until no further improvements are obtained for high values of N.

To determine an appropriate value for N experimentally, we consider several benchmark circuits. We apply the static test compaction procedure to their test sets using $N = 1, 2, \cdots, 7$. The results are shown in Table I as follows.

Column *init* shows the size of T after the first application of forward-looking reverse order fault simulation. Column $N = N_0$, for $1 \le N_0 \le 7$, shows the size of T after applying the static test compaction procedure using $N = N_0$ for test vector improvement.

The smallest test set size is marked with an asterisk. In addition, if a value of N does not produce the highest number of detected faults, it is marked with an x. As discussed at the end of Section III, the compacted test set T always detects all the faults detected by the initial test set. However, different numbers of additional faults may be

TABLE I. Selecting N

circuit	init	N=1	N=2	N=3	N=4	N=5	N=6	N=7
s208	40	39	36	35	37	36	*33	35
s298	44	39	37	37	37	*35	38	37
s344	43	36	31	28	29	*27	*27	*27
s382	51	41	40	38	37	37	38	*33
s386	93	91	86	87	84	*81	*x81	*x81
s420	84	74	66	66	62	62	59	*58
s510	96	78	78	82	79	79	*77	82
s526	89	75	x78	x71	x71	71	*x68	70
s641	76	50	43	37	*35	37	37	36
s820	149	137	129	x127	*118	129	125	128
s953	147	124	119	x112	111	114	*x107	109
s1196	212	174	169	158	154	152	*144	149
s1423	107	x83	x65	x64	62	65	x55	x51
average	94.69	80.08	75.15	72.46	70.46	71.15	68.38	68.92

detected for different values of N. The last row of Table I shows the average test set size in each column.

From Table I it can be seen that the smallest test set size is obtained most often for $N = 6$. The lowest average test set size is also obtained for $N = 6$. However, there are also cases where $N = 6$ does not result in the highest number of detected faults. The same applies to $N = 7$, which yields the second lowest average number of tests. We selected to use $N = 4$, which yields the third lowest average number of tests.

The results using $N = 4$ are shown in Tables II and III. The test sets in Table II are the same as the ones used for Table I. They are generated by a procedure that selects, for every fault, the test type that is more likely to exist. For comparison, the test sets in Table III are biased towards including as many skewed-load tests as possible.

Column *initial* shows information for the initial test set T after the first application of forward-looking reverse order fault simulation. Subcolumn *tot* shows the total number of tests in the test set. Subcolumn *brd* shows the number of broadside tests. Subcolumn *skw* shows the number of skewed-load tests. Subcolumn *f.c.* shows the transition fault coverage of the test set.

Column *skw→brd* shows the number of skewed-load tests in the initial test set such that the static test compaction procedure decided to add to the compacted test set broadside tests based on them. Column *brd→skw* shows the number of broadside tests in the initial test set such that the static test compaction procedure decided to add to the compacted test set skewed-load tests based on them.

Column *final* shows information for the test set T after static test compaction. Subcolumn *tot* shows the total number of tests in the test set. Subcolumn *%red* shows the reduction in the number of tests as a percentage compared with the initial test set. Subcolumn *brd* shows the number of broadside tests. Subcolumn *skw* shows the number of skewed-load tests. Subcolumn *f.c.* shows the transition fault coverage of the test set. Subcolumn *ntime* shows the normalized run time of the static test compaction procedure. For normalization we divide the run time of the procedure by the run time for simulating the initial test set. Since static test compaction is achieved by fault simulation, this gives an indication of the computational effort expended by the procedure.

From Tables II and III it can be seen that the static test compaction procedure replaces the types of significant numbers of tests. In most cases, it replaces broadside tests by skewed-load tests. However, there are also cases where it replaces skewed-load tests by broadside tests, especially when the initial test set is biased to contain as many skewed-load tests as possible.

The reduction in test set size is significant, and in several cases, it goes together with a small increase in fault coverage. This is possible when the initial test set does not detect all the detectable faults.

V. CONCLUDING REMARKS

We described a static test compaction procedure for test sets consisting of both broadside and skewed-load tests. A unique feature of the procedure is that it allows the type of a test to be modified (from broadside to skewed-load or from skewed-load to broadside) if this contributes to test compaction. For every test in the initial test set, the procedure attempts to obtain a broadside test, and separately, a skewed-load test, that detect as many faults as possible. The test with the largest number of detected faults is included in the compacted test set.

Experimental results were presented using transition faults under available mixed test sets to demonstrate that the procedure replaces the type of a significant number of tests, and is able to reduce the test set size significantly. It also increases the fault coverage in cases where the given test set does not detect all the detectable faults.

REFERENCES

[1] J. Savir and S. Patil, "Scan-Based Transition Test", IEEE Trans. on Computer-Aided Design, Aug. 1993, pp. 1232-1241.

[2] J. Savir and S. Patil, "Broad-Side Delay Test", IEEE Trans. on Computer-Aided Design, Aug. 1994, pp. 1057-1064.

[3] N. Ahmed, M. Tehranipoor, C. P. Ravikumar and K. M. Butler, "Local At-Apeed Scan Enable Generation for Transition Fault Testing Using Low-Cost Testers", IEEE Trans. on Computer-Aided Design, May 2007, pp. 896-905.

[4] G. Xu and A. D. Singh, "Scan Cell Design for Launch-on-Shift Delay Tests with Slow Scan Enable", IET Computers & Digital Techniques, May 2007, pp. 213-219.

[5] S. Wang, X. Liu and S. T. Chakradhar, "Hybrid Delay Scan: A Low Hardware Overhead Scan Based Delay Test Technique for High Fault Coverage and Compact Test Sets", in Proc. Design Autom. and Test in Europe Conf., 2004, pp. 1296-1301.

[6] I. Pomeranz and S. M. Reddy, "Effectiveness of Scan-Based Delay Fault Tests in Diagnosis of Transition Faults", IET Computers & Digital Techniques, Sept. 2007, pp. 537-545.

[7] I. Park and E. J. McCluskey, "Launch-on-Shift-Capture Transition Tests", in Proc. Intl. Test Conf., 2008, pp. 1-9.

[8] L. N. Reddy, I. Pomeranz and S. M. Reddy, "COMPACTEST-II: A Method to Generate Compact Two-Pattern Test Sets for Combinational Logic Circuits", in Proc. Intl. Conf. on Computer-Aided Design, 1992, pp. 568-574.

[9] I. Hamzaoglu and J. H. Patel, "Compact Two-Pattern Test Set Generation for Combinational and Full Scan Circuits", in Proc. Intl. Test Conf., 1998, pp. 944-953.

[10] N. Tendolkar, R. Raina, R. Woltenberg, X. Lin, B. Swanson and G. Aldrich, "Novel Techniques for Achieving High At-Speed Transition Fault Test Coverage for Motorola's Microprocessors Based on PowerPC(TM) Instruction Set Architecture", in Proc. VLSI Test Symp., 2002, pp. 3-8.

[11] Y. Shao, I. Pomeranz and S. M. Reddy, "On Generating High Quality Tests for Transition Faults", in Proc. Asian Test Symp., Nov. 2002, pp. 1-8.

[12] W. Qiu, J. Wang, D. M. H. Walker, D. Reddy, X. Lu, Z. Li, W. Shi and H. Balachandran, "K Longest Paths Per Gate (KLPG) Test Generation for Scan-Based Sequential Circuits", in Proc. Intl. Test Conference, 2004, pp. 223-231.

[13] Z. Chen, D. Xiang and B. Yin, "The ATPG Conflict-Driven Scheme for High Transition Fault Coverage and Low Test Cost", in Proc. VLSI Test Symp., 2009, pp. 146-151.

[14] P. Goel and B. C. Rosales, "Test Generation and Dynamic Compaction of Tests", in Proc. Test Conf., 1979 pp. 189-192.

[15] I. Pomeranz, L. N. Reddy and S. M. Reddy, "COMPACTEST: A Method to Generate Compact Test Sets for Combinational Circuits", in Proc. Intl. Test Conf., 1991, pp. 194-203.

[16] J.-S. Chang and C.-S. Lin, "Test Set Compaction for Combinational Circuits", in Proc. Asian Test Symp., 1992, pp. 20-25.

[17] Y. Matsunaga, "MINT -An Exact Algorithm for Finding Minimum Test Sets", IEICE Trans. Fundamentals., vol. E76-A, No. 10, Oct. 1993, pp. 1652-1658.

[18] S. Kajihara, I. Pomeranz, K. Kinoshita and S. M. Reddy, "Cost-Effective Generation of Minimal Test Sets for Stuck-at Faults in Combinational Logic Circuits", IEEE Trans. on Computer-Aided Design, Dec. 1995, pp. 1496-1504.

[19] I. Hamazaoglu and J. H. Patel, "Test Set Compaction Algorithms for Combinational Circuits", in Proc. Intl. Conf. on Computer-Aided Design, 1998, pp. 283-289.

[20] M. Abramovici, M. A. Breuer and A. D. Friedman, *Digital Systems Testing and Testable Design*, IEEE Press, 1995.

[21] L. N. Reddy, I. Pomeranz and S. M. Reddy, "ROTCO: A Reverse Order Test COmpaction Technique", in Proc. EURO-ASIC, 1992, pp. 189-194.

[22] I. Pomeranz and S. M. Reddy, "Forward-Looking Fault Simulation for Improved Static Compaction", IEEE Trans. on Computer-Aided Design, Oct. 2001, pp. 1262-1265.

[23] X. Lin, J. Rajski, I. Pomeranz and S. M. Reddy, "On Static Test Compaction and Test Pattern Ordering for Scan Designs", in Proc. Intl. Test Conf., 2001, pp. 1088-1097.

[24] A. H. El-Maleh and Y. E. Osais, "Test Vector Decomposition-Based Static Compaction Algorithms for Combinational Circuits", ACM Trans. on Design Automation of Electronic Systems, Oct. 2003, pp. 430-459.

[25] I. Pomeranz and S. M. Reddy, "On Test Generation with Test Vector Improvement", IEEE Trans. on Computer-Aided Design, March 2010, pp. 502-506.

[26] C. Barnhart, V. Brunkhorst, F. Distler, O. Farnsworth, B. Keller and B. Koenemann, "OPMISR: The Foundation for Compressed ATPG Vectors", in Proc. Intl. Test Conf., 2001, pp. 748-757.

[27] J. Rajski, J. Tyszer, N. Kassab, N. Mukherjee, R. Thompson, K.-H. Tsai, A. Hertwig, N. Tamarapalli, G. Mrugalski, G. Eide and J. Qian, "Embedded Deterministic Test for Low Cost Manufacturing Test", in Proc. Intl. Test Conf., 2002, pp. 301-310.

978-1-61284-657-6/11 $26.00 © 2011 IEEE

TABLE II. Experimental results

circuit	N	initial				skw→brd	brd→skw	final					ntime
		tot	brd	skw	f.c.			tot	%red	brd	skw	f.c.	
s208	4	40	25	15	94.47	2	6	37	7.50	22	15	94.47	154.00
s298	4	44	25	19	94.97	4	11	37	15.91	17	20	94.97	64.67
s344	4	43	28	15	97.97	4	7	29	32.56	18	11	97.97	114.33
s382	4	51	29	22	93.98	2	9	37	27.45	14	23	93.98	102.25
s386	4	93	43	50	95.85	7	13	84	9.68	40	44	96.11	82.43
s420	4	84	47	37	94.17	2	14	62	26.19	30	32	94.17	173.43
s510	4	96	64	32	96.86	9	24	79	17.71	38	41	96.86	195.40
s526	4	89	29	60	93.06	6	8	71	20.22	26	45	93.25	117.10
s641	4	76	53	23	100	0	14	35	53.95	11	24	100	251.20
s820	4	149	105	44	92.99	2	41	118	20.81	66	52	93.11	150.75
s953	4	147	118	29	98.74	5	27	111	24.49	71	40	98.79	261.45
s1196	4	212	209	3	100	0	41	154	27.36	114	40	100	169.91
s1423	4	107	56	51	98.24	5	12	62	42.06	23	39	98.66	304.38
s5378	4	408	239	169	97.85	8	25	244	40.20	113	131	97.86	596.86
s9234	4	673	360	313	92.91	18	31	381	43.39	177	204	93.22	447.98
s13207	4	849	420	429	96.61	15	38	492	42.05	219	273	96.80	1304.44
s15850	4	683	250	433	94.91	15	26	396	42.02	150	246	95.11	1105.05
s35932	4	115	66	49	89.78	3	13	34	70.43	14	20	89.78	4616.67
b03	4	50	36	14	99.74	2	11	31	38.00	13	18	100	164.33
b04	4	101	79	22	99.52	1	16	49	51.49	27	22	99.78	292.72
b05	4	130	33	97	95.63	7	8	78	40.00	18	60	96.03	99.98
b07	4	90	26	64	95.56	8	12	58	35.56	15	43	98.09	170.77
b08	4	72	32	40	96.56	4	8	57	20.83	20	37	97.39	144.40
b09	4	41	21	20	99.12	4	5	30	26.83	13	17	99.41	146.50
b10	4	81	39	42	97.59	4	21	51	37.04	17	34	99.31	143.33
b11	4	109	47	62	98.42	9	17	89	18.35	33	56	98.85	158.61
b14	4	372	204	168	80.88	13	24	197	47.04	97	100	85.75	567.99
b15	4	774	224	550	67.04	60	42	353	54.39	125	228	92.16	573.80
b20	4	614	157	457	86.37	14	18	261	57.49	28	233	90.23	837.05
b21	4	500	217	283	75.82	18	34	204	59.20	73	131	83.39	869.81

TABLE III. Experimental results

circuit	N	initial				skw→brd	brd→skw	final					ntime
		tot	brd	skw	f.c.			tot	%red	brd	skw	f.c.	
s208	4	46	10	36	94.23	7	1	33	28.26	12	21	94.23	-
s298	4	42	15	27	94.97	8	5	36	14.29	15	21	94.97	-
s344	4	40	13	27	97.67	4	1	26	35.00	12	14	97.82	-
s382	4	48	8	40	93.98	6	1	35	27.08	11	24	93.98	-
s386	4	91	31	60	95.73	16	2	84	7.69	35	49	96.11	-
s420	4	92	22	70	94.17	10	2	65	29.35	17	48	94.17	-
s510	4	90	22	68	96.86	26	2	75	16.67	32	43	96.86	176.00
s526	4	90	22	68	93.16	7	3	73	18.89	25	48	93.35	113.50
s641	4	79	10	69	100	11	0	37	53.16	11	26	100	566.00
s820	4	146	32	114	92.50	27	6	123	15.75	43	80	93.05	188.90
s953	4	148	32	116	98.74	22	4	110	25.68	41	69	98.79	227.86
s1196	4	207	3	204	100	89	2	155	25.12	86	69	100	174.35
s1423	4	108	22	86	98.14	6	1	58	46.30	15	43	98.66	329.00
s5378	4	399	148	251	97.83	16	10	246	38.35	103	143	97.86	646.01
s9234	4	678	222	456	92.87	31	12	383	43.51	146	237	93.24	472.35
s13207	4	862	293	569	96.67	28	23	499	42.11	185	314	96.78	1279.79
s15850	4	680	182	498	94.80	28	15	396	41.76	142	254	95.10	1087.77
s35932	4	117	0	117	89.78	10	0	32	72.65	10	22	89.78	4729.61
b03	4	50	11	39	100	6	1	31	38.00	12	19	100	-
b04	4	86	2	84	99.82	6	1	41	52.33	7	34	99.82	365.60
b05	4	119	4	115	95.87	3	1	73	38.66	4	69	96.03	104.78
b07	4	99	7	92	96.44	9	0	64	35.35	13	51	98.81	190.80
b08	4	68	12	56	97.27	11	0	59	13.24	20	39	97.39	105.00
b09	4	39	5	34	99.12	4	2	30	23.08	6	24	99.71	-
b10	4	79	7	72	98.74	5	2	51	35.44	9	42	98.97	120.00
b11	4	115	15	100	98.74	8	5	84	26.96	17	67	98.96	153.50
b14	4	340	21	319	80.83	22	3	173	49.12	27	146	85.57	564.21
b15	4	758	28	730	67.62	71	24	363	52.11	77	286	91.61	530.63
b20	4	611	12	599	86.35	23	0	278	54.50	23	255	90.86	816.57
b21	4	469	25	444	75.95	25	1	212	54.80	28	184	84.25	786.88

Efficient and Product-Representative Timing Model Validation

Eun Jung Jang[1], Anne Gattiker[2], Sani Nassif[2], and Jacob A. Abraham[1]

[1]Computer Engineering Research Center, The University of Texas Austin, {ejang,jaa}@cerc.utexas.edu
[2]Austin Research Lab, IBM, Austin, TX 78758, {gattiker,nassif}@us.ibm.com

Abstract—Timing analysis is a key sign-off step in the design of today's chips, but as technology advances, it becomes ever more challenging to create timing models that accurately reflect real timing-related behavior. Complex dependencies on second order phenomena, such as pattern density and stress/strain make it very difficult to develop device models and simulation tools that accurately predict the timing behavior that will be seen in actual product silicon. As a result, it is necessary to validate timing models in silicon. Traditional ways to validate timing models use ring oscillators or perform delay testing but both approaches have significant drawbacks. Ring oscillators lack diversity in circuit structure and present layout configurations that are not typical of real products. Delay test can be expensive to apply and provides directly only path delay information not individual gate delays.

To address these limitations, we explore the potential of a new test structure-based method of timing model validation. The proposed approach combines benefits of a ring oscillator and path delay testing while addressing their limitations. Specifically, the test structure is composed of circuits that are physically synthesized and therefore product-representative, but configures the devices under test into oscillating paths so that measurement is easy and inexpensive. Path delay test ATPG is used to generate test patterns whose oscillation frequencies provide measures of path delays. Gate delays are deduced from those path delays using a matrix that codes the delay elements comprising each path in a careful way that overcomes overdetermination problems in the matrix algebra. Results show that RMS errors can be maintained under 5% for all gate types using a chosen circuit.

I. INTRODUCTION

The study of mismatches between timing models used during chip design for timing sign-off and actual silicon timing behaviors is gaining more attention with device scaling. Extracting individual gate delays from path delays will make validation of the timing model much easier. However, working backwards to get individual gate delays from path delays is a difficult problem. Here we take advantage of the fact that for timing model validation, as opposed to delay testing, we have a new degree of freedom to create our own circuits from which to deduce gate delays from path delay measurements. The important thing in those circuits is that they are product-representative, which we can achieve by creating them through the typical ASIC circuit synthesis process.

The reason silicon validation of timing models is important is that there are so many subtle effects in today's circuits (e.g., layout effects like pattern density or stress/strain causing changes to device performance) that characterization based just on circuit simulation is bound to suffer from inaccuracies. From the standpoint of accuracy, it would be desirable to characterize real paths in the circuit. However, the difficulty of applying delay tests, and the challenges associated with deconvolving gate delays from measured path delays are serious obstacles to practical application. Here we make progress toward relaxing the deconvolution obstacle by creating our own test structure circuits that are representative of product physical layout configurations, but can be tailored for information extraction, unlike real product delay paths. We overcome the measurement difficulty obstacle by configuring the circuits into an oscillator that allows for easy frequency-based measurement. The typical ring-oscillator-based approach for timing model validation shares the advantage of easy measurement of our approach, but suffers from non-representative layout/circuit structure and typically requires a separate ring for each configuration we wish to study. A method that configures real product delay paths into oscillating configurations was reported in [13] and is nicely product-representative, but requires modification of the circuit and, given its critical-path characterization objective, limits itself to real paths. Since our interest is in timing model validation (presumably generally for a library) rather than critical path characterization as in [13], we have the freedom to create our own circuit. Doing so avoids the need to make changes in the product and allows tailoring the circuit for ease of information extraction. It also allows the strategy to be implemented on a test chip or in the scribe line, which in turn also makes possible things like characterization of multiple instances of the structure to capture spatial variability.

The contributions of this paper are as follows.

1) Our test structure addresses the limitation of traditional ring oscillators. Since the test structure is a product-representative physically synthesized circuit, the timing effect coming from the physical design can be analyzed.

2) While traditional delay testing provides the information of path delays only, we provide a method to determine the gate delays from path delays, which directly provides a feedback from silicon to the timing model validation procedure. Direct measurement of individual gate delays is almost impossible because they are too short.

3) The delay paths are not limited to the critical and near-critical paths. This is important because short paths can also provide useful timing information.

4) By creating our own test structure circuits, we ease the difficulty of deducing gate delays from measured path delays.

II. RELATED WORK

There are several commonly used methods to collect timing information and to validate a timing model during the post-silicon stage. One is using on-chip monitors such as ring oscillators [9], [1], [4], [5]. Ring oscillators have many advantages. (i) Since they have small area, they can be built on the scribe lines. (ii) They can be measured easily with minimal errors. (iii) They do not require expensive testers [7]. However, ring oscillators have some limitations. (i) They exercise the gate of interest at a limited combination of input slew and output capacitance. (ii) They can oscillate too fast, resulting in incomplete logic swings. Lastly, (iii) they may not reflect the effects caused by irregular structures of layout of circuits since they have very simple and regular structures. As a result, ring oscillators may not reflect the timing behavior of regular circuity accurately.

Another method of obtaining timing information is performing delay testing. Applying delay tests using traditional delay testing, however, is difficult. Expensive high-speed external test equipment is required to apply test vectors and to observe responses at the system operating speed. Even if the test vectors can be applied at slower speeds, the response must be captured at-speed [12]. Also, this approach provides only measured path delays, not individual gate delays.

[11] proposed a correlation method using path delay testing. The authors take path delay testing to study the mismatch between design and silicon with a non-parametric learning approach. In [3], the authors suggest a statistical diagnosis method to rank potential sources of mismatch. While they suggest using a path-delay-based method, the limited number of paths that can be observed during the post-silicon stage remains as a problem. Because we create our own test structure, we can mitigate this problem.

Delay testing using oscillation alleviates the difficulty of applying delay tests and observing responses. Delay testing using oscillation has been studied [13], [14], [2]. [14] and [2] proposed oscillation delay test methods which consist of sensitizing a path under test and then incorporating it in a ring oscillator. While our work uses a test structure, they require modification of the actual product circuit to detect delay and stuck-at faults in the paths under test and to find the maximum operating speed. Wang, et al., proposed the Path-RO structure [13], which makes potential critical paths of the circuitry oscillate. Modified scan flip-flops at both ends of the critical and near-critical paths enable path oscillation. This scheme makes the measurement of critical paths easier and it is useful for speed binning. However, path delay measurements are only possible for those paths that have modified scan flip-flops on both ends. Also, modification of the actual product circuit is required. Our method has a different objective, measures the path delays of non-critical paths as well, and does not require modification of product circuits.

III. OVERVIEW

A. Background

Our method uses path delay measurement results. To exercise a path, a transition should be propagated along the path. Each test pattern consists of two vectors. The first vector, v1, initializes the path. And then, the second vector, v2, will sensitize the path. A transition will occur at the input of the path and propagate to the output of the path. If a signal is on the path, it is called an *on-input*. On the other hand, if a signal is an input of a gate on the path but it is not on the path, it is called an *off-input*. There are two types of path-delay faults based on the direction of transitions. They are a rising fault and a falling fault.

There are different types of sensitization of a path. If all the off-input of a path can be set to non-controlling values, we say the path is *statically sensitizable*. Otherwise, the path is *statically unsensitizable*. When no transition can propagate from the input to the output of a path, the path is a *false path*. If the values of the off-inputs of a path can be set to non-controlling values in both vectors, v1 and v2, the path is *single-path sensitizable*. For the purpose of timing model validation, we must be able to tell which one of the paths the measured path delay comes from. If there are multiple paths sensitized for a measured path delay, we cannot tell which one of those is the measured one. Theoretically, the path that has the longest delay will be the one. If the expected delays of the sensitized paths are significantly different from each other, it will be easy to find the measured path. However, with variability, it is hard to guarantee that the expected longest path is the true longest path. Moreover, there can be many near-critical paths. Thus, it is desirable to have only one path sensitized at a time so that we can guarantee that the measured path delay comes from the expected path. Consequently, we will apply the *single-path sensitizable* constraint.

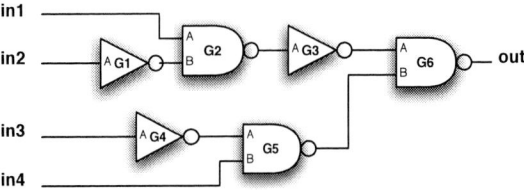

Fig. 1. Example Circuit

Path	linear sum. of segments
Path 1	G2/A/R + G3/A/F + G6/A/R
Path 2	G2/A/F + G3/A/R + G6/A/F
Path 3	G1/A/R + G2/B/F + G3/A/R + G6/A/F
Path 4	G1/A/F + G2/B/R + G3/A/F + G6/A/R
Path 5	G4/A/R + G5/A/F + G6/B/R
Path 6	G4/A/F + G5/A/R + G6/B/F
Path 7	G5/B/R + G6/B/F
Path 8	G5/B/F + G6/B/R

TABLE I

LIST OF PATHS OF FIG. 1

B. Matrix Representation

The proposed scheme utilizes path delay information to estimate gate delays. The circuit in Figure 1 has four paths from primary input, in1, in2, in3, and in4 to primary output out. Each path can have either a rising or a falling transition. Thus, there are eight different path transitions. Table I shows those eight paths. Figure 2 illustrates the matrix representation of Figure I. The path information matrix, PI, represents the path transition information, and it is a binary matrix. For example, P1 is the rising transition from in1 to out. It passes through G2, G3, and G6. In detail, if we name the upper input of 2-input gate and the input of inverter A, P1 passes through G2/A (rise), G3/A (fall), and G6/A (rise). These three gate transitions are mapped to the third, eighth, and fifteenth columns, respectively. In the PI matrix, the first row represents P1. Therefore, the third, eighth, and fifteenth columns of the first row are set to one, while other columns have zeroes. In the same manner, P2 represents the falling transition from in1 to out, etc. Assume that we can generate all eight different path transitions and measure the path delays. Then, we can generate the PI matrix along with a path delay vector, PD. Each row of PD is the path delay of each path transition, and each row in the gate delay vector, GD, is the gate delay value. The first row of GD is the delay from input A to

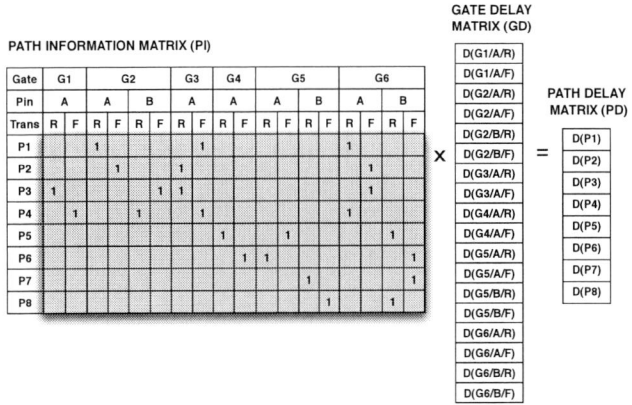

Fig. 2. Matrix Representation of Table I

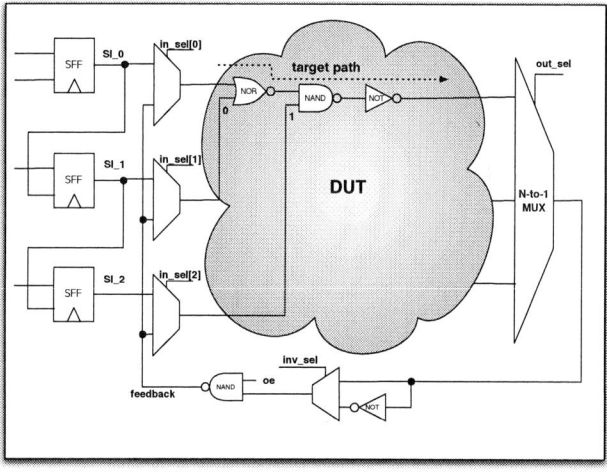

Fig. 3. Test Structure

the output of G1 gate when A transits from 0 to 1. The second row is the falling transition delay of the same gate as the first row. Here an equation, $PI \cdot GD = PD$ holds. Equally, $GD = PI^{-1} \cdot PD$ holds. Thus, if it is possible to solve the second equation, we can deduce all the gate delays of the circuit.

C. Test Structure

For easy, frequency-based measurement of path delays, a path delay ring oscillator is utilized. The scheme is to select one target path at a time, make the path oscillate, and repeat this process for all target paths. This way, the period of the target path can be easily found by observing the oscillation frequency. Figure 3 shows the test structure. It consists of the following parts. The first part is a design under test (DUT). Any synthesizable combinatorial circuit can be utilized as a DUT. We are using a synthesized circuit as DUT so that it is product-representative. The second part is a scan chain, which sets side input values of a DUT, in_sel, out_sel and inv_sel properly. M 2-to-1 multiplexers are required for the M-bit primary input of the DUT. Also, an N-to-1 multiplexer for the N-bit primary output of the DUT is necessary. The test structure has a small control circuit as well.

First of all, the side input of the on-path should be set as non-controlling values to test the target path so that one-to-one mapping of the delay of sensitized path to the measured path delay is possible. In other words, the path should be single path sensitizable. To set the side input values properly, a scan chain is used.

There are two operation modes, *scan mode* and *oscillation mode*. During scan mode, scan data is scanned in to set the side input of on-path as non-controlling values which are determined by the ATPG (Automatic Test Pattern Generation) tools. In this example, scan sequence "1-0-X" will be scanned in. During scan mode, control input values are scanned in as well, which are in_sel, inv_sel and out_sel. The scan chain for the control signal is not shown in Figure 3. in_sel selects either scan_in or feedback. in_sel is a one-hot-code that has the same number of bits as M. In this example, to select the target path of the DUT, in_sel will be set as "001". By setting in_sel[2] and in_sel[1] as 0, SI_1 and SI_2 will be connected to the primary input of the DUT. However, in_sel[0] is 1, thus, the input of target path will be connected to feedback through the multiplexer. out_sel selects one of the output that will be connected to the returning path. The target path can have either an odd or an even number of inverting

gates. In Figure 3, the target path has three inverting gates, making the output phase opposite to the input phase. Thus, inv_sel is set as 0. If it were to have an even number of inverting gates, inv_sel should be set as 1 to make the target path oscillate. Once the scan shift operation ends, the side input of NAND gate, SI_2, is set as 1, and the side input of NOR gate, SI_1, is set as 0, respectively. Since in_sel, out_sel and inv_sel are set as proper values, a loop composed of the target path and the returning path is made. The returning path is the path from the primary output of the DUT to the primary input of DUT. The returning path passes through the N-to-1 multiplexer, the 2-to-1 multiplexer below the DUT, and the 2-to-1 multiplexer connected to primary input. During scan mode, OE stays 0.

Once scan mode ends, OE is asserted to 1, and oscillation mode starts. During this mode, the feedback signal will oscillate. All the primary input values of the DUT will remain the same during this time except the target primary input, which is connected to the feedback signal. This means that the two vectors for delay testing, v1 and v2, have only one bit difference. Therefore, we do not need to worry about logic hazards (glitches) [10], [8]. If the frequency of feedback is f_{p1}, it indicates

$$f_{p1} = \frac{1}{T_{P1,rise} + T_{P1,fall}}.$$

Here, $T_{P1,rise}$ is the rising transition delay of the loop including P1, and $T_{P1,fall}$ is the falling transition delay of the same loop. By observing the duty cycle, we can also estimate the rising and the falling transition delays of the path separately. After measuring the frequency and the duty cycle of path P1, the test structure will go into scan mode again to set the values to measure path P2, and so on.

Since all the multiplexers and the control circuitries are synthesizable, the measured path will be the entire path that includes the target path, and the returning path. The interconnect delay of a path is a small fraction of the overall path delay, and it can be treated as constant since all the paths share one returning interconnect line. For simplicity, we take advantage of the fact that for our small-area structure, interconnect delay is a small fraction of the overall delay and ignore it. Each gate transition will have rising and falling transitions together, because the period is the summation of both the rising path transition and the falling path transition.

The test structure eliminates the need for expensive high-speed test equipment. The speed of scan shift does not matter here. Also, it does not require at-speed capture of responses. It can also be implemented with process monitors to account for PVT variations. Multiple replicated test structures can be built on a single test chip to study intra-die variation.

D. Collapsed Matrix

To get reliable gate delay values, it is important to obtain enough number of linearly independent rows in the PI matrix. Otherwise, the matrix will be under-determined. In this case, there will be no unique solution. The maximal number of linearly independent columns of a matrix is the *column rank* of the matrix. Likewise, the maximal number of linearly independent rows of a matrix is the *row rank* of the matrix. Since the column rank and the row rank of a matrix are always equal, they are called the *rank* of the matrix. The issue is that it is hard to get high row ranks with the PI matrix as it is. The number of columns in the PI matrix is approximately the same as the number of gate delays in the test structure. The row rank will be the same as at most the total number of paths of the circuit, but in practice will be lower as explained below. The

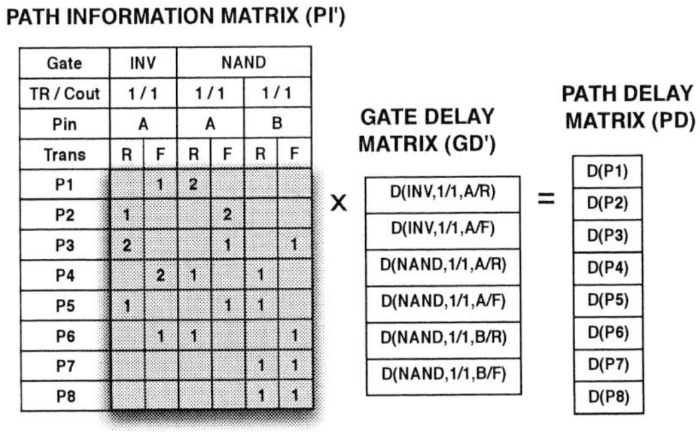

PATH INFORMATION MATRIX (PI')

Gate	INV		NAND			
TR / Cout	1/1		1/1		1/1	
Pin	A		A		B	
Trans	R	F	R	F	R	F
P1		1	2			
P2	1			2		
P3	2			1		1
P4		2	1			1
P5	1				1	1
P6		1	1			1
P7					1	1
P8					1	1

GATE DELAY MATRIX (GD')

D(INV,1/1,A/R)
D(INV,1/1,A/F)
D(NAND,1/1,A/R)
D(NAND,1/1,A/F)
D(NAND,1/1,B/R)
D(NAND,1/1,B/F)

PATH DELAY MATRIX (PD)

D(P1)
D(P2)
D(P3)
D(P4)
D(P5)
D(P6)
D(P7)
D(P8)

Fig. 4.   Collapsed Matrix

number of gate delays of an inverter is 2 including rising and falling delays, whereas the number of gate delays of a 2-input NOR or a 2-input NAND is 4 including the combination of rising/falling and from the input pin A/B to the output pin. The number of single-path sensitizable paths of the DUT is only a fraction of the total number of paths. Furthermore, among the single-path sensitizable paths, many are linearly dependent to each other. If one path is a span of existing paths, the rank of matrix PI will not increase when the path is added to the matrix.

For the reasons explained above, it is hard to obtain high row ranks of PI to solve $GD = PI^{-1} \cdot PD$ and to obtain reliable gate delay values as it is. Thus, it will be necessary to transform the matrix so that it is possible to solve the GD matrix.

One way of doing such transformation is to collapse the columns of matrix PI. Instead of unique gate delays, the columns can be grouped based on certain characteristics. Traditionally, the gate delay is a function of the input transition time and the output capacitance. Thus, for the same type of gates, it is reasonable to categorize gate delays based on those two. Here, the same type of gates means the same standard cells. Even for the same gate, the delays from input A to output and from input B to output will be different. Also, the delays of the rising transition and the falling transition will not be the same. Therefore, input pins and the direction of transition should be considered when categorizing gate delays as well.

In sum, we are considering five factors that affect gate delays: type of gates, input net transition times, output net capacitances, input pins, and the directions of transition. In other words, if there are multiple gates that are mapped to the same standard cells and have the same direction of transition, similar input transition times and similar output net capacitance, we can assume those gates will have very similar gate delay values. Therefore, they can be grouped together. Among those five factors, input net transition times can be obtained using timing models. The other four factors do not rely on timing models. Once the physical design is set, they can be easily found. However, input net transition times can only be estimated with timing models. Thus, the proposed method assumes that the inaccuracy of timing models is within reason. Also, the entire procedure of timing model validation will be iterative except for the measurement step. In other words, the timing model validation flow will be repeated for the calibration of the timing model.

Figure 4 shows a collapsed matrix representation of Figure 1, assuming that the transition time of every input net is 1 and the output capacitance of every gate is 1 for simplicity. There are less columns in the path information matrix PI' compared to matrix PI in Figure 2. It has reduced from 18 to 5. Now it is more solvable. PI' is defined as the collapsed matrix of PI.

IV. IMPLEMENTATION

Figure 5 shows the implementation flow. First of all, the design is written in HDL. To obtain the gate-level design, the design is synthesized using Synopsys DFT Compiler using three standard cells - inverters, 2-input NAND, and 2-input NOR gates. While we limit the types of gates to those three commonly used ones here, the same procedures can be applied for other standard cells as well. However, it is desirable not to use too many different standard cells for a design because it will lower the chance of solving matrices with higher accuracies.

To select paths to test, an STA (Static Timing Analysis) tool, Synopysys PrimeTime, is used. For each input and output combination of the DUT, the STA tool extracts paths. The information of extracted paths is read in by an ATPG tool, Synopsys TetraMAX. In this step, a path delay fault testing is used to find single path sensitizable paths among all the extracted paths and gives test patterns that exercise those single path sensitizable paths.

The generated test patterns are loaded into the test structure through scan chains using automatic test equipment (ATE) during scan mode. Then, the test structure goes into oscillation mode. During this mode, an output pin of the test structure starts oscillating at a specific frequency and duty cycle. External equipment can measure these frequency and duty cycle, and the target path delay can be deduced from these two values. This procedure is repeated for all the target paths.

After measuring path delays, the path information matrix, PI', and the path delay vector, PD, can be built. As explained previously, it is hard to solve the path delay information matrix, PI, as is. Thus, collapsing the matrix is necessary. If we have the full rank of the final path information matrix, PI', then we can assume that we have enough extracted paths. In case we do not have the full rank and there are more paths to extract, we can go back to the path extraction step. The STA will extract more paths and the same procedures will be performed until either we have the full rank or there are no more paths to extract. If we do not have the full rank and there are no more paths to extract, the path information matrix ends up either underdetermined or overdetermined. We reject underdetermined matrices due to the reasons explained previously. Overdetermined matrices can still be solved. An overdetermined PI' matrix has more rows than columns, and the problem can be viewed as an overdetermined least square problem.

Once the matrices are formulated, they are solved using MATLAB. We adapt the Moore-Penrose pseudo-inverse method to approximately solve overdetermined least square problems.

One thing to note is that we should consider how to handle input net transition time and output net capacitance. To formulate a matrix, we need to bin them. There are trade-offs between accuracy and the rank of the path delay information matrix depending on how we bin the input net transition times and the output net capacitance, as further discussed below.

V. EXPERIMENTS

A. Experiment Setup

We have complete freedom on the selection of the DUT. In this work, we used some of ISCAS-85 benchmark circuits [6] and some other circuits such as adders and decoders to show how well the method works for different circuits. As mentioned, any combinatorial circuit can be a candidate DUT, although the performance may vary. All benchmarks were mapped to IBM 45nm technology library. Experiments were conducted following the steps of implementation flow in Figure 5 except that path delay

978-1-61284-657-6/11 $26.00 © 2011 IEEE

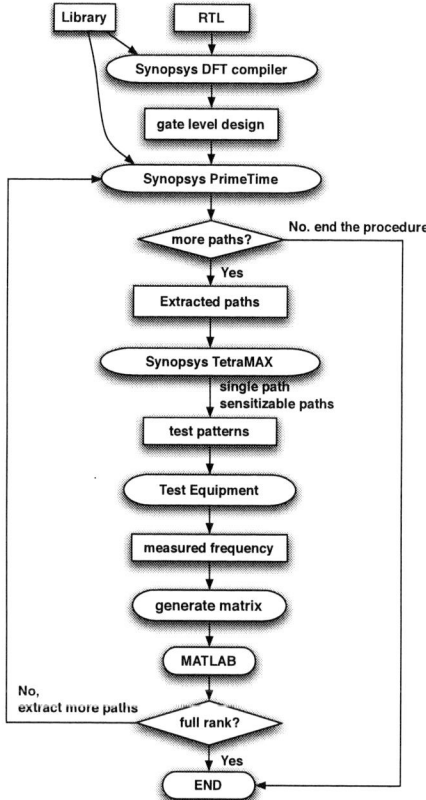

Fig. 5. Implementation Flow

circuit	STA	ATPG	rank	gate delays
c17	18	18	16	26
c432	51562	3158	303	669
c499	204656	38688	262	771
c880	25758	15891	317	850
c1355	211312	40200	178	602
c1908	162836	7575	450	1280
c2670	33380	24467	663	1614
c3540	218406	7622	583	1916
c5315	255164	95196	514	1631
c6288	190054	1987	526	1618
c7552	350918	80803	940	2823
FA	34	29	21	43
RCA4	274	200	81	175
RCA16	3394	2151	322	704
BIN_GRAY8	58	58	58	102
GRAY_BIN8	1498	1498	58	114
BCD_BIN16	87086	1071	132	336
BIN_BCD16	68764	488	202	594

TABLE II

NUMBER OF PATHS FOR EACH STEP (STA, ATPG), RANK, AND
NUMBER OF GATE DELAYS OF *PI*

circuit	rank	g. delays	circuit	rank	g. delays
c17	9	14	FA	11	22
c432	50	51	RCA4	22	27
c499	**77**	**77**	RCA16	22	27
c880	**88**	**88**	BIN_GRAY8	9	14
c1355	78	79	GRAY_BIN8	22	37
c1908	93	95	**BCD_BIN16**	**86**	**86**
c2670	**87**	**87**	BIN_BCD16	68	73
c3540	161	163	–	–	–
c5315	**52**	**52**	–	–	–
c6288	**51**	**51**	–	–	–
c7552	25	26	–	–	–

TABLE III

RANK, AND NUMBER OF GATE DELAYS OF *PI'*

information obtained from PrimeTime (STA) was used instead of measured frequency. PrimeTime gives sensitized path information, path delays of the sensitized paths, and the gate delays on the path based on the timing library. Path delay values from PrimeTime were used instead of measurement since we do not have the real path delay measurement. The gate delays extracted from PrimeTime were used to represent the timing library.

B. Experimental Results

Table II shows the number of paths for each step of Figure 5 and the number of gate delays in the DUT. It shows why collapsing matrices is necessary. The second column specifies the number of extracted paths using a STA tool. The third column represents the number of single path sensitizable paths, which is the number of paths that can be monitored. The fourth column shows the rank, which is the same as the number of linearly independent rows of matrix, *PI*. The fifth is the number of gate delays in the DUT. They are the same as the number of columns of the path information matrix *PI*. For a matrix to have a full rank, the rank must be equal to the number of gate delays.

We can see that none of benchmark circuits gives matrix *PI* full ranks. This means we cannot obtain accurate gate delay values using matrix *PI*. Since it is hard to obtain sufficiently accurate gate delay values with *PI*, we used a collapsed matrix *PI'*. Table III shows the rank and the number of gate delays of *PI'* matrix. We can clearly see that many of the benchmark circuits provide full ranks with *PI'* matrix (such circuits are displayed in bold). The ranks of matrices depend on the circuit structure and the binning. Specifically, we found that for the benchmarks that we used, the dependency on the given range of input transition time was much smaller than

the dependency on the given range of output net capacitance. Thus, we binned the transition time by 2, while binning the output net capacitance by 20. We chose a bin strategy that provided an acceptable trade-off between accuracy and solvability. Among the benchmark circuits that give full ranks, we chose "BCD_BIN16" for the rest of the experiments.

After solving the collapsed matrices, we obtained the gate delay values using least square methods. Figure 6(a) shows the rise delay of inverter-type gate which is obtained from our approach together with the golden timing model numbers we wish to match. In this case, we get those golden numbers as reported by the STA tool. Gate delay is a function of the input transition time and the output capacitance. Therefore, we presented 3-dimensional plots here. The plot lying flat at the bottom shows the difference between the two. We can see that the results from our approach and the gate delays from the STA correlate well. The delay plots of other types of gates will not be presented here since they show similar trends to that of rise delay of the inverter. RMS errors between the timing model and the matrix results were under 5% for any gate type for the circuit. The RMS errors for the other types of gates are shown in Table IV.

To replicate the mismatches between timing models and silicon measurements, we conducted a separate experiment where we assumed the actual rise delay of an inverter type gate is 10% smaller than the timing model. We subtracted 10% of the rising inverter delay from path delays whenever the paths passed through

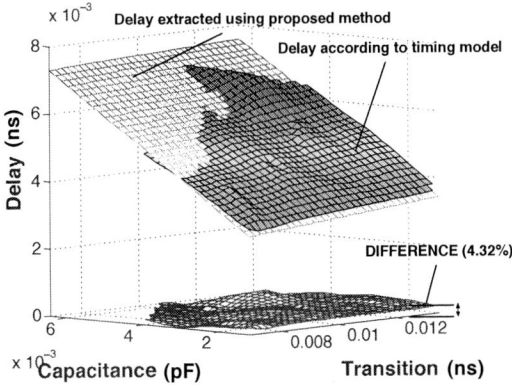

(a) Gate Delay - Rise Transition of Inverter

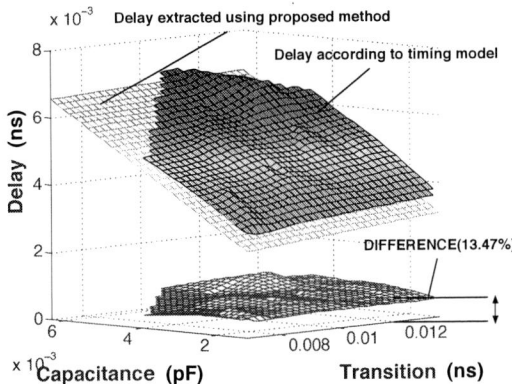

(b) Gate Delay - Rise Transition of Inverter with 10% mismatch

Fig. 6.   Experiment Results

TYPE	pin	direction	RMS error
INV	A	rise	4.32%
INV	A	fall	4.01%
NAND	A	rise	3.71%
NAND	A	fall	2.71%
NAND	B	rise	3.47%
NAND	B	fall	2.44%
NOR	A	rise	2.16%
NOR	A	fall	4.83%
NOR	B	rise	2.77%
NOR	B	fall	3.28%

TABLE IV

RMS ERROR FOR EACH GATE TYPE

an inverter and the transition of the inverter was rising. Figure 6(b) shows the delay of the inverter when we assumed the 10% of mismatches. We can see the difference between matrix results and the library became larger than when there was no mismatch. RMS error of rise delay of inverter became 13.47% while RMS error of other gate types remained under 5%. These results correctly indicate that there are mismatches between the timing model and the silicon results and the source of timing mismatches is mis-modeling of the inverter. This case study shows what type of analyses may be possible with our method.

These results show we can estimate gate delays based on measured path delays (in this case, imitated by PrimeTime) with small error and identify intentionally inserted mismatches between timing model and "actual" silicon behavior. Different structures of DUT

give different results. For improvements in error, future work will concentrate on a better bin strategy and a DUT design providing greater diversity in order to get a large number of data points.

VI. CONCLUSIONS

In this paper, we proposed a new method of timing model validation using silicon results. Our method uses a test structure composed of circuits that are physically synthesized and therefore product-representative. Our method also allows for easy, frequency-based measurement by configuring the circuits into an oscillator. Thus, it combines the benefits of ring oscillators and path delay testing while addressing their limitations. Moreover, our method provides a way to extract gate delay values using matrices with the obtained delay testing results, which enables direct validation and calibration of timing models. With the freedom to create our own circuit, we show that RMS errors can be maintained under 5% for all gate types for a chosen circuit. As different designs of DUT give different quality of results, future work will include the research on identifying properties that make a circuit more suitable as a DUT and developing a circuit that has those properties.

REFERENCES

[1] K. Agarwal and S. Nassif. Characterizing process variation in nanometer CMOS. In *Proceedings of the 44th annual Design Automation Conference*, page 399. ACM, 2007.

[2] K. Arabi, H. Ihs, C. Dufaza, and B. Kaminska. Digital oscillation-test method for delay and stuck-at fault testing of digital circuits. In *Proceedings of the 1998 IEEE International Test Conference*, page 100. IEEE Computer Society, 1998.

[3] P. Bastani, N. Callegari, L. Wang, and M. Abadir. Statistical diagnosis of unmodeled systematic timing effects. In *Proceedings of the 45th annual Design Automation Conference*, pages 355–360. ACM, 2008.

[4] M. Bhushan, A. Gattiker, M. Ketchen, and K. Das. Ring oscillators for CMOS process tuning and variability control. *IEEE Transactions on Semiconductor Manufacturing*, 19(1):10–18, 2006.

[5] D. Boning, S. Nassif, A. Gattiker, and F. Lui. Test structures for delay variability. In *Proceedings of the 85th ACM/IEEE International Workshop on Timing Issues in the Specification and Synthesis of Digital Systems*, page 109. ACM, 2002.

[6] F. Brglez and H. Fujiwara. A neutral netlist of 10 combinational benchmark circuits and a target translator in Fortran. In *Proc. of International Symposium on Circuits and Systems*, volume 663, page 698, 1985.

[7] M. Ketchen, M. Bhushan, and D. Pearson. High speed test structures for in-line process monitoring and model calibration. In *IEEE Proc. ICMTS*, pages 33–38, 2005.

[8] S. Nowick and D. Dill. Exact two-level minimization of hazard-free logic with multiple-input changes. *IEEE Transactions on Computer-Aided Design of Integrated Circuits and Systems*, 14(8):986–997, 1995.

[9] H. Onodera and H. Terada. Characterization of WID Delay Variability Using RO-array Test Structures. *ASIC, 2009. ASICON '09. IEEE 8th International Conference on*, pages 658 – 661, Sep 2009.

[10] A. Virazel, R. David, P. Girard, C. Landrault, and S. Pravossoudovitch. Delay fault testing: Choosing between random SIC and random MIC test sequences. *Journal of Electronic Testing*, 17(3):233–241, 2001.

[11] L. Wang, P. Bastani, and M. Abadir. Design-Silicon Timing Correlation A Data Mining Perspective. In *Design Automation Conference, 2007. DAC'07. 44th ACM/IEEE*, pages 384–389, 2007.

[12] W. Wang and S. Gupta. Weighted random robust path delay testing of synthesized multilevelcircuits. In *12th IEEE VLSI Test Symposium, 1994. Proceedings.*, pages 291–297, 1994.

[13] X. Wang, M. Tehranipoor, and R. Datta. Path-RO: a novel on-chip critical path delay measurement under process variations. In *Proceedings of the 2008 IEEE/ACM International Conference on Computer-Aided Design*, pages 640–646. IEEE Press, 2008.

[14] W. Wu, C. Lee, M. Wu, J. Chen, and M. Abadir. Oscillation ring delay test for high performance microprocessors. *Journal of Electronic Testing*, 16(1):147–155, 2000.

978-1-61284-657-6/11 $26.00 © 2011 IEEE

Special Session: Multifaceted Approaches for Field Reliability

Session Organizer: Yasuo Sato, Kyushu Institute of Technology
Session Chair: to be decided

Field Reliability is a so complex problem that a single approach is not enough. Three distinguished researchers address their various state-of-the-art approaches, which include debug, soft-error and field test. They will also raise an issue of testing.

"Programmability based approach to post-silicon debug and rectification"

Masahiro Fujita, Tokyo University, Japan

Despite the intensive efforts of pre-silicon verification, logical bugs can escape and appear only after chips have been fabricated. In order to realize post-silicon patches, some sorts of programmability is essential. In this paper, we discuss about the two techniques we have been working on. The first approach is to synthesize circuits considering changes of specification as well as faults of some hardware components. Instead of synthesizing circuits from a single design description, variants of design descriptions are automatically generated and circuits which can perform most of them by re-programming micro-controllers are synthesized. The second approach is to introduce look-up-table (LUT) in the hard wired logic in such a way that functionality of the circuit can be adjusted in the field if necessary. We discuss about the usefulness of the two approaches with some experimental results.

"An EDA tool chain for soft-error tolerant VLSI design"

Yusuke Matsunaga, Kyushu University, Japan

This talk will introduce a research project on EDA technologies concerning soft-error tolerant VLSI design. Under 60nm technology or beyond, the influence of soft-errors occurred in logic circuits is increase drastically, so that support of EDA tool will be inevitable.

The main objective of the project is to establish a complete tool-chain, not only to develop each technology on particular design level. The bottom level of the tool-chain is to characterize soft-error related parameters of cell libraries. The second step is to calculate the propagation probability of a soft-error induced pulse at logic gate level. The third level is to calculate the soft-error rate (SER) of RT level circuit, which is composed of control (FSM) part and data-path part. As well as the brief summary of existing techniques, novel techniques under developing will be discussed.

"Accurate and Efficient SoC Field Test for Failure Prediction"

Michiko Inoue[*], Nara Institute of Science and Technology, Japan
Seiji Kajihara[*], Kyushu Institute of Technology, Japan
*Japan Science and Technology Agency CREST, Japan

With miniaturization of CMOS devices, transistor aging which results in performance degradation becomes a crucial issue in highly reliable systems. Reliability-aware design could resolve this problem where some timing margin is added according to aging estimation. However, the worst case estimation to ensure the reliability would add excessive margin and incur appreciable performance degradation. On-line testing to predict or detect failures is a promising approach to avoid a sudden system down and guarantee high system reliability. We propose a novel failure prediction system that predicts failures based on accurate delay measurements for system-on-chips (SoC)s in the field.

The proposed system realizes accurate and efficient failure prediction with combination of a delay measurement at each core and a SoC test controller that schedules and manages core delay measurements and predicts failures in cores. Major challenges of the proposed system are (1) *accuracy* - delay measurement should be accurate under temperature and voltage variations, (2) *coverage* - delay measurement should cover a large number circuit elements, (3) *limited test application-time* - there are several test chances during a long lifetime, but a period of each test chance is quite short, and (4) *priority* - cores or parts of cores to be tested should be selected according to their degradation degrees. Tackling these issues, the proposed system realizes truly accurate and efficient failure prediction for SoCs.

Advanced Methods for Leveraging New Test Standards

Organizer: Mike Laisne, Qualcomm

Abstract: This session explores how newly introduced standards are supporting innovative industry practices. First, a new update for 1149.1, including new instructions, extensions, and a new procedural language, is reviewed. Next, techniques for implementing concurrent test using P1687 are described. Finally, a method is examined, using P1581, for connectivity testing of ICs with no boundary scan.

"Innovative practices with the new IEEE P1149.1-2011 JTAG update"
By: CJ Clark* (Intellitech Corp.)

"Test Concurrency using P1687"
By: Songlin Zuo* (Qualcomm) and Michael Laisne (Qualcomm)

"IEEE P1581 - Simplifying Connectivity Tests for Complex Memories and other Non-Boundary Scan Devices"
By: Heiko Ehrenberg* (Goepel)

Session 4

Special Session 4A: New Topics
Parametric Yield and Reliability of 3D Integrated Circuits: New Challenges and Solutions

Siddharth Garg, University of Waterloo (s6garg@ecemail.uwaterloo.ca)

Diana Marculescu, Carnegie Mellon University (dianam@ece.cmu.edu)

Abstract

3D integration is a promising new technology that offers numerous potential benefits including reduced wire length, high tier-to-tier bandwidth and low latency, and the possibility for heterogeneous integration of disparate technologies. As a result, 3D integrated circuits (IC) are being aggressively investigated as a potential replacement for conventional planar ICs in both academia and industry.

While the benefits of 3D integration are numerous, as outlined above, a number of challenges that can potentially hinder the adoption of this technology remain to be addressed. Higher on-die temperatures in 3D ICs have already been pin-pointed as a major issue and numerous solutions have been proposed to address this challenge. Recently, we have demonstrated that the impact of process variations on the performance (in terms of clock frequency or cycle time) of 3D ICs is another critical challenge – in particular, a 3D implementation has *lower parametric yield* for any given timing specification vis-à-vis an equivalent planar 2D implementation[1]. Intuitively, this is because the clock frequency of a 3D IC is constrained by the impact of process variations on critical paths within each tier (within-die variations) and across tiers (die-to-die variations).

A proposed solution to the process variation problem involves speed-binning manufactured die *before* 3D assembly and packaging[2][3]. The speed-bin information for bare, unpackaged die is used to determine an optimal 3D assembly strategy that maximizes parametric yield. As a simple example, one 3D assembly strategy could be to bind "slow" die with "slow" die, and "fast" die with "fast" die. Economically obtaining accurate speed-bin information for bare, unpackaged die represents a new *test challenge* for which innovative solutions are required. We will discuss some promising solution to this challenge based on *inferring* the speed-bin information from other, more readily available test data. These might include leakage current measurements[3], or detailed test measurements from a small number of die on a wafer[4].

References:

[1] Garg, S. and Marculescu, D. "3D-GCP: An analytical model for the impact of process variations on the critical path delay distribution of 3D ICs," in proceedings of IEEE Symposium on Quality of Electronic Design (ISQED), 2009.

[2] Ferri, C., Reda, S. and Bahar, R.I. "Strategies for improving the parametric yield and profits of 3D ICs," in proceedings of the IEEE/ACM International Conference on Computer-Aided Design (ICCAD), 2007.

[3] Garg, S. and Marculescu, D. "System-level process variability analysis and mitigation for 3D MPSoCs," in proceedings of the Design Automation and Test in Europe (DATE) Conference and Exhibition, 2009.

Security-Aware SoC Test Access Mechanisms

Kurt Rosenfeld
Google, Inc.
New York, NY
kuro@google.com

Ramesh Karri
Polytechnic Institute of NYU
Brooklyn, NY
rkarri@duke.poly.edu

Abstract—Test access mechanisms are critical components in digital systems. They affect not only production and operational economics, but also system security. We propose a security enhancement for system-on-chip (SoC) test access that addresses the threat posed by untrustworthy cores. The scheme maintains the economy of shared wiring (bus or daisy-chain) while achieving most of the security benefits of star-topology test access wiring. Using the proposed scheme, the tester is able to establish distinct cryptographic session keys with each of the cores, significantly reducing the exposure in cases where one or more of the cores contains malicious or otherwise untrustworthy logic. The proposed scheme is out of the functional path and does not affect functional timing or power consumption.

I. INTRODUCTION

SoC design cycles are getting shorter and designs are getting more complex. Pressure for more productivity per designer per day has led to reuse of design modules and, in many cases, obtaining modules from external sources of intellectual property (IP). It is impractical, and sometimes impossible, for SoC designers to manually assess the security of each of the cores they use. Modular design methodologies hide the internals of the modules, which boosts productivity, but unfortunately it shifts designers from *knowing* the logic of the chip to merely *hoping* that each piece actually operates according to its interface specifications. Continued progress in VLSI requires that we not let complexity and short design cycles undermine our ability to produce chips that are trustworthy. In the long run, we need to work toward designing in such a way that the security failure of one module does not result in the security failure of the entire system. One step in that direction is the topic of this paper, to reduce the risk of cascading security failure in the test subsystems of chips we design.

Test access mechanisms (TAMs) are present in every complex chip and are used for a large and growing variety of purposes. Their original purpose was to enable efficient structural testing of digital systems instead of the notoriously inefficient process of black-box functional testing. As system design has evolved, the role of TAMs has been extended to include invoking BIST, programming nonvolatile memory, debugging embedded microprocessors, initializing the configuration of FPGAs, initializing volatile run-time configuration registers, and enabling and disabling system components including the TAM itself. A security-aware TAM improves upon conventional TAM concepts by protecting the data that traverses it, thus reducing the risks arising from untrustworthy cores.

A. Assumptions

The threat model addressed in this paper is that one or more untrustworthy cores intercept or modify the test data that passes between the tester and a core. We make the following assumptions:

1) The SoC contains many cores, some of which are untrustworthy.
2) The SoC contains trustworthy inter-core functional wiring.
3) The SoC contains trustworthy test access wiring.
4) Cores are connected to the tester by shared wiring. This can be either a daisy-chain scheme or a bus scheme.
5) Some or all cores, including their test wrappers, are opaque.
6) The SoC design is known to the attacker as much as it is known to the designer. If a crypto key is embedded in the design, the key is known to the attacker.

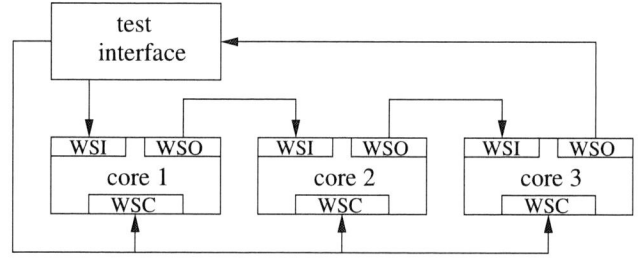

Fig. 1. Data sent from the test controller to core 2 passes through core 1, giving core 1 an opportunity to intercept it. Likewise, data passing from core 2 back to the tester passes through core 3, giving core 3 an opportunity to intercept it.

Some examples of TAM-related threats in a daisy-chained TAM architecture, as shown in Figure 1, include the following:

1) Core 1 sniffs data passing from tester to core 2. Core 1 leaks the data. Core 2 can be an FPGA or microprocessor, and the data can be its bitfile or executable program. This attack results in leakage of intellectual property which can result in monetary losses or expose the leaked data to reverse engineering and additional security exposure.
2) Core 1 modifies data passing from the tester to core 2. Core 2 can be a crypto core and the data can be a key. This attack can create a back door into the system.

B. Constraints

One way to reduce TAM-related risk is to use a star topology for test data. The star topology avoids the threats listed above by avoiding the placement of two cores on the same test wiring where they might interfere with each other. This topology is seldom used, since it results in high wiring cost. As the number of cores in SoC designs increases, star-topology TAM wiring becomes less and less practical. Instead, we reject that technique in favor of cryptography, which allows us simultaneously have the low cost of the bus topology and the good security of the star topology.

C. Core Test Wrappers

To facilitate design automation, modular design, and efficient reuse of cores, the cores are often delivered to SoC integrators in a *wrapped* form, which means that the complexity of their internal test structures is hidden behind some wrapper logic, which exposes a simplified, standardized interface to the outside. IEEE 1500 [1] defines this interface. Although wrapping can improve the productivity of SoC integrators, it requires the SoC integrator to trust the wrapper implementations provided by the core vendors. Wrapped versus unwrapped cores is a complex decision with security ramifications.

II. PRIOR WORK

The threat of malicious inclusions in chip designs has been discussed at a technical level in the security and VLSI communities, and at a policy level in defense communities. DARPA has led US DoD research in the area, funding the TRUST program [2], which aims to ensure that critical government operations are able to source trustworthy chips. King [3] shows the construction of malicious modifications to a CPU design that give the attacker the flexibility to choose the details of the attack at run time, in the field, after deployment. Wang [4] presents a taxonomy of malicious hardware. Kim [5] examines the risks malicious modules in an SoC abusing their bus-master capability. They provide a security-enhanced bus arbiter that traps the bus transactions of rogue modules and inhibits them from further action by cutting power to the offending module. The risks of having malicious chips on a JTAG [6] chain were studied in [7], and the authors developed countermeasures.

The SoC TAM security problem differs significantly from the JTAG security problem. Primarily, the difference is that all cores of an SoC are fabricated together. The JTAG security solutions proposed in [7] assume that each chip exists first in isolation, gets packaged and tested, and then shipped to the customer. While in isolation, before being shipped, keys can be set up for use by the tester for secure communication over the untrustworthy test bus in the field. Since this isolated stage does not exist for SoC cores, the TAM security solutions developed for JTAG do not apply to SoCs.

III. PROPOSED APPROACH

We propose TAM enhancements that leverage the SoC designer's control over the inter-core wiring. The result is

that untrustworthy cores are prevented from sniffing communication on the test bus. As with most communication security schemes, key exchange is a pivotal issue. Many cores are to a single test bus, and to single out a target core for communication, the TAM must provide the tester with a mechanism for securely distinguishing the target core from the rest of the cores. We propose that the SoC integrator construct a scan chain outside of any of the wrapped cores, and connect each core to the output of one of the cells in the scan chain. The foundation of the security of our system are the assumptions that:

1) A core, however malicious and devious it may be, cannot affect the intercore wiring of the SoC.
2) A core cannot control where in the intercore wiring its terminals are connected.

Thus, the tester securely distinguishes each core by which scan cell it connects to, as shown in Figure 2.

As stated in Section I-B, TAMs are under significant pressure to minimize their cost. For this reason, SoCs use shared wiring to connect the tester with the cores. During testing, the target core is addressed using any of various schemes, while the other cores are expected to be passive. We describe three scenarios where untrusted cores are placed on the shared TAM wiring. For each scenario, we discuss how the economy of shared wiring can be retained while protecting sensitive test data from sniffing as it passes by or through malicious cores. The first scenario is where a wrapped cores is obtained already containing the security-aware TAM enhancements described in this paper. The second scenario is where a wrapped cores is obtained without any TAM security enhancements, and has sensitive test data. The third scenario is where a core is obtained that is untrusted but no sensitive test data will be exchanged with it.

A. Security-enhanced Test Wrapper

We enhance the security of the standard core test wrapper by using cryptography to protect the data. Standard crypto primitives are used and the details of their design and implementation are outside the scope of this paper. Here, we focus on the practical aspects of key exchange and issues specifically relevant to the SoC test problem.

A common technique for session key establishment is to use the Diffie-Hellman [8] protocol. In Diffie-Hellman, each party generates a random number, sends a message, and does some arithmetic, and the result is that the two parties agree on a secret that was never explicitly transmitted over the wire. Although Diffie-Hellman is a powerful building block, for our application, we can obtain better security at lower cost by taking advantage of practical constraints on what a malicious core can do. Keys can be generated by the test controller or external tester, and distributed to each core at test initialization time as shown in Figure 2. Cheap and secure, this is our preferred key distribution scheme.

The SoC designer can choose between on-chip key generation and off-chip key generation. If done on-chip, the external tester (i.e., ATE) does not have access to keys, which

978-1-61284-657-6/11 $26.00 © 2011 IEEE

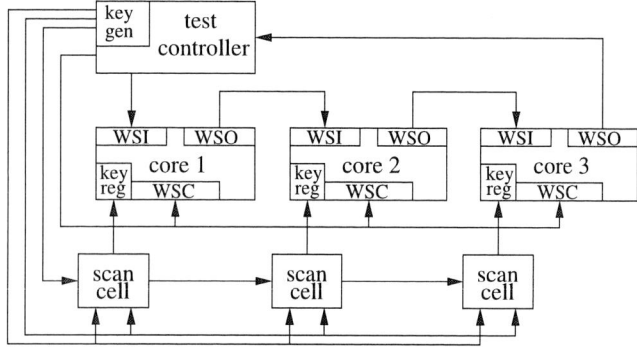

Fig. 2. A chain of scan cells is used for distributing keys to each of the cores. The scan cells are configured not to expose key bits at their outputs while they are being shifted.

improves security in some situations. However, for reasons of cost, as discussed in Section IV, in most circumstances we expect SoC designers to prefer off-chip key generation and cryptography. Key generation entails selecting a key for each core that has a security-enhanced wrapper. The most important characteristic of the keys is that no core should be able to learn the key of another core. As stated in Section I-A, we assume that the design is known to the attacker. This precludes the possibility of hard-coding the keys. Instead, for on-chip key generation, a hardware random number generator is used. Holleman [9] reported a hardware number generator requiring 0.031 mm^2 of die area in 0.35 μm four-metal, double-poly CMOS. If implemented in a current fabrication process with smaller feature sizes, the die area for the circuit would be correspondingly smaller.

When key bits are transmitted to the cores during key setup, they are also stored by the test controller or external tester. The storage of key bits or cipher state is necessary because it is the basis for encrypted communication. However, the SoC designer has a choice of whether to maintain crypto sessions when accessing other cores. For example, if the test schedule involves communicating with core 1 and then with core 2, and then with core 1 again, the question is whether the test controller should maintain the crypto session (cipher state) associated with core 1 while accessing core 2. If it does, then it can resume communications with core 1 without the delay of reinitializing the cipher state associated with core 1. On the other hand, if the on-chip test controller is performing the crypto, then registers must be added to the test controller to store the cipher state, which increases the die area overhead. In the case of off-chip key generation and crypto, storing session state is not a problem. Offloading the crypto to the ATE is consistent with our goals and with the threat model stated in Section I-A.

The scan cells in the key setup scan chain, though very simple, provide two properties that are very important for security. First, as shown in Figure 3, they accept an output inhibit signal, O_INH*, which forces the output to zero when it is pulled low. This has the effect of blocking cores from

observing other cores' key bits while they are being shifted in. Second, the logic gates are, for all intents and purposes, unilateral. There is no way for a core to actively force a value onto the flip-flop to affect the key bits that are received by other cores.

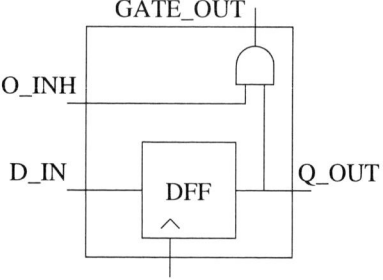

Fig. 3. The key setup scan chain conveys data from the test controller to the core wrapper key registers without allowing it to be sniffed or modified by other cores. Other than the basic distributed shift register functionality, the only extra functionality we require of our scan cell is an output inhibit input (O_INH) to ensure that the key is not leaked during shifting. After the tester has the key bits shifted to their intended location, the tester deasserts the output inhibit signal so that the cores receive their key data.

Communication requirements of cores vary over a wide range, and optimal test access design involves allocation of test buses and scheduling of tests [10]. Cores using BIST exclusively may be interfaced using only a serial test interface. Cores exposing extensive internal scan chains to the tester typically make use of a parallel test bus to increase test speed. In either case standard low-cost symmetric cryptographic modules are placed between the test interface and the core, as shown in Figure 4. The costs of these modules are discussed in Section IV.

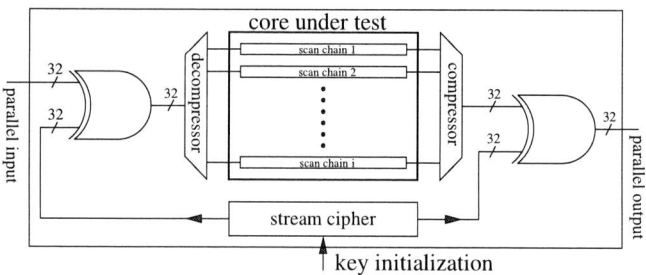

Fig. 4. In a typical security-enhanced wrapped core, a word of compressed test data arrives via the parallel data input, is decrypted instantaneously, decompressed and applied to the inputs of the core's scan chains. The outputs are compressed, encrypted, and sent out. Standard wrapper components are not shown, such as the parallel bypass.

If an untrusted core was obtained in an unwrapped state, the test wrapper described in this subsection should be added by the SoC integrator. Purely from a security standpoint, it is good to obtain cores in an unwrapped state. However, it adds significant work for the SoC integrator, losing the productivity benefits of test integration standards like IEEE 1500. Obtaining cores prewrapped with security-aware wrappers is probably the option most SoC integrators would prefer.

An untrustworthy core can be provided to the SoC integrator prewrapped with a security-enhanced wrapper. The presence of the security-enhanced wrapper does not make the core trustworthy, even if the wrapper itself is free of malicious features. However, the untrustworthy core, even if its wrapper contains malicious features, cannot undermine the TAM security of the SoC. This is in stark contrast with the conventional daisy-chain TAM architecture, where a single untrustworthy core breaks the security of all other cores on the chain.

B. A Security Overwrapper for Prewrapped Cores

In cases where an important module is only available as a conventionally wrapped core, a security *overwrapper* can be used. The functionality provided by the overwrapper, when combined with the core's included wrapper, is equivalent to that of the of the security-enhanced wrapper discussed in Section III-A. The security overwrapper contains functional blocks for decrypting input test data and encrypting output test data.

C. Interoperability with Noncompliant Cores

The TAM security scheme described here does not require that all cores in an SoC comply. Noncompliant wrapped cores can be used as they are, provided that no security-critical data passes over their test interfaces. The presence of noncompliant cores on the test bus does not undermine the security guarantees provided to the compliant cores.

IV. Costs

The SoC TAM security enhancements presented in this paper were designed to be efficient in terms of die area, test time, and effort for the SoC integrator.

A. Die Area Cost

The security enhancements contribute to die area in three ways. To a certain extent, these costs can be traded off, and the optimal choice depends on economics.

1) Wiring Cost: The extra wiring cost of the security enhancements is the cost of three extra wires on the test bus. In a typical SoC with serial control of the core wrappers and a 32-bit path for test data, there is a minimum of 40 wires without the enhancements, and 43 with the enhancements, a 7.5% overhead.

2) Core Wrapper Area: Each core whose test data the SoC integrator decides to protect needs to have its own crypto hardware, whether provided by the core supplier or by the SoC integrator. Assuming a 32-bit parallel test data path, the wrapper must decrypt 32 bits of input test data while encrypting 32 bits of output test data. To implement this with no additional latency, a stream cipher is used. 64 bits of keystream are required for each cycle of the test clock. This is achieved by using a keystream generator that produces multiple bits per clock cycle. The main additional hardware requirements for the security-enhanced core wrapper are:

- 32 XOR gates to decrypt the input test stimulus
- 32 XOR gates to encrypt the output test response

- a stream cipher that generates 64 bits of keystream per cycle of test clock

The Trivium [11] stream cipher meets the requirements. In its 64-bit form, it is equivalent to 5504 NAND gates. Assuming the XOR gates are equivalent to 2.5 NAND gates, the XOR gates used in each core wrapper are equivalent to 160 NAND gates. The total area overhead is therefore approximately 5700 NAND gates. For an SoC with n cores with security-enhanced wrappers, the gate count cost is

$$overhead = n \times 5700 \qquad (1)$$

The percentage gate count overhead is

$$\left(\frac{\sum_1^n (g_i + 5700)}{\sum_1^n g_i} - 1 \right) \times 100\% \qquad (2)$$

where g_i is the number of NAND gate equivalents in core i. For example, for an SoC an average of 100,000 NAND equivalent gates per core, the area overhead is 5.7%.

3) Test Controller Area: The cryptography can be handled by the test station or by the on-chip test controller. If it is handled by the test controller, it must have hardware for generating random key bits, and for storing them. It also needs the encryptor and decryptor blocks. If the cryptography is handled by the test station, the test controller's complexity is essentially the same as without the security enhancements.

If the designer prefers to perform the cryptography and key generation using the on-chip test controller, then there are three cases. In the first case, the test controller only maintains information about a single core at a time. This requires a key setup whenever addressing a new core. For this, the die area overhead in the test controller associated with the security-enhanced TAM is the area of the stream cipher and XOR gates, the equivalent of 5700 NAND gates. In cases where all of the key setup is done at initialization time, the die area overhead is $5700 + 12nk$ NAND gates, where n is the number of cores and k is the key length. The third case is where cipher state is maintained by the test controller for each security-enhanced wrapped core. Essentially, this means keeping a copy of the state register of the stream cipher, which is almost the same as the test controller simply having separate instances of the stream cipher module for each security-enhanced wrapped core with which it will communicate. The die area overhead of this is approximately $5700n$ NAND gates. This minimizes test time, but is the most expensive option in terms of die area.

B. Test Time

The security enhancements do not affect the test clock speed, test duration, or test scheduling. However, the security-enhanced cores need to have their key registers initialized before testing can commence. The worst-case test time overhead is the time to program all key registers, back-to-back. For a k-bit key and n cores in the SoC, the worst case key initialization time is

$$t_{init} = kn \qquad (3)$$

The key setup clock frequency can be assumed to be the same as the test clock frequency. The percentage test time overhead is

$$\left(\frac{\sum_1^n t_i + k}{\sum_1^n t_i} - 1 \right) \times 100\% \qquad (4)$$

where t_i is the number of test clock cycles required to test core i. For example, assume an 80-bit key length and 50 cores in the SoC, the worst-case key initialization time is 4000 cycles of the key setup clock. If the average core requires more than 8000 bits of test data, the the test time overhead of the security enhancements is less than 1%.

C. Effort for SoC Integrator

Under the proposed scheme, cores fall into three categories:

1) no security-enhancements,
2) cores shipped with security-enhanced wrappers as described in Section III-A, and
3) cores that were shipped with non-security-aware wrappers and had the security overwrapper added as described in Section III-B.

The effort for the SoC integrator for category-1 cores is zero. The effort for using category-2 cores is very low. The signaling of the cores is consistent, so the per-core additional effort is just assigning the signals to connect the core to the key-setup scan chain. Category-2 is the preferred category, achieving all of the security benefits with minimal effort. Category-3 cores are slightly more effort for the SoC integrator, but still minimal work. The difference is that category-2 cores come with the crypto modules already in place whereas category-3 cores require the SoC integrator to add the crypto module between the core and the test bus terminals.

V. Conclusion and Future Work

We presented a scheme that eliminates the risk of a malicious SoC core sniffing test data. The essential contribution is a straightforward way of establishing cipher keys without any hard-coded secrets in the design. The area overhead is under 6% and the test time overhead is under 1%. For typical chips where a minority of the cores have secrecy-sensitive test data, only those cores need the security-aware wrapper, and the total area overhead can be correspondingly reduced to 3% or less. Additional area savings are available when the functional clock is more than 8 times higher in frequency than the test clock. In such cases, the Trivium stream cipher can be used in a configuration that produces 8 bits at a time instead of 64 bits at a time, and it can be clocked by the functional clock instead of by the test clock. This reduces the size of the main source of area overhead, the stream cipher, by 32%. We showed how the SoC integrator can use his control over the inter-core wiring to maintain the security of the test data. Future work using the same key setup scheme will provide not only secrecy guarantees for the test data, but integrity guarantees as well. As the number of cores in SoCs increases, and cores are obtained from an increasingly wide variety sources, limiting the damage done by a single rogue core has become an important concern with a practical solution.

References

[1] F. DaSilva, Y. Zorian, L. Whetsel, K. Arabi, and R. Kapur, "Overview of the IEEE P1500 standard," vol. 1, sep. 2003, pp. 988 – 997.

[2] D. Dean. R. Collins, "Trust, a proposed plan for trusted integrated circuits," http://www.dtic.mil/cgi-bin/GetTRDoc?AD=ADA456459.

[3] S. T. King, J. Tucek, A. Cozzie, C. Grier, W. Jiang, and Y. Zhou, "Designing and implementing malicious hardware," *USENIX Workshop on Large-Scale Exploits and Emergent Threats*, 2008.

[4] X. Wang, M. Tehranipoor, and J. Plusquellic, "Detecting malicious inclusions in secure hardware: Challenges and solutions," *IEEE International Workshop on Hardware-Oriented Security and Trust*, pp. 15 –19, jun. 2008.

[5] L.-W. Kim and J. D. Villasenor, "A system-on-chip bus architecture for thwarting integrated circuit trojan horses," *IEEE Transactions on Very Large Scale Integration (VLSI) Systems*, vol. PP, no. 99, pp. 1 –5, 2010.

[6] IEEE Std 1149.1-2001, Test Access Port and Boundary-Scan Architecture.

[7] K. Rosenfeld and R. Karri, "Attacks and defenses for jtag," *Design Test of Computers, IEEE*, vol. 27, no. 1, pp. 36 –47, jan. 2010.

[8] W. Diffie and M. Hellman, "New directions in cryptography," *IEEE Transactions on Information Theory*, vol. 22, no. 6, pp. 644 – 654, nov. 1976.

[9] J. Holleman, B. Otis, S. Bridges, A. Mitros, and C. Diorio, "A 2.92 µW hardware random number generator," *Proceedings of the 32nd European Solid-State Circuits Conference*, pp. 134 –137, sep. 2006.

[10] K. Chakrabarty, "Optimal test access architectures for system-on-a-chip," *ACM Transactions on Design Automation of Electronic Systems*, vol. 6, pp. 26–49, 2001.

[11] C. D. Canniere and B. Preneel, "Trivium specifications," *ECRYPT Stream Cipher Project*, 2006.

Design and Analysis of Ring Oscillator based Design-for-Trust technique

Jeyavijayan Rajendran, Vinayaka Jyothi
ECE Department
Polytechnic Institute of
New York University
{jrajen01,vjyoth01} @students.poly.edu

Ozgur Sinanoglu
Computer Eng. Department
New York University
Abu Dhabi
ozgursin@nyu.edu

Ramesh Karri
ECE Department
Polytechnic Institute of
New York University
rkarri@poly.edu

Abstract—Due to the increasing opportunities for malicious inclusions in hardware, Design-for-Trust (DFTr) is emerging as an important IC design methodology. In order to incorporate the DFTr techniques into the IC development cycle, they have to be practical in terms of their Trojan detection capabilities, hardware overhead, and test cost. We propose a non-invasive DFTr technique, which can detect Trojans in the presence of process variations and measurement errors. This technique can detect Trojans that are inserted in all or a subset of the ICs. It is applicable to both ASICs and FPGA implementations. Circuit paths in a design are reconfigured into ring oscillators[1] (ROs) by adding a small amount of logic. Trojans are detected by observing the changes in the frequency of the ROs. An algorithm is provided to secure all the gates, while reducing the hardware overhead. We analyzed the coverage, area and test time overhead of the proposed DFTr technique. To demonstrate its effectiveness in the real world, the proposed technique had been validated by a red-team blue-team approach.

I. INTRODUCTION

Globalization of the IC design flow is creating opportunities for rogue elements within the supply chain to corrupt the IC design [1]. It is possible for an attacker to gain access and control a target IC any time in its life. To establish trust during fabrication, using trusted foundries for fabrication have been proposed [2]. However, they are not economically feasible and go against the globalization trend. In an alternate approach to establish trust, the design is hardened before fabrication by inserting DFTr infrastructure, and the trustworthiness of the fabricated IC is verified using the inserted infrastructure.

One DFTr technique to detect Trojans[2] in an IC is to create identity for a design; any alteration in the design should change this identity. A design's circuit path delays of an can be used as an identity of that design. Path delay measurement-based Trojan detection technique has been proposed in [3].

We propose to configure *functional paths into ROs* to detect inserted Trojans. Previously, ROs have been used to give a unique identity to individual ICs [4] where the goal was to identify individual chips that implement the same design. We use the frequencies of ROs to embody each design with a unique identity. The frequency of the RO depends upon the components in the circuit and changes with any modification in the design. Changes in the frequency of an RO due to the inserted Trojan alter the identity of the design and

thus enable the detection of Trojans. The frequency of each RO is calibrated such that the effects of process variations and measurement errors are minimized. The proposed DFTr technique is non-invasive and can detect Trojans even if the Trojans are inserted in only some of the fabricated ICs. An *algorithm is proposed to configure the circuit paths into ROs* such that the number of secured gates is maximized and the hardware overhead is minimized.

While it is important to incorporate the DFTr techniques into the IC design flow, it is essential to analyze the impact of the DFTr techniques on the IC area and on the IC development time. We analyze the proposed DFTr technique using traditional Design- for-Test (DFT) metrics such as area, test time and coverage. In addition, we perform *a red-team blue-team evaluation* where outside attackers try to compromise the hardened design.

A. Previous work

Most DFTr techniques detect Trojans by analyzing the power sidechannel. Based on the assumption that Trojans consume additional power, measurement of IC power dissipation can be used to detect Trojans [5]. One can measure the power consumed in specific parts of the chip, by increasing the switching activity in those parts. Input patterns can be crafted to increase switching activity in the targeted region and hence maximizing the power consumption of that targeted region [6]. To overcome the effect of process variations, statistical techniques have been proposed in [7]. Since Trojan circuits draw extra current from the power supply, measurement of the current flowing through power ports of the chip may also detect Trojans [8]. Power analysis based Trojan detection methods may not work if the Trojans are power-gated.

Trojans can also be detected by activating them and observing their malicious responses [9], [10]. Since it is likely that Trojans are inserted in the hard-to-excite nodes in a design, applying input patterns and making the hard-to-excite nodes easily testable is another approach to detect Trojans [10]. Each Trojan activation method assumes a model for every Trojan that it targets. But in reality, the intentions of an attacker as well as his/her Trojans cannot be modeled.

Trojans are also detected based on their impact on path delays. These methods do not assume any Trojan model, similar to the proposed technique. In [3], test patterns are generated to excite the paths in the design and statistical

---

[1] A ring oscillator consists of an odd number of inverting elements connected in a ring.

[2] Trojans are deliberate and malicious changes that are made to an IC design.

978-1-61284-657-6/11 $26.00 © 2011 IEEE

techniques are applied to overcome the effect of process variations. Operating the IC at its critical speed or greater than its critical speed, and checking for violations in their behavior is another technique used to detect Trojans [11]. Since the inserted Trojans might impact at least one of the sidechannels, measuring multiple sidechannels can detect Trojans [12].

II. THREAT MODEL AND MOTIVATION

A. Threat model

Malicious changes can be made at any phase of the IC design such as specification, design, fabrication, testing, and packaging [13]. The proposed DFTr technique targets Trojans that are inserted during fabrication. We focus on Trojans in the form of addition or modification of existing logic gates. While the proposed technique does not assume any Trojan model, we use the smallest gate-level Trojan, the non-inverting buffer, as an example. While the previous detection methods assumed that Trojans are inserted in all the fabricated ICs, we aim at coping with the more challenging case of Trojans that are inserted in only some of the fabricated ICs.

B. A motivational example

Consider the hardened C17 circuit shown in Figure 1 wherein the circuit paths are arbitrarily configured into two ROs ($RO1$ and $RO2$). The blue (darkly shaded) components indicate RO1 and the brown (lightly shaded) components indicate RO2.

RO1 includes the gates $G1$, $G2$, $G4$, and $G6$ and the two-extra muxes $M1$ and $M2$, and an inverter $Inv1$. This additional inverter makes RO1 inverting (as there are only 4 NAND gates in this RO). RO1 is activated by enabling $TrE1$ signal HIGH and by applying the pattern 00111.

RO2 includes the gates $G3$ and $G5$ and an additional mux $M3$, and an inverter $Inv2$. RO2 is activated by enabling $TrE2$ signal HIGH and by applying the pattern 01000. Only one RO is activated at a time by applying the corresponding pattern and enabling the corresponding TrE signal. We implemented the hardened design on a Xilinx Spartan XC3S500E FPGA board. The golden frequencies of the two ROs are $F_{golden,RO1}$ = 107.91 MHz and $F_{golden,RO2}$ = 176.52 MHz.

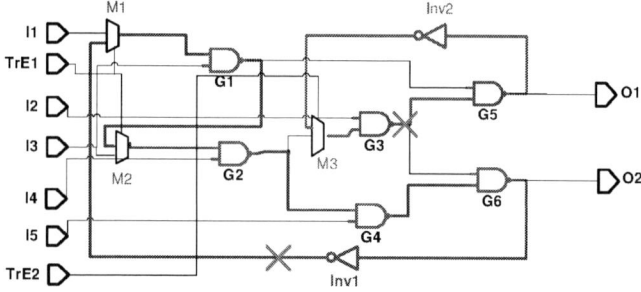

Fig. 1. Gate level diagram of the C17 circuit with two ROs – RO1 and RO2 to protect all the gates. Blue (darkly shaded) wires indicate the path of RO1 and the brown (lightly shaded) wires indicate the path of RO2.

Recently the effects of both inter-die and intra-die variations are experimentally studied [14]. Hundreds of ring oscillators were implemented on tens of FPGAs (Xilinx Spartan3E S500)

and the effect of both inter-die and intra-die variations are studied. The maximum inter-die variation was found to be 6.6%. One can estimate the change in the frequency of RO1 due to process variations $\triangle F_{pv,RO1} = \pm6.6\% \times 107.91$ MHz which is 7.13 MHz. Hence, due to process variations, the frequency of RO1 can vary between 100.78 MHz (= 107.91 - 7.13) and 115.04 MHz (= 107.91 + 7.13). Similarly, the frequency of RO2 can vary within the range 164.87 MHz - 188.17 MHz.

Let us now insert Trojans into the hardened design. The cross-points shown in Figure 1 indicate the location of Trojans.

Trojan 1: An inverting buffer is at the output of gate G3. Now the frequency of RO1 is 0 MHz as there are an even number of inverting elements in the RO1 path. Trojan is detected as RO1 is not oscillating.

Observation 1: Only the frequency of RO1 changes and the frequency of RO2 remains the same. One RO may not be enough to detect the changes in all the parts of a circuit.

Trojan 2: A non-inverting buffer is inserted between O1 and M3. Now the frequency of RO1 is 90.50 MHz and the frequency of RO2 remains the same. Trojan is detected due to a change in RO1 frequency and this change is less than the change due to process variations.

Trojan 3: A non-inverting buffer is inserted between O2 and M1. Now the frequency of RO1 is 110.27 MHz and the frequency of RO2 is 123.62 MHz. Trojan is detected due to changes in the frequencies of RO1 and RO2

Observation 2: Some Trojans are detected by changes in the frequencies of multiple ROs.

III. RING OSCILLATOR BASED DFTR

A. The idea

Any additional gate (loading gate) that is connected to the output of a gate (loaded gate) increases the fan-out delay of the loaded gate and the delays of the paths wherein the loaded gate resides. If the loaded gate is on the critical path or on a near-critical path, then the increased delay may violate the functional specification[3].

An attacker can insert malicious gates in non-critical paths in such a way that the modified path does not violate the critical path constraint. In our technique, the paths in the design are reconfigured into ROs such that the additional delays caused by Trojans can still be measured as changes in ROs' frequencies. The test time is significantly reduced as one ROs can simultaneously give information about the delay of multiple gates.

B. Effect of process variations

We use the frequency of ROs as the Trojan detection metric. In order to detect a Trojan in the presence of process variations, one has to embed a RO in such a way that the change in frequency of the RO due to Trojan insertion ($\triangle F_{Trojan}$) is greater than the change in the frequency due to process variations ($\triangle F_{PV}$). If $\triangle F_{Trojan}$ is greater than $\triangle F_{PV}$, then it results in a successful detection. On the other

---

[3]An attacker chooses not to insert the Trojan on critical or near-critical paths as delays in these paths can be explicitly measured by conventional delay testing.

978-1-61284-657-6/11 $26.00 © 2011 IEEE   106

```
SecuredGates = φ
ConditionList = φ
While (SecuredGates ≠ AllGates)
    SelectedPath = Path with maximum number of unsecured gates
    Embed_Ring_Oscillator(SelectedPath)
    Pattern = ComputeExciteCondition(SelectedPath)
    ConditionList = ConditionList U {Pattern}
    SecuredGates = SecuredGates U {GatesInSelectedPath}
```

Fig. 2. Algorithm for securing gates – a greedy algorithm for embedding of ROs into a design

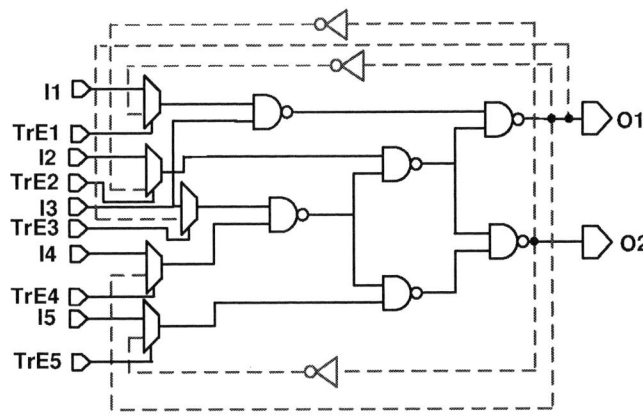

Fig. 3. The hardened C17 circuit uses five ROs generated by the proposed algorithm. The dotted lines show the ROs and the 3 inverters which are inserted to create the ROs.

hand, if $\triangle F_{Trojan}$ is less than $\triangle F_{pv}$, the inserted Trojan cannot be differentiated from process variation, and hence, remains undetected.

The ability of the proposed method depends on the measurement of $\triangle F_{pv}$. To calculate the effect of process variations on these frequencies we used the results from [14]. This study has revealed that the frequency of the ROs deviates up to 6.6% from their golden value due to inter-die variations and up to 1.1% due to intra-die variations. In this work, we will consider only the dominant variation factor, i.e., the inter-die variation.

Even though the C17 circuit shown in Figure 1 can detect Trojans, it is hardened by arbitrarily embedded ROs.

C. Algorithm for securing gates

To ensure the detection of an inserted Trojan, one has to embed ROs by covering all the gates. However, every additional RO entails an overhead of at least a mux and potentially an inverter.

The proposed algorithm is shown in Figure 2. Each iteration of the algorithm selects the path with the maximum number of unsecured gates, embeds an RO to secure the gates in that path, and computes the input pattern[4] that sensitizes the embedded RO. All the sensitized patterns are stored in the *ConditionList* The algorithm terminates when all the gates are secured. The RO paths originate from primary or pseudo-primary inputs and terminate at primary or pseudo-primary outputs [5]. Embedding a RO necessitates the insertion of a mux, and possibly an inverter if the path is non-inverting. Each mux is controlled by a TrE signal. If one has a constraint on the number of ROs that can be embedded, then the termination condition can be modified accordingly, but still a maximum number of gates can be secured.

D. DFTr methodology

The proposed DFTr methodology includes three steps:

1) Step 1: Embed ring oscillators: See Section III-C.

2) Step 2: Determine golden frequencies: Upon the insertion of ROs, the designer determines the frequency of the ROs F_{golden}, based on simulation results in case of an ASIC and based on implementation results in case of an FPGA. The designer can also determine the effect of process variations on

this golden frequency from the data provided by the fabrication company.

3) Step 3: Test procedure: The designer/tester performs the following steps on each fabricated IC, as dictated by our assumption that Trojans may have been inserted in only some or all of the ICs. The designer applies each of the input patterns in the ConditionList and measures the corresponding frequency $F_{measured}$. Then, the amount of deviation of the $F_{measured}$ from that of F_{golden} is determined. If this change is greater than the change due to process variations for any pattern, then the IC has a *Trojan*. Otherwise, the chip is *Trojan free*.

4) Application of the proposed DFTr technique to the motivational example: While the hardened C17 circuit in Figure 1 has arbitrarily embedded ROs, Figure 3 shows the hardened C17 circuit using the proposed algorithm. In this new hardened C17 circuit, 5 ROs are embedded using 5 additional muxes and 3 inverters. Modification to harden the circuit are only made at the inputs in the form of adding muxes and inverters if necessary. The delay penalties of the muxes at the inputs can be reduced by using a modified scan flip flop design, explained below.

E. Integrating the DFTr into DFT flow

As can be seen, the proposed DFTr flow is similar to the DFT flow and hence it can be readily incorporated into the IC design flow. Also one can incorporate the proposed DFTr infrastructure into the already existing DFT infrastructure by modifying the scan flip flop or the scan latch. Figure 4(a) shows the original scan latch where the muxes for the RO are added to the Data (D_{IN}) line of the scan latch. This might increase the critical path delay. One can re-organize the scan latch to incorporate the mux for the RO path so that modified scan latch supports both DFT and DFTr operations. This modified scan latch is shown in Figure 4(b). An additional multiplexer and an OR gate along with a TrE pin are added. In a conventional scan latch, when TE pin goes HIGH, the test inputs are latched into the scan latch. In this modified scan latch, when the TE pin and the TrE pin of the corresponding scan cell are HIGH, then the RO path is enabled, and the scan

---

[4]The pattern that sensitizes the ring oscillator is computed by invoking an ATPG tool (Synopsys Tetramax) with the proper constraints. Details are omitted due to space constraints.

[5]Long paths, wherein the effect of process variations and a Trojan cannot be differentiated, can be divided into sub-paths, each of which is configured into a RO.

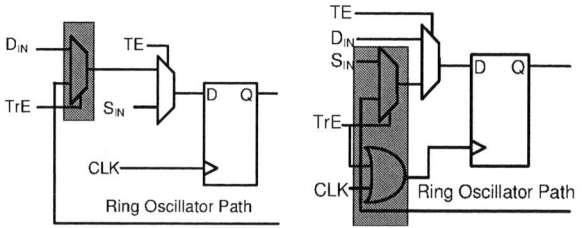

Fig. 4. (a) Conventional scan latch with DFTr logic inserted in the Data In(D_{IN}) line. (b) A Modified scan latch with DFTr architecture inserted in the Scan In (S_{IN}) line. The shaded region shows the proposed DFTr logic.

latch is turned into a transparent latch by feeding a constant-HIGH, instead of the clock signal. By using such a modified latch, the delay of the additional multiplexers and inverters can be hidden.

IV. ANALYSIS OF RO BASED DFTR TECHNIQUE

The proposed ring-oscillator based DFTr method is evaluated using the ISCAS-85 benchmark circuits.

A. Experimental setup

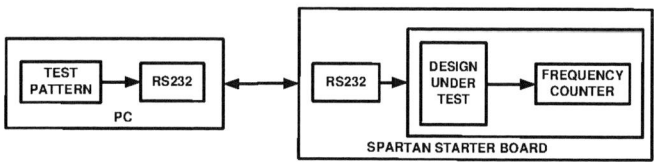

Fig. 5. Experimental setup

The RO embedding algorithm was implemented in C and the ROs were embedded into six ISCAS-85 combinational benchmark circuits. The patterns used to sensitize the ROs were computed by using the Synopsys Tetramax ATPG tool. The hardened design-under-test and a frequency counter were synthesized using Xilinx tools and implemented on a Xilinx Spartan XC3S500E FPGA board. A Perl script applied patterns to sensitize the selected ROs in the hardened benchmark on the FPGA board via the RS232 port, and to instruct the frequency counter to measure the selected frequency. The frequency was measured by the frequency counter with respect to the 50 MHz clock of the FPGA. This experimental setup is shown in Figure 5.

B. Gate Coverage of ROs and Trojan detection capabilities

Figure 6 shows the number of ROs required to secure a given percentage of gates for the ISCAS-85 benchmarks using the proposed algorithm. The golden frequencies of the ROs in a hardened design are first measured.

The plot resembles the conventional fault coverage curve. For example in C1908, only 23 ROs are sufficient to secure the first 50% of gates (440 gates), while 50 additional ROs are necessary to increase the percentage of secured gates from 90% to 100% (792 gates to 880 gates).

For each of the hardened benchmarks, we generated 30 variants by inserting a non-inverting buffer Trojan at a random location. A Trojan is detected if the frequency of at least one of the embedded ROs falls outside the golden range. This

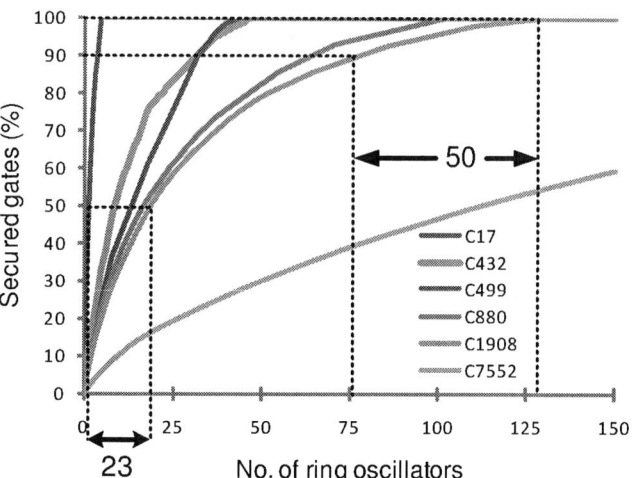

Fig. 6. Number of ring oscillators required to secure a given percentage of gates for the different ISCAS-85 benchmarks. For C1908, 23 ROs secure 50% of the gates, and 50 additional ROs are needed to improve the coverage from 90% to 100%.

verification step when applied to every hardened benchmark was able to detect the Trojan is inserted into the design. Of course, for small benchmarks the locations of the inserted Trojans covered the entire design, while for large benchmarks they only covered a small percentage of the design. Most of the Trojans were detected by more than one RO. Since the ISCAS benchmarks are relatively small, each of the selected paths was covered by a single RO.

C. Area cost

Let 'G' be the number of gates in a design and let $G_{path,min}$ be the minimum number of gates in any path of the design. In the worst case, each shortest path is embedded with a RO, which results in embedding N_{RO} ROs.

$$N_{RO} = \left\lceil \frac{G}{G_{path,min}} \right\rceil$$

As each RO necessitates a mux and possibly an inverter, the area cost includes N_{RO} muxes and at most N_{RO} inverters in the worst case. The actual number of muxes and inverters for some of the benchmarks are listed in Table I.

Benchmark	No. of gates	No. of ROs	No. of extra Muxes	No. of extra Inverters
C17	6	5	5	3
C432	160	47	47	16
C499	202	44	44	44
C880	383	103	103	51
C1908	880	128	128	80
C7552	3512	671	671	0

TABLE I
SUMMARY OF THE AREA COST ASSOCIATED WITH THE RO-BASED DFTR METHOD TO COVER ALL POSSIBLE TROJANS

Column 1 in Table I lists the number of gates in the original benchmark and column 2 lists the number of ROs necessary to secure all the gates in the hardened benchmark. Columns 3 and 4 tabulate the additional multiplexers and inverters introduced by the proposed DFTr scheme. Almost half of the paths that are reconfigured as ROs do not require extra inverters because they contain an odd number of inverting elements already.

978-1-61284-657-6/11 $26.00 © 2011 IEEE

D. Test time overhead

The test time overhead of the proposed method depends upon the frequencies of the ROs. Higher RO frequencies translate into lower test time.

Let F_{RO} be the frequency of an RO and it is represented by B bits and the frequency resolution F_{RE} is given as, $\frac{F_{RO}}{2^B}$. One can successfully detect the Trojan in the presence of both process variations and resolution error if,

$$\triangle F_{Trojan} \geq |\triangle F_{PV} + \triangle F_{RE}|$$

Hence, to increase the Trojan detection capability of the proposed DFTr technique, F_{RE} can be reduced by increasing B. The value of B impacts the test time. The maximum F_{RE} that can be tolerated without compromising the Trojan detection capability is limited by the value of $\triangle F_{PV}$. Hence the minimum number of bits B_{min}, required to represent the RO frequency is given as,

$$B_{min} = log_2 \frac{\triangle F_{PV}}{F_{RO}}$$

Since $\triangle F_{PV}$ is 6.6%, in this experiment we used 5 bits to represent the RO frequency. The maximum number of clock cycles $N_{F_{clk,max}}$ depends on the RO with the minimum frequency $F_{RO,min}$ and it is given as,

$$N_{F_{clk,max}} = \frac{F_{RO,min}}{F_{clk}} \times 2^B$$

Hence, the total test time for a design embedded with N_{RO} ROs is $N_{RO} \times N_{F_{clk}}$.

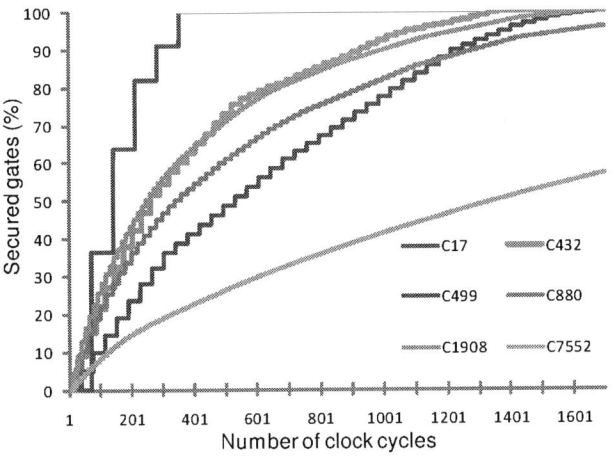

Fig. 7. Test time for different ISCAS-85 benchmarks using the proposed algorithm.

Figure 7 shows the test time overhead to secure a particular percentage of gates, for different benchmarks. Similar to the trend in Figure 6, the number of clock cycles required to verify the last 10% of the gates is much higher when compared to the first 50%. Furthermore, test time for different benchmarks differ even when the same number of ring oscillators are used, as $F_{RO,min}$ varies for these benchmarks.

E. Red-team Blue-team validation

Although we validated the proposed DFTr technique using a traditional DFT metric of gate coverage, in practice a 100% gate coverage does not guarantee that all attacks can be detected.

A DFTr technique is only as good as its effectiveness against the last attack. We tested the effectiveness of this method in a real-world scenario by participating in the 2009 Embedded Systems Challenge [15], [16]. In this challenge, each team submitted a hardened design (Blue team activity) and the other teams attempted to compromise it by inserting Trojans (Red team activity).

The design that was used for this challenge was a cryptographic hardware and it is shown in Figure 8. This cryptographic hardware implemented a Trivium stream cipher that can communicate with a PC either by using the RS232 protocol or using the JTAG protocol on a RS232 channel. The cryptographic hardware uses 1564 gates. We hardened this cryptographic hardware design using a single RO. This single RO was embedded at arbitrary places and not embedded based on the proposed algorithm. Four teams inserted Trojans into this hardened design. The Trojans from different teams are described below:

Red-Team Trojan 1: This Trojan was inserted in the Trivium stream cipher module and it leaked the secret key through the RS232 channel. But, this Trojan changed the frequency of the ring oscillator.

Red-Team Trojan 2: This team figured out the hardening mechanism. When they tried to insert Trojans, they realized that the frequency of the RO varied. Yet, this team inserted a counter and the Trojan changed in the RO frequency.

Red-Team Trojan 3: This Trojan was inserted in the Trivium stream cipher module and it leaked the secret key through the RS232 channel. The inserted Trojan changed the RO frequency. This team figured out the hardening mechanism and hence they hardcoded the RO frequency.

Red-Team Trojan 4: This Trojan was inserted in the RS232 module. It made the plaintext to bypass the Trivium module and returned the plaintext. This Trojan also changed the RO frequency.

As can be seen above, all the four Trojans altered the golden frequency indicating the presence of a Trojan. However, at the time of the challenge, we detected only two of the Trojans. The other two Trojans hardcoded the frequency of the single RO and presented this as the response to the sensitizing pattern. However, the hardcoding can be detected by varying the supply voltage and by checking for a change in the RO frequency. If the measured frequencies remain constant, it indicates a hard code attack.

F. Comparison with previous work

We compared the proposed DFTr technique with path delay measurement [3] and Trojan activation [10]. In [3], a random number of paths are excited to detect Trojans. The excited paths' delays are measured with the Automatic Test Equipment (ATE). Although randomly generated path delay patterns fail to cover all the gates, the path delay method can be utilized in conjunction with the proposed algorithm thereby lowering the exponential test time cost of the path delay method down to a linear test time cost (2N).

In [10], Trojans are activated by applying a specific number of input patterns through the scan flip flops. Additional scan flip flops are inserted to increase controllability and observ-

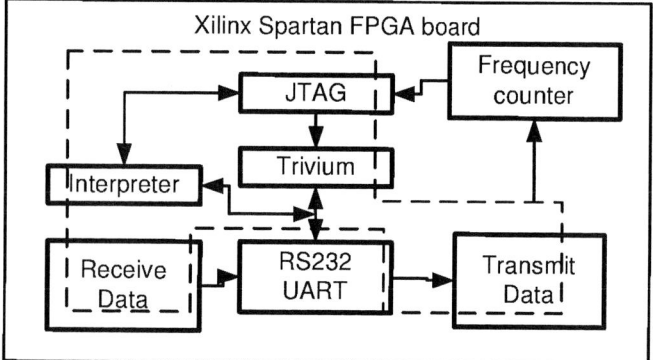

Fig. 8. The hardened Trivium stream cipher with one ring oscillator which is used for the Embedded Systems Challenge. The dotted line represents the single ring oscillator that spans the entire design.

Method Trojans	An inverter	3-bit counter	10-bit counter
RO+Proposed Alg.	1665	1665	1665
Path delay [3]+ Proposed Alg.	66	66	66
Trojan activation [10]	66	528	67584

TABLE II

NUMBER OF CLOCK CYCLES REQUIRED BY THREE DFTR METHODS TO DETECT TROJANS

ability of some nodes. The best-case scenario for this method is where there are no additional flip flops except at the PIs and POs and all the internal nodes are easily testable. Let us consider three Trojan models – an inverter, a 3-bit counter and a 10-bit counter. The test times using three different techniques for C1908 are provided in Table II. It can be seen that the combination of path delay method and the proposed algorithm yields the lowest test time, although such a method necessitates the use of an ATE. The performance of Trojan activation method increases exponentially with the size of the Trojan.

V. CONCLUSION AND FUTURE WORK

The proposed technique is applicable to both ASIC and FPGAs. In the context of an ASIC, the golden frequencies of the ROs can be estimated by considering the effect of process variations using simulation. ROs can be crafted such that the effect of process variations on their frequencies is minimized. For post-fabrication testing, a low-cost ATE is sufficient to apply the input patterns. In the context of FPGAs, the attacker is assumed to be capable of modifying the FPGA bit stream. The designer implements the hardened design on the FPGA and measures the golden frequencies of the ROs. After the FPGA is deployed on the field, the designer applies input patterns in an off-line mode and measures the frequencies of the ROs to detect any malicious change to the design.

We proposed a DFTr technique that reconfigures functional paths into Trojan-detecting ROs. We showed that the technique could successfully detect Trojans. The test cost analysis had shown that this approach scales linearly with the number of gates in the design. While we presented a delay measurement based technique, a similar analysis can be conducted by measuring the power consumption of the IC, when a particular RO oscillates. The increased switching activity in the RO

path magnifies the power consumption of the gates in that particular path, enabling the detection of Trojans. To overcome the variations in RO frequencies due to temperature, one can either test the ICs in a controlled temperature environment or calibrate the RO frequencies taking the effect of temperature into account. ROs embedded by the proposed technique can be re-used as reliability and temperature sensors [17], [18]. Furthermore, the test time can be reduced by embedding ROs in such a way that two or more ROs can oscillate concurrently.

VI. ACKNOWLEDGMENTS

This work was supported by National Science Foundation grants CNS 0621856, CNS 0831349, and CNS 0958510. We also thank Kurt Rosenfled for his valuable comments.

REFERENCES

[1] http://www.acq.osd.mil/dsb/reports/ADA435563.pdf.
[2] "http://www.dmea.osd.mil/otherdocs/accreditedsuppliers.pdf."
[3] Y. Jin and Y. Makris, "Hardware Trojan detection using path delay fingerprint," *IEEE International Workshop on Hardware-Oriented Security and Trust*, pp. 51–57, Jun 2008.
[4] G. Suh and S. Devadas, "Physical Unclonable Functions for Device Authentication and Secret Key Generation," *Design Automation Conference, 2007. DAC*, pp. 9 –14, 4-8 2007.
[5] D. Agrawal, S. Baktir, D. Karakoyunlu, P. Rohatgi, and B. Sunar, "Trojan detection using ic fingerprinting," *IEEE Symposium on Security and Privacy*, pp. 296–310, May 2007.
[6] M. Banga and M. Hsiao, "A Novel Sustained Vector Technique for the Detection of Hardware Trojans," *IEEE International Conference on VLSI Design*, pp. 327–332, Jan. 2009.
[7] R. M. Rad, X. Wang, M. Tehranipoor, and J. Plusquellic, "Power supply signal calibration techniques for improving detection resolution to hardware Trojans," *IEEE/ACM International Conference on Computer-Aided Design*, pp. 632–639, 2008.
[8] X. Wang, H. Salmani, M. Tehranipoor, and J. Plusquellic, "Hardware Trojan Detection and Isolation Using Current Integration and Localized Current Analysis," *IEEE International Symposium on Defect and Fault Tolerance of VLSI Systems*, pp. 87–95, Oct. 2008.
[9] S. Jha and S. Jha, "Randomization Based Probabilistic Approach to Detect Trojan Circuits," *IEEE High Assurance Systems Engineering Symposium*, pp. 117–124, Dec. 2008.
[10] H. Salmani, M. Tehranipoor, and J. Plusquellic, "New design strategy for improving hardware Trojan detection and reducing Trojan activation time," *IEEE International Workshop on Hardware-Oriented Security and Trust*, pp. 66–73, Jul. 2009.
[11] J. Li and J. Lach, "At-speed delay characterization for ic authentication and Trojan horse detection," *IEEE International Workshop on Hardware-Oriented Security and Trust*, pp. 8–14, Jun. 2008.
[12] S. Narasimhan, D. Dongdong, R. Chakraborty, S. Paul, F. Wolff, C. Papachristou, K. Roy, and S. Bhunia, "Multiple-parameter side-channel analysis: A non-invasive hardware Trojan detection approach," *IEEE International Symposium on Hardware-Oriented Security and Trust*, pp. 13–18, Jun. 2010.
[13] J. Rajendran, E. Gavas, J. Jimenez, V. Padman, and R. Karri, "Towards a comprehensive and systematic classification of hardware Trojans," *IEEE International Symposium on Circuits and Systems*, pp. 1871–1874, May 2010.
[14] A. Maiti, J. Casarona, L. McHale, and P. Schaumont, "A large scale characterization of RO-PUF," *IEEE International Symposium on Hardware-Oriented Security and Trust*, pp. 94–99, Jun. 2010.
[15] "Securing FPGA design using PUF-chain and exploitation of other Trojan detection circuits," http://isis.poly.edu/~kurt/s/esc09_submissions/reports/Polytechnic_JV.pdf.
[16] "http://spectrum.ieee.org/semiconductors/design/creative-winners-in-hardware-trojan-contest."
[17] M. Ketchen, M. Bhushan, and R. Bolam, "Ring Oscillator Based Test Structure for NBTI analysis," *IEEE International Conference on Microelectronic Test Structures*, pp. 42–47, Mar. 2007.
[18] S. Park, C. Min, and S.-H. Cho, "A 95nw ring oscillator-based temperature sensor for RFID tags in 0.13 μm CMOS," *IEEE International Symposium on Circuits and Systems*, pp. 1153–1156, May 2009.

978-1-61284-657-6/11 $26.00 © 2011 IEEE

The Buck Stops with Wafer Test: Dream or Reality?
Organizer: Suriyaprakash Natarajan, Intel Corporation, suriyaprakash.natarajan@intel.com
Moderator: Arani Sinha, Advanced Micro Devices, arani.sinha@amd.com

In the industry today, testing packaged chips achieves the outgoing DPPM (defective parts per million) requirements. Usually, functional and structural test patterns are used at wafer sort, followed by functional/structural testing with packaged parts, and then by functional system level testing, each subsequent stage significantly more expensive than the previous one. By and large, wafer test have not been used for performance binning and reliability screening. Also, packaged parts are tested in burn-in chambers and on load boards, using either the same structural patterns used at wafer sort or with functional test patterns.

As design complexity has gone up significantly over the past decade, the test cost has grown disproportionately. If a die is found to be faulty at a stage after wafer sort, then the design house incurs the cost of packaging, and for subsequent testing. In one business model, the fabless design house buys dies from the foundry that pass wafer sort, and needlessly pays for dies that are later found to be bad after packaging. Furthermore, this problem can be severe for dies that go into multi-chip modules or stacked ICs, as all the dies in the packaged chip have to be thrown away even if one constituent die is found to be bad. Since it is increasingly more expensive to test chips down-stream (using functional testers or on a system) and since there is downward pressure on product costs with the advent of inexpensive SoCs, it becomes important to achieve maximum test quality in terms of defectivity and binning, at wafer sort.

Known good die (KGD) refers to dies which have been tested to the same quality and reliability levels as their packaged counterparts. Even though the KGD problem has been discussed since the mid 90's, the problem becomes more and more difficult with every new process node and new design requirement. There is need to develop high quality tests, on-die DFX instrumentation, and reliability screens that can be applied at wafer sort, given pin constraints. In addition to addressing failure mechanisms of a bare die, such as gross defects, small delay defects, cross-talk, and variations due to photolithography, a good wafer sort test methodology and associated tests should not only be enough to reject bad bare dies, but also characterize the dies enough to enable estimation of the performance/power of packaged parts on a system.

Presentations:

"A Case for Known Good Die"
Karim Arabi, Qualcomm

"Challenges with Achieving KGD for High Performance Products"
T. M. Mak, Intel Corporation

"Mixed Wafer Test Strategies for the Automotive, 0 DPPM Market"
LeRoy Winemberg, Freescale Semiconductor

Session 5

978-1-61284-657-6/11 $26.00 © 2011 IEEE

Apprentice – VTS Edition: Season 4

Organizers:
Kee Sup Kim, Samsung, kee.sup.kim@samsung.com
Rob Roy, Atrenta, robroy@atrenta.com

The main objective of this active "panel" is to increase technical interaction among attendees .Team leaders will recruit participants to their team. Each team will try to clearly articulate the problems and come up with ways to solve this problem in the form of new business proposal. The teams will present their findings and business proposals in front of judges later during the conference. The winning team will be announced during the social event
.

Forming the Teams
Each team will be required to have members from three different continents and at least two students to promote interaction among people who may not usually interact at VTS. The teams that do not meet this requirement can still participate but won't be eligible for the award.

Developing the Plan
Each team will have a working discussion session during the panel to define the problem and convincing enough approach to address the problem. Teams are encouraged to carry on conversations outside of the panel sessions also to further develop their ideas. Each team needs to come up with five minute presentation for the judging session.

Team Leaders.
The Team Leaders will be announced at the beginning of the conference.   As an option, anyone can be a team leader as long as one more participant will join the team.

Judging
The judging session will be scheduled during for Wednesday Panel session. Each team will have five minute to present and five minutes for questions from judges and audience. The audience will have a chance to vote on winning proposals. The winning team announcement and the award delivery will be at the end of the judging session.

Blake Winchell, Partner Ventures, Managing Director,

Laura Oliphant: Intel, Director of Intel Capital

Yervant Zorian: Synopsys, VP

Award Sponsors: Syntest has provided the awards for the wining team. First ten people to join the winning team will receive the awards.   Atrenta will also provide a special award to the leader of the first place team.

Special Session 5B: Panel

How Much Toggle Activity Should We Be Testing With?

Organizer: *Xiaoqing Wen*, Kyushu Institute of Technology, Japan

Moderator: *Mohammad Tehranipoor*, University of Connecticut, USA

Panelists:

Rohit Kapur, Synopsys, Inc., USA

Anand Bhat, Texas Instruments India, Pvt. Ltd., India

Amitava Majumdar, Advanced Micro Devises, Inc., USA

LeRoy Winemberg, Freescale Semiconductor, Inc., USA

Abstract:

Power dissipation of an LSI circuit during scan testing, especially at-speed scan testing, can be several times higher than that during functional operations. Excessive test power causes hot spots and/or severe IR drop that may lead to chip damage, undue yield loss, or reliability degradation, especially for low-power LSI circuits. As a result, it is becoming increasingly important to reduce test power by lowering test-induced toggle activity in order to make scan test "power-safe". However, with the stress on reducing toggle activity during scan test one might question: Have we gone too far? Should we reduce toggle activity below functional levels? Should we even plan for many test sets with different toggle activities? Can the test power problem be solved by existing DFT and ATPG solutions? What's missing in today's solutions? What's next for low-power testing? This panel provides an interactive forum to discuss these critical questions with industry experts from both semiconductor and EDA companies. It helps practitioners and researchers alike in their quest for more effective and more efficient solutions to the test power problem.

Session 6

978-1-61284-657-6/11 $26.00 © 2011 IEEE

A Novel Mechanism for Speed Characterization During Delay Test

Amitava Majumdar, Arani Sinha, Nehal Patel, Ramamurthy Setty, Yan Dong, Shu-Hsuan Chou

AMD, Inc.

1 AMD Place, Sunnyvale CA 94088, USA

E-mail: {amitava.majumdar, arani.sinha, nehal.patel, ramu.setty, yan.dong, shu-hsuan.chou}@amd.com

Abstract— The impact of di/dt noise and static IR drop on at-speed scan testing has been reported in literature. Delays of paths can be impacted during delay testing by IR drop and di/dt noise in ways that change the delay ordering of paths. This, in turn, affects the ability of such tests to catch certain delay defects and impairs its use for speed binning. It is important, therefore, to address IR drop during delay tests. This paper proposes an instrumentation methodology for a design based on launch-off-capture delay tests to control the time interval between shift and capture cycles. This mechanism improves speed characterization of devices and achieves a higher capture frequency compared with traditional methods. It also addresses reduction of di/dt noise during capture. The proposed scheme is realized by (i) pipelined scan-enable and (ii) a deterministic launch-and-capture method for the tile under test. This mechanism has been implemented on silicon and experimentally observed to increase the speed of a device between 8% and 24% relative to traditional methods.

Keywords - di/dt noise; launch-off-capture; pipelined scan enable; IR drop.

I. INTRODUCTION

Well before power-grid noise and static IR drop became test issues, the authors highlighted the need for control of di/dt during scan-based delay tests. Researchers have since observed IR drop and di/dt noise and their impact on test responses [3].

It can be said that the goal of scan-based delay tests is to test for delay defects in an electrical environment (power-profile) akin to normal functional operation. But traditional scan-based methods, developed over the past three decades, are often incapable of satisfying all these requirements -- not because of any inherent deficiency of scan, but because this was never a requirement. It is only today -- with delay defects becoming more prevalent and insidious, with recent increases in circuit complexity, with supply voltage falling below 1V and further eroding signal-to-noise margins, and reductions in device sizes making devices more vulnerable to parasitics -- that power-grid noise has become an issue and restrictions on power-profile during test, a requirement.

Four aspects of power grid noise are relevant to test:

1. **Peak current draw** during test can often exceed amounts for which power grids are normally designed. This can happen during both shift and capture.

2. As a direct consequence of peak current draw, a **static resistive drop** is observed during both shift and capture.

3. An **instantaneous inductive drop** (di/dt) that is observed during capture.

4. Depending on the actual electrical event, it is not uncommon to experience **inductive-capacitive-resistive oscillations** in the power grid, affecting FF equilibria and, hence, their stability. This usually occurs during capture.

Due to these phenomena, increasingly prevalent in new technology nodes, scan patterns shifted in and captured responses shifted out can have errors that invalidate these tests. Numerous researchers have looked into the effect of power-grid noise on delay test [1], [5], [9], [10], [11]. Power-grid noise can also cause clocks to behave abnormally. An example abnormality is a general slowing of the clock, a phenomenon known as clock stretching [2], which can prevent detection of delay defects.

In this work, we devise a method that allows us to explore the transitionary phases between scan shift and capture with different parameters (such as latency, activity, etc.) and a reasonably high level of granularity. For this purpose, we use the launch-off-capture delay test scheme. It has been observed that shift pulses and capture pulses induce both static power drop and instantaneous inductive noise through power and ground. Whereas during shift, both types of power-grid noise can be absorbed by reducing shift frequency, noise during capture can invalidate tests. The objective of this work is to apply the capture pulses in a manner that achieves a better speed characterization of the device using delay tests on an ATE.

We do not want to apply capture pulses too soon after shift pulses because there is a high resistive drop due to peak current draw of shift patterns. At the same time, we would like to reduce instantaneous noise at the start of the capture pulses to eliminate any error in observed response during capture. The inductive transients occur at the start of the capture pulses, then die out as capture is in progress. Since a current builds up during scan shift, it is possible to reduce the instantaneous current during capture by controlling the time interval between shift and capture. Thus, by limiting the incremental change in the magnitude of the current, di, during the initial capture pulse, we limit the magnitude of di/dt noise. A mechanism is designed to exert fine-grain control over when the capture pulses are applied. This scheme is realized by means of (i) pipelined scan enable and (ii) a deterministic launch-and-capture mechanism that controls, among other things, the interval between the last shift pulse and the first capture pulse. The scheme is presented later in this work.

Ideally, the chip should be tested at frequencies close to its functional frequency. Researchers have addressed the problem of correlating operating frequencies in the functional mode with that during scan capture [8]. If the impact of IR drop and di/dt noise is reduced, then the frequency at which the chip operates is expected to increase. In other words, the passing frequency of the chip during the capture phase is expected to correlate better with the functional frequency. We have implemented this scheme on silicon and observed that the passing frequency during capture goes up between 8% and 24%.

In the rest of this paper, Section II discusses background work. In Section III, we describe the implementation of this scheme. In Section IV, we describe the experimental results.

II. BACKGROUND

At-speed test techniques can cause power variations that are not observed during functional operation. During shift, this is caused by the unusually high toggle content of the scan-shift patterns, even though scan shift occurs at low frequencies. During capture, the power variations are caused by the toggle content as well as the high frequency of capture. There are two distinct causes: static power dissipation, which averages out during the application of the pattern, and instantaneous power variation, typically seen at the beginning of a pattern sequence. The static type is caused by the current through the resistive elements of the power-ground grid. The instantaneous, or di/dt, variation is caused by the current through the power network and package pins that activates the inductance of those two circuit components.

As stated earlier, this work addresses the impact of IR drop and di/dt noise on delay test, and proper speed characterization of devices. We will briefly discuss the existing solutions for reducing di/dt during capture phase.

One of the solutions for reducing di/dt noise is to reduce the rate of change of voltage. However, it will not work for at-speed delay testing. Another solution, proposed by LogicVision (currently Mentor Graphics), applies a series of false capture pulses in "burst mode" before the actual capture pulses are applied [5]. The initial capture pulses ensure the di/dt noise dies down. This requires all scan chains to be modified so they can be configured as circular registers. During the initial capture pulses, the scan flops are in scan mode and the values are rotated in the scan chain. This methodology is intrusive.

Another proposed solution is to apply a number of capture pulses and gradually increase the frequency of the pulses [1]. The authors showed with oscilloscope measurements that this scheme dramatically reduced di/dt noise. However, the fast capture pulses during at-speed test are normally derived from the PLL, and this requires a complicated clocking mechanism.

Whereas the work in [1] proposes progressively increasing the capture frequencies, we apply the capture pulses after the shift pulses in such a manner that neither the remnant IR drop from shift nor the di/dt during capture can impact its speed markedly. The scan-shift frequency is slower than capture frequency; however, the amount of toggle-in scan data creates a large current during shift. We want to apply the capture pulses so the impact of IR drop from shift is low, yet the change in current di is small at the start of capture, which ensures the di/dt value is a small quantity during capture. This scheme does not require any modification to the underlying test generator or the clocking mechanism.

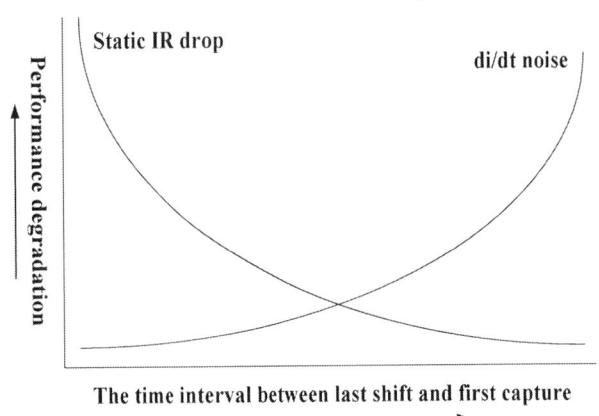

Figure 1. The application of capture pulses.

III. IMPLEMENTATION

A. Proposed Idea

We describe the proposed idea using Figure 1. Scan patterns are applied to the circuit using a number of shift pulses, and then the response from the circuit is captured by two (or more) capture pulses applied at scan-capture speed. Ideally, scan-capture speed should be either equal to or close to functional speed even if it is not achieved. A static IR drop builds up when the shift pulses are applied. After the shift pulses end, the current slowly dies. This is shown by the curve labeled "static IR drop" in Figure 1. Also, when we apply the capture pulses, there is an instantaneous change in current (as well as static IR drop due to the capture pulses). This instantaneous current is higher if there is a large time interval between the shift and the capture pulses. This is shown by the curve labeled "di/dt noise" in Figure 1.

We theorize we can reduce the magnitude of the instantaneous current by reducing the time interval between shift and capture pulses. Since there is a remnant current from the shift phase, the instantaneous change in current di is small. As Figure 1 shows, there is a region in which both static IR drop and di/dt noise are low. We propose the capture pulses be applied in this region so proper speed characterization of a device can be achieved. We also propose the instrumentation necessary to control the application of the capture pulses. This helps achieve a higher delay test frequency, which is closer to the functional frequency.

B. Pipelined Scan Enable

This section describes the pipelined scan enable mechanism. In the past, different variants of pipelined scan-enable have been proposed for controlling the scan-enable signal in the narrow time window between shift and capture in the launch-off-last-shift delay test scheme [4]. We, too, pipeline the scan-enable signal for essentially the same reason (i.e., for fine control of the de-assertion between shift and capture pulses). This sub-section describes the implementation.

We test each tile in isolation. The pipelined scan-enable signal is daisy chained through all tiles and the targeted tile or tile under test (TUT) gets the correct value. There is a pipeline flop in each tile on the daisy chain. We shift in the scan-enable value 0, for use in capture mode, through the pipeline using the slow scan clock while scan shift is in progress in the scan chains.

It should be noted that the scan-enable signal is 1 in the targeted tile while scan shift along the daisy chain is in progress. This is because scan shift is in progress in the targeted tile. The change in scan-enable value to a 0 in TUT corresponds to the end of scan-shift cycles. The scan-enable signal is routed through a central tile (henceforth referred to as the DFT tile). If the TUT is the nth tile in the daisy chain, then the shift of the scan-enable value 0 starts n cycles before the end of shift cycles so the scan-enable signal reaches the targeted tile after the last shift pulse. This describes the de-assertion of the scan-enable signal. Please refer to Figure 2. The input of the multiplexor labeled 1 is selected at this time.

When external scan-enable pin is asserted, the tile is set in the scan-shift mode. The assertion is asynchronously achieved from the scan-enable pin at the chip boundary. Figure 2 shows the inputs of the multiplexors labeled 0 are selected during this time.

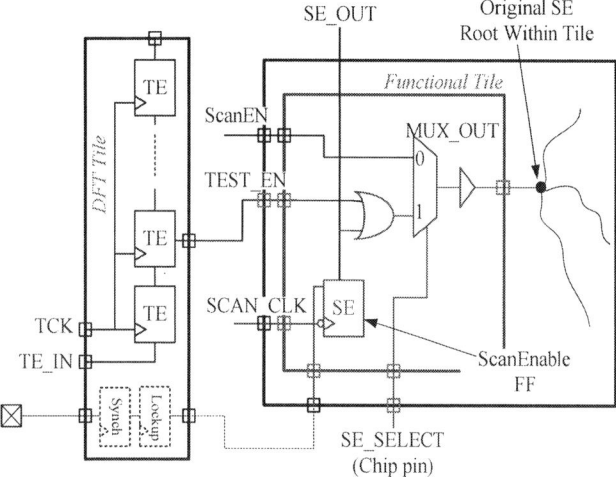

Figure 2. The pipelined scan-enable mechanism.

We use the chip pin SE_SELECT, or the scan-enable select signal, to control the multiplexor. As already

described, each tile has a pipeline register that is labeled SE in Figure 2. The scan-enable signal in the targeted tile is de-asserted synchronously using SE and the OR gate in the diagram. At this time, the output of SE is 0, the TEST_EN signal is set to 0, and the SE_SELECT signal is 1. Therefore, the inputs of the multiplexors labeled 1 are selected.

If TEST_EN is 1, the tile stays in shift mode. The TEST_EN can be used to keep the power grid in adjacent tiles alive during transition from shift to capture mode. We can program TEST_EN to 1 for all the other tiles, which are not targeted for delay test. This would always keep those adjacent tiles in shift mode, and help control the switching activity in the design to maintain a desired level. We want to control the switching activity in the other tiles to mimic the switching activity during the functional mode.

C. Launch-and-capture Mechanism

If a tile has a position n in the daisy chain, it would take n slow-speed scan clock cycles to program the pipeline flop (SE in Figure 2) to 0. Due to n pipelined stages, the SCAN_EN signal needs to be de-asserted n shift cycles before going to the capture mode. The capture pulse-generation logic uses the SE-distance counter. When tile number n is targeted, the SE-distance counter is also programmed to the value n.

Once this counter reaches 0, it triggers the capture-latency counter. The capture-latency counter helps us control the distance between shift and capture, and is set to a value equal to the routing delay on the scan-enable signal within the tile. The value programmed in the capture-latency counter is an estimate of the within-tile routing delay in terms of the number of clock pulses.

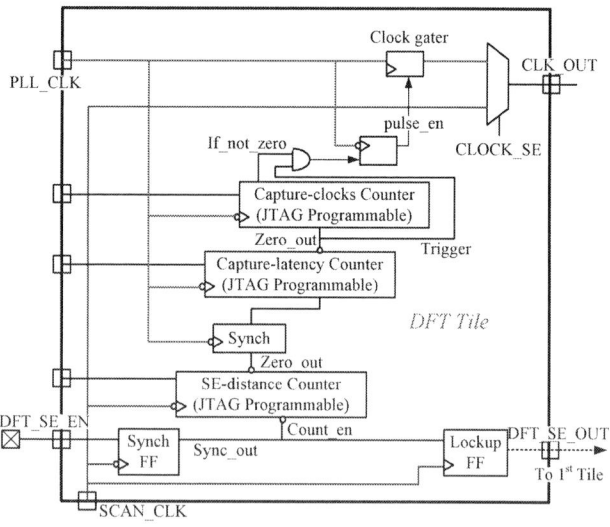

Figure 3. Launch-and-capture mechanism.

After the capture-latency counter decrements to a value of 0, the high-speed capture pulses are triggered. The capture-clocks counter is programmed to a value equal to the number of pulses required. More than two capture pulses may be desired because it has been observed that the underlying finite state machine closely resembles the functional behavior with application of multiple pulses [6]. All three registers are programmed using the JTAG test access ports [7].

SE-distance counter: The scan-enable signal is daisy chained from the DFT tile to all tiles, as already described. When tile number n is targeted, the SE-distance counter is programmed to the value n. While the scan-enable value of 0 (as required during capture) propagates to the targeted tile n through the daisy chain, the SE-distance counter decrements. Once the scan-enable value of 0 reaches the tile n, the SE-distance counter also reaches 0. The value of 0 on the SE-distance counter triggers the capture-latency counter. This counter is running on the negative edge of the scan clock to avoid any race condition.

Capture-latency counter: The purpose of this counter is to delay the high-speed pulse trigger to accurately control the duration between shift and capture. The latency counter should be programmed to such a value that the total latency (in terms of the total number of PLL pulses) is greater than scan-enable latency within the tile. Since the pipelined scan-enable signal is used, we care only about the scan-enable routing delay inside the tile. The trigger for this counter has to be synchronized because there is a clock domain crossing from the slow scan clock to the fast PLL clock. This counter is on the negative edge of the PLL clock to avoid any race condition. The latency of the scan-enable signal within the tile is determined by static timing analysis. If the latency is high, the capture pulses will be delayed, which defeats the purpose of this work because the delay between shift and capture will increase. Therefore, we define an upper limit on this latency and meet that limit during physical optimization of every tile.

Capture-clocks counter: The purpose of this counter is to control the total number of high-speed clock pulses during the capture mode. This counter is clocked by the negative edge of the PLL clock. We need a minimum of two high-speed capture pulses to exercise transition delay faults.

Figure 3 shows the counters and the synchronizers in this launch-and-capture mechanism. The shift clock is applied externally on the ATE equipment and the capture clock, as just described, is availed by programming the PLL.

The node DFT_SE_IN is connected to the external scan-enable pin. The synchronizer is needed on the scan-enable signal since it is clocked by the ATE clock and needs to be synchronized to the system clock. The signal DFT_SE_OUT is the daisy-chained signal that goes to all the tiles under test. The daisy chain from DFT_SE_OUT is shown explicitly in Figure 2. Since both the synchronizer and the pipeline flop shown as SE in Figure 2 are on the negative edge of the scan clock, a lockup latch is inserted between them. When the scan-enable signal goes low at the output of the synchronizer, the SE-distance counter is enabled. When the SE-distance counter reaches the value 0, the capture-latency counter is enabled. The capture-latency counter is on the negative edge of the PLL clock, so another synchronizer is placed between the SE-distance counter and the capture-latency counter. When the capture-latency counter reaches value 0, the capture-clock counter is enabled. When the

capture-clock counter reaches value 0, the signal labeled pulse_en goes low. While pulse_en signal is high, we observe the PLL clock at the CLK_OUT pin.

D. Waveforms

The waveforms for scan-enable are presented in Figure 4. This is presented for a case in which the SE distance is 2, the latency is 3, and the number of capture pulses is 3. It takes one scan clock pulse for the scan-enable to be synchronized. It takes two scan clock pulses for the scan-enable signal to reach the TUT. Also, three capture pulses are triggered after capture latency equal to three capture pulses.

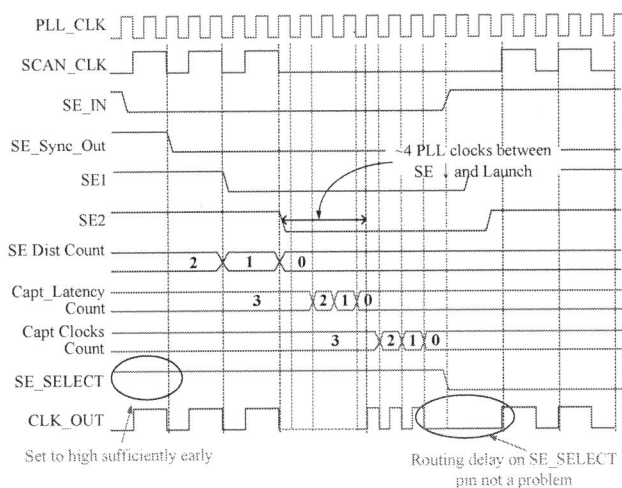

Figure 4. Scan-enable de-assertion.

Figure 4 shows the free-running PLL clock and the scan clock applied on the tester (respectively labeled PLL_CLK and SCAN_CLK). The next signal, SE_IN, is the scan-enable signal at the chip pin. When it is de-asserted, the synchronized signal, SE_Sync_Out that is synchronized with the shift clock, is shown next. One shift cycle after the SE_Sync_Out signal is de-asserted, the scan-enable signal in the first tile, SE_1, goes down. Another shift cycle later, the scan-enable signal in the second tile, SE_2, goes low. Thus, the de-asserted value is being shifted through the pipeline registers, labeled SE, in Figure 2.

While the de-asserted value of the scan-enable signal is in propagation through the daisy chain, the SE-distance counter is decrementing at every shift pulse. It goes to 0 simultaneously as the shifted scan-enable value reaches the targeted tile. The capture-latency counter and the counter that determines the number of capture pulses are labeled respectively Capt_Latency_Count and Capt_Clocks_Count. During this time, the SE_SELECT signal from Figure 2 is high and it goes low after the capture pulses are applied so the scan-enable assertion can occur asynchronously.

Note that the time interval between the last capture pulse and first shift pulse can be made long without any impact on the scheme. Therefore, the routing delay (if any) on the SE_SELECT signal, which has to propagate from a chip to a tile in the core of the chip, is not an issue during assertion of the scan-enable signal. In other words, we can give enough time to the SE_SELECT signal to propagate from the chip pin to the multiplexor in the targeted tile. Next, we can assert the scan-enable signal. Thus, the commencement of shift can be done independent of routing delay after completion of capture.

On the other hand, we can assert the SE_SELECT signal to 1 to choose the pipelined scan-enable sufficiently early. This will switch the mux to the input labeled 1 earlier than the last shift pulse; however, the scheme will work because the output of the register labeled SE will change to 0 only after the last shift pulse. This will address any concern with routing delay on the SE_SELECT signal during scan-enable de-assertion. This signal does not have to meet any timing constraint, and static timing analysis is not required for this signal.

IV. EXPERIMENTAL RESULTS

The new ac scan methodology is implemented in silicon for a graphic processor chip. The observations on silicon are presented in this section.

Tables I, II, and III show the increase in ac scan-capture frequency obtained by our methodology. Five different values of scan-capture latency are used – these values are 0, 3, 7, 15, and 31. Three different values of Vdd are used = 0.8V, 1.0V and 1.2V. The experimental results are shown for 5 different tiles. These tiles are geographically spread across the chip. Each measurement is done on 5 chips and the average number is shown. The numbers shown in these tables are the percentage improvements in capture frequency due to the new method. The percentage increase is calculated as:

$$\frac{\left(freq\_new\_method - frequency\_legacy\_method\right)}{frequency\_legacy\_method} * 100\%$$

At 0.8V, the percentage improvement ranges between 13.6% and 23.8%. For a tile, the highest increase in ac scan frequency among all values of capture latency is considered. At 1.0V, the increase is between 11.7% and 20.5%. At 1.2V, the percentage increase is between 8% and 12.2%.

TABLE I. INCREASE IN CAPTURE FREQUENCY AT 0.8V

SE latency (Time interval)	Tile1	Tile2	Tile3	Tile4	Tile5
0	15.3%	21.9%	18.0%	20.0%	13.0%
3	15.4%	21.0%	20.8%	23.6%	12.5%
7	16.8%	22.9%	21.9%	17.5%	12.3%
15	17.2%	23.8%	18.8%	24.3%	13.6%
31	13.5%	22.1%	21.7%	21.4%	12.1%

TABLE II. INCREASE IN CAPTURE FREQUENCY AT 1.0V

Capture latency (Time interval)	Tile1	Tile2	Tile3	Tile4	Tile5
0	16.2%	13.5%	11.2%	13.5%	8.7%
3	12.0%	11.0%	11.5%	10.0%	7.6%
7	13.7%	11.0%	10.5%	11.3%	6.5%
15	18.7%	18.4%	20.5%	15.5%	10.4%
31	18.7%	12.2%	12.0%	15.4%	11.7%

TABLE III. INCREASE IN CAPTURE FREQUENCY AT 1.2V

SE latency (Time interval)	Tile1	Tile2	Tile3	Tile4	Tile5
0	9.2%	8.4%	6.9%	11.0%	5.6%
3	10.0%	10.9%	11.2%	10.5%	6.8%
7	8.1%	7.0%	7.4%	8.9%	4.7%
15	11.9%	8.3%	6.1%	12.2%	8.0%
31	9.3%	10.0%	10.5%	10.2%	5.4%

Overall, the new method achieves at-speed scan testing closer to the functional frequency, even though the functional frequency is not achieved. Other phenomena need to be investigated and analyzed for exercising the circuit at near-functional condition. After using the new scheme, the scan-capture frequency achieved is between 75% and 95% of the functional frequency.

One thing we are interested in is the frequency improvement curve as a function of the latency counter. In theory, IR drop will gradually reduce during the dead cycles between shift and capture, but the effect of di/dt noise will progressively increase. A sweet spot exists where the fastest passing frequency can be achieved. Figure 5 shows the frequency improvement curve as functions of capture latency and voltage. This is plotted for both the traditional (or legacy) ac scan method (labeled by TAC) and the improved ac scan method (labeled by IAC). In this figure, SE0 refers to the situation when latency value is 0, and similarly for SE3, SE7, SE15, and SE31. We see that the maximum frequency is achieved for capture latency values of 7 and 15. We see that this is true for both legacy and new methods. Even though the %increase, relative to latency value of 0, is higher for legacy method, the actual value of passing frequency is higher for the new method.

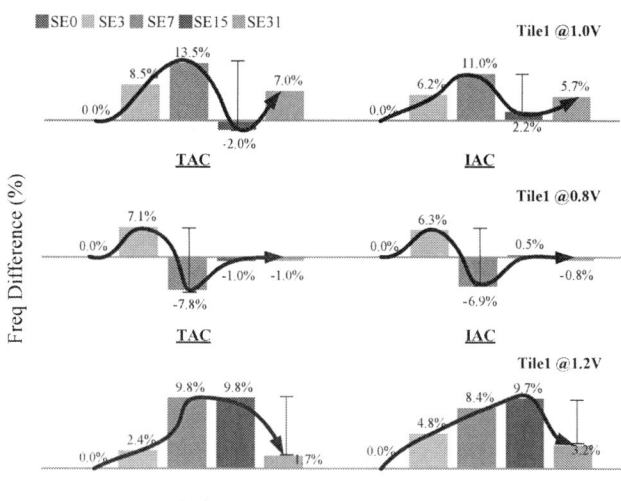

Figure 5. Frequency curve as a function of latency value between shift and capture.

Next, we would like to study the performance of the targeted tile with toggles in the surrounding tiles. We introduce toggle in the scan chains in the surrounding tiles through both the scan and capture periods of the targeted tile. By introducing the toggles, we try to capture the activity in a surrounding tile during actual functional operation. In general, the higher toggle rate provides higher current and exercises higher IR drop. In this experiment, we use toggle rate values of 12.5% (tog12.5), 25% (tog25), and 50% (tog50). In one case, we toggle only one surrounding tile and in a second scenario, we toggle three different surrounding tiles. Table IV shows the degradation in capture frequency is between 1.5% and 6.9% when 3 surrounding tiles are toggled. It should be noted that the degradation is more pronounced for capture latency values of 0 and 3 where IR drop is high to begin with. On the other hand, the impact is negligible if only one adjacent tile is toggled as shown in Table V. We observe that the worst degradation in this case is 2.3%. We see that there is an isolated situation where the capture frequency actually increases by 2.2%. There is another instance when SE latency is 15 and the applied voltage is 1V that we see that the degradation in capture frequency is worse with 1 tile toggling as compared to the situation when 3 tiles are toggling.

TABLE IV.   NEW METHOD WITH 3 ADJACENT TOGGLING TILES

SE latency	1.0V			1.2V		
	tog12.5	tog25	tog50	tog12.5	tog25	tog50
0	-4.5%	-6.9%	-4.5%	-3.3%	-3.3%	-4.9%
3	-4.3%	-4.3%	-4.3%	-1.5%	-3.2%	-1.5%
7	-2.0%	-4.1%	-4.1%	0.0%	-1.6%	-1.6%
15	2.2%	0.0%	-4.5%	0.0%	0.0%	-1.6%
31	0.0%	-2.2%	-2.2%	0.0%	-1.7%	-1.7%

TABLE V.   NEW METHOD WITH 1 ADJACENT TOGGLE TILE

SE latency	1.0V			1.2V		
	tog12.5	tog25	tog50	tog12.5	tog25	tog50
0	0.0%	0.0%	-2.3%	0.0%	-1.6%	-1.6%
3	-2.1%	0.0%	-2.1%	0.0%	0.0%	0.0%
7	-2.0%	-2.0%	-2.0%	0.0%	0.0%	0.0%
15	0.0%	-2.3%	0.0%	0.0%	0.0%	0.0%
31	0.0%	0.0%	0.0%	0.0%	0.0%	0.0%

Next we will present the degradation numbers graphically. For these graphs, please refer to Figure 6 and Figure 7. We show performance variations where five chips are sampled. With 3 surrounding toggling tiles, there are larger variations. We also observe different values for different chips. This is because of process variation, and IR drop and di/dt noise can be impacted by process variation. The impact of process variation calls for characterization of the design and decide on one or two values of scan-capture latency, which should be used to test the chip in production.

V.   CONCLUSION

We have a presented a methodology for delay testing and speed characterization. The proposed method aims to control the time interval between the last scan-shift pulse and the first capture pulse. There is a region in this interval in which the maximum speed of a part is achieved. This can be used in both defect screening and frequency binning. We describe a launch-and-capture mechanism and pipelined scan-enable

design to achieve the same. We observe by measurements on silicon implementation of this scheme that the AC scan-capture frequency increases by a magnitude of between 8% and 24%.

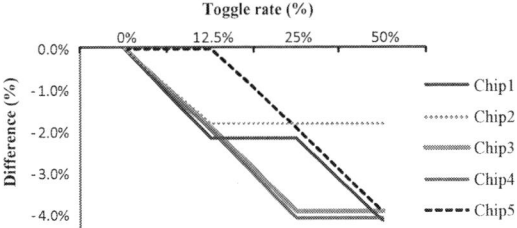

Figure 6.   Variation due to 3 adjacent toggling tiles.

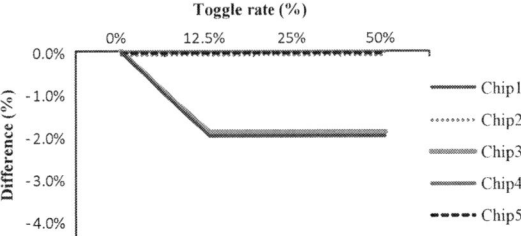

Figure 7.   Variation due to 1 adjacent toggling tile.

REFERENCES

[1] P. Pant and J. Zelman, "Understanding Power Supply Droop During At-speed Scan Testing," Proc. of VLSI Test Symposium, pp. 227-232, 2009.

[2] J. Rearick and R. Rodgers, "Calibrating Clock Stretch During AC Scan Testing," Proc. of International Test Conference, pp. 266-273, 2005.

[3] J. Saxena, K. Butler, V. Jayaram, S. Kundu, N.V. Arvind, P. Sreeprakash, and M. Hachinger, "A Case Study of IR-Drop in Structured At-Speed Testing," Proc. of International Test Conference, pp. 1098-1104, 2003.

[4] N. Ahmed, M. Tehranipoor, C.P. Ravikumar, and K. Butler, "Local At-Speed Scan Enable Generation for Transition Fault Testing Using Low-Cost Testers," IEEE Transactions on Computer-Aided Design of Integrated Circuits and Systems (CAD/ICAS), vol. 26, no. 5, pp. 896-906, May 2007.

[5] B. Nadeau-Dostie, K. Takeshita, and J.-F. Côté, "Power-Aware At-Speed Scan Test Methodology for Circuits with Synchronous Clocks," Proc. of International Test Conference, paper 9.3, 2008.

[6] J. Rearick, "Too much delay fault coverage is a bad thing," Proc. Of International Test Conference, pp. 624-633, 2001.

[7] K.P. Parker, The Boundary Scan Handbook, Springer, 2003.

[8] J. Zeng, M.S. Abadir, A. Kolhatkar, G. Vandling, L.-C. Wang, and J.A. Abraham, "On Correlating Structural Tests with Functional Tests for Speed Binning of High Performance Design," Proc. of International Test Conference, pp. 31-37, 2004.

[9] C. Liu, Y. Wu, and Y. Huang, "Effect of IR-Drop on Path Delay Testing Using Statistical Analysis," Proc. of Asian Test Symposium, pp. 245-250, 2007.

[10] N. Ahmed, M. Tehranipoor, and V. Jayaram, "Transition Delay Fault Test Pattern Generation Considering Supply Voltage Noise in a SOC Design," Proc. of Design Automation Conference, pp. 533-538, 2007.

[11] Z. Abuhamdeh, B. Hannagan, A. L. Crouch, and J. Remmers, "A Production IR-drop Screen on a Chip," IEEE Design and Test of Computers, vol. 24, issue 3, pp. 216-224, 2007.

An Efficient Method to Screen Resistive Opens under Presence of Process Variation

Seongmoon Wang

s2.wang@samsung.com, Samsung Austin Semiconductor, SARC, Austin, TX 78754

Abstract

In this paper, a cost efficient test methodology to screen chips that have resistive open defects under the presence of process variation is proposed. The proposed test methodology is based on small delay defect testing. The entire test session is divided into several subsessions. In each subsession, test patterns are applied with a different frequency of test clock, all of which are faster than the rated clock. Unlike others, different test patterns are generated and applied in each subsession to reduce test application time. A simple three step screening method is also proposed. The first step identifies scan outputs that can fail only if there are defects under the possible worst case process variation. In the second step, we assume that a chip failed due to a defect if the number of faulty scan outputs in a test pattern is much larger than that of faulty scan outputs of a typical defect free chip. Finally, the third step screens defective chips by comparing the number of fail patterns. Among 10 benchmark circuits used for the experiments, the proposed method was able to screen successfully more than 90 % of defective chips for 8 circuits.

1 Introduction

Partially broken interconnect wires can be resistive open defects. Incompletely filled vias and contacts behave also as weak resistive opens. Resistance opens increase delays of paths where they are located but will not change the function of the chip. Resistance opens located on short paths (paths whose delays are much shorter than those of critical paths) manifest themselves as small delay faults. They escape traditional transition delay test patterns since paths they are located at have large slacks. However these defects can be a serious reliability problem. Electromigration will keep carving defective wires (vias, or contacts) increasing their resistances and the chip will fail in the system eventually. Hence, chips that have resistive opens should be eliminated before shipping to maintain the reliability of the product.

Process variation is the deviation of parameters from expected characteristics due to the limited process controllability. Deep submicron chips are more sensitive to process variation [7]. Since process variation can increase (or decrease) delays of circuit paths, screening chips that have resistive opens is even more difficult under the presence of process variation. Even if the size (the amount of delay increase) of a delay defect is fairly large, if the defect is on a wire that resides only on short paths, the defect will not be detected by at-speed delay test patterns. Detecting such defects can be possible only by applying the test clock at a faster speed than the rated clock [1, 16]. Chips that fail delay test merely due to excessive process variation should be distinguished from chips that have resistive opens and not be thrown away to avoid unnecessary loss of yield. The main difference is that unlike resistive opens, delays increased due to process variation will not grow by electromigration. On the other hand, if a chip fails any test pattern applied at its system clock speed, it should be discarded regardless of the cause of failure.

Many researchers have attempted to tackle this problem by statistical test methods. As a popular statistical test approach, using small delay defect testing for outlier screening has been proposed by several researchers [12, 5, 15, 16]. Sato et al. proposed statistical delay quality model (SDQM) [8, 9] to estimate the probability of detecting small delay defects based on detectable delay defect sizes. Devta-Prasanna et al. proposed a method to accurately measure small delay defect coverage [4]. Timing-aware test pattern generation techniques that specifically target small delay defects are proposed in [6]. The three-tier foundation of statistical testing to screen latent defects is presented in [13]. Tayade at al. determine whether the deviation in circuit delay is due to random process parameters or the presence of a latent defect using delay variance in [12]. Several techniques improve the efficiency of detecting small delay defects by observing the circuit's response to a delay test at multiple sample times [16, 18]. Outlier screening methods based on Machine Learning are proposed in [15, 5]

This paper [1] proposes a cost efficient test method, which is also based on small delay defect testing, to screen chips that have latent defects (resistive opens) in the presence of process variation. Our objective is to simplify screening procedures so that statistical testing can be done without significantly increasing test application time even with low cost testers, which have limited capabilities. Machine Learning based screening methods [5, 15, 16] and variance based screening method [12] may increase test application time significantly and not be applicable to test flows that use low cost testers. Another important goal is to use an existing transition delay ATPG (automatic test pattern generator), which does not use any timing information, to minimize ATPG run time overhead.

The rest of this paper is organized as follows: Section 2 describes the test generation method used for the proposed test method. In Section 3, three different methods we use to screen resistive opens during test application are described. Section 4 describes the simulation setup for the experiments conducted to demonstrate the feasibility of our method and presents simulation results. Section 5 gives conclusions.

2. Test Generation

Although the proposed method is applicable to non-scan designs, in the rest of this paper, we assume that the circuit under test (CUT) employs full scan and both test generation and test application utilize scan without loss of generality. Like typical scan based delay testing, we assume that no primary outputs are observed (only scan outputs are observed) and the primary input part does not change between the initialization cycle and the launch cycle in any test pattern [11]. Hence, no primary inputs can trigger transitions during launch cycles.

Like others [16, 17], the proposed screening technique also applies test patterns at several different test clock frequencies, which are faster than the rated clock frequency, to detect small delay defects. However, unlike [16, 17], where the same set of test patterns is applied repeatedly at different test clock frequencies, in the proposed technique, a different set of test patterns is applied at each test clock frequency, targeting a spe-

---

[1] This work was done when the author was with NEC Labs., America.

978-1-61284-657-6/11 $26.00 © 2011 IEEE

cific part of the design. The entire test session is divided into g subsessions, $S_1, S_2, \ldots, S_g$. In subsession S_i, the period between the launch clock and the capture clock of each two pattern test is set to tp_i. S_1 is conducted first, followed by S_2, and so on (tp_1 is the shortest and tp_g is the longest).

Compared to applying the same set of test patterns repeatedly at different test clock frequencies, generating a different set of test patterns for each test clock frequency can reduce test application time significantly, especially, when test patterns are applied at many different test clock frequencies. If a set of T test patterns is applied g times (at g different test clock frequencies) repeatedly like [16, 17], the test application time required is simply gT. However, since the same T test patterns are applied g times, some test patterns may not detect any new fault at test clock frequencies applied later, wasting test application time. In contrast, in this paper, since specific faults are targeted for each test clock frequency (subsession), if there are no more faults that can be detected in the targeted circuit region, which is determined by the test clock frequency and the circuit structure, then the ATPG stops generating test patterns for the subsession and moves on the next subsession.

In [16, 17], since the same T test patters are repeatedly applied, the number of test patterns that are stored in the ATE memory is only T. In contrast, in the proposed method, since different test patterns are generated for each test clock frequency, the total number of test patterns generated by the proposed method may be larger than T. However, according to our experiments (see Section 4), test data volumes of the proposed method are only slightly larger or even smaller for some circuits than those of the method where test patterns are generated all together without grouping scan outputs.

The performance (the ratio of defective chips that are successfully screened) of the proposed screening method depends on not only the frequency of each test clock but also the number of test clock frequencies. In general, using larger number of test clock frequencies leads to better performance at the expense of longer test application time and larger test data volume. Although finding optimal test frequencies and the optimal number of test clock frequencies is an important issue, it is beyond the scope of this paper. We are currently investigating an algorithmic approach to find the optimal number of test clock frequencies and optimal test clock frequencies for a given number of test clock frequencies. Since all test clocks used are faster than the rated clock speed, hazards may be captured at some scan outputs (although the chip does not to produce any hazard during its normal operation). Hazards at scan outputs can deviate test results as pointed in [14]. However, we do not consider hazards in this paper since we use a regular transition delay ATPG tool with little modification.

2.1. Grouping Scan Outputs

Since test patterns are applied at faster clock speed than the rated clock, some scan outputs may capture errors even if there are no defects or large process variation in the CUT. Hence, those scan outputs that are driven by paths whose nominal delays are longer than the test clock period should be masked to avoid discarding defect free chips. When test patterns are applied with a faster clock period, more scan outputs should be masked. Hence, scan outputs are organized into several groups according to delays of paths that drive them.

The path whose delay is the longest among all paths that end at scan output s_j is called the *longest path* of scan output s_j and denoted by $LP(s_j)$ and its nominal delay is denoted by $DLP(s_j)$. Once periods of all test clocks are determined, scan outputs in the design are organized logically (this organization does not involve any design change) into g

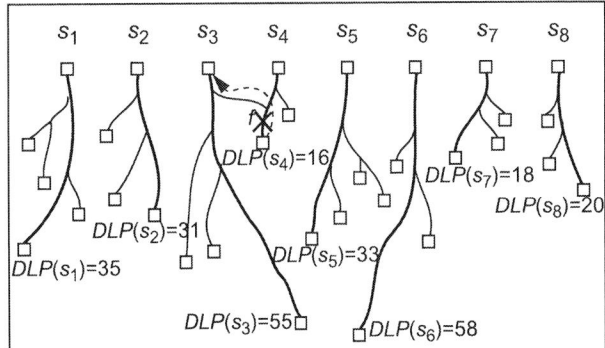

test clock periods: $tp_1 = 21$ ns, $tp_2 = 36$ ns, $tp_3 = 60$ ns
$C_1 = \{s_4, s_7, s_8\}$, $C_2 = \{s_1, s_2, s_5\}$, $C_3 = \{s_3, s_6\}$
Figure 1. Grouping Scan Cells by Path Delays

groups, $C_1, C_2, \ldots, C_g$, according to delays of their longest paths. During subsession S_i, only the scan outputs that belong to scan group C_i and the scan outputs whose longest paths have shorter delays than tp_i, i.e., scan groups C_k, where $k = 1, 2, \ldots, i - 1$, are observed. Response values captured in the other scan outputs are masked (not observed to make pass/fail decisions for CUTs). Also during test generation for S_i, only scan outputs that belong to $C_1, C_2, \ldots, C_i$ are used by ATPG as observation points.

Figure 1 shows eight scan outputs, $s_1, s_2, \ldots, s_8$, and their paths. Assume that the entire test session is divided into three subsessions S_1, S_2, and S_3, and periods of test clocks are respectively 25 ns, 38ns, and 60 ns, i.e., $tp_1 = 25$ ns, $tp_2 = 38$ ns, and $tp_3 = 60$ ns. Since delays of all paths that drive scan outputs s_4, s_7, and s_8 are shorter than 25 ns, group C_1 contains s_4, s_7, and s_8 as its elements. During S_1, only scan outputs that belong to C_1 are observed. Scan outputs s_1, s_2, and s_5 among the remaining scan outputs belong to S_2 because delays of their longest paths are smaller than 38 ns. During S_2, the scan outputs that belong to C_2 as well as C_1 are observed. Finally, the remaining scan outputs s_3 and s_6 comprise C_3. All scan outputs, $s_1, \ldots, s_8$, are observed during the last subsession S_3. Hence, unless a fault is untestable, it will be eventually detected by the proposed test generation scheme.

Grouping scan outputs according to delays of their longest paths requires timing information of all design elements (gates and wires). In this paper, delay descriptions of wires and gates in the design are obtained from the standard delay format (SDF) file of the design, which is extracted from the physical design data. The rising delay of a gate (described in an SDF file) is usually different from its falling delay. Although it is not required, in this paper, we assume that the rising delay and the falling delay of each gate are the same for the sake of simplicity and clarity of illustration. However, the proposed technique is applicable to the more realistic case where the rising delay is different from the falling delay of the same gate (indeed, in the experiments shown in Section 4.2, we allowed different falling and rising delays for the same gate).

Before grouping scan outputs according to delays of their longest paths, we identify the longest path of every scan output. Note that if a path is the longest path of a scan output, then every subpath to the scan output must also be the maximum delay subpath. This property allows finding the longest path of each scan path in linear time. A linear time algorithm, which is based on dynamic programming [3], to find the longest path of each scan output is described in Figure 2. Once the longest path of every scan output is identified, scan outputs are organized into g groups according to delays of their longest paths. Compared to time complexity of test generation with delay data [4, 2], time complexity of grouping is negligible.

978-1-61284-657-6/11 $26.00 © 2011 IEEE

```
longest_path()
    for every scan_output so_i
        trace_path(so_i);
end

trace_path(g_i)
    if g_i.mark = 1 /* gate g_i has already been visited */
        return 1;
    else if g_i.pi = 1 /* gate g_i is a primary input */
        gate.delay ← 0.0;
    else if g_i.si != 1 /* gate g_i is not a scan input */
        for every fanin f_j of g_i
            trace_path(f_j);
        end for;
        max ← 0.0;
        max_gate ← NULL;
        for every fanin f_j of g_i
            if f_j.max_delay > max
                max ← f_j.max_delay;
                max_gate ← f_j;
            end if;
        end for;
        g_i.max_delay ← max + g_i.delay;
    else /* gate g_i is a scan input */
        gi.max_delay ← g_i.delay;
    end if;
    g_i.mark ←1;
end
```

Figure 2. Pseudo-code for Identifying Longest Path

2.2. Test Generation for Divided Subsessions

If a fault is detected by a test pattern during test generation, it is dropped from the fault list to reduce test generation and fault simulation time. Ordering test subsessions from S_1 to S_g during test generation (also test application) improves the performance of the proposed method by detecting faults always at the fastest possible frequency. For example, in Figure 1, assume that fault f can be observed at both scan output s_4, which belongs to C_1, and s_3, which belongs to C_3. Since test patterns for S_1 are generated earlier than test patterns for S_3, the ATPG will generate a test pattern that detects delay fault f at s_4 while generating test patterns for S_1 and then drop f from the fault list. Hence, all delay defects on the longest path of s_4 that manifest themselves as f and whose delay defect sizes are larger than 5 ns ($tp_1 = 21$ ns and the path where f is located has 16 ns delay) will be detected at s_4 during S_1. On the other hand, if process variation across the chip can increase path delays up to 10 %, then the longest path of s_4 will have 17.6 (16×1.1) ns delay in the worst case if there are no defects in the chip. Since it is still much shorter than tp_1, defect free chips will not produce an error at s_4 in the presence of process variation that increases path delays by up to 10 %.

Now, assume that S_3 is conducted before S_1 and accordingly test patterns for S_3 are generated before S_1. The ATPG will generate a test pattern that detects f at s_3 while generating test patterns for S_3 and drop f from the fault list. Since f is already dropped, the ATPG will not attempt to generate a test pattern for f in any of the following test subsessions. Hence, unless a test pattern among the test patterns that the ATPG generate for S_1 later fortuitously detects f, f will not be detected in S_1. If the path (highlighted with a dotted arrow in Figure 1) of s_3 where f is located has only 18 ns nominal delay, then that path has 42 ns slack (tp_3 is 60 ns) during S_3.

3. Screening Methods

Recently, several publications show that applying Machine Learning techniques [15, 16] and statistical screening techniques [12] to fail logs obtained from small delay defect testing can achieve good outlier screening results. This implies

that most chips that have latent defects behave differently from defect free chips during small delay defect testing. From this, we derive two reasonable hypotheses: 1) increases in path delays caused by typical resistive opens are larger than those in path delays caused by most process variations, 2) typical resistive opens are local delay defects while process variation is distributed across the chip. There may be very subtle resistive opens so that their delay defect sizes are close to delay increases caused by process variation. However, if the size of a defect is very small, then the attrition rate at the defect location caused by electromigration will also be slow. Hence, this chip will survive throughout its expected life time.

While applying test patterns generated by the procedure described in Section 2, we employ three different screening criteria in sequence for each failing chip to identify chips that have latent defects. The proposed three screening methods, which are described in the following section, are based on the above two hypotheses. Since one of our objectives is to reduce test application time, we apply the Abort-on-Fail strategy during test application, i.e., when a CUT is found to be defective by any test pattern, we stop applying more test patterns to the failed CUT and move on to the next chip.

3.1. Screening with FODVs

The first screening method exploits the first one of the two hypotheses described above. Assume that process variation can increase nominal path delays by up to 5 % (although delay fluctuations of individual gates and wires due to process variation can be larger than 5 %, increases in path delays will be much smaller since delays of some wires and gates will be even smaller than their nominal delays in the presence of process variation). Consider conducting subsession S_i whose clock period is tp_i. Let the nominal delay of the longest path of scan output s_j, which is observed during S_i, be d, i.e., $DLP(s_j) = d$. Process variation can increase $DLP(s_j)$ to $d \times 1.05$ in the worst case. Assume that tp_i is larger than $d \times 1.05$. Since $d \times 1.05 < tp_i$, no errors will be captured at scan output s_j during S_i if a CUT is defect free. In other words, if a CUT fails a test pattern at s_j during subsession S_i, then it can be explained only by a defect.

To screen defective chips by this method, we generate an additional vector, called the *fail-only-by-defect vector (FODV)*, for each test subsession. The FODV for subsession S_i is denoted by $FODV_i$. If there are g test subsessions, then we need g FODVs. The FODV value for each scan output is determined from the delay of its longest path and the period of test clock applied in subsession S_i. The FODV value of scan output s_j for subsession S_i is defined as follows:

$$FODV_i(s_j) = \begin{cases} X & \text{if } s_j \text{ is not observed in } S_i \\ 1 & \text{if } DLP(s_j)(1.0 + MPV + GB) > tp_i \\ 0 & \text{otherwise,} \end{cases}$$

$$(1)$$

where MPV is the maximum path delay increase rate due to process variation and GB is a guard band added to minimize the chance of discarding good chips.

Figure 3 explains how to determine FODV values for subsession S_2 of the circuit shown in Figure 1. The scan outputs, s_1, s_2, s_4, s_5, s_7, and s_8, which belong to C_1 and C_2, are observed in S_2. The other scan outputs, which are not observed during S_2, are assigned X's in $FODV_2$. Assume that path delays can increase up to 5 % by process variation. We use a guard band to minimize the chance of discarding good chips due to large process variation. If 3 % is used as a guard band, we may assume that path delays can increase up to 8 % instead of 5 % by process variation. Delays of the longest paths of s_1

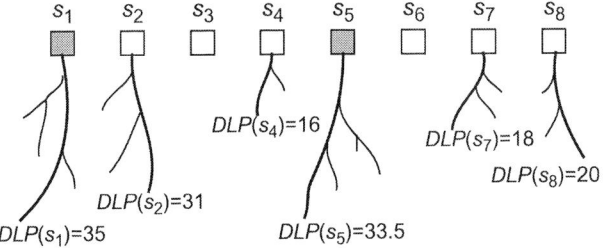

expected max process variation = 5 %   guard band = 3 %

$tclk_2$ = 36 ns   FODV<s1, s2, ..., s8>=1,0,X,0,1,X,0,0

Figure 3. Determining FODV Values

and s_5 are larger than $tp_2 = 36$ when multiplied by 1.08, i.e., $DLP(s_1) \times 1.08 = 37.8$, $DLP(s_5) \times 1.08 = 36.7$. Hence, s_1 and s_5 are assigned 1's and the other observed scan outputs, s_2, s_4, s_7, and s_8, are assigned 0's in $FODV_2$.

If s_j is assigned a 0 in $FODV_i$ and captures an error during S_i, then the CUT must have a defect. In contrast, if no faulty scan outputs, which capture errors, are assigned 0 in any FODV, then we apply the next screening criterion, which is described in the following section, to determine whether the CUT failed due to a defect or large process variation, Since only one FODV is required for each test subsession, the increase in overall test data volume due to FODVs is very small.

3.2. Screening by Number of Faulty Scan Outputs

The second method screens defective chips by finding chips that produce too many faulty scan outputs in any test pattern. If process variation increases and also decreases delays of gates and wires across a chip with equal probabilities, then the number of paths whose delays increase significantly due to process variation will be small. Hence, if there are no defects in a chip and that chip fails due to process variation, the number of faulty scan outputs will be small in most test patterns.

In contrast, a resistive open is a local defect that increases the delay of a wire (a via, or a contact). A resistive open at a wire affects all paths that pass through that wire. Hence, a defect can cause much more scan outputs to fail than process variation. For example, in Figure 4, if a defect is located at signal l_a, it affects delays of all paths from l_a to s_3, s_4, and s_5. If there is a test pattern that activates the defect at l_a and propagate the activated fault effect to all s_3, s_4, and s_5, then s_3, s_4, and s_5 will capture errors. The major limitation of this screening method is that it is effective only for defects at wires that drive many scan outputs. For example, in Figure 4, if a defect is at wire l_b, which drives only s_3, then it will be difficult to distinguish its behavior from excessive process variation.

To employ this screening method, we need to collect statistical data, i.e, the number (a typical, an average, and/or a worst case) of scan outputs that capture errors for each test pattern, from a large number of defect free chips during test application. Since these data should be obtained from known defect free chips, they may not be available, especially in the early stage of production. If that is the case, then we can use data that are collected from simulations instead of real chips and count the number of scan outputs that capture errors in each test pattern. The data collected from the simulations can be used to sort out a few defect free chips during real test application. As more defect free chips are collected, we can update the original data (collected from simulations) with data collected from the real defect free chips.

3.3. Screening by the Number of Fail Patterns

The third screening method makes the screening decision with accumulated results of previously applied test patterns. Since the increase in delay is larger (the first hypothesis), de-

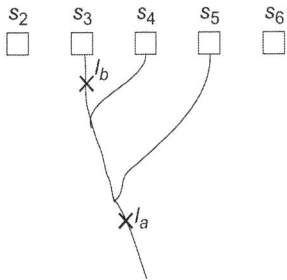

Figure 4. Defect Locations and Failing Scan Outputs

fective chips may fail test patterns that defect free chips do not usually fail. If the number of test patterns a chip fails is much larger than that of test patterns typical defect free chips fail after so many test patterns are applied, then we suppose that the chip is defective, discard it, and move on the next chip.

4. Experiments

4.1. Simulation Setup

Since chips that have resistive opens were not available, we solely relied on simulations to study the feasibility of the proposed technique. We assume that at-speed transition delay testing is conducted before small delay defect testing and removes all chips that fail any pattern. We defined the rated clock speed of each benchmark circuit as the delay of the longest path of the scan output that has the largest delay among all scan outputs, i.e., $max_{\forall s_j}\{DLP(s_j)\}$. For each circuit, we divided the test session into three subsessions, S_1, S_2, and S_3. The period of test clock for S_3, tp_3, was defined as $0.85 \times$ *the_rated_clock_period*, $tp_2 = 2tp_3/3$, and $tp_1 = tp_3/3$. Test patterns were generated with an existing transition delay ATPG tool for three subsessions separately as described in Section 2 by using the broadcast (launch-off-capture) scheme [10].

In the experiments, every chip was modeled by an SDF file. We first extracted a *reference SDF* file from physical design data of each circuit. The delays described in the reference SDF file were assumed to be nominal delays of the design. Then we generated 100 SDF files to represent 100 defect free chips, which have process variation, from the reference SDF file by increasing or decreasing the delay of every wire and gate described in the reference SDF file by a random amount within a predetermined range (± 5 % in the all experiments shown in Table 1 and ± 10 % in the all experiments shown in Table 2). In the rest of paper, an SDF file that represents a defect free chip (with process variation) is called a *defect free SDF file* and an SDF file that represents a defective chip is called a *defect SDF file*.

Then we made another set of 100 defect free SDF files to make 100 defect SDF files. Each defect SDF file was made by injecting a delay defect (by adding extra delay into the delay description of a wire) into one of these 100 defect free SDF files, which have random process variation. The size of each injected delay defect was half the delay of the longest path of the scan output that has the smallest longest path delay among all scan outputs in the design, i.e., $0.5 \times min_{\forall s_j}\{DLP(s_j)\}$. To select the 100 faults to be injected to make the 100 defect SDF files, we randomly selected 34 faults from the faults that are detected by test patterns generated for S_1, 33 faults from the faults that are detected by test patterns generated for S_2, and finally 33 faults from the faults that are detected by test patterns generated for S_3.

4.2. Simulation Results

We conducted simulations with large ISCAS 89 and ITC 99 benchmark circuits. Simulation results are shown in Table 1

Table 1. Simulation Results (5 % Process Variation)

CKT name	TL(grp) total(S_1, S_2, S_3)	TL (whole)	DLP max (ns)	DLP min (ns)	clock periods tp_1, tp_2, tp_3 (ns)	# screened > max + 1 GB 5	# screened > max + 1 GB 2	# screened > max + 2 GB 5	# screened > max + 2 GB 2
s1423	191(73, 73, 45)	183	16.68	1.42	4.73, 9.45, 14.18	22(11,11,0)	22(17, 5, 0)	21(11,3,7)	22 (17, 2, 3)
s5378	306(12, 125, 169)	481	5.956	0.55	1.69, 3.38, 5.07	91 (37,51,3)	91(44,44, 3)	43(37,0,6)	49(44, 0, 5)
s9234	519(19, 211, 289)	745	13.28	1.032	3.76, 7.53, 11.29	100(23,77,0)	100(23, 77, 0)	30(23,0,7)	30(23, 0, 7)
s13207	535(124, 110, 301)	993	12.96	0.497	3.93, 7.87, 11.80	100(10,90,0)	100(10, 90, 0)	99(10,88,1)	99(10,88, 1)
s15850	460(100, 88, 272)	312	18.33	0.538	5.13,10.26, 15.40	85(0,85,0)	85(0, 85, 0)	4(0,4,0)	5 (0, 5, 0)
s38417	309(78, 501, 330)	784	10.75	0.494	3.07, 6.15, 9.22	100(3,97,0)	100(100, 0, 0)	100(3,97.0)	100(100, 0, 0)
s38584	1202(240, 698, 264)	2718	12.92	0.527	3.775,7.55, 11.32	100(1,99,0)	100(5, 95, 0)	96(1,95,0)	96(5, 91, 0)
b20s	1573(231,850, 492)	961	37.69	2.239	10.41,20.83,31.24	15(8, 7, 0)	15(12, 3, 0)	12(8, 2, 2)	14(12, 0, 2)
b21s	1932(311,688, 933)	1330	38.49	2.186	10.71,21.43,32.14	100(5,95,0)	100(5, 95, 0)	100(5,95 0)	100(5, 95, 0)
b22s	2978(274,1004,1700)	2202	38.41	2.150	10.88,21.77,32.64	100(2,98,0)	100(2,98,0)	100(2, 98, 0)	100(2, 98,0)

and Table 2. Since the test clock period of each subsession is determined from the delay of the longest path of each scan output, if a chip is defect free, then it rarely failed any test pattern even under fairly large process variation (typically, only one or two chips among 100 chips failed just only one test pattern). Hence, to increase the number of failing defect free chips, we applied test clock at 10 % faster frequency during simulations with SDF back-annotation, which models test application, than tp_1, tp_2, and tp_3. For example, tp_1, tp_2, and tp_3 for s1423 are respectively 4.73, 9.45, and 14.18 ns. However, test clock periods for S_1, S_2, and S_3 we applied during simulations were respectively 4.25, 8.50, and 12.76 ns.

The column labeled *TL(grp)* shows the number of test patterns generated for the proposed method (see Section 2); the total number of test patterns generated (the first number in the entry) and the number of test patterns for each test subsession (three numbers separated by commas in the parenthesis). The column labeled *TL (whole)* shows the number of test patterns generated without grouping scan outputs. The column *DLP max* gives the delay of the longest path of the scan output whose longest path is the largest among all scan outputs while the column *DLP min* gives the delay of the longest path of the scan output whose longest path is the smallest among all scan outputs. The column labeled *clock periods* gives the test clock periods used for grouping scan outputs.

The number of successfully screened defective chips is shown in the columns labeled *# screened > max + 1* and *# screened > max + 2*. We experimented with two different guard bands when computing FODVs for the first screening method: 5 % (columns labeled *GB 5*) and 2 % (columns labeled *GB 2*). The first number in each entry in these columns shows the total number of successfully screened defective chips and the following three numbers in the parenthesis are respectively the number of successfully screened defective chips by each of the three screening methods (Section 3). While simulating 100 defect free SDF files, for every test pattern applied, we recorded the number of faulty scan outputs of the SDF file that has the largest number of faulty scan outputs among all 100 defect free SDFs. We used the recorded largest numbers of faulty scan outputs to define thresholds for the second screening method (see Section 3). For the results shown in the columns labeled *# screened > max + 1* (*# screened > max + 2*), we determined a CUT (an SDF file) is defective if the number of faulty scan outputs of the CUT is at least 2 (3) scan outputs larger than the recorded largest number of faulty scan outputs in the response to the corresponding test pattern. Likewise, the threshold for the third screening method was defined with the number of fail patterns of the defect free SDF file that failed the largest number of test patterns among all 100 defective free SDF files. If a CUT failed more test pat-

Table 2. Simulation Results (10 % Process Variation)

CKT name	# screened > max + 1 GB 5	# screened > max + 1 GB 2	# screened > max + 2 GB 5	# screened > max + 2 GB 2
s1423	16(7,9,0)	16(11, 5, 0)	16(7,0,9)	16(11,0,5)
s5378	32 (12,20,0)	32(12,20, 0)	23(12,0,11)	23(12,0,11)
s9234	99(16,83,0)	99(16, 83,0)	99(16,0,83)	99(16, 0,83)
s13207	75(73, 2,0)	75(73, 2, 0)	73(73, 0,0)	73(73, 0,0)
s15850	2(0, 2,0)	2(0, 2, 0)	2(0,0,2)	2 (0, 0, 2)
s38417	100(0,100,0)	100(97,3,0)	100(0,0,100)	99(97,2,0)
s38584	100(1,99,0)	100(5, 95, 0)	96(1,36,63)	100(5, 90,5)
b20s	100(95, 5, 0)	100(100, 0, 0)	100(95, 5, 0)	100(100,0,0)
b21s	100(5,95,0)	100(5, 1, 94)	100(5,95 0)	100(5,0,95)
b22s	100(2,98,0)	100(4,96,0)	100(2,98,0)	100(4,96,0)

terns than that number, we assumed that the CUT is defective.

Table 1 shows that the total number of test patterns generated by the proposed test generation method (see Section 2) was smaller than that of test patterns generated without grouping scan outputs (the column labeled *TL (whole)*) for five circuits out of ten circuits. (Note that we did not optimize either of the ATPG tools to generate the minimum test pattern set. Hence, if they are optimized, results may differ.) The simulation results are overall very promising. When the threshold for the second screening method is set to the maximum number of faulty scan outputs + 2 (the columns labeled *# screened > max + 1*), the proposed method was able to successfully screen most defective chips except s1423 and b20s. The largest number of defective chips were screened by the second screening method among the three screening methods for most benchmark circuits except s38417 and s5378 for which the first screening method screened the largest number of defective chips. Only a few or no defective chips were screened by the third method. This is mainly due to the fact that the third screening method is employed last among the three screening methods. As expected, when a smaller guard band is used (more scan outputs will be assigned 0's in FODVs), more defective chips are screened by the first screening method.

Table 2 shows results with 10 % process variation, i.e., 10 % of random variation was given to every gate and wire in the reference SDF file to model process variation. Overall, results with 10 % process variation are little worse than those with 5 % process variation shown in Table 1. However, both results follow a similar trend; high success rates for larger benchmark circuits and low success rates for smaller benchmark circuits with a few exceptions such as s15850 and b20s. Compared with 5 % process variation, while results for s15850 are very poor with 10 % process variation, results for b20s are much better with 10 % process variation. Note that for some circuits such as s1423 and s9234, all defective chips that were screened by the second screening method when the threshold is set to the maximum faulty scan outputs + 2 (columns labeled *# screened > max + 1*) were screened by the third screening method when the threshold is set to the maximum faulty scan outputs + 3 (columns labeled *# screened > max + 2*). This implies that the third screening method can effectively screen defective chips that escape the second screening method.

Although the simulation results shown above seem promising, there are still several issues to be addressed. First, we need to experiment with real chips. Some assumptions we made during simulations may not hold true for real chips. According to our extensive experiments, test clock frequencies are very important to achieve good results with the proposed test method. An efficient algorithm to find the optimal number of test clock frequencies and optimal test clock frequencies for a given number of test clock frequencies will be necessary. In this paper, we used a regular transition ATPG tool to generate test patterns with almost no modification. Although this is one of our goals in this paper, to further improve the performance, we need to optimize test pattern generation by propagating the fault effect of each fault to the maximum number of scan outputs and generating hazard free patterns.

5. Conclusions

In this paper, a test methodology to screen chips that have resistive open defects in the presence of process variation is proposed. The proposed test methodology is based on small delay defect testing. The test session is divided into several subsessions. In each subsession, test patterns are applied at a different frequency of test clock, all of which are faster than the rated clock. Unlike others [16, 17], where the same set of test patterns are repeatedly applied with different frequencies of test clock, in the proposed method, different test patterns are generated and applied with each frequency of test clock. This scheme can reduce test application significantly. Scan outputs in the design are organized into the same number of scan groups as the number of subsessions according to delays of paths that end at each scan output. While test patterns are generated for a subsession, only scan groups whose scan outputs' longest path delays are shorter than the period of test clock of the subsession are observed. An algorithm that can identify the longest path of each scan output in linear time is also proposed.

The proposed methodology uses three screening methods, which are based on two hypotheses: 1) increases in path delays caused by typical resistive opens are larger than those in path delays caused by most process variations, (2) typical resistive opens are local delay defects while process variation is distributed across the chip. The first screening method exploits the first hypothesis to identify which scan outputs cannot fail if there are no defects under the possible worst case process variation. If any of these scan outputs captures an error in any test pattern, we assume that the CUT is defective. The second screening method, which is based on the second hypothesis, determines that a CUT is defective if the number of faulty scan outputs in a test pattern is much larger than that of faulty scan outputs that most defect free chips produce in the corresponding pattern. Finally, the third screening method compares the number of fail patterns. If the number of fail patterns of a CUT is larger than that of fail patterns of typical defect free CUTs, then we assume that the CUT is defective. To demonstrate the feasibility of the proposed test method, we conducted experiments with SDF files generated from physical design data. Simulation results with SDF files show that the proposed test method can distinguish resistive open defects from large process variation for most large ISCAS 89 and ITC 99 benchmark circuits. Among ten benchmark circuits, the proposed method was able to screen successfully more than 90 % defective chips for eight circuits.

Acknowledgment

The author would like to thank Wenlong Wei of Marvell for help with developing the simulation environment.

References

[1] N. Ahmed and M. Tehranipoor. A Novel Faster-Than-at-Speed Transition-Delay Test Method Considering IR-Drop Effects. *IEEE Trans. on Computer-Aided Design of Integrated Circuit and System*, Vol. 10:1573–1582, October 2009.

[2] N. Ahmed, M. Tehranipoor, and V. Jayaram. Timing-Based Delay Test for Screening Small Delay Defects. In *Proceedings IEEE-ACM Design Automation Conference*, pages 320–325, 2006.

[3] P. H. Bardell, W. H. McAnney, and J. Savir. *Built-In Test for VLSI: Pseudorandom Techniques*. John Wiley & Sons, 1987.

[4] N. Devta-Prasanna, S. K. Goel, A. Gunda, M. Ward, and P. Krishnamurthy. Accurate Measurement of Small Delay Defect Coverage of Test Patterns. In *Proceedings IEEE International Test Conference*, pages 1–10, 2009.

[5] D. Drmanac, B. Bolin, L.-C. Wang, and M. S. Abadir. Minimizing Outlier Delay Test Cost in the Presence of Systematic Variability. In *Proceedings IEEE International Test Conference*, pages 1–10, 2009.

[6] S. K. Goel, N. Devta-Prasanna, and R. P. Turakhia. Effective and Efficient Test Pattern Generation for Small Delay Defect. In *Proceedings VLSI Testing Symposium*, pages 111–116, 2009.

[7] P. Nigh and A. Gattiker. Test Method Evaluation Experiments & Data. In *Proceedings IEEE International Test Conference*, pages 454–463, 2000.

[8] Y. Sato, S. Hamada, T. Maeda, A. Takatori, and S. Kajihara. Evaluation of the Statistical Delay Quality Model. In *Proceedings IEEE Asian and South Pacific Design Automation Conference*, pages 305–310, 2005.

[9] Y. Sato, S. Hamada, T. Maeda, A. Takatori, Y. Nozuyama, and S. Kajihara. Invisible Delay Quality – SDQM Model Lights Up What Could Not Be Seen. In *Proceedings IEEE International Test Conference*, pages 1–10, 2005.

[10] J. Savir. Broad-side Delay Test. *IEEE Trans. on Computer-Aided Design of Integrated Circuit and System*, Vol. 13(8):1057–1064, August 1994.

[11] J. Saxena, K. M. Butler, J. Gatt, R. R, S. P. Kumar, S. Basu, D. J. Campbell, and J. Berech. Scan-Based Transition Fault Testing - Implementation and Low Cost Test Challenges. In *Proceedings IEEE International Test Conference*, pages 1120–1129, 2002.

[12] R. Tayade, S. Sundereswaran, and J. Abraham. Small-Delay Defect Detection in the Presence of Process Variations. In *Proceedings IEEE International Symposium on Quality Electronic Design*, pages 711–716, 2007.

[13] R. P. Turakhia, W. R. Daasch, J. Lurkins, and B. Benware. Changing Test and Data Modeling Requirements for Screening Lantent Defects as Statistical Outliers. *IEEE Design & Test of Computers*, pages 100–109, March-April 2006.

[14] J. Wang, H. Li, Y. Min, X. Li, and H. Liang. Impact of Hazards on Pattern Selection for Small Delay Defects. In *Proceedings IEEE Pacific Rim International Symposium on Dependable Computing*, pages 49–54, 2009.

[15] S. H. Wu, D. Drmanac, and L.-C. Wang. A Study of Outlier Analysis Technique for Delay Testing. In *Proceedings IEEE International Test Conference*, pages 1–10, 2008.

[16] S. H. Wu, B. N. Lee, L.-C. Wang, and M. S. Abadir. Statistical Analysis and Optimization of Parametric Delay Test. In *Proceedings IEEE International Test Conference*, pages 1–10, 2007.

[17] H. Yan and A. D. Singh. Experiments in Detecting Delay Faults Using Muliple Higher Frequency Clocks and Results from Neighboring Die. In *Proceedings IEEE International Test Conference*, pages 105–111, 2003.

[18] H. Yan and A. D. Singh. On the Effectiveness of Detecting Small Delay Defects in the Slack Interval. In *Proceedings IEEE International Workshop on Defect Based Testing*, pages 49–54, 2004.

On Clustering of Undetectable Transition Faults in Standard-Scan Circuits

Irith Pomeranz
School of Electrical & Computer Eng.
Purdue University
W. Lafayette, IN 47907, U.S.A.

Abstract - **Transition faults are used for modeling delay defects. A comparison between transition faults and single stuck-at faults indicates that many more transition faults than single stuck-at faults in standard-scan circuits are undetectable. Furthermore, this paper shows that undetectable transition faults in benchmark circuits appear in larger clusters than single stuck-at faults, where a cluster consists of several undetectable faults that are included in the same connected subcircuit. This implies that test sets for transition faults do not cover delay defects uniformly across the circuit. The paper studies the clustering of undetectable transition faults in standard-scan benchmark circuits by considering exhaustive as well as deterministic test sets. It defines double transition faults that provide targets for improving the coverage of subcircuits with undetectable transition faults, and presents the results of test generation.**

Keywords: Double faults, scan circuits, transition faults, undetectable faults.

I. INTRODUCTION

Gate-level fault models are used for representing the behavior of defects at the gate level, where test generation is tractable. A test set generated for a target gate-level fault model is expected to be effective in detecting defects even if the fault model does not capture the behavior of defects accurately. For example, a test set generated for single stuck-at faults is expected to detect static defects associated with the sites of stuck-at faults. A test set generated for transition faults is expected to detect localized delay defects associated with the sites of transition faults.

When a target fault is undetectable, it leaves an uncovered site in the circuit. Considering single stuck-at faults, it was demonstrated in [1] that undetectable single stuck-at faults cluster in subcircuits of benchmark circuits. This implies that certain subcircuits are uncovered, or less covered than others, by a test set for single stuck-at faults. The clustering effect for transition faults in standard-scan circuits is expected to be more severe than that of stuck-at faults, for the following reason.

A test for a single transition fault also detects a corresponding single stuck-at fault [2]. Conversely, the existence of an undetectable single stuck-at fault implies the existence of an undetectable transition fault. Therefore, clustering effects that exist for stuck-at faults also exist for transition faults. With enhanced-scan [3], every two-pattern test can be applied to the circuit. As a result, if an enhanced-scan circuit has no undetectable stuck-at faults, all its transition faults are detectable as well. The situation is different with standard-scan. Standard-scan allows skewed-load [4] or broadside [5] tests to be applied. In a skewed-load test, a scan operation is followed by an additional scan shift cycle and a functional capture cycle that define the two patterns of the test. In a broadside test, a scan operation is followed by two functional capture cycles that define the two patterns of the test. With both types of tests, there are two-pattern tests that cannot be applied to the circuit. As a result, the set of undetectable transition faults is significantly larger under standard-scan than under enhanced-scan, and it is significantly larger than the set of undetectable single stuck-at faults. In particular, circuits with no undetectable single stuck-at faults have undetectable transition faults under standard-scan. With more undetectable faults, clustering can be more severe.

Motivated by these observations, this paper studies the clustering effect of undetectable transition faults under skewed-load tests and under broadside tests. The results indicate that significant clustering occurs in both cases.

Another observation that is important to this discussion is that the effects of delay defects can accumulate along a path, causing a path to fail only after it accumulates a sufficient number of extra delays. This is the motivation for using the path delay fault model [6]. It also motivated the development of delay fault models that involve subsets of lines, such as the double transition fault model from [7], the segment delay fault model from [8], and the transition path delay fault model from [9].

These models support a solution, similar to the one used in [1] for single stuck-at faults, where multiple transition faults are used to provide additional coverage for subcircuits that contain undetectable single transition faults. The procedure proposed in this paper considers double transition faults that consist of an undetectable transition fault and an adjacent detectable transition fault in order to improve the coverage of subcircuits containing undetectable transition faults. The motivation for considering such double faults is the following.

Based on [10], if an undetectable fault f_{j1} occurs in the circuit without being detected and a second fault f_{j2} occurs, a test set that detects f_{j2} when it is present alone may not detect the double fault that consists of f_{j1} and f_{j2}. When f_{j2} is a detectable transition fault that is adjacent to an undetectable transition fault f_{j1}, the likelihood that the presence of f_{j1} will invalidate the tests for f_{j2} is higher than when the faults are in different parts of the circuit. Considering the double fault explicitly, it is possible to ensure that f_{j2} will be detected when f_{j1} is present. This is important for improving the coverage of delay defects around the sites of f_{j1} and f_{j2}.

This paper performs the study of clustering of undetectable transition faults in two parts. The first part of the study uses exhaustive test sets for finite-state machine benchmarks. These test sets provide complete information about undetectable transition faults. The second part of the study uses deterministic broadside test sets for ISCAS-89 benchmarks. In this case, undetected faults are treated in the same way as undetectable faults. This represents the more practical situation where it is not possible to detect every fault or prove that it is undetectable.

The paper is organized as follows. Section II defines adjacency between undetectable faults. It describes the partitioning of the set of undetectable faults, denoted by U, into subsets U_0, U_1, $\cdots$, U_{m-1} of adjacent faults. It then defines corresponding subcircuits G_0, G_1, $\cdots$, G_{m-1} where undetectable faults cluster. Section III considers finite-state machine benchmarks under exhaustive test sets for transition faults. Section IV considers ISCAS-89 benchmarks under deterministic broadside test sets for transition faults. Section V considers the possibility of increasing the coverage of subcircuits with undetectable faults by targeting double transition faults.

II. ADJACENT UNDETECTABLE FAULTS

We denote the $v \to v'$ transition fault on a line g by $g/v \to v'$. We say that two transition faults $f_{j1} = g_{j1}/v_{j1} \to v'_{j1}$ and $f_{j2} = g_{j2}/v_{j2} \to v'_{j2}$ are adjacent if one of the following conditions is satisfied. (1) g_{j1} and g_{j2} are the same line. (2) For a gate γ, g_{j1} is the input of γ and g_{j2} is the output of γ, or vice versa. (3) For a gate γ, g_{j1} and g_{j2} are inputs of γ. (4) g_{j1} is a fanout stem and g_{j2} is one of its fanout branches, or vice versa. (5) For a fanout stem g, g_{j1} and g_{j2} are fanout branches of g.

Let $U = \{f_0, f_1, \cdots, f_{n-1}\}$ be a set of undetectable faults. The procedure for partitioning U into subsets U_0, U_1, $\cdots$, U_{m-1} of adjacent faults proceeds as follows. Initially, $U_i = \{f_i\}$ for $0 \le i < n$. The procedure repeats the following step in order to merge pairs of subsets that contain adjacent faults until no additional merging is possible.

For every pair of subsets U_{i1} and U_{i2} such that $i1 < i2$, the procedure checks whether U_{i1} and U_{i2} contain faults f_{j1} and f_{j2}, respectively, such that f_{j1} and f_{j2}

are adjacent. If so, it adds the faults from U_{i2} to U_{i1}, and removes U_{i2}.

For illustration we show in Figure 1 part of finite-state machine benchmark *bbara*. We consider the undetectable transition faults under skewed-load tests. The notation $v \to v'$ above a line indicates that the $v \to v'$ transition fault on the line is undetectable by skewed-load tests. All the undetectable faults in Figure 1 are included in a single subset of adjacent undetectable faults.

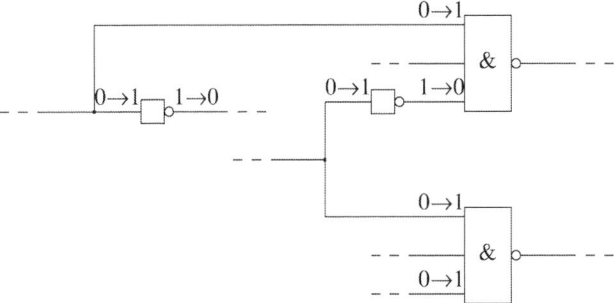

Figure 1. Part of *bbara*

Let $U_i = \{f_j : 0 \le j < p\}$. Let $f_j = g_j / v_j \to v'_j$, for $0 \le j < p$ (f_j is the $v_j \to v'_j$ transition fault on line g_j). U_i defines a subcircuit $G_i = \{g_j : f_j \in U_i\}$ such that the lines in G_i have undetectable faults from U_i. Since every line in G_i has at least one undetectable transition fault, the transition fault coverage for G_i is at most 50%. This is typically significantly smaller than the overall transition fault coverage for a circuit.

III. EXHAUSTIVE TEST SETS

This section considers exhaustive test sets of the following types. E_{skw} is an exhaustive skewed-load test set. E_{brd} is an exhaustive broadside test set. These test sets provide complete information about detectable and undetectable transition faults in standard-scan circuits.

For every test set E_{typ}, where $typ = skw$ or brd, fault simulation of the set of transition faults F under E_{typ} yields a set D_{typ} of detectable faults, and a set U_{typ} of undetectable faults. For $typ = skw$ and brd, we partition U_{typ} into subsets as described in Section II. The results for finite-state machine benchmarks [11] are shown in Table I in the following format.

Row typ correspond to U_{typ}, for $typ = skw$ and brd. In every case, column *und* shows the number of undetectable faults in U_{typ}, column *sub* shows the number of subsets obtained by partitioning U_{typ} based on adjacency relations between faults, column *max* shows the size of the largest subset obtained, and column *ave* shows and the average size of a subset.

Most of the circuits in Table I do not have any undetectable single stuck-at faults. When they do, the number of undetectable single stuck-at faults is small. The number of undetectable single transition faults is significantly higher than the number of undetectable sin-

TABLE I. Exhaustive test sets

circuit	typ	und	sub	max	ave
bbara	skw	8	2	7	4.00
bbara	brd	24	3	12	8.00
bbsse	skw	25	12	6	2.08
bbsse	brd	37	12	8	3.08
beecount	skw	7	2	6	3.50
beecount	brd	17	5	8	3.40
cse	skw	55	14	18	3.93
cse	brd	66	12	40	5.50
dk14	skw	15	5	8	3.00
dk14	brd	19	7	6	2.71
dk15	skw	2	1	2	2.00
dk15	brd	5	4	2	1.25
dk16	skw	93	13	36	7.15
dk16	brd	51	17	11	3.00
dk17	skw	7	3	3	2.33
dk17	brd	9	6	3	1.50
dk512	skw	19	5	11	3.80
dk512	brd	22	6	9	3.67
donfile	skw	44	6	33	7.33
donfile	brd	22	9	7	2.44
ex2	skw	40	4	26	10.00
ex2	brd	17	10	6	1.70
ex3	skw	16	5	7	3.20
ex3	brd	4	2	2	2.00
ex4	skw	32	3	25	10.67
ex4	brd	27	7	16	3.86
ex5	skw	14	5	5	2.80
ex5	brd	22	6	15	3.67
ex6	skw	19	4	8	4.75
ex6	brd	22	10	4	2.20
ex7	skw	18	2	15	9.00
ex7	brd	19	3	15	6.33
mark1	skw	42	2	40	21.00
mark1	brd	44	4	31	11.00
opus	skw	19	7	8	2.71
opus	brd	7	5	3	1.40

gle stuck-at faults due to the restrictions of standard-scan. The following points can be seen from Table I.

Both skewed-load and broadside tests have significant numbers of undetectable transition faults, which show clear clustering effects. The faults are partitioned into a small number of subsets, and the maximum as well as average subset size is high compared with the number of undetectable faults. For example, *mark*1 has 42 undetectable faults under skewed-load tests. These faults are partitioned into two subsets, the largest of them containing 40 undetectable faults. The circuit has 44 undetectable faults under broadside tests. These faults are partitioned into four subsets, the largest of them containing 31 undetectable faults.

IV. DETERMINISTIC BROADSIDE TEST SETS

This section repeats the experiment of Section III using a broadside test set T_{brd} for transition faults. The test set is generated by a combination of simulation-based and deterministic test generation procedures. The deterministic test generation procedure is based on the sequential test generation procedure from [12].

As in Section III, we perform fault simulation of the set of transition faults F under T_{brd} to obtain a set D_{brd} of detected faults, and a set U_{brd} of undetected faults. In this case, we do not refer to the faults in U_{brd} as undetectable. This is due to the fact that the test generation procedure that produces T_{brd} may abort on some faults. However, without tests for these faults, the fault sites remain uncovered, and the faults should be treated in the same way as undetectable faults for the discussion of clustering.

We partition U_{brd} into subsets as described in Section II. The results for ISCAS-89 benchmarks are shown in Table II in the same format as Table I.

TABLE II. Deterministic broadside test sets

circuit	typ	und	sub	max	ave
s208	brd	95	7	55	13.57
s298	brd	109	5	96	21.80
s344	brd	38	4	25	9.50
s382	brd	165	7	107	23.57
s386	brd	161	1	161	161.00
s420	brd	222	14	77	15.86
s510	brd	103	22	32	4.68
s526	brd	372	1	372	372.00
s641	brd	50	11	28	4.55
s820	brd	324	3	321	108.00
s953	brd	102	13	42	7.85
s1196	brd	2	2	1	1.00
s1423	brd	326	56	64	5.82
s5378	brd	841	34	590	24.74
s9234	brd	3009	142	600	21.19
s13207	brd	4868	301	2573	16.17
s15850	brd	8985	233	3958	38.56
s35932	brd	9191	578	7344	15.90
s38417	brd	2263	440	156	5.14
s38584	brd	10264	489	6881	20.99

The smaller circuits in Table II do not have any undetectable single stuck-at faults. Nevertheless, most of the circuits have significant numbers of undetected transition faults. Table II demonstrates significant clustering effects for the undetected transition faults. The faults are partitioned into small numbers of subsets, and the largest as well as average subset size is typically high.

V. TEST GENERATION FOR DOUBLE TRANSITION FAULTS

The results of Sections III and IV indicate that undetectable (or undetected) transition faults tend to cluster in certain subcircuits. This section defines sets of double transition faults whose detection will improve the coverage of subcircuits containing undetectable (or undetected) transition faults. We focus on the motivation given by [10] for considering double faults, namely, the possibility that a test for a detectable fault will be invalidated by the presence of an undetectable fault. After defining the set of target double faults, this section describes a test generation process for these faults, and presents experimental results.

Starting from a given test set T for single transition faults, the proposed process obtains the set of detected faults D and the subsets of undetected faults $U_0, U_1, \cdots,$

U_{m-1}. We assume that the subsets U_0, U_1, $\cdots$, U_{m-1} are ordered such that $|U_i| \geq |U_{i+1}|$ for $0 \leq i < m-1$. For $i = 0, 1, \cdots, m-1$, the process obtains a subset of double transition faults $F_{2,i}$ based on U_i. It then performs test generation for the faults in $F_{2,i}$. If new tests are generated based on $F_{2,i}$, they are added to T. Thus, tests generated for U_i are available when U_{i+1} is considered, for $0 \leq i < m-1$.

A. Double faults

The goal of defining the set $F_{2,i}$ of double transition faults is to provide a target for improving the coverage of the subcircuit G_i that contains the subset of undetected faults U_i. In particular, the goal is to ensure that the detectable faults in (and around) G_i will continue to be detected in the presence of the undetected faults in U_i. Due to their proximity, the presence of faults from U_i may affect values of lines in and around G_i. These values may be necessary for the detection of detectable faults in and around G_i. As a result, tests for the detectable faults in and around G_i are more likely to be invalidated by the presence of the faults in U_i. Therefore, double faults involving faults from U_i and detectable faults adjacent to them are important to consider.

For $U_i = \{g_j/v_j \rightarrow v'_j : 0 \leq j < p\}$, we have that $G_i = \{g_j : f_j \in U_i\}$. We also define a subcircuit H_i that contains lines, which are adjacent to G_i, as follows.

For every $g_j \in G_i$ and every line h_k that is adjacent to g_j, if $h_k \notin G_i$, we include h_k in H_i.

For illustration we consider ISCAS-89 benchmark circuit $s27$ shown in Figure 2. Considering a broadside test set, we find that the subcircuit $G_i = \{4, 16, 19\}$ contains undetected transition faults. The lines included in this subcircuit are marked with squares in Figure 2. The subcircuit $H_i = \{14, 17, 18, 20\}$ is defined based on G_i as follows. Lines 14 and 17 are added to H_i since they are adjacent to line 16. Lines 18 and 20 are added to H_i since they are adjacent to line 19. The lines included in H_i are marked with circles in Figure 2.

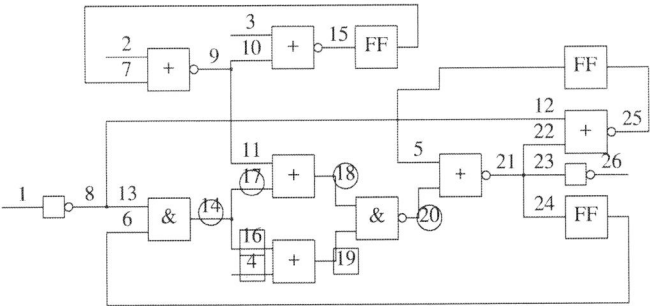

Figure 2. ISCAS-89 benchmark $s27$

It is possible to define $F_{2,i}$ based on G_i or based on $G_i \cup H_i$. We consider the case of $G_i \cup H_i$. To obtain the case where H_i is not used it is possible to set $H_i = \phi$.

For every pair of lines $g_{j1} \in G_i \cup H_i$ and $g_{j2} \in G_i \cup H_i$ such that $j1 < j2$, for every $v_{j1} \in \{0,1\}$, and for every $v_{j2} \in \{0,1\}$, the double fault $(g_{j1}/v_{j1} \rightarrow v'_{j1}, g_{j2}/v_{j2} \rightarrow v'_{j2})$ is included in $F_{2,i}$ if the following conditions are satisfied.

(1) $g_{j1}/v_{j1} \rightarrow v'_{j1} \in U_i$ or $g_{j2}/v_{j2} \rightarrow v'_{j2} \in U_i$ (that is, one of the faults is undetected).
(2) $g_{j1}/v_{j1} \rightarrow v'_{j1} \in D$ or $g_{j2}/v_{j2} \rightarrow v'_{j2} \in D$ (that is, one of the faults is detectable).

Some of the faults in $F_{2,i}$ may be detected by the given test set T. Fault simulation with fault dropping of $F_{2,i}$ under T identifies such faults. The remaining faults are targeted during test generation.

B. Test generation

For the faults in $F_{2,i}$ that are not detected by T, it is possible to perform deterministic test generation by modifying a test generation procedure for single transition faults [13]-[18] to consider double transition faults. For the experiment reported in the next subsection, the test generation procedure obtains additional tests for the faults in $F_{2,i}$ by introducing small numbers of bit changes to the tests in T. Changing small numbers of bits is used by simulation-based test generation procedures and built-in self-test methods to obtain new tests.

The procedure used here considers broadside tests. A broadside test consists of a scan-in state s, and two primary input vectors a_1 and a_2. After s is scanned in, a_1 and a_2 are applied in two consecutive functional clock cycles. We represent a broadside test as a single vector $t = sa_1a_2$. For a circuit with K state variables and N primary inputs, the bitwise representation of t is $t = t(0)\, t(1) \cdots t(K+2N-1)$.

For $M = 1, 2$ and 4, the procedure considers tests that differ in M bits from the tests included in T. For every value of M, the procedure performs M iterations over the tests in T. When a test $t \in T$ is considered, the procedure first includes in a set I the indices of all the bits of t by setting $I = \{0, 1, \cdots, K+2N-1\}$. It generates modified tests based on t as long as I contains at least M indices, as follows.

It selects a subset $I_{curr} \subseteq I$ of M indices randomly. It defines a modified version of t by complementing $t(i)$ for every $i \in I_{curr}$. It then removes I_{curr} from I.

This procedure considers $(K+2N-1)/M$ modified versions of t in every iteration. Since it performs M iterations, it considers approximately $K+2N-1$ modified versions of t for every value of M.

For illustration we consider a circuit with three state variables and one primary input. Let $t = 000\ 0\ 0$ as shown in Table III(a). With $M = 1$ the procedure considers t once, and modifies t one bit at a time. Suppose that the bits are selected such that $I_{curr} = \{1\}, \{3\}, \{2\}, \{4\}$ and finally $\{0\}$. This results in the tests shown in Table III(b).

With $M = 2$ the procedure considers t twice, and modifies t two bits at a time. Suppose that the bits in the first iteration are selected such that $I_{curr} = \{2,3\}$ and then $\{0,4\}$. Suppose that the bits in the second iteration are selected such that $I_{curr} = \{1,3\}$ and then $\{2,4\}$. This results in the tests shown in Table III(c). With $M = 4$ the procedure considers t four times, and modifies t four bits at a time. The result may be the tests shown in Table III(d).

TABLE III. Example of new tests

(a)	000 0 0
(b)	010 0 0
M=1	000 1 0
	001 0 0
	000 0 1
	100 0 0
(c)	001 1 0
M=2	100 0 1
	010 1 0
	001 0 1
(d)	111 0 1
M=4	011 1 1
	110 1 1
	111 1 0

The procedure simulates the faults in $F_{2,i}$ under every test t_k obtained based on every test $t \in T$. Suppose that a new test t_k detects a fault $(f_{j1}, f_{j2}) \in F_{2,i}$. Then t_k is added to T and (f_{j1}, f_{j2}) is removed from $F_{2,i}$.

New tests added to T in this process are also used for generating new tests. This continues until no additional tests are obtained, which detect faults from $F_{2,i}$.

C. Experimental results

We defined double faults and generated additional tests for the circuits in Table II. We used the transition fault test sets used for Table II after static test compaction. For every subset U_i of undetected faults, we first found the set of double faults $F_{2,i}$ using $H_i = \phi$. Only if the number of double faults was lower than 100000, we recomputed $F_{2,i}$ using the set of adjacent lines H_i as defined in Subsection V.A. This was done in order to limit the numbers of double faults and obtain moderate increases in test set size. The results are reported in Tables IV and V as follows.

The first row for every circuit shows the number of tests in T before adding any tests based on the subsets U_i. Next, there is a row for every subset U_i that resulted in the generation of new tests. A circuit is omitted from Tables IV and V if no tests were added based on any one of the largest subsets U_i. For $s38417$ we show only the first 30 subsets that resulted in the generation of new tests.

A row corresponding to a subset U_i shows the index i under column i. Column Hi has a 1 if H_i is used for defining double faults as described in Subsection V.A, or a 0 if $H_i = \phi$ is used. Column und shows the number of faults in U_i. Column $double$ shows the number of double faults included in $F_{2,i}$ based on U_i, followed by the number of double faults detected after fault simulation

TABLE IV. Double transition faults

circuit	i	Hi	und	double total	detect	tests
s208						33
s208	0	1	55	3986	3936	37
s298						36
s298	0	1	96	27608	27574	44
s382						45
s382	0	1	107	27668	27641	49
s386						58
s386	0	1	161	65430	65388	69
s420						66
s420	1	1	55	3986	3936	70
s420	2	1	55	3986	3936	71
s526						70
s526	0	1	372	190388	190258	81
s820						133
s820	0	1	321	284582	284512	147
s1423						136
s1423	0	1	64	14400	14395	137
s1423	1	1	23	1576	1568	141
s1423	34	1	4	78	77	142
s1423	45	1	2	42	42	143
s1423	46	1	2	38	37	144
s1423	47	1	2	46	46	145
s5378						403
s5378	0	1	590	316128	316128	452
s5378	4	1	23	930	930	458
s5378	7	1	8	264	264	463
s9234						685
s9234	0	0	600	148552	147996	783
s9234	2	1	305	113692	113691	784
s9234	3	0	265	106164	105658	792
s9234	10	1	46	6530	6512	793
s9234	14	1	38	2348	2290	795
s9234	16	1	33	2320	2282	796
s9234	60	1	5	162	162	797
s9234	70	1	3	48	46	798
s9234	72	1	3	38	38	799
s9234	77	1	3	90	90	801
s9234	78	1	3	90	90	803
s9234	79	1	3	36	34	804
s9234	80	1	3	36	34	805
s13207						691
s13207	0	0	2573	3150700	3149288	1293
s13207	2	1	152	27020	26978	1295
s13207	4	1	100	13382	13371	1297
s13207	177	1	1	10	10	1298
s13207	178	1	1	10	10	1299
s15850						573
s15850	2	0	690	263362	263337	574
s15850	17	1	46	3102	3102	582
s15850	51	1	9	166	166	583
s15850	52	1	9	166	166	584

and test generation. Column $tests$ shows the number of tests in the test set T after U_i is considered.

It should be noted that U_0 is the largest subset of undetected faults, and the subset size decreases as i is increased. The size of G_i, and the size of $G_i \cup H_i$, may not decrease monotonically with i since it depends on the locations of the faults included in U_i and on the circuit structure around these locations.

Most of the circuits in Table IV have smaller subsets of undetected faults than the ones reported in Table IV. However, these subsets do not contribute additional tests since the faults defined for them are either already

TABLE V. Double transition faults

circuit	i	Hi	und	double total	detect	tests
s38417						1469
s38417	0	1	156	33296	33277	1479
s38417	1	1	101	22646	22578	1480
s38417	3	1	56	5790	5769	1486
s38417	4	1	55	5522	5501	1492
s38417	5	1	55	5960	5942	1498
s38417	6	1	46	4292	4279	1504
s38417	7	1	46	4292	4279	1513
s38417	9	1	41	4516	4513	1518
s38417	10	1	40	4126	4123	1522
s38417	11	1	40	11320	11320	1550
s38417	12	1	40	9080	9080	1583
s38417	13	1	37	2828	2812	1586
s38417	14	1	36	2498	2485	1590
s38417	15	1	31	6262	6262	1606
s38417	16	1	29	2614	2612	1610
s38417	17	1	22	1304	1303	1618
s38417	19	1	21	4032	4032	1632
s38417	21	1	19	730	722	1634
s38417	22	1	18	956	956	1638
s38417	23	1	17	652	647	1641
s38417	24	1	17	652	647	1643
s38417	25	1	17	688	680	1644
s38417	26	1	17	854	853	1649
s38417	27	1	16	568	560	1650
s38417	28	1	15	484	480	1653
s38417	29	1	14	496	491	1654
s38417	30	1	14	496	491	1655
s38417	31	1	14	496	491	1656
s38417	32	1	14	496	491	1657
s38417	33	1	14	438	437	1659

detected or undetectable. Most of the additional tests are generated based on the largest subsets of each circuit, where detectable faults that are not already detected are defined. Moreover, in the circuits for which no additional tests were obtained for any subset, the largest subset size is relatively small. An example is $s344$, for which the largest subset contains 25 undetected faults.

From Tables IV and V it can be seen that the numbers of tests added based on double faults are significant (although substantially smaller than the number of double faults considered). This indicates that there are significant numbers of cases where undetected transition faults invalidate tests for adjacent detectable faults. Considering the double faults thus provides an improvement to the delay defect coverage around the undetected faults.

Finally, test generation can stop when a constraint on test set size is reached.

VI. CONCLUDING REMARKS

This paper studied the clustering of undetectable transition faults in standard-scan benchmark circuits using exhaustive as well as deterministic test sets. Transition faults in standard-scan circuits may be undetectable due to the type of tests applicable to the circuit (skewed-load or broadside tests), as well as due to the structure of the circuit. The results demonstrated that clustering effects are significant with both types of standard-scan test sets. The paper defined double transition faults to improve the cov-

erage of subcircuits with undetected faults. In particular, it addressed the possibility that tests for detectable transition faults will be invalidated in the presence of adjacent undetected faults. It also presented the results of test generation for these faults. The results indicated that there are significant numbers of cases where undetected transition faults invalidate tests for adjacent detectable faults. Therefore, considering the double faults provides an improvement to the delay defect coverage around the undetected faults.

REFERENCES

[1] I. Pomeranz and S. M. Reddy, "On Clustering of Undetectable Single Stuck-At Faults and Test Quality in Full-Scan Circuits", IEEE Trans. on Computer-Aided Design, July 2010, pp. 1135-1140.

[2] J. Waicukauski, E. Lindbloom, B. Rosen and V. Iyengar, "Transition Fault Simulation", IEEE Design & Test, April 1987, pp. 32-38.

[3] S. Dasgupta, R. G. Walther, T. W. Williams and E. B. Eichelberger, "An Enhancement to LSSD and Some Applications of LSSD in Reliability, Availability and Serviceability", in Proc. Fault-Tolerant Computing Symp., 1981, pp. 880-885.

[4] J. Savir and S. Patil, "Scan-Based Transition Test", IEEE Trans. on Computer-Aided Design, Aug. 1993, pp. 1232-1241.

[5] J. Savir and S. Patil, "Broad-Side Delay Test", IEEE Trans. on Computer-Aided Design, Aug. 1994, pp. 1057-1064.

[6] G. L. Smith, "Model for Delay Faults Based Upon Paths", in Proc. Intl. Test Conf., 1985, pp. 342-349.

[7] I. Pomeranz, S. M. Reddy and J. H. Patel "On Double Transition Faults as a Delay Fault Model", in Proc. Great Lakes Symp. on VLSI, 1996, pp. 282-287.

[8] K. Heragu, J. H. Patel and V. D. Agrawal, "Segment Delay Faults: A New Fault Model", in Proc. VLSI Test Symp., 1996, pp. 32-39.

[9] I. Pomeranz and S. M. Reddy, "Transition Path Delay Faults: A New Path Delay Fault Model for Small and Large Delay Defects", IEEE Trans. on VLSI Systems, Jan. 2008, pp. 98-107.

[10] M. Abramovici, M. A. Breuer and A. D. Friedman, Digital Systems Testing and Testable Design, IEEE Press, 1995.

[11] 1991 MCNC International Workshop on Logic Synthesis.

[12] X. Lin, I. Pomeranz and S. M. Reddy, "MIX : A Test Generation System for Synchronous Sequential Circuits", in Proc. VLSI Design Conf., 1998, pp. 456-463.

[13] L. N. Reddy, I. Pomeranz and S. M. Reddy, "COMPACTEST-II: A Method to Generate Compact Two-Pattern Test Sets for Combinational Logic Circuits", in Proc. Intl. Conf. on Computer-Aided Design, 1992, pp. 568-574.

[14] I. Hamzaoglu and J. H. Patel, "Compact Two-Pattern Test Set Generation for Combinational and Full Scan Circuits", in Proc. Intl. Test Conf., 1998, pp. 944-953.

[15] N. Tendolkar, R. Raina, R. Woltenberg, B. Lin, B. Swanson and G. Aldrich, "Novel Techniques for Achieving High At-Speed Transition Fault Test Coverage for Motorola's Microprocessors Based on PowerPC(TM) Instruction Set Architecture", in Proc. VLSI Test Symp., 2002, pp. 3-8.

[16] Y. Shao, I. Pomeranz and S. M. Reddy, "On Generating High Quality Tests for Transition Faults", in Proc. Asian Test Symp., 2002, pp. 1-8.

[17] W. Qiu, J. Wang, D. M. H. Walker, D. Reddy, X. Lu, Z. Li, W. Shi and H. Balachandran, "K Longest Paths Per Gate (KLPG) Test Generation for Scan-Based Sequential Circuits", in Proc. Intl. Test Conf., 2004, pp. 223-231.

[18] Z. Chen, D. Xiang and B. Yin, "The ATPG Conflict-Driven Scheme for High Transition Fault Coverage and Low Test Cost", in Proc. VLSI Test Symp., 2009, pp. 146-151.

Designing a Fast and Adaptive Error Correction Scheme for Increasing the Lifetime of Phase Change Memories

Rudrajit Datta and Nur A. Touba

Computer Engineering Research Center
University of Texas at Austin, Austin, TX 78712
rudrajit.datta@mail.utexas.edu, touba@ece.utexas.edu

Abstract

This paper proposes an adaptive multi-bit error correcting code for phase change memories that provides a manifold increase in the lifetime of phase change memories thereby making them a more viable alternative for DRAM main memory. A novel aspect of the proposed approach is that the error correction code (ECC) is adapted over time as the number of failed cells in the phase change memory accumulates. The operating system (OS) monitors the number of errors corrected on a memory line, and when the number of errors on a line begins to exceed the strength of the ECC present, the ECC strength is adaptively increased. As this happens, the performance of the memory system gracefully degrades because more storage is taken up by check bits rather than data bits thereby reducing the effective size of a cache line since less data can be brought to the cache on each read operation to the PCM main memory. Experimental results show that the lifetime of a phase change memory can be significantly extended while keeping the fraction of data to check bits as high as possible at each stage in the lifetime of the phase change memory.

1. Introduction

Memory technology scaling drives increasing density, increasing capacity, and falling price-capability ratios. Storage mechanisms in prevalent memory technologies require inherently un-scalable charge placement and control. This in turn has put memory scaling, a first-order technology objective, in jeopardy. Dynamic Random Access Memory (DRAM) has been used as the main memory in computer systems for decades due to its high-density, high-performance and low-cost. However, DRAM technologies, facing both scalability and power issues, will be difficult to scale down beyond 50nm [Zhang 09] due to various limitations associated with device leakages and retention time.

Phase change memory (PCM) provides a non-volatile storage mechanism amenable to process scaling. Phase change memories function by alternating between low resistance crystalline and high resistance amorphous states. The thermally induced phase transition is brought about by injecting current into the storage material during writes. The state of the cell is then detected during reads with the

high resistance state being interpreted as a zero and the low resistance state as one. PCM, relying on analog current and thermal effects, does not require control over discrete electrons. As technologies scale and heating contact areas shrink, programming current will also scale linearly. PCM scaling mechanism has been demonstrated in a 20nm device prototype and is projected to scale to 9nm [Lee 09]. As a scalable DRAM alternative, PCM could provide a clear roadmap for increasing main memory density and capacity.

However, one major challenge that needs to be addressed for PCM is its limited write endurance. PCM writes induce thermal expansion and contraction within the storage element, degrading injection contacts and limiting endurance to hundreds of millions of writes per cell at current processes. In current devices, a PCM cell typically supports around 10^7 writes [Ferreira 10]. Thus, PCM will wear-out quickly if used as a main memory.

This is a significant limitation and a prime reason why PCM is not yet a ready substitute for DRAM main memory. Current PCM prototypes are not designed to mitigate PCM endurance. One major challenge in designing ECC for PCM based systems is that the number of cell failures is a monotonically increasing function of memory writes. In order to mitigate the time-dependant nature of failures, this paper proposes a novel adaptive error correction technique that increases PCM endurance several times. The core idea here is to start with a nominal ECC, depending on experimentally determined error rates for PCM, and then adaptively boost the ECC strength to keep up with increasing failure rates of PCM. The dynamic control over the ECC is achieved by involving the underlying operating system (OS). The OS monitors the maximum number of errors corrected per PCM line and compares this against the strength of the ECC currently in place. When number of errors corrected on a memory line read approaches the capacity of the existing error code, the ECC strength is increased. This can be done on the next reboot or done by writing main memory to disk and reconfiguring the ECC when it is paged back in. The increase in ECC strength is achieved by breaking up a memory line into segments and then implementing separate ECC for each segment. Taken together, the combined effect of the segmented ECCs can correct up to tens of bits per memory line.

978-1-61284-657-6/11 $26.00 © 2011 IEEE

Note that as the ECC of the memory is increased, the performance of the memory system gracefully degrades because more storage is taken up by check bits rather than data bits. However, this strategy is much better than using a worst-case ECC which would give worst-case performance throughout the lifetime of the system. The degradation comes in the form of reducing the effective size of a cache line since less data can be brought to the cache on each read operation to the PCM main memory. Note that if the cache itself is implemented with more reliable SRAM, then the number of check bits stored in the cache does not need to be increased. So the total cache capacity remains the same. If the cache line size is reduced, then the cache can store more lines. Strategies for designing the cache to accommodate graceful degradation of the line size is discussed as well as the performance impact which is highly dependent on the number and locality of memory references for an application.

2. Related Work

Various approaches have been adopted to counter the limited write endurance of phase change memories. [Zhang 09] presents a hybrid PRAM/DRAM memory architecture that uses an OS level paging scheme to improve PRAM write performance and lifetime. They use a 7-error correcting BCH code for the ECC. BCH codes provide the desired level of reliability but require increasing number of cycles for correcting multi-bit errors [Lin 83]. [Xu 10] proposes a novel sensing mechanism for multi-level PCM structures to address the reliability issue using either BCH or LDPC codes, for both of which the decoding time scale with number of errors being detected.

[Ferreira 10] shows how PCM writes can be minimized thereby increasing their lifetime. Note that this methodology could be used on top of the methodology proposed in this paper. [Lee 09] uses buffer reorganization and partial write techniques to mitigate high energy PCM writes but improves PCM lifetime to only about 5.6 years.

Traditional schemes like using spare rows and columns as well as bit interleaving, as shown in [Stapper 92], are likely to prove insufficient because of the prohibitively high error rate in PCM systems.

While the PCM reliability issue has primarily been addressed from an architecture standpoint, solutions using novel ECC have yet to be fully explored. PCM differs from standard DRAM in a fundamental way in that the number of failures for a PCM cell is a function of time or more accurately a function of the number of writes/cell. The following section presents a detailed overview of the proposed scheme.

3. Overview of Proposed Approach

Given that bit failure rates for phase change memories increase with continuous usage, the proposed approach of adaptively increasing the strength of the ECC to keep up

with the increasing failure rate is a good strategy. Note that this strategy requires involving the operating system (OS).

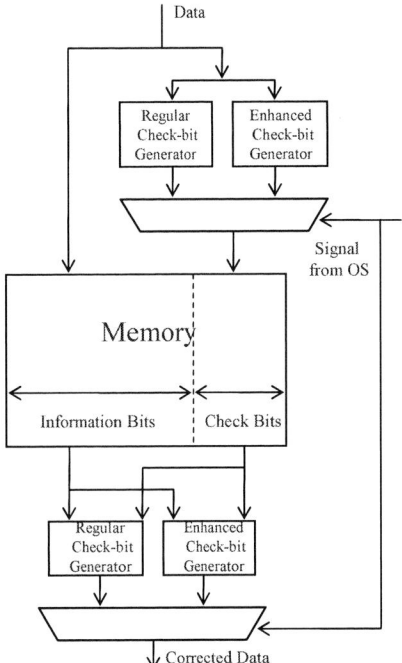

Figure 1. Adaptive ECC implementation

ECC decoding and correction is performed by the memory controller after the data and check bits have been read from the memory chip and are in the processor core. The memory controller can be programmed as to what granularity of ECC is to be implemented. In the proposed method, every time the memory controller reads a memory line, the number of errors corrected is compared against a threshold which depends on the existing strength of the ECC in place. When the number of errors corrected for a particular read begins to approach the given threshold for the implemented ECC, the OS switches to a stronger ECC. This is accomplished by writing all of the main memory to disk and paging the memory back into physical memory with the stronger ECC. Note that this process of reconfiguring the ECC occurs very infrequently (on the order of months or years). Another way to implement it would be to record the need for an ECC reconfiguration and then perform it on the next reboot.

For the memory controller to be able to switch to an ECC of greater strength, the hardware has to be in place from before. Depending on up to how many levels the ECC will be stepped up, the full hardware to perform the necessary encoding-decoding will have to be implemented in the memory controller from the beginning. The memory controller will then choose which of the existing encoding-decoding schemes to employ using information received from the OS.

978-1-61284-657-6/11 $26.00 © 2011 IEEE   135

Figure 1 shows a possible implementation of the scheme that can switch between two modes of ECC. The OS signals when the memory controller needs to switch from the basic ECC scheme to the advanced scheme. In the general case where several different levels of ECC hardware are implemented, the OS will signal the memory controller when to step up the strength of ECC.

Strengthening the ECC requires increasing the redundancy of the code, which means more check bits and fewer data bits can be stored in each line of the memory. When a cache miss occurs, fewer data words can be brought into the cache in each main memory access. For example, if a line in the memory initially contained 1024 data bits (i.e., 16 64-bit words), but then if the redundancy of the ECC is increased by 25%, then each line would only have 768 data bits (i.e., 12 64-bit words).

Note that if the cache is implemented with more reliable SRAM rather than PCM technology, then it is not necessary to store as many check bits in the cache as is needed in a PCM main memory, nor is it necessary to scale up the number of check bits in the cache over time. So as the number of check bits stored in the phase change main memory is increased over time, the total storage capacity of the cache is not affected. What is affected is the bandwidth coming into the cache. However, the reduction in system performance would be very application dependent and would depend on the locality of the data and number of required memory accesses. Experimental results are shown in Sec. 6 exploring the impact on performance.

There are a number of different options for how the cache is implemented to accommodate a gracefully degrading line size coming from the main memory.

One would be to have multiple valid bits for each original full size line in the cache. As the effective bandwidth is reduced when the ECC is strengthened in the phase change main memory, then each read from the main memory will only partially fill a line in the cache which would be indicated with the appropriate subset of valid bits for the line. If for example, the number of data bits was reduced by 50%, then each memory access would fill half of the cache line. Another way to think about it is that a cache with 1024 lines where each line originally stored 16 words would be effectively transformed to a cache with 2048 lines where each line stores only 8 words. The cache capacity doesn't change, only the effective line size changes.

Another way to implement the cache would be to adjust the associativity when the bandwidth from the PCM main memory is reduced. For example, a two-way set associative cache could be converted to a four-way set associative cache when the line size is reduced in half.

In either of these cases, the impact of reducing the line size will depend on the locality and frequency of memory references in the application. The reduction in line size is partially offset by the increase in either the associativity or the number of lines.

This gives an overview of the approach which is general and could be used for any type of ECC. Next, one way for implementing the ECC with this scheme will be proposed which utilizes OLS codes. The advantage of using OLS codes over traditional multi-bit ECC such as BCH, LDPC codes, is that correction can be performed in a single clock cycle and the amount of time required is independent of the number of errors.

4. Orthogonal Latin Square Codes

OLS codes, as the name suggest, are based on Latin squares. A Latin square [Hsiao 70] of order (size) m is an m x m square array of the digits $0, 1, \ldots, m - 1$, with each row and column a permutation of the digits $0, 1, \ldots, m - 1$. Two Latin squares are orthogonal if, when one Latin square is superimposed on the other, every ordered pair of elements appears only once.

As explained in [Datta 10], a t-error correcting majority decodable code works on the principle that $2t + 1$ copies of each information bit are generated from $2t + 1$ independent sources. One copy is the bit itself received from memory or any transmitting device. The other $2t$ copies are generated from $2t$ parity relations involving the bit. By choosing a set of h Latin squares that are pair-wise orthogonal, one can construct a parity check matrix such that the number of 1's in each column is $2t = h + 2$. The orthogonality condition ensures that for any bit d, there exists a set of $2t$ parity check equations orthogonal on d_i, and thus makes the code self-orthogonal and one-step majority decodable. One-step majority decoding is the fastest parallel decoding method. The t-error correcting codes generated by OLS codes [Hsiao 70] have m^2 data bits and $2tm$ check bits per word.

5. Adaptive Error Correction Code

The principle behind the proposed method is that as the number of permanent errors keeps increasing over time, the ECC needs to be increased in strength. The trade off comes in the form of using up more of the memory for storing the ECC check bits.

If for k data bits in a memory line, a t-error correcting code is used, then this is sufficient for all errors $\le t$. To mitigate more than t errors, more check bits are required. They cannot be added without reducing the number of data bits per line because for off-chip main memory, the n-bit bus width transferring data and check bits from the memory to the processor is fixed such that

$$n = data\ bits + check\ bits$$

A straightforward approach for strengthening the ECC for an OLS code would be to simply directly increase t for the whole line. However, rather than doing that, it is more efficient to divide the line into fragments and increase the number of errors corrected in each fragment as will be shown in this section.

If an n-bit line consists of k data bits, it can be broken up into fragments each of size k_i, and r_i bits of ECC are separately implemented for each data fragment k_i such that

$$\sum_i \left(k_i + r_i\right) = n$$

Consider the case where all k bits have a t-error correction OLS code implemented on it. Then the total number of bits, data plus check bits, would be

$$k + 2t\sqrt{k} \quad\dots\dots\dots\dots\dots\dots(5)$$

Now if the line is broken up into fragments, each of size $k_i = k/f$ and a $t/\sqrt{f}$ -error correcting OLS code is implemented for each fragment. The total number of bits still remains

$$\left\{k/f + 2\left(t/\sqrt{f}\right)\sqrt{k/f}\right\} * f = k + 2t\sqrt{k} \quad\dots(6)$$

Although the total number of bits is the same in both the cases, (5) and (6), the error correction capacity is different for both. In case (5), the line can withstand all error patterns affecting up to t-bits. In case (6), each fragment can handle up to $t/\sqrt{f}$ errors. But overall the line can handle all error patterns affecting up to $t/\sqrt{f}$ bits in each fragment and *some* error patterns affecting up to $t\sqrt{f}$ bits on the entire line. As is shown later in section 6, for randomly occurring errors, the property to correct some error patterns of size greater than the individual capacity of the ECC in each fragment is significant and as simulations show, a fair number of error patterns can be tolerated using this property.

So the overall idea is the following. An initial code over across all n bits is selected to protect the PCM memory based on characterization tests. Then during the course of operation as the number of failed cells accumulates over time, the strength of the ECC is increased by implementing ECC on increasingly smaller fragments.

Consider a numerical example to illustrate the scheme. Consider a memory line with 256 data bits. Initially a 3-error correcting OLS code is employed. Thus the total number of bits in the line is,

$$256 + 2 * 3 * \sqrt{256} = 352$$

In an enhanced ECC mode, 25% of the memory line is used to store extra check bits. Hence the total number of data bits per line now becomes 192. The rest

$$352 - 192 = 160$$

bits are used for storing ECC. But instead of implementing ECC over the entire 192 data bits at a time, the line is broken up into fragments of size 64, 64, 16, 16, 16 and 16 bits. Next a 3-error correcting OLS code is implemented on each of the 64-bit fragments and a 2-error correcting OLS code on each of the 16-bit fragments, bringing the total number of bits to

$$\left(64 + 2 * 3 * 8\right) * 2 + \left(16 + 2 * 2 * 4\right) * 4 = 352$$

But now instead of being able to correct only 3-error patterns all 2-error patterns, 99.97% of all 3-error patterns, 99.73% of all 4-error patterns and so on, up to a small fraction of 14-bit errors can be corrected. Thus the approach of breaking up a line into fragments and using separate ECC for each fragment is more efficient in terms of error correcting capacity than implementing a single ECC on the whole line.

The selection of fragment sizes and their respective ECC bits is a combinatorial problem. In the cases where there is more than one possible way to break up a memory line into identical division of data and check bits, the combination which can correct maximum number of errors is chosen.

6. Experimental Results

The bit error rate for a memory is defined as the number of failed bits divided by the total size of the memory (or in other words, the probability that each bit has failed). The bit error rate for PCM memories starts very small and grows over time as more cells fail. Figure 3 compares the proposed adaptive error correction scheme with the approach in [Zhang 09] where a 7-error BCH code is used and can tolerate bit error rates of up to 0.145%. The adaptive scheme discussed in this paper was implemented with a line size of 1024 and starts with an initial correction capability of 3-errors. As the OS detects that the bit error rate is exceeding the strength of the ECC, 25% of the data bits are converted to check bits by dividing the memory into fragments and adding check bits for each fragment. This process is repeated as the error rate continues to increase. As can be seen in Fig. 3, the proposed scheme can tolerate bit-error rates from 0.08% to a significantly higher 1.2%.

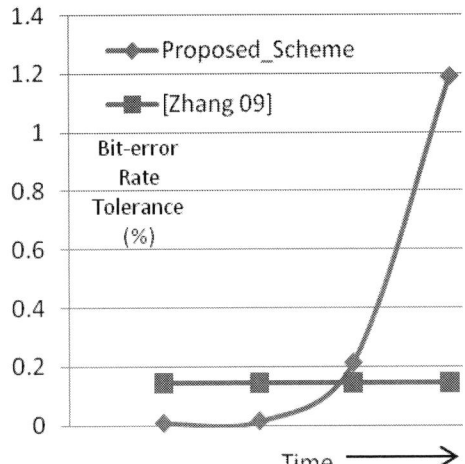

Figure 3. Adaptive fault tolerance

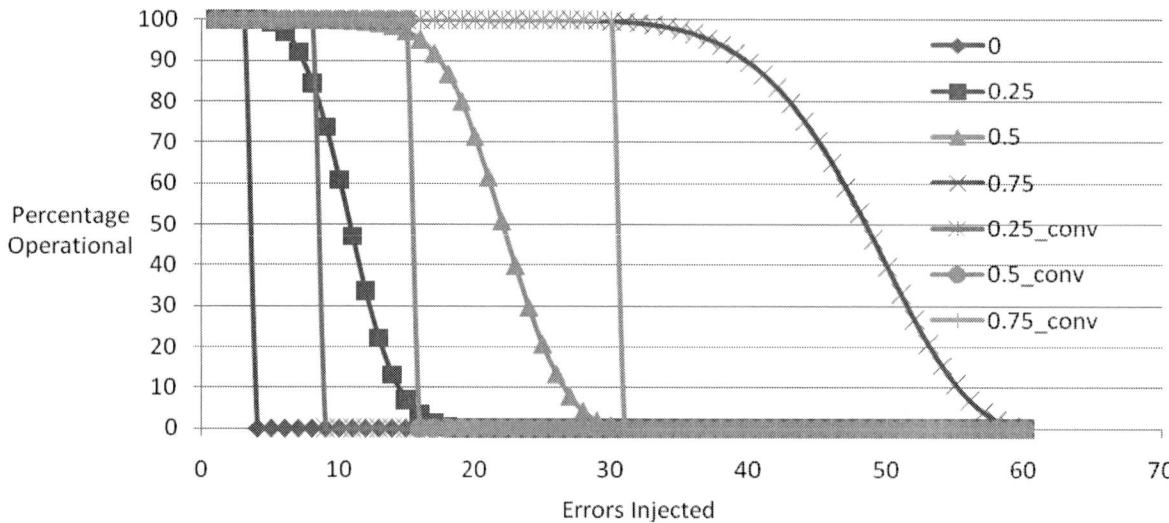

Figure 4. Percentage of operational cache lines versus number of errors injected (out of 100,000 experiments)

Table 1 shows how the error tolerance varies with size of the memory as the number of check bits is increased. As can be seen from the data, very high error rates of around 10^{-2} can be tolerated in an adaptive manner.

Table 1. Error Tolerance (no. of errors / no. of bits * 100) for varying line sizes

Memory Size	Fraction of Memory Used for Storing Extra Check-bits			
	0.0	0.25	0.5	0.75
128MB	0.008	0.015	0.213	1.190
256MB	0.006	0.042	0.205	1.117
1GB	0.005	0.026	0.154	0.989
4GB	0.003	0.020	0.125	0.916

The monotonic decrease in tolerance with increasing size can be explained by the fact that as the number of lines increase, the difference

$$n * E[one\ line] \sim E[n\ lines]$$

is likely to increase, where n is the number of lines, *E[one line]* is expected error tolerance of a single line and *E[n lines]* is the error tolerance of n lines. This mirrors a likely scenario where errors will accumulate faster on some lines than others unless the data read/write pattern is absolutely random, which is not the usual case due to data locality.

Another aspect of evaluating the proposed scheme is to study the distribution of error tolerance in each line for different ECC configurations. Figure 4 shows results for a 1024 bit line in which the initial stating point is a 3-bit error correcting ECC and then 25% of the data bits were converted to check bits, and then 50%, and finally 75%. Errors were injected at random, and Fig. 4 shows the percentage of lines that have not failed across 100,000 experiments. The *x*-axis corresponds to the number of errors injected in the line, and the *y*-axis corresponds to the percentage of lines that were able to tolerate that many errors for different configurations of the ECC.

Results are shown for both the ECC scheme described in Sec. 5 which breaks up the line into fragments and implements ECC separately for each fragment, and the conventional case where the ECC was implemented across all the data bits at once. As can be seen from the results, the fragmented ECC scheme can easily tolerate more errors than the conventional method. The former is able to tolerate 20% more errors per line at 90% probability, than the conventional method, when half the line is used for storing check bits. When 3/4th of the line is used for storing check bits, the fragmented scheme can tolerate 33% more errors at 90% probability.

Figures 5 and 6 show the effect of reducing memory line size to accommodate extra check bits. One way to implement this, as was described in Sec. 3, is to adjust the associativity and line size of the cache. In both Figs. 5 and 6, the *y*-axis plots cycles per instruction (CPI) for a set of SPEC2006 [Spec 06] benchmarks across different cache configurations defined as follows:

cache_config1 – line size 512B, associativity 4
cache_config2 – line size 256B, associativity 8
cache_config3 – line size 128B, associativity 16

The CPI was calculated assuming single cycle for all non-memory instructions and five cycles [Wulf 95] for instructions that caused a cache miss and needed to access main memory. As can be seen from the figures, there is little degradation on performance for reduced line sizes and increasing associativity. Moreover, the fact that these changes are expected at an interval of every few years lessens the performance impact.

The simulations were done using Pin [Pin 04], a dynamic instrumentation tool. Pin was used to obtain memory traces of the benchmarks for various cache configurations. These memory traces were then used as an input to the DineroIV [Dinero IV] cache simulator to generate cache miss rates.

978-1-61284-657-6/11 $26.00 © 2011 IEEE

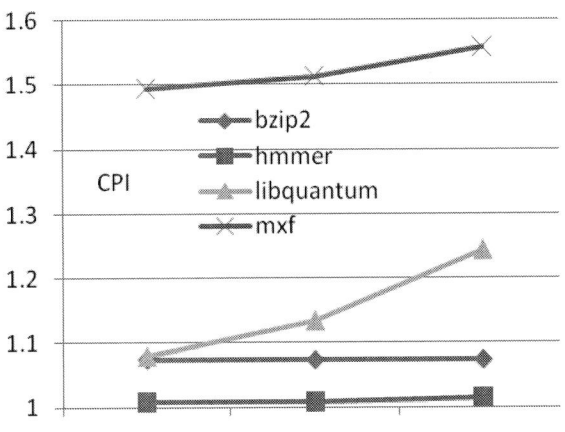

Figure 5. Variation of CPI for different cache configurations for four different SPEC2006 benchmarks for 64KB cache

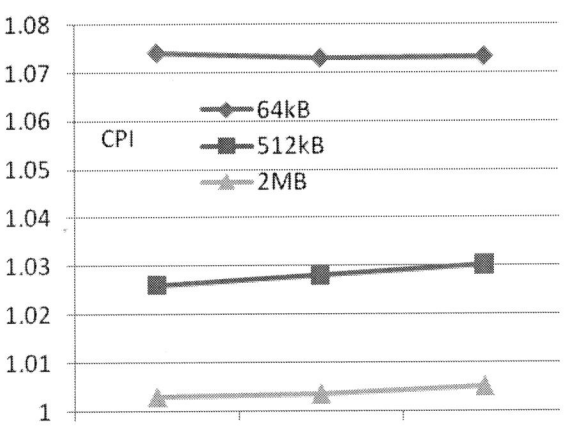

Figure 6. Variation of CPI for different cache configurations across different cache sizes for the SPEC2006 benchmark *bzip2*

7. Conclusions

This paper described an adaptive error tolerance scheme that can extend to 8x more error tolerance than that of [Zhang 09] for similar initial redundancy. As the PCM memory degrades to the point where it exceeds the capability of the method in [Zhang 09] to continue operation, the proposed method can continue operation by adaptively increasing the ECC. The performance of the memory will gracefully degrade due to reducing the effective line size for each memory read to service a cache miss. However, the impact of this can be minimized by careful cache design since the total cache storage capacity as a whole is not impacted.

Acknowledgments

This research was supported in part by the National Science Foundation under Grant No. CCR-0426608.

References

[Datta 10] R. Datta, N. Touba, "Post-Manufacturing ECC Customization Based on Orthogonal Latin Square Codes and Its Application to Ultra-Low Power Caches", *Proc. of International Test Conference*, Paper 7.2, 2010.

[Dinero IV] J. Edler, M. D. Hill, "DineroIV – Trace Driven Uniprocessor Cache simulator for memory references," http://www.cs.wisc.edu/~markhill/DineroIV/

[Ferreira 10] A.P. Ferreira, M. Zhou, S. Bock, B. Childers, R. Melhem, D. Mosse, "Increasing PCM Main Memory Lifetime", *Design Automation and Test in Europe*, pp. 914-919, 2010,.

[Hsiao 70] M.Y. Hsiao, D.C. Borren, R.T. Chien, "Orthogonal Latin Square Codes", *IBM Journal of Research and Development*, Vol. 14, No. 4, pp. 390-394, July 1970.

[Lee 09] B. C. Lee, E. Ipek, O. Mutlu, D. Burger, "Architecting Phase Change Memory as a Scalable DRAM Alternative", *Proc. of International Symposium of Computer Architecture*, pp. 2-13, 2009.

[Lin 83] S. Lin, D. Costello, *Error Control Coding: Fundamentals and Applications*. Prentice Hall, 1983.

[Pin 04] V.J.Reddi, A. Settle, D.A.Connors, R.S.Cohen, "PIN: A Binary Instrumentation Tool for Computer Architecture Research and Education", *Proc. of Workshop on Computer Architecture Education*, June 2004.

[Spec 06] Standard Performance Evaluation Corporation. SPEC CPU2006 Benchmarks. http://www.spec.org/cpu2006/.

[Stapper 92] C.H. Stapper, Hsing-san Lee, "Synergistic Fault-Tolerance for Memory Chips", *Proc of IEEE Transactions on Computers*, Vol. 41, No. 9, pp 1078-1087, Sep. 1992.

[Wulf 95] W.A. Wulf, S.A. McKee, "Hitting the memory wall: implications of the obvious", *ACM SIGARCH Computer Architecture News*, Vol. 23 Issue 1, Mar.1995.

[Xu 10] W. Xu, T. Zhang, "Using Time-Aware Memory Sensing to Address Resistance Drift Issue in Multi-Level Phase Change Memory", *Proc. of International Symposium of Quality Electronic Design*, pp. 356-361, 2010.

[Zhang 09] W. Zhang, T. Li, "Exploring Phase Change Memory and 3D Die-Stacking for Power/Thermal Friendly, Fast and Durable Memory Architectures", *Int. Conference on Parallel Architectures and Compiler Techniques*, pp. 101-112, 2009.

Programmable Extended SEC-DED Codes for Memory Errors

Valentin Gherman, Samuel Evain, Fabrice Auzanneau, Yannick Bonhomme

CEA, LIST, Embedded Systems Reliability Laboratory,
Point Courrier 94, 91191 Gif-sur-Yvette CEDEX, FRANCE
firstname.lastname@cea.fr

Abstract—Redundant memory columns are an essential ingredient of memory design for yield and reliability. They are used either as spare columns for the replacement of completely defective regular columns or to store check-bits for error detection and correction codes. Column replacement allows to mask isolated malfunctioning storage cells as well. Unfortunately, the number of columns with defective storage cells that can be masked in this way cannot exceed the number of spare columns which is usually quite low. Here, we propose a way to increase the capacity of masking memory columns with isolated defective storage cells using spare memory columns. For this purpose, single error correction and double error detection (SEC-DED) codes already available for the protection against soft errors are extended such that all double-bit errors which affect a fixed sub-set of bit positions in the code words can be corrected. The cardinality of this sub-set is significantly higher than the number of spare columns. A bit-swapper is employed to map the bit positions that are protected by the extended SEC-DED code against double-bit errors to the memory columns with defective storage cells. In this way, single-bit soft-errors affecting any bit position can be corrected simultaneously with single-bit hard errors induced by any sub-set of memory columns. The bit-swapper can be dynamically reconfigured based on status information that designates the memory columns with defective storage cells. This facilitates the integration into built-in self-repair (BISR) schemes.

Keywords- yield; memory repair; BISR; error correction

I. INTRODUCTION

Manufacturing and wear-out induced defects are identified as major threats for the yield and the reliability of memories produced with advanced scaled-CMOS technologies [2][5]. In parallel, soft-error rates at chip and system levels remain essentially unchanged or they increase, as is the case with the SRAM memories [1].

Memory protection against soft errors is usually ensured with the help of single error correction and double error detection (SEC-DED) codes [4][9]. In such cases, redundant memory columns are necessary to store the check-bits of the SEC-DED codes. Additional redundant memory columns, called spare columns, are required to replace completely malfunctioning regular columns affected by manufacturing or wear-out induced defects [10][13][17][19].

In memory units with a large number of banks, the majority of banks will not have completely defective columns and the available spare columns can be used to mask isolated defective storage cells [7]. In this way, one can reduce the pressure on other memory repair strategies that can handle defective storage cells such as the employment of spare words [17][19].

The number of single-bit hard errors that can be masked with memory column replacement grows linearly with the number of bits in each memory word that can be stored in spare columns [10][13][17][19], in other words the number of spare columns that can intersect a memory word. Recently, programmable restricted error correction codes have been introduced to enable the correction of an exponential number of single-bit hard errors with respect to the number of spare columns per memory word [7]. Unfortunately, this method is not appropriate for a memory that is already protected by an error detection and correction code.

In this paper, a memory protection scheme is proposed based on the extension of a systematic SEC-DED code with a number of check-bits equal to the number of spare columns available in a memory bank. This extension enables the correction of all double-bit errors that affect at least one bit position from a fixed sub-set of bit positions in the extended code words. Any double-bit error in which these bit positions are not involved remains detectable. The cardinality of the sub-set of better protected bit-positions is significantly higher than the number of spare columns per memory word. Any single-bit hard error affecting these bit positions can be corrected simultaneously with any single-bit soft error. The requirement is that at most one of the storage cells where the bits of each code word are stored is defective.

With the proposed extended SEC-DED (E-SEC-DED) code, already the presence of one single spare column per memory word allows to mask out two different single-bit hard errors instead of only one as is the case with column replacement techniques [10][13][17][19]. Furthermore, spare memory columns with defective storage cells can still be used to mask out defective storage cells in regular memory columns as long as each memory word contains at most one defective storage cell. The proposed E-SEC-DED code has a hierarchical structure [8] and can be easily reduced to the original SEC-DED code or to another E-SEC-DED code with fewer supplementary check-bits.

A bit-swapper is proposed to rearrange the bits of the E-SEC-DED code words before they are stored such that the memory columns with defective storage cells receive bits protected against any double-bit error which might affect them. The bit-swapper can be dynamically programmed based on test result information that indicates the columns with defective storage cells in the accessed memory bank. This further improves the defect masking capacity [16] and facilitates the integration into a memory built-in self-repair (BISR) scheme.

In the resulting memory repair scheme, the spare memory columns can be used either for column replacement or to store additional check-bits. A similar approach was also proposed in [6] where, besides column replacement, the redundant columns are employed to reduce the miscorrection probability of multiple-bit soft errors.

A brief review of systematic linear block codes is given in Section II. The E-SEC-DED codes are introduced in Section III. A bit-swapper design and a way to program it are shown in Section IV. Section V presents an adaptation of the proposed scheme for memories that are insensitive to soft errors. The paper achievements are summarized in Section VI.

II. SYSTEMATIC LINEAR BLOCK CODES

The data protected with linear block codes is organized in code words of length n that contain k data-bits and $r=n-k$ check-bits [12][14]. A binary matrix H, also called parity-check matrix, can be defined such that each code word V fulfills the relation below [4]:

$$\bigoplus_{i=0}^{n-1}\left(H_j^{\ i} \wedge V_i\right)=0; \quad 0 \le j < r, \tag{1}$$

where the symbols '$\oplus$' and '$\wedge$' denote the exclusive disjunction (xor) and conjunction operators (and), respectively. Each column in the H-matrix corresponds to a particular bit position in the code words V.

In the case of systematic linear block codes, the matrix columns that correspond to the check-bit positions are linearly independent and usually form an identity or triangular matrix. Consequently, the H-matrix can be brought to the form $[G, I_r]$, where G is an $r \times k$ binary matrix and I_r is the $r \times r$ identity matrix.

In order to ensure the SEC-DED property, the columns H_i of the H-matrix must be different from the all-zero vector, from each other and from the bitwise xor-sum of any two H-matrix columns, as illustrated below ($0 \le i,p,q < n$):

$$H_i \neq H_{p\neq i} \neq 0; \tag{2}$$

$$H_i \neq H_p \oplus H_q \tag{3}$$

Fast detection of double-bit errors is allowed by the fact that all SEC-DED code words have a fixed parity, e.g. the Hamming and the Hsiao SEC-DED codes [9][12]. Usually, the code words of these SEC-DED codes have an even parity:

$$\bigoplus_{i=0}^{n-1}V_i = 0$$

Upon a read operation of a memory word V' previously stored as a code word V, syndrome bits S_j ($0 \le j < r$) are calculated according to the following expression:

$$S_j = \bigoplus_{i=0}^{n-1}\left(H_j^{\ i} \wedge V'_i\right)=\bigoplus_{i=0}^{n-1}\left[H_j^{\ i} \wedge \left(V'_i \oplus V_i\right)\right]$$

Due to the compliance of the stored code word V with relation (1), the syndrome S is independent of V and only depends on the error vector $V \oplus V'$. If S is an all-zero vector, the read code word V' is assumed to be error-free ($V'=V$). Otherwise, the syndrome S is used to correct or detect the occurred errors. A single-bit error generates a syndrome identical to the column of the H-matrix that corresponds to the corrupted bit. For the ith bit position of the code word V', the formula below is used to check whether the syndrome S matches the ith column of the H-matrix:

$$BitFlip_i = \bigwedge_{j=0}^{r-1}\left(H_j^{\ i} \overline{\oplus} S_j\right); \quad 0 \le i < n, \tag{4}$$

where the symbol '$\overline{\oplus}$' stands for the negated exclusive disjunction (xnor) operator. A match ($BitFlip_i=1$) indicates that

an error has occurred at the ith bit position. This error can be corrected with the expression below:

$$V_i = V'_i \oplus BitFlip_i; \quad 0 \le i < n$$

In case of SEC-DED codes with even parity, the occurrence of a double-bit error is indicated if the following expression becomes true:

$$\bigvee_{j=0}^{r-1} S_j \wedge \overline{\bigoplus_{i=0}^{n-1} V'_i}, \tag{5}$$

where the symbol '$\vee$' stands for the disjunction (or) operator.

III. EXTENDED SYSTEMATIC SEC-DED CODES

Here, we consider the extension of a systematic linear block SEC-DED code with s additional check bits that enables (a) the correction of all single-bit errors, (b) the correction of all double-bit errors which affect at least one bit position from a fixed sub-set of f check-bit positions and (c) the detection of the remaining double-bit errors. The parameter f is taken as the minimum between the total number of check-bits $r+s$ and 2^s+s-1. This is equivalent to the extension of an $r \times n$ SEC-DED H-matrix with s additional columns and s additional lines such that the columns of the resulting $(r+s) \times (n+s)$ H^*-matrix satisfy the following relations:

$$H^*_i \neq H^*_{p\neq i} \neq 0; \tag{6}$$

$$H^*_i \oplus H^*_{i\neq l} \neq H^*_{p\neq l} \oplus H^*_{q\neq l}; \tag{7}$$

$$H^*_i \neq H^*_p \oplus H^*_q; \tag{8}$$

$$0 \le i,p,q < n+s; \quad n+s-f \le l < n+s$$

In the relations above and all-over this paper, the first $k=n-r$ columns in the H^*-matrix ($0 \le i < k$) correspond to the data-bits, the following r columns ($k \le i < n$) to the check-bits of the original SEC-DED code and the last s columns ($n \le i < n+s$) to the additional check-bits.

The structure of an H^*-matrix can be selected as shown in Figure 1. One can identify four additional blocks besides the H-matrix of the initial SEC-DED code: an $r \times s$ all-zero block, an $s \times (n+s-f)$ all-zero block, an $s \times s$ identity matrix and an $s \times (f-s)$ H°-matrix with all columns different from the all-zero vector and from each other. The H°-matrix has the same properties as the parity-check matrix of a SEC Hamming code [9].

The resulting extended SEC-DED (E-SEC-DED) code is systematic, since the last $r+s$ columns of the H^*-matrix are linearly independent. This is due to the $r \times s$ all-zero block and the $s \times s$ identity matrix on the last s H^*-matrix columns and to the fact that the last r columns of the H-matrix are linearly independent as the initial SEC-DED code is systematic.

The resulting E-SEC-DED code is also a *hierarchical* code [8] since each E-SEC-DED code word contains a SEC-DED code word. This is a consequence of the $r \times s$ all-zero block in the H^*-matrix due to which the relationship between the k data-bits and the initial r check-bits is not affected by the introduction of the new s check-bits. An example of hierarchical code is the Hamming SEC-DED code obtained by the addition of an overall parity bit to the SEC code words [9].

If at least one index is smaller than n (6) and (8) are fulfilled, since the columns of the H-matrix, and implicitly of the H^*-matrix, must satisfy (2) and (3) while the first r values of the last s H^*-matrix columns are equal to zero. These relations

978-1-61284-657-6/11 $26.00 © 2011 IEEE

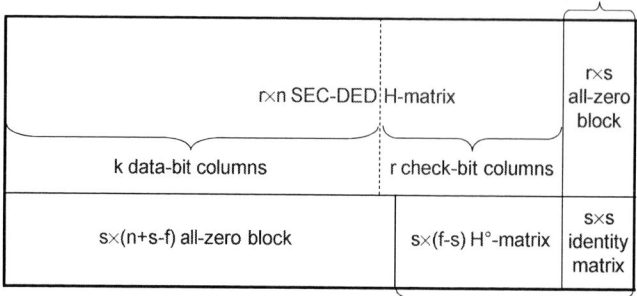

matrix columns corresponding to the s additional check-bit positions

Figure 1. Block structure of a (r+s)×(n+s) E-SEC-DED H^*-matrix obtained by the extension of an r×n SEC-DED H-matrix.

are also fulfilled when all indices are larger or equal to n, since the columns of the $s \times s$ identity matrix are linearly independent.

The relation (7) is fulfilled by the columns of the H^*-matrix in the following cases:

- all indices are larger or equal to n, since the columns of the $s \times s$ identity matrix are linearly independent,
- at least one index is smaller than n and at least another index is larger or equal to n, since the H^*-matrix columns which correspond to the indices larger or equal to n have the first r values equal to zero while the H^*-matrix columns which correspond to the indices smaller than n should satisfy (2) and (3),
- all indices are smaller than n and at most one of the indices i, p and q is larger or equal to $n+s-f$, since the H°-matrix columns are different from the all-zero vector and from each other while the first $n+s-f$ columns of the H^*-matrix have the last s values equal to zero,
- all indices are smaller than n and larger or equal to $n+s-f$, since this case involves only linearly independent H-matrix columns that correspond to the check-bit positions of a systematic SEC-DED code.

It comes out that (7) is automatically implied by the construction of the H^*-matrix except for the case when all indices are smaller than n and only one index, say i, is smaller than $n+s-f$ such that the xor-sum of the H^*-matrix columns corresponding to the remaining three indices (H^*_i, H^*_p, H^*_q) generates a vector with the last s values equal to zero. This exception can be handled with the help of an additional constraint imposed to the SEC-DED H-matrix.

In order to understand this constraint consider the example in Figure 2 where a 3×7 H°-matrix is used to extend a 7×39 H-matrix to a 10×42 H^*-matrix as shown in Figure 1. In this H°-matrix, there are 7 triplets of columns whose xor-sum generates an all-zero vector. If the columns of the H°-matrix are indexed from left to right starting with the index 1, these triplets can be identified as follows: (1, 2, 5), (1, 3, 6), (2, 3, 4), (4, 5, 6), (1, 4, 7), (2, 6, 7) and (3, 5, 7). The extension of these triplets over the entire H^*-matrix defines 7 triplets of check-bit columns in the H-matrix. The xor-sums of the columns in each one of these H-matrix triplets give 7 non-zero vectors that must be different from any other column in the H-matrix such that (7) can be fulfilled.

For example, the columns of the H-matrix of a Hsiao SEC-DED code have one single non-zero value in the check-bit columns and larger odd numbers of non-zero values in the data-bit columns. Consequently, the xor-sum of 3 check-bit columns gives a vector with 3 non-zero values and 7 different triplets of check-bit columns define 7 distinct vectors with 3 non-zero values. So in Figure 2, the number of 7-bit vectors with 3 non-zero values which can be selected for the data-bit columns of a Hsiao H-matrix is $\binom{7}{3}$-7=28. For the rest of the 32-28=4 data-bit columns, one must select vectors with 5 non-zero values.

Consider now the same 3×7 H°-matrix used to extend an 8×72 H-matrix of a Hsiao SEC-DED code to an 11×75 H^*-matrix. In this case, the 64 data-bit columns of the Hsiao H-matrix can be constituted of $\binom{8}{3}$-7=49 vectors with 3 non-zero values and 64-49=15 vectors with 5 non-zero values.

matrix columns corresponding to the 3 additional check-bit positions

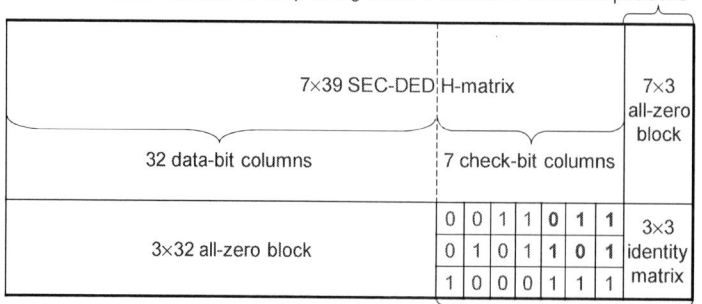

f=10 matrix columns corresponding to the check-bit positions involved only in correctable double-bit errors

Figure 2. Block structure of a 10×42 E-SEC-DED H^*-matrix obtained by the extension of a 7×39 SEC-DED H-matrix.

The bold characters in Figure 2 define a 2×3 H°-matrix that can be used to extend a 7×39 SEC-DED H-matrix to a 9×41 H^*-matrix if only two additional check-bits are available. The xor-sum of the columns in the only possible triplet of H°-matrix columns produces an all-zero vector. As a consequence, the xor-sum of the corresponding check-bit columns in the SEC-DED H-matrix gives a non-zero 7-bit vector that should be different from any other column in the H-matrix. In case of a Hsiao H-matrix, $\binom{7}{3}$ -1=34 distinct vectors with 3 non-zero values remain available for the selection of the 32 data-bit columns.

Similar considerations can be made in the case of H-matrices that correspond to Hamming SEC-DED codes [9].

In the extreme case when only one spare column is available, a 1×1 H°-matrix ($H^{\circ l}{}_i$=1) can be used to extend a 7×39 H-matrix to an 8×40 H^*-matrix. In this case, the H-matrix can be selected without any restriction. f=2 distinct single-bit hard errors can be masked out instead of a single one as is the case with the conventional column replacement even if the last check-bit of the E-SEC-DED code is a replica of the last check-bit of the extended SEC-DED code.

Spare memory columns with defective storage cells can still be used to correct double-bit errors as long as at most one of the bits of each E-SEC-DED code word is stored in a defective storage cell.

Fast correction of the data-bit errors is possible since each data-bit position in the E-SEC-DED code words can be affected by maximum $f\leq r+s$ correctable double-bit errors. Consequently, for these data-bit positions ($0\leq i<k$) (4) can be adapted as follows:

$$BitFlip_i = \overset{r+s-1}{\underset{j=0}{\wedge}}\left(H^*{}_j^i\,\overline{\oplus}S_j\right)\vee\overset{n+s-1}{\underset{l=n+s-f}{\vee}}\left[\overset{r+s-1}{\underset{j=0}{\wedge}}\left(H^*{}_j^i\oplus H^*{}_j^l\right)\overline{\oplus}S_j\right]\quad(9)$$

This expression cannot be applied to the last f check-bit positions since they can be involved in $n+s-1$ correctable double-bit errors. One way to correct the check bits is to regenerate them out of the corrected data bits. Usually, the correction of check-bit errors is only necessary if one needs to avoid the accumulation of soft errors in the memory. This can be done during scrubbing campaigns performed while the memory is idle [15][18].

A benefit of the hierarchical structure of the proposed E-SEC-DED code is the fast detection of any uncorrectable double-bit error. If the SEC-DED code words inside the E-SEC-DED code words have an even parity, the detection of double-bit errors can be ensured with the help of the expression below which has one additional factor as compared to (5):

$$\overset{r-1}{\underset{j=0}{\vee}}S_j\wedge\overline{\overset{n-1}{\underset{i=0}{\oplus}}V'_i}\wedge\overline{\overset{r+s-1}{\underset{j=r}{\vee}}S_j}\quad(10)$$

The correctable double-bit errors are filtered out with the help of the last factor in (10) when at least one of the last s syndrome bits is different from zero. The only correctable double-bit errors that cannot be handled in this way affect two bit positions that correspond to H^*-matrix columns with identical values on the last s lines. Due to the structure of the H^*-matrix, these correctable double-bit errors affect one bit position that corresponds to a column in the H°-matrix and another bit position that corresponds to an identical column in the $s\times s$ identity matrix. Consequently, these correctable double-bit errors affect one and only one of the first n bit positions and

are filtered out by the second factor in (10).

If the last s check-bits of an E-SEC-DED code word are ignored, the remaining n bits receive a SEC-DED protection independently of any single-bit or multiple-bit error that may affect the last s check-bits. This can be achieved if the last s syndrome bits S_j ($n\leq j<n+s$) in (9) and (10) are forced to zero during the error checking and correction phase. Consequently, (9) and (10) become equivalent to (4) and (5), respectively.

The hierarchical structure of the E-SEC-DED code concerns any sub-set of the check-bits added to a SEC-DED code. Due to the $r\times s$ all-zero block and the $s\times s$ identity matrix on the last s H^*-matrix columns in Figure 1, any sub-set of the last s check-bits can be ignored without affecting the relationship between the data-bits and the remaining check-bits. In Figure 2 for example, the 10×42 H^*-matrix (f=10) can be reduced to a 9×41 H^*-matrix (f=5) if the last row and column of the 10×42 H^*-matrix are ignored and the first four columns of the 3×7 H°-matrix are masked to zero. Ignoring the last row and column of the 9×41 H*-matrix is equivalent to neglecting the last check-bit by forcing to zero the corresponding syndrome bit in (9) and (10).

IV. MEMORY BISR BASED ON E-SEC-DED CODES AND BIT-SWAPPING

The bit positions of the E-SEC-DED code words which are better protected against the double-bit errors can be mapped to the memory columns with defective storage cells based on bit-swapping, as illustrated in Figure 3. A schematic view of a bit-swapper is presented in Figure 4. The jth bit position of the E-SEC-DED code words V is mapped to the ith memory column C_i with the help of multiplexors controlled by the M_i^j control signals. Each of the first $n+s-f$ multiplexors from the left hand side receives inputs from the last f code word bit positions. These inputs can only be selected if the driven memory column has defective storage cells. This is performed via bit-swapping: if the jth bit position in the E-SEC-DED code words is mapped to the ith memory column, then i^{th} bit position in the E-SEC-DED code words is mapped to the jth memory column. In order to enable bit-swapping, each of the last f multiplexors receives inputs from the first $n+s-f$ code word bit positions. Whenever bit-swapping is not required, an additional input is selected which comes from the code word bit position that has the same index as the driven memory column.

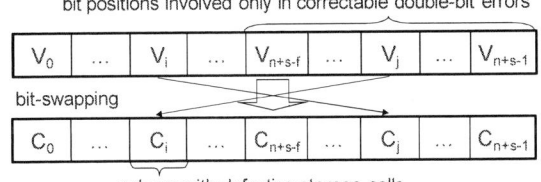

Figure 3.   The bit-swapper maps bit positions of the E-SEC-DED code words to memory columns.

If a repair scheme needs to be reconfigured dynamically depending on the accessed memory bank or memory segment, the reduction of the amount of configuration information becomes crucial [16]. The M_i^j control signals can be encoded with the help of the test result bits T_i ($0\leq i<n+s$) that indicate whether the ith column in a memory bank has defective storage cells or not. The M_i^j control signals that are used to select the inputs of

978-1-61284-657-6/11 $26.00 © 2011 IEEE

the first $n+s\text{-}f$ multiplexors from the left hand side $(0\leq i<n+s\text{-}f)$ in Figure 4 can be calculated with the help of the test result bits T_i $(0\leq i<n+s\text{-}f)$ and T_j $(n+s\text{-}f\leq j<n+s)$ as follows:

$$M_i^{\,i} = \overline{T_i}$$

$$M_i^{\,j} = T_j \wedge \overline{T_j} \wedge \left(\overline{\bigvee_{t=j+1}^{n+s-1} M_i^{\,t}}\right) \wedge \left(\overline{\bigvee_{s=0}^{i-1} M_s^{\,j}}\right); \qquad (11)$$

$$0 \leq i < n+s-f; \quad n+s-f \leq j < n+s$$

The third factor in (11) indicates whether a check-bit position larger than j has already been mapped to the ith memory column. The last factor in (11) is used to verify if j^{th} check-bit position has been mapped to a memory column with an index smaller than i.

The $M_i^{\,j}$ control signals used to select the inputs of the last f multiplexors $(n+s\text{-}f\leq i<n+s)$ in Figure 4 can be calculated as follows:

$$M_i^{\,i} = \overline{\bigvee_{s=0}^{n+s-f-1} M_s^{\,i}}; \qquad (12)$$

$$M_i^{\,j} = M_j^{\,i};$$

$$n+s-f \leq i < n+s; \quad 0 \leq j < n+s-f$$

In order to ensure fast logic implementations of (12) without increasing the amount of configuration information, one can use the configuration bits $T_i^{\,\prime} = M_i^{\,i}$ $(n+s\text{-}f\leq i<n+s)$ instead of the test result bits T_i $(n+s\text{-}f\leq i<n+s)$. The $T_j^{\,\prime}$ bits can be calculated off-line after the production test or during in-field BISR. In this case, $T_j^{\,\prime}$ should replace T_j $(n+s\text{-}f\leq j<n+s)$ in (11) and the following relation should be used on-line instead of (12):

$$M_i^{\,i} = T_i^{\,\prime}; \quad n+s-f \leq i < n+s$$

As mentioned in Section III, an $(n+s)$-bit E-SEC-DED code can be reduced to a $(n+s\text{-}s^{\prime})$-bit E-SEC-DED code $(0<s^{\prime}\leq s)$ if the last $s^{\prime}$ check-bits are ignored. This may be helpful in the case of memory banks that contain $s^{\prime}$ completely defective columns. With the remaining $s\text{-}s^{\prime}$ check-bits, it is still possible to mask the presence of $f=$minimum$(r+s\text{-}s^{\prime},\ 2^{s\text{-}s^{\prime}}+s\text{-}s^{\prime}\text{-}1)$ memory columns with isolated defective storage cells. The third factor in (11) guarantees that the last s check-bits are the first ones to be assigned to the memory columns with defects. If a $(n+s\text{-}s^{\prime})$-bit E-SEC-DED code is needed, then the very last $s^{\prime}$ check-bits can be ignored by programming to logic 1 the configuration bits $T_j^{\,\prime}$ $(n+s\text{-}s^{\prime}\leq j<n+s)$. Column replacement based on bit-shifting [16] can be applied after bit-swapping in order to mask out the $s^{\prime}$ completely defective columns.

We implemented the proposed repair scheme and a conventional scheme, based on a SEC-DED code and column replacement [16], for 32-bit memories and different numbers of spare columns. The repair logic of each protection scheme was implemented as a pure combinational block. Synthesis results obtained with Synopsys Design Compiler and a 45nm standard cell library (TSMC N40LP CMOS) are reported in Table 1. The larger area of the proposed scheme can be considered as acceptable due to the fact that the repair logic is shared by the whole memory. Burst access latencies are significantly smaller than random access latencies and are roughly similar for both protection schemes. The random access latencies appear only when a protection scheme needs to be reconfigured due to a change of the accessed memory bank or memory segment. Obviously, the proposed repair scheme outperforms the conventional scheme with respect to the number of correctable single-bit hard errors.

V. UTILIZATION OF BIT-SWAPPING FOR SOFT-ERROR INSENSITIVE MEMORIES

Soft-error protection may not always be needed [3][11]. In the memory banks where each memory word contains maximum one defective storage cell, the bit-swapping memory protection scheme can be adapted to ensure only single-error protection if the E-SEC-DED code is replaced by a Hamming SEC code [9] with the check-bit number equal to the number of spare columns s. This Hamming code is defined by an $s\times(2^s\text{-}1)$ H-matrix in which all columns are different from each other and from the all-zero vector. The number of columns of this H-matrix is equal to $2^s\text{-}1$, the maximum number of non-zero s-bit vectors. The resulting SEC code words contain $2^s\text{-}1\text{-}s$ data bits and s check-bits.

As the number of memory spare columns and implicitly the number of check bits in the Hamming SEC code is usually small, only a sub-set of bits in the $(n+s)$-bit memory words can be encoded as a SEC code word. The bit-swapper in Figure 4, with $f=2^s\text{-}1$, can be used to map the bit positions protected against the single-bit errors to the memory columns with isolated defective storage cells. As before, this approach works only if a memory word can be affected by at most one defective storage cell.

A comparison of the defect masking scheme presented here with schemes proposed in [7] and [16] for 32-bit memories is given in Table 2. Memories with one single spare column $(s=1)$ are not considered since in their case the column replacement schemes perform better with respect to all parameters. Only the memory repair scheme proposed in [7] allows to mask the same number $(f=2^s\text{-}1)$ of single-bit hard errors based on the concept of programmable restricted SEC codes. Nevertheless, the scheme

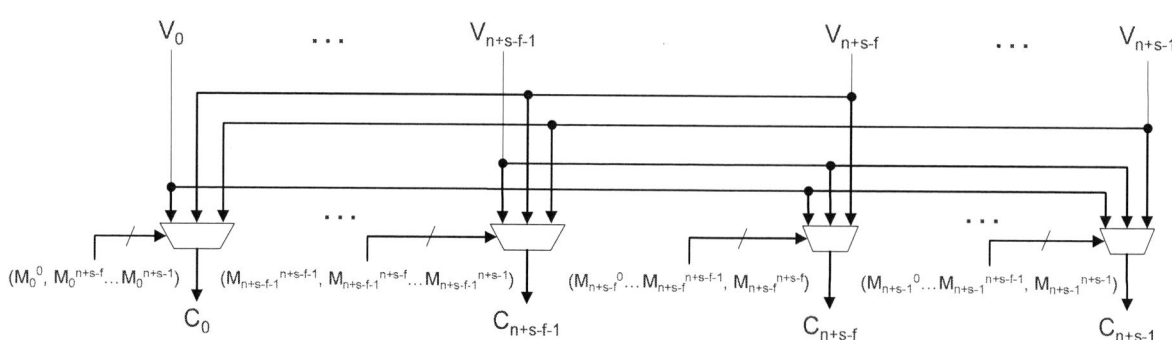

Figure 4. Mapping of the E-SEC-DED code word bits to memory columns via bit-swapping and column replacement.

proposed here has a slightly lower logic overhead and burst accesses latencies between 2 and 5.5 times smaller than the method in [7]. Performance overhead can be further reduced by (a) bypassing the repair logic when error-free memory banks are accessed (b) using column replacement whenever the number of memory columns with defects does not exceed the number of spare columns.

VI. CONCLUSIONS

Spare memory columns normally aimed at replacing completely defective memory columns can also be used to mask out malfunctioning storage cells. Unfortunately, conventional column replacement has limited repair efficiency for this kind of defects. Here, a memory protection scheme was introduced to ensure the correction of both hard and soft errors based on extended SEC-DED (E-SEC-DED) codes and bit-swapping. In this way, the number of columns where the defective storage cells can be masked is significantly increased with respect to solutions based on conventional SEC-DED codes and column replacement. Even in the case when only one spare column is available, two distinct columns with defective storage cells can be masked out instead of a single one as is the case with conventional columns replacement. A generic approach has been proposed to construct the parity-check matrices of E-SEC-DED codes that enable the correction of a large number of double-bit errors. The proposed E-SEC-DED codes have a hierarchical structure and can be easily reduced to the original SEC-DED code or to E-SEC-DEDs with a lower number of supplementary check-bits. This allows their application to memory banks with an arbitrary number of completely defective columns. We have shown how to configure the proposed memory repair scheme out of (built-in) test result bits that indicate the memory columns with damaged storage cells. We have also presented how to apply this repair approach to memories that do not require soft error protection and we have evaluated the improvements with respect to state-of-the-art memory repair schemes.

REFERENCES

[1] R. Baumann "Soft Errors in Advanced Computer Systems," IEEE Design and Test of Computers, pp. 258-266, 2005.

[2] S. Borkar "Designing Reliable Systems from Unreliable Components: The Challenges of Transistor Variability and Degradation," IEEE Micro 25,

No. 6, pp. 10–16, 2005.

[3] G. Cellere et al. "Can Atmospheric Neutrons Induce Soft Errors in NAND Floating Gate Memories?" IEEE Electron Device Letters, vol. 30, no. 2, pp. 178-180, 2009.

[4] C.L. Chen, M.Y. Hsiao "Error-Correcting Codes for Semi-conductor Memory Applications: A State of the Art Review," IBM J. Res. Develop., vol. 28, no. 2, pp. 124-134, March 1984.

[5] C. Constantinescu "Trends and Challenges in VLSI Circuit Reliability," IEEE Micro, Volume 23, Issue 4, pp. 14–19, July 2003.

[6] R. Datta, N.A. Touba "Exploiting Unused Spare Columns to Improve Memory ECC," Proc. IEEE VLSI Test Symposium, pp. 47-51, May 2009.

[7] S. Evain, Y. Bonhomme, V. Gherman "Programmable Restricted SEC Codes to Mask Permanent Faults in Semiconductor Memories," IEEE International On-Line Testing Symposium, 2010.

[8] B. Godard, J.-M. Daga, L. Torres, G. Sassatelli "Hierarchical Code Correction and Reliability Management in Embedded nor Flash Memories," IEEE European Test Symposium, pp. 84-90, 2008.

[9] R.W. Hamming "Error Correcting and Error Detecting Codes," Bell Sys. Tech. Journal, Vol. 29, pp. 147-160, April 1950.

[10] M. Horiguchi "Redundancy techniques for high-density DRAMS," IEEE International Conference on Innovative Systems in Silicon, pp. 22-29, October 1997.

[11] M. Hosomi, H. Yamagishi, T. Yamamoto "A novel nonvolatile memory with spin torque transfer magnetization switching: spin-ram," International Electron Devices Meeting, 2005, pp. 459–462.

[12] M.Y. Hsiao "A Class of Optimal Minimum Odd-weight-column SEC-DED codes," IBM Journal of Research and Development, Vol. 14, pp. 395-401, 1970.

[13] I. Kim et al. "Built In Self Repair for Embedded High Density SRAM," International Test Conference, pp. 1112-1119, 1998.

[14] S. Lin, D. J. Costello "Error Control Coding: Fundamentals and Applications," Prentice-Hall, Inc., Englewood Cliffs, NJ, 1983.

[15] S. Mukherjee et al. "Cache scrubbing in microprocessors: myth or necessity?" 2004.

[16] M. Nicolaidis, N. Achouri, L. Anghel "A Diversified Memory Built-In Self-Repair Approach for Nanotechnologies," IEEE VLSI Test Symposium, pp. 313-318, 2004.

[17] S.E Schuster "Multiple word/bit line redundancy for semiconductor memories," IEEE J. Solid-State Circuits, No.5, pp.698- 703, 1978.

[18] M. Spica, T.M. Mak "Do we need anything more than single bit error correction (ECC)?" Proc. IEEE International Workshop on Memory Technology, Design, and Testing (MTDT), 2004.

[19] C.H. Stapper, H.-S. Lee "Synergistic fault-tolerance for memory chips," IEEE Transactions on Computers, Volume 41, Issue 9, pp.1078 – 1087, Sept. 1992.

TABLE I. SYNTHESIS RESULTS OF REPAIR SCHEMES BASED ON E-SEC-DED + BIT-SWAPPING VS. SEC-DED + DISTANT COLUMN REPLACEMENT

# spare columns	E-SEC-DED + bit-swapping				SEC-DED + distant column replacement [16]			
	# correctable hard errors	Random access latency [ns]	Burst access latency [ns]	area [nand2]	# correctable hard errors	Random access latency [ns]	Burst access latency [ns]	area [nand2]
s=3	f=10	10.82	2.08	2074	f=3	9.16	2.02	733
s=2	f=5	9.72	1.80	1292	f=2	7.89	1.73	596
s=1	f=2	8.72	1.80	747	f=1	6.55	1.73	484

TABLE II. SYNTHESIS RESULTS OF REPAIR SCHEMES BASED ON BIT-SWAPPING VS. SOLUTION IN [7] VS. DISTANT COLUMN REPLACEMENT

# spare columns	Bit-swapping + restricted SEC				Programmable restricted SEC [7]				Distant column replacement [16]			
	# correctable hard errors	Random access latency [ns]	Burst access latency [ns]	area [nand2]	# correctable hard errors	Random access latency [ns]	Burst access latency [ns]	area [nand2]	# correctable hard errors	Random access latency [ns]	Burst access latency [ns]	area [nand2]
s=3	f=7	6.92	1.04	460	f=7	6.73	2.25	593	f=3	5.87	0.51	220
s=2	f=3	8.07	0.60	1023	f=3	8.87	3.43	1032	f=2	6.72	0.39	331

Training-Based Forming Process for RRAM Yield Improvement

Hsiu-Chuan Shih, Ching-Yi Chen
and Cheng-Wen Wu

Department of Electrical Engineering
National Tsing Hua University
Hsinchu, Taiwan
E-mail: {hcshih, cychen95, cww}@larc.ee.nthu.edu.tw

Chih-He Lin and Shyh-Shyuan Sheu

Electronics and Optoelectronics Research Laboratories
Industrial Technology Research Institute
Hsinchu, Taiwan
E-mail: {chihhelin, sssheu}@itri.org.tw

Abstract—**Over the past decade, the resistive memory device known as RRAM has been studied extensively in many ways, and many of its problems have been identified, discussed, and some solved. It is time to move from material, process, and device to circuit design and yield, in order to commercialize RRAM. However, as we move from resistive device to memory circuit, new problems do appear, partly because the operating conditions of resistive devices on real RRAM circuit differ from those in an experimental environment for single devices. In this paper, an over forming problem has been identified from our analysis, and we propose a solution based on training sequence. As a result, by solving the over forming problem, RRAM yield can be improved significantly.**

RRAM; forming process; training sequence; yield improvement; non-volatile memory; memory testing

I. INTRODUCTION

RRAM is a new type of non-volatile random access memory that uses resistive devices as storage elements. Its attractive characteristics of low power, high speed, high density, and compatibility with CMOS technology make it a potential candidate to replace the widely-used flash memories. The resistive device has a typical metal-insulator-metal structure, in which the state of the insulator layer determines the resistance of the device. The resistive switching between the high resistance state (HRS) and low resistance state (LRS) within the device is thereby used to represent the logic value transition.

There are many resistive device research works over the past decade, and most of them focus on the performance of resistive devices of various materials [1-7] and the interpretation of the resistive switching mechanism [8-10]. The resistance switching mechanism has been explained in many ways, such as the alteration of the bulk insulator, the modification of metal/insulator interface, and the formation of localized metal atom chains, yet there is no fixed model so far. The technologies are moving from material and device to circuit design, in spite of the unsolved issue about the resistance switching mechanism. The first and the only work so far that achieved the RRAM chip design can be found in [11].

Forming is a process that initializes an RRAM device. The initial resistance of an RRAM device after fabrication is extremely high. In the forming process, a dedicated set operation (i.e., rising transition) by high voltage is applied on cell arrays for a long time to decrease the device's resistance level to a normal set state (i.e., LRS), and then the cell state can be switched between HRS and LRS by the peripheral read/write circuit. The forming voltage level is high, which is typically proportional to the thickness of the oxide layer in an RRAM device. Take the HfO_2 based RRAM device in [12] as an example, the thickness of the HfO_2 layer of 20 nm to 60 nm requires the forming voltage of 10V to 25V. Therefore, an external tester is usually required for supplying the high forming voltage, and the forming process is consequently time-consuming and expensive. Recently, certain works are focusing on forming-free cell development [13-16]. In addition, it is anticipated that an HfO_2 based RRAM cell can be forming-free if the thickness of the HfO_2 layer can be scaled down to 3 nm [17]. However, most of these forming-free devices suffer from unstable switching or insufficient endurance.

In this paper, we will present the failure analysis of some 128Kb HfO_2 based RRAM circuits. The circuits are developed based on the resistive device proposed in [17], and the RRAM cells have the typical one-transistor-one-resistive-device (1T-1R) structure [11]. The RRAM cell performs well at the cell device level; however, the fabricated RRAM chip has a low yield of 38% originally. By analyzing the cells' resistance distribution, we discovered that the RRAM chip has an over forming problem, which results in temporary stuck-at-faults in cells for a certain period. To solve such a problem, we propose a solution based on training sequence.

II. RESISTIVE RANDOM ACCESS MEMORY (RRAM)

The RRAM cell in the chip under measurement is composed of an HfO_2 based resistive device and a transistor serving as the switch of the cell [11]. The resistive device has the $TiN/Ti/HfO_2/TiN$ composition, in which the Ti layer in the device can receive oxygen from the HfO_2 layer and turn the HfO_2 layer into $HfO_{1.4}$. The oxygen vacancies generated in the HfO_2 layer are considered a probable physical mechanism leading to the resistance change of the device. The oxygen vacancies will form a conducting path between the top and bottom electrode layers during a set operation, and the resistance is thus reduced. On the other hand, the conducting path will break partially during a reset operation, which results in a high resistance device. The device can be repeatedly switched between the high and low resistance

978-1-61284-657-6/11 $26.00 © 2011 IEEE

sates. The initial resistance of the RRAM device is extremely high, and a forming process is applied on the memory cell arrays to pull the device down to a low resistance state. Forming requires an external tester for supplying the high forming voltage, and the forming process is consequently time-consuming and expensive.

Each of the chips in our experiment has 32 individual 128 Kb-RRAM cores. The RRAM core has a typical RAM organization as shown in Fig. 1. There are word-lines (WLs) and bit-lines going through the cell array. During a write operation, a set/reset voltage is applied on the selected BL and the cross voltage on the selected cell changes its resistance state. As to the read operation, a small current is applied on the selected cell. The read circuit as shown in Fig. 2 compares the resulting cross voltage on the selected cell with a reference voltage to determine the data value. The reference voltage is provided by a reference cell of certain resistance. However, considering the potential read failure caused by inaccurate reference resistance, which is widely discussed in other non-volatile RAMs such as MRAM or PCM, the read circuit can alternatively have an external reference voltage.

To help separate the HRS and the LRS distributions, a verification mechanism is adopted for the write operation of the chip. As the verification mechanism is enabled, successively increased voltage pulses and verify-read operations are applied during a write operation, until the resistance of the target cell reaches the threshold resistance of the other state. In the write circuit, the set and reset voltage levels applied on the selected BL, i.e., V_m and V_p, are dynamically adjusted by controlling V_{ref}. The set and reset operations use different reference resistance values for the verify-read operation, which are the margins of the corresponding state. The verification mechanism tends to move the distributions behind the state margins, and mitigate the overlap between the HRS and the LRS distributions. In addition, increasing voltage pulses will tighten the distribution of both states. Such technique is commonly-used in flash memories, and it has been adopted in a previous RRAM work [15]. It is shown that the verification mechanism with increasing voltage pulses can help stabilize

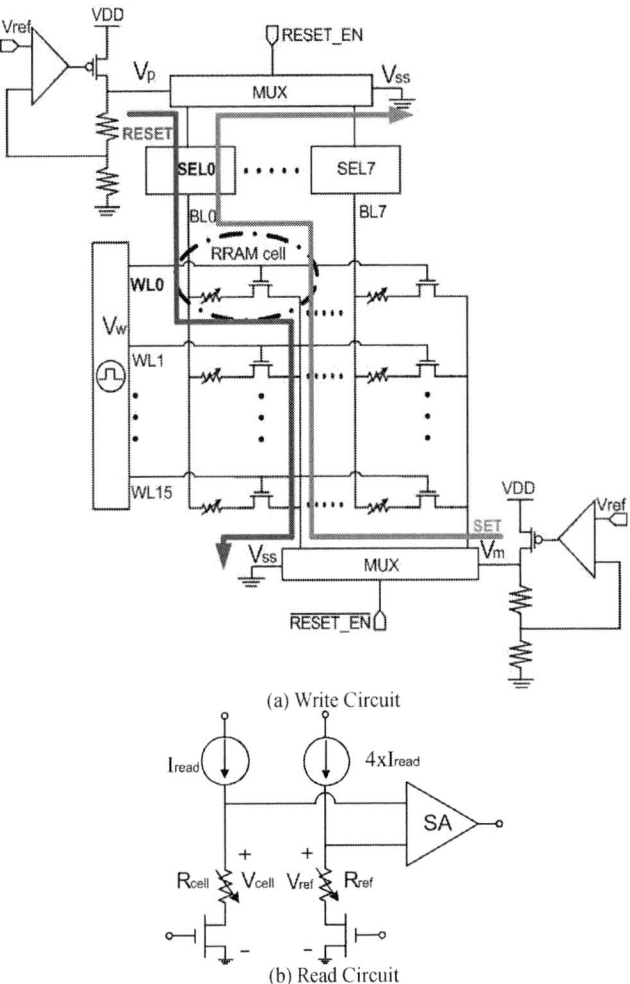

(a) Write Circuit

(b) Read Circuit

Figure 2.   Write circuit and read circuit [11].

the cell state and result in improved endurance.

III.   OVER FORMING

We applied the MSCAN test on a 128Kb RRAM core manufactured by a commercial 0.18 um CMOS technology.

$$\{\updownarrow(W0); \updownarrow(R0); \updownarrow(W1); \updownarrow(R1)\}. \qquad (1)$$

The raw yield of the chip is only 38%, in which most failed cells show the stuck-at-zero symptoms. However, these cells are not permanent failed. Most of them behave correctly and the failure rate decreases after an exercise of hundreds of set and reset operations.

To analyze the failure mechanism, we measured the resistance distribution of the cells in the RRAM cores. The result is shown in Fig. 3. In reality we did not get the exact resistance value of all cells, since we can only achieves the data output signals. In the read circuit of the core, the read operation is performed by voltage comparison between the voltage difference across the selected cell and a reference voltage level, which is applied externally. An output logic-1 indicates the cross voltage of the accessed cell is higher than

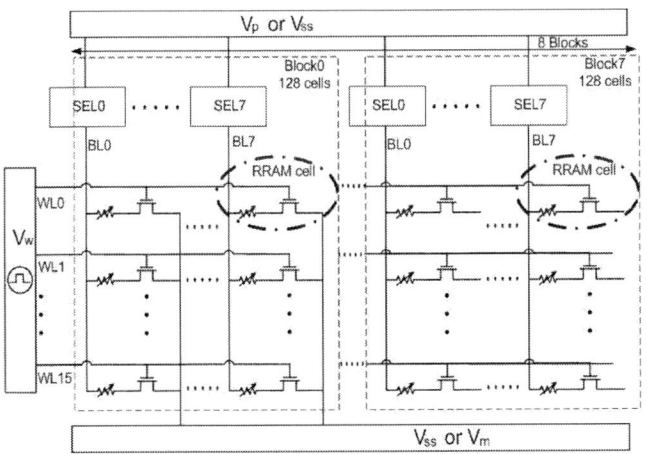

Figure 1.   Organization of RRAM cell array [11].

978-1-61284-657-6/11 $26.00 © 2011 IEEE   147

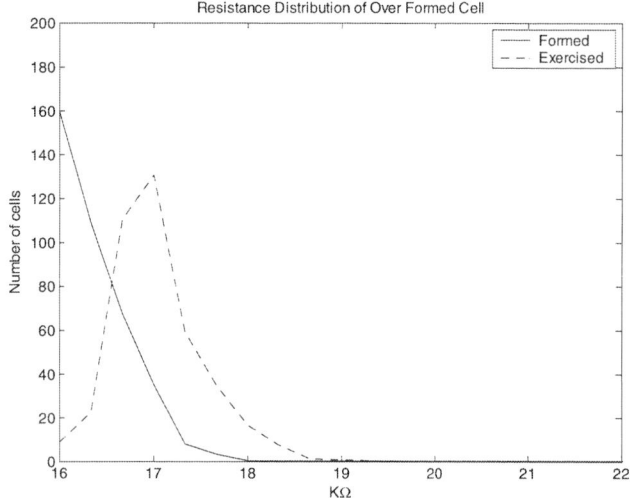

Figure 3. The change in the resistance distribution of a 128Kb RRAM core.

the reference voltage. By measuring the data value of cells for different reference voltage levels, we can derive the voltage range of a cell's cross voltage. Then the cell resistance can be derived by the cross voltage value. In the target case, 0.30 volt can be realized as approximate 20 KΩ. Accordingly, we employ the reference voltage to derive the distribution of cells' resistance.

In Fig. 3, two curves respectively present the distribution of cells' resistance at the forming process finished, and after the exercise of hundreds of set/reset operations. The result shows that most cells have resistance of less than 16KΩ after the forming process. However, the set operation for the core is designed for pulling the cell resistance to lower than 20 KΩ, and the strength of reset operation is therefore design for cells of slightly lower than 20 KΩ. The over-formed cells of less than 16 KΩ require a stronger reset stimulus for reset, and the operation region of these cells may be mismatched with the ability of the write circuit. The fact explains the result of a great number of stuck-at-zero cells in the test result.

The cell distribution after hundreds of set and reset operations in Fig. 3 shows the set state of cells move to around 17 KΩ. For the time the failure rate has decreased, which should be on account of the cells recovered from temporary stuck faults.

IV. TRAING SEQUENCE

To solve the over forming problem, a probable solution is to adjust the strength and the length of forming. However, as the distribution shown in Fig. 3, cells in a core may have different sensibility to the forming process and it therefore results in a wide range of initial cell resistance. If the forming strength is reduced, it could bring the initial resistance of certain cells beyond the normal range.

From the fault recovery phenomenon, we realize the effect of a sequence of write operations on the HRS and LRS

distribution of RRAM cells. Accordingly, we proposed a method to improve the yield of RRAM chips.

A. Training in the Forming Process

For the over forming problem, we suggest an additional training sequence after the forming operation:

$$\updownarrow(W1, W0)^n. \qquad (2)$$

The sequence expressed by the March notations is a series of repeated set and reset operations over the whole cell arrays, where n is the length of the training sequence. The training repeats the March element for n times.

As introduced in Sec. 2, the verification mechanism is adopted for the write operation of the chip. A set/reset operation successively applies increasing voltage pulses and verifies the cell resistance, until the resistance of the target cell reaches the margin of another state. The technique helps separate and tighten the HRS and the LRS distributions. However, certain cells could still be over-set or over-reset if they are too sensitive to the write operations. The initial set/reset voltage results in an extreme change in their resistance state. Cells consequently perform unstable switch behavior, and it manifests as the temporary failed cells in the chip.

Moreover, unbalanced strength between set and reset operation also results in unstable switch behavior. If the strength of reset operation is too strong to certain cell, the cell resistance would be pushed over the normal HRS range. After that, the strength of a subsequent set operation would be insufficient to pull down its resistance. Thus there is a strong dependence between the success of a write operation and the cell state.

The training sequence has the ability to concentrate reset state and set state of cells into a range approximate to the margin of normal HRS and LRS region. By repeated switches between HRS and LRS, the two states inside the devices are imperceptibly formed, and the cells would change between two fixed resistance levels. Though the exact physical mechanism of the resistive device is not confirmed now, the effect of training sequence shown in Fig. 3 is consistent with above analysis.

B. Decision of the Training Sequence Length

The length of the training sequence n is a key factor of the training effect. Insufficient training cannot achieve the concentrated distribution and the stable switching between states. However, excessive training consumes the endurance of devices. Since the physical mechanism of the resistive device operations is still a mystery, we comprehend the influence of training length on the resistive devices by realistic measurement.

The cell resistance distributions of set state and reset state for different n are shown in Fig. 4 and Fig. 5 respectively, which are measured according to

$$\{(\text{Forming}); (R_{dis}); \uparrow(W1); (R_{dis}); \uparrow(W0); (R_{dis});$$
$$\uparrow(W1, W0)^{10}; \uparrow(W1); (R_{dis}); \uparrow(W0); (R_{dis});$$
$$\uparrow(W1, W0)^{100}; \uparrow(W1); (R_{dis}); \uparrow(W0); (R_{dis});$$
$$\uparrow(W1, W0)^{1000}; \uparrow(W1); (R_{dis}); \uparrow(W0); (R_{dis});\}. \quad (3)$$

In the above patterns, R_{dis} is the procedure to trace the distribution of cell resistance, which contains a sequence of read operations with different reference voltage levels for the read circuit. We apply the training sequences in length of ten, of a hundred, and of 1K, and observe the resistance distribution after each training sequence by the $\{\uparrow(W1);$ $(R_{dis}); \uparrow(W0); (R_{dis});\}$ pattern. By such steps the observed distributions are, in fact, for n of 11, 112, and 1113 respectively, instead of exact ten, a hundred, and 1K. Nevertheless, we append sequences of a round number for simplification, since we focus on the trend for n in different scale. In addition, we apply the pattern in an ascending sequence in the experiment, while the direction does not affect the result.

Fig. 4 clearly shows the change in the set state distribution. It continuously moves toward 20 KΩ as n increases, which is the margin of LRS designed for the write circuit. And only few of cells have a resistance exceeding the margin. Compared with n of hundred, the distribution after training of 1K tends to expand. Since the result of a reset operation is highly related to the original state in the accessed device, a chip of less concentrated distribution would behave more unstably. The set distribution might extend due to finite ability of training sequence or a declined reset state distribution.

Fig. 5 shows the reset state distributions for different n. In the chip under measurement, the verification mechanism for reset operation is disabled due to certain voltage mismatch problems. The training sequence is therefore carried out by the verify-set operations and the primitive reset operations. Thus, the reset state distribution has no distinct margin as the set state does, and it leans toward the margin of set state 20 KΩ. Accordingly, it has no alternative to set the reference of read operation as the set state margin for this RRAM chip, for the better yield.

In Fig. 5 the number of cells within the resistance range is much smaller, and the curve is therefore represented in different scale with Fig. 4. Note that, most of cells have a resistance larger than 33 KΩ, and the main distribution falls outside the range shown in Fig. 5. In the target RRAM chip, the cross voltage of the resistive device is designed to be clamped under 0.5 volt to avoid disturb the cell data during a read operation. With such constraint, the device of resistance larger than 33 KΩ appears the same cross voltage during read operation. As a result, we cannot conduct the exact resistance distribution above 33 KΩ and the figure shows distribution in a limited range.

Nevertheless, the distribution within the measurable range still reveals certain information. For the reset state, a cell of resistance below 20 KΩ is a faulty cell. The distribution for cells before any training sequence contains a number of such faulty cells. These cells behave similar to the stuck-at-zero. It should be due to the over-formed effect. After a hundred times of training, the stuck-at-zero cells are

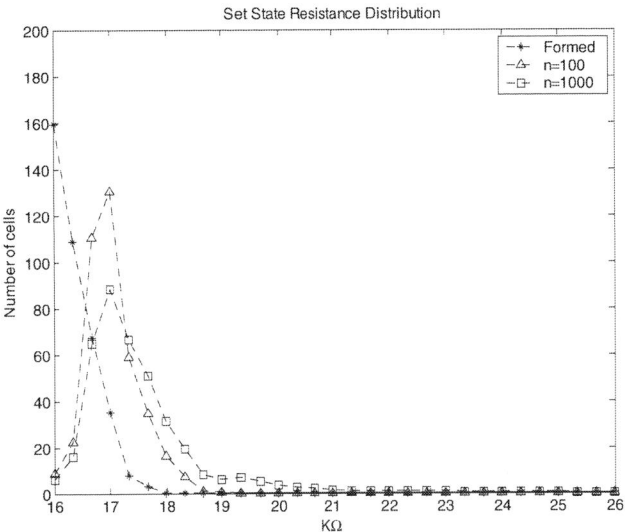

Figure 4.   Effect of the training sequence on the set state.
μ_Formed=16.3481, σ_Formed=0.3917;
μ_100=17.0594, σ_100=0.6534;
μ_1000=17.6342, σ_1000=1.4436

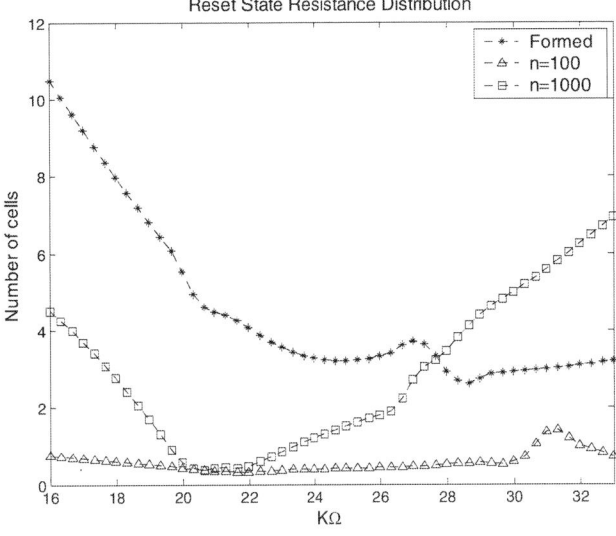

Figure 5.   Effect of the training sequence on the reset state.

greatly decreased. However, after another training sequence of 1K, faulty cells increase again.

To realize the cause of this phenomenon, we applied one more 1K-training sequence on the cell arrays. As a result, the number of faulty cells keeps growing. We present the distribution of cells that are kept at the same resistance level after the last training sequence in Fig. 6. The result shows almost the entire of faulty cells results from the first 1K-training sequence remains failed after another 1K-training sequence. That is, these cells have been destroyed, which results in permanent fault behaviors. The poor endurance is inconsistent with the performance demonstrated during the cell device level. There might be certain problems occurs in

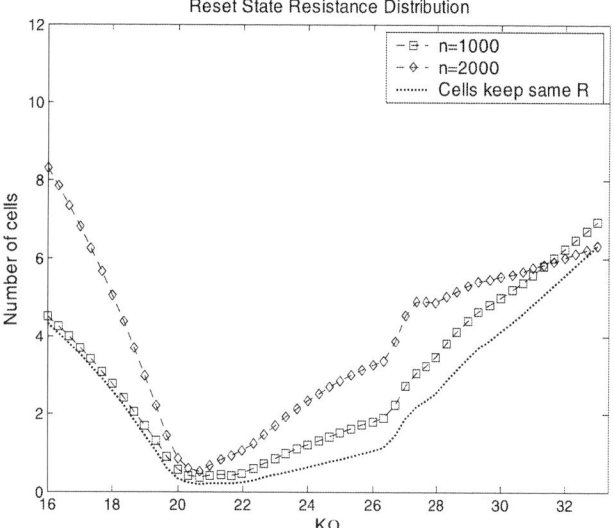

Figure 6. The faulty cells remain failed after another 1K-training.

the write circuit design and result in a mismatch between cells' operating region and the generated write strength. The endurance problem would be improved as the technologies get matured. It is much less perplexing than the unstable switch problem with the present technologies.

As a result, n of 100~150 is appropriate for training RRAM cells in the case. The set state distribution is concentrated in a short range, while the reset state is pushed to a high level beyond 33 KΩ. The resistance distributions of two states are distinct and stable during switch between states. More training tends to decline the effect of stabilization, while the sequence can be a stress on devices and the cells lack of endurance could be destroyed permanently.

The length of the training sequence should be dependant on the device characteristics, and n can be different to different devices. Nevertheless, it can be determined by measurement during the development stage. In case of variation occurs during manufacturing, the length can be further designed to be reconfigurable in the circuit and changed according to chip's characteristics.

C. Self-Forming

Based on the training sequence's ability of bringing the device state to be close to the margins of HRS and LRS, we proposed a self-forming method that forms the RRAM devices by the training sequence alone, without the ATE process. According to the measurement result, the training sequence shows a superior effect than the ATE forming process. We therefore replace the conventional ATE forming process by the training sequence.

The training sequence is simply a repeated set and reset operations performed by the write circuit, which does not require high voltage sources. It can be realized by a simple built-in circuit or utilize an existing built-in self-test (BIST) circuit to accomplish the sequence, which would greatly reduce the forming time and the cost for tester equipments.

TABLE I.   EXPERIMENTAL RESULT

	ATE Forming	Forming w/ training sequence	Self-forming
Reset fail rate	61.79%	01.28%	09.63%
Set fail rate	00.20%	04.76%	02.78%
Yield	38.01%	93.96%	87.99%
Time (sec)	450	454.9	4.9

V.   EXPERIMENTAL RESULT

We respectively apply the training sequence to the RRAM cores that have finished the ATE forming process and apply the self-forming to the circuits without the ATE forming. After that, we apply the MSCAN test to check the training effect. The RRAM circuits are manufactured by 0.18 um CMOS technology. Each of the cores is 128Kb and the I/O width is 8. The self-forming at the time is realized by the FPGA board.

Corresponding time cost is evaluated according to the following equation:

$$T_{Forming} = T_{ATE} + (T_{Set} + T_{Reset}) * N_{Addr} * n. \qquad (4)$$

It includes the time for the high voltage forming procedure provided by ATE, T_{ATE}, and the time for training sequence of length n, as the product term shows. T_{ATE} averages 7.5 minutes for the target 128Kb RRAM core. The set operation and the reset operation are 2.8 us and 375 ns respectively, performed word-oriented. Hence the training time is accumulated by number of word, N_{Addr}, which is 16K for the chip in our experiment.

The test result and the forming time are shown in Table 1. Conventionally, the forming procedure is the high voltage process by ATE, whose result is listed in the first column. The chip formed by ATE results in a poor yield of 38%. The result of appending the training sequence into the conventional forming process is shown in the second column, in which the training sequence is 100 in length. Compared to ATE forming, the proposed method has greatly improved the yield by 56% in additional 4.9 seconds.

As to the proposed self-forming technique by training sequence, the result is shown in the last column. In this case the reference resistance is 26.67 KΩ, and the length of training sequence for self-forming is 100. The distribution of formed cells is shown in Fig. 7, where the set distribution is concentrated around 24 KΩ and the reset distribution is beyond 33 KΩ. The results shows that without the high voltage process by ATE forming, the training sequence also provides a great forming effect, which results in about 50% yield improvement. Moreover, required time for forming a 128Kb RRAM core is significantly reduced to less than 5 seconds.

The forming time grows rapidly with the memory capacity. A 4Mb RRAM chip requires a conventional forming process of four hours by an ATE, and it would make a great cost for mass production. On the other hand, the proposed self-forming by training sequence method can be realized by a simple built-in circuit without an external tester resource. Consequently, the few forming time can be further reduced by parallel forming for multiple cores.

978-1-61284-657-6/11 $26.00 © 2011 IEEE

Corresponding cost can be well controlled for mass production.

VI. CONCLUSIONS

The RRAM devices in a chip may have different sensitivities to the forming process due to immature process or process variation. Thus the over forming problem is likely to occur, and results in temporary failed cells. The proposed training sequence technique has a great effect on forming the cells into stable states. The training sequence adopted after a conventional ATE forming process improves the yield from 38% to approximately 94%. Moreover, it can even replace the long-time ATE forming process and achieve an 88% yield in less than 5 seconds for a 128Kb RRAM core. The verification mechanism is disabled for reset operations in the chips in our experiment, and the capability to tighten states only works for the set state. With the full verification mechanism for both states, the set and rest states can be well separated. In that case the training sequence is expected to have a greater effect on yield improvement.

ACKNOWLEDGEMENT

This work is supported in part by the National Science Council, ROC, under contract NSC 98-2221-E-007-082-MY3.

REFERENCE

[1] C.Y. Liu, P.H. Wu, A. Wang, W.Y. Jang, J.C. Young, K.Y. Chiu, and T.Y. Tseng, "Bistable Resistive Switching of A Sputter-Deposited Cr-Doped SrZrO3 Memory Film," IEEE Electron Device Lett., vol. 26, no. 6, pp. 351-353, June 2005.

[2] M. Fujimoto, H. Koyama, S. Kobayashi, Y. Tamai, N. Awaya, Y. Nishi, and T. Suzuki, "Resistivity and Resistive Switching Properties of Pr0:7Ca0:3MnO3 Thin Films," Appl. Phys. Lett., vol. 89, no. 24, p. 243504, Dec. 2006.

[3] H. Sim, D. Choi, D. Lee, S. Seo, M.J. Lee, I.K. Yoo, and H. Hwang, "Resistance-Switching Characteristics of Polycrystal-line Nb2O5 for Nonvolatile Memory Application," IEEE Electron Device Lett., vol. 26, no. 5, p. 292-294, May 2005.

[4] M. Fujimoto, H. Koyama, M. Konagai, Y. Hosoi, K. Ishihara, S. Ohnishi, and N. Awaya, "TiO2 Anatase Nanolayer on TiN Thin Film Exhibiting High-Speed Bipolar Resistive Switching," Appl. Phys. Lett., vol. 89, no. 22, pp. 223509, Nov. 2006.

[5] S. Seo, M.J. Lee, D.H. Seo, E.J. Jeoung, D.S. Suh, Y.S. Joung, I.K. Yoo, I.R. Hwang, S.H. Kim, I.S. Byun, J.S. Kim, J.S. Choi, and B.H. Park, "Reproducible Resistance Switching in Polycrystalline NiO Films," Appl. Phys. Lett., vol. 85, no. 23, pp. 5655-5657, Dec. 2004.

[6] D. Lee, H. Choi, H. Sim, D. Choi, H. Hwang, M.J. Lee, S.A. Seo, and I.K. Yoo, "Resistance Switching of the Nonstoichiometric Zirconium Oxide for Nonvolatile Memory Applications," IEEE Electron Device Lett., vol. 26, no. 10, pp. 719-721, Oct. 2005.

[7] H.Y. Lee, P.S. Chen, C.C. Wang, S. Maikap, P.J. Tzeng, C.H. Lin, L.S. Lee, and M.J. Tsai, "Low-Power Switching of Nonvolatile Resistive Memory Using Hafnium Oxide," Jpn. J. Appl. Phys., vol. 46, no. 4B, pp. 2175-2179, Apr. 2007.

[8] J.J. Yang, M.D. Pichett, X. Li, D.A.A. Ohlberg, D.R. Stewart AND R.S. Willeams, "Memristive Switching Mechanism for Metal/Oxide/Metal Nanodevices," Nature Nanotechnology, vol. 3, no. 7, pp. 429-433, July 2008.

[9] B. Gao, S. Yu, N. Xu, L.F. Liu, B. Sun, X.Y. Liu, R.Q. Han, J.F. Kang, B. Yu, and Y.Y. Wang, "Oxide-based RRAM switching mechanism: A new ion-transport-recombination model," in Proc. IEEE Int. Electron Devices Meeting (IEDM), Dec. 2008, pp. 1-4.

[10] C. Cagli, D. Ielmini, F. Nardi, and A. L. Lacaita, "Evidence for threshold switching in the set process of NiO-based RRAM and physical modeling for set, reset, retention and disturb prediction" in Proc. IEEE Int. Electron Devices Meeting (IEDM), Dec. 2008, pp. 1-4.

[11] S.S. Sheu, P.C. Chiang, W.P. Lin, H.Y. Lee, P.S. Chen, Y.S. Chen, T.Y. Wu, F.T. Chen, K.L. Su, M.J. Kao, K.H. Cheng, M.J. Tsai, "A 5ns Fast Write Multi-Level Non-Volatile 1 K Bits RRAM Memory with Advance Write Scheme," in Symp. on VLSI Circuit, June 2009, pp. 82-83.

[12] Ch. Walczyk, Ch. Wenger, R. Sohal, M. Lukosius, A. Fox, J. Dabrowski, D. Wolansky, B. Tillack, H.J. Mussig, and T. Schroeder, "Pulse-Induced Low-Power Resistive Switching in HfO2 Metal-Insulator-Metal Diodes for Nonvolatile Memory Applications," J. Appl. Phys., vol. 105, no. 11, June 2009.

[13] L. Goux, J.G. Lisoni, X.P. Wang, M. Jurczak, and D.J. Wouters, "Optimized Ni Oxidation in 80-nm Contact Holes for Integration of Forming-Free and Low-Power Ni/NiO/Ni Memory Cells," IEEE Trans. on Electron Device, vol. 56, no. 10, Oct. 2009.

[14] H. B. Lv, M. Yin, Y. L. Song, X. F. Fu, L. Tang, P. Zhou, C. H. Zhao, T. A. Tang, B. A. Chen, and Y. Y. Lin, "Forming Process Investigation of CuxO Memory Films," IEEE Electron Device Letters, vol. 29, no. 1, pp. 47-49, Jan. 2008.

[15] H. B. Lv, M. Yin, P. Zhou, T. A. Tang, B. A. Chen, Y.Y. Lin, A. Bao, and M. H. Chi, "Improvement of Endurance and Switching Stability of Forming-Free CuxO RRAM," in Proc. Non-Volatile Memory Technology Symp., May 2008, pp. 52-53.

[16] L. Courtade, Ch. Turquat, Ch. Muller J.G. Lisoni, L. Goux, and D.J. Wouters, "Improvement of resistance switching characteristics in NiO films obtained from controlled Ni oxidation," in Proc. Non-Volatile Memory Technology Symp., Nov. 2007, pp. 1-4.

[17] H.Y. Lee, P.S. Chen, T.Y. Wu, Y.S. Chen, C.C. Wang, P.J. Tzeng, C.H. Lin, F. Chen, C.H. Lien, and M.J. Tsai, "Low Power and High Speed Bipolar Switching with a Thin Reactive Ti Buffer Layer in Robust HfO2 Based RRAM," IEDM, Dec. 2008, pp. 1-4..

The Bang for the Buck with Resiliency: Yield or Field?
Organizer: Arani Sinha, AMD (arani.sinha@amd.com)
Moderator: Suriyaprakash Natarajan, Intel Corporation
(Suriyaprakash.natarajan@intel.com)

Today's electronic systems and those envisioned for the near future exploit significant integration of devices on a chip with incredibly shrinking device/interconnect geometries. Such systems operate very close to their power/performance margins to achieve maximum profitability. The increase in the design complexity of such systems that now include digital and analog components, coupled with the race to reach the market faster severely constrains the resources and time to validate and test them. Furthermore, products can also fail to operate correctly in the field prior to their expected end-of-life due to transient errors or aging.

This scenario requires that the designs be resilient and adaptive by construction allowing automated identification of operating points and elimination of defective parts. Building resiliency into designs can increase cost in terms of design time, power, area, and in some cases, degraded performance. However, products come in different price ranges and with different life expectancies. For example, customers who buy a server product may pay a premium price and expect it to operate for long periods of time without an outage, while buyers of smart phones may pay little, expect more features, and care less about reliable operation of features other than phone calls. Hence, the benefit of resilient features can be realized in different products in different ways. In some products they could ensure reliable field operation, while in others, they could reduce the silicon bring-up time, and increase manufacturing yield by identifying and circumventing defective but redundant parts of a design with some potential loss of performance.

This session explores the state of the art resilient techniques for digital and mixed-signal designs that aid improving manufacturing yield or operating in the field, and the cost/benefit trade-offs involved.

Presentations:

"Resilient Circuits for Improving Microprocessor Performance and Energy Efficiency"
Carlos Tokunaga, Intel Corporation

"Resiliency: An IP Design Perspective"
Vikas Chandra, ARM

"High Performance Mixed-Signal/RF: Yield Recovery via Post-Manufacture Tuning"
Abhijit Chatterjee, Georgia Tech

Session 7

978-1-61284-657-6/11 $26.00 © 2011 IEEE

Modified Flip-flop Architecture to Reduce Hold Buffers and Peak Power during Scan Shift Operation

Prakash Narayanan, Rajesh Mittal, Sumanth Poddutur, Vivek Singhal, Puneet Sabbarwal

Texas Instruments India Pvt. Ltd, Bangalore, India.

{prakash.narayanan, rajmittal, sumanthp, viveks, puneet.sabbarwal}@ti.com

Abstract − Hold timing closure and scan power are major concerns for any design. Hold closure for scan shift operation generally causes addition of buffers in the data path between flip-flops. This results in increased gate count that will toggle during the functional mode of operation thereby resulting in an increase in functional power. Scan operation also causes higher switching activity due to high toggling in a given test cycle. There are two components of power i.e. peak power and average power. Peak power increases IR-drop in the design, thereby reducing the voltage across the transistor and can lead to failure. In this paper we will present a modified flip-flop architecture that will serve two purposes i.e. enabling hold timing closure across process, voltage, temperature and reducing peak power during scan shift operation with minimal impact to functional timing and area. The modified flip-flop will introduce a half cycle delay in the data path invariant of process, voltage, temperature thereby easing hold closure. Test time and coverage are not impacted by the same. Existing ATPG tool generated pattern can be applied with this scheme. This approach reduces peak power close to 50% and reduces hold buffer area close to 40% in a given design.

Keywords-Hold timing closure, scan shift power, slack profile, etc

I. INTRODUCTION

In present day ultra-deep sub-micron technology multi-million gate SOCs (Systems on chips), scan tests are mandatory to catch the manufacturing defects. Scan test involves scan shift-in, capture and shift-out operation. In scan shift, vectors are pumped into all the flip-flops to put the design into a known state [1]. During capture, the required number of clock pulses is given to the design and the state of the design is captured into flip-flops. The captured data is then shifted out to the tester for comparing it with the expected vector. Scan chains are formed by stitching together flip-flops one after the other. Q of the first flip-flop is connected to the scan-in pin of the next flip-flop, etc.

The scan chain is just daisy chaining of flip-flops so as to shift-in and shift-out the data and there is no combinatorial logic, the issue of hold violation is more prevalent in scan shift mode [2]. Despite all the efforts put to achieve clock balancing across all the flip-flops there will be clock skew between them. In order to ensure that there is no hold-timing related issue between any two flip-flops, either their clocks need to be balanced accurately or buffers need to be inserted in the data path. Addition of buffers in the data path results in increased gates toggling during functional operation as well.

It is well known that power consumption in test mode is higher than that in normal mode [3] because test patterns,

compared to functional ones, are less spatially and temporally correlated. In a full scan circuit almost entire design will be exercised during scan shift operation. All the nodes of the circuit are toggled and hence it causes very high power dissipation. Both average and peak power increases due to high toggling activity in scan test. Peak power is of higher concern than the average power. Average power indicates the requirement of battery current to be supported by tester and excessive burn out of chip. ATE (Automatic Test Equipment) can support the average power as required. Further more, average power can be reduced by reducing the frequency and excessive burn out can be avoided. On the other hand, peak power is independent of frequency. Higher peak power results in a higher IR drop which would in turn result in increase in delay of the logic cells resulting in timing closure issues. Increase in peak power would lead to device failing at a higher than an intended voltage. A device that fails testing at low voltage could be classified as faulty even though it would perform functionally resulting in lower yield due to power droop and cross talk [4].

Issues caused due to high peak power are generally not visible in simulation and are difficult to model accurately. Such issue goes through prolonged debug cycle and delay the time to market of the device. In today's world time to market is very critical, so we need to have something that is correct by construction and should not invest time in doing unwanted debug. The ramp to production should be fast. The proposed approach is an attempt in that direction.

Power consists of two parts, namely leakage power and dynamic power. Leakage power is more due to the circuit design and CMOS properties. This component is omnipresent from the time the power supply is provided to the chip. Dynamic power is the power consumption of cells when their outputs are transitioning from $1 \rightarrow 0$ or $0 \rightarrow 1$. Lowering the toggling activity at any instant of time reduces the dynamic power and hence also the peak power of the chip.

The objective of this paper is to share the new flip-flop architecture variants and the associated methodology that has been developed to ensure the following:

- Hold timing closure across process, voltage and temperature with minimal impact to area and power.

- Peak power reduction by way of skewing the data ensuring robustness of tests.

The reminder of the paper has been organized as follows. Section II does discuss prior art that has been followed with regards to both scan shift hold-timing closure and peak power reduction during scan shift. The modified scan flip-flop architecture along with complete flow depicting the proposed architecture has been discussed in Section III. The buffer reduction and peak power reduction are described in Section IV. In Section V, some conclusive remarks are drawn.

II. PREVIOUS WORK

Most of the previous work targets either hold buffer reduction or peak power reduction as given below. The benefit of proposed modified flip-flop architecture is to target both.

In [2] a new class of flip-flops is introduced for which the scan input hold requirement and data input setup time are small. The purpose of this is to ease scan chain hold violation while meeting the data path frequency. However there is increased in power consumption during scan mode where scan test are run at higher speed so as to reduce test time. The increased power consumption requires more robust power grid design and package support. In this paper a modified hold-friendly flip-flop is proposed to address the scan mode power problem and easing hold buffer requirement as observed in [2].

In [5] a hold friendly scan flip-flop is introduced whose scan pin hold characteristic has improved while data pin timing and power are left intact. This characteristic helps to resolve scan chain hold problem while meeting the maximum frequency in data path. This solution can reduce the number of buffers inserted in the scan chain to fix hold violations. The new flip-flop can save up to 27% area and approximately 15% power as compared to the usage of normal flip-flops combined with hold-fixing buffers. Another similar circuit technique for reducing the requirements on hold buffering is proposed in [6], where the scan input data is delayed before it is latched by the flip-flop. A circuit technique to reduce the hold buffer requirement has been proposed in [7], where a delayed and isolated (based on scan enable) version of the output "DQ" is available in "SO". All these technique can be used to reduce hold buffer during shift mode only and hold closure across PVT (Process, Voltage and Temperature) need to be guaranteed. There is no significant improvement in peak power. In proposed approach delay added in flip-flop output is very large that prevents hold violation across PVTs and result in reduction of peak power as well.

There has been several prior works done in order to reduce scan shift power. The approach of gating the output of a flip-flop has been proposed in [8]. This will prevent all the switching or toggle at the output of the flip-flops from propagating to the combinatorial logic. And results in power savings with regards to combinatorial logic and related interconnect switching power. This approach has a couple of drawbacks. The first being, that it adds unnecessary delay at the flip-flop output during the normal functional operation as well. The second drawback is the area increase by way of this gating circuit being present at the output of the flip-flops. Power reduction by blocking circuitry varies from design to design and governed by the extent of blocking circuitry. Proposed scheme has very minimal impact to area of the design and no adverse impact to timing in functional mode.

Traditionally most common method used is non-zero clock skew in order to reduce peak power across synchronous logic as given in [9]. An approach to reduce the scan shift power by skewing the clocks going to different scan chains, so that the edges of all the clocks do not occur at the same time has been proposed in [10]. Different phase shifted versions of the scan clock is provided to the different flip-flops in the design. Here clock skew varies across PVT condition. This approach is feasible only after CTS (clock tree synthesis). The average power increases due to addition of more delay buffer. The reduction in peak power is in order of 10-15%. With the proposed approach, skew of half clock period is guaranteed which can result in close to 50% peak power reduction irrespective of any PVT with no additional buffer. The original approach of multi-phase shifting to reduce scan power has been introduced in [11]. The drawbacks to the approaches described in [10] and [11] are that additional clock phasing logic needs to be implemented in the clock path thereby complicating the design process.

The low power technique as given in [12] talks about shifting in constant value into the scan chains in order to reduce scan shift power. The logic is configured to load constant values from the decompressor onto the scan chains and shifted into the core. Random values in scan chains are replaced with constant values as they are shifted through the core. A repeat fill heuristic is used to generate the constant values. Implementation of approach proposed in [12] results in pattern count inflation with respect to power saving. Almost 80% increase in the pattern count has been seen on 1.2M gate design with 10-20% power reduction option given. This option can only be used in cases where the additional test time overhead is acceptable as against a design revision to fix the power issues. Other pattern-based optimization has also been researched in several prior works which involve techniques like X-fill [13] [14], preferred-fill [15], constant-fill [16], etc. The proposed architecture and method is pattern agnostic

In [17] an adaptive scan architecture approach is presented. In this, scan chains are partitioned into multiple segments and when scan data is being shifted in a particular segment then the other segments clock can be gated off to save power. In case of scan chains where the flip-flops are getting clocked from the same functional clock domain, additional clock gating logic needs to be introduced to separate the clocks for different segments. This introduces complexity in the clock path and will affect the insertion delay and skew in the functional mode of operation as well. This technique results in area overhead in terms of the trigger circuits, clock division logic, delay circuits, etc. In [18] improvements has been made with regards to scan power reduction while not complicating the design, the details of the contribution is not explained in detail here for a lack of space. The proposed circuit technique and method is much simpler since the clock path is not changed or altered in any way, and the delay introduced is only in the data path.

III. MODIFIED SCAN FLIP-FLOP ARCHITECTURE

The original scan flip-flop consists of 2 latches i.e. a master and slave latch. A typical D flip-flop has the following inputs "D" and "CLK" and the output "Q". A scan flip-flop has an extra multiplexer in the "D" path. It has two more inputs "SD" and "Scan_en", where "SD" stands for the scan data. Scan chains are constructed by connecting the flip-flops one after the other by connecting the "Q" output of a previous flip-flop to the "SD" input of the subsequent flip-flop, etc as shown in Figure 1. In some cases, a scan flip-flop could have an additional data output called "SQ" which is the scan data output. "SQ" is generally buffered "Q" output which will be used to drive scan data. The separate "SQ" helps in closing scan path independent of functional path. The data that arrives at the "SD" pin of a flip-flop is available at the output port "Q" or "SQ" some delay after the clock edge.

In case the clock skew between two adjacent flip-flops happens to be greater than the "CLK" to "Q/SQ" delay there will be a need to add hold buffers to prevent race conditions. The addition of hold buffers needs to be done across all PVT corners. The number of buffers inserted will be higher in the case where the "CLK" to "Q/SQ" delay is very low. There is no combinatorial logic between "Q" and "SD" during scan operation so only net delay will come into picture and there is higher possibility of hold violation. These buffers that have been inserted to meet shift hold-timing will unnecessarily toggle during normal functional operation resulting in increased functional power. The high power i.e. instantaneous power is also caused due to most of the flip-flops toggling during the positive edge.

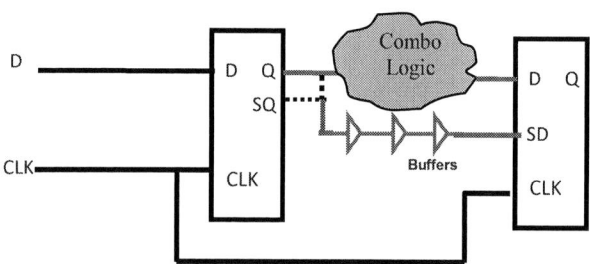

Figure 1 Scan chain hookup

The proposed modified scan flip-flop architecture is shown in the Figure 2. This introduces a half cycle delay in the output of the scan flip-flop resulting in the data being available after half a clock cycle. There are two variants of this modified scan flip-flop that we have considered here. The difference lies in whether the delay needs to be introduced in the "Q" path or "SQ" path. The output stage which introduces the half cycle delay could be as simple as using a "multiplexer" with the select line being the same "CLK". The multiplexer has been integrated into new library cell and area penalty is just 1.5 micron sq per existing flop.

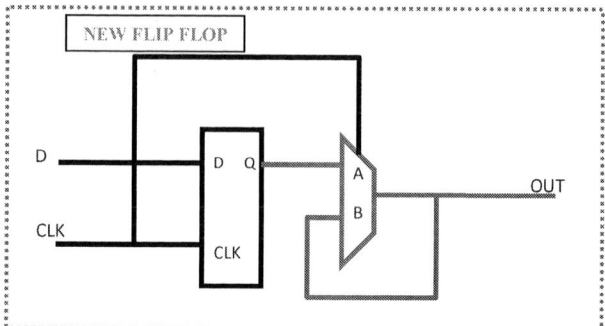

Figure 2 New flip-flop architecture with multiplexer

It could also be designed by using a latch at the output that is transparent only when the "CLK" is low. This circuit implementation is as shown below in Figure 3. However preference is to use multiplexer structure as latch consumes more area and power as compared to a multiplexer.

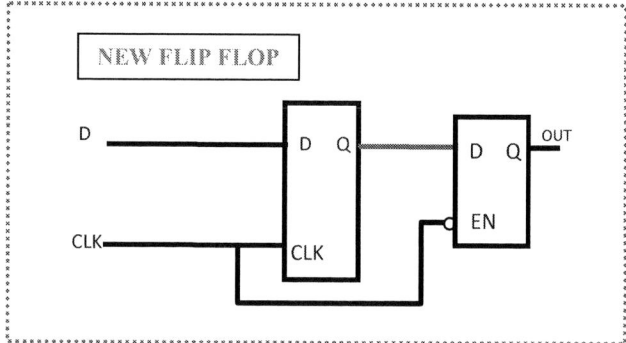

Figure 3 New flip-flop structure with latch

The proposed implementation works with identification of setup timing slack in the "Q" path. The setup timing slack at a particular functional frequency is governed by combinatorial logic which causes extra delay between "Q/SQ" and "D" path. The paths where slack is greater than half cycle are points of interest. The first analysis is done on "Q" path as proposed circuit modification on "Q" path help hold-timing closure in both functional mode and scan mode of operation. The "Q" path where slack is more than half cycle is delayed without causing any new timing violation. The analysis on whether there is enough slack available is done on a couple of TI (Texas Instruments) SOCs and IPs that were taped out in the last year or so. The slack profile of some of our SOCs is given in Table 1 and Table 2. It can be observed that there are at least 60% of the paths which have greater than 50% of period as slack across all the corners. As highlighted in Table 1, a 65nm design consisting of 101k flops has 67936 flop where slack is greater than half cycle.

978-1-61284-657-6/11 $26.00 © 2011 IEEE   156

DESIGN 1 (101K flops, 3M gates) , Functional Clock Frequency 30Mhz		
S.No	Slack (ps)	START POINT (at Q pin)
1	0-3000	2453
2	3000-6000	586
3	6000-9000	1502
4	9000-12000	1126
5	12000-15000	12725
6	15000-18000	17276
7	18000-21000	**36720**
8	21000-24000	**20564**
9	>24000	**10652**

Table 1 65nm design-1 slack report

As highlighted in Table 2, 65nm design consisting of 32k flops has 19523 flop where slack is greater than half cycle.

DESIGN 2 (32K flops, 200k gates) , Functional Clock Frequency 20Mhz		
S.No	Slack (ps)	START POINT (at Q pin)
1	0-3000	1982
2	3000-6000	3200
3	6000-9000	1752
4	9000-12000	232
5	12000-15000	1455
6	15000-18000	2171
7	18000-21000	1018
8	21000-24000	6
9	>24000	**19523**

Table 2 65nm design-2 slack report

In case where this is not feasible based on functional path and slack is less than half cycle and hold violation exist during shift, the scan flip-flop which introduces the half cycle delay in the "SQ" path will be considered. The complete flow is shown in Figure 4. With the flow we achieve hold closure in shift mode across PVT and also reduce peak power of the design during scan shift, scan capture and functional mode of operation. The power savings obtained is not exactly 50% even though half of the flip-flops are converted since the clock tree power increases with the addition of multiplexer.

The approach presented above is very much applicable for designs where the functional operating frequency of most of the design is lower than the scan shift frequency that will be targeted. In cases of designs, where the functional operating frequency is significantly higher we would not have a slack profile along the lines of tables 1 and 2. Such timing critical designs will have very less slack and hence modifying the flip-flops in those designs or logic will not be possible in the "Q" output. In such cases, all the flip-flops could be replaced with the new flip-flops that have both "Q" output (used for functional operation) and "SQ" (used for scan shift). And the multiplexer could be introduced in the "SQ" output so that a half cycle delay could be introduced only in the scan shift path. This way the functional path and timing is undisturbed. Hold timing during scan shift operation can be guaranteed, and the scan shift power can be reduced significantly.

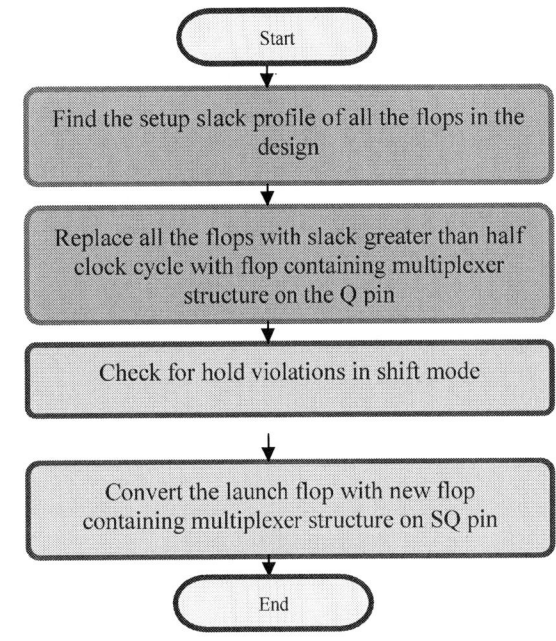

Figure 4 Flow depicting proposed methodology

IV.   RESULTS

This section illustrates the benefits in terms of both the area savings due to hold buffer relaxation as well as the power savings.

A. Hold buffer area saving

The table 3 illustrates the savings in hold buffer area that was achieved for 4 designs. Column 2 in the table represents the number of flip-flops that required hold buffering. Column 3 indicates the actual hold buffer area and column 5 indicates the savings we get when the new flip-flop structure is used.

Design	Flio-flops that required hold buffering in shift mode	Buffer area required for hold buffering(in micron2)	Area impact with proposed solution (1.5 micron2 extra area per flop)	% Savings
Design1	2629	6497	3943.5	39.30
Design2	1632	4117	2448	40.54
Design3	1436	4242	2154	49.22
Design4	3372	12046.44	5058	58.01

Table 3: Hold buffer area saving with proposed flop

B. Power Reduction with proposed technique

In design1 having 3 million gates, consisting 101K flip-flops, we calculated the setup slack profile in functional mode as explained in Figure 4 and conservatively changed 44% of the flip-flops (even though greater than 60% had more than half cycle slack) in the design with the proposed flip-flop and did a VCD based dynamic IR drop analysis.

In Figure 5 (in the last page), the IR drop plot for scan shift mode of the design generated using [17] is shown. The plot on the left side corresponds to the original design without any modification to the flip-flops. The plot on the right side indicates the reduction in IR drop when 44% of the flip-flops were converted with the new proposed scan flip-flop in the paper. As the scan flip-flop with multiplexer on the Q pin was used to replace the existing normal scan flip-flop, which resulted in half clock cycle delay in all the modes, the power reduction is achieved not only in scan shift mode but also functional and scan capture modes.

Figure 6 represents the current waveform during one cycle of scan shift. In Figure 6, the waveform in green color represents the current profile during scan shift mode for the original design. The waveform in red represents the current profile when flip-flops were converted to proposed new flip-flops. One can note the reduction of peak power from 1.2mA to 0.8mA from the graph.

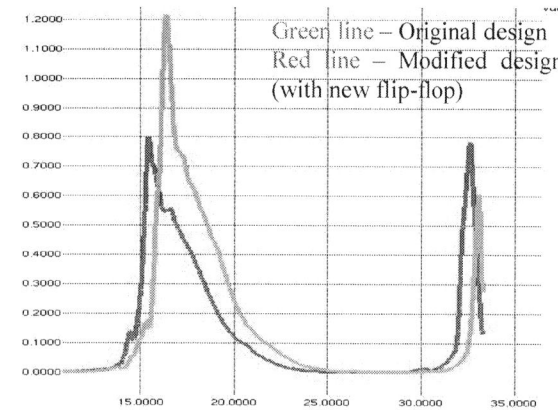

X– axis – time in nano seconds, Y – axis – Current in mA

Figure 6 Current profiles of original design and modified design

Table 4 summarizes all the results for peak power reduction explained above. The area overhead mentioned in the table takes into account the area saving obtained due to hold buffering in the design.

Design flip-flop count		101K	
Gate count		3 million	
% flip-flops converted		44%	
Area overhead incurred due to conversion		0.053 mm^2	
	Before flip-flop conversion	After flip-flop conversion	% reduction
peak current	1.2 mA	0.8 mA	33.30%

Table 4: Power savings after modification of flip-flops

V. CONCLUSION

This paper described a simple modified flip-flop architecture that has the following benefits.

- An introduction of half-cycle delay ensures that across process, voltage and temperature conditions there are no hold-timing violations as long as the clock skew between adjacent flip-flops is not greater than half a cycle.

- It has been shown that if the modified flip-flop architecture is used for just the flip-flops that have hold violations then the area incurred due to the addition of the multiplexer at the output is actually lesser than the area taken by hold buffers (inserted based on typical backend flows), resulting in area savings.

- In cases of designs where the overhead of adding the multiplexer at the output of the flip-flops is possible for 50% of the flip-flops, significant power savings can be achieved during scan shift operation.

- In cases where the functional setup slack is more than half a cycle, if the "Q" output or functional output is modified with this new flip-flop then we can save on not only the scan shift power, but also capture power and functional mode power as well.

REFERENCES

[1] H.J.Wunderlich, Abramovici M, Breuer MA, Friedman , "Digital Systems Testing and Testable Design", 1990, John wiely & Son Inc.

[2] R. Ahmadi, "A hold-friendly flip-flop for area recovery", ISCAS 2007

[3] Y. Zorian, "Distributed BIST Control Scheme for Complex VLSI Devices", VTS, 1993

[4] J.T. Tudu, E. Larssont, V. Singh, V.D. Agarwal, "On Minimization of Peak Power for Scan Circuit during Test," ETS, 2009

[5] R.Ahmadi, "A power efficient hold-friendly flip-flop." Circuits and Systems and TAISA Conference, 2008.

[6] W-J. Cheng, C-T. Fan, C-I. Huang, "Mux scan cell with delay circuit for reducing hold-time violations", US Patent 6,895,540 B2

[7] L-M Jin, "Scan cell including a propagation delay and isolation element", US Patent 6,412,098,B1

[8] H.J. Wunderlich & Gerstendorfer, "Mimimized power consumption for scan based BIST", ITC, 1999.

[9] Andreev. Alexander E. Scepanovic. Ranko,"Method of decreasing instantaneous current without affecting timing", US Patent 6795954, 2004.

[10] J. Zhang, T. Zhang, Q. Zuo, "Multi-phase Clock Scan Technique for Low Test Power", Proc. of High Desnsity Packaging and Microssytem Integration, pp 1-5,. 2007.

[11] R.Sankaralingam, N.A.Touba, "Multi-Phase Shifting to Reducing Instantaneous Peak Power During Scan", Proc. IEEE Latin American Workshop, pp-78-83, 2003.

[12] D. Czysz, M. Kassab, X. Lin, G. Mrugalski, J. Rajski, J. Tyszer "Low power EDT shift and capture in EDT enviornment", ITC, 2008.

[13] T-T. Chen, W-L. Li, P-H. Wu, J.-C. Rau, "A New Scheme for Reducing Shift and Capture Power using the X-Filling Methodology", ATS, 2009.

[14] X. Wen, K. Miyase, T. Suzuki, Y. Yamato, S. Kajihara, L.-T. Wang, K. Saluja, "A Highly-Guided X-Filling Method for Effective Low-Capture-Power Scan Test Generation", ICCD, 2006.

[15] S. Remersaro, X. Lin, Z. Zhang, S. M. Reddy, I. Pomeranz, J. Rajski, "Preferred Fill: A Scalable Method to Reduce Capture Power for Scan Based Designs", ITC, 2006.

[16] S. Ravi, R. Parekhji, A. Sabne, R. Tiwari, A. Shrivatsava, "A Generic Low Power Scan Chain Wrapper for Designs Using Scan Compression", VTS, 2010.

[17] L. Whetsel, "Adapting Scan Architectures for Low Power Operation", ITC, 2000.

[18] L.Whetsel, K. Butler, J. Saxena, "An Analysis of Power Reduction Techniques in Scan Testing", ITC 2001.

[19] RedHawk tool from Apache; http://www.apache-da.com.

[20] R. Sankaralingam, N.A. Touba "Inserting Test Points to Control Peak Power During Scan Testing", International Symposium on Defect and Fault-Tolerance in VLSI Systems, 2002.

[21] S. Kajihara, K. Ishida, K. Miyase, "Test Vector Modification for Power Reduction during Scan Testing", VTS, 2002.

[22] K. M. Butler, J. Saxena, T. Fryars, G. Hetherington, A. Jain, and J. Lewis, "Minimizing power consumption in scan testing: pattern generation and DFT techniques," ITC, 2004.

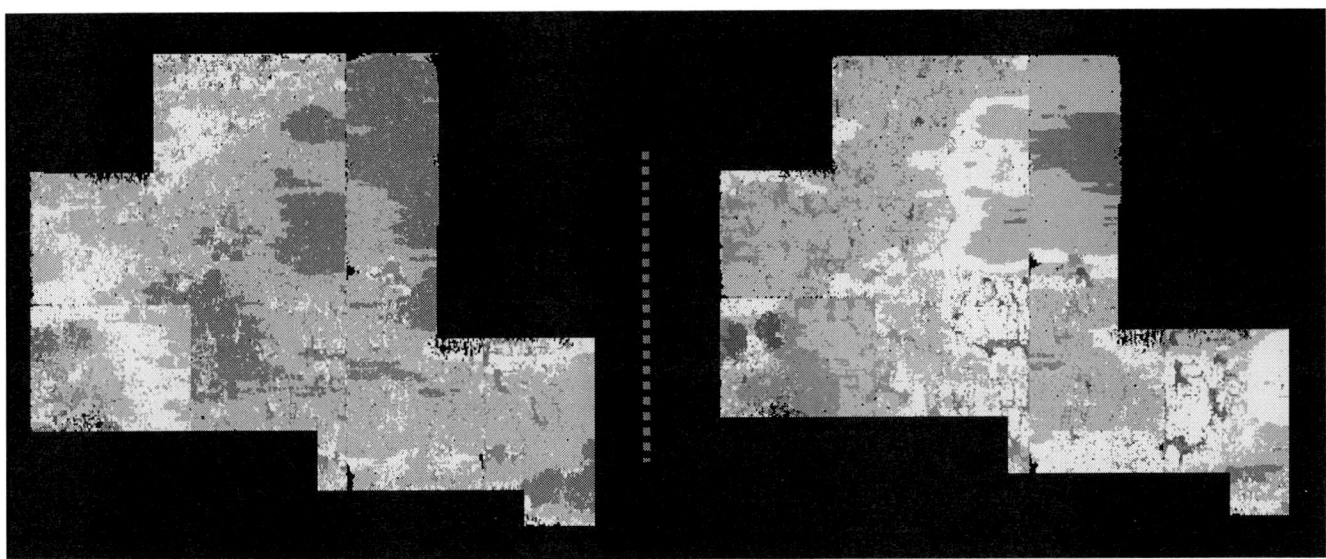

The gradient of IR drop ranges from Red -> Orange -> Yellow -> Green -> Blue, with Red being the worst to Blue being ideal

Figure 5 IR drop plots - before modification and after modification

2011 29th IEEE VLSI Test Symposium

Power-Safe Test Application Using An Effective Gating Approach

Considering Current Limits

Wei Zhao[1], Mohammad Tehranipoor[1], and Sreejit Chakravarty[2]

[1]ECE Department, University of Connecticut, {wzhao,tehrani}@engr.uconn.edu

[2]LSI Corporation, sreejit.chakravarty@lsi.com

Abstract—**Freezing scan cell outputs can block transitions to the combinational components thus reduce shift power. The extra logic introduces area overhead, reduces timing margin and increases power in capture mode. This paper proposes a partial gating flow that calculates instance toggling probability to identify power sensitive cells. The toggling rate reduction tendency is demonstrated to be useful in estimating how much extra logic is needed to achieve a desired shift power reduction rate for a design. To ensure power safety across entire test session, the toggling rate metric is enhanced to consider the effect of capture power increase. A complementary pair of weights can adjust the power change in shift and capture modes, thus achieve an overall balanced power safety. The toggling probability metric along with the proposed flow provide a flexibility that benefits various practical power requirements when considering current limits of both circuit and tester.**

Keywords-**low power test, scan cell gating, shift power, capture power, tester probe, weighted switching**

I. INTRODUCTION

Power consumption not only has become a critical concern in deep sub-micron design phrase, but also in test stage. Excessive switching activity occurs during scan chain shifting while loading test stimuli and unloading test responses, as well as in launch and capture cycles using functional clocks [1]. As test procedures and test techniques do not necessarily have to satisfy all power constraints defined in the design phase, the higher switching activity causes higher power supply currents and higher power dissipation, which can result in several issues that may not exist in functional operation, for example, high temperature, performance degradation and power supply noise, or even irredeemable damage to circuit under test (CUT) or tester.

Due to the high capital cost of automatic test equipments (ATEs), it is vital for them to work in extremely safe condition. One of the greatest concerns is to keep practical test current and power within their delivery capabilities. As technology scales, the allowable current during wafer test falls behind functional operation of packaged chips, in which many chips today already consume tens of amperes of current [2]. Designing prob cards with thousands of probe contacts may not only be achievable, but also introduces significant inductance [3][4], which increases power supply noises. Much work has already been done to address power issues from perspective of CUT, but this does not guarantee all final test patterns are power-safe

during actual wafer test [5]. To ensure power-safety in test, it is necessary not only to be aware of CUT's functional limitation, but also current limitations of tester probes, especially the commonly used low-cost testers. We define the major goal of this work to be using a low-power technique to keep test power under a predefined threshold that suits both circuit and tester. Another consideration is the associated cost. It is well understood that many existing low-power test techniques have some trade-offs, for example, in circuit performance, die size or test time. It is necessary to evaluate the cost of any low-power endeavors before they are conducted in chip design or silicon test. Another goal of this work is to make power consumption maneuverable. That is, our efforts can be used as a guide for DFT engineers to meet various test power requirements optimally.

There are numerous existing low-power test techniques to mitigate power issues, which can be classified into two major categories: 1. DFT-based solutions, which rely on modifications of scan structure, for example, scan clock blocking [6], scan chains segmentation [7], and scan-cells gating [8][9][10]; 2. ATPG-based solutions, which rely on analysis and adjustment of test pattern contents, for example, various filling methods [11][12], primary inputs control [13], etc. We notice that most of these work placed emphasis on addressing either shift power or capture power issue. They are unaware of current limitations imposed by CUTs and testers.

This work is a development of scan-cells gating methodology, thus it is one of the DFT-based solutions and requires hardware change. It has been demonstrated in [9][14] that by means of inserting extra logic on the outputs of scan cells, the transitions to be propagated to the combinational logic can be frozen, thus shift power can be reduced dramatically. Even with a portion of test points insertion, i.e. partial gating, the authors observed prominent shift power reduction. Critical paths were considered to avoid timing violations [9]. Nonetheless, due to the diversity of VLSI designs, not all circuits can be observed to have the same amount of shift reduction claimed by the authors even with full gating scheme. There is no golden rule to determine a fixed gating ratio that suits all kind of designs. What percentage of gating should be applied to their designs still remains a dilemma for circuit designers or DFT engineers. In addition, the missing part of most previous gating works is that they ignored the impact of gating elements on capture power, though in a relatively smaller range than it has on shift power. In many situations, the impact becomes non-negligible. To remedy the incomplete part of gating methodologies, our work distinguishes from previous

---

* This work was supported in part by grants from LSI Corporation and NSF CCF 0811632.

978-1-61284-657-6/11 $26.00 © 2011 IEEE

research in several aspects:

1) Evaluating the effectiveness of partial gating methodology and estimating a gating ratio for a desired shift power reduction rate.
2) Considering capture power increase, which is one of the byproducts of gate insertion, and incorporating it with shift power reduction to devise an enhanced metric for evaluating a balanced power during entire test session.
3) Considering current limitation imposed by both CUT and tester probes. Our developed strategy is not sheer power reduction oriented, but rather ensuring power-safety in test application.
4) Addressing power issues of all kinds of test patterns, i.e. transition delay faults, path delay faults, stuck-at faults, etc.

The remainder of the paper is organized as follows. Section II introduces scan cell blocking elements, current limitations and a strategy of achieving overall power safety. Section III describes a metric as well as its enhanced version in identifying power-sensitive scan cells considering the impact on shift power as well as capture power. Section IV presents an integrated power analysis flow for evaluating the effectiveness of gating methodology and metrics. In Section V, experimental results and analysis are presented. Finally, the concluding remarks and future work are given in Section VI.

II. PRELIMINARIES

A. Gating Elements

An implementation of a frozen scan cell is shown in Figures 1(a) and 1(b), with the former frozen at logic 0 and the latter at 1. An extra AND gate is inserted between flip-flop Q output and combinational logic, with the inversion of scan enable as the other input. During test mode, scan enable is high, the extra inverter outputs zero which is then fed to the AND gate, thus combinational logic always receives a logic zero from the sequential cell. When CUT switches to capture mode, the AND gate becomes transparent. Likewise, an extra OR gate is able to freeze sequential output at value 1 as in Figure 1(b). Using this gating implementation, each inserted gating module increases total transistor count by 8 for freezing at logic 0, or 6 for freezing at logic 1.

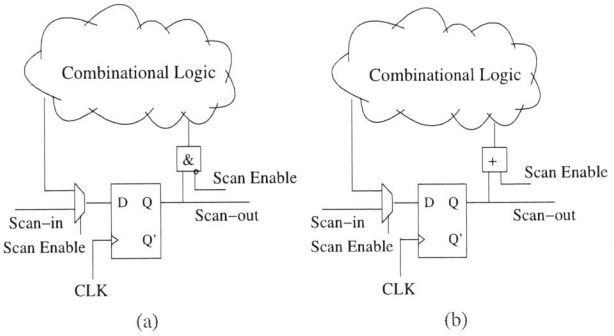

Fig. 1. Scan cell with extra logic at the output, frozen at (a) logic 0 and (b) logic 1.

B. Current Limitations

Shift power is observed to be larger than capture power in many cases. Figure 2(a) is an example of weighted switching activity (WSA) [8][15] plot during entire test pattern application for benchmark b19 (see size in Table I in Section V). Each test cycle is associated with a WSA dot in the plot. We can roughly estimate in this example that the average shift WSA is 2.5 times more than that of capture. Scan output freezing can significantly reduce shift power [9][14], but would have a negative impact on the capture power, since these additional elements become completely redundant in capture mode, the switching of which draw non-negligible current from power supply, especially when scan-cell-to-gate (STG) ratio is large. Clearly, capture power increase rate is dependent on circuit topology. For some designs, capture power increase is relatively small and should not be a major concern. However, it is entirely possible for a certain design that has capture power increasing beyond a limit that would cause at-speed power issues, which has not been taken into account in previous work using gating methodology.

Now consider Figure 2(b). Assume P_s and P_c are the original power level for shift and capture power in the non-gated design. $\bar{P}_s$ is the desired optimum power-safe level in shift mode considering power capability of tester prob, while $\bar{P}_c$ is the desired power-safe level in capture mode for CUT working properly. $\bar{P}_s$ and $\bar{P}_c$ are not necessarily a same value due to different test frequencies. These parameters can be quantified through many approaches, for example, power constraints of CUT, early stage power analysis, power specification of tester, etc. After they are determined, we give the definition of $\langle \bar{\Delta}s, \bar{\Delta}c \rangle$ as:

- $\bar{\Delta}s = (P_s - \bar{P}_s)$, power reduction goal for partial gating.
- $\bar{\Delta}c = (\bar{P}_c - P_c)$, capture power increase margin.

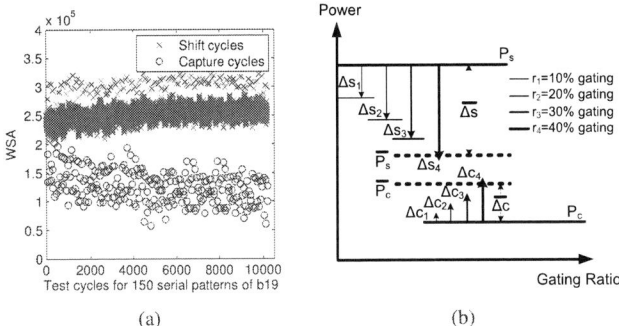

Fig. 2. Gating strategy: (a) Shift power is observed to be much greater than capture power. (b) Power safety for both shift and capture power.

Suppose four gating ratios $r_{i=1,2,3,4} = \{10\%, 20\%, 30\%, 40\%\}$ are implemented respectively, with shift and capture power change rate as $\langle \Delta s_i, \Delta c_i \rangle$ in each case. A higher r_i usually implies a larger Δs_i and Δc_i. However, there are two possible outcomes for adopting a larger gating ratio: (1) $\Delta s_i > \bar{\Delta}s$; (2) $\Delta c_i > \bar{\Delta}c$. It is not power-safe to consider solely (1) while ignoring (2). To ensure power safety in both shift and capture modes, there should be a trade-off on the practical gating ratio selection so that Δc_i does not go up beyond the margin. A criterion for capture power consideration is specified as Equation 1 shows, where μ_{thr} is a predefined threshold for whether considering capture power or not.

$$\bar{\Delta}c \begin{cases} > \mu_{thr}\bar{\Delta}s, \text{no need for capture power analysis.} \\ \leq \mu_{thr}\bar{\Delta}s, \text{consider capture power during gating.} \end{cases} \quad (1)$$

Equation 1 can be understood as follows: when $\bar{\Delta}c$ is estimated to be greater than $\mu_{thr}\bar{\Delta}s$, capture margin is wide enough, there is no risk on P_c value change, thus we will not consider capture power increase as a drawback during implementing scan cell gating. Then the problem is reduced to regular gating scheme as in [14]. Otherwise, capture power needs to be taken into consideration as Section III does. Generally, a smaller μ_{thr} indicates that more importance should be assigned on controlling P_c to keep at-speed power at a safe level.

Note that in nowadays tests, P_c can be already above $\bar{P}_c$. Previous low-power techniques can reduce P_c. Our previous work in [5] can also obtain a launch-to-capture low-power TDF pattern set that keeps P_c within a threshold. So the assumption in Figure 2(b) is that by means of other techniques, P_c is already safe when there is no gating, and it should not break the power-safe level after applying gating methodology in this work.

III. POWER-SENSITIVE SCAN CELL SELECTION

It has been observed that some scan cells have a much larger impact on toggle rates of combinational logic than other scan cells. These scan cells are called power-sensitive scan cells, by freezing the output of which, a same number of extra gates can reduce more power. Though further reduction can be achieved by gating more scans, in is not practical in use due to the impact on area and timing slack. Since we are not always aware of a specific gating ratio that suits the design, we design the partial gating goal here to be finding a set of scan cells, by gating which, P_s can be reduced below the safe level, $\bar{P}_s$.

In order to identify power sensitivity of scan cells, we first calculate a sum of toggling rate of all instances constituting the combinational part in the normal design, i.e. no gating on any scan cell. Then the calculation process is iterated for modified designs by gating each scan cell at logic 0 or logic 1 and compare each time the outcome rate with that of normal design. Those scan cells with larger toggling rate reduction on combinational logic can be regarded as power-sensitive ones. Note that the power-sensitive cell identification process is a static analysis based on circuit topology that the selection result is completely pattern independent and it needs to be run only once.

In order to evaluate the toggling rate of each logic gate, we consider the toggling probability of all nets first, since once the toggling probability of output pins are determined, that gate's switching probability is defined. The toggling probability of a net i, TP_i, is defined as Equation (2), where $P_i(0)$ is the probability for net i being logic 0, $P_i(1)$ for being 1. For the entire circuit that contains M logic gates, the toggling rate of CUT, TR_{comb}, is given as Equation (3). The coefficient k_m is the power weight for each $gate_m$, that is, a more power consuming gate will be assigned a larger k. Such information can be extracted from cell library and technology.

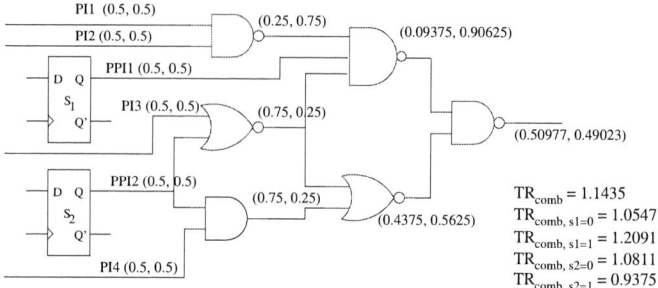

Fig. 3. An example of net toggling probability and instance toggling rate calculation.

$$TP_i = P_i(0) \times P_i(1) \quad (2)$$

$$TR_{comb} = \sum_{m=1}^{M} k_m \cdot TP_{output\_pin\_of\_gate_m} \quad (3)$$

For each scan cell s_j, its toggling rate reduction (TRR) is calculated twice for being gated at either 0 or 1, as Equation 4 shows. The larger value among the two is adopted, which meanwhile determines the type of logic, i.e. AND or OR gate should be inserted onto the output of s_j. To simplify the process, we do not consider freezing the $\bar{Q}$ pin of flip-flop, though for some scan cells this pin is also connected to combinational logic.

$$TRR_{s_j} = max \begin{cases} (TR_{comb} - TR_{comb,s_j=0})/TR_{comb}, \text{for 0 gating.} \\ (TR_{comb} - TR_{comb,s_j=1})/TR_{comb}, \text{for 1 gating.} \end{cases} \quad (4)$$

Initially, all primary inputs (PIs) and pseudo-primary inputs ($PPIs$) are assigned a toggling probability of (0.5, 0.5) with the first value in the bracket as probability of being logic 0 while the latter for logic 1. All other nets are initially assigned (-1,-1) indicating their probability has not been determined. If probabilities of all input pins of an instance are defined, its output pin's toggling probability is also determined, which recursively triggers the probability calculation and determination of its fan-out gates. A TRR calculation process terminates after no more nets can be updated through this topology traversal. There could still be a few nets or instances undetermined in the end, which we assign (0.5, 0.5) to them.

An example of TRR calculation is given in Figure 3. In this example, suppose $k = 1$ for all instances. TR_{comb}=1.1435. Considering gating D flip-flop s_1 at first, PPI_1 probability becomes (1, 0) or (0, 1), then by updating probability on the other nets, we get $TR_{comb,s_1=0}$=1.0547. $TR_{comb,s_1=1}$=1.2091 Then PPI_1 recovers to half-half probability, and start to gate s_2 using similar process, and get $TR_{comb,s_2=0}$=1.0811. $TR_{comb,s_2=1}$=0.9375. Finally, TRR of s_1 and s_2 can be obtained: TRR_{s_1}=0.0888/1.1435=7.77%, TRR_{s_2}=0.2060/1.1435=18.0%. Thus, scan cell s_2 is more power sensitive than s_1, and should be inserted an OR gate at the Q pin to be frozen at logic 1.

For a large synthesized circuit, its topology can be understood from netlist. We also start from PIs and $PPIs$ and calculate probabilities of as many nets as possible. After scan cell gating iteration is finished, all TRR_{s_j} are determined, then sorted in descending order, with top $x\%$ identified as power-sensitive scan cells. The x value can be adjusted to meet different desired $\langle \Delta s_x, \Delta c_x \rangle$. Greedy algorithm is introduced in [14] to consider correlation between scan cells. They picked up several top scan cells from the first result, assume they are already gated, then calculate TRR for the rest ones and sort again, etc. The correlation handling process is more time consuming, and we did not observe distinct discrepancy on the quality, i.e. Δs of power sensitive cells between whether considering scan cells correlation or not. It is considered only when CPU runtime is not a critical concern.

Now let us consider the impact of inserted logic on capture power. D pin of each flip-flop is fed by combinational logic, and its value will impact the transition on next arriving cycle. Hence reducing the toggling rate on all D pins can offset the increased capture power in some degrees, as the launch-to-capture cycle immediately follows the last shift cycle. Thus, to take capture power into consideration, each scan cell is associated with a new toggling rate reduction value, TRR_D, which benefits power safety during at-speed test cycles. In order to achieve a balance between shift power reduction and capture power increase for the gating methodology, we propose a metric, TRR_{BL}, to evaluate power-sensitive scan cells in Equation 5. α and β are two positive adjustable weights assigned based on $\langle \bar{\Delta s}, \bar{\Delta c} \rangle$, or simply μ_{thr}. If μ_{thr} is relatively small, i.e. capture power margin is stringent during test, a larger β needs to be assigned in this case. All previous work can be seen as an extreme condition with $\alpha = 1$ and $\beta = 0$.

$$TRR_{BL} = \alpha \cdot TRR_{comb} + \beta \cdot TRR_D, \\ 0 \leq \alpha, \beta \leq 1, (\alpha + \beta = 1). \qquad (5)$$

IV. Validation Flow

We illustrate a flow in Figure 4 that validates the effectiveness of power-sensitive cell selection proposed in Section III. After synthesizing a RTL description to a gate level netlist, a stand alone power-sensitivity detection routine sorts all scan cells based on their TRR_{BL} value using a pre-defined pair of (α, β) weight, thus a complete scan cell list can be obtained. Static timing analysis (STA) is done to determined critical paths, and those power sensitive cells on the critical paths will be removed from the list.

Firstly, we select a top $x\%$ flip-flop from this list for freezing. The original and modified netlists are fed to physical synthesis tool for placing and routing, respectively. We use original design to generate ATPG patterns, which will be used as stimuli in logic simulation for evaluating dynamic power of both original and modified netlists. During serial pattern simulation, WSA for each clock cycle is recorded, which will then be used to determine both peak and average WSA for that design after simulation finishes. The flip-flop freezing process is iterated for other gating ratios, for example, $2x\%$, $3x\%$, ... till 100%, a.k.a, full gating. Again, shift and capture WSA are collected from all clock cycles for those designs. Comparison will be made among different gating ratios, as well as among

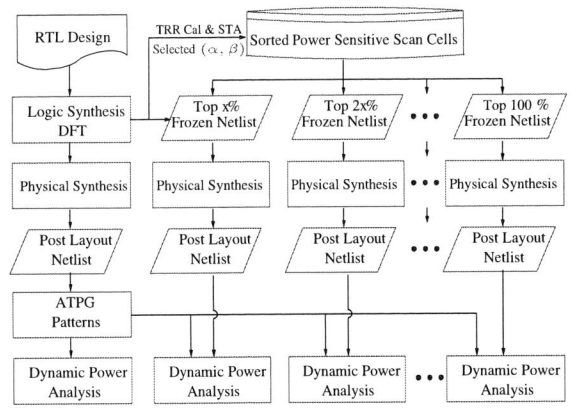

Fig. 4. Flow diagram of validating power-sensitive scan cell selection.

different (α, β) pairs. We would like to determine: 1) the effectiveness of power-sensitive cell selection; 2) an optimum ratio for a specific design; 3) balance between shift and capture power.

V. Experiment Results

The proposed flow is implemented on three benchmarks with different sizes and STG ratio listed in Table I. The RTL descriptions were logically synthesized in Synopsys DC Compiler in *flattened mode with area optimization*. An in-house tool was developed to calculate probability and TRR in each design to identify power-sensitive cells with CPU runtime listed in the last column of Table I which were obtained on a Linux desktop with 2.4GHz CPU and 2GB RAM. Gate insertion in netlist was handled by another in-house tool developed in C. The resulting netlists were then placed and routed using Cadence SOC Encounter. Transition delay fault (TDF) patterns were generated using Synopsys TetraMax. Pattern simulation and WSA calculation were performed with Synopsys VCS with the PLI procedures implemented in C. For relatively larger benchmarks like wb_conmax and b19 with a large pattern count, test cycles to be simulated are selected uniformly from all cycles in the entire pattern set, so as to save simulation time without getting biased result.

A. Gating Ratio and TRR Analysis

In this part, we set $(\alpha, \beta)=(1,0)$, i.e. consider TRR_{comb} only. So shift power reduction is the target. We performed scan cell freezing in s38417 from top 3% till 100%. The maximum TRR_{comb} was 43.7%, as shown in Figure 5(a). In addition, great linearity is observed between gating ratio and TRR_{comb}. WSA is measured for all clock cycles in s38417 when shifting TDF patterns. Peak WSA results are given in Figure 5(b). It is observed that shift WSA reduction in s38417 was saturated at TRR_{comb} around 23%, equivalently 47% gating ratio. Freezing more scan cells cannot reduce shift power any further.

Meanwhile, peak capture WSA increase can reach as high as 18%. Even at shift saturation point, capture power still has 10% increase. If we focus on shift linear part in Figure 5(b), that is, when no more than 47% gating needs to be considered, a desired $\bar{\Delta s}$ can be mapped to a TRR_{comb} value easily. Then a gating ratio can be estimated from Figure 5(a) accordingly.

978-1-61284-657-6/11 $26.00 © 2011 IEEE

TABLE I
BENCHMARKS CHARACTERISTICS.

Benchmark	# of Scan	# of Gates	STG	Shift Cycles	Capture Cylces	TRRs CPU Runtime
s38417	1564	4673	1:3	9646	182	4s
wb_conmax	770	18640	1:25.4	6250	250	2min
b19	5868	57400	1:9.8	13394	181	42min

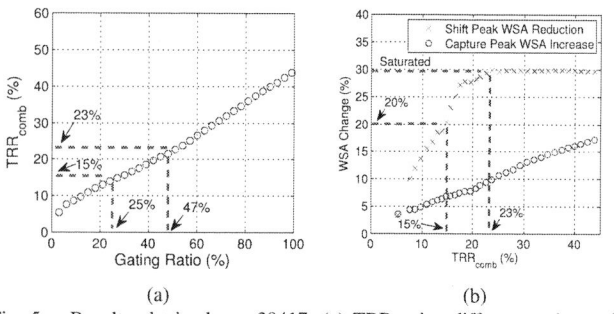

(a) (b)

Fig. 5. Results obtained on s38417: (a) TRR using different gating ratios, (b) Shift and capture peak WSA change with different TRRs.

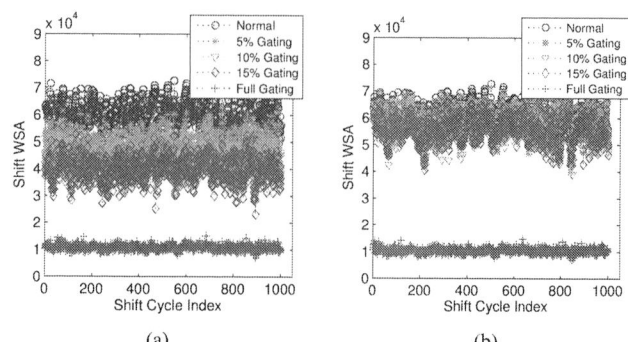

(a) (b)

Fig. 6. Shift WSA plot for (a) Deterministic scan cell selection vs. (b) Random scan cell selection.

Different circuits may not necessarily have the same TRR characteristic and saturation ratio as in s38417. We can do similar analysis on a design. The information obtained in this stage can be used to estimate a primitive gating ratio in the beginning. A hypothetic example is given at the end of Subsection V-B.

B. Evaluation of Power-Sensitive Scan Cells

To demonstrate the effectiveness of power-sensitive cell identification process, we selected top 5%, 10% and 15% scan cells for freezing respectively and observed their WSA reduction rates which were compared with randomly selected 5%, 10% and 15% scan cells. Figure 6(a) shows WSA plot for 1000 shift cycles from 40 TDF patterns in wb_conmax (scan chain length 25). As gating ratio is increased, shift power is reduced further. With 15% gating, the average WSA is reduced by 36%, which is near to half the effect full gating can achieve: an 82% reduction in this benchmark. However, none of the randomly selected ratio has noticeable shift power reduction, as shown in Figure 6(b). We believe a randomly selection of larger number of scan cells could become effective in shift power reduction, but it will add cost to silicon area, as well as capture power increase.

Table II gives more detailed area overhead data, as well as WSA result in shift and capture cycles for wb_conmax. 0% represents normal design, i.e. no gating, and 100% represents full gating design. The WSA data were collected for all simulated clock cycles. As a 0 or 1-gated insertion introduces different number of transistors, a netlist generated after random selection does not necessarily have the same number of gates with that the deterministic selection using a same gating ratio. However, shift WSA reduction ratios are quite distinct among these selection schemes. For example, a 15% random selection achieved only 6.7% reduction compared to 36% by using our flow. Therefore, the proposed flow in this work will be very effective in achieving the shift power reduction goal.

Moreover, if scan gating budget is stringent, we can figure out an optimum gating ratio for a specific power reduction objective. Take this hypothetical example. Suppose a chip is designed to consume k amperes of current during normal operation, which after fabricated, will be tested on a low-cost tester that provides source and measure currents within $\bar{P}_s = 2 \cdot k$ amperes. Partial gating is considered to be applied during design-for-test to avoid possible power issue in silicon test. A gating ratio is needed to be determined. Firstly, two groups of patterns, both test and functional are simulated. Peak WSA data for shift and functional cycles are collected respectively. Assume in this case, peak shift WSA is obtained to be 2.5 times larger than its functional mode. Thus, peak current during shift operation is supposed to be $P_s = 2.5 \cdot k$ amperes. And $\bar{\Delta}s = (P_s - \bar{P}_s) = 0.5 \cdot k$ amperes, a 20% shift power reduction requirement. Similar TRR and power analysis is done as in Subsection V-A. Suppose the plots we get are same with s38417 in Figures 5(a) and 5(b). We first obtain from Figure 5(b) that a 20% shift WSA reduction requires a 15% TRR. While TRR is directly proportional to gating ratio, we can obtain from Figure 5(a) that a 25% gating ratio is able to achieve the goal of $\bar{\Delta}s$.

C. Capture Power Consideration

Though WSAs in capture cycles are observed to be smaller than that of shift cycles, the negative impact of power increase during at-speed cannot be neglected. Worsely, the two Capture WSA columns in Table II show that deterministic selection increases higher capture power than random selection. Thus, considering TRR_{comb} alone cannot address capture power issue. As stated in Subsection II-B, it is required for a design to keep capture power in a safe threshold, while adding extra logic can possibly break the at-speed safety line. We proposed a balanced TRR calculation method at the end of Section III. In this part, we used two extreme combinations of (α, β) on b19 and observed shift and capture power change. They are: only consider TRR_{comb}, i.e. $(\alpha, \beta) = (1,0)$, which we believe has major impact on shift power, or only TRR_D, i.e. $(\alpha, \beta) = (0,1)$, which would impact capture power.

Figure 7(a) show that if we only consider freezing D pins of scan cells, $(\alpha, \beta) = (0,1)$, the shift power reduction is less effective than considering mere freezing instance output pins, $(\alpha, \beta) = (1,0)$, but it introduced less capture power increase when gating ratio is below 50%, as shown in Figure 7(b). Note that, gating ratio greater than 50% is not recommended since in many situations shift power reduction becomes saturated at this rate, as exemplified in Figure 5(b).

The impact of (α, β) change on shift and capture power is also observed in wb_conmax, as shown in Figures 8(a) and

TABLE II
CHARACTERISTICS OF WB_CONMAX WITH DIFFERENT GATING RATIO, EITHER DETERMINISTIC OR RANDOM.

Gating Ratio	Power-Sensitive Cells Selection						Random Scan Cell Selection					
	# of Gates	Core Area μm^2	Shift WSA		Capture WSA		# of Gates	Core Area μm^2	Shift WSA		Capture WSA	
			Max	Avg	Max	Avg			Max	Avg	Max	Avg
0%	18460	1296951	74959	59558	45653	38180	18460	1296951	74959	59558	45635	38180
5%	18515	1299178	65088	47803	46337	38597	18515	1299133	73392	58137	45907	38278
	0.2% ↑		13.2% ↓	19.7% ↓	1.5% ↑	1.1% ↑		0.2% ↑	2.1% ↓	2.4% ↓	0.56% ↑	0.26% ↑
10%	18577	1301551	56768	41018	46448	38990	18571	1301371	71034	56080	45721	38596
	0.4% ↑		24.3% ↓	31.1% ↓	1.7% ↑	2.1% ↑		0.3% ↑	5.2% ↓	5.8% ↓	0.15% ↑	0.83% ↑
15%	18647	1304160	52886	38145	46737	39269	18627	1303592	70259	55569	46496	38814
	0.6% ↑		29.4% ↓	36.0% ↓	2.4% ↑	2.9% ↑		0.5% ↑	6.3% ↓	6.7% ↓	1.8% ↑	1.7% ↑
100%	19730	1345907	15183	10771	49805	42380	19730	1345907	15193	10771	49805	42380
	4% ↑		80% ↓	82% ↓	9.1% ↑	11% ↑		4% ↑	80% ↓	82% ↓	9.1% ↑	11% ↑

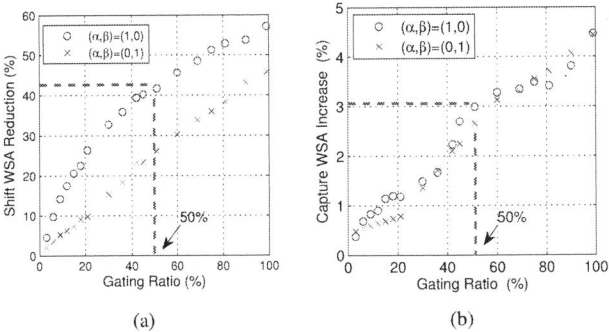

Fig. 7. Results obtained on b19: (a) shift WSA reduction, (b) capture WSA increase, based on different power sensitivity calculated by $(\alpha, \beta)=(1,0)$ and $(\alpha, \beta)=(0,1)$.

8(b). Two other weight pairs are also included. In (a) we did not see difference on shift WSA change when gating ratio is less than 20%. When greater than 20% and less than 50%, (0.5, 0.5) and (0, 1) pairs have less shift power reduction rate than both (1, 0) and (0.8, 0.2), which is expected. However, they brought down capture power increase rate from 10% increase to 4%. If capture power is a stringent requirement during at-speed test, the later two (α, β) would ensure more power safety during at-speed test.

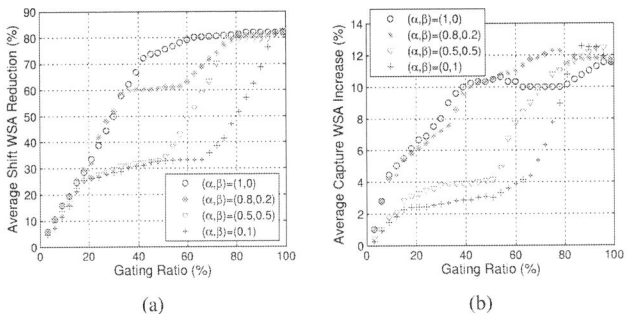

Fig. 8. Results obtained on wb_conmax result (a) shift WSA reduction, (b) capture WSA increase, based on different (α, β) pairs.

VI. CONCLUSIONS AND FUTURE WORK

We have presented a novel power-sensitive scan identification metric and flow. We demonstrated its effectiveness of power reduction during shift, as well as ensuring power safety in capture mode. The results showed that capture power increase rate can be controlled without compromising much effectiveness in shift power reduction. The parameters in the new metric can be adjusted accordingly to meet different shift and capture power requirement during silicon test. Meanwhile,

the linear relationships among gating ratio, TRR and shift power reduction rate we have observed can be used to estimate how much extra logic should be added to achieve the safe power goal. In the future, we are considering improving the efficiency of TRR calculation routine, running experiments on non-flattened hierarchical circuits as well as industry designs, and observing low-power efforts achieved on industry circuits based on our methodology in collaboration with our industry liaison.

REFERENCES

[1] L.T. Wang, C. E. Stroud and N. A. Touba, *System-on-Chip Test Architectures: Nanometer Design for Testability (Systems on Silicon)*, Chapter 7, 2009

[2] S. Kundu, T.M. Mak and R. Galivanche, *Trends in manufacturing test methods and their implications*, in Proc. *IEEE International Test Conference (ITC'04)*, 2004.

[3] P. Girard, *Survey of low-power testing of VLSI circuits*, in *IEEE Design & Test of Computers*, vol. 19, no. 3, pp. 80-90, May-June 2002.

[4] M. Tehranipoor and K.M. Butler, *Power Supply Noise: A Survey on Effects and Research*, in *IEEE Design & Test of Computers*, vol. 27, no. 2, pp. 51-67, March-April 2010.

[5] W. Zhao, J. Ma, M. Tehranipoor and S. Chakravarty, *Power-Safe Application of Transition Delay Fault Patterns Considering Current Limit during Wafer Test*, to appear in *IEEE Asia Test Symposium*, 2010.

[6] R. Sankaralingam, B.Pouya and N.A. Touba, *Reducing power dissipation during test using scan chain disable*, in Proc. *VLSI Test Symposium (VTS'01)*, 2001.

[7] P. Rosinger, B.M. Al-Hashimi and N. Nicolici, *Scan architecture with mutually exclusive scan segment activation for shift and capture-power eduction*, in *IEEE Transactions on CAD of Int. Cir. and Systems*, vol. 23, no. 7, pp. 1142- 1153, July 2004.

[8] S. Gerstendorfer and H.J. Wunderlich, *Minimized power consumption for scan-based BIST*, in Proc. *IEEE International Test Conference (ITC'99)*, 1999.

[9] M. Elshoukry, M. Tehranipoor and C. P. Ravikumara, *A critical-path-aware partial gating approach for test power reduction*, in *ACM Transactions on Des. Autom. Electron. Syst. 12, 2*, Apr. 2007.

[10] R. Sankaralingam and N.A. Touba, *Inserting test points to control peak power during scan testing*, in Proc. *IEEE International Symposium on Defect and Fault Tolerance in VLSI Systems (DFT'02)*, 2002.

[11] X. Wen, Y. Yamashita, S.Kajihara, L.T. Wang, K.K. Saluja and K. Kinoshita, *On low-capture-power test generation for scan testing*, in Proc. *VLSI Test Symposium (VTS'05)*, 2005.

[12] S. Remersaro, X. Lin, S.M. Reddy, I. Pomeranz and J. Rajski, *Low Shift and Capture Power Scan Tests*, in Proc. *IEEE International Conference on VLSI Design (VLSID'07)*, 2007.

[13] T.C. Huang and K.J. Lee, *Reduction of power consumption in scan-based circuits during test application by an input control technique*, in *IEEE Transactions on CAD of Int. Cir. and Systems*, vol. 20, no. 7, pp. 911-917, Jul 2001.

[14] T.C. Huang and K.J. Lee, *Scan Shift Power Reduction by Freezing Power Sensitive Scan Cells*, in *Journal of Electronic Testing*, vol. 24, no. 4, pp. 327-334, 2008.

[15] J. Lee, S Narayan, M. Kapralos and M. Tehranipoor, *Layout-Aware, IR-Drop Tolerant Transition Fault Pattern Generation*, in Proc. *Design, Automation and Test in Europe (DATE'08)*, 2008.

Power-Aware Test Generation with Guaranteed Launch Safety for At-Speed Scan Testing

X. Wen[1], K. Enokimoto[1], K. Miyase[1], Y. Yamato[1], M. A. Kochte[1,2], S. Kajihara[1], P. Girard[3], and M. Tehranipoor[4]

[1] Kyushu Institute of Technology, Iizuka, Japan
[2] University of Stuttgart, Stuttgart, Germany
[3] LIRMM, Montpellier, France
[4] University of Connecticut, Storrs, USA

Abstract - At-speed scan testing may suffer from severe yield loss due to the ***launch safety problem***, where test responses are invalidated by excessive ***launch switching activity*** (LSA) caused by test stimulus launching in the at-speed test cycle. However, previous low-power test generation techniques can only reduce LSA to some extent but cannot guarantee launch safety. This paper proposes a novel & practical power-aware test generation flow, featuring ***guaranteed launch safety***. The basic idea is to enhance ATPG with a unique two-phase (***rescue & mask***) scheme by targeting at the real cause of the launch safety problem, i.e., the excessive LSA in the neighboring areas (namely *impact areas*) around long paths sensitized by a test vector. The ***rescue phase*** is to reduce excessive LSA in impact areas in a focused manner, and the ***mask phase*** is to exclude from use in fault detection the uncertain test response at the endpoint of any long sensitized path that still has excessive LSA in its impact area even after the rescue phase is executed. This scheme is the first of its kind for achieving guaranteed launch safety with minimal impact on test quality and test costs, which is the ultimate goal of power-aware at-speed scan test generation.

Keywords – test generation; test power; at-speed scan testing; power supply noise; launch safety.

I. INTRODUCTION

Shrinking feature sizes and increasing clock frequencies have made timing-related defects a major cause for failing integrated logic circuits. At-speed scan testing is thus required so as to achieve sufficient product quality. At-speed scan test vectors are usually generated by *automatic test pattern generation* (ATPG) based on either the transition delay fault model and/or the path delay fault model.

Compared with slow-speed scan testing, at-speed scan testing is confronted with more severe challenges in terms of ***test quality*** and ***test costs***. These challenges have been successfully mitigated to some extent by such approaches as timing-aware ATPG [1] for test quality improvement and test compression [2] for test cost reduction. However, in recent years, at-speed scan testing is starting to suffer more and more from a new challenge, namely ***test power*** [3].

The test power problem is caused by the ever-growing gap between *functional power* (lower) and *test power* (higher), which has widened from about 2X to 5X [4] by ever-shrinking functional power (due to aggressive low-power design practices such as power gating/clock gating) and ever-increasing test-mode power (due to fault/block parallelism and functional circuit/clocking constraints being ignored for test efficiency). This means that function-mode-oriented wafer/package level heat management and *power supply network* (PSN) design are getting relatively weaker with respect to potentially excessive test power [5]. As a result, ***heat-related test safety*** (i.e., over-heat may damage circuits) and ***power-supply-noise-related test safety*** (i.e., power supply noise may invalidate test responses) have become serious problems in at-speed scan testing [5,6].

Generally, heat-related test safety relies on average shift power, which can be effectively and predictably reduced below a safety level by practical techniques, e.g., *scan chain segmentation* [7]. On the other hand, power-supply-noise-related test safety largely depends on the ***launch switching activity*** (LSA) caused by test stimulus launching at the beginning of the at-speed test cycle. As illustrated in Fig. 1 based on the *launch-on-capture* (LOC) clocking scheme, the first capture C_1 may cause excessive LSA, resulting in IR-drop and $L\,di/dt$ that reduce effective power supplies to cells, leading to increased path delay, and finally timing failures at the second capture C_2. This paper will focus on power-supply-noise-related test safety, referred to as ***launch safety*** hereafter, since *launch switching activity* (LSA) is its determining factor. Clearly, a launch-safe test vector is one that will not cause excessive-LSA-induced timing failures.

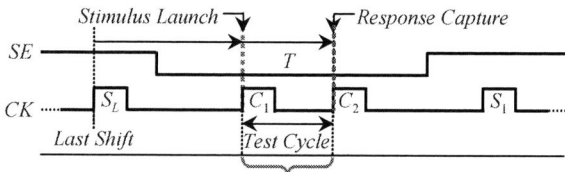

Figure 1. Launch safety in LOC-based at-Speed scan testing.

Achieving launch safety requires sufficient reduction of excessive LSA. However, while previous low-LSA test generation techniques (ATPG, test compaction, and X-filling) [5] can reduce LSA to some extent, they may not guarantee launch safety [8] due to the following problems:

• **Problem-1 (*Unfocused Effect*)**: Previous techniques only reduce the total LSA for the whole circuit in an unfocused manner. However, the real cause for timing failures (i.e., excessive LSA in neighboring areas around long sensitized paths) often remains [6,8]. In addition, unfocused LSA reduction constrains too many logic values in a test vector, causing test quality degradation and test data inflation.

• **Problem-2 (*Unguaranteed Sufficiency*)**: Previous low-LSA techniques can reduce LSA to some extent. However, none of them can guarantee sufficient LSA reduction for all test vectors of any circuit. This is unacceptable in industry since yield loss risk remains even if only one test vector obtained by low-LSA test generation is still launch-unsafe.

978-1-61284-657-6/11 $26.00 © 2011 IEEE

In this paper, we propose a novel and practical scheme to achieving guaranteed launch safety with minimal impact on test quality and test costs in power-aware test generation. The basic idea consists of three integral parts as follows:

(1) *Risky Path Identification*: A *risky path* P of a test vector V is a path sensitized by V and has excessive *launch switching activity* (LSA) in its *impact area* (composed of the cells whose LSA impacts the delay of the path P). V is said to be *launch-risky* if it has at least one risky path.

(2) *Risky Path Reduction*: Focused LSA reduction is conducted for the impact areas of risky paths in order to effectively reduce risky paths. This may turn a launch-risky test vector into a launch-safe one. Even if it cannot, it usually reduces the number of remaining risky paths.

(3) *Risky Path Masking*: Since the value at the endpoint of any remaining risky path is uncertain due to excessive LSA in its impact area, it is excluded from use for fault detection. This is done by placing an X at the expected test-response-vector bit corresponding to that endpoint. As a result, no yield loss will occur. Note that this is data masking, without any performance penalty / additional circuit overhead.

Clearly, *risky path masking* is the core part to achieving guaranteed launch safety. Although being simple and straightforward, this part only becomes feasible under three critical conditions: (**CC1**) risky paths are identified; (**CC2**) the number of risky paths is small; and (**CC3**) an ATPG flow is devised to recover the lost fault detection capability due to masking. Only when **CC1 ~ CC3** are all satisfied can *risky path masking* achieve guaranteed launch safety with minimal impact on test quality and test costs.

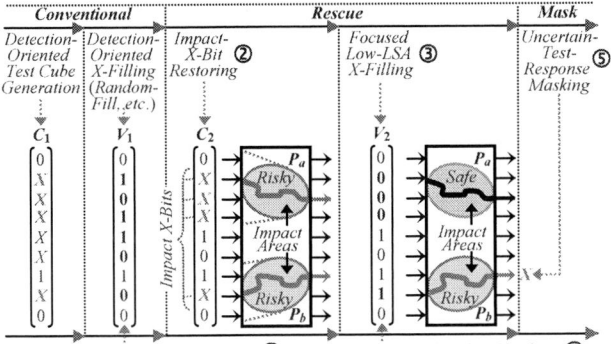

Figure 2. Basic idea for guaranteeing launch safty.

In this paper, we try to satisfy the three critical conditions with a unique two-phase ATPG scheme. As illustrated in Fig. 2, a test cube C_1 (with Xs) is generated and then turned into a test vector V_1 (without Xs) by detection-oriented X-filling (usually *random-fill* for high test quality and small vector count). Conventional ATPG ends here, but the new scheme continues with two more phases as follows:

● **Rescue**: *LSP-based launch safety checking* (①) identifies all *risky paths* of V_1 by checking the LSA in the impact area of each *long sensitized path* (LSP) under V_1. Suppose that P_a and P_b are found to be risky paths. Then, *impact-X-bit restoring* (②) identifies those bits in V_1 that are originally

X-bits in C_1 (before X-filling) and can reach the impact areas of P_a and P_b, and turns them back into X-bits (*impact-X-bits*) to create a new test cube C_2. After that, *focused low-LSA X-filling* (③) is conducted to turn C_2 into V_2 with reduced LSA in the impact areas of P_a and P_b.

● **Mask**: *LSP-based launch safety checking* (④) identifies that P_a is now safe but P_b is still risky under V_2. In this case, *uncertain-test-response masking* (⑤) is conducted to place an X at the endpoint (FF input) of P_b in the test response to V_2. This makes the uncertain value observed by the FF to be ignored in test response comparison, thus avoiding yield loss. Note that this masking needs no additional circuitry.

The advantages of the proposed flow are as follows:

● ***Focused LSA Reduction***: LSA is reduced only for necessary vectors (*launch-risky vectors*) and only in necessary areas (*impact areas*). That is, there is no over-reduction of LSA for launch-safe test vectors or in areas with low or timing-failure-non-causing LSA. This not only greatly improves the effectiveness of risky path reduction but also avoids unnecessary test quality degradation.

● ***Guaranteed Launch Safety***: Masking any uncertain test response guarantees launch safety as the last resort. This is made possible by focused LSA reduction, which makes the number of remaining risky paths small, if any.

● ***Minimal Impact on Test Quality & Test Costs***: Focused LSA reduction only uses necessary resources (i.e., *impact-X-bits*) but keeps original logic values at other bits already optimized by detection-oriented X-filling (e.g., *random-fill*). Furthermore, masking-induced loss in fault detection capability is mostly recovered by test vectors generated in subsequent ATPG runs. Therefore, the original test quality is preserved and severe test data inflation is avoided.

The rest of the paper is organized as follows: Sect. II describes the background. Sect. III presents the novel scheme for achieving guaranteed launch safety. Sect. IV shows experimental results, and Sect. V concludes the paper.

II. BACKGROUND

A. Launch Safety Checking

Ideal launch safety checking is time-consuming and memory-intensive due to timing-accurate logic simulation, IR-drop analysis, and delay calculation. This fact makes it necessary to use simplified metrics. As a result, *toggle count* (TC) and *weighted switching activity* (WSA) for FFs, the whole circuit, or regions in a total or instantaneous manner are often used for estimating LSA [5]. However, these metrics are not targeted at long sensitized paths that are most susceptible to the impact of LSA. To address this issue, the *critical capture transition* (CCT) metric assesses LSA around critical paths [9], and the *critical area targeted* (CAT) metric estimates LSA around the longest sensitized path of a test vector [8].

In this paper, we use an improved metric based on the CAT metric [8] for launch safety checking. The CAT metric is extended to check all long sensitized paths for higher accuracy. Details will be presented in Subsection III.B.

B. Low-LSA Test Generation

Three typical approaches to reduce LSA are available [5]:

- **ATPG**: Low-LSA test vectors can be generated with reversible capture-transition-triggered backtracking as well as clock-disabling with pre-calculated (e.g., *default values*, *clock control cubes*, etc.) or dynamically-calculated values.

- **Low-LSA Test Compaction**: This can be done as dynamic compaction by properly selecting secondary faults or as static compaction by properly selecting test cubes to be merged. Both of them try to avoid concentrated LSA.

- **Low-LSA X-Filling**: X-bits, directly left in a partially-specified test cube or obtained from a fully-specified test vector by test relaxation [10], can be filled with proper logic values so as to reduce LSA [11]. There are three types of such techniques: (**1**) *FF-Oriented*: Transitions at FF outputs are reduced by input-output equalizing (e.g., *preferred-fill* [12], *JP-fill* [13], *iFill* [14], etc.) or clock disabling (e.g., *CTX-fill* [5]); (**2**) *Node-Oriented*: Transitions inside a circuit are directly reduced (e.g., *PWT-fill* [5]); and (**3**) *Critical-Area-Oriented*: Transitions in specific areas inside a circuit are reduced (e.g., *CAT-fill* [8], *CCT-fill* [9], etc.).

However, these previous techniques can only reduce LSA to some extent, but cannot guarantee launch safety. As discussed in Sect. I, the reason comes from two problems: (**i**) **unfocused effect** (i.e., most of them only reduce total LSA for the whole circuit but excessive LSA may still remain in neighboring areas around long sensitized paths) and (**ii**) **unguaranteed sufficiency** (i.e., they cannot guarantee sufficient LSA reduction for all test vectors). In this paper, we propose a two-phase (**rescue** & **mask**) scheme to achieve guaranteed launch safety effectively.

III. New Power-Aware Test Generation Scheme

A. Test Generation Flow

As shown in Fig. 3, conventional test generation (**A~E**) starts from initial fault list generation (**A**). A partially-specified test cube C_1 is generated to detect a primary fault and dynamic compaction is then conducted (**B**). Here, any transition, path, or small-delay ATPG can be used. Then, detection-oriented X-filling is conducted to turn C_1 into a fully-specified test vector V_1 (**C**). *Random-fill* is often used in industry for this purpose since its fortuitous detection capability greatly improves unmodeled-defect detection (thus *higher test quality*) and reduces test data volume (thus *lower test costs*). After that, fault simulation is conducted to update the fault list (**D**), and the termination condition is checked to decide whether to continue test generation (**E**).

This conventional ATPG flow is enhanced to guarantee launch safety by adding a new two-phase scheme (① ~ ⑤):

- **Phase-I (Rescue)**: This phase consists of ① ~ ③. *LSP-based launch safety checking* (①) is to identify all *long sensitized paths* (LSP) under V_1 and check the *launch switching activity* (LSA) in the neighboring area (called *impact area* to be defined in III.B) of each LSP. If the impact area of an LSP has excessive LSA, the LSP is called

a *risky path*. V_1 is *launch-risky* if it has at least one risky path. In this case, **impact-X-bit restoring** (②) is conducted to restore those logic bits in V_1 to X-bits (called *impact-X-bits* to be defined in III.C) if they are originally X-bits in C_1 and can reach the impact area of at least one risky path. This way, a new test cube C_2 is obtained efficiently without using time-consuming test vector relaxation [10]. After that, **focused low-LSA X-filling** (③) is conducted for the impact-X-bits to reduce LSA in the impact areas of the risky paths in a focused manner. This results in a new test vector V_2.

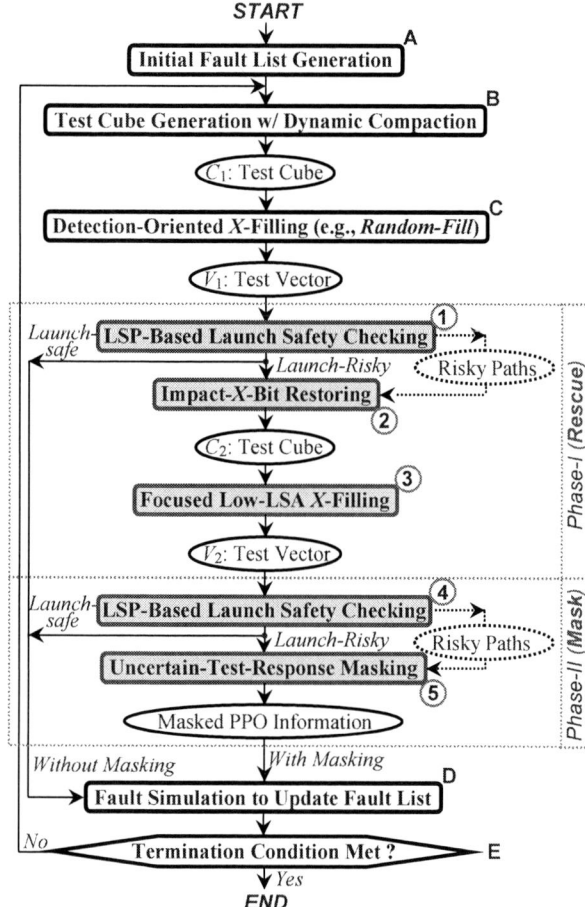

Figure 3. Test generation scheme with guaranteed launch safety.

- **Phase-II (Mask)**: This phase consists of ④ and ⑤. First, *LSP-based launch safety checking* (④) is conducted on V_2. If V_2 is found to be launch-risky, an erroneous test response may appear at the endpoint (a FF input or a pseudo primary output (PPO) in the circuit model) of a risky path. To avoid this risk of yield loss in production test, **uncertain-test-response masking** (⑤) is conducted by placing an X as the test response at the endpoint PPO of any remaining risky path for V_2 in production test data. This masking incurs no circuit overhead. In addition, fault simulation with masked PPOs is conducted to update the fault list so that masked-PPO-induced change in fault detection capability is properly reflected in the result of the current ATPG run (**E**).

B. LSP-Based Launch Safety Checking

Since it is a *long sensitized path* (*LSP*) that is the most susceptible to the impact of excessive LSA, we conduct *LSP-based launch safety checking* as follows:

Definition 1: The *aggressor region* of a gate G, denoted by $AR(G)$, is composed of aggressor cells (gates and FFs) whose transitions strongly impact the supply voltage of G.

In an LSI chip, the current flows through C4 pads to cells through a *power grid* composed of alternate metal lines of *VDD* and *GND* in each layer. The metal layers are connected by vias, and cells are connected to lower-level (e.g., M2 through M4) vias. Thus, the aggressor region of a gate G can be identified as follows [15]: First, identify the *powering via* for G, by which G is directly powered. Then, identify all current-sink cells for the powering via of G, and these cells are the aggressor cells for G. A simplified example (*GND* wires ignored) is shown in Fig. 4, where the aggressor region of G_1 consists of G_2, G_3, and G_4. Note that G_2 is a stronger aggressor than G_3 and G_4 that are farther away.

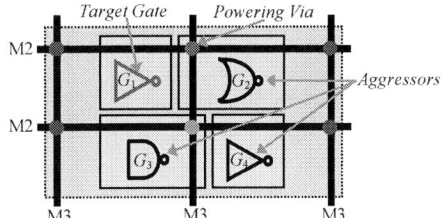

Figure 4. Aggressor region.

Definition 2: The *impact area* of P, denoted by $IA(P)$, consists of the aggressor regions of all on-path gates (G_1, G_2, ..., G_n) of P. That is, $IA(P) = AR(G_1) \cup AR(G_2) \cup ... \cup AR(G_n)$, as illustrated in Fig. 5.

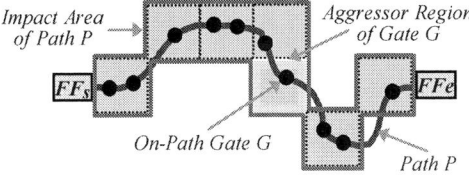

Figure 5. Impact area.

Definition 3: A path P is said to be a *risky path* under a test vector V if (1) P is sensitized by V and (2) the LSA in the impact area of P under V is excessive (w.r.t. a threshold).

Definition 4: A test vector V is said to be a *launch-risky test vector* if V has at least one risky path.

Based on the above definitions, LSP-based launch safety checking can be conducted by the following procedure:

LSP-Based_Launch_Safety_Checking ()
{ *Input*: test vector V, design data (netlist, layout, power supply network), path length threshold α, WSA level threshold β
   Output: status of V, risky paths of V
OP-1: Identify all LSPs, i.e., paths that are sensitized by V and whose lengths are greater than α.
OP-2: Identify the impact area of each LSP.
OP-3: Run logic simulation and calculate the WSA (*weighted switching activity*) for the impact area of each LSP.

Op-4: Identify all risky paths of V by checking if the WSA for the impact area of each LSP is greater than β. Output all risky paths of V.
Op-5: Output the status of V. V is launch-risky if it has at least one risky path; otherwise, V is launch-safe. }

In this procedure, a and β are the *path length threshold* and *WSA threshold*, respectively. In our experiments, they were set to 70% of the length of the longest structural path and 30% of the maximum WSA in the impact area, respectively. In practice, they can be set by test engineers.

The major advantages of LSP-based launch safety checking over previous techniques are as follows:

● *High Accuracy*: The procedure targets at long sensitized paths (LSPs), whose delay increase is the dominant cause of timing failures in the test cycle as shown in Fig. 1.

● *High Resolution*: The procedure identifies all risky paths, not just reporting whether a test vector is launch-risky or not. It is this detailed information on risky paths that makes it possible to effectively conduct focused LSA reduction in Phase-1 and guarantee launch safety in Phase-II.

C. Impact-X-Bit Restoring

As shown in Fig. 3, if a test vector is identified by LSP-based launch safety checking (①) as launch-risky, *rescue* is then conducted in Phase-I by reducing the excessive LSA in the impact area of each risky path as much as possible. This goal is realized by *impact-X-bit restoring* (②) for obtaining necessary X-bits, and *focused low-LSA X-filling* (③) for filling those X-bits with proper logic values so as to reduce LSA. This subsection describes impact-X-bit restoring.

Two previous approaches are available for obtaining X-bits needed for low-LSA X-filling. One is *test cube preservation* [3], in which ATPG is forced to leave X-bits in a deterministically-generated test cube by disabling *random-fill* or other detection-oriented X-filling processes. Another is *test vector relaxation* [10], in which a fully-specified test vector set is turned into a partially-specified test cube set while preserving its original fault coverage.

However, *test cube preservation* suffers from significant test quality degradation and test vector count inflation since the fortuitous-detection-capability of detection-oriented X-filling (e.g., *random-fill*) is not used. For example, our experiments on a 600K-gate industrial circuit block showed a 51% increase in test vector count when *random-fill* was disabled. On the other hand, *test vector relaxation* can be conducted on a compact test set and maintains its size. However, this approach is a static post-ATPG process for a complete test set, which is hard to apply in a dynamic ΛTPG flow for a single test vector.

To preserve the test quality benefit of detection-oriented X-filling as much as possible while obtaining X-bits in a dynamic manner for individual test vectors, we propose a new technique, namely *impact-X-bit restoring*, as follows:

Definition 5: Let V_1 be a launch-risky test vector, obtained by detection-oriented X-filling from a test cube C_1. An X-bit in C_1 that can reach the impact area of at least one risky path of V_1 is called an *impact-X-bits* for V_1.

Impact-X-bit restoring turns every logic bit b in a fully-specified test vector V_1 to X if b corresponds to an impact-X-bit for V_1. The result is a new partially-specified test cube C_2.

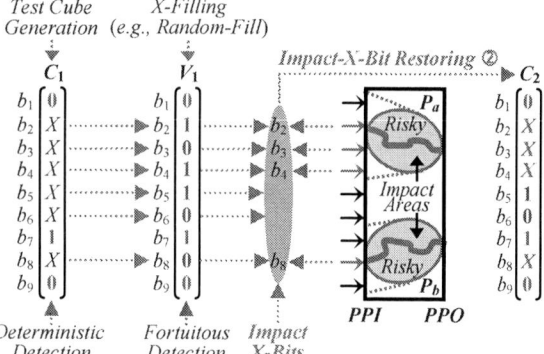

Figure 6. Impact-*X*-bit restoring.

Fig. 6 illustrates *Impact-X-bit restoring*. Here, C_1 is a test cube generated deterministically, with logic values at b_1, b_7, and b_9 for detecting targeted faults. All other bits in C_1 are filled by *random-fill* for fortuitous detection, resulting in V_1. If V_1 has two risky paths (P_a, P_b), the impact-X-bit set for V_1 is $S = \{b_2, b_3, b_4, b_8\}$. Turning the bits in V_1 corresponding to S back into X-bits results in C_2 that detects all faults detected by C_1. Note that *random-fill*-assigned logic values remain unchanged at b_5 and b_6, helping preserve some fortuitous-detection-capability of *random-fill*. Thus, *impact-X-bit restoring* not only obtains X-bits directly related to the LSA in the impact areas of risky paths but also helps preserve test quality and avoid severe test vector count inflation.

D. Focused Low-LSA X-Filling

In Fig. 3, after X-bits are obtained by *impact-X-bit restoring* (②), *focused low-LSA X-filling* (③) is conducted on those X-bits. The term "*focused*" means that all of the X-bits are directly related to the impact areas of risky paths.

In this paper, we apply an improved form of *JP-fill* [13] that uses assignment / justification / multi-pass probability calculation to fill X-bits for equalizing PPI and PPO values at **candidate PPI-PPO bit-pairs** (i.e., bit-pairs in the form of $<0/1, X>$, $<X, 0/1>$, or $<X, X>$). Since the effectiveness of X-filling depends on the filling order, we improve *JP-fill* with a new weight to order candidate PPI-PPO bit-pairs.

Definition 6: Let $bp = <x, y>$ be a candidate PPI-PPO bit-pair, whose PPI-bit x can reach risky paths: P_1, P_2, ..., P_m. The **weight** of bp, denoted by *weight* (bp), is defined as

$$weight(bp) = \sum_{i=1}^{m} reached(x, P_i) / all(P_i)$$

where *reached*(x, P_i) is the number of nodes reachable from the PPI-bit x in the impact area of P_i, and *all*(P_i) is the number of all nodes in the impact area of P_i.

Clearly, *weight*(bp) indicates the impact of reducing a transition at the candidate PPI-PPO bit-pair bp on reducing LSA in the impact areas of risky paths. We use this weight to determine the order of processing candidate PPI-PPO bit-pairs so as to achieve more effective LSA reduction.

E. Uncertain-Test-Response Masking

As shown in Fig. 3, *focused low-LSA X-filling* (③) in the **rescue phase** (Phase-I) results in a test vector V_2. Generally, V_2 may still be launch-risky although it often has fewer risky paths. In this case, the **mask phase** (Phase-II) is executed as the last resort for guaranteeing launch safety.

In Phase-II, if *LSP-based launch safety checking* (④) finds V_2 to be launch-risky, **uncertain-test-response masking** (⑤) is then conducted as illustrated in Fig. 7.

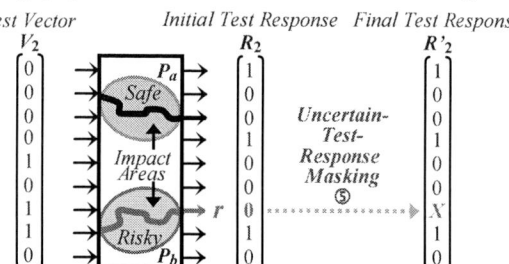

Figure 7. Uncertain-test-response masking.

In Fig.7, test vector V_2 has two long sensitized paths: a safe path P_a (due to focused low-LSA X-filling) and a risky path P_b (due to insufficient LSA reduction) whose endpoint corresponds to the output bit r. Here, although R_2 (the initial test response to V_2 obtained by logic simulation) has 0 at r, excessive LSA in the impact area of P_b may cause a timing error at r, falsely making 1 to appear at r in test mode. To avoid possible yield loss, **uncertain-test-response masking** (⑤) is conducted by replacing 0 with X at r in the final test response R'_2. This instructs the tester not to compare at r in production test, thus avoiding any possible power-supply-noise-induced yield loss in at-speeding scan testing.

Note that masking incurs no area/performance overhead as it just puts an X (unknown) at r in the final test response. Although faults detected by V_2 only at r become undetected by V_2 due to masking, r is only masked for V_2 but available for fault detection by other test vectors. That is, the fault list is updated by fault simulation with masked PPO information (**D** in Fig. 3), and ATPG continues in which initially r is not masked. This way, the lost fault detection capability at r for V_2 can be recovered by subsequent ATPG runs, only at the cost of slightly more test vectors.

F. Extension to Compressed Scan Testing

The proposed rescue-&-mask scheme for achieving guaranteed launch safety can be readily extended to any test compression environment. For example, in broadcast-scan-based test compression [2], the constraints imposed by the combinational decompressor on inputs can be embedded into an integrated combinational circuit model [16], on which *LSP-based launch safety checking*, *impact-X-bit restoring*, and *focused low-LSA X-filling* can be directly applied. The major concern is with test response compaction, where Xs introduced by *uncertain-test-response masking* may disturb fault detection. However, such Xs are sparse and their number is very small for a test vector. This makes their impact on test quality and test costs minimal.

978-1-61284-657-6/11 $26.00 © 2011 IEEE

IV. EXPERIMENTAL RESULTS

The power-aware test generation flow of Fig. 3 was implemented with TetraMAX® as the base ATPG, and the rescue-&-mask scheme for guaranteeing launch safety was coded in C. Six large ITC'99 benchmark circuits were synthesized and physically designed with a power supply network. Experiments were conducted on a workstation (Intel Xeon® 3.33GHz with 64GB main memory). Table I shows circuit statistics and experimental results.

In our experiments, test costs were evaluated by test vector count (*# of Vectors*), while test quality was evaluated not only by transition fault coverage (*FC*) but also by *bridging coverage estimate* (*BCE*) [17] and *statistical delay quality level* (*SDQL*) [18] for more comprehensive evaluation. *BCE* is used in industry to assess the capability of detecting unmodeled structural defects (especially, bridging defects), and *SDQL* is used in industry to assess the capability of detecting unmodeled small-delay defects.

First, the conventional ATPG flow was executed, and its results are shown under "*Conventional Flow*" in Table I as baseline values. Next, the proposed ATPG flow was executed, and its results are shown under "*Proposed Flow*" in Table I with four parts: (**1**) Changes in test vector count and three test quality metrics (*FC, BCE, SDQL*) are shown under "*ATPG Results %Change*"; (**2**) the average number of long sensitized paths per vector (*Ave. # of LSPs / Vec.*), the average number of risky paths per vector (*Ave. # of Risky Paths / Vec.*), and the percentage of risky vectors (*% of Risky Vec.*) are shown under "*Launch Safety Checking*"; (**3**) the percentage of impact-X-bits (*% of Impact-X-Bits*) and the ratio of focused low-LSA X-filling making risky paths into safe paths (*Resuce Ratio (%)*) are shown under "*Rescue*"; (**4**) the number of masked PPOs (*# of Masked PPOs*) and the average number of masked PPOs per vector (*Ave. # of Masked PPOs / Vec.*) are shown under "*Mask*".

Due to the nature of the proposed ATPG flow, it always guarantees launch safety. From Table I, it can be seen that the impact of this new ATPG flow on test quality and test costs is minimal since there is little change in test vector count, *FC, BCE,* and *SDQL*. This is because of **the nature of masked PPOs**, i.e., masked PPOs are sparse and the number of masked PPOs is extremely small, as indicated by "*Ave. # of Masked PPOs / Vec.*" in Table I.

Note that the above nature also holds in any test compression environment since it is only related to the combinational portion and independent of the decompressor and the compactor. This indicates that the proposed rescue-&-mask scheme also works for compressed scan testing.

V. CONCLUSIONS

This paper has addressed the fundamental issue in power-aware test generation for at-speed scan testing, i.e., how to guarantee launch safety instead of merely reducing launch switching activity to some extent. A novel two-phase scheme has been proposed to guarantee launch safety with minimal impact on test quality and test costs. The ***rescue phase*** is to reduce excessive LSA around long sensitized paths, and the ***mask phase*** is to exclude any uncertain test response from being used for fault detection. Experimental results have validated this novel approach to guaranteeing launch safety.

Future work includes conducting evaluation experiments by using a commercial-grade test compression tool.

REFERENCES

[1] X Lin, et al., "Timing-Aware ATPG for High Quality At-Speed Testing of Small Delay Defects," *Proc. ATS*, pp.139-146, 2006.
[2] N.A. Touba, "Survey of Test Vector Compression Techniques," *IEEE Design & Test Magazine*, Vol. 23, Issue 4, pp. 294-303, 2006.
[3] J. Saxena, et al., "A Scheme to Reduce Power Consumption during Scan Testing," *Proc. ITC*, pp. 670-677, 2001.
[4] S. Sde-Paz and E. Salomon, "Frequency and Power Correlation between At-Speed Scan and Functional Tests," *Proc. ITC*, Paper 13.3, 2008.
[5] P. Girard, et al., *Power-Aware Testing and Test Strategies for Low Power Devices*, Springer, 2009.
[6] C. P. Ravikumar, et al., "Test Strategies for Low-Power Devices," *J. of Low Power Electronics*, Vol. 4, No.2, pp. 127-138, 2008.
[7] L.Whetsel, "Adapting Scan Architectures for Low Power Operation," *Proc. ITC*, pp. 863-872, 2000.
[8] K. Enokimoto, et al., "CAT: A Critical-Area-Targeted Test Set Modification Scheme for Reducing Launch Switching Activity in At-Speed Scan Testing," *Proc. ATS*, pp. 99-104, 2009.
[9] X. Wen, et al., "Critical-Path-Aware X-Filling for Effective IR-Drop Reduction in At-Speed Scan Testing," *Proc. DAC*, pp.527-532, 2007.
[10] K. Miyase, et al., "XID: Don't Care Identification of Test Patterns for Combinational Circuits," *IEEE TCAD*, 23-2, pp. 321-326, 2004.
[11] X. Wen, et al., "On Low-Capture-Power Test Generation for Scan Testing," *Proc. VTS*, pp. 265-270, 2005.
[12] S. Remersaro, et al., "Preferred Fill: A Scalable Method to Reduce Capture Power for Scan Based Designs," *Proc. ITC*, Paper 32.2, 2006.
[13] X. Wen, et al., "A Novel Scheme to Reduce Power Supply Noise for High-Quality At-Speed Scan Testing," *Proc. ITC*, Paper 25.1, 2007.
[14] J. Li, et al., "On Capture Power-Aware Test Data Compression for Scan-Based Testing," *Proc. ICCAD*, pp. 67-72, 2008.
[15] J. Lee, et al., "Layout-Aware, IR-Drop Tolerant Transition Fault Pattern Generation," *Proc. DATE*, pp. 1172-1177, 2008.
[16] K. Miyase, et al., "A Novel Post-ATPG IR-Drop Reduction Scheme for At-Speed Scan Testing in Broadcast-Scan-Based Test Compression Environment," *Proc. ICCAD*, pp. 97-104, 2009.
[17] B. Benware, et al., "Impact of Multiple-Detect Test Patterns on Product Quality," *Proc. ITC*, pp. 1031-1040, 2003.
[18] Y. Sato, et al., "Invisible Delay Quality - SDQM Model Lights Up What Could Not Be Seen," *Proc. ITC*, Paper 47.1, 2005.

TABLE I. EVALUATION RESULTS

Circuit	# of Gates	# of FFs	Conventional Flow (with possible launch risk)				Proposed Flow (with guaranteed launch safety)										
			ATPG Result (Baseline)				ATPG Result (% Change)				Launch Safety Checking			Rescue		Mask		CPU (Sec.)
			# of Vectors	FC	BCE	SDQL	Δ # of Vectors	Δ FC	Δ BCE	Δ SDQL	Ave. # of LSPs / Vec.	Ave. # of Risky Paths / Vec.	% of Risky Vec.	% of Impact-X-Bits	Resuce Ratio (%)	# of Masked PPOs	Ave. # of Masked PPOs / Vec.	
b17	21,235	1,317	970	73.0	66.1	16.9	+0.05	+0.02	−0.01	+0.01	0.13	0.01	0.19	29.18	25.00	2	0.01	471
b18	64,009	3,064	1,933	65.4	64.0	204.6	+0.01	+0.04	−0.01	+0.03	0.45	0.21	0.90	27.86	8.90	28	0.01	1,777
b19	128,576	6,130	2,806	66.5	63.9	303.8	−0.06	+0.02	−0.01	+0.01	0.25	0.01	0.11	45.29	96.00	1	0.01	4,607
b20	20,271	430	1,496	93.8	63.3	116.5	+0.01	+0.01	−0.04	+0.01	0.25	0.15	0.93	14.70	0.02	8	0.01	343
b21	20,148	430	1,503	94.1	57.1	137.6	+0.04	+0.01	−0.01	+0.01	0.33	0.30	1.22	12.79	0.02	23	0.02	342
b22	29,926	645	1,919	94.4	62.3	140.5	+0.08	+0.01	−0.01	+0.03	0.05	0.01	0.15	11.37	48.00	3	0.01	597

SLIDER : **A Fast and Accurate Defect Simulation Framework**

Wing Chiu Tam and R. D. (Shawn) Blanton
ECE Department, Carnegie Mellon University, Pittsburgh, USA
{wtam,blanton}@ece.cmu.edu

Abstract—As integrated circuit (IC) manufacturing entered the nano-scale era, defect observability has greatly diminished. As a result, test-fail data diagnosis and mining are playing an indispensable role in providing feedback for yield learning. Accurate simulation of defect behavior is vital to this process but, unfortunately, cannot be achieved with simulation at the logic-level alone. This work proposes a framework to enable fast and accurate defect simulation, by making use of existing and well-developed mixed-signal simulation technology (traditionally used for design verification). While previous work has considered this topic before, the innovation here centers on two aspects: (i) accuracy resulting from defect injection taking place at the layout level, (ii) speedup resulting from careful and automatic partitioning of the circuit into digital and analog domains for mixed-signal simulation, and (iii) complete automation that involves defect injection, design partitioning, netlist extraction, mixed-signal simulation, and test-data extraction. The mixed-signal framework developed can be applied in a variety of settings that include diagnosis resolution improvement, defect localization, fault model evaluation, and virtual failure data creation. Experiments demonstrate that the proposed framework is scalable to handle large designs efficiently. A second set of experiments demonstrates how defect localization can be dramatically improved (> 53%) by more accurate defect simulation.

Keywords- mixed-signal simulation; defect modeling; volume diagnosis; yield learning; layout analysis.

I.   INTRODUCTION

The ever-increasing complexity of nano-scale IC manufacturing introduces new defect types and associated failure mechanisms. The traditional approach of using test structures to understand defects, while still very useful, has greatly diminished in applicability because of the limited feedback [1] it can provide. This makes yield learning increasingly difficult. IC testing, conventionally used to find failing parts, now also serves an added function of aiding the discovery and understanding of the root-cause of failures. Through diagnosis, valuable information can be extracted [2-4]. When a sufficiently large number of failures are analyzed, one can potentially discover commonalities that underlay many of the failures. Armed with the information extracted from testing, the design/process can be compensated to improve yield while test itself can be adjusted to drive product quality improvement [2-8].

Failure diagnosis involves basically two steps: (1) using software that accepts a circuit description, tests, and the tester response to deduce possible locations, structural changes, and associated behaviors of the defect(s) that cause failure (traditionally referred to as logic diagnosis), and (2) the physical analysis of the failed chip to identify the culprit process step that lead to the defect (traditionally referred to as physical failure analysis). The ability to simulate a defect

(*i.e.*, evaluate the output response of a circuit for the given input stimulus in the presence of a given defect) is essential for any logic diagnosis method. Since diagnosis is not perfect in that it typically produces several seemingly-equivalent candidates, it is important that the ranking criteria indicate how closely a defect's simulated response matches the observed tester response. Since the logic level offers a good tradeoff between accuracy and efficiency, most testing tasks and diagnosis techniques (*e.g.*, [9-13]) operate at the logic level. The erroneous behavior of a defect therefore must also be abstracted to the logic level and is typically known as a fault model. While this approach has the advantages of abstracting away unnecessary details and providing tremendous speedup, it likely discards important information for ensuring accuracy. For example, it is well-known that a Byzantine bridge can confound logic-level diagnosis [14]. Another example is an open defect whose faulty behavior changes with its position along the interconnect network [15-16]. A traditional logic-level fault simulator would not be able to capture/simulate these behaviors.

At the other extreme, one can represent the defect as a geometric change at the layout level. Specifically, a defect is injected into the design layout, a circuit is then extracted from the modified layout, and then circuit-level simulation of the extracted circuit is performed [17]. This approach exhibits tremendous accuracy but is not scalable since it employs circuit-level simulation to the entire circuit. Thus, its usage is limited to situations when the circuits under consideration are small (typically no more than a thousand gates). In [17], the authors create a plethora of virtual failures for a number of small circuits to validate diagnosis techniques and layout-based defect localization techniques.

A.   Motivation

The simulation speed-vs-accuracy tradeoff is nothing new. It is present in any task involving computer simulation/modeling. For example, in IC design, both analog and digital blocks are utilized. Clearly, the analog blocks cannot be simulated in the digital domain. On the other hand, it is inefficient (and often infeasible) to simulate the entire design in the analog domain. The solution developed by the design community is to simulate the analog blocks with circuit-level accuracy and the digital blocks with logic-level accuracy along with appropriate signal conversion at interfaces that interconnect the two types of blocks.

The same concept can be applied to defect simulation. To accurately capture defect behavior, only the defect site and a few of its subsequent logic stages need to be simulated accurately. This is possible since any intermediate voltage level that results due to the defect often becomes logically well-defined (either logic 1 or logic 0) in subsequent logic stages, and therefore does not require circuit-level simulation accuracy from that point onwards. Thus, simulation of the remaining portions of the circuit can be carried out completely in the digital domain. Figure 1 illustrates this concept. Due to the presence of a bridge defect (represented as a resistor inside the red circle), the pull-up network in the NOR gate and the pull-down network of NAND gate are in contention for the driver input values applied, resulting in the unknown logic value X for both wires involved in the bridge. However, all signals have well-defined, valid logic values after two stages of logic. In other words, only the region inside the red dotted rectangle requires simulation with circuit-level accuracy.

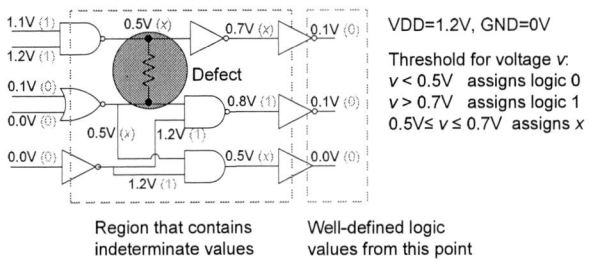

Figure 1: Circuit region that contains indeterminate logic value (X) due to the presence of a bridge defect. All intermediate values become well-defined logic values after two stages of logic.

It is certainly possible to build a mixed-signal defect simulator from scratch based on this concept. However, it is far more efficient and robust to use existing tools for mixed-signal simulation, which is a mature technology with associated modeling languages (*e.g.*, Verilog-AMS [18]) and commercial simulators (*e.g.*, Cadence AMS Designer [19]). Using these tools requires that the design be carefully partitioned into digital and analog blocks. Specifically, all regions that require high simulation accuracy are typically consolidated into a single analog block. Changes can then be made to the analog block to represent the presence of one or more defects. The rest of the design remains as a digital block. Once the design is properly divided, the mixed-signal simulator can be directly used.

B. Our Contribution

Several prior works have explored the possibility of accurate defect modeling/simulation. The inductive fault analysis tool CARAFE [20] identifies likely faults in a circuit and creates corresponding circuit-level netlists, taking into account the layout, the defect characteristics, and the fabrication technology. CODEF [21-23] is a process simulator that was applied to gates/cells or small pieces of the layout in order to predict the impact of a particulate contamination on the layout geometry. The impact is then mapped to a circuit-level defect which is then simulated for improving diagnosis [21] and understanding defect behavior [22-23]. In [24], a 4-bit ALU is used as a base design for investigating a diverse set of realistic defects (modeled as changes in the circuit-level netlist) that are crafted specifically to generate responses that are used to benchmark diagnosis techniques. The work in [17] is a significant extension of [24], where defect generation and injection is completely automated and performed at the layout level to remove its dependence on a specific design. Work in [25] simulates transistor-level defects using SPICE and abstracts

the resulting behavior to logic level by using a new algebra. Work in [26-27] uses mixed-level fault simulation to simulate a defect more accurately while maintaining a tolerable simulation speed. Unfortunately, the work in [17, 21-24] depends on simulating the entire circuit at the transistor level and is therefore not scalable. The work in [25-27], while scalable, does not consider layout and therefore sacrifices accuracy. In addition, the variety of defects considered in [25-27] is quite limited. By making use of a commercial mixed-signal simulator and performing defect injection at the layout-level, the framework described here solves the scalability problem without loss of accuracy and thus enables new applications as the result of the efficiency improvement. In fact, this work is a completely re-designed version of the framework in [17], with scalability enhancement through mixed-signal simulation and support for multiple defect injection. The newly designed framework is called SLIDER (Simulation of Layout-Injected Defects for Electrical Responses).

While it is true that many defect behaviors can be mapped to the logic-level [28] to gain scalability, the mapping process often requires pre-characterization of the defect behaviors [25]. In addition, logic-level abstraction will not be able to capture/simulate defect behavior that is both a function of the tests applied and their location of occurrence. In contrast, SLIDER (automatically) injects defects at the layout level and extracts a netlist of the affected area.

It should be emphasized that, in contrast to [17], SLIDER is a very general framework that enables/supports many testing tasks. One obvious application of SLIDER is to improve diagnosis by further analyzing diagnosis candidates to infer more information. For example, in diagnosis it is typically the case that several equally-ranked candidates are reported. SLIDER can be used to experiment with different defect-injection scenarios at the suspect failure locations to resolve this ambiguity. Even in the case when there is no ambiguity in the diagnosis (*i.e.*, only one candidate reported with a "100% match"), it is possible to use SLIDER to characterize the defect (*e.g.*, identify the range of resistance of an open or a bridge) which can help to further localize the defect, yielding additional information for any follow-on analyses.

Another application of SLIDER is to measure fault model and test metric effectiveness. Tests generated using a given fault model can be simulated with SLIDER to evaluate its effectiveness in defect detection. Tests generated/selected using a given test metric (*e.g.*, Physically-aware N-detect [29]) can be subject to the same analysis to evaluate the effectiveness of the test metric.

Because of its efficiency, the SLIDER framework can be used to create a large population of virtual failures for other uses. First, it can be used to validate/evaluate *any* test and yield learning techniques that are based on analyzing test fail data. Consider the evaluation of a systematic defect identification methodology. SLIDER can be used to create test data for both systematic and random defects of different proportions. After applying the systematic defect identification methodology to the virtual failure data, the result can be used to evaluate how much "noise" (*i.e.*,

978-1-61284-657-6/11 $26.00 © 2011 IEEE

proportion of random defects) can be tolerated while still providing the correct answer.

Second, SLIDER can aid the "debugging" of test and yield learning techniques. It can be used to identify the boundary cases for which a technique begins to fail. This exposes incorrect assumptions or software bugs in the technique by providing a counter-example. By understanding when and how the technique fails, it makes improvement of the technique possible.

Finally, the virtual failure population can play an active role in yield-learning methodologies. Since both the failure data and the defect characteristics are available, various machine-learning techniques can be applied to learn valuable information from the virtual data. Lessons learned can then be applied to real-world data. This idea is explored in [30], in which a decision forest [31] (a type of data classifier) is trained using the virtual failure population to separate bridge and non-bridge defects. The decision forest is then applied to classify real failure data from an Nvidia GPU. Unfortunately, the experiment performed in [30] uses circuit-level simulation of small circuits to generate the training data and therefore its applicability to large designs can be quite limited. The SLIDER framework resolves this issue.

The rest of this paper is organized as follows: Section II describes the details of the SLIDER framework. This is followed by a run-time experiment carried out using SLIDER for 12 placed-and-routed benchmarks [32-33], and an example application of the framework for diagnosis resolution improvement and defect localization in Section III. Finally, the work is summarized in Section IV.

II. SIMULATION FRAMEWORK

In this section, we describe the structure and flow of the SLIDER framework.

A. Framework Structure

Figure 2 shows the key components of SLIDER. It is composed of an analog block creation unit (ABCU), a digital block creation unit (DBCU), a mixed-signal simulation engine (MSSE), and a result parser (RP). The shaded blocks are implemented in SLIDER. The MSSE used in SLIDER is Cadence AMS designer [11]; however any other mixed-signal simulator can be easily used instead.

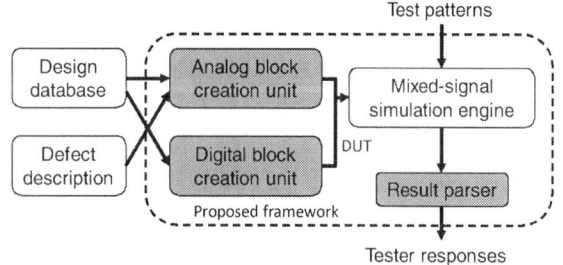

Figure 2: Block diagram of the SLIDER framework.

B. Framework Flow

The input to the framework is the design database, a description of the defects to be injected into the layout, and the test patterns. The design database must contain both the layout and the logical netlist. The defect description specifies the geometry of one or more defects (*i.e.,* coordinates, size, polygon, layer, *etc.*). The supported defect types include: open, resistive via, signal bridge, supply bridge, and various kinds of cell defects that include nasty polysilicon defect [34], transistor stuck-open, and transistor stuck-closed. The test patterns mimic the tester stimulus that would be applied to the circuit during production testing.

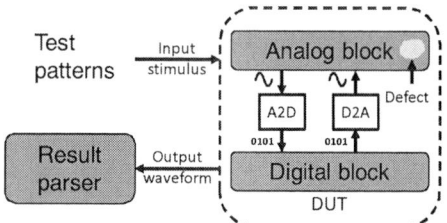

Figure 3: Block diagram that illustrates the top-level testbench created by SLIDER for mixed-signal simulation.

Using these inputs, the SLIDER framework creates a testbench that can be understood directly by the MSSE to perform mixed-signal simulation for the defects of interest, as illustrated in Figure 3. The flow starts by automatically partitioning the design into its respective analog and digital blocks, which is accomplished by ABCU and DBCU in Figure 2, respectively. The analog block contains all instances and nets that are required to produce accurate circuit-level simulation results for the defect sites and their influence regions, while the digital block contains everything that remains. Defects are then injected into the analog block at the layout level. The analog block, the digital block, and their connections form the design under test (DUT). To capture the DUT's response, a Verilog module, namely the RP, is defined to sample the output waveform at the frequency defined by the test clock. Once the testbench is created, the MSSE performs a transient analysis for a duration required for application of all the test patterns. For example, if the test pattern count is 288 and the test clock period is 10ns, then the duration for transient analysis is 2880ns. During simulation, the MSSE automatically takes care of the cross-domain signals through the use of connection modules, an analog-to-digital (digital-to-analog) convertor for a signal that crosses from the analog (digital) domain to the digital (analog) domain. The module can be completely specified by the user in the Verilog-AMS language [35].

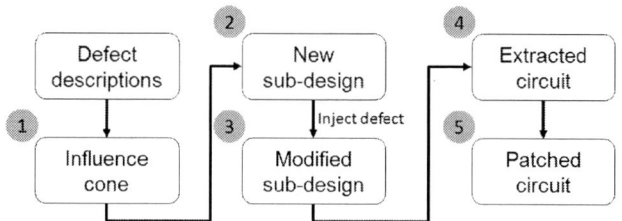

Figure 4: Flow diagram that illustrates the analog block creation process.

This process is repeated automatically for each simulation run, where a run can consist of multiple defects. It should be noted that because the design database can be very large, copying the database is intentionally avoided. Instead,

978-1-61284-657-6/11 $26.00 © 2011 IEEE   174

the design database is modified in place to output the testbench for MSSE. Changes to the original design database are immediately discarded afterwards. The design is purged and re-opened for the next iteration.

C. Analog Block Creation

Figure 4 illustrates the flow for analog block creation. For each defect, its layout location of injection is used to determine the *affected nets*. A net is considered affected when its corresponding layout polygons have a non-zero-area overlap with the defect polygon. For each affected net, an *influence cone* is identified which consists of all the nets that can influence/control the voltage of the affected net, or whose voltage can be rendered indeterminate by the affected net. Thus, the influence cone typically consists of signals in the transitive fan-in and fan-out of the affected net. Table 1 summarizes the types of nets in the influence cone. Column 1 gives the net type while column 2 provides a description. Figure 5 illustrates the various nets in an influence cone. In this example, an open (shown as a red circle) is injected into net N_1. The transitive fan-in of N_1 for one logic stage is $\{N_4\}$. (Refer to column 3 of Table 1 for other net types). The influence cone is the union of all the nets and is therefore $\{N_1, N_2, N_4 - N_7\}$ for $d = 260nm$, $i = 1$, and $j = 2$. Parameters d, i, and j are all specified by the user for each defect to be injected. The process of influence-cone determination is repeated for each defect to be injected in the circuit. The influence cones for all the defects are consolidated into a single cone using a set union operation.

It should be emphasized that the influence cone is intentionally defined to have a well-defined logical boundary (*i.e.*, a signal net or a standard cell is never split) to simplify the design partitioning process.

Net type	Description	Example (see Fig. 5)
transitive fanin(x,i)	The fanin logic cone of x for i logic stages.	fanin(N_1,1)=$\{N_4\}$
transitive fanout(x,j)	The fanout logic cone of x for j logic stages.	fanout(N_1,1)=$\{N_6\}$ fanout(N_1,2)=$\{N_6, N_7\}$
neighbors(x,d)	The nets within a physical distance d from the net x in the layout.	neighbor(N_1, 260nm)= $\{N_5\}$
side inputs(x)	The nets driving the side inputs of the receiver instances driven by x.	sideinput(N_1)= $\{N_2, N_5\}$

Table 1: Various nets that can be in the influence cone of a net x.

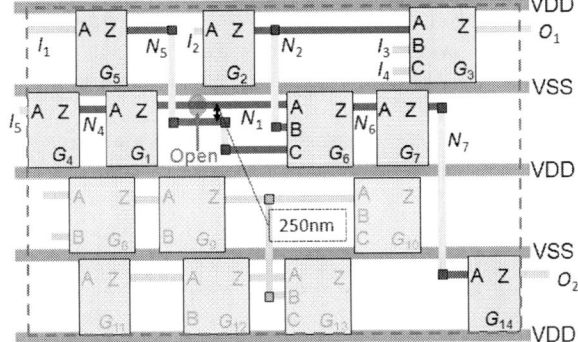

Figure 5: Illustration of the influence cone for an open defect (in red).

In the second step, a new, layout-level sub-design is created using all the layout polygons that belong to the nets in the consolidated cone. The driving cells (*e.g.*, G_1, G_2, G_4–

G_7 in Figure 5) and the receiver cells (*e.g.*, G_1, G_3, G_6, G_7, G_{14} in Figure 5) of these nets are also included in the sub-design. The resulting polygons in the sub-design now defines a bounding box (*e.g.*, red dotted rectangle in Figure 5), which is used to clip the polygons that belong to the global nets, such as VDD and GND. The clipped polygons of the global nets are added to the sub-design. Input and output pins are then added to represent the I/O of the sub-design. Using the same example in Figure 5, the input pins added are $I_1 - I_5$, the output pins added are $O_1 - O_2$, and the global signals included are VDD and GND. Note that the sub-design creation is an inexpensive operation because the sub-design layout is very sparse.

The third step modifies the sub-design layout to inject the defect. Table 2 summarizes the layout modification(s) made to the sub-design for various defect types. Column 1 of Table 2 shows the defect type, while columns 2 – 3 show the layout before and after the modification for each defect type, respectively. The method for generating cell defects is not listed in Table 2 since they can be generated by applying the same methods described in Table 2 at the standard-cell level instead of at the interconnect level.

Defect type	Original layout	Modified layout	Modification to the extracted netlist
open	N_1	N_{1a} N_{1b}	Add a resistor to the nodes (N_{1a}, N_{1b})
resistive via	N_1	N_{1a} N_{1b}	Add a resistor to the nodes (N_{1a}, N_{1b})
signal bridge	N_1 N_2	N_{1a} N_{1b} N_{2a} N_{2b}	Add three resistors to the nodes (N_{1a}, N_{1b}), (N_{2a}, N_{2b}), (N_{1a}, N_{2a})
supply bridge	N_1	N_{1a} N_{1b}	Add two resistors to the nodes (N_{1a}, N_{1b}), (Vdd/Vss, N_{1a})

Table 2: Illustration of layout-level modifications supported and the associated circuit-level modifications.

In the fourth step, a circuit-level netlist is extracted from the "defective" (modified) sub-design layout using the Cadence extraction engine [36]. Since the layout geometry is changed before the extraction, the extraction engine would reflect the change in the form of modified connectivity and parasitic components (*e.g.*, coupling to adjacent lines, additional load, *etc.*) within the netlist. This ensures that the defect is accurately represented.

The final step (step 5) patches the extracted circuit-level netlist. This step adds flexibility for the user to modify the netlist to more accurately represent the defect. For example, in the open defect injection, it might be desirable to change the resistance of the open node in order to model a partial open instead of a complete open. The patching process for each defect type is described in column 4 of Table 2. The patched circuit is directly used in the mixed-signal simulation.

D. Digital Block Creation Flow

During the formation of the analog block, the digital block is formed in the background. In fact, it is the design database that remains after the polygons for the analog block are removed. Since nets and cells are removed from the original design to form the analog block, there is a

978-1-61284-657-6/11 $26.00 © 2011 IEEE   175

Circuit name	No. of fan-in levels	No. of fan-out levels	Neighbor distance (μm)	No. of gates	No. of gates in cone	No. of patterns	Fault coverage (%)	Test Period (ns)	Simulation time (s)	Total time (s)	Simulation time per pattern (s)	Total time per pattern (s)
C432	1	2	0.5	172	33.62	49	98.93	10.00	48.21	53.55	0.98	1.09
C1355	1	2	0.5	582	24.13	101	100.00	10.00	45.15	50.65	0.45	0.50
C3540	1	2	0.5	1743	37.15	139	100.00	10.00	90.43	96.29	0.65	0.69
C7552	1	2	0.5	3813	27.92	102	99.97	10.00	65.48	72.11	0.64	0.71
S13207	1	2	0.5	8483	22.69	270	99.98	10.00	100.97	109.07	0.37	0.40
S15850	1	2	0.5	9940	24.60	135	100.00	10.00	118.96	129.74	0.88	0.96
S35932	1	2	0.5	16353	42.16	23	100.00	10.00	113.62	128.09	4.94	5.57
S38417	1	2	0.5	22609	34.42	108	100.00	10.00	123.58	135.93	1.14	1.26
S38584	1	2	0.5	20193	48.09	151	99.97	10.00	162.69	175.15	1.08	1.16
B14	1	1	0.4	10033	55.87	746	99.98	10.00	362.74	372.16	0.49	0.50
B18	1	1	0.4	116840	63.94	1212	99.90	18.00	1750.39	1803.42	1.44	1.49

Table 3: Summary of the runtime experiment results to demonstrate the scalability of the SLIDER framework.

connectivity change in the database. Therefore, new input and output ports are needed to reflect this change. Using the same example in Figure 5, the output ports added include $I_1 - I_5$, and the input ports added are $O_1 - O_2$.

After both the digital block and the analog block are created, the DUT includes instantiations of these two blocks and connects them together accordingly using analog-to-digital and digital-to-analog converters as shown in Figure 3.

III. EXPERIMENT

It has already been demonstrated in [17] that representing defects at the layout-level is accurate and can create a realistic failure population that exhibits characteristics that include location-dependent responses (*i.e.*, injecting the same type of defect at different locations along the same wire segment results in different responses). We therefore do not repeat this analysis here. Instead, experiments are performed to demonstrate scalability. Twelve benchmark circuits of much larger sizes than those used in [17] are placed-and-routed using Cadence Encounter [37]. For each circuit, 100 random defects taken from Table 2 are injected and simulated. The results are averaged and summarized in Table 3. Column 1 shows the circuit name, while columns 2 - 4 give the transitive fan-in and fan-out levels, and physical distance parameter *d*, respectively. The total number of gates for each circuit is given in column 5, while column 6 gives the number of gates in the influence cone which is an approximate size measure of the analog block. The number of test patterns, the corresponding fault coverage, and the test-clock period (in ns) are provided in columns 7 – 9, respectively. These test patterns are generated using a commercial ATPG tool. The simulation time and the total runtime are summarized in columns 10 and 11, respectively. Comparing columns 10 and 11, it is clear that the circuit partitioning adds negligible overhead to the total runtime, which is dominated by the simulation time. It should be emphasized that, more often than not, only a subset of test patterns needs to be simulated. For example, it is likely only necessary to simulate test patterns that sensitize the affected net which is typically a very small percentage of the total number of test patterns. Therefore, the simulation time and the total runtime are normalized by the pattern count and are shown in columns 12 and 13, respectively. Note that the simulation time is not further normalized by the test-clock period since the circuit has to be fully simulated for the entire test-clock period to obtain meaningful results.

To demonstrate the utility of the SLIDER framework, a second experiment uses SLIDER to improve diagnosis. In this experiment, virtual failures from the first experiment are diagnosed using a commercial diagnostic tool. Particular attention is paid to ambiguous diagnosis results. SLIDER is then used to resolve this ambiguity. Specifically, defects are injected into the suspect failure sites and simulated to identify a defect that best matches the observed response. It should be emphasized that this is a blind experiment in that the identified defect is checked against the injected defect at the *end* of the experiment. In this experiment, two examples are given to illustrate: 1) diagnostic resolution improvement, and 2) defect localization.

Candidate name	Fault matching score	Fault type	Response match?
N7671	100%	sa01	No
FIXED_FANIN_N7671_0	100%	sa01	Yes

Table 4: Characteristics of diagnosis candidates for example one.

The first example uses a diagnosis result for the ISCAS-85 benchmark circuit C7552. The commercial tool reports two fault candidates whose characteristics are summarized in Table 4. Both candidates have the type "sa01" (*i.e.*, the candidate exhibits stuck-at-0 for some failing test patterns and stuck-at-1 behavior for other failing test patterns). Both candidates match all the failing patterns and therefore they have a 100% matching score, which is calculated using the formula [38-39]:

$$\text{Fault Matching Score} = \frac{tfsf}{tfsf + tfsp + tpsf}$$

where *tfsf* is the number of times when both the tester and the simulator produce failing responses, and *tfsp* (*tpsf*) is the number of times when the tester failed (passed) but the simulator passed (failed).

Since a sa01 tester response is a strong indication of an open, an open defect is injected at various locations along the nets that correspond to the two diagnosis candidates. However, only defects injected into the second candidate produce a mixed-signal simulation response that exactly matches all the passing and failing test patterns. Therefore, the candidate N7671 is eliminated, leaving only the second candidate, which, upon verification, is the actual site used to create the original (virtual) test response. This clearly demonstrates that diagnostic resolution can be improved using the SLIDER framework.

The second example uses a diagnosis result for the ISCAS-85 benchmark circuit C3540. In this example, defect localization for the diagnosis candidate is performed. For this run, the commercial diagnosis tool returns a single candidate (N317) of type sa01 with a 100% matching score. This net has four fan-outs (G1 – G4), as shown in Figure 6. (This figure is drawn based on the topology of the actual layout but is not drawn to scale.) Again, because of the sa01 tester response, open defects are injected at various locations (red circles with labels) along the layout-level net that corresponds to the candidate. A simulation response match occurs only when the defect is injected at location 1 shown in Figure 6. In other words, the defect has been localized to the region defined by the red dotted line (an improvement of 53.33%), which again has been verified to be correct by examining the analog block used to create the test response used for diagnosis.

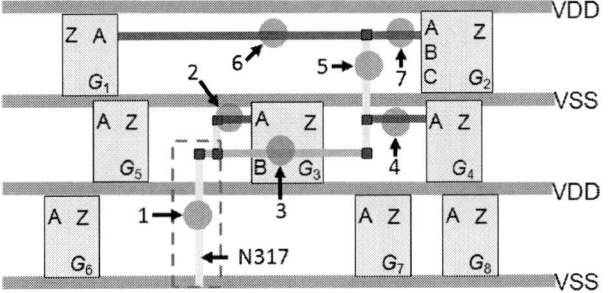

Figure 6: Injection of an open defect (red circle) in various locations in a diagnosis candidate (N317) for better defect localization.

IV. SUMMARY

This paper describes the SLIDER framework for performing accurate and scalable defect simulation. SLIDER makes use of mature, commercial mixed-signal technology. It is therefore capable of handling large designs and poised to benefit from any future improvements in mixed-signal simulation environments. The framework enables/supports many testing tasks that include diagnosis resolution improvement, defect localization, test metric/fault model evaluation, and virtual failure data creation. Experiments have demonstrated the utility of the framework to improve diagnosis. Specifically, diagnosis resolution improvement and defect localization are achieved through more accurate defect simulation with SLIDER.

REFERENCES

[1] B. Kruseman, *et al.*, "Systematic Defects in Deep Sub-Micron Technologies," *International Test Conference* pp. 290-299, 2004.

[2] J. E. Nelson, W. Maly, and R. D. Blanton, "Diagnosis-Enhanced Extraction of Defect Density and Size Distributions from Digital Logic ICs," *SRC TECHCON*, 2007.

[3] T. Huaxing, *et al.*, "Analyzing Volume Diagnosis Results with Statistical Learning for Yield Improvement," *European Test Symposium*, pp. 145-150, 2007.

[4] K. Martin, *et al.*, "A Rapid Yield Learning Flow Based on Production Integrated Layout-Aware Diagnosis," *International Test Conference*, pp. 1-10, 2006.

[5] X. Yu, *et al.*, "Controlling DPPM through Volume Diagnosis," *VLSI Test Symposium*, pp. 134-139, 2009.

[6] M. Sharma, *et al.*, "Efficiently Performing Yield Enhancements by Identifying Dominant Physical Root Cause from Test Fail Data," *International Test Conference*, pp. 1-9, 2008.

[7] J. Jahangiri and D. Abercrombie, "Value-Added Defect Testing Techniques," *IEEE Design & Test of Computers*, vol. 22, pp. 224-231, 2005.

[8] R. Turakhia, *et al.*, "Bridging DFM Analysis and Volume Diagnostics for Yield Learning - A Case Study," *VLSI Test Symposium*, pp. 167-172, 2009.

[9] R. Desineni, O. Poku, and R. D. Blanton, "A Logic Diagnosis Methodology for Improved Localization and Extraction of Accurate Defect Behavior," *International Test Conference*, pp. 1-10, 2006.

[10] S. D. Millman, E. J. McCluskey, and J. M. Acken, "Diagnosing CMOS Bridging Faults with Stuck-at Fault Dictionaries," *International Test Conference*, pp. 860-870, 1990.

[11] S. Venkataraman and S. B. Drummonds, "POIROT: a Logic Fault Diagnosis Tool and its Applications," *International Test Conference*, pp. 253-262, 2000.

[12] T. Bartenstein, *et al.*, "Diagnosing Combinational Logic Designs Using the Single Location At-a-Time (SLAT) Paradigm," *International Test Conference*, pp. 287-296, 2001.

[13] D. B. Lavo, I. Hartanto, and T. Larrabee, "Multiplets, Models, and the Search for Meaning: Improving Per-Test Fault Diagnosis," *International Test Conference*, pp. 250-259, 2002.

[14] D. B. Lavo, T. Larrabee, and B. Chess, "Beyond the Byzantine Generals: Unexpected Behavior and Bridging Fault Diagnosis," *International Test Conference*, pp. 611-619, 1996.

[15] Y. Sato, *et al.*, "A Persistent Diagnostic Technique for Unstable Defects," *International Test Conference*, pp. 242-249, 2002.

[16] R. Rodriguez-Montanes, *et al.*, "Diagnosis of Full Open Defects in Interconnecting Lines," *VLSI Test Symposium*, pp. 158-166, 2007.

[17] W. C. Tam, O. Poku, and R. D. Blanton, "Automated Failure Population Creation for Validating Integrated Circuit Diagnosis Methods," *Design Automation Conference*, pp. 708-713, 2009.

[18] "The Verilog-AMS Language Reference Manual," 2.3 ed. Napa, CA.: Accellera Organization, Inc., 2008.

[19] "The Virtuoso AMS Designer Environment User Guide," ed. San Jose, CA: Cadence Design Systems Inc., 2009.

[20] A. Jee and F. J. Ferguson, "Carafe: an Inductive Fault Analysis Tool for CMOS VLSI Circuits," *VLSI Test Symposium*, pp. 92-98, 1993.

[21] J. Khare and W. Maly, "Rapid Failure Analysis Using Contamination-Defect-Fault (CDF) Simulation," *IEEE Transactions on Semiconductor Manufacturing*, vol. 9, pp. 518-526, 1996.

[22] J. Khare and W. Maly, "Inductive Contamination Analysis (ICA) with SRAM Application," *International Test Conference*, pp. 552-560, 1995.

[23] J. Khare, W. Maly, and N. Tiday, "Fault Characterization of Standard Cell Libraries Using Inductive Contamination Analysis (ICA)," *VLSI Test Symposium*, pp. 405-413, 1996.

[24] T. Vogels, *et al.*, "Benchmarking Diagnosis Algorithms with a Diverse Set of IC Deformations," *International Test Conference*, pp. 508-517, 2004.

[25] P. Banerjee and J. A. Abraham, "A Multivalued Algebra For Modeling Physical Failures in MOS VLSI Circuits," *IEEE Transactions on Computer-Aided Design of Integrated Circuits and Systems*, vol. 4, pp. 312-321, 1985.

[26] W. Meyer and R. Camposano, "Fast Hierarchical Multi-Level Fault Simulation of Sequential Circuits with Switch-Level Accuracy," *Design Automation Conference*, pp. 515-519, 1993.

[27] M. B. Santos and J. P. Teixeira, "Defect-Oriented Mixed-Level Fault Simulation of Digital Systems-on-a-Chip Using HDL," *Design, Automation and Test in Europe*, pp. 549-553, 1999.

[28] R. D. Blanton, K. N. Dwarakanath, and R. Desineni, "Defect Modeling Using Fault Tuples," *IEEE Transactions on Computer-Aided Design of Integrated Circuits and Systems*, vol. 25, pp. 2450-2464, 2006.

[29] Y.-T. Lin, *et al.*, "Physically-Aware N-Detect Test Pattern Selection," *Design, Automation and Test in Europe*, pp. 634-639, 2008.

[30] J. E. Nelson, W. C. Tam, and R. D. Blanton, "Automatic Classification of Bridge Defects," *International Test Conference*, p. 10.3, 2010.

[31] L. Rokach and O. Maimon, Data Mining With Decision Trees, 1 ed.: World Scientific, 2008.

[32] F. Brglez and H. Fujiwara, "A Neutral Netlist of 10 Combinational Benchmark Circuits and a Target Translator in Fortran," *International Symposium on Circuits and Systems*, pp. 1929-1934, 1985.

[33] F. Brglez, D. Bryan, and K. Kozminski, "Combinational Profiles of Sequential Benchmark Circuits," *International Symposium on Circuits and Systems*, pp. 1929-1934, 1989.

[34] R. D. Blanton, *et al.*, "Fault Tuples in Diagnosis of Deep-Submicron Circuits," *International Test Conference*, pp. 233-241, 2002.

[35] K. Kundert and O. Zinke, The Designer's Guide to Verilog-AMS, 1 ed.: Springer, 2004.

[36] "The Diva Reference Manual," ed. San Jose, CA: Cadence Design Systems Inc.

[37] "The Cadence Encounter Reference Manual," ed. San Jose, CA: Cadence Design Systems Inc.

[38] "The Tetramax User Guide," ed. Mountain View, CA.: Synopsys Inc., 2009.

[39] "The Cadence Encounter Test User Guide," ed. San Jose, CA: Cadence Design Systems Inc.

An Industrial Case Study of Analog Fault Modeling

Ender Yilmaz*, Anne Meixner**, Sule Ozev*
Arizona State University USA*
Intel Corporation. Hillsboro, OR, USA **

Abstract- *Analog fault modeling (AFM) provides a quantitative measure of quality and insight into defective device behavior. However, the high computational burden typically associated with fault simulation makes it unappealing for industrial applications. We propose an efficient methodology to reduce computational burden of the AFM method by exploiting the hierarchical nature of process variation. We apply the proposed methodology on an industrial SerDes TX Driver circuit and achieve 98% simulation time reduction. We quantify defect impact with a defect severity measure.*

1. Introduction

Today's assessment of analog test coverage consists of ad-hoc checklists of analog tests and exercising all the analog circuitry in some manner. We have no knowledge as to the importance of one test over another. As analog designs change to adapt to process constraints and product requirements, evaluating test coverage during product definition and pre-silicon validation adds value to the design and test process.

A very promising method, analog fault modeling (AFM), which enables such coverage assessment, has been proposed nearly two decades ago. Early work on AFM focused on parametric defects and circuit level defects. [1] proposed an approach with process variation; while [2] investigated circuit level defects. Parametric faults are typically simulated with out of tolerance deviations [3, 4], while open and short defects are simulated via injecting respectively a large and small resistance [5, 6, 7, 8]. Early work on defect modeling did not include the masking effect of process variation [9, 10], which is becoming increasingly prominent with more advanced processes. Process variation is incorporated in AFM [11, 12] at a cost of increasing computational complexity. Although AFM provides great insight in defective behavior and test coverage, it has remained an academic research topic due to its extensive computational requirements.

In this paper, we provide a feasible implementation methodology for defect-oriented simulations that substantially reduces the computation requirements. We exploit the hierarchical structure of process variation to split the simulation process in multiple steps and apply a pruning algorithm to eliminate unnecessary steps. In the first step we analyze the impact of die-to-die variations. This first hand analysis gives us defects that have an effect on performance. In the second step, we include within die (WID) variation for the defects that have a possibility of resulting in specification violation. By judiciously using these simulations, the overall computational burden can be greatly reduced.

By demonstrating the AFM approach on a commercial device, we show that fault simulation is both feasible and provides invaluable information for test quality optimization and yield improvement.

The results of this defect analysis can be used for many purposes. In this work, we analyze the coverage of the manufacturing tests based on this defect analysis. We also present a methodology to identify sensitive nodes in the circuit, so that they can be targeted in the layout step for yield improvement.

2. Analog Fault Modeling Approach

We focus on structural defects and use resistive opens and shorts which are commonly used defect models. The inclusion of frequency related defect models as discussed in [13] was deemed unnecessary for this evaluation. Since defect size can vary, we investigate multiple resistance values for each defect location to avoid incorrect conclusions on defect coverage.

2.1. Defect oriented simulation flow

Defect oriented simulation can be separated into four steps: defect list generation, defect injection, simulation, and defect assessment.

1. Defect list is generated examining the schematic or the layout of the circuit. This list includes possible defects generated during manufacturing process due to various defect mechanisms, such as extra metal deposition.
2. Assuming single defect model, defects are injected to the circuit one at a time.
3. Defective circuits are simulated and circuit performance parameters are obtained.
4. Performance parameters are analyzed for specification or test result violations and a coverage table is generated.

For this study we generated the defect list via the schematic method. Generating a defect list via a layout analysis [2] results in the most accurate defect list and offers a ranking of defects based process defect statistics. However as it is an established technique we wanted to focus on the simulation challenges of including process variation. In addition for test coverage analysis waiting for a layout does not enable improvement of test coverage with early feedback. The results from performing AFM at the schematic level can be used to improve defect robustness.

2.2. Process variation model

Although circuits are designed to tolerate process variation, when combined with defects, process variation can result in unpredictable consequences. Therefore, it is necessary to incorporate the effect of process variation.

Process variation consists of several layers. We group these layers into high-level and low-level variation such that we separate correlated and uncorrelated variation. We define high level variation as the overall effect of lot-to-lot, wafer-to-wafer, and die-to-die variation. Low-level variation is defined as the effect of WID (aka mismatch) variation. High-level variation is typically higher and varies parameters (e.g. transistor Vt) of the same nature at the same rate; hence, it is common mode variation. Low-level variation can be considered as independent variation.

Splitting the variation enables us to conduct analysis in two consecutive steps. First, the circuit is subjected to high-level variation through sampling device parameters from the statistical distribution model of high level variation. In this work, a skew based statistical model is employed. Conceptual representation of the model is provided in Figure 1. The figure shows fast and slow corners of the silicon for important process level parameters. High level variation is represented with a number of skew points that bound the most likely region of the distribution. In the second step, Monte Carlo simulation on selected transistors is conducted for each skew point to simulate for the effect of WID variation.

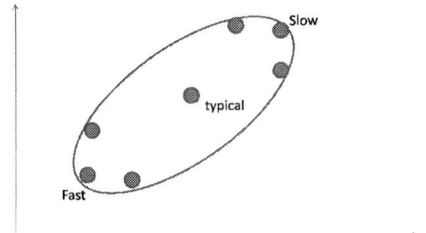

Figure 1 Skews Model for Process Variation

Considering that there may be many possible defects and simulating for each defective circuit for process variation requires sampling a large number of device instances, it is not surprising that defect oriented approach is computationally expensive. In this work, we show how we can mitigate the computation burden and reduce simulation time substantially by exploiting several observations. These will be described in the next subsection.

2.3. Computationally efficient defect simulation strategy

Simulation process of our approach consists of three consecutive steps;
1. Simulating the typical behavior of defective circuits
2. Simulating high level variation for each defective circuit
3. Simulating low level variation for each defective circuit

The third step is computationally the most expensive. Therefore, skipping this step reduces the simulation time appreciably. To this end, we make use of four observations.

• *Impact of high level variation differs for fault-free and faulty devices.* Process variation for defective devices cannot be predicted from the fault free simulation results. We illustrate this observation through a sample defect. Figure 2 shows the response of a short defect we have analyzed. X-axis shows the resistance values of the defect models ranging from 1 to 5k ohms. Y-axis shows the response in terms of one sample performance parameters. Red dashed lines indicate the range of fault free response in presence of process variation. Vertical blue bars show the range of defective response in presence of process variation for six different defect models. Analysis of several defect responses reveals that response variation depends on the defect model and it cannot be estimated from fault free response. Therefore, it is not possible to skip the second step of simulations.

• *Hierarchical process model reduces computational burden.* Process variation is typically emulated by sampling from a parametric distribution and this approach requires a large number of device instances to be sampled. Since skew parameters are much smaller in number, an exhaustive evaluation of the high level process space is possible by separating the WID variation from high level variation. The total number of simulated skews is 7.

• *The effect of WID variation is similar for faulty and fault free circuits.* Although high-level variation differs for fault-free circuits and faulty circuits, WID variation results in similar tolerance range for both fault free and faulty circuits (see Figure 3).

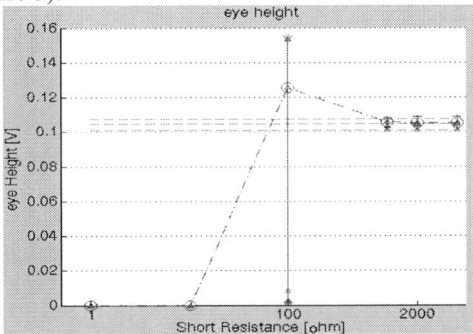

Figure 2: Performance parameters in presence of a short between the output node and ground

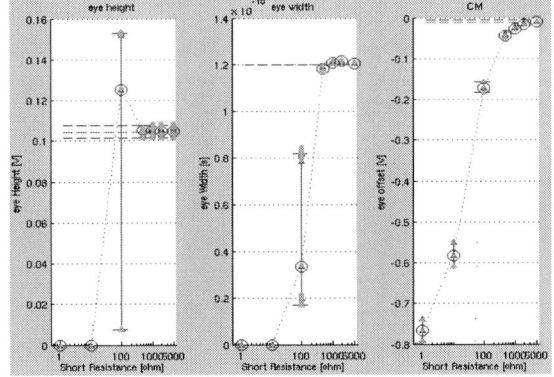

Figure 3 WID impact: Short between output node and ground

Therefore, within die variation level obtained for fault-free circuits can often be used as a guide to estimate WID variation for defective circuits. In cases where WID variation can be estimated with high confidence, mismatch simulation can be avoided. Therefore, this observation enables us to reduce to number of 3rd step simulations.

• *Analysis of the circuit architecture enables further reduction in simulation time.* Defect extraction effort and simulation time can be significantly reduced by isolating sub-blocks from the circuit and simulating them.

3. Case Study: PCI-Express TX Driver

We demonstrate our feasible AFM method on a PCI-Express TX analog front-end circuit in a 65nm process.

3.1 Circuit Background: PCI-Express TX Driver circuit

We chose to analyze the analog driver circuit, which is the outmost circuit interfacing the channel. The driver circuit is a

978-1-61284-657-6/11 $26.00 © 2011 IEEE

5 bit source series terminated digital to analog converter [14]. High level diagram of the driver is illustrated in Figure 4. Each incoming bit is connected to a set of identical cell circuits. Each bit is connected to 2^n cells, where n is the significance number of the bits. The driver generates signal levels proportional to the number of the active cells. Note this is design for a 65nm process.

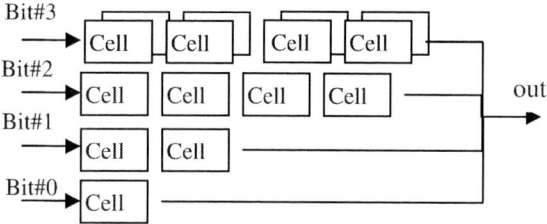

Figure 4: High-level Representation of the Tx Driver

3.1.1. Simulation Setup

High level representation of the simulation setup is shown in Figure 5. We included the test environment model to obtain realistic responses. Pseudo random input is applied at the input, coded according to the PCI express standard. We evaluated the TX Driver performance with respect to three commonly used test parameters for high speed links: eye height, eye width, and eye offset [15]. Eye height and width measure the vertical and horizontal opening of the eye diagram, while eye offset measures vertical shift of the eye. Eye height is typically determined via voltage margining [16] method by introducing a common mode shift until the signal becomes undetectable. Similarly, eye width is measured through time margining; introducing a shift in time domain. Eye offset can be determined by computing the average integral of the differential input.

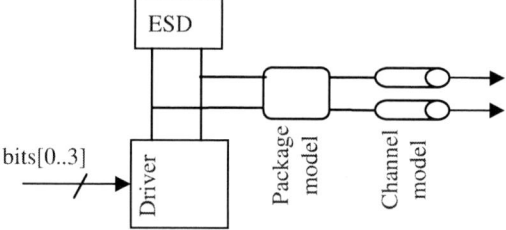

Figure 5: Simulation Setup

3.2. Defect List Generation and Defect Equality

Defect list is generated by analyzing the circuit to find possible defect locations. We exploit the modularity of the driver to reduce the defect list extraction effort. Since all cells are identical, examining only one cell suffices to generate the complete defect list. The generated defect list can be duplicated for the other cell circuits in the driver. Also, the cell circuit has a symmetric structure enabling further effort saving through examining only one half of the cell circuit. However, identical defects in various building blocks do not necessarily result in identical behavior.

Even though all cells are identical, their response may differ due to differing input bit patterns. We group the cells with identical inputs. Simulation for each group is necessary to obtain an accurate overall response. For this design, cells are

organized into 4 groups depending on input bits connected. A further reduction can be obtained through symmetric architecture of the cells. The overall simulation savings using architectural information are discussed in the results section.

A defect list of 17 shorts and 15 opens is generated through examining the cell schematic using practical heuristics for opens and shorts as follows:

- All nodes can be shorted to either VCC or Vss
- For all Transistors Gate to Drain, Gate to Source and Drain to source shorts included
- Opens introduced at all likely junctions

3.3 Defect Injection

Defect-oriented simulation requires knowledge on the circuit level model of the defects. Shorts and Opens are both modeled with resistors. Traditionally, short defects are assigned 1 ohm and open defects are assigned a large 1M ohms. However, depending on the defect mechanism and the defect size, defect models can assume a wide range of resistance values. In this work, a realistic range of defect resistances for the 65nm process was chosen for simulation as follows:

- Shorts: 1, 100, 1K, 2k, 5K Ohms
- Opens: 1k, 2K, 5K, 10K, 100K, Infinity ohms.

3.4. Simulation Flow

Figure 6 shows the simulation flow. We start by simulating the fault-free circuit which yields the fault-free response in the presence of high-level and low-level process variation. These results serve as a reference to assess defective responses.

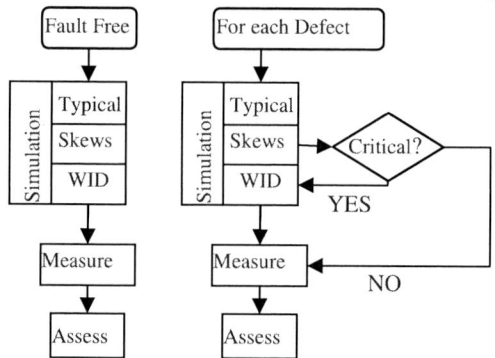

Figure 6: Simulation Flow

Defective circuits go through a similar flow; however, we apply a pruning method to reduce simulation time. First two steps of the simulation are necessary for all defects. The third step may be dropped in cases where WID variation of a defective device is guaranteed not to change the pass/fail criteria. The proposed pruning algorithm decides whether a defective device needs to be simulated for WID variation.

3.5 Pruning Algorithm

Once we obtain the typical response and high-level variation response of defective circuits through first two simulation steps, we can estimate worst case response of WID variation utilizing WID variation results obtained for the fault-free circuit. Worst case variation limits are defined using the equations listed below.

$$pp_{i,min} = \min(pp_{i,skew\{j\}}) - 6std(pp_{i,FF,WID}) \quad (1)$$
$$pp_{i,max} = \max(pp_{i,skew\{j\}}) + 6std(pp_{i,FF,WID}) \quad (2)$$

978-1-61284-657-6/11 $26.00 © 2011 IEEE

$$spec_{i,min} < pp_{i,min} < pp_{i,max} < spec_{i,max} \quad (3)$$

$pp_{i,skew\{j\}}$ is the ith performance parameter for j^{th} skew, where j is the skew index. $Std(pp_{FF,WID})$ is the standard deviation of the response of ith performance parameter for WID variation. Since WID variation does not change considerably for fault-free and defective circuits (observation #3), we use fault-free WID variation amount as a guide to estimate the worst case scenario for defective responses. To increase the confidence of the approach, we chose 6σ window. Equations 1 and 2 yield a minimum and a maximum worst case point for the performance parameter of defective circuit.

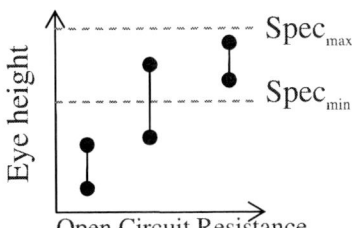

Figure 7: Defect Pruning critical Defect

We define a defect as critical if it does not satisfy Equation 3 for any of the defect models. Critical defect concept can be explained via Figure 7 which illustrates the response obtained for a critical defect for 3 defect models. Red dashed lines show the acceptable region confined by minimum and maximum specification limits. Vertical lines show the worst case response range, where solid circles at the bottom and the top of the lines are $pp_{i,min}$ and $pp_{i,max}$ points respectively. According to our pruning strategy, if any of the behaviors fall outside of the acceptable region, a defect is considered as critical. In Figure 7, defective response violates the criteria given in Equation 3 for the first two defect resistance values; hence it is identified as critical.

This algorithm identifies the defects that are susceptible to WID variation. For example, the response range for the second defect model in Figure 7 intersects with one of the specification limits. Hence, there may be a set of device instances for this particular defect, whose performance parameters fall very close to the boundary. These instances can easily be pushed to the other side to the limit by WID variation. Pruning algorithm guarantees to capture those cases in the specified defect resistance range.

3.6 Fault Coverage Assessment

Fault coverage assessment and pruning algorithm requires an acceptable region definition of the performance parameters. These specifications are given by the designer. For the circuit in this study, the specifications are as follows:

- Eye Height (min): 80mV
- Eye Width (min): 118ps
- Offset: +/- 200mV

Depending on the fault location and the defect size, each fault may not result in a specification violation. Our goal is to compute the probability of specification of violation for each fault and the probability of detecting this violation by one of the three tests defined above. We use these two measures to calculate fault coverage. Fault coverage is defined below as:

$$FP_{Fi} = Pr\left(\frac{device\ with\ fault\ F_i}{fails\ at\ least\ one\ spec}\right)$$

$$DP_{F_i}^{T_j} = Pr\left(\frac{device\ with\ fault}{F_i\ fails\ test\ T_j}\right)$$

$$FaultCoverage_{F_i}^{T_j} = 100\ Pr\left(\begin{array}{c} device\ with\ fault\ F_i \\ fails\ test\ T_j\ and\ at \\ least\ one\ specification \end{array}\right)$$

$$FaultCoverage_{F_i}^{T_j} = 100\ \frac{DP_{F_i}^{T_j}}{FP_{Fi}} \quad (4)$$

The above equation is valid if the test (T_j) set consists of direct specification based tests.

4. Analysis of Design Robustness with Respect to Defects

The impact of defects on performance may be substantially different; some defects degrade the performance severely and deserve more attention. *A priori* knowledge of such sensitive defect locations can be used by designers to build defect-robust circuits. For instance, if the drain of a transistor is sensitive to an open defect, placing multiple connectors (vias) at the drain will make the design more robust against this defect.

In order to assess the impact level of a defect, we define the defect severity (DS) metric as described by Equation 5.

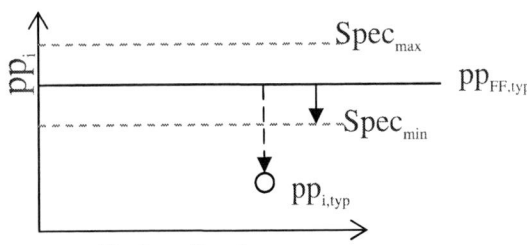

Figure 8: Normalized defect severity indicates the defect impact

$$DS(R) = \begin{cases} \frac{pp_{i,typ}(R) - pp_{FF,typ}}{spec_{max} - pp_{FF,typ}}, & if\ pp_{i,typ} > pp_{FF,typ} \\ \frac{pp_{i,typ}(R) - pp_{FF,typ}}{pp_{FF,typ} - spec_{min}}, & else \end{cases} \quad (5)$$

Where, $pp_{i,typ}$ is the ith performance parameter for of the typical skew and $pp_{FF,typ}$ is performance parameter of the fault-free response for typical skew. This formulation defines the deviation of the response from the nominal in terms of allowed deviation. Defect severity metric is illustrated in Figure 8. The vertical dashed line shows the deviation of a performance parameter for a defect with a particular resistance model. DS metric is simply the ratio of the length of the vertical dashed line to the length of the vertical solid line for this particular example. DS is a function of defect resistance value.

We define another metric, DSS (Defect Severity Score), to estimate the severity conditioned on defect resistance distribution in Equation 5. This metric incorporates the statistical distribution of defect resistance value.

$$DSS = \int_0^\infty DS(R)pdf(R)dR \quad (6)$$

978-1-61284-657-6/11 $26.00 © 2011 IEEE   181

where Pdf(R) is the probability distribution function of defect resistance distribution. DSS indicates the severity given a particular defect takes place in the device. However, the probability of that particular defect may be low, which would obviate the attempt to alter the design to make circuit robust for that particular defect. Or the probability may be high which would make it important to alter the design for improved robust operation. Expected value of DSS can be estimated by incorporating probabilities of the defect occurrences through weighing DSSs with defect occurrence probabilities. Expected defect severity score (EDSS) metric is defined in Equation 6.

$$EDSS_k = \Pr\,(Defect_k)DSS_k \qquad (7)$$

where, k is the defect index and Pr(Defect$_k$) is the probability of occurrence of kth defect. Ordering the defects with respect to EDSS enables us to assess the realistic impact of the defects and devote resources such as redesign effort and die area more efficiently.

5. Results

The proposed simulation flow was applied to the PCI express driver circuit. Simulations were run on 3GHz Quad core machines using multiple threads. We report simulation time in terms of equivalent CPU time of a single core machine. Table 1 lists the duration of each simulation step. In the first two steps, all defects are simulated, while in last step only the 18 defects that are identified as critical in the pruning step are simulated. The most significant contributor is the WID variation simulation.

Table 1: Simulation Time

	Time [Hr]
Sim. Step #1, Typical	12
Sim. Step #2, Process skew	96
Sim. Step #3, WID	1,296

Table 2: Simulation time savings

	Final Simulated	No pruning	No Defect equality
Time [Hr]	1,404	9,468	71,010
Saving		85.2%	98%

Results in Table 2 show that simulation time without pruning is 9468 CPU hours, while it is 1404 CPU hours with the proposed pruning algorithm, saving 85.2% of the simulation time.

Defect equality approach enables simulation time reduction as well. 4 out of 15 cells are simulated due to the identical structure of the cells, and only one half of the defects per cell are injected thanks to symmetry. Hence, the overall saving of the pruning and defect equality approach is 98%. Without defect equality, the required simulation time is 71010 (extrapolated) hours.

5.1. Fail Probability and Per-Test Fault Coverage

32 defects for 4 different cells and 2 defects that are common to all cells are simulated for 6 defect models. Out of 130 defect types, only 18 of them proved to be critical. This suggests that not all defects result in a specification violation. Defect coverage table is shown in Table 3. The leftmost column shows in which cell the defects are located. The second column indicates the type of the defects, the 3rd column shows fail probability, and the rest of the columns show the coverage of faults for individual performance parameters.

Table 3: Fail Probability and Fault Coverage Table

Bit#	type	FP_{Fi}	Fault Coverage		
			Eye height [%]	Eye width [%]	Eye offset [%]
all	short	0.50	80.96	100.00	66.66
all	short	0.50	42.86	100.00	19.04
2	short	0.33	100.00	0.00	0.00
2	short	0.50	100.00	0.00	0.00
2	short	0.33	100.00	21.42	0.00
2	short	0.50	100.00	0.00	0.00
2	short	0.50	100.00	33.34	0.00
2	short	0.36	100.00	0.00	0.00
2	short	0.67	100.00	0.00	0.00
2	short	0.50	100.00	33.34	0.00
2	open	0.67	100.00	21.43	0.00
2	open	0.74	100.00	12.90	0.00
2	open	0.26	100.00	0.00	0.00
2	open	0.67	100.00	21.43	0.00
3	short	0.14	0.00	100.00	0.00
3	short	0.33	0.00	100.00	0.00
4	short	0.14	0.00	100.00	0.00
4	short	0.17	0.00	100.00	0.00

Results show that faults are fully covered by eye height and eye width test and eye offset test does not improve the coverage. This information enables us to optimize test process by dropping redundant tests and ordering tests to reduce the expected test time. For instance, we can drop eye offset test, and scheduling eye height test before eye width test will reduce the expected test time for fails. This assumes defects are equally likely. For this case study the analysis is straightforward to conclude by manual observation for this case study, algorithms can be developed for test optimization when the number of specifications is much higher.

5.2. Analysis of Design Robustness

Based on the fault simulation results, only 18 out of 130 defects result in failure. Therefore, schematic based analysis showed that the circuit can be considered robust with respect to most of the defects. In order to further improve robustness, we only need to focus on these 18 sensitive defects.

Robust design techniques can be applied at the layout level by locally laying devices out and routing wires to reduce the probability of a failure at a potential cost of area. Defect severity measure enables one to optimize the cost of this effort by prioritizing via defect impact.

Equations 4-6 can be used to obtain the severity of the defects to improve the design for defect robust operation. Defect resistance distribution, pdf(R), is considered uniform for 65nm process based upon previous internal analysis. In this work we assumed equal occurrence probability for all defects (due to schematic based defect generation). These two assumptions reduce equations 6-7 to equation 5. Hence, we used only equation 5 to evaluate the severity of the defects.

Figure 9 shows defect severity plot of the 18 critical defects. X-axis shows the number each defect and y-axis shows severity in terms of DS parameter. The first 14 of the bars are for short defects and the last 4 bars are for open defects. The contribution of each defect resistance model is represented with a different color. Height of the bars indicates which defect is more important, providing useful information to improve defect robustness.

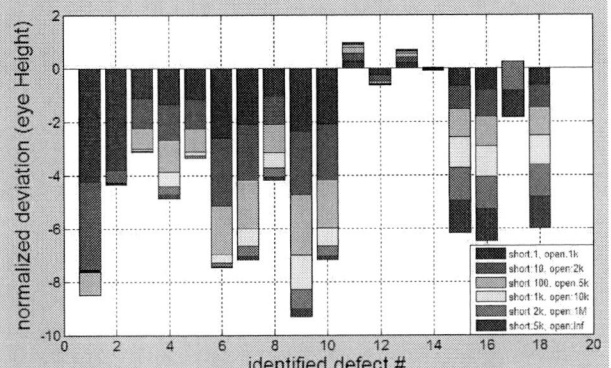

Figure 9: Defect Severity

According to these results, the most influential defect is the 9th, therefore it should be addressed first. The 1st and the 6th defects are next in the priority list. Defect priority list enables the layout engineer to devote the limited chip area to important defects more efficiently

6. Conclusions and Future Work

We presented an efficient implementation methodology for AFM technique on an in industrial SerDes driver circuit using an industry process and its associated process variation. Results show 98% simulation time reduction compared to the traditional AFM implementation. Given the trend in VLSI design to rely on IP blocks that would be used for multiple products, the simulation investment in assessing an IP's analog fault coverage is worthwhile. The AFM assessment permitted a realistic assessment of the manufacturing tests. The resulting 18 defective circuits had any significant impact and that eye height and width together provide complete fault coverage. In addition, we suggested improving design robustness by defining a defect severity measure and assessing it on all 18 defects. This measurement enabled us to rank their impact and hence, to improve the circuit layout.

There exist several directions for future work. The most immediate is to follow our case study with actual silicon results. A simple experiment would be to assess the overlap between eye height and width measurements. Next would be to apply this methodology to more complex analog circuits e.g. a clock data recovery circuit. Out of the four basic observations that we utilize to make AFM efficient, the first two are general and apply to all circuits. Utilization of last two observations which employ circuit specific information (symmetry and WID variation) while not necessary, will significantly help to boost the efficiency as demonstrated in this paper. As analog circuits often have multiple identical elements and differential circuits are required in SerDes like devices it is not unrealistic to presume these can be used to advantage. Finally the focus of this work has been on analog

fault coverage so another direction to apply this work is to the much harder problem of analog yield prediction.

7. Acknowledgements

The authors would like to acknowledge Paul Newman for his expert advice on process variation simulation strategy; Steve Cove and Mark Seidel for design knowledge of the TX circuit design and Hyung Soo Kim for generating the external loopback interconnect model.

8. References

[1] L. Milor and V. Visvanathan. Detection of catastrophic faults in analog integrated circuits. IEEE Transactions on Computer-Aided-Design, 8(2):114–130, Feb 1989.

[2] Meixner, A.; Maly, W. Fault Modeling for the Testing of Mixed Integrated Circuits. Test Conference, 1991, Proceedings., International , vol., no., pp.564, 26-30 Oct 1991

[3] A.J. Bishop and A. Ivanov. Fault simulation and testing of an OTA biquadratic filter. In IEEE International Symposium on Circuits and Systems, May 1995

[4] M. Soma. An experimental approach to analog fault models. In IEEE Custom Integrated Circuits Conference, May 1991

[5] F. Azais, Y. Bertrand, M. Renovell, A. Ivanov, and S. Tabatabaei. An all-digital DFT scheme for testing catastrophic faults in PLLs. IEEE Design and Test of Computers, pages 60–67, January-February 2003

[6] C. Sebeke, J.P. Teixeira, and M. J. Ohletz. Automatic fault extraction and simulation of layout realistic faults for integrated analogue circuits. In IEEE European Design and Test Conference, March 1995.

[7] Soon-Jyh Chang and Chung Len Lee and Jwu E. Chen," Structural Fault Based Specification Reduction for Testing Analog Circuits", Kluwer J. Electron. Test., vol. 18 no.6, pp.571-581, 2002

[8] H. Stratigopoulos, S. Mir, E. Acar, and S. Ozev, "Defect Filter for Alternate RF Test", IEEE European Test Symposium, pp.101-106, 2009

[9] R. Voorakaranam, S. Chakrabarti, J. Hou, A. Gomes, S. Cherubal, A. Chatterjee, and W. Kao. Hierarchical specification-driven analog fault modeling for efficient fault simulation and diagnosis. In Proceedings of the International Test Conference, pages 903–12, 1997.

[10] N. Nagi and J. A. Abraham. Hierarchical fault modeling for analog and mixed-signal circuits. In IEEE VLSI Test Symposium, pages 96–101, May 1992

[11] C. Y. Chao, H. J. lin, and L. Milor. Optimal testing of vlsi analog circuits. IEEE Transactions on Computer Aided Design of Integrated Circuits and Systems, 16(1):58–77, January 1997.

[12] L. Milor and A. L. Sangiovanni-Vincentelli. Minimizing production test time to detect faults in analog circuits. IEEE Transactions on Computer-Aided Design of Integrated Circuits and Systems, 13(6):796–813, June 1994

[13] E. Acar and S. Ozev, .Defect-oriented testing of RF circuits,. IEEE Transactions on Computer-Aided Design of Integrated Circuits and Systems, vol. 27, no. 5, pp. 920.931, May 2008

[14] Menolfi, Christian, et.al. "A 16 Gb/s Source-Series Terminated Transmitter in 65nm CMOS SOI" ISSC 2007 pages 446-447.

[15] Mak, T.M.; Tripp, M.; Meixner, A.;"Testing Gbps interfaces without a gigahertz tester" Design & Test of Computers, IEEE Volume 21, Issue 4, July-Aug. 2004 Page(s):278 – 286

[16] Meixner, A.; Kakizawa, A.; Provost, B.; Bedwani, S., "External Loopback Testing Experiences with High Speed Serial Interfaces," IEEE International Test Conference, 2008.

978-1-61284-657-6/11 $26.00 © 2011 IEEE

A New Methodology for Realistic Open Defect Detection Probability Evaluation under Process Variations

Jesus Moreno, Victor Champac
Dept. of Electronic Engineering
National Institute for Astrophysics, Optics
and Electronics-INAOE, Puebla, Mexico.
jmoreno(champac)inaoep.mx

Michel Renovell
LIRMM-Universite de Montpellier II
161 rue Ada 34392 Montpellier, France
renovell@lirmm.fr

Abstract—**CMOS IC scaling has provided significant improvements in electronic circuit performance. Advances in test methodologies to deal with new failure mechanisms and nanometer issues are required. Interconnect opens are an important defect mechanism that requires detailed knowledge of its physical properties. In nanometer process, variability is predominant and considering only nominal value of parameters is not realistic. In this work, a model for computing a realistic coverage of via open defect that takes into account the process variability is proposed. Correlation between parameters of the affected gates is considered. Furthermore, spatial correlation of the parameters for those gates tied to the defective floating node can also influence the detectability of the defect. The proposed methodology is implemented in a software tool to determine the probability of detection of via opens for some ISCAS benchmark circuits. The proposed detection probability evaluation together with a test methodology to generate favorable logic conditions at the coupling lines can allow a better test quality leading to higher product reliability.**

I. INTRODUCTION

As the use of electronic components increases, the expectation of lower cost, better accuracy and higher reliability increases. Lower cost and better accuracy are achieved by putting more transistors per unit of silicon, using design automation, increasing device operation speed and reducing its power consumption. However, these design steps cannot guarantee reliability. In fact, as the circuit density increases, the probability of a manufacturing defect increases. A higher expectation of reliability can only be met by more thorough and comprehensive testing. Due to the complexity of IC technological process, many physical defects occur during the manufacturing of any system. The typical defects encountered in today technologies and modeled in yield simulators are the so-called spot defects that may cause shorts and/or opens at one or more of the different conductive levels of the devices. Open defects are a major defect mechanism in modern integrated circuits [1] [2] and have been found to be an important contributor to test escapes [3]. In damascene-copper process interconnect opens have become an important factor in product yield loss. An open defect may affect the material

of a connecting line due to metal migration for example but open defects mainly affect contacts and vias [4]. Note that the number of vias is becoming greater than the number of transistors in some large IC's such as microprocessors [5]. In today process, it is consequently crucial to have a very realistic and accurate determination of the coverage of these via open defects. The objective of this work is to propose a method to compute the realistic coverage of via open defects. A full interconnect open separates the original node into two independent nodes: the node staying connected and still electrically controlled by the driving gates and the floating node connected to the driven gates. The electrical behavior of this floating node is determined by several parameters such as the electrical parameters of the disconnected transistors (L, W, Cox,..), the charges trapped on the floating gate of these disconnected transistors and, more important, a number of capacitances coupling the floating node to its environment [6] [7] [8] [9] [10] [11] [12]. In this work we assume that the nominal values of all these parameters are known:

- Locations of via open defects are extracted from the layout,
- Transistor parameters (L, W, Cox) are given by the library description,
- Technological parameters are obtained from the process,
- Capacitance parameters (area) are extracted from the layout.

Classically, the nominal values of these electrical parameters are used to compute with a dedicated floating node model [6] [7] [9] [10] [12], the equivalent floating node voltage which determines the state of the transistors and the logic value at the output of the driven gate. This way, each via open defect can be declared detected or not detected and the global open coverage is determined for a given test sequence. In today nanometer processes, variability is predominant and considering only nominal value of parameters is not realistic. The originality of this work is to propose a model for computing a realistic coverage of via open defect taking into account the process variability. The remainder of this paper is organized as follows.

In Section II, the detection conditions for via opens defects are stated. In Section III, a probability model for via open defects under parameter variability is proposed. In Section IV, the via open defect probability is evaluated. Finally, in Section V, the conclusions of this work are presented.

II. VIA OPEN DEFECT DETECTION CONDITIONS

A. Defect Modeling

In this work full interconnect opens are considered. This means that there is non-significant influence from the input signal over the floating line. The behavior of the defective gate(s) is determined by the induced voltage at the floating line (V_{if}). In Figure 1 the main factors influencing the behavior of the interconnect open are shown. C_0^r (C_1^r) is the part of the routing capacitance with one terminal connected to V_{GND} (V_{DD}). The loading gates influence the floating line with the parasitic overlap capacitances (C_{gdon}, C_{gson}, C_{gdop}, C_{gsop}), intrinsic gate charge (Q_{GT}) [9] of the transistors connected to the floating line and poly-bulk/well capacitances (C_{pb}, C_{pw}) of each transistor connected to the floating line. The voltages at the drain/source terminals of the transistors (V_{sn}, V_{dn}, V_{sp}, V_{dp}) connected to the floating line need to be determined. These voltages depends on the actual test pattern and the gate topology. Furthermore they may depend on the history of the gate [9]. C_{C_1} to C_{C_n} are the the capacitances coupled to the floating line.

Their terminal voltages V_{C_1} to V_{C_n} depend on the actual test pattern. The behavior of a gate with an interconnect open also depends on the trapped gate charge deposited during fabrication whose value is difficult to predict. The trapped gate charge builds up a voltage at the floating gate called "trapped gate voltage" [8] [9].

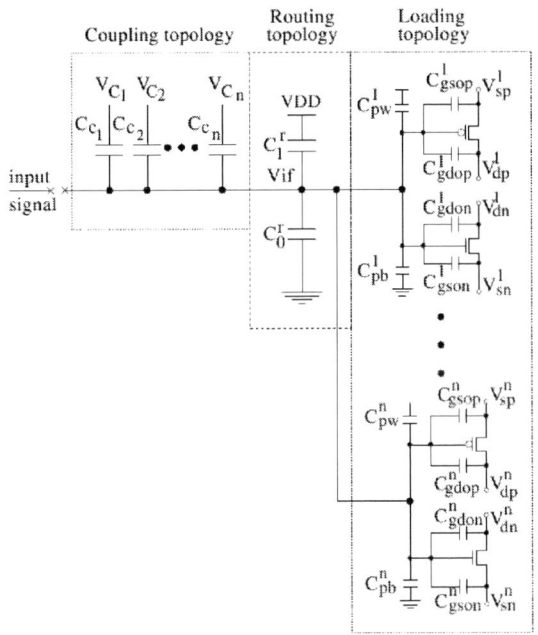

Figure 1. Electrical model of an interconnect open considering several loading gates

B. Detecting Conditions under parameter nominal value

The logic threshold is normally used as a reference to determine faulty/fault-free behavior. However, in advanced technologies parameter variations and noise may invalidate the gate threshold reference. In this work, an interconnect open having a voltage at the floating line lower (greater) than V_{TN} ($V_{DD} - |V_{TP}|$) is considered assured logic testable by a stuck-at 0 (1) vector. The detection of an interconnect open is evaluated for two conditions: a) assured logic testable for a stuck-at 0 vector, and b) assured logic testable for a stuck-at 1 vector.

1) Assured logic testable for a stuck-at 0 vector: The necessary condition(s) to assure an induced voltage at the floating line non-greater than the threshold voltage of the n-channel transistor (V_{TN}) will be obtained. This condition assures than the interconnect opens is detected by a stuk-at 0 vector. Taking into account the law of charge conservation at the floating node, an explicit expression can be obtained to estimate the minimum value of capacitance to ground of the floating line (C_0) to have at most an induced voltage of V_{TN} at the floating line. According to this, the detection condition for an interconnect open is given by,

$$
\begin{aligned}
C_0 \; & \geq \; -C_{FD} + \frac{V_{DD}(C_{pw}^T + C_1^r)}{V_{TN}} + \frac{Q_{GT}^T}{V_{TN}} \\[2mm]
& + \frac{\sum_{i=1}^{n} C_{gsop}^i V_{sp}^i}{V_{TN}} + \frac{\sum_{i=1}^{n} C_{gson}^i V_{sn}^i}{V_{TN}} + \frac{\sum_{i=1}^{n} C_{gdon}^i V_{dn}^i}{V_{TN}} \\[2mm]
& + \frac{\sum_{i=1}^{n} C_{gdop}^i V_{dp}^i}{V_{TN}} + \frac{\sum_{i=1}^{n} C_{C_i} V_{C_i}}{V_{TN}} \quad\quad (1)
\end{aligned}
$$

Where:

$$
C_0 = C_0^r + C_{pb}^T
$$

$$
\begin{aligned}
C_{FD} \; = \; & \sum_{i=1}^{n} C_{gson}^i + \sum_{i=1}^{n} C_{gdon}^i + \sum_{i=1}^{n} C_{gdop}^i \\[2mm]
& + \sum_{i=1}^{n} C_{gsop}^i + \sum_{i=1}^{n} C_{pw}^i + \sum_{i=1}^{n} C_C^i + C_1^r
\end{aligned}
$$

If for unsensitized gates some of the drain/source voltages are not determined by the actual input vector then worst case conditions are considered.

2) Assured logic testable for a stuck-at 1 vector: In a similar way an expression to estimate the minimum value of the capacitance to V_{DD} of the floating line (C_1) to have at least an induced voltage of $V_{DD} - |V_{TP}|$ can be obtained. This condition assures that the interconnect open behaves as a stuck-at 1 fault.

978-1-61284-657-6/11 $26.00 © 2011 IEEE

C. Detecting conditions under process variability

In nanometer technologies, the process parameters may suffer significant variations from the nominal values due to imperfections during the manufacturing process [13] [14] [15]. For instance, channel length variation may occur due to a combination of photolithography, gate etching, ion implant, spacer formation and thermal processing effect [13]. Variation in channel length has the greatest effect on circuit performance [14]. Other important parameters are effective gate oxide thickness variation [14] and random doping fluctuations [13]. Because of this the influence of the parameter process variations on the detecting conditions of interconnect opens must be taken into account.

A simplified equation of 1, in terms of its parameters, is given by

$$C_0 = f(P_1, \ldots, P_n) \qquad (2)$$

where $P_1, \ldots, P_n$ are parameters subject to process variations. These variations can be approximated by a normal distribution even thought they are not strictly normal random variables [15]. The advantage of this approach is that the characterization of the multivariate normal function (mean μ_{C_0} and standard deviation σ_{C_0}) includes the correlation information between different variables.

In the equation 2, the process parameters $(P_1, \ldots, P_n)$ are normal random variables with mean $\mu_{P_1}, \ldots, \mu_{P_n}$ and standard deviation $\sigma_{P_1}, \sigma_{P_2}, \ldots, \sigma_{P_n}$. The mean (μ_{C_0}) and the standard deviation (σ_{C_0}) of the random variable C_0 can be estimated by multivariable Taylor-series expansion [16].

$$
\begin{aligned}
\mu_{C_0} = {} & f(\mu_{P_1}, \ldots, \mu_{P_n}) \\
& + \frac{1}{2} \sum_{i=1}^{n} \frac{\partial^2 f(P_1, \ldots, P_n)}{\partial P_i^2} \sigma_{P_i}^2 \\
& + \sum_{i=1}^{n-1} \sum_{j=i+1}^{n} \frac{\partial^2 f(P_1, \ldots, P_n)}{\partial P_i P_j} \rho_{P_i P_j} \sigma_{P_i} \sigma_{P_j} \quad (3)
\end{aligned}
$$

$$
\begin{aligned}
\sigma_{C_0}^2 = {} & \sum_{i=1}^{n} \left(\frac{\partial f(P_1, \ldots, P_N)}{\partial P_i} \right)^2 \sigma_{P_i}^2 \\
& + 2 \sum_{i=1}^{n-1} \sum_{j=i+1}^{n} \frac{\partial f(P_1, \ldots, P_n)}{\partial P_i} \frac{\partial f(P_1, \ldots, P_n)}{\partial P_j} \rho_{P_i P_j} \sigma_{P_i} \sigma_{P_j}
\end{aligned}
$$
$$(4)$$

where $\rho_{P_i P_j}$ is the correlation coefficient between two process parameters. The distributions of these process parameters is provided by the silicon foundry (the corner models), but the correlation between these parameters are in many cases confidential or not well known. The correlation coefficients have been obtained based in statistical theoretical equations [16] and using analytical expressions describing the transistor operation. Table 1 shows the obtained correlation coefficients for some parameters in TSMC CMOS technology 0.18μm.

Table I
CORRELATION COEFFICIENTS

	Tox	Vtn	Vtp	XW	XL	Cgon	Cgop
Tox	1						
Vtn	0.81	1					
Vtp	-0.56	-0.31	1				
XW	0	-0.01	-0.00	1			
XL	0	-0.09	-0.09	0	1		
Cgon	-0.08	-0.15	0.28	0	0	1	
Cgop	-0.09	-0.20	0.32	0	0	1	1

The normal distribution function for C_0 can be expressed as,

$$C_0 \sim N\left(\mu_{C_0}, \sigma_{C_0}\right) \qquad (5)$$

In a similar way, the normal distribution function for the actual extracted routing capacitance to ground C_0^{ext} can be expressed as,

$$C_0^{ext} \sim N\left(\mu_{C_0^{ext}}, \ \sigma_{C_0^{ext}}\right) \qquad (6)$$

where $\mu_{C_0^{ext}}$ and $\sigma_{C_0^{ext}}$ are media and standard deviation of the normal distribution of the layout extracted capacitance of the floating line to ground C_0^{ext}.

Then, the statistical detection condition for an interconnect open is stated as,

$$C_0^{ext} \sim N\left(\mu_{C_0^{ext}}, \sigma_{C_0^{ext}}\right) \geq C_0 \sim N\left(\mu_{C_0}, \sigma_{C_0}\right) \qquad (7)$$

In other words, the previous expression gives the detection condition for an interconnect open considering process variations.

III. VIA OPEN DEFECT PROBABILITY MODEL

A joint probability density function $(f(C_0, C_0^{ext}))$ [16] for the two continuos random variables C_0 and C_0^{ext} that reflects the behavior of these two variables can be obtained (See Fig. 2). The joint probability distribution of the two random variables provides a method for calculating the probability such that C_0 and C_0^{ext} take a value in the detectability region of two-dimensional space (bottom plane in Fig. 2). Furthermore, the detectability region must satisfy the condition $C_0^{ext} \geq C_0$ as stated by equation 8. This region corresponds to the left side of the line bisecting the bottom plane in Fig. 2. The probability $(P\left[C_0^{ext} \geq C_0\right])$ that takes a value in the detectability region is equal to the volume of the shaded region.

The probability of detection of an interconnect open can be analytically estimated making a double integral of $f(C_0, C_0^{ext})$ over the detectability region, and it is given by,

$$
P\left[C_0^{ext} \geq C_0\right] =
$$
$$
\int_{\mu_{C_0} - 3\sigma_{C_0}}^{\mu_{C_0} + 3\sigma_{C_0}} \left(\int_{C_0}^{\mu_{C_0^{ext}} + 3\sigma_{C_0^{ext}}} f(C_0, C_0^{ext}) \, dC_0^{ext} \right) dC_0
$$
$$(8)$$

978-1-61284-657-6/11 \$26.00 © 2011 IEEE

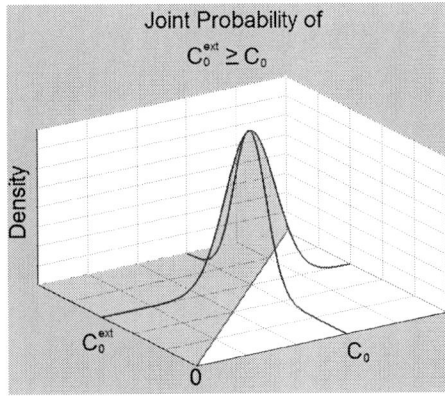

Figure 2. Region to estimate fault coverage of an interconnect open

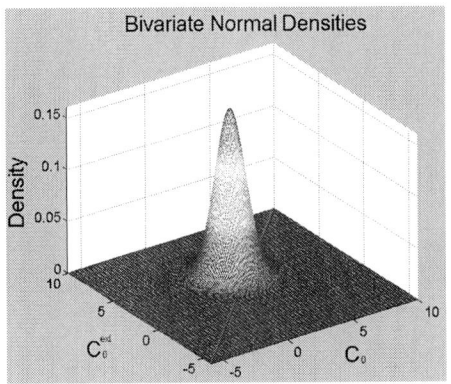

(a) Volume of $f(C_0, C_0^{ext})$ with the detectability region.

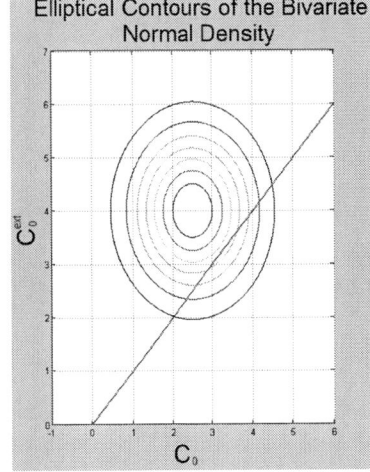

(b) Elliptical countours of the volume of the joint probability density function.

Figure 3. Volume and elliptical contours for a case with a high probability of detection.

As an example, let us consider $C_0^{ext}(3,1)$ and $C_0(5,1)$ as the normal distribution functions. Using MAPPLE, the bivariate function of these two variables is plotted in Figure 3a. The elliptical contours of the volume under the surface are shown in Figure 3b. The detection condition of equation 8 is also indicated in Figures 3a and 3b by a dividing line. Only the region $N\left(\mu_{C_0^{ext}}, \sigma_{C_0^{ext}}\right) \geq N\left(\mu_{C_0}, \sigma_{C_0}\right)$ is considered to estimate the probability of detection. This region is located to the left of the dividing line (See Figure 3). As a large portion of the volume of $f(C_0, C_0^{ext})$ is over the detectability region, the probability of detection for this case is 87%.

Figure 4 shows the volume of the joint probability density function of C_0 and C_0^{ext} for a case of low probability of detection (4.4%). In this case, $C_0^{ext}(4,1)$ and $C_0(6.4,1)$ are the normal distribution functions of the random variables C_0 and C_0^{ext}. The low probability of detection is due to the fact that only a small portion of the volume of $f(C_0, C_0^{ext})$ is over the detectability region.

IV. VIA OPEN DETECTION PROBABILITY EVALUATION

A. Detection condition optimization

Figure 5 shows the test flow to enhance the detectability of interconnect opens. This is mainly composed of: a) an enhanced test generation strategy (OPVEG [18]), and b) open detection probability evaluation. OPVEG obtains a enhanced test vector set. Favorable test logic conditions at the coupled signals of maximal neighbors are attempted to be obtained during test process generation. The detection probability for interconnect open is evaluated for the test set generated by OPVEG. If the detection probability is not enough, then OPVEG will attempt to generate a new test vector set putting more favorable logic conditions at the coupled signals. Otherwise, the final test set vectors is obtained. Effort in test pattern generation for interconnect opens has also been devoted by other authors [19] [20] [21].

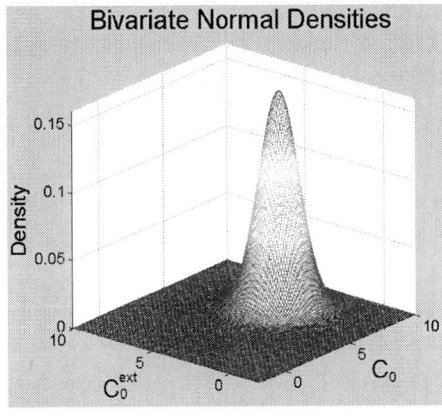

Figure 4. Volume of $f(C_0, C_0^{ext})$ with the detectability region for a case with low probability of detection.

B. Simulation

The proposed methodology to evaluate de detection probability of interconnect opens under process variations has been

978-1-61284-657-6/11 $26.00 © 2011 IEEE

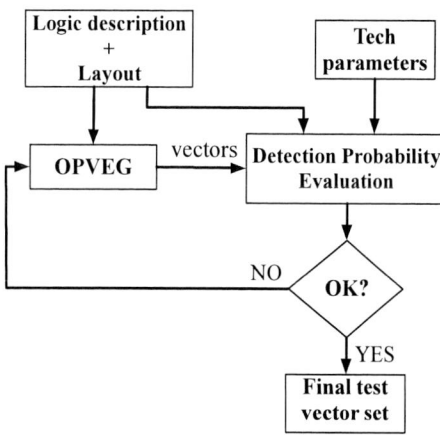

Figure 5. Test framework for interconnect opens

Table II
DETECTION PROBABILITY FOR C432

F_{CL}	Victim Line Factor				
	400% S0:54	200% S0:133	100% S0:166	50% S0:178	10% S0:178
400%	89.47%	89.27%	88.06%	87.57%	87.57%
200%	90.09%	91.35%	91.32%	90.61%	90.61%
100%	90.78%	91.86%	91.99%	92.57%	92.57%
50%	94.36%	92.48%	92.73%	93.33%	93.33%
10%	95.45%	93.01%	93.21%	93.62%	93.62%

Table III
DETECTION PROBABILITY FOR C499

F_{CL}	Victim Line Factor				
	400% S0:27	200% S0:85	100% S0:166	50% S0:183	10% S0:186
400%	69.58%	78.96%	80.39%	80.39%	80.39%
200%	78.52%	88.26%	84.71%	84.71%	84.20%
100%	80.14%	90.49%	90.74%	90.77%	89.80%
50%	80.48%	91.62%	92.10%	92.77%	93.01%
10%	80.50%	92.31%	93.88%	95.78%	95.84%

implemented in a software tool. This tool uses layout information and circuit logic description. Test vector generation of opens considering all the signals at the neighbor lines is a computationally intensive process. Because of this only those opens (vias) locations susceptible (**Selected Opens-SO**) to be significantly influenced by their coupled signals are selected to enhance the detectability of via opens using OPVEG [18]. Possible opens in vias/contacts located nearest to the gate driving the interconnect are analyzed. However, the favorable logic conditions also benefits other open locations in the stem. The total capacitance of a line (C_L) is composed by its capacitance to ground and V_{DD}. An open-i is selected when its line has at least one coupled capacitance C_c^i equal or greater than the capacitance C_L multiplied by a **Victim Line Factor** F_{VL}. This condition is stated by

$$C_c^i \geq F_{VL}(C_{GND} + C_{VDD}) \qquad (9)$$

The previous condition determines the set of Selected Opens (SO).

The **Coupling Line Factor** F_{CL} is used to select those coupled signals of the Selected Opens that OPVEG attempts to set for favorable logic conditions. Signals with coupling capacitance values (C_c^i) equal or higher than the total capacitance of the line (C_L) multiplied by F_{CL} are selected for test process generation using OPVEG. This condition is stated by

$$C_c^i \geq F_{CL}(C_{GND} + C_{VDD}) \qquad (10)$$

Tables II-V show the simulation results for some ISCAS85 benchmark circuits using the flow diagram shown in Figure 5. A zero value for the trapped gate charge has been considered. The Victim Line and the Coupling Line Factors are independently varied between 400% (coupling capacitance four times higher than C_L) and 10%. The number of Selected Opens (SO) increases for a lower value of the F_{VL}. For the same F_{VL}, the detection probability increases as F_{CL} decreases. In other words, forcing favorable logic conditions at more coupled signals improves via open detectability.

C. Impact of spatial correlation

In this section, spatial correlation effects of the parameters for those gates connected to the floating line is analyzed. In the previous discussion, different threshold voltages have been assumed for the transistors affected by the open. This is valid when considering the effect of the random dopant fluctuations. However, the threshold voltage of the affected transistors located at different space positions are correlated due to systematic effects intra-die variations. We have also studied the effect of the channel length and gate oxide thickness on the probability of detection of interconnect opens. The effect of the spatial correlation is considered by extending the terms of equations 3 and 4 that consider the correlation coefficients between the two parameters of interest. Figures 6 and 7 show the density distributions of the required condition to detect the open with a SAO vector C_0 for different correlation coefficient

Table IV
DETECTION PROBABILITY FOR C1908

F_{CL}	Victim Line Factor		
	50% S0:29	25% S0:153	10% S0:238
400%	75.34%	63.72%	62.28%
200%	75.34%	63.72%	62.28%
100%	75.34%	63.72%	62.28%
50%	78.32%	64.53%	63.49%
10%	92.44%	81.69%	78.95%

Table V
DETECTION PROBABILITY FOR C3540

F_{CL}	Victim Line Factor			
	100% S0:34	50% S0:226	25% S0:742	10% S0:896
400%	72.57%	61.66%	61.03%	65.98%
200%	75.51%	61.92%	61.09%	66.04%
100%	83.21%	64.34%	61.83%	66.58%
50%	85.56%	67.18%	62.69%	67.51%
10%	87.92%	79.01%	77.40%	82.52%

978-1-61284-657-6/11 $26.00 © 2011 IEEE

values of the gate oxide parameter (TOX) and the threshold voltage of the Nmos transistor V_{TN}, respectively. The degree of correlation of TOX has little influence on the density distribution of C_0 (See Fig. 6). However, the correlation coefficient value of parameter VTN has a significant influence on the density distribution if C_0 (See Fig. 7). The correlation influence of the transistor channel length parameter is even more significant. Hence, the open defect detection probability depends on the location of the gates tied to the floating gate node.

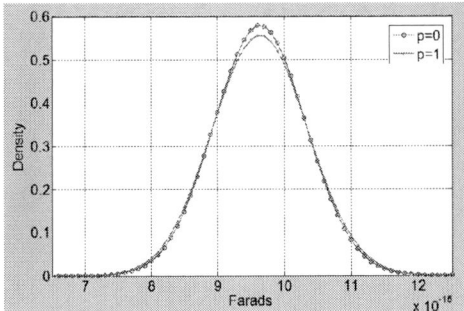

Figure 6. Spatial correlation impact of parameter TOX on the density distribution of C_0, ρ is the correlation coefficient for TOX.

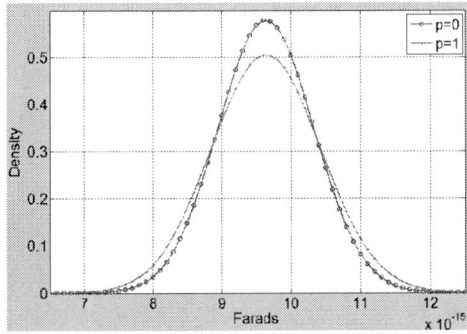

Figure 7. Spatial correlation impact of parameter V_{TN} on the density distribution of C_0, ρ is the correlation coefficient for V_{TN}.

V. CONCLUSION

In this work a new methodology for computing a realistic coverage of via open defect taking into account the process variability has been proposed. This has been made by obtaining the normal distributed values of the detection conditions for the open. The correlation between the parameters has been considered. Furthermore, spatial correlation of the parameters for those gates tied to the defective floating node can also influence the detectability of the defect. The open defect probability detection depends on the location of the gates tied to the floating gate node. The proposed methodology has been implemented in a software tool to determine the probability of detection of via opens for some ISCAS benchmark circuits. The proposed detection probability evaluation together with a test methodology to generate favorable logic conditions at the coupling lines can

allow a better test quality leading to higher product reliability.

ACKNOWLEDGMENT - The work has been partially supported by CONACYT (México) through the PhD scholarship number 160492.

REFERENCES

[1] Y. Sato, I. Yamazaki, H. Yamanaka T. Ikeda, and M. Takura. "A persistent diagnostic technique for unstable defects", *International Test Conference*, pp. 242-249, 2002.

[2] R. Blish, T. Dellin, S. Huther, M. Johnson, J. Maiz, B. Likins, N. Lycoudes, J. McPherson, Y. Peng, C. Peridier, A. Preussger, G. Propkop, and L. Tullos. *Critical Reliability Challenges for the International Technology Roadmap for Semiconductors*, ITRS, 2003.

[3] W. Needham, C. Prunty, and E.H. Yeoh, "High Volume Microprocessor Test Escapes, an Analysis of Defect Our Test are Missing", *Proceedings International Test Conference 1998*, pp. 25-34, 1998.

[4] R. Rodriguez-Montañez, J. Pineda, P. Volf , "Resistance Characterization for Weak Open Defects", *IEEE Design & Test*, pp. 18-26, 2002.

[5] K.M. Thompson, "Intel and the myths of test", *In the Keynote Address of ITC*, 1995.

[6] C. L. Henderson, J. M.Soden and C. F. Hawkins, "The Behavior and Testing Implications of CMOS IC Logic Gate Open Circuits", *International Test Conference*, pp. 302-310, 1991.

[7] M. Renovell and G. Cambon, "Electrical Analysis and Modeling of Floating-Gate Fault", *IEEE Transactions on Computer-Aided Design*, Vol. 11, No. 11, pp. 1450-1458, November 1992.

[8] S. Johnson, "Residual Charge on the Faulty Floating Gate MOS Transistors", *International Test Conference*, pp. 555-561, 1994.

[9] Haluk Konuk, "Voltage- and Current-Based Fault Simulation for Interconnect Open Defects", *IEEE Transactions on Computer-Aided Design of Integrated Circuits and Systems*, Vol 18, No. 12, pp. 1768-1779, December 1999.

[10] Antonio Zenteno, Victor H. Champac, "Detectability Conditions of Full Opens in the Interconnections", *Journal of Electronic Testing: Theory and Applications*, Vol 17, pp 85-95, 2001.

[11] Daniel Arumi, Rosa Rodriguez-Montanes, and Joan Figueras, "Experimental Characterization of CMOS Interconnect Open Defects", *IEEE Transactions on Computer-Aided Design*, Vol. 27, No. 1, January 2008.

[12] Sudhakar M. Reddy, Irith Pomeranz, Chen Liu, "On Tests to Detect Via Opens in Digital CMOS Circuits", *Design Automation Conference*, pp. 840-845, June 2008.

[13] Jaume Segura, Charles F Hawkins, "CMOS Electronics How it Works, how it fails", *Wiley Inter Science*, 2004.

[14] K. Bernstein, K. Carrig, C. Durham, P. Hansen, D. Hogenmiller, E. Nowak, and N. Rohrer, "High Speed CMOS Design Styles"*Kuwer Academic Publishers*, 1998.

[15] Sani R. Nassif, "Modeling and Analysis of Manufacturing Variations", *IEEE Custom Integrated Circuits Conference*, pp. 223-228, 2001.

[16] Athanasios Papoulis, "Probability, Random Variables and Stochastic Processes", *Mc-Graw-Hill, Inc.*, Third Edition, 1991.

[17] http://www.maplesoft.com.

[18] Roberto Gomez, Alejandro Giron, Victor Champac, "A Test Generation Methodology for Interconnection Opens Considering Signals at the Coupled Lines", *Journal of Electronic Testing: Theory and Applicattions*, Kluwer Academic Publishers, Volume 24 , Issue 6, Pages: 529-538, December 2008.

[19] H. Takahashi, Y. Higami, T. Kikkawa, T.Aikyo, Y.Takamatsu, K. Yamazaki, T. Tsutsumi, H.Yotsuyanagi, M. Hashizume, "Test Generation and Diagnostic Test Generation for Open Faults with Considering Adjacent Lines", *IEEE International Symposium on Defect and Fault Tolerance in VLSI Systems*, 2007.

[20] Stefan Spinner, Ilia Polian, Piet Engelke, Bernd Becker, Martin Keim, Wu-Tung Cheng, "Automatic Test Pattern Generation for Interconnect Open Defects", *IEEE VLSI Test Symposium*, 2008.

[21] Xijiang Lin, Janusz Rajski, "Test Generation for Interconnect Opens" *International Test Conference*, pp. 1-7, 2008.

Session 8

978-1-61284-657-6/11 $26.00 © 2011 IEEE

Impact of the Application Activity on Intermittent Faults in Embedded Systems

Julien Guilhemsang, Olivier Héron,
Nicolas Ventroux, Olivier Goncalves
CEA, LIST, 91191, Gif-sur-Yvette CEDEX, France
e-mail: julien.guilhemsang@cea.fr

Alain Giulieri
LEAT
Université de Nice-Sophia Antipolis, CNRS, 250, rue
Albert Einstein, 06560, Valbonne, France
e-mail: giulieri@polytech.unice.fr

Abstract—Future embedded systems are going to be more sensitive to hardware faults. In particular, intermittent faults are going to appear faster in future technologies. Understanding the occurrence of faults and their impact on systems and applications can help to improve the fault-tolerance of systems. However, there is no study on their effects in more complex digital circuits. We propose an experimental platform for accelerating and catching the occurrence of intermittent faults in complex digital circuits. We experimentally show that intermittent faults can appear during the lifetime of the circuit, very early before the wear-out period. We studied the impact of processor activity on intermittent faults rate. We conclude that a continuous usage of circuits causes the occurrence of intermittent faults earlier than a low usage under identical operating conditions. We show that applications do not have the same sensitivity to intermittent faults.

Intermittent faults, aging, embedded processor cores, FPGA

I. INTRODUCTION

The reliability of embedded systems is an important issue, and continuous advances in integration technologies have a negative impact on it. Devices are going to be more sensitive to transient faults, likewise intermittent and permanent faults are going to appear earlier in future technologies. Understanding the occurrence of faults and their impact on systems and applications can help to improve the fault-tolerance level of systems.

Transient and permanent faults have been studied for long time and past studies proposed a variety of solutions to tolerate them. Alternatively, intermittent faults arbitrarily occur over time several times and moreover, they can appear in *bursts* from few nano-seconds to several seconds [1]. Compared to aging related faults, they may appear earlier and may be predominant in future systems, whereas there is no published papers on present technologies and particularly in embedded systems. Intermittent faults are mainly due to process variations and aging [2][3] combined with dynamic variations of supply voltage and junction temperature [4].

Process variations bring up both random and systematic defects during the fabrication process. These defects have an impact on the circuit behavior and can induce distortion of the output signal for analog circuits and variable delays for digital circuits. At transistor level, the process variations have an impact on the threshold voltage V_T, the body factor γ and the current factor β [5]. There are two principal types of

process variations: Die-To-Die variations that induce parameter deviations between the dies in a same wafer and With-In-Die variations that induce parameter deviations between the transistors of a same die. Relatively to With-In-Die variations, it results in different working frequencies for different parts of a same chip [6]-[8].

Moreover, the degradation mechanisms such as Electromigration (EM), Time-Dependent Dielectric Breakdown (TDDB), Negative Bias Temperature Instability (NBTI), Hot Carrier Injection (HCI) and Stress Migration (SM) will decrease timing margins over time and thus cause timing problems [5][9][10]. Under nominal operating conditions, HCI and EM are activated by a dynamic stress i.e. when transistors switch, while NBTI and TDDB are activated by a constant stress (no transistor switching). All of these physical phenomena are closely related to junction temperature. A high temperature level accelerates the activation of NBTI, EM, SM, TDDB, while HCI is activated by a low temperature level. An increase of temperature level will accelerate circuit aging, will decrease timing margins and then will promote the occurrence of intermittent faults [4]. In our case study, we apply a high operating temperature level to accelerate the activation of these aging phenomena. As a consequence, it is not trivial to say which activity profile (i.e. dynamic or constant switching stress) and operating conditions will accelerate which failure mechanism and hence will lead to the first circuit failure and highest failure rate.

After analyzing the State-of-the-Art, we can say that the intermittent fault problem was mainly discussed based on knowledge of physics and results obtained from accelerated stress tests on devices (single transistor), but there is no published study that discuss on their effects on more complex digital circuits (processor core, memories, peripherals). One approach would be to build a macro-level failure model of the entire circuit by combining the different models existing at device or library level (bottom-up modeling approach), as done in [9] [22]. One would perform simulations at different corners and would conclude with statistical results. Alternatively, several campaigns of accelerated stress test would be applied to a set of identical circuit dies issued from different lots/wafers (empirical approach). Both approaches still remain a big challenge.

As a starting point, this paper makes the following contributions:

- A generic experimental platform to accelerate the occurrence of intermittent faults and catch the resulting errors in complex digital circuits (here, IBM PowerPC440 processor, bus and peripherals). The die is encapsulated in a standard packaging with no heat sink and no fan.
- For a chosen technology and circuit design (Xilinx Virtex5FX in 65nm with a *hard* PPC440 core), we show experimentally that errors can appear frequently (*bursts*) during the lifetime period of the circuit and very early before the fatal failure (loss of functionality).
- We show that a high usage (relatively to embedded applications) causes the occurrence of errors earlier than a low usage of the circuit under identical operating conditions. Moreover, the former situation causes a higher number of errors than the latter one.

Section 2 describes the major aging failures that affect the processor lifetime and their activation condition. This section allows understanding why it is not trivial to compare the lifetime of two systems, even if their respective activities are very different. Sections 3 and 4 present the experimental platform and the experiment. Section 5 discusses on the results and section 6 concludes the paper.

II. AGING INDUCED FAILURE MECHANISMS

Failures in chips are mainly related to assembly, mounting, handling and wafer processes [10]. In this paper, we only focus on failure related to wafer process. After analyzing the State-of-the-Art [5], [9]-[14], [17] and [18], we can say that the five following aging-induced failure mechanisms become a major issue for processor lifetime in current and future technology nodes: Time-Dependent Dielectric Breakdown (TDDB), Negative Bias Temperature Instability (NBTI), Hot Carrier Injection (HCI), Electromigration (EM) and Stress Migration (SM).

The use of thin gate oxides in deep submicron technologies combined with the non-ideal scaling of voltage supply increases electrical field stress and thus, accelerates the activation of NBTI, HCI and TDDB failure mechanisms [9]. In contrast to previous failure mechanisms that occur in the transistor, EM and SM failure mechanisms occur in vias, contacts and along long metal wires. Advances in circuit speed, device miniaturization and density increase the current density and thus accelerate failure activation.

Transistor switching activates EM, SM and HCI, while transistor idle states activate TDDB and NBTI. EM depends on the current density which reaches the maximum value when the transistor is switching. A current pulse induces a temperature pulse and leads to the dilatation of wires. Then, the thermo-mechanical stress experienced by the wires activates SM. HCI is a complex phenomenon that occurs when a current flows in the canal. Leakage current in the dielectric of the gate causes TDDB. The electric field intensity in the dielectric gate influences NBTI.

An important remark concerns the exponential dependence of most failure mechanisms to the junction temperature (Arrhenius model). Except HCI, the activation of other mechanisms is accelerated when the temperature increases. A high temperature level stress of the chip above the limit will accelerate the activation of NBTI, TDDB, EM and SM. The platform we propose applies an over temperature stress to the chip under test in order to accelerate the experiment.

To eliminate the *weak* circuits that do not verify the expected reliability requirements before being shipped to the user, Accelerated Life Tests (ALT), also known as IC burn-in, [19] are applied along the manufacturing process, from wafer-level process to packaging-level process. The technique consists in stressing the circuit at its limits in order to extimate the Failure In Time (FIT) of the circuit (number of errors per billion of hours). FIT values are generally computed with the aid of complex statistical formulas [10]. In that way, the experiment necessarily targets a large set of the same circuit.

Relatively to the behavior of the different failure mechanisms mentioned above, it is not easy to determine if a complex system like a processor will age faster if the processor is running an application or doing nothing (idle state). The experimental platform will try to provide an answer to this question.

III. PRESENTATION OF THE EXPERIMENTAL PLATFORM

The purpose of our platform is to accelerate the occurrence of faults caused by aging phenomenon and to catch the resulting errors in a circuit under test. By this mean, we attempt to say if intermittent faults exist and if yes, understand their behavior. This is done by stressing the processor with a high temperature and by applying a set of stimuli on the different entries of the processor under test.

Given a technology node, an external environment, an instruction set architecture and a design, one issue is to compare the impact of two activity levels on the intermittent error rate in a processor. More precisely, this platform will determine how the intermittent errors are affected by the processor state: active or idle. During an active state, the processor is powered and executes different instructions. During an idle state, the processor is powered but only executes a *NOP* instruction.

Our platform is divided into three main elements: the Circuits Under Test (CUT), the External Stress Manager (ESM) and the Internal Stress Manager (ISM). The CUT can represent one or several systems consisting in a processor, a memory and communication peripherals. The ESM controls the circuit temperature and the power supply. Finally, the ISM loads the benchmark sets in the processor memories and gathers the responses (and so the errors). Fig. 1.a illustrates a typical composition of the platform with more than one CUT.

A. Circuit Under Test (CUT)

The Circuit under Test is a Virtex5-FX FPGA that contains an IBM PowerPC 440 processor, implemented in hardware. The circuit under test is mounted into a board, as illustrated in Fig. 1.b. The board is an Avnet AESV5FXT board [20]. In our experiment, the platform is composed of six boards.

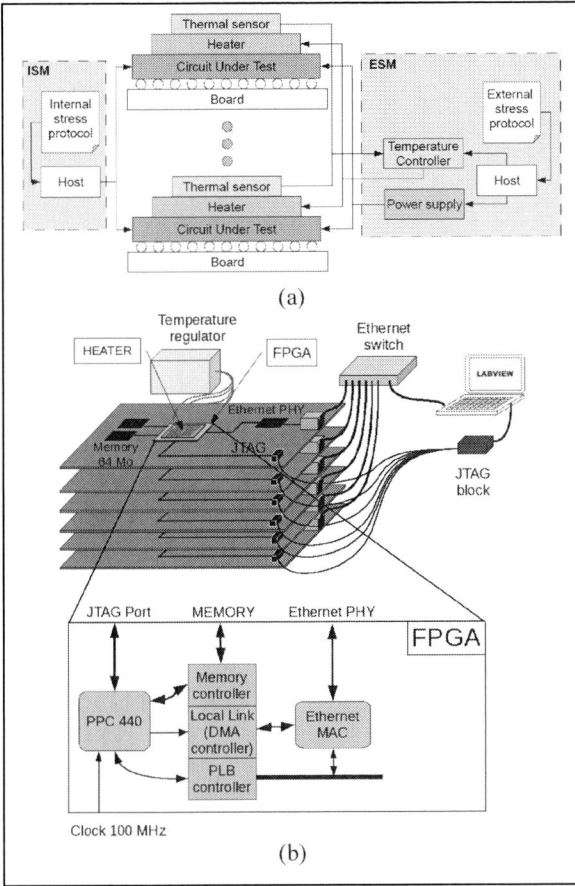

(a)

(b)

Figure 1. (a) Platform functional diagram and (b) Platform schematic.

Circuit I/O pads are connected to the following elements: a clock generator, a power supply, a JTAG port (IEEE1149.1 standard), a peripheral transceiver and connectors. The JTAG port allows the connection between the host computer and the processor core, and will be used to download programs into the system. The CUT offers to the processor a communication structure based on a R/W memory, a communication bus and a peripheral. The memory stores both program (sequence of instructions) and data. The peripheral enables communications of data from the platform to the external host computer.

B. External Stress Management (ESM)

Aging analysis can be done through over temperature stress and/or over power supply stress. Both temperature and power supply are controlled by the host computer through a specific bus. ESM controls the start of the experiment; sets the temperature and voltage values from a stress protocol file; records measurements; and checks continuously the

electrical values according to the safety limits. The recorded values are the instantaneous circuit voltages, currents and junction temperature. A temperature sensor is connected to the bottom side of the board under the circuit. If the circuit temperature or current grows above a predefined threshold, the experiment automatically stops and requests maintenance. The program used to control the ESM is implemented with the commercial tool LabView.

A stress protocol file describes the temperature and the voltage values over the time. Relatively to temperature, the protocol lists the temperature value and pitch at each period.

To increase the junction temperature of a circuit, the circuit is generally put in an oven. The temperature stress affects the entire board - including solders and connectors - which can disturb the process. To prevent such situation, a flexible heater [21] is mounted to the top side of the CUT packaging. An acrylic pressure sensitive adhesive that can support high temperatures links the heater to the packaging. To reach high junction temperatures (200 °C) and prevent the variations of the ambient temperature, the CUT is put in an oven in which the ambient temperature is roughly constant and equal to 50 °C. note that the thermal resistance of recent devices becomes very low due to the use of efficient thermal sink and spreader. Hence, the temperature junction of the device is almost equal to the heater temperature. A PID-based controller (Proportional Integral Derivative) controls the heater temperature, as shown in Fig. 1.b.

A programmable power generator controls the board power supply. To allow a voltage stress, the power supply generator must be connected directly to the CUT pad and hence, it must bypass the voltage regulator on the board. In our case (Avnet's board), the voltage stress is not possible. A current monitor is serially inserted between the power supply and the board so that the instantaneous current variations of the board can be monitored and recorded during the experiment.

C. Internal Stress Management (ISM)

The ISM controls the stimuli applied to the CUT, and especially in the processor core. One objective of our experiment is to compare the impact of two activity levels on intermittent error rate in a processor core. Fig. 2 shows two different profiles of internal stress that imply two different activity levels.

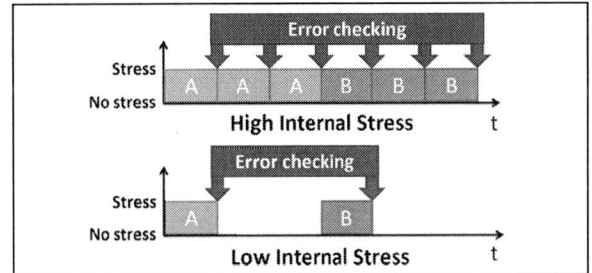

Figure 2. Internal stress examples. A and B blocks represent one execution of two different applications. Errors are checked at the end of each execution of each application.

978-1-61284-657-6/11 $26.00 © 2011 IEEE

We use a set of applications to create an activity in the different parts of the processor, and to detect errors in these parts. The activity level depends on the ratio between the active state and the idle state. Fig. 2 shows two different levels of internal stress that imply two different activity levels.

The ISM runs on the host computer with the LabView tool. It reads an internal stress protocol file that lists the application collection to be run at different periods. At the end of the execution of each application, it first gathers the recorded data in the CUT memory through Ethernet communication and next checks the presence of errors in the data. The following steps show how the different applications are loaded in the CUT and how results are checked:

a) The first step consists in configuring the FPGA by downloading the bitstream. This step is done only once at the beginning of the experiment.

b) Then, the application binary is downloaded in the external memory and the processor is started (JTAG commands). This step is also done when the ISM changes the application e.g. when changing from application (A) to application (B) (Fig. 2).

c) The Ethernet communication between CUT and host computer is checked (done for each CUT). If there is no link, the processor of failed CUT is resetted and Step *b)* is done again. If failed again, the processor is considered as permanently defective.

d) The application starts when the processor receives the "start" command from ISM. In that way, the ISM can synchronize the different CUTs.

e) After one application iteration, the processor stores the application results in the external memory and computes a signature (~one byte).

f) The host computer sends a command to request the results and signature. The processor sends them consecutively three times to prevent communication errors.

g) The processor returns to the step *d)* and waits (again) for the "start" command.

h) The ISM compares the collected data with golden values. If a difference appears, the ISM logs the difference for post-analysis. The comparison considers the data size and content. If no result is received by the ISM, the experiment automatically stops and ISM requests maintenance.

IV. DESCRIPTION OF THE EXPERIMENT

The previous section describes the different components of the platform. This section describes the different stress protocols used to make our analysis.

In order to study the impact of different activities on the error rate of a CUT, we choose to compare two sets of applications and boards that induce two different activity values. In each set, three CUTs run the same application at each period. Among the three CUTs, one CUT will not be under a temperature stress (witness board). In the first set,

the processors run an application during 100% of the time, while the processors allocated to the second set remain in idle state during 99.7% of the time. By comparing the number of errors and the time instant when they occur, we will be able to determine which activity level has the greater impact on the processor reliability.

A. Stress protocols

As mentioned in section III, two different stresses are considered: internal stress and external stress.

The internal stress protocol is composed of 7 applications: *QuickSort* is a part of MiBench collection [23]. *DCT, Quant* and *FIR* are classical embedded applications used in signal processing field. *TestFunct, TestFunct2* and *TestFunct3* are Software-Based Self-test (SBST) programs generated by ATPG [22]. Table I shows the characteristics of each application. The 1st column shows the average rate of load and store operations. The 2nd column shows the average instruction rate. The last column is the program duration.

TABLE I. APPLICATIONS CHARACTERISTICS

Application	#ls./s.	#ins./s.	Duration (s)
Qsort	8,567,619.96	25,157,627.08	4.64
FIR	5,113,756.02	12,600,864.70	3.80
Quant	3,982,256.49	10,202,307.42	5.74
Testfunct2	1,323,048.81	6,084,324.06	0.76
Testfunct	925,207.17	3,544,534.22	1.09
DCT	27,544.99	180,340.58	0.99
Testfunct3	110.17	9,531,901.79	33.92

The first set of CUTs (high internal stress) runs each application continuously during 1 hour with no interruption. The 7 applications are re-executed once again after 7 hours. On the contrary, the second set of CUTs (low internal stress) runs each application once time during the time frame of 1 hour. In the remaining time, it executes NOP operations (Fig. 2).

External stress only involves the temperature control. ESM applies an over stress on 4 CUTs (a first CUT pair is under high internal stress and the second pair is under low internal stress). The two other processors run at ambient temperature. No error should be observed on these two processors. They will allow to validate the results.

B. Observation Mechanism

In the platform, an error can occur at different stages of the experiment. We enumerate four types of errors:

- *Bitstream loading error:* such error can occur during when the configuration file (bitstream) is loaded into the FPGA. In our case, this error only happened when a CUT already failed.
- *Program loading error:* such error can occur during when the program is loaded into the main memory of

the board. We only experienced this type of error after the fatal failure of the board.

- *Communication error*: such error occurs when the Labview's program tries to g with the boards in order to get the results. The communication is done through an Ethernet port with the TCP/IP protocol. This error can be caused by a fault in the processor or DMA controller or Ethernet controller or memory controller. We are not able to identify the root cause. Such error can be caused by the occurrence of an intermittent fault.
- *Computation error*: such error appears when a permanent or an intermittent fault occurs during a computation step in the processor. In this case, the results will differ from the expected ones. Computation error was the major category in our experiments.

Our objective is to catch computation errors and avoid other error sources. The following section only refrs to computation errors.

V. RESULTS

The experiment was performed in two phases. In a first experiment phase, the CUT under temperature stress run during six months at 145 °C. No errors were observed in the 6 different CUTs. Therefore, we can assume that any error observed after this first phase is not due to the hardware or software design. Since the chips were stressed, the CUTs begin the second phase with a non-zero aging level. During the second phase, the CUTs under stress temperature run during 250 hours at 160°C. During this second period, computation errors were observed, as follows (no errors in the witness CUTs).

A. Emphasis on intermittent errors

Fig. 3.a shows the number of times an application fails over the time for the two CUT under high internal stress. Each point accounts for the total of fails observed in the time frame of 7 hours. Firstly, these results show that intermittent errors exist at system level and very early before the fatal system failure. Actually, the system fails with some applications while no failures were observed for others. The system can continue after starting a new application. Thus, we can develop methods to detect if a system is about to be faulty and methods for system recovery. Secondly, the number of intermittent errors seems to increase over the time before the fatal failure. Well, the monitoring of the number of intermittent errors can be used to predict the fatal failure of the system.

Fig. 3.b shows the accumulation of errors observed on one CUT under high internal stress with the program *TestFunct3*. As a reminder, the program runs during 1 hour every 7 hours; this program does not fail at each execution time. It only fails 5 times among the 30 execution occurrences.

Once an error appears, if no action is taken to stop the program, the program can continue its execution but by producing errors. Here, the error is corrected by stopping the processor and next loading a new program. Therefore, this result confirms that suspension techniques to recover from intermittent errors, as proposed in [1], are good candidates.

B. Impact of internal stress on observed intermittent errors

Here, we compare the number of observed errors in the response between the two application sets. As a reminder, a CUT under low internal stress runs an application once time every 1 hour. In both application sets, we only account for the errors observed during the first application occurrence.

Fig. 3.c shows the number of computation errors in the 4 CUTs (2 CUTs under high internal stress and 2 CUTs under low internal stress). A total of 25 errors over the time is reported for CUTs under high internal stress and only 2 errors for CUTs under low internal stress. The mean error rate for CUTs under high internal stress is equal to 0.08 errors per hour while the mean is equal to 0.006 errors per hour for others. CUTs under high internal stress are 12 times more subject to intermittent errors than CUTs under low internal stress.

C. Impact of intermittent errors on applications

Fig. 3.d shows the number of computation errors detected in the various applications regarding the two CUTs under high internal stress. We can see that the number of intermittent errors is not evenly distributed among them. We note that the applications with the highest number of errors are those with the highest number of load/store operations per second (Table I). The number of load/store operations per second denotes the traffic between the processor and the external memory. For this design, applications that cause a high traffic with the memory are the most likely to fail.

VI. CONCLUSION

Before the fatal failure of a system appears, the system can experience intermittent errors. Our experiment shows that intermittent errors can be observed at system level. Moreover, we show that if no action is taken, the error still occurs intermittently, but a simple re-start or shutdown of the processor seems to be sufficient for recovery. Until now, no published study has analyzed the effect of intermittent errors on an embedded system in a current technology. In this paper, we presented a generic experimental platform able to stress an embedded system and generate intermittent faults.

Our platform can be adapted to any digital test vehicles design, technology and chip assembly. For the same external stress, several experiments can be conducted with different conditions, such as the activity level. Therefore, it can help us to determine which system parameters have the greatest impact on the activation of intermittent faults.

In our case study, we show that a PPC440 (embedded in a Xilinx FPGA) will fail intermittently more often over time when it is continuously used than a PPC440 under low usage. From that, we can tailor a detection technique for intermittent and permanent errors relatively to this behavior.

REFERENCES

[1] P. M. Wells, K. Chakraborty, and G. S. Sohi, "Adapting to intermittent faults in multicore systems," in International conference on Architectural Support for Programming Languages and Operating Systems (ASPLOS). New York, NY, USA: ACM, 2008, pp. 255–264.

[2] C. Constantinescu, "Impact of deep submicron technology on dependability of vlsi circuits," in International Conference on Dependable Systems and Networks (DSN). Bethesda, MD, USA, 2002, pp. 205–209.

[3] S. Borkar et al., "Parameter variations and impact on circuits and microarchitecture," in Design Automation Conference (DAC). New York, NY, USA: ACM, 2003, pp. 338–342.

[4] C. Constantinescu, "Impact of intermittent faults on nanocomputing devices," Proc. IEEE/IFIP DSN (Supplemental Volume), Edinburgh, UK, pp. 238–241, 2007.

[5] G. Gielen et al., "Emerging yield and reliability challenges in nanometer cmos technologies," Design, Automation and Test in Europe (DATE), pp. 1322–1327, 2008.

[6] S. Duvall, "Statistical circuit modeling and optimization," in IEEE International Workshop on Statistical Metrology (IWSM), 2000, pp. 56– 63.

[7] K. Bowman, S. Duvall, and J. Meindl, "Impact of die-to-die and withindie parameter fluctuations on the maximum clock frequency distribution for gigascale integration," IEEE J. Solid-State Circuits, vol. 37, no. 2, pp. 183 –190, feb 2002.

[8] K. Bowman et al., "Impact of die-to-die and within-die parameter variations on the clock frequency and throughput of multi-core processors," IEEE Trans. VLSI Syst., vol. 17, no. 12, pp. 1679 –1690, dec. 2009.

[9] J. Srinivasan et al., "The impact of technology scaling on lifetime reliability," in DSN, Florence, Italy, 2004, pp. 177–186.

[10] Renesas Technology, "Semiconductor reliability handbook," Renesas Technology, Tech. Rep. Rev 1.01, Nov. 2008.

[11] V. Reddy et al., "Impact of negative bias temperature instability on digital circuit reliability," in Reliability Physics Symposium Proceedings, 2002. 40th Annual, 2002, pp. 248–254.

[12] Jedec Publication, "Failure mechanisms and models for semiconductor devices," Jedec, Tech. Rep. JEP122C, Mar. 2003.

[13] S. Lin et al., "Impact of off-state leakage current on electromigration design rules for nanometer scale cmos technologies," in 2004 IEEE International Reliability Physics Symposium Proceedings, 2004. 42nd Annual, 2004, pp. 74–78.

[14] M. White et al., "Product Reliability Trends, Derating Considerations and Failure Mechanisms with Scaled CMOS," in 2006 IEEE International Integrated Reliability Workshop Final Report, 2006, pp. 156–159.

[15] B. Greskamp, S. Sarangi, and J. Torrellas, "Threshold voltage variation effects on aging-related hard failure rates," in IEEE International Symposium on Circuits and Systems, 2007. ISCAS 2007, 2007, pp. 1261–1264.

[16] W. Goes and T. Grasser, "Charging and discharging of oxide defects in reliability issues," in IEEE IIRW, 2007, pp. pp. 27–32.

[17] Jedec Publication, "Foundry process qualification guidelines," Jedec, Tech. Rep. JP001.01, May 2004.

[18] AVNET, "AES-V5FXT-EVL30-G Evaluation Kit." [Online]. Available: http://www.em.avnet.com

[19] Minco, "Flexible heaters design guide." [Online]. Available: http://www.minco.com/

[20] M. Guthaus et al., "MiBench: A free, commercially representative embedded benchmark suite," in IEEE International Workshop on Workload Characterization, 2001, pp. 3–14.

[21] N. Kranitis et al., "Software-based self-testing of embedded processors," Computers, IEEE Transactions on, vol. 54, no. 4, pp. 461–475, April 2005.

[22] T. Gupta et al., "RAAPS: Reliability Aware ArchC based Processor Simulator," To appear in Proceeding of IEEE Int. Integrated Reliability Workshop, 2010.

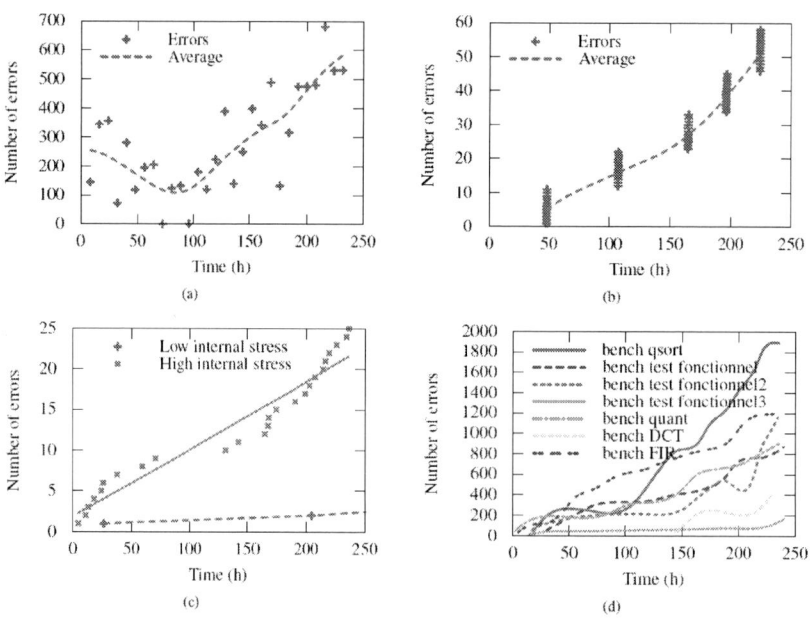

Figure 3. (a) Total number of program errors observed at 160°C. Each point represents the total number of errors during 7 hours.
(b) Number of errors for one board under high internal stress with the program *TestFunct3*.
(c) Number of errors detected during the first run of each application (every one hour) for boards under high and low internal stresses.
(d) Number of errors detected for each application for boards under high internal stress.

An Analytical Method for Estimating SET Propagation

Sreenivas Gangadhar, Spyros Tragoudas
Department of Electrical and Computer Engineering
Southern Illinois University Carbondale
Carbondale, IL 62901
{gangadha, spyros}@engr.siu.edu

Abstract—**In sub-micron technology, a small inaccuracy in computing the probability of occurrence of a soft error results into an unacceptable chip failure rate. A method to estimate the probability of SET propagation to the output gate at any time instant within the latching window is proposed. Its accuracy is evaluated using Monte Carlo simulations.**

I. INTRODUCTION

Technology scaling challenges the reliability of the digital systems. Increase in the clock frequency and decrease in the supply voltage raise concerns on the effect of Single Event Transients [SETs]. Generation of an SET is due to a strike by high-energy neutrons from cosmic radiation or alpha particles from radioactive contaminants in the packaging material. The electron-hole pair produced by such a strike separates promptly in the presence of electric field creating an inversion layer. This temporary inversion layer created under the gate of a transistor produces a short voltage pulse, the SET. If this generated SET propagates and latches at a flip-flop, it causes Single Event Upset [SEU] or, otherwise called a soft error. Thus, a soft error is generated if the SET arrives within a given time period (also called the latching window) and this requires time related calculations.

The soft error rate [SER] of a circuit is defined as product of the rate of effective particle hit and probability of the soft error from the particle hit. Almost all models in the literature [9], [12], [16], [19], [20], [22] among others propound logical masking, electrical masking, and latching window filtering for the accurate estimation of SER.

Another measure, the one followed in this paper, is the SET effect at an output which is the product of P_{gen} the probability of generating an SET at a gate at time t and P_{prop} the probability to propagate the generated SET to an output within the latching window. The SET generation probability P_{gen} will be calculated using [4]. The SET generation model of [4] eliminates restrictions for P_{gen} that were assumed in [1], [9], [7], [17]. Such details are given in [4] and the focus of this paper is to calculate the probability P_{prop}. The proposed method to calculate the probability of the SET propagation

This material is based upon work supported by the National Science Foundation under Grant No. 0702628. Any opinions, findings, and conclusions or recommendations expressed in this material are those of the author(s) and do not necessarily reflect the views of the National Science Foundation.

to the output P_{prop} is based on Boolean functions. We chose to manipulate these functions using Binary decision diagrams (BDDs) but Boolean expression diagrams (BEDs) can also be used to eliminate memory related issues of BDDs at the expense of higher execution time. The proposed method uses the concept of disjoint covers [5]. Disjoint covers take structural correlations into consideration for accurate probability calculations. This is an improvement over [1]. SET propagation by [1] traverses the netlist but does not take structural correlations into consideration. Moreover, disjoint covers of a Boolean function assure the certainty of logical masking effects.

Binary Decision Diagrams were proposed in [21][22] for SER analysis. These methods are not able to calculate the probability of SET to propagate to the output at any time instant. None of the existing work in error propagation benefits from the timing related information at the gates and interconnects.

The circuit expansion method [6] along with timing related information is used by [7], [17] for SET analysis. A limitation of the circuit expansion method is that only the earliest time that an SET can propagate to the output can be reported. Furthermore, this model cannot be applied to calculate the SET propagated to an output within the latching window but goes beyond the clock cycle. The basis of the probabilistic model in [17] is the Bayesian network that eliminates probabilistic in-dependencies in a circuit. The disadvantage of this model is that for a circuit with large re-convergent paths the Bayesian network may explode and is also tedious task to generate Bayesian networks for different particle strikes. The proposed approach overcome all the above limitations of [7], [17]. The error propagation probability is initially calculated without considering electrical attenuation. Subsequently, it is shown how to modify propagation probability in the presence of electrical attenuation.

Most of the work in literature [4][20][21] assumes that every pulse propagated to the output due to SET has width that completely overlaps with the latching window. However, it is possible that a large pulse width is shifted and only a part of it is within the latching window. Thus, only a portion of the latching window exhibits erroneous behavior whereas another portion exhibits error-free behavior. In reality, the latching window is not much larger than the minimum gate delay. For example, in $180nm$ technology the latching window is shown to be no more than three times the delay of a buffer [14]. For simplicity, the method is explained assuming that the latching window is twice the buffer delay but it can be generalized to any constant. The method is initially presented assuming that there is no electrical attenuation. It is shown in [11][12] that SETs generated with small widths practically do not propagate to an output. Thus, there is no reason to apply the proposed theory. However, [11][12] show that the SET of width two or three times the gate delay is not impacted by electrical attenuation, and the presented results apply in this case. We show, however, that the derived probabilities still need to be modified in order to take into consideration the small width pulses that are generated in internal lines due to reconvergencies which will not make to the output because of electrical attenuation.

The accuracy of the proposed analytical method is evaluated using Monte Carlo simulations. For a circuit with huge number of inputs(flops), Monte Carlo simulation is impractical. Scalability of the proposed analytical method is ensured by storing and manipulating the functions using BDDs.

The paper is organized as follows. Section II describes the SET effect assuming that the SET is responsible for a pulse that reaches before the latching window and terminates after the latching window. This special condition has been followed so far in the literature to calculate the SET effect at an output with an assumption that the pulse overlaps the latching window completely. This is a reasonable assumption, since typically pulses that propagated to the output are wider than the latching window [20], [21]. However, in general the pulse propagated to the output may overlap partially with the latching window. Section III describes how to cope with this condition and calculate the SET effect appropriately. Section III also describes how to modify the propagation probabilities to consider electrical masking at the circuit gates. Section IV validates the proposed approach using Monte Carlo based simulations. Section V concludes.

II. A SPECIAL CASE AT THE LATCHING WINDOW

For simplicity, the method is described assuming that all gates have unit delay and a latching window with a width of two time units. Let the latching window start at time instance t and end at time instance $t+2$. We sample for potential errors due to SETs at time instances $t+1$ and $t+2$ i.e., we assume that the two time intervals $(t, t+1)$, and $(t+1, t+2)$ denote the latching window. The special case where an error propagates to output no later than time t and terminates no earlier than time $t+2$ is described. This means that the error is through out the latching window. We distinguish this case based on whether the error free behavior of the output is 0 (as in Figure 1a) or 1 (as in Figure 1b).

The framework for SET propagation uses symbolic simulation. Functions and their disjoint covers are generated at each gate g at particular time t. This will be used to calculate the probability of the gate being zero or one at that particular

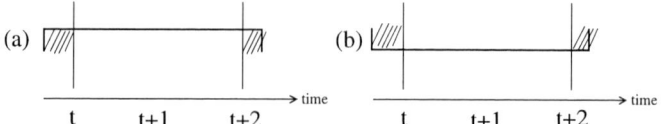

Figure. 1. Without electrical attenuation, Pulse starts no later than time t and finishes no earlier than time $t+2$ when the error free value at the output line is either zero as in (a) or one as in (b).

time t. In this work, a different function is defined at the output of each gate g at each different time point t. Each input I is associated with several variables I_t defined at times $t \geq 0$. Functions at the output of the gates are formed at each time point depending on such variables I_t.

Symbolic simulation of a circuit produces Boolean functions at each gate of the circuit. We use the same symbol g to denote a gate g and the function at the output of gate g. Let g_t denote the gate g or the function at the output of gate g at time t. Let $P(g)$, denote the probability that function g evaluates to binary value 1 and let $P(\overline{g})$ denote the probability that function g evaluates to binary value 0. We denote gate g that particle strikes as g^S and time of strike as t^S.

Signal probabilities at each time instance at a given gate are generated as described in [4]. The method uses Boolean functions stored using BDDs. Further, disjoint covers are used in order to take reconvergencies into consideration. It also describes how to derive probability P_{gen} for the SET to be generated at a gate due to particle strike at time t^S. The remainder of this Section describes how to compute the probability for the SET to propagate to the output and completely overlaps with the latching window.

We want to calculate the probability of the SET propagated to the latching window under the special condition that it has propagated no later than time t and having a duration of at least two units. We distinguish between the two error propagation probabilities. One kind of error propagation probability is when the error free behavior of the line is 0 and the other is when the error free behavior of the line is 1. The overall error propagation probability is the sum of these two error propagation probabilities. The following will discuss and computes these two error propagation probabilities.

Particle strike may occur at any gate, at any time and initiates the SET (error). If the SET is generated at any internal gate then the approach operate on a modified circuit where the gate of strike g^S is considered as an input. This requires standard graph theoretic manipulation where any path from inputs to gate g^S is removed. This is done with graph traversals which are not described due to space limitations.

The probability by which the error at gate g^S can be observed at the output is defined as the propagation probability (P_{prop}). To compute the propagation probability we use Boolean difference method. Consider the Boolean function $O(i[1], i[2]...i[n])$ at the output O where $i[1], i[2]...i[n]$ are the n inputs. Our method allows the particle strike at any internal gate. For simplicity, in the description of the method,

978-1-61284-657-6/11 $26.00 © 2011 IEEE   198

we assume that the particle strike S happens at input $i[1]$ which is denoted by $i^S[1]$. The Boolean difference of output O w.r.t input $i[1]$ is defined as

$$d(O)/d(i[1]) = O(1, i[2]...i[n]) \oplus O(0, i[2]...i[n]) \quad (1)$$

The Boolean difference of output O w.r.t input $i[1]$ indicates whether O is sensitive to changes in $i[1]$. This means that if $d(O)/d(i[1])=1$ then the value at output O changes as a result of the error at $i[1]$. If $d(O)/d(i[1])=0$ then the error at $i[1]$ is said to be unobservable.

The Boolean difference in Equation (1) is extended in order to consider gate delays in the circuit. Assume a particle strike at time t^S on input $i[1]$. In this case the Boolean function at output O at time $t^S+\triangle$ is $O_{t^S+\triangle}(i_{t^S}^S[1], i_{t_2}[2]...i_{t_n}[n])$ where $i_{t^S}^S[1], i_{t_2}[2]...i_{t_n}[n]$ are n inputs at n different times and $\triangle$ denotes the maximum combinational delay between input $i^S[1]$ and output O.

The Boolean difference at output O considering the gate delays is

$$d(O_{t^S+\triangle})/d(i_{t^S}^S[1]) = O_{t^S+\triangle}(1, i_{t_2}[2]...i_{t_n}[n]) \oplus$$

$$O_{t^S+\triangle}(0, i_{t_2}[2]...i_{t_n}[n]) \quad (2)$$

The expression $d(O_{t^S+\triangle})/d(i_{t^S}^S[1])=1$ implies that the output O at time $t^S + \triangle$ is sensitive to input $i^S[1]$ at time t^S. The probability of this function in Equation (2) being one, i.e., $P(d(O_{t^S+\triangle})/d(i_{t^S}^S[1]))$, gives the propagation probability of an error generated at $i^S[1]$ at time t^S propagate to output O at time $t^S + \triangle$ assuming an error free value of 0 at output O.

The latching window of two units is defined at time instances t, $t + 1$, $t + 2$. The error has to be observed at both the time instants $t + 1$ and $t + 2$. However, we must insist that output O must have identical error free behavior at both the time instances $t + 1$, $t + 2$. Let $PP\_1\_(t + 1)$ denote the probability of observing an error at output O at time $t + 1$ provided that the error free state of output O is 1 defined in Equation (3). Likewise we define $PP\_0\_(t+1)$ in Equation (4).

$$PP\_1\_(t + 1) = P(d(\overline{O_{t+1}})/d(i_{t^S}^S[1])) \cdot P(O_{t+1}) \quad (3)$$

$$PP\_0\_(t + 1) = P(d(O_{t+1})/d(i_{t^S}^S[1])) \cdot P(\overline{O_{t+1}}) \quad (4)$$

Similarly, we define $PP\_1\_(t + 2)$, $PP\_0\_(t + 2)$. Using Equations (3),(4), we conclude that the probability P_{prop} to observe the pulse at the latching window at time instances $t+1$ and $t + 2$ is:

$$P_{prop} = PP\_1\_(t + 1) \cdot PP\_1\_(t + 2)+$$

$$PP\_0\_(t + 1) \cdot PP\_0\_(t + 2) \quad (5)$$

The probability of an error propagated P_{error} to output O due to strike at gate g^S is the error propagation probability as defined in Equation (5) times the SET generation probability

P_{gen}. Thus, the probability of an error P_{error} at the latching window is

$$P_{error} = P_{gen} \cdot P_{prop} \quad (6)$$

III. THE GENERAL CASE AT LATCHING WINDOW

As in Section II, we assume a latching window of two time units and that we sample for an error at times $t + 1$, $t + 2$. In general, we have more cases for an error to propagate within a latching window. The pulse propagated to the output O may finish at time instance $t + 1$ and there is no error for the remaining of the latching window. We distinguish between such cases based on whether the error free behavior of the output is 0 (as in Figure 2a) or 1 (as in Figure 2b). Furthermore, the pulse propagated to the output O may start at time instance $t + 1$ without any error prior to that during the latching window. We distinguish between such cases based on whether the error free behavior of the output is 0 (as in Figure 2c) or 1 (as in Figure 2d).

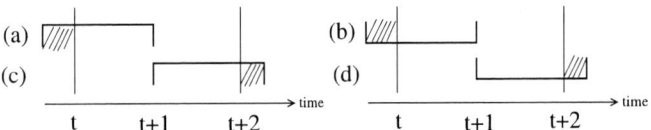

Figure. 2. Without electrical attenuation, Pulse finishes at time $t + 1$ when the error free value at the output line is either zero as in (a) or one as in (b). Pulse starts at time $t + 1$ when the error free value is either zero as in (c) or one as in (d)

In order to ensure that the pulse propagated finishes by time $t + 1$, we need to observe the error during the time $(t, t+1)$. See also Figures 2a,2b. However prior to time instance t, the error may or may not be present but not required for the calculations. This is a don't care condition.

On the other hand, the error after time instance $t + 2$ may or may not be present but not required for the calculations. This is a don't care condition. See also Figures 2c,2d

Considering all the four cases as shown in Figure 2, the error propagated P_{prop} can be calculated using Equations (3),(4) as shown below in Equation (7). Each product term in Equation (7) calculates the error propagation probability in each of the cases defined above for the general case of latching window without any electrical attenuation. The error propagation probability P_{error} can be calculated as in Equation (6).

$$P_{prop} = PP\_0\_(t + 1) \cdot PP\_1\_(t + 2)+$$

$$PP\_1\_(t+1) \cdot PP\_0\_(t+2) + PP\_1\_(t+1) \cdot PP\_1\_(t+2)+$$

$$PP\_0\_(t + 1) \cdot PP\_0\_(t + 2) \quad (7)$$

In the presence of electrical attenuation, the signal probability at the output of a gate at a given time cannot be accurately described even if the signal probability is given for each input at any time. This is the topic of current research, [8] among others. Therefore we are reluctant to propose error propagation methods that assume accurate signal probabilities for the lines. Instead, we consider the following approach.

Considering electrical attenuation we observed that the presented theory applies only when the SET width is of at least two times the max gate delay along any path to the output. Let us assume for simplicity that all gates have unit delay and that the SET is of width two units. We generate the error propagation probability assuming that any pulse of unit length arriving within the latching window will be eliminated due to electrical attenuation. Such a pulse can only be generated by pulses shifted by one unit at the input of the output gate as illustrated at gate G in the Figure 3.

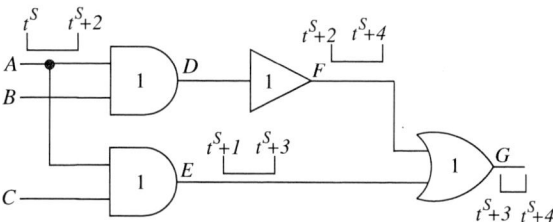

Figure. 3. SET propagation with electrical masking

In Figure 3, the numbers within each gate represent delay. Consider the particle strike at gate A^S at time t^S which initiates the SET 1-0-1 say of width two units (double the gate delay). It is observed that the SET generated at gate A^S is propagated to output G through two different re-convergent paths. The corresponding waveforms at each gate are also shown in Figure 3. Although this probability at output G can be calculated precisely with the technique of this paper, we have experimentally found in Section IV that it does not need to be taken into consideration in our calculations.

The latching window of two time units is defined by $(t, t+1)$ and $(t+1, t+2)$. The remaining show how to modify the proposed method in the presence of electrical attenuation will observe only the pulses of width at least two units propagated to an output due to SET. Therefore, we need two more time units $(t-1,t)$ and $(t+2,t+3)$ to consider the pulses of 2 units for this general case as shown in Figure 4.

Consider a pulse of width two units propagated to output O that finishes at time instance $t+1$ and that there is no error for the remaining of the latching window. We distinguish between such cases based on whether the error free behavior of the output is 0 (as in Figure 4a) or 1 (as in Figure 4b). Furthermore, the pulse of width two units propagated to the output O may start at time instance $t+1$ without any error prior to that during the latching window. We distinguish between such cases based on whether the error free behavior of the output is 0 (as in Figure 4c) or 1 (as in Figure 4d).

In Figure 4a,4b, to ensure that the pulse propagated is of at least two units, we need to observe the error during the time $(t-1,t+1)$. However, prior to time instance $t-1$, the error may or may not be present but not required for the calculation so is a don't care condition. Similarly for the other two cases in Figure 4c,4d, the error after time instance $t+3$ may or may not be present but not required for the calculation so is a don't care condition.

Considering all the four cases as shown in Figure 4, the error propagated P_{prop} can be calculated using Equations (3),(4) as shown below in Equation (5). Each product term in Equation (5) calculates the error propagation probability in each of the cases of Figure 4 defined above for the general case of latching window without any electrical attenuation.

$$P_{prop} = PP\_0\_(t-1) \cdot PP\_0\_(t+1) +$$
$$PP\_1\_(t-1) \cdot PP\_1\_(t+1) + PP\_0\_(t+1) \cdot PP\_0\_(t+3) +$$
$$PP\_1\_(t+1) \cdot PP\_1\_(t+3) \qquad (8)$$

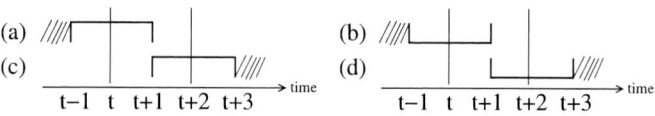

Figure. 4. With electrical attenuation, Pulse finish at time $t+1$ when error free value is (a) zero (b) one, Pulse start at time $t+1$ when error free value is (c) zero (d) one

The performance of this heuristic is evaluated in Section IV using Monte Carlo simulations which can incorporate electrical attenuation with more accuracy even without relying on spice simulation and also more accurate than the signal probabilities.

IV. EXPERIMENTAL RESULTS

The proposed method is implemented in the C language. We present experimental results on the ISCAS'85 combinational circuits [23]. Experiments were performed on a 750MHz Sun Blade 1000 workstation with 1-GB RAM. In our study, a standard cell library in the $TSMC$ $180nm$ CMOS technology is used for fixed delays of the gates. For comparison with Monte Carlo simulations, the precision of the fixed delays are kept intact.

The probability of the SET strike, likelihood ratio, gate that SET strikes are user defined. In all the experiments, the likelihood ratio, the uncorrelated primary inputs probability and the probability of the SET strike on any gate are assumed to be having a probability of 0.5.

In order to verify the validity of the proposed error propagation approach, we devised a Monte Carlo simulation. In the Monte Carlo, randomly generated input vectors are fed to the circuit primary inputs. Consider a particle strike at gate g which generates the SET. For each random input vector, we observe if the SET generated is propagated to the output within a latching window. If propagated then it is a success. We calculate the probability P_{MC} from the Monte Carlo simulations. P_{MC} is defined as the ratio of success to the number of random input vectors applied. Two conditions to terminate the Monte Carlo simulations are 1. P_{MC} must be almost constant and, 2. the number of zero's and one's applied at each input should be equi-probable. This means that the ratio of number of zeros or ones applied at each input to the total number of random input vectors applied is approximately 0.5.

TABLE I
EXPERIMENTAL RESULTS FOR UNIT FIXED DELAY MODEL WITHOUT ELECTRICAL ATTENUATION

Circuit	SET gate	Output gate	Special Case					General Case				
			Analytical		Monte Carlo		% diff	Analytical		Monte Carlo		% diff
			P_{error}	CPU time (seconds)	P_{MC}	CPU time (seconds)		P_{error}	CPU time (seconds)	P_{MC}	CPU time (seconds)	
C2670	305	2970	0.1523	80.13	0.1517	1876.29	0.39	0.12295	94.62	0.1224	1882.29	0.45
C5315	234	6877	0.1905	68.97	0.18938	2583.46	0.57	0.13934	72.18	0.13842	2573.46	0.73
C7552	632	10711	0.14219	122.39	0.14162	2893.58	0.4	0.118	120.86	0.11731	2904.71	0.58

TABLE II
EXPERIMENTAL RESULTS FOR VARIABLE FIXED DELAY MODEL WITHOUT ELECTRICAL ATTENUATION

Circuit	SET gate	Output gate	Special case					General case				
			Analytical		Monte Carlo		% diff	Analytical		Monte Carlo		% diff
			P_{error}	CPU time (seconds)	P_{MC}	CPU time (seconds)		P_{error}	CPU time (seconds)	P_{MC}	CPU time (seconds)	
C2670	305	2970	0.0625	79.64	0.062313	1874.32	0.29	0.0472	87.34	0.04683	1881.72	0.78
C5315	234	6877	0.064516	68.32	0.064	2585.46	0.8	0.04189	69.87	0.0417	2582.61	0.45
C7552	632	10711	0.047205	123.25	0.046875	2891.58	0.7	0.0391	122.85	0.03865	2902.98	0.64

TABLE III
EXPERIMENTAL RESULTS FOR GENERAL CASE IN THE PRESENCE OF ELECTRICAL ATTENUATION

Circuit	SET gate	Output gate	Unit Fixed Delay Model				% diff
			Analytical		Monte Carlo		
			P_{error}	CPU time (seconds)	P_{MC}	CPU time (seconds)	
C2670	305	2970	0.09797	87.56	0.0963	1654.32	1.7
C5315	234	6877	0.186153	70.15	0.18392	2432.34	1.2
C7552	632	10711	0.03514	138.32	0.034816	2688.13	0.94

We have done Monte Carlo simulations assuming that SET width, gate delays and latching window are one unit wide and there is no electrical attenuation. This was done in order to validate the Boolean difference based method that forms the basis of the proposed framework. ISCAS'85 benchmarks are selected so that the proposed method can be validated by comparing with the Monte Carlo simulations. In reality, scanned sequential circuits can have large number of inputs and thus Monte Carlo simulations cannot be performed on such circuits. The results in Table I show that the presented analytical method for the unit delay model is within 0.5% of Monte Carlo on average. Also the CPU time required for the proposed method has a speed-up of at-least 30 on average.

Then we considered fixed gate delays from $TSMC$ $180nm$ CMOS technology whose description(precision) is 6 digits. We injected the SET with double the max gate delay. Latching window was considered to be equal to the buffer delay and no electrical attenuation is considered. Although the latching window cannot be in reality as small as one buffer delay [14]. Under these assumptions we implemented Monte Carlo simulation to validate the probability derived using the proposed analytical method for the special case on Section III. The results listed in Table II show that the presented analytical method for the variable fixed delay model is also within 0.5% of Monte Carlo on average. The proposed analytical method is at-least 30 times more faster than the Monte Carlo simulations.

Column 1 in Tables I,II, lists the benchmark circuit. We considered three large circuits from ISCAS'85 to ensure the scalability of the proposed method. Column 2 reports the gate at which particle strikes. The output gate to which SET propagates is reported in Columns 3. Columns 4-8 in Table II report the results for the experiments under special case with no electrical attenuation. Columns 4,5 report the error propagation probability P_{error} (as in Equation (6)) and the CPU time for the proposed analytical method. Columns 6,7 report the probability P_{MC}, and CPU time for the Monte Carlo simulations. Column 8 reports the % difference between P_{error}, P_{MC}. Columns 9-13 report the experimental results for the general case using variable fixed delay model. Columns 9,10 report the error propagation probability P_{error} (as in Equation (6)) and the CPU time for the proposed analytical method. Columns 11,12 report the probability P_{MC}, and CPU time for the Monte Carlo simulations. Column 13 reports the % difference between P_{error}, P_{MC}.

It is interesting to observe that our approach can handle 6-digit precision in gate delay descriptions. Details on how to implement the analytical method to handle the large precision gate delays are not listed here due to space limitations.

The remainder focuses on the general case in Section III, assuming electrical attenuation. In our method, as well as in Monte Carlo simulations we assumed that a pulse with width of double or more the gate delay is propagated through the gate without any attenuation and that a pulse of gate delay width is eliminated at the gate. The Monte Carlo simulations also incorporated observations from [8] where pulse on more than one input at a gate may result into a pulse whose width is larger than any of the input pulse widths.

Table III reports the results for the proposed general case in the presence of electrical attenuation. In Table III, Column 1 lists the circuit. The gate at which particle strikes and the output gate to which SET is propagated is reported in Columns 2,3 respectively. Columns 4,5 report the error propagation probability P_{error} (as in Equation (6)) and the CPU time for the proposed analytical method. Columns 6,7 report the probability P_{MC}, and CPU time for the Monte Carlo simulations. Column 8 reports the % difference between P_{error}, P_{MC}. The results in Table III show that the proposed method is always within 2% of the Monte Carlo simulations and therefore the assumption in Section III for simplifying the required probability calculation is acceptable. The proposed analytical method is at least 30 times more faster than the Monte Carlo simulations.

V. Conclusions

This is the first method that calculates SET effects for a pre-specified latching window size. We presented a generalized method to compute the error propagation probability when the SET propagated to the output overlaps partially with the latching window. We also proposed a method to compute the error propagation probability in the presence of electrical attenuation at circuit gates. The accuracy of the proposed method is evaluated using Monte Carlo simulations.

References

[1] G. Asadi, M. B. Tahoori, *An Accurate Estimation Method Based on Propagation Probability*, in Proceedings of Design, Automation and Test in Europe, 2005.

[2] R. E. Bryant, *Graph Based Algorithms for Boolean Function Manipulation*, in IEEE Transanctions on Computer Aided Design, pp. 677691, August 1986.

[3] D. R. Cox, H. D. Miller, *The Theory of Stochastic Processes*, John Wiley & Sons Inc., 1968.

[4] S. Gangadhar, S. Tragoudas, *Probabilistic Method for the Impact of an SET in Combinational Logic*, in Proceedings of International On-Line Test Symposium, 2010.

[5] A. Ghosh, S. Devadas, K. Keutzer, J. White, *Estimation of Average Switching Activity in Combinational and Sequential Circuits*, in Proceedings of Design Automation Conference, 1992.

[6] S. Manich, J. Figueras, *Maximizing the Weighted Switching Activity in Combinational CMOS Circuits under the Variable Delay Model*, in Proceedings of European Design and Test Conference, 1997.

[7] V. Massimo, *Accurate Single-Event-Transient Analysis via Zero-Delay Logic Simulation*, in IEEE Transactions on Nuclear Science, Vol. 50, No. 6, 2003.

[8] M. Skoufis, S. Tragoudas, *Coping with Delays and Hazards in Buses and Random Logic in Deep Sub-Micron*, Dissertation at Southern Illinois University Carbondale, Carbondale, IL, August 2009.

[9] N. Miskov-Zivanov, D. Marculescu, *MARSC-Modeling and Reduction of Soft Errors in Combinational Circuits*, in Proceedings of Design Automation Conference, pp. 767-772, 2006.

[10] F. Najm, *Transition Density, A Stochastic Measure of Activity in Digital Circuits*, in Proceedings of the 28th Design Automation Conference, 1991.

[11] B. Narasimham, B. L. Bhuva, *Characterization of Heavy-ion, Neutron and Alpha Particle-Induced Single Event Transient Pulse Widths in Advanced CMOS Technologies*, Dissertation at Graduate School of Vanderbilt University, Nashville, TN, December 2008.

[12] M. Omana, G. Papasso, D. Rossi, C. Metra, *A Model for Transient Fault Propagation in Combinational Logic*, in Proceedings of International On Line Test Symposium, 2003.

[13] A. Papoulis, *Probability, Random Variables and Stochastic Process*, 2nd Edition, McGraw-Hill Book Co., 1984.

[14] E. Salman, E. G. Friedman, *Reducing Delay Uncertainty in Deeply Scaled Integrated Circuits Using Interdependent Timing Constraints*, in Proceedings of ACM International Workshop on Timing Issues in the Specification and Synthesis of Digital Systems (TAU), March 2010.

[15] A. Sanyal, S. Kundu, *On Derating Soft Error Probability Based on Strength Filtering*, in Proceedings of International On Line Test Symposium, 2007.

[16] P. Shivakumar, M. Kistler, W. Keckler, D. Burger, L. Alvisi, *Modeling the Effect of Technology Trends on the Soft Error Rate of Combinational Logic*, in Proceedings of Dependable Systems and Networks, 2002.

[17] R. Thara, B. Sanjukta, *A Timing-Aware Probabilistic Model for Single-Event-Upset Analysis*, in IEEE Transactions on VLSI Systems, Vol. 14, No. 10, 2006.

[18] C. Yu, C. Zhuo, *Soft Error Verification for Sequential Circuits*, Department. of EECS, University of Michigan, Ann Arbor.

[19] N. M. Zivanov, D. Marculescu, *Soft Error Rate Analysis for Sequential Circuits*, in Proceedings of Design, Automation and Test in Europe, 2007.

[20] M. Zhang, N, R. Shanbhag, *A Soft Error Rate Analysis (SERA) Methodology*, in Proceedings of International Conference on Computer-Aided Design, 2004.

[21] B. Zhang, M. Orshansky, *Symbolic Simulation of the Propagation and Filtering of Transient Faulty Pulses*, in Workshop on System Effects of Logic Soft Errors, Urbana Champion, IL, April 2005.

[22] Zhang, W. Shen, M. Orshansky, *FASER: Fast Analysis of Soft Error Susceptibility for Cell-Based Designs*, in Proceedings of International Symposium on Quality Electronic Design, 2006.

[23] www.heera.engr.siu.edu/usr/local/benchmarks

[24] http://vlsi.colorado.edu/ fabio/CUDD/cuddExtDet.html

Adaptive Error-Prediction Flip-flop for Performance Failure Prediction with Aging Sensors

C. V. Martins, J. Semião
Univ. of Algarve / INESC-ID Lisbon
Faro, Portugal
{cvmartins, jsemiao}@ualg.pt

J. C. Vazquez, V. Champac
INAOE
Puebla, Mexico
{jcvazquez, champac}@inaoep.mx

M. Santos, I. C. Teixeira, J. P. Teixeira
INESC-ID Lisbon / IST-UTL
Lisbon, Portugal
{marcelino.santos, isabel.teixeira, paulo.teixeira}@ist.utl.pt

Abstract—**This paper presents a new approach on aging sensors for synchronous digital circuits. An adaptive error-prediction flip-flop architecture with built-in aging sensor is proposed, performing on-line monitoring of long-term performance degradation of CMOS digital systems. The main advantage is that the sensor's performance degradation works in favor of the predictive error detection. The sensor is out of the signal path. Performance error prediction is implemented by the detection of late transitions at flip-flop data input, caused by aging (namely, due to NBTI), or to physical defects activated by long lifetime operation. Such errors must not occur in safety-critical systems (automotive, health, space). A sensor insertion algorithm is also proposed, to selectively insert them in key locations in the design. Sensors can be always active or at pre-defined states. Simulation results are presented for a balanced pipeline multiplier in 65 nm CMOS technology, using Berkeley Predictive Technology Models (PTM). It is shown that the impact of aging degradation and/or PVT (Process, power supply Voltage and Temperature) variations on the sensor enhance error prediction.**

Keywords: aging sensor; performance failure prediction; NBTI; delay insertion.

I. INTRODUCTION

In nanometer technologies, variability is becoming one of the leading causes for chip failures and delayed schedules [1]. In general, the root causes can be (1) Process variability (e.g., V_{th} variations, thickness variations [2], gate L (Length) and W (Width) variations, etc.), (2) Operational variability (multiple design modes, Dynamic Voltage and Frequency Scaling [3], power gating, etc.), (3) Operation-dependent variability (e.g., PVT (Process, power-supply Voltage, and Temperature) variations [4], crosstalk [5], IR-drop [6], etc.) and (4) Long-term variability (aging [7] effects, namely due to NBTI (Negative Bias Thermal Instability) [8], radiation, etc.). Variability causes increasing uncertainty in system behavior, namely on its performance.

Moreover, variability also decreases circuit dependability, i.e., its ability to deliver the correct functionality within the specified time frame. Hence, lower circuit's dependability and reliability [9][10] are to be expected, when moving to low nanometer range. Clearly, PVTA (PVT and Aging) variations degrade circuit's dependability. Conservative approaches, with wider relative circuits' time slacks, are required when technology is down scaled [11]. Time slack (τ_{slack}) is the time period exceeding the longest propagation delay time between memory elements, to accommodate PVT variations.

Semiconductor aging [7] causes long-term performance degradation, and may activate physical defects latent in production. For instance, MOS threshold voltage (V_{th}) parametric variations may be induced either due to holes trapped in the thin gate oxide (SiO_2), or due to Si/SiO_2 surface change. The first effect tends to decrease V_{thN} and to increase $|V_{thP}|$ (NMOS and PMOS V_{th}, respectively). The second effect, under high radiation, tends to increase V_{thN} and $|V_{thP}|$. Hence, V_{th} shifts will modify the timing performance of digital circuits.

NBTI has been identified as the dominant long-term effect in nanometer CMOS technologies [8][12]. It primarily affects PMOS transistors, increasing $|V_{thP}|$ along the time. This will ultimately cause a delay fault. For safety-critical (e.g., automotive) and mission-critical applications (e.g., space), this must not occur. Performance failure *prediction* must be implemented.

The purpose of this paper is to present an adaptive error prediction flip-flop (AEP-FF), to be used in a performance failure prediction methodology for safety-critical, high-performance systems. This methodology is based on monitoring long-term performance degradation of CMOS digital systems. In particular, key flip-flops (FF) are chosen to include the aging sensor functionality and monitor FF data input late transitions. The proposed AEP-FF performs locally all the monitoring procedure, reducing to a minimum the global interconnections for aging sensor insertion. The observation (or guard-band) interval, t_g, at the end of the clock cycle, is defined by design, so the time period in which abnormal delays are observable is user's defined. However, with the novel sensor's architecture, its sensitivity (measured by t_g) increases with its PVTA variations. This way, the sensor FF will adapt and increase the guard-band, as circuit variability increases with aging.

The paper is organized as follows. Section II briefly reviews previous work on aging and its impact on system performance, and on aging sensors. In section III, the proposed flip-flop architecture is presented. Section IV describes the aging sensor methodology, including sensor insertion procedure and circuit reconfiguration. In section V, simulation results are presented. Finally, conclusions are summarized in section VI.

II. PREVIOUS WORK

The impact of NBTI on digital circuit performance has been under a significant research effort [8][12][13]. NBTI modeling in static and dynamic operation has been proposed [13]. Solutions to enhance tolerance to V_{thP} degradation have also been presented [7][14], and [12] shows how design optimization can restrict NBTI performance degradation. Moreover, various aging sensor topologies have been presented [15][16][17][18] to globally detect circuit's performance degradation.

A different aging sensor approach is the circuit failure prediction technique proposed by M. Agarwal et al. in [19]. The underlying idea is to anticipate system failure, before it really occurs. Their major application was to reduce the pessimistic worst-case delay to accommodate PVT variations, which significantly limits system performance.

More recently, an aging sensor methodology focusing a different application was presented in [11][20][21][22]. Here, aging monitoring is performed during product lifetime under heterogeneous VT (power supply Voltage and Temperature) variations. The methodology includes (a) programmable aging sensor design, resilient to aging and showing low sensitivity to PVT variations, and (b) a monitoring procedure, with automatic sensor insertion.

Nevertheless, the proposed methodology has limitations, namely (1) sensor must have better performance and lower sensitivity to PVT variations than the CUT (Circuit Under Test); (2) methodology is based in monitoring time interval (guard-band) that needs to be synchronous with the clock; (3) sensors are active in pre-defined periods during which the CUT critical paths may not be activated. The present paper presents a novel FF architecture with built-in aging sensor, based on the previous work done in [21]. However, a different approach is used, in order to deal with the reported limitations of previous aging sensors topologies.

III. ADAPTIVE ERROR-PREDICTION FLIP-FLOP

In a synchronous digital system, data signals processed in combinational modules are registered in storage elements (FF). However, if a signal propagation delay is large enough (e.g., due to PVTA variations), it will induce a de-synchronization effect due to the increased difference between the critical path propagation delay, T_{pd}, and the clock distribution network delay, τ. Consequently, an aging sensor must monitor performance degradation locally, at key memory cells, where synchronization errors start to occur.

We refer these FF as CME (Critical Memory Elements). Aging sensors must be integrated in FF terminating critical or near-critical paths (CP). The sensor monitors late transitions at the FF's data input.

The topology and principles of operation of the proposed Adaptive Error-Prediction Flip-flop (AEP-FF) are shown in Figures 1 and 2, respectively. The Delay Element (DE) delays data signals captured at the Master Latch output, during CLK low state. The Stability Checker (SC) analyzes data transitions during CLK high state. This way, the DE propagation delay is the effective observation (or guard-band) interval, t_g, used by the sensor. Late transitions at FF data input (propagated to the Master Latch output) will be identified by the SC. As it will be shown later, SC has on-state retention logic, to discard the use of an additional latch to store the aging sensor output signal (AS_OUT).

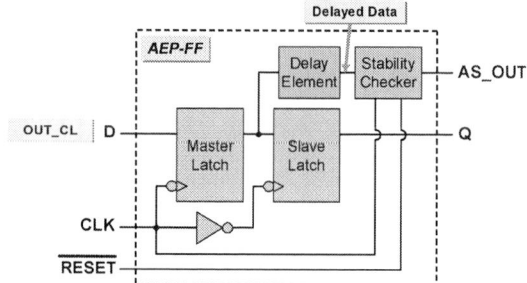

Figure 1. Adaptive Error-Prediction Flip-flop topology.

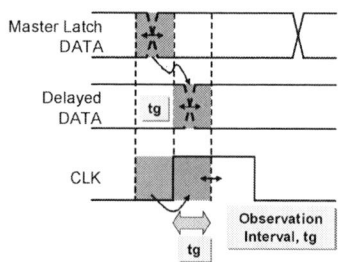

Figure 2. Aging sensor flip-flop principles of operation.

Using the flip-flop data signal at the output of the Master Latch to drive the DE, instead of the FF input data signal, D, as in previous aging sensor architectures [19][20], simplifies DE design. Basically, the new delay element is a simple buffer that introduces a delay to create a virtual guard-band where late transitions at FF data input are signalized. The loading effect of the sensor is inside the FF; hence, it does not explicitly impact the signal path. Moreover, if PVTA variations occur in DE, this virtual guard-band will increase accordingly. It is possible to create a power-on state for the DE, and activate it in short periods to maintain its propagation delay approximately constant when compared with the circuit's performance degradation. Even when the sensing operation is always ON, the increased workload of the DE will cause sensor's guard-band to increase as aging effects cumulatively degrade DE performance. This way, the sensor's sensitivity is adapted with the cumulative aging

978-1-61284-657-6/11 $26.00 © 2011 IEEE

degradation of the circuit. Another advantage is that the guard-band signal does not need to be distributed as a second balanced clock to the sensing FFs, as in [19][20].

Figure 3 presents three typical DE architectures. These buffers have different delay capability and different aging performance degradations. The DE architecture should be chosen according to the technology used: e.g., architecture (a) does not work in nanometer technologies, (b) and (c) do; the time slack margin, τ_{slack}/T_{clk}, is also relevant. In fact, for high performance circuits (low τ_{slack}/T_{clk}), architecture (c) is preferred. Moreover, changing W/L transistors ratios also change the sensor's effective guard-band, t_g. Unlike in [20][21][22], t_g is not programmable, it is defined at design time. However, t_g is adaptive with PVTA variations, enhancing sensor's detection sensitivity.

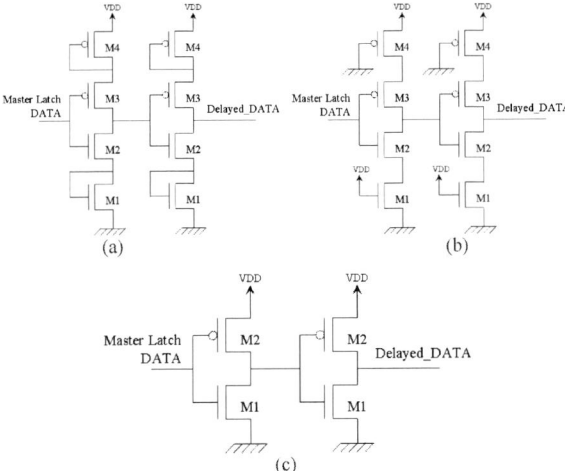

Figure 3. Delay element typical architectures.

The novel Stability Checker (Figure 4) is implemented with dynamic CMOS logic and has built-in on-retention logic. During CLK low state, and considering that AS_out signal is low, X and Y nodes are pulled up (making AS_out to stay low). When CLK signal changes to high state, M3 and M4 are OFF, and according to Delayed_DATA signal, one of the nodes X or Y changes to low. If, during the high state of the CLK, a transition in Delayed_DATA occurs, the high X or Y node is pulled down by transistor M2 or M5, respectively, driving AS_out to go high. From now on, M9 transistor is OFF. Hence, X and Y nodes are not pulled up during CLK low state, unless the active low RESET signal is active. X and Y nodes remain low, helped by transistors' M3 and M4 activation during AS_out high state. For the RESET signal to restore the cell's sensing capability, it must be active, at least during the low state of one clock period.

The proposed SC architecture, with the on-retention logic implemented with transistors M3, M4, M8 and M9, does not need an additional latch to retain the SC output signal when it's active. However, additional research must be pursued to analyze the SC operation in the presence of cross-talk noise, especially when CLK signal is activated, as false-positive errors may be signalized. Moreover, only FF's internal clock

signal is triggering the beginning of the observation interval, t_g. As mentioned, guard-band interval is the DE propagation delay, ultimately limited by half the clock cycle (when CLK signal is high). We refer it as a *virtual guard-band* because there is no signal explicitly representing the observation interval. Each sensing FF will have its own unique PVTA-dependent guard-band (each local DE may age differently).

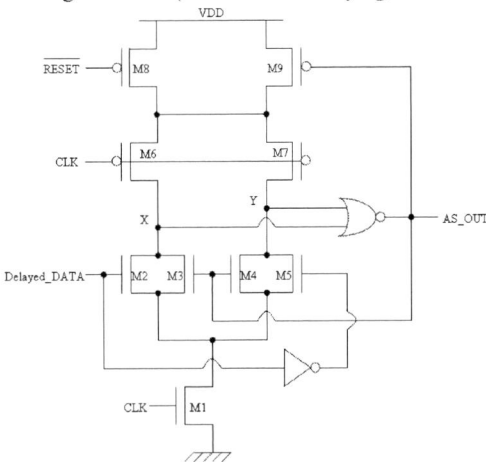

Figure 4. Stability checker architecture with on-retention logic.

IV. AGING MONITORING PROCEDURE

The monitoring procedure comprises (1) aging sensors (AS) characteristics, (2) overhead attributes, (3) monitoring effectiveness, and (4) sensor insertion criteria, algorithm and tool. Valued AS characteristics are: low power consumption, area overhead and performance degradation, and sensitivity to PVTA variations (increasing t_g as variability increases). Due to DE input location (at the Master Latch output), the performance degradation of the AEP-FF is negligible.

Overhead attributes are hardware overhead. In terms of hardware, one metric is defined: global overhead, GO (GO=n_{AS}/m, where n_{AS} is the number of inserted AS and m is the total number of memory elements in the CUT).

In terms of monitoring effectiveness, two issues are critical: looking at the right place, and long enough to uncover abnormal delays (the activation of the critical paths is an important task, but is out of the scope of this work). Hence, sensors must be inserted at the end of critical (CP) and near-critical delay paths. How many near-critical paths should be sensed? A CP delay variation margin (α) is user's defined, in order to allow the decision on how many and which near-critical paths should be sensed.

The following sensor insertion criteria are considered:
1. CP identification, and signal path delay map building (using a Statistical Timing Analysis tool, e.g., PrimeTime™). The path with the longest propagation delay time (t_{pdmax}) is identified, and all signal paths are ranked, according to their delay times.
2. CME identification, according to user's defined (α) monitoring effectiveness. Selected CME end all signal

978-1-61284-657-6/11 $26.00 © 2011 IEEE   205

paths whose propagation delay exceeds [$\alpha.t_{pdmax}$]. Typically, α=90%, to consider PVTA variations.

3. Circuit reconfiguration at logic-level, to automatically replace critical flip-flops (CME) with AEP-FFs.

An algorithm to apply the three criteria and perform sensor insertion is embedded in a proprietary tool, DyDA (Dynamic Delay Analyser) [4]. DyDA (1) collects Prime Time™'s output data, (2) generates path delay distribution and potential CME insertion, and (3) performs netlist reconfiguration (upon user's decision (α)), replacing few FF by AEP-FFs. To help user's decision on defining α, DyDA tool generates a ranked list of all CMEs, based on their input cone's critical path and a graph with the distribution of the memory elements' input cone CP and all paths distribution.

Figure 5 depicts path delay and CME distribution curves for a 2-stage, 4-bit Pipeline Multiplier (PM) (AMS 0.35μm), used as CUT1. PM is a 36 FF, 52 gate structure, 9 (8 data, 1 CLK) inputs, 8 outputs (Fig. 6). As shown, for the longest paths, with α=90%, 5 CME are needed. If a larger safety margin is required (α<90%), 4 additional CME are needed. Note that more than one large delay path converge to the same FF. In fact, in this case, 23 long signal paths converge in 5 CME of the 20 FF at the end of pipeline stage 1.

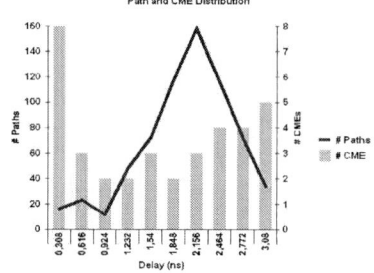

Figure 5. Dyda's path delay and CME distribution for CUT1 (PM, 0.35μm).

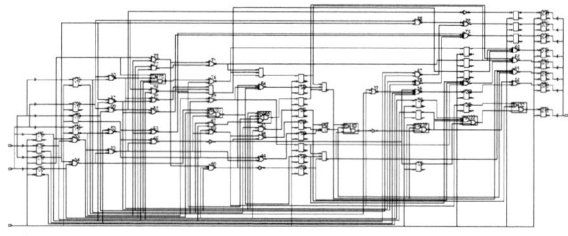

Figure 6. CUT1 (2-stage, 4-bit pipeline multiplier (PM)).

V. SIMULATION RESULTS

Simulation results are presented for CUT1 (PM) (0.35μm and PTM 65nm [23]) and CUT2 (inverter chain creating a CP example) (figure 7) (PTM 65nm). A 1V nominal V_{DD} is considered, with variations from 1.2V to 0.8V. T variations are restricted from 27°C to 150°C. WC (worst case) VT conditions are V_{DD}=0.8V, T=150°C. Aging fault injection, by V_{thP} modification, is performed by Spice VTHO parameter modulation. NBTI aging is assumed to equally degrade V_{thP} of all PMOS transistors.

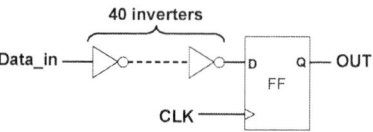

Figure 7. CUT2 (40 inverters' chain, creating a typical CP example).

Considering 65nm PTM CUT2 simulations with HSPICE, the propagation delay time of CUT2 CP at nominal conditions (NC) leads to 328ps (L-H transition). WC VT conditions lead to t_{pLH}=805ps, which represents a serious degradation, as compared to nominal t_{pLH}. The clock period is defined as $T_{clk}=\tau_0+\tau_{slack}$, where $\tau_0=t_{pLH}+t_{SU}$, and t_{SU} is the FF set-up time, namely t_{SU}=39ps. Hence, τ_0= 844ps. In order to accommodate process variations, a time slack was set as 30% of τ_0. Therefore, T_{clk} = 1.1ns will guarantee correct operation under worst case PVT conditions.

Simulations were carried out to compute propagation delay degradation due to NBTI-induced aging. The impact of up to 30% variations in $|V_{thP}|$ of the PMOS in the CUT on % variations of t_{pLH} and t_{pHL} was evaluated (Fig. 8): 30% V_{thP} variation of leads to $\Delta t_{pl\,H}$= 16% at VT NC. Worst case VT scenario leads to higher delay degradation (35%).

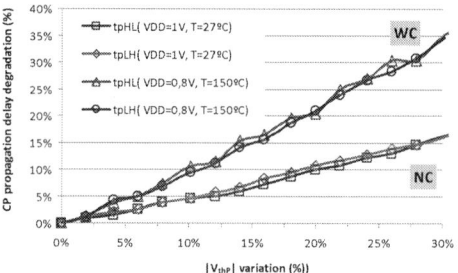

Figure 8. CP's propagation delay variations (Δt_{pLH} and Δt_{pHL}) in CUT2 with aging (ΔV_{thP}), for nominal (NC) and worst-case (WC) conditions.

For CUT2, AS insertion was carried out, with selective AEP-FF insertion, implemented with the DE shown in fig. 3-c) (with minimum-size PMOS transistors). The AEP_FF performance degradation was evaluated: only 5% set-up time degradation was measured ($t_{SU\_AEP\_FF}$ = 41ps) for WC conditions (when compared to T_{CLK}, it's irrelevant), while no hold-time degradation was measured ($t_{Hold\_AEP\_FF}$ = 9ps).

Error-prediction by the AEP-FF starts to occur for 10% degradation in V_{thP} and for 9% degradation in the critical path propagation delay (L-H transition). The sensor continues its error-prediction until the error occurs (when t_{slack} is violated). The sensor's effective virtual guard-band, t_g, was computed under VT variations (Fig. 9) (only L-H transition is shown, as H-L transitions' results are similar) and under V_{thP} aging variations (Fig. 10). Moreover, when analyzing t_g variation with VTA variations (Figs. 9 and 10), unlike in [19][20], our proposed sensor increases the guard-band interval as variability in the circuit increases with V reduction and T and $|V_{thP}|$ increase. This is an important result, because V_{thP} aging degradation, or V depletion, or T increase, unlike other aging sensor solutions, work in favor of the error-prediction methodology. In fact, the AS becomes

978-1-61284-657-6/11 $26.00 © 2011 IEEE

more sensitive to CUT delay variations, as VTA variability increases. Previous works [19][20] put a heavy burden on AS design, to guarantee that the AS sensitivity to VTA variations does not reduce significantly the detectability window of the sensor.

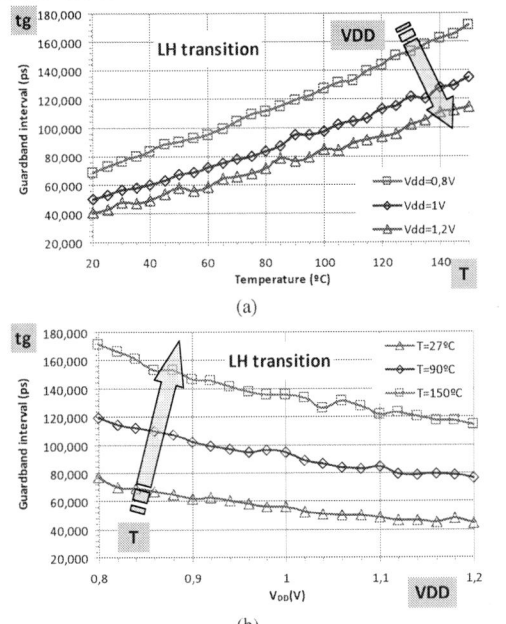

(a)

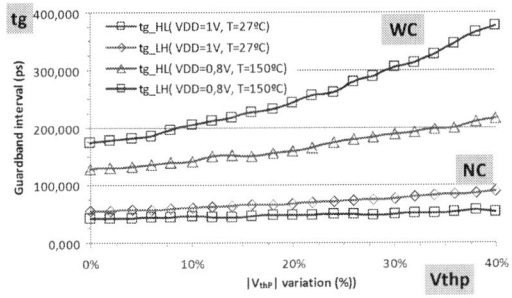

(b)

Figure 9. AS effective guard-band variation (L-H transition). (a) under temperature variations; (b) under power-supply voltage variations.

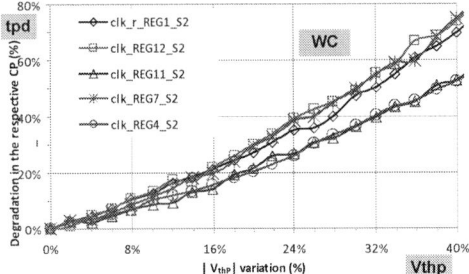

Figure 10. AS effective guard-band variation (L-H and H-L transitions) with aging (V_{thP} variation), for nominal and WC conditions in CUT2.

Furthermore, in [19][20][21] SC operation significantly reduces the effective t_g of the sensor. In high performance circuits, the SC must also react fast to maintain the predictive error detection capability. In the proposed AEP-FF, the data signal that drives the SC is always stable, due to the fact that Delayed_DATA signal is analyzed at the output of the master latch of the FF. Hence, there is no limitation imposed by the eventually slow operation of the SC.

For the CUT1 Pipeline Multiplier (PM, Fig. 6), implemented with 65nm PTM, the 5 most critical memory elements are the same as the ones identified by DyDA on the

0.35µm version (Fig. 5). Fig. 11 depicts their WC propagation delay variation with NBTI-induced aging.

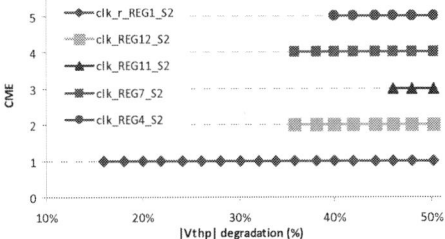

Figure 11. Propagation delay degradation of 5 CP ending at 5 CME in CUT1, under % variations in the PMOS V_{thP}, under worst case VT conditions.

After PM circuit reconfiguration, replacing the 5 CME with the AEP-FF, HSPICE simulations were performed with T_{clk}=780ps, where 60% corresponds to the propagation delay of the CP under WC conditions, with τ_{slack}=40% T_{clk} to accommodate WC PVT variations. Fig. 12 shows, for PM (CUT1) and WC VT conditions, the abnormal delays detection ranges at the inputs of the 5 CME (slowest HL or LH transitions), showing how effective the aging sensor is. The sensor can detect from 16% to 40% V_{thP} degradation.

Figure 12. Effective detection ranges in percent variations of V_{thP} degradation variations, for CUT1 under NC VT conditions at the 5 CME

For safety-critical applications, the sensor must provide reliable operation also in the presence of P (Process) variations. Monte Carlo (MC) simulations under WC VT conditions were carried out. In all MC simulations, a Gaussian distribution with +/-3σ variation of +/-10% of the nominal values is assumed [24] for 3 MOSFET parameters: L_{eff}, t_{ox} and V_{th}. MC simulations were run varying MOSFET parameters in the CUT and in the sensor's transistors. For each set of MC simulations, 30 runs were performed. Under P variations, detection on fast samples requires high V_{thP} degradation. We define the Detection Probability (DP) as the percentage of the 30 MC runs that detect the abnormal delay. Fig. 13 presents MC results for DP in the 5 CME of CUT1.

The monitoring methodology was used to define sensor insertion in a set of benchmark circuits (Table 1): ITC'99 benchmarks, 2-stage 4-bit Pipeline Multiplier (PM 4-2), 4-stage, 16-bit PM (PM 16-4) and a simple PIC controller. Results show that few AEP-FF need to be inserted. Of course, for sequential circuits it is less rewarding. However, propagation delay distribution depends on circuit synthesis,

978-1-61284-657-6/11 $26.00 © 2011 IEEE

which can be used to restrict the global overhead (GO), making less circuit nodes prone to unsafe time response.

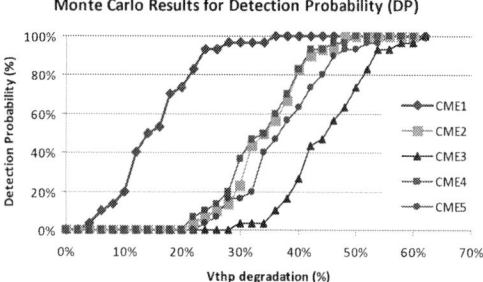

Figure 13. Monte Carlo statistic results for first detections at the 5 CME for percent variations of V_{thP} degradation variations, in CUT1 under WC VT conditions.

TABLE I. AGING SENSOR INSERTION USING DyDA TOOL (AMS 0.35μM)

Circuit	# FF	# AEP-FF	GO (%)
B01	5	2	40.0
B03	30	22	73.3
B06	9	3	33.3
B09	45	23	51.1
Bal. PM 4-2	36	5	13.9
Simple PIC	66	16	24.2
Bal. PM 16-4	480	7	1.5

VI. CONCLUSIONS

In this paper, a new Adaptive Error-Prediction Flip-flop (AEP-FF) was presented, to be used in a performance failure prediction methodology for safety-critical applications. It was shown that sensor's guard-band increases with V_{DD} depletion, T increase and NBTI-induced aging. The AEP-FF includes new DE and a new SC architecture with on-retention logic feature. AEP-FF architecture reduces the performance degradation overhead due to sensor insertion, as compared to previous solutions. Moreover, SC operation does not limit the effective detection window of the sensor, as in previous AS topologies. As sensors age, sensor's performance degradation works in favor of the predictive error detection. The methodology includes aging sensor design and insertion, using DyDA tool. Simulation results demonstrate AEP-FF robustness on detecting late transients at FF input data signal, even in the presence of its own aging.

Future work includes cell library design, for complete methodology automatic insertion, workload-dependent NBTI effects, analysis of sensor's robustness in the presence of cross-talk noise and more extensive MC analysis to account for process variations, and will be reported in the future.

ACKNOWLEDGMENT

The work has been partially supported by ENIAC SE2A Project, and by FCT (INESC-ID multiannual funding) through the PIDDAC Program.

REFERENCES

[1] Mentor Graphics Corp., P&R Whitepaper, "Design for Variability: Managing Design, Process, and Manufacturing Variations in Physical Design",http://www.mentor.com/resources/techpubs/upload/mentorpaper_43548.pdf, Oct., 2008.

[2] A. Asenov, S. Kaya, J. Davies, S. Saini, "Oxide thickness variation induced threshold voltage fluctuations in decanano MOSFETs: a 3D density gradient simulation study", Superlattices and Microstructures, Vol. 28, pp. 507-515, 2000.

[3] S. Das, D. Roberts, S. Lee, S. Pant, D. Blaauw, T. Austin, T. Mudge, K. Flautner, "A self-tuning DVS Processor using Delay Error Detection and Correction", IEEE J. Solid-Sate Circuits, vol. 41, n° 4, pp. 792-804, April 2006.

[4] J. Semião et al., "Time Management for Low-Power Design of Digital Systems", ASP Journal of Low Power Electronics (JOLPE), Vol. 4, N° 3, pp. 410–419, December 2008.

[5] M. Cuviello, S. Dey, X. Bai, Y. Zhao. "Fault Modeling and Simulation for Crosstalk in System-on-Chip Interconnects". Int. Conf. on Computer Aided Design, pp. 297-303, Nov. 1999.

[6] H. Chen, L. Wang. "Design for Signal Integrity: The New Paradigm for Deep-Submicron VLSI Design". Proc. Int. Symp. on VLSI Technology, pp. 329-333, June 1997.

[7] S.V. Kumar, C.H. Kim, S. Sapatnekar, "Adaptive Techniques for Overcoming Performance Degradation due to Aging in Digital Circuits", Proc. IEEE ASP-DAC, pp. 284-289, Jan. 2009.

[8] V. Reddy, et al., "The Impact of NBTI on the Performance of Combinational and Sequential Circuits", Proc. ACM/IEEE Design Automation Conf. (DAC), pp. 364-369, 2007.

[9] J.W. McPherson, "Reliability Challenges for 45 nm and Beyond", Proc. ACM/IEEE Design Autom. Conf. (DAC), pp. 176-181, 2006.

[10] B. C. Paul et al., "Temporal Performance Degradation under NBTI: Estimation and Design for Improved Reliability of Nanoscale Circuits", Proc. DATE, pp. 780-785, 2006.

[11] J. Semião et al., "Delay-Fault Tolerance to Power Supply Voltage Disturbances Analysis in Nanometer Technologies", Proc. IEEE Int. On-Line Testing Symp. (IOLTS), pp. 223-228, 2009.

[12] K. Kang, S. Gangwal, S. Phil Park, and K. Roy, "NBTI Induced Performance Degradation in Logic and Memory Circuits: How Effectively Can We Approach a Reliability Solution?", Proc. Asia / South Pacific Design Autom. Conf. (ASP-DAC), pp. 726-731, 2008.

[13] R. Vattikonda, W. Wang, Y. Cao, "Modeling and Minimization of PMOS NBTI Effect for Robust Nanometer Design", Proc. DAC, pp. 1047-1052, 2006.

[14] W. Wang et al., "An Efficient Method to Identify Critical Gates under Circuit Aging", Proc. ICCAD, pp. 735-740, 2007.

[15] J. Tschanz, et al. "Adaptive Frequency and Biasing Techniques for Tolerance to Dynamic Temperature-voltage Variations and Aging," Proc. IEEE Int. Solid-State Circ. Conf. (ISSCC), pp. 292-293, 2007.

[16] C. R. Gauthier, P. R. Trivedi, G. S. Yee, "Embedded Integrated Circuit Aging Sensor System", Sun Microsystems, US Patent 7054787, May 30, 2006.

[17] D. Kim, J. Kim, M. Kim, J. Moulic, H. Song, "System and Method for Monitoring Reliability of a Digital System", IBM Corp., US Patent 7495519, Feb. 24, 2009.

[18] J. Keane, T. Kim, C. Kim, "An on-chip NBTI sensor for measuring PMOS threshold voltage degradation", Proc. Int. Symp. on Low Power Electronics and Design (ISLPED), pp. 189-194, 2007.

[19] M. Agarwal, et al., "Circuit Failure Prediction and Its Application to Transistor Aging". Proc. VLSI Test Symp. (VTS), pp. 277-286, 2007.

[20] J. C. Vazquez et al., "Built-In Aging Monitoring for Safety-Critical Applications", Proc. IEEE Int. On-Line Test Symp. (IOLTS), pp. 9-14, 2009.

[21] J. C. Vazquez et al., "Low-sensitivity to Process Variations Aging Sensor for Automotive Safety-Critical Applications", Proc. IEEE VLSI Test Symposium (VTS), pp. 238-243, 2010.

[22] J.C. Vazquez, et al., "Predictive Error Detection by On-line Aging Monitoring", Proc. IEEE Int. On-Line Test Symp. (IOLTS), 2010.

[23] Predictive Technology Model (PTM), http://www.eas.asu.edu/~ptm/.

[24] Int. Technology Roadmap for Semiconductors, http://www.itrs.net/

Calibrated high-efficiency testing and modelling methodologies for concentrated multi-junction solar cells

Jeffrey F. Wheeldon, SUNLab, School of Information Technology and Engineering, University of Ottawa, Ottawa, ON, Canada

Abstract

Multi-junction solar cells (MJSC) under concentrated sunlight are leading the way forward to record efficiencies (>40%, under AM1.5D solar spectrums) and are ideally suited for utility-scale energy production. MJSC are able to achieve high efficiencies by partitioning the solar spectrum using multiple n-p junctions composed of III-V and IV group semiconductors. To reduce the cost associated with their more complex epitaxial growth processes, MJSC are coupled with concentrator optics (typical concentrations of 500-1000 suns). The University of Ottawa's SUNLab focuses its research on improving the efficiency of concentrated high-efficiency MJSC systems. This research takes place in three crucial areas: laboratory characterization of MJSC, numerical modelling of thermal, electrical and optical properties of cells and systems, and outdoor testing of concentrated photovoltaic systems. Laboratory testing of MJSC are performed under controlled temperature and spectral irradiance conditions, which allow accurate efficiency comparisons of different solar cell designs. Flash and continuous solar simulators illuminate the solar cells with intensities of up to 1000 suns. Flash solar simulators are the standard method of characterizing solar cells under concentration, whereas testing under continuous concentrated solar radiation provides critical additional information regarding thermal reliability (see Fig. 1(a)). MJSC are composed of subcells connected electrically in series, making their performance sensitive to changes in the solar spectrum. Spectral sensitivity is quantified by performing quantum efficiency measurements that determine a subcell's ability to extract an electron for a given incident photon. For each experimental measurement, numerical modelling assists in determining the limiting elements within a design. These numerical simulations cover the critical areas, such as the concentrating optics with complex raying tracing, current-voltage characteristics of the MJSC using semiconductor device simulators, and finite element method thermal models of the MJSC on carrier. Outdoor testing of a solar system measures the efficiency under real-world conditions. The University of Ottawa's outdoor test site has three solar tracking stations that measure the power output of a photovoltaic system while monitoring the irradiance and spectrum of the sun, which determine overall system efficiencies (see Fig. 1(b)).

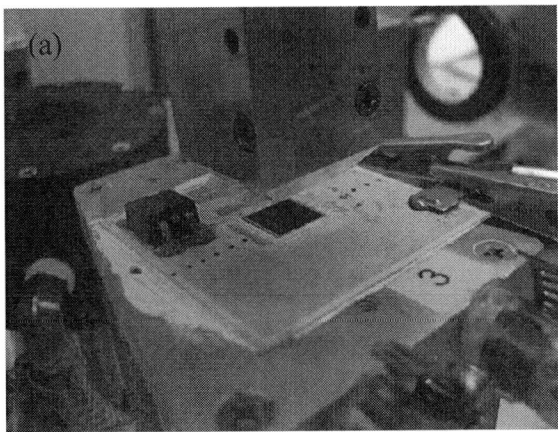 

Figure 1: (a) Laboratory characterization of a MJSC under continuous concentrated solar radiation. (b) University of Ottawa's solar tracker station, in which the performance of concentrated photovoltaic systems are measured over long periods of time under real-world conditions.

Session 9

978-1-61284-657-6/11 $26.00 © 2011 IEEE

Coverage Closure in SoC Verification: Are we chasing a mirage?

Organizer/Moderator: Shobha Vasudevan, Associate Professor at University of Illinois at Urbana Champaign, USA

Abstract:

With over 78% of designs being heterogeneous integrations of diverse components, SoCs are ubiquitous. This integration, though, brings with it the malaise of challenges in verification and validation. SoC verification has a number of unique challenges beyond traditional ASIC type of designs. The typical SoC flow consists of the following development phases: System Design, Software Design, HW/SW Integration, SoC HW Integration and HW IP design.

The typical verification/validation platforms include **Virtual Prototyping, RTL Simulation, FPGA Prototyping/Emulation, Silicon validation.** This panel will have specialists in all the 4 verification/validation platforms targeted at the various development phases.

Coverage closure, or complete confidence in the design, is a dream that every SoC developer, vendor, IP provider and user nurtures. As the number of integrated components increases, with different levels of IP protection and reuse, verifying the whole is much harder than the sum of its parts. How do contemporary industries "let go" of the design in the absence of coverage closure? Is a warm, fuzzy feeling about the design enough or do we need to get serious about achieving coverage closure? How realistic is the notion of achieving coverage closure?

Watch the panelists take on these serious questions that plague the SoC design industry and hear expert points of view.

Session 10

A Scan Cell Architecture for Inter-Clock At-Speed Delay Testing

Kyoung Youn Cho and Rajagopalan Srinivasan
NVIDIA
2701 San Tomas Expressway
Santa Clara, CA 95050
Email: {kcho, rasrinivasan}@nvidia.com

Abstract—**At-speed delay testing is inevitable for improving the test quality of modern high-speed semiconductor chips. This paper presents a scan cell architecture for at-speed testing of delay faults in inter-clock logic. The technique utilizes commercially available ATPG tools for test pattern generation and internal PLL clocks for test pattern application. The hardware modification is contained within the scan cells and no additional global routing is required. Simulation results using three industrial designs demonstrate that the technique is effective in detecting delay faults in inter-clock logic.**

Keywords-**design for testability (DFT); scan cell architecture; delay testing; at-speed testing; inter-clock logic**

I. INTRODUCTION

As the semiconductor chip fabrication technology advances, delay defects become more important defects than static or permanent defects that can be modeled as single stuck-at faults (SSFs) [Mak 04]. In order to generate test patterns for detecting delay defects, the transition fault model [Waicukauski 87], the gate delay fault model [Carter 87], the segment delay fault model [Heragu 96], and the path delay fault model [Smith 85] can be used. Among the delay fault models, the transition fault model has been used most frequently in production testing because of its advantages over other fault models: SSF automatic test pattern generation (ATPG) tools can be easily modified to generate transition fault test patterns and the number of faults increases linearly with the number of gates [Wang 06]. The *transition fault model* represents defects that delay transitions at gate inputs and outputs, consisting of two types: the slow-to-rise type and the slow-to-fall type.

For at-speed transition fault testing in high-speed semiconductor chips, the launch-capture clocks are usually generated by internal phase-locked loops (PLLs) rather than primary input clock signals driven by automatic test equipments (ATEs). This is because ATEs may not provide at-speed clocks to the input pins of device under tests (DUTs). One issue with using internal PLL clocks is that current ATPG tools assume that clock signals are controlled by primary input pins. Therefore, during test pattern generation it is required to modify the circuit model such that clock signals are driven by primary input pins, while during test pattern application on an ATE launch-capture clocks are derived from PLLs. In [Lin 03], at-speed clock sequences

were integrated into an ATPG algorithm to generate test patterns utilizing clock sequences generated by PLLs.

These days, industrial chips contain multiple clock domains. Testing delay faults in inter-clock logic remains a difficult problem, whereas that in intra-clock logic is relatively easy. In order to improve delay defect coverage, the delay faults in inter-clock logic need to be detected in at-speed delay testing.

This paper presents a scan cell architecture for improving the test quality of delay faults in inter-clock logic. The two clocks considered in this research are synchronous clocks derived from a single clock source such as a PLL; the clock frequency of one clock is assumed to be 2 times higher than that of the other clock. In the technique, muxed-D scan design [Williams 73] is used, and test patterns are generated utilizing currently available transition fault ATPG tools using launch-on-capture technique [Savir 94]. The scan cell architecture can also be used for testing other delay fault models such as the path delay, gate delay, and segment delay fault models.

This paper is organized as follows: Section II provides the background of this paper; Section III presents the test application technique presented in this paper; Section IV describes hardware implementation of the technique; Section V reports the experimental results that support the effectiveness of the presented scan cell architecture; Section VI concludes this paper.

II. BACKGROUND

In this paper, the definitions from [Furukawa 06] are used. A *clock domain* is a portion of a circuit clocked by the same clock or synchronous clocks. Two clocks are called *synchronous clocks* if one of them is derived from the other clock; the derivation includes inversion or clock frequency division. An *Intra-clock logic block* is a combinational logic block of which the inputs and outputs are connected to flip-flops that are clocked by the same clock. An *Inter-clock logic block* is a combinational logic block of which the inputs and outputs are connected to flip-flops clocked by two synchronous clocks.

Let us consider the circuit in Fig. 1. In the example circuit, it is assumed that clock signals CLK1 and CLK2 are synchronous clocks. There is an intra-clock path from

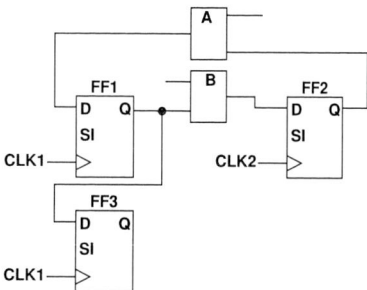

Figure 1. Circuit to explain inter-clock and intra-clock paths

FF1 to FF3; the clock of the source scan cell is the same as that of the destination scan cell. There is also a signal path from flip-flop FF2 to flip-flop FF1 through gate A; FF2 is clocked by CLK2 while FF1 is clocked by CLK1. Therefore, this path is an inter-clock logic path.

In at-speed delay fault testing utilizing PLL clocks, the clock waveforms during test application on ATE are different from those assumed by ATPG tools. Due to the difference of clock waveforms, perfect parts may fail test patterns, making the patterns inapplicable.

Several techniques can be used to test semiconductor designs with inter-clock logic. One simple technique to avoid false failures is to mask the related scan cells during ATPG. *Masking* a scan cell means that the logic value of the scan cell is replaced by a logic-X when the scan cell captures a new logic value. If an expected value is a logic-X, the corresponding response is not compared. For example, if flip-flop FF1 in Fig. 1 is masked during ATPG, the false failures on an ATE caused by the inter-clock logic path through gate A can be avoided. However, the delay faults in the inter-clock logic path through gate A cannot be detected, reducing the test coverage of the test set. In addition to that, the delay faults on the signal path from flip-flop FF1 to flip-flop FF3 cannot be detected even though the signal path is an intra-clock logic path; due to the masking of FF1, a logic-X is propagated from FF1 to FF3. The results of the presented technique will be compared with those of this masking technique.

In [Furukawa 06], a test clock generation scheme for inter-clock at-speed testing was presented. In the technique, each inter-clock logic is tested separately by pulsing a launch clock on the source scan cells and pulsing a capture clock on the destination scan cells. The main issue with this technique is that it requires new ATPG tools that are not commercially available currently.

III. INTER-CLOCK DELAY FAULT TESTING

Figure 2 illustrates an example clock domain. The clock signals CLK1 and CLK2 are generated by an internal PLL; in this paper, CLK2 is derived from CLK1 by dividing the clock frequency by two. There are two intra-clock logic blocks (logic blocks A and C) and one inter-clock logic

block (logic block B). In this paper, the concept is explained using the clock domain in Fig. 2.

Figure 3 illustrates the relationship between CLK1 and CLK2. The head of an arrow represents the capture clock edge, while the tail indicates launch clock edge. Table I reports the launch and capture clocks for each clock cycle time: T1, T2, T3, and T4. For example, T3 is the clock cycle time for inter-clock logic B, in which CLK2 is the launch clock while CLK1 is the capture clock.

In the presented technique, test sets are generated using a currently available transition fault ATPG tool. For test pattern generation, netlists are modified such that the two clock signals are controlled by primary input pins. Therefore, ATPG tools assume that the two clocks have the same waveform as illustrated in Fig. 4.

Test pattern generation is conducted using the following steps.

1) Mask the scan cells driven by inter-clock logic B and clocked by CLK2
2) Generate a transition fault test set: TS_1
3) Write undetected faults: F_{undet}
4) Unmask scan cells masked in step 1;
5) Generate a transition fault test set targeting F_{undet}: TS_2

In step 1, the scan cells are masked to remove false failures on an ATE; without the scan cell masking, perfect chips will fail the generated patterns on an ATE because the clock waveforms assumed during the test pattern generation are different from those during the test application. Figure 5 illustrates the clock timing waveforms for test set TS_1 during the test application on an ATE. This clock waveform is known as *launch aligned double-capture* [Wang 06]. In this clock waveforms, the scan cells driven by inter-clock logic B and clocked by CLK2 may capture wrong values if the transitions triggered by the capture clock of CLK1 are propagated to the scan cells. The mismatches are caused because it is assumed that the capture clock edges of CLK1

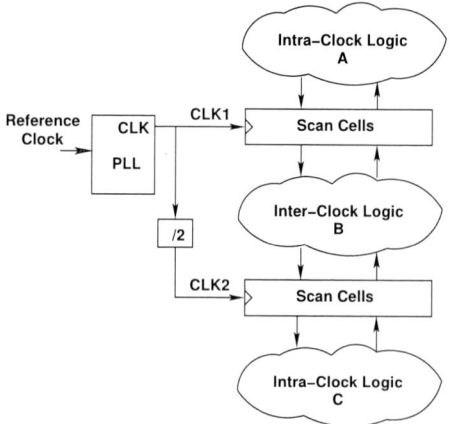

Figure 2. An example clock domain

978-1-61284-657-6/11 $26.00 © 2011 IEEE   214

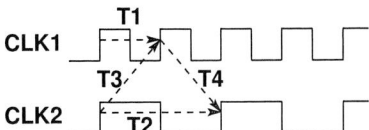

Figure 3. Synchronous clocks: CLK1 and CLK2

Table I. Launch/capture clock

Clock cycle	Launch clock	Capture clock	Logic block
T1	CLK1	CLK1	Intra-clock logic A
T2	CLK2	CLK2	Intra-clock logic C
T3	CLK2	CLK1	Inter-clock logic B
T4	CLK1	CLK2	Inter-clock logic B

and CLK2 are aligned during the test pattern generation. These false failures are avoided by masking the related scan cells. Therefore, test set TS_1 targets the transition faults in intra-clock logic A (clock cycle T1), the faults in intra-clock logic C (clock cycle T2), and the faults in inter-clock logic B that are captured on the scan cells clocked by CLK1 (clock cycle T3).

The undetected faults (F_{undet}) include the faults in inter-clock logic B that were not covered by TS_1 because of the masking of scan cells and the faults in intra-clock logic A and C that were not detected by TS_1. Figure 6 illustrates the clock timing waveform for test set TS_2 on an ATE. This clock waveform is known as *capture aligned double-capture* [Wang 06]. One issue with the clock timing waveform in Fig. 6 is that the logic values to be captured on the launch clock of CLK1 can be changed because the signal transition through the flip-flops triggered by the launch clock of CLK2 may propagate different logic values to the input of the scan cells. In conventional technique, the scan cell is masked to

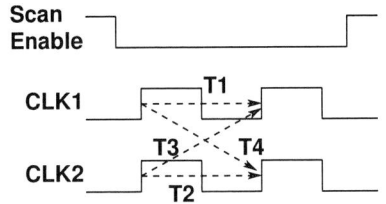

Figure 4. Clock timing waveform during test pattern generation

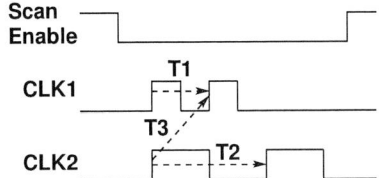

Figure 5. Clock timing waveform for test set TS_1 on an ATE: launch aligned double-capture

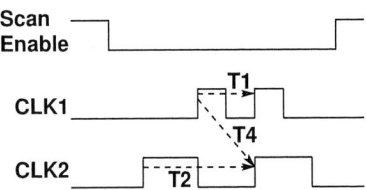

Figure 6. Clock timing waveform for test set TS_2 on an ATE: capture aligned double-capture

avoid the mismatches on an ATE. This masking technique, however, increases the number of logic-X in the expected values and reduces test coverage; it also increases the burden of ATPG process. In order to overcome this issue, the traditional scan cell is modified as illustrated in Fig. 7. The dark area is the circuit added to the traditional scan cell architecture. The input and output ports of the scan cell are not changed; this means that currently available electronic design automation (EDA) tools can be used without any modification. Only the scan cells that are clocked by CLK1 and driven by inter-clock logic block need to be replaced by the modified scan cell.

Figure 8 shows the details of enable_not (EN) signal generator utilized in Fig. 7. The EN signal generator can be implemented using three two-input NOR gates. Figure 9 illustrates the waveform of EN signal. The tri-state buffer in Fig. 7 is enabled during the scan shift operation. Once the scan enable (SE) signal is disabled, the tri-state buffer is disabled until the first pulse on CLK signal. Therefore, the modified scan cells capture the logic values assigned after the scan shift-in operation rather than those propagated by the launch clock of CLK2. This technique is possible because the disabled tri-state buffer keeps the logic value in CMOS technology. The feasibility and application to test techniques of keeping logic values using disabled tri-state buffers were demonstrated in [Datta 04].

One issue with the modified scan cell is hold time violations because the EN signal is triggered by the clock input

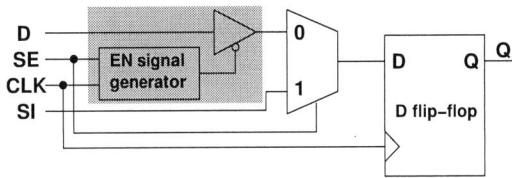

Figure 7. Modified scan cell

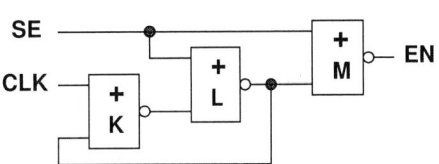

Figure 8. Enable signal generator

978-1-61284-657-6/11 $26.00 © 2011 IEEE   215

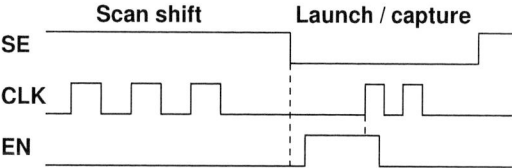

Figure 9. Waveform of enable_not (EN) signal generator in Fig. 8

of the scan cell. This issue can be resolved by using slow (high voltage threshold) NOR gates such that the transition happens after the hold time violation limit of the scan cell.

Let us consider the slow-to-rise transition fault in the circuit of Fig. 10. The transition fault is on inter-clock logic from CLK1 to CLK2; therefore, it cannot be detected by test set TS_1 as indicated in Fig. 5. Therefore, the transition fault is targeted during the generation of test set TS_2. Figure 10(a) illustrates the logic states during test pattern generation. Note that only traditional scan cell model is used during the ATPG process. The first logic value on the node is the state after the scan shift-in operation; the second is after launch clock; the last is after capture clock. In order to detect the fault, an ATPG tool will assign a logic-0 to both of the flip-flops FF1 and FF2 through the scan shift-in operations. After the scan shift-in operations, a logic-0 is assigned to the fault site and a logic-1 is assigned to the D input of FF1. A launch clock will create rising transition at the fault site and a capture clock captures the response on FF2. On an ATE, however, this does not happen. Let us look at Fig. 10(b). The first logic value on the node is the state after the scan shift-in operation; the second is after the launch clock of CLK2; the third is after launch clock of CLK1; the last is after capture clock. Note that a launch clock of CLK2 is earlier than that of CLK1 as illustrated in Fig. 6. A launch clock of CLK2 propagates a logic-0 to the D input of FF1 and a launch clock of CLK1 propagates a logic-0 to the fault site. This propagation does not create rising transition on the fault site. Furthermore, FF2 capture logic-1 rather than logic-0, causing false failures on ATE. In order to avoid this wrong value propagation through FF1, FF1 is masked during ATPG in conventional masking technique. Figure 10(c) illustrates the logic states when scan cell FF1 is replaced by the modified scan cell presented in Fig. 7. By replacing scan cell FF1 with the modified scan cell, the transition fault can be tested without false failures.

IV. IMPLEMENTATION OF TEST LOGIC

The hardware implementation of test clock waveform generation is not unique. In order to make this paper self-contained, one implementation is presented. Figure 11 illustrates the block diagram of the test clock generation logic. TEST_MODE selects either the test mode or the normal mission mode.

Figure 12 describes the circuit for generating delayed scan enable signal. SE_delay is synchronized to the falling edge

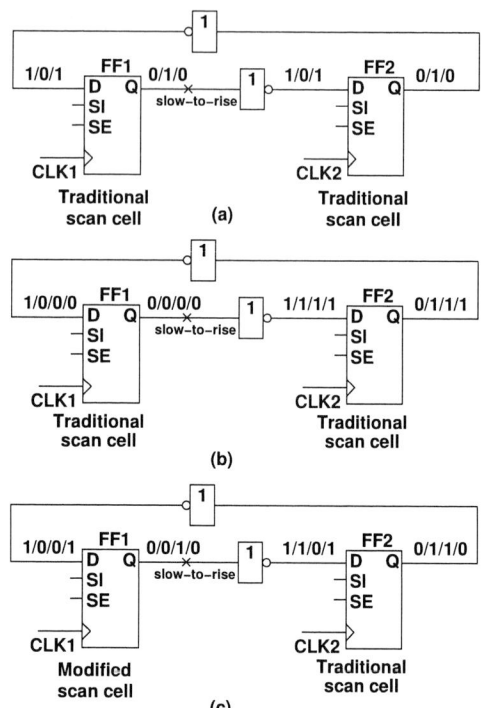

Figure 10. Example: (a) During pattern generation; (b) During ATE testing using traditional scan cell; (c) During ATE testing using modified scan cell

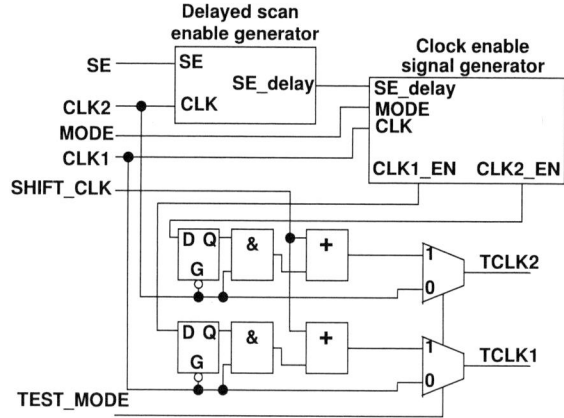

Figure 11. Block diagram of test clock generation

of CLK2 while SE may not. This synchronization is required for the correct operation of the clock enable signal generator. Figure 13 describes clock enable signal generator while Table II reports the clock mode control signal. The signal waveforms of the clock generation block are illustrated in Fig. 14.

The presented scan cell can also be used to detect SSFs even if a hold time violation exists on the capturing scan cell. This is possible because the disabled tri-state buffer blocks the logic transition that causes the hold-time violation.

978-1-61284-657-6/11 $26.00 © 2011 IEEE

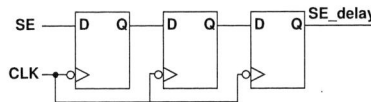

Figure 12. An implementation of delayed scan enable signal generator

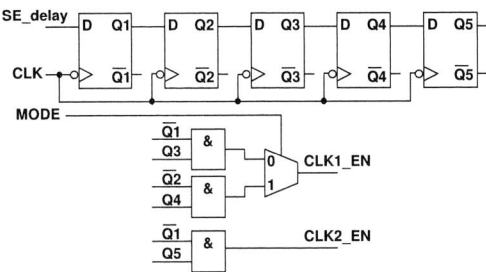

Figure 13. Clock enable signal generator

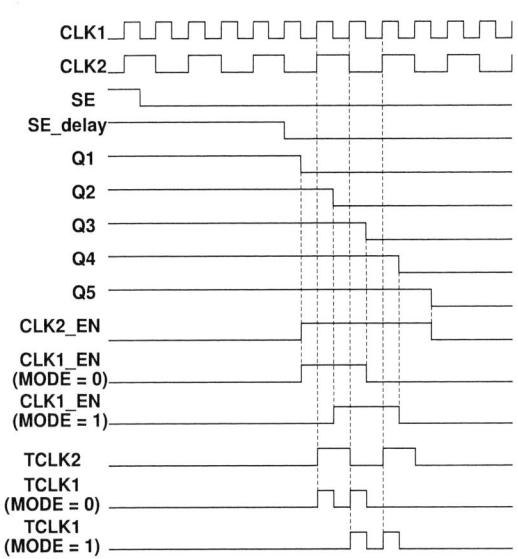

Figure 14. Clock timing waveform

The added logic in Fig. 8 does not affect the normal system operation. In the normal system operation, scan-enable signal is set to a logic-0. After one system clock is applied to the modified scan cells, the logic-1 of CLK assigns a logic-0 to the output of gate K independently of the logic value on the other combinational feedback input of the gate. This logic-0 assigns a logic-1 to the output of gate L because SE is set to a logic-0 in the normal operation. Therefore, a logic-0 is assigned to EN, enabling the tri-state buffer. In addition to that, the logic-1 at the output of gate L assigns logic-0 to the output of gate K independently of the logic value on CLK. Therefore, the power consumption of the added logic during the normal system operation is minimized because the outputs of the NOR gates does not change the logic states.

In this paper the technique is applied to a circuit in which one clock is two-times faster than the other, while the technique can also be applied to general inter-clock logic circuits with different clock frequency ratios.

In muxed-D scan cell based designs, launch-on-shift [Savir 93] or launch-on-capture techniques are used for delay fault testing. The presented technique cannot be used with launch-on-shift technique. However, this does not impair the effectiveness of the presented technique because launch-on-shift technique is not used in our production testing because of the disadvantage of the technique: the scan enable signal must toggle at the rated system clock speed; state transition may not be possible in the functional mode, causing non-functional behavior [Wang 05].

The hardware cost of the presented scan cell is three NOR

Table II. Mode control signal

Clock alignment	MODE
Launch aligned double-capture	0
Capture aligned double-capture	1

gates and one tri-state buffer. The added logic increases the D to Q delay of the modified scan cell and capacitive loading to the clock signal; these increases need to be taken into consideration during the logic and clock tree synthesis. However, all the modification is restricted within scan cells without any additional global routing of signals.

V. EXPERIMENTAL RESULTS

To demonstrate the effectiveness of the presented scan cell architecture, three generations of NVIDIA motherboard graphics processing unit (mGPU) with integrated media communications processor or media control processor (MCP) [NVIDIA] were used. In this research, a commercial ATPG tool was used to generate transition fault test sets. Table III reports the information of the designs: the number of transition faults and the number of scan cells for the three designs. Table IV shows the number of scan cells that need to be replaced by the presented scan cells depicted in Fig. 7; these scan cells need to be masked when the conventional masking technique is used. In the designs, 1.64%, 1.66%, and 1.30% of the scan cells need to be replaced by the presented scan cell for designs A, B, and C, respectively

In this research, 10,000 patterns were generated for each of test set TS_1 and TS_2. Table V presents the pattern generation results for TS_1. The transition fault test coverages of designs A, B, and C are 71.20%, 81.41%, and 86.05%,

Table III. Design information

Design A	Num. of transition faults	125,195,071
	Num. of scan cells	3,647,025
Design B	Num. of transition faults	73,906,760
	Num. of scan cells	2,141,449
Design C	Num. of transition faults	66,058,772
	Num. of scan cells	1,814,486

Table IV. Number of scan cells that need to be replaced with the presented scan cells

Design	Number of scan cells to be replaced
Design A	59,780 (1.64%)
Design B	35,477 (1.66%)
Design C	23,576 (1.30%)

respectively. Note that test patterns for designs A and B were generated using distributed ATPG with multiple processors while those for design C were generated using a single processor.

After the test set TS_1 was generated, the undetected faults (F_{undet}) were targeted in a new ATPG session to generate test set TS_2. The total numbers of transition faults targeted in generating TS_2 are 34M, 14M, and 9M for designs A, B, and C, respectively. Table VI compares the results of the presented technique to those of the conventional capture masking technique. The results demonstrate that more transition faults were detected by utilizing the presented technique: compared to the capture masking technique, 1,312,878 (14.57%), 232,042 (7.53%), and 43,612 (5.86%) more transition faults were detected for designs A, B, and C, respectively. Test generation times did not show consistent trends for the designs; CPU time was reduced for designs B and C, while it increased for design A.

VI. CONCLUSIONS

This paper presented a scan cell architecture for at-speed testing of delay faults in inter-clock logic blocks. The technique uses currently available ATPG tools for test pattern generation; it utilizes the launch and capture clocks from an internal PLL for test pattern application on ATE. Experimental results using commercial designs demonstrated that the technique is effective in detecting the delay faults in inter-clock domain logic.

REFERENCES

[Carter 87] Carter, J. L., V. S. Iyengar, and B. K. Rosen, "Efficient Test Coverage Determination for Delay Faults," *Proc. Intl. Test Conf.*, pp. 418–427, 1987.

Table V. Experimental results: launch aligned double-capture

Design		
Design A	Patterns	10K
	Test coverage	71.20%
	CPU Time	436,179 sec$^{\&}$
Design B	Patterns	10K
	Test coverage	81.14%
	CPU Time	66,065 sec*
Design C	Patterns	10K
	Test coverage	86.05%
	CPU Time	265,697 sec$^{\#}$
$^{\&}$ Distributed ATPG with 10 processors		
* Distributed ATPG with 8 processors		
$^{\#}$ ATPG with 1 processor		

Table VI. Experimental results: capture aligned double-capture

Design		Capture mask technique	Presented technique
A	Faults	34,188,321	34,188,321
	Patterns	10K	10K
	Detected faults	9,012,983	10,325,861
	CPU time	311,041 sec$^{\&}$	323,320 sec$^{\&}$
B	Faults	14,386,722	14,386,722
	Patterns	10K	10K
	Detected faults	3,083,400	3,315,442
	CPU time	68,446 sec*	61,165 sec*
C	Faults	9,329,485	9,329,485
	Patterns	10K	10K
	Detected faults	744,223	787,835
	CPU time	167,214 sec$^{\#}$	166,136 sec$^{\#}$
$^{\&}$ Distributed ATPG with 10 processors			
* Distributed ATPG with 8 processors			
$^{\#}$ ATPG with 1 processor			

[Datta 04] Datta, R., R. Gupta, A. Sebastine, J. A. Abraham, and M. d'Abru, "Tri-Scan: A Novel DFT Technique for CMOS Path Delay Fault Testing," *Proc. Intl. Test Conf.*, pp. 1118–1127, 2004.

[Furukawa 06] Furukawa, H., X. Wen, L.-T. Wang, B. Sheu, Z. Jiang, and S. Wu, "A Novel and Practical Control Scheme for Inter-Clock At-Speed Testing," *Proc. Intl. Test Conf.*, Paper 17.2, 2006.

[Heragu 96] Heragu, K. J. H. Patel, and V. D. Agrawal, "Segment Delay Faults: A New Fault Model," *Proc. VLSI Test Symp.*, pp. 32–39, 1996.

[Lin 03] Lin, X., et al., "High-Frequency, At-Speed Scan Testing," *IEEE Design & Test of Computers*, vol. 20, no. 5, pp. 17–25, 2003.

[Mak 04] Mak, T. M., A. Krstic, K. Cheng, and L. Wang, "New Challenges in Delay Testing of Nanometer, Multigigahertz Designs," *IEEE Design & Test of Computers*, vol. 21, no. 3, pp. 241–247, 2004.

[NVIDIA] NVIDIA, http://www.nvidia.com

[Savir 93] Savir, J. and S. Patil, "Scan-Based Transition test," *IEEE Trans. on CAD*, vol. 12, no. 8, pp. 1232–1241, Aug. 1993.

[Savir 94] Savir, J. and S. Patil, "On Broad-Side Delay Test," *Proc. VLSI Test Symp.*, pp. 284–290, 1994.

[Smith 85] Smith, G. L., "Model for Delay Faults Based upon Paths," *Proc. Intl. Test Conf.*, pp. 342–349, 1985.

[Waicukauski 87] Waicukauski, J. A., E. Lindbloom, B. K. Rosen, and V. S. Iyengar, "Transition Fault Simulation," *IEEE Design & Test of Computers*, vol. 4, no. 2, pp. 32–38, 1987.

[Wang 05] Wang, L.-T., X. Wen, P.-C. Hsu, S. Wu, and J. Guo, "At-Speed Logic BIST Architecture for Multi-Clock Designs," *Proc. Intl. Conf. on Computer Design*, pp. 475–478, 2005.

[Wang 06] Wang, L.-T., C.-W. Wu, and X. Wen (Eds.), *VLSI Test Principles and Architectures*, Morgan Kaufmann Publishers, San francisco, CA, 2006.

[Williams 73] Williams, M. J. Y. and J. B. Angell, "Enhancing Testability of Large-Scale Integrated Circuits via Test Points and Additional Logic," *IEEE Trans. on Computers*, vol. C-22, no. 1, pp. 46–60, Jan. 1973.

978-1-61284-657-6/11 $26.00 © 2011 IEEE

Design and Implementation of A Time-Division Multiplexing Scan Architecture Using Serializer and Deserializer in GPU Chips[1]

Amit Sanghani[‡], Bo Yang[‡], Karthikeyan Natarajan[‡] and Chunsheng Liu[§]

[‡] DFT Engineering, NVIDIA Corp.
2701 San Tomas Expy, Santa Clara, CA 95050, USA
{asanghani, byang, knatarajan}@nvidia.com

[§] Test Development, Altera Corp.
101 Innovation Drive, San Jose, CA 95134
cliu@altera.com

Abstract

We present the design and implementation details of a time-division demultiplexing/multiplexing based scan architecture using serializer/deserializer. This is one of the key DFT features implemented on NVIDIA's Fermi family GPU (Graphic Processing Unit) chips. We provide a comprehensive description on the architecture and specifications. We also depict a compact serializer/deserializer module design, test timing consideration, design rule and test pattern verification. Finally, we show silicon data collected from Fermi GPUs.

1 Introduction

Scan-based Design-for-Test (DFT) technique has been widely adopted in Integrated Circuit(IC) manufacturing test due to its simplicity and high fault coverage [4]. In test mode, during shift operation, test stimuli are shifted from Automatic Test Equipment (ATE) test channels into scan chains, and test responses are shifted out back to ATE for comparison. Each scan element can be viewed as a primary input or primary output, greatly enhancing the controllability and observability of internal nodes of the device.

The number of total scan chains and the longest chain in a scan-based design are determined by the total number of scan IOs of the chip and test channels available on ATE. Generally, an ATE with more test channels is also more expensive. Test application time of scan-based patterns is proportional to both the longest chain and the shift speed. Since the size of scan-based patterns is usually dominant in a test suite, one of the most important objectives of various scan architectures is to reduce test application time of scan patterns using inexpensive ATEs, i.e., using ATEs with limited number of channels and/or low frequency channels to reduce test expense without compromising test quality. A lot of work has been conducted to

target this problem via different scan chain manipulations and methods such as bandwidth matching and test resource partitioning [1, 8, 9, 10, 13].

Another dimension in test time reduction is through test data compression, which reduces the amount of test data that must be stored on ATE for a deterministic test set. This is usually done through additional on-chip hardware to decompress the test stimuli from the ATE and to compact the responses into the ATE for comparison [3, 5, 7, 11]. With test compression techniques, a small number of ATE channels can often drive a large number of internal scan chains, hence the depth of each ATE channel is minimized, which can reduce ATE test time significantly [2, 3, 6].

Low cost ATEs can generally drive test channels at up to several hundred Mhz, e.g. 200MHz used in this work. The scan IOs can also operate at this frequency during scan shift. However, internal scan chains typically operate at a much lower frequency ranging from 10Mhz to 50Mhz due to chip timing and power constraints, especially for large designs such as Graphic Processing Unit (GPU). Test compression logic further slows down the internal shift. The gap between ATE bandwidth and internal scan frequency causes inefficient use of ATE, hence higher test expense.

A solution to this problem is to insert time division logic, usually through the use of serializer and deserializer, between high speed IO and internal chains to match the bandwidth difference. This will not only allow more efficient use of ATE capacity to reduce test time, but also lead to reduced number of ATE channels. Some work has been reported on the use of bandwidth matching [1, 8, 9, 10, 12, 13]. However, no real industrial implementations have been reported in the literature. It is especially a challenge on a GPU because of its complexity and various stringent constraints.

In this paper, we present the design of a time-division demultiplexing/multiplexing based scan architecture for bandwidth matching. This is implemented on NVIDIA's Fermi family GPU chips. We show that with such an implementation, ATE resource can be efficiently utilized and test time significantly reduced without exceeding timing and power

---

[1]The work of C. Liu was conducted when he was with NVIDIA Corp.

978-1-61284-657-6/11 $26.00 © 2011 IEEE

limitations. In Section 2, we illustrate the overall design architecture. In Section 3, we show the design of the serializer/deSerializer modules. We then discuss some design rules in Section 4 and show the flow for pattern generation and verification in Section 5, respectively. Finally, we present some related data from a real GPU in Section 6.

2 Architecture

In this section, we propose a comprehensive design for a scalable scan architecture using serializer/deserializer. It fits into current scan and test compression flow smoothly with extremely small area overhead. The term "deserializer" refers to the time-division demultiplexing logic that connects a small number of high speed input IO to a large number of low speed internal scan chains. The term "serializer" refers to the time-division multiplexer on the other side of the chains. For test compression, this architecture is optimized for the Adaptive Scan scheme of Synopsys, where a decompressor/compressor pair (or CODEC) is used. This architecture satisfies the requirements from both regular scan without test compression (which is bypassed) and scan with test compression (for production test). It can be configured to interface with scan data pipeline flops (added to meet scan shift timing of up to 200Mhz shift frequency on ATE) or IO pads directly.

Since commercial EDA tools do not support such a scan scheme, we develop a methodology to modify the gate level netlists to make them compatible with ATPG tool for pattern generation. The generated test patterns are then manipulated to match the original netlists that are with serializer/deserializer logic. Such patterns can be used in pattern verification and post-silicon bring-up.

The top level scan architecture is depicted in Figure 1. We use "S/D" to denote serializer/deserializer for the sake of simplicity. On the left side of the figure, we indicate various clock domains of signals in each node in the architecture. ATE/IO shift speed is targeted at 200MHz (S/D_fastclk) and internal shift speed is targeted at 50Mhz (shift_slowclk or bypass test clock). Therefore, a fastclk/slowclk ratio of 4 is available in this case. In practice, the ratio is dependent on the ratio of ATE/IO shift speed over internal shift speed and targeted ATE configurations.

The test data from ATE are shifted in through scan data pipeline flops (used to meet scan timing), at the speed of $T \times f$, into the SSI (S/D-Scan-In) input of a deserializer module. T is the fastclk/slowclk ratio and f is the operating frequency (shift speed) of the internal chains. Each SSI can serially feed data at the speed of $T \times f$ into T internal PSIs (Pseudo-Scan-In) at speed f. If there are replicated design modules, i.e. the modules with identical logic such as stream processors (SM), the SSI will be broadcasted to all of the replicated modules. As a result, for non-replicated

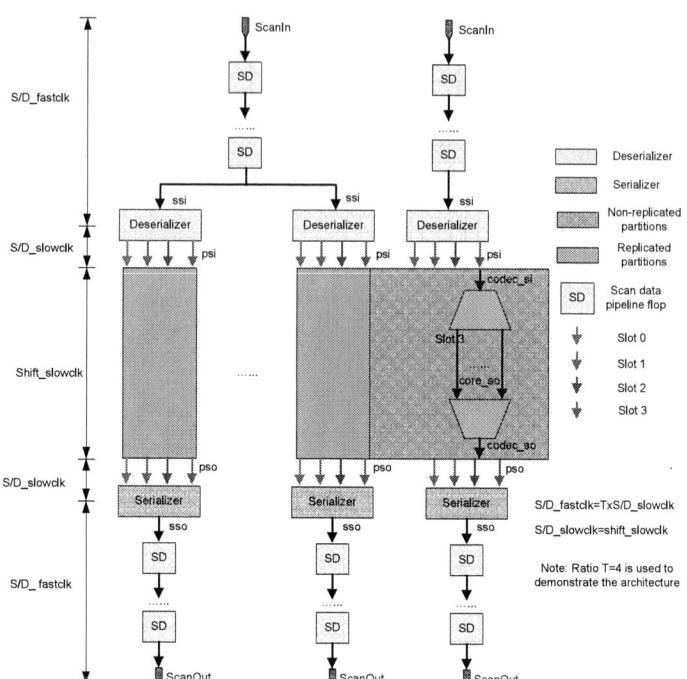

Figure 1. Time-division multiplexing scan architecture.

structure, one chip scan-in can drive T internal regular scan chains or input ports on the decompressors (referred to as codec_si in the figure). For replicated structure one chip scan-in can support $T \times n$ (n is the number of replicated copies) codec_si signals. In replicated structure, T deserializers are used for one chip scan-in instead of a single deserializer to achieve better scan timing.

Similarly on the scan-out side, for both replicated and non-replicated structures, T internal PSO (Pseudo-Scan-Out) at the speed of f are multiplexed to feed port SSO (S/D-Scan-Out) at the speed of $T \times f$. The SSO data are shifted out through scan data pipeline flops to chip scan-out and then to ATE. To make this architecture more efficient, for regular scan mode without compression, the number of internal chains should be the multiple of T; and with test compression the number of codec_si and codec_so ports on compressors/decompressors should be the multiple of T.

The number of internal clocks (slowclk) is dependent on functional clock domains, which is determined by the clock design. For smaller chip, one global $T \times f$ shift clock can drive all the internal clock domains. Serializer and deserializer design is able to handle the data transfer from $T \times f$ clock domain to internal test clock domains smoothly, taking into account the insertion delays. For a large chip, however, multiple $T \times f$ clocks should be used to facilitate the $T \times f$ clock tree balancing.

For both regular scan (without test compression) and scan with test compression, S/D and Non-S/D modes are

978-1-61284-657-6/11 $26.00 © 2011 IEEE   220

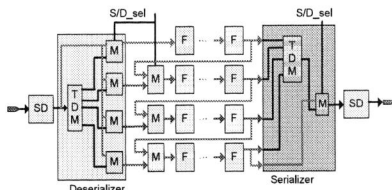

Figure 2. Regular Scan-Non-S/D mode vs. Scan-S/D mode.

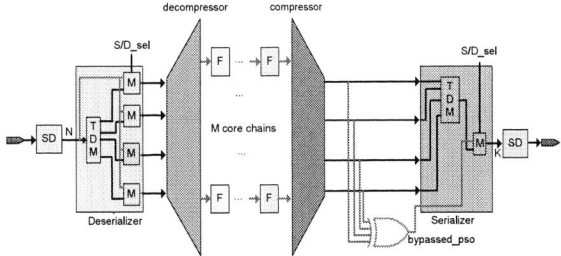

Figure 3. Test compression Scan-Non-S/D mode vs. Scan-S/D mode.

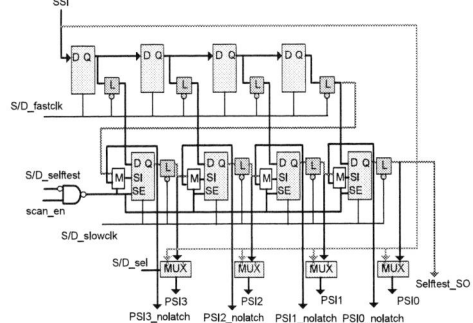

Figure 4. Deserializer design with $T = 4$.

provided. Non-S/D are modes with serializer and deserializer bypassed. They are useful for ATE debug and can serve as a backup if S/D logic does not work properly.

The Scan-Non-S/D and Scan-S/D structures without test compression are shown in Figure 2. We omit the clock signal names on each node of the data flow, see Figure 1. The figure shows the same architecture as in Figure 1, but depicts more details inside the serializer and deserializer. TDM represents a Time-Division-Multiplexing unit that distributes serial scan-in data onto multiple internal chains. Without S/D logic, the scan design is optimized as N balanced chains with the longest chain length being L. In Scan-S/D mode, i.e. when S/D_sel = 1, each scan chain is sliced into T internal chains (Scan_S/D chains) and these T shorter chains are connected to the serializer modules as illustrated in Figure 2. A multiplexer M is inserted before each Scan_S/D chain. When S/D_sel = 0, indicating a Scan-Non-S/D or bypass mode, T Scan-S/D chains are stitched into a long scan chain (Scan-Non-S/D chain) and both deserializer and serializer modules are bypassed. Hence in this mode scan shift works just as without S/D logic. Note that in Non-S/D mode, in order to optimize the scan chain balance, the T internal chains between a pair of deserializer and serializer do not need to be stitched into the same Scan-Non-S/D chain. They can be slice-diced with chains of other serializer/deserializer pairs to form optimized chains.

The S/D design with test compression is shown in Figure 3. In both Non-S/D and S/D modes, the configuration of internal scan chains and CODECs remain unchanged. When S/D_sel is 1 (S/D mode), the decompressor takes $T \times N$ PSI signals from deserializer module and feeds $T \times M$ internal scan chains, also referred to as core chains, as scan-in (core_si). The compressor compresses data from the $T \times M$ core chains scan-out (core_so) to feed $T \times K$ PSO signals, which are then multiplexed into K-bit chip scan-out. When S/D_sel = 0 (Non-S/D mode), each bit in the N-bit chip scan inputs is broadcasted to the T codec_si ports (inputs to the decompressor, see Figure 1), which are driven by the T PSIs from the same deserializer. On the output side, the PSOs feeding the TDM logic in the serializer is XORed to generate the bypassed_pso signal. This signal is selected in the serializer module and then becomes chip scan-out. Such XOR gates can be viewed as an extra level of compressor logic. In stead of stitching T chains in S/D mode to a sin-

gle chain, the use of fixed internal core chains with extra level test compression logic avoids scan chain reconfiguration when switching between different modes. It eliminates the need for a large number of reconfiguration multiplexors by compromising the controllability and observability. Since the Non-S/D mode is a downgraded backup mode, such a scheme does not hurt test quality.

3 Serializer/deserializer modules design

In this Section, we present more details on the design of S/D modules.

3.1 Deserializer design

A deserializer design with $T = 4$ is illustrated in Figure 4. It takes SSI data at the speed of $T \times f$ from one scan data pipeline flop or chip scan-in port, and feeds four PSIs operating at the speed of f. The inputs of the MUX(M) are from the output of the flop. When S/D_sel is 0, the deserializer module is bypassed and the SSI is broadcasted to all four PSIs. When S/D_selftest = 1, the SE signals on the lower four scan flops are controlled by scan_en and a chain can be extracted between selftest_so and SSI. The S/D modules will be flattened after they are inserted into the design. Hence for modules whose PSI_nolatch signals don't have sinks, these floating nets will be optimized away. The operation modes of the deserializer are described in Table 1. Note that in capture mode, the flops in deserializer modules should maintain their values. This is done by inserting loop back MUX(M).

978-1-61284-657-6/11 $26.00 © 2011 IEEE   221

Table 1. Deserializer operation modes

Modes	S/D_sel	S/D_selftest	scan_en
Shift	0	0	1
	TimeDeMuxing SSI to T PSIs		
	S/D_fastclk and S/D_slowclk toggle		
	internal scan clocks toggle		
Capture	0	0	0
	4 flops on S/D_slowclk hold data		
	S/D_fastclk and S/D_slowclk stop		
	internal_shiftclk clock may toggle		
Selftest	0	1	X
	Shift from PSI to selftest_so		
	S/D_fastclk runs at f		
Bypass	1	X	X
	SSI broadcasts to PSIs		
	S/D_fastclk runs at f		

3.2 Serializer design

Serializer shown in Figure 5 takes T-bit PSO data at the speed of f from internal scan chains in regular scan mode, or codec_so ports on compressors in test compression mode. The data are then time-multiplexed to SSO at the speed of $T \times f$. The $T \times f$ speed SSO data are then shifted out to chip scan-out port through scan data pipeline flops. When S/D_sel = 0, serializer is bypassed and SSO is driven by bypassed_PSO.

During scan capture mode (scan_en=0), the T-bit one-hot counter (marked as green in the lower part of Figure 5) is set to 1110, resulting in parallel_load set to 0. The four flops in the middle of Figure 5 load data from the four flops on top in parallel. During load_unload (scan_en=1), the middle four flops form a shift register running at S/D_fastclk, and data will be shifted out to SSO. It can be seen that in load_unload, parallel_load is set to 0 every T cycles.

Since the internal scan chains are on scan clock domain while the flops capturing PSO data are on S/D clock domain, these capture flops are designed as negative edge triggered flops so that the paths between the last flop in scan chains and these capture flops are half cycle paths, which can help solve the potential hold violation problem. An alternative solution is to use positive triggered flops and add a lockup latch after the last flop of each scan chain. However, this is infeasible with test compression in a complex design, because the number of internal scan chains can be extremely large, which can incur huge area overhead. In test compression mode, global signals such as mask and unload_enable also have deserializer inserted and their outputs interface with serializer through combinational compressor logic. To address the possible hold violation problem, PSI_nolatch signals are used.

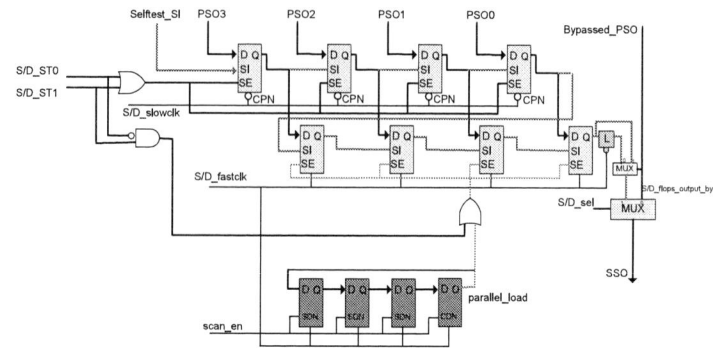

Figure 5. Serializer design with $T = 4$.

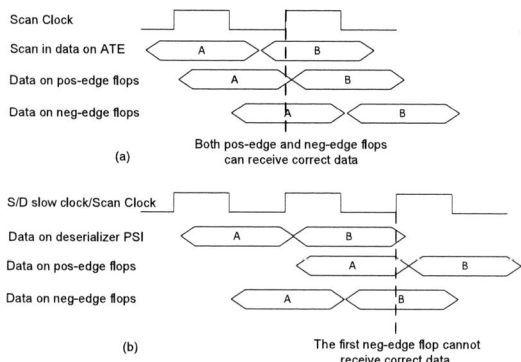

Figure 6. Data slip in scan chains starting with a neg-edge flop

4 DFT design rules

To ensure the correct data transfer from scan inputs or inbound scan data pipeline flops to internal scan chains, and then to outbound scan data pipeline flops or scan outputs, the following design rules need to be followed:

- Rule 1: No scan chains should start with negative edge triggered flops.

- Rule 2: No scan chains should end with negative edge triggered flops.

In a design using Synopsys Adaptive Scan, violating the above rules only compromises test time reduction because it makes shift timing marginal. For example, if a compressor violates Rule 2, the valid data on scan outs are available from a full cycle to a half cycle. In a regular scan design without test compression, optimizing test timing on scan pin can resolve this problem. However, in a S/D based design, violating these rules will result in the break of shift functionality, as shown in Figure 6. Figure 6 (a) shows the data capturing of the flops in a scan chain without S/D. In each test cycle, both pos-edge and neg-edge flops can load data

978-1-61284-657-6/11 $26.00 © 2011 IEEE   222

from the same cycle. In S/D design as shown in (b), the neg-edge flops are not able to retain the data from the same cycle as pos-edge flops, causing incorrect load operations.

In the serializer design, the capturing flops on the S/D_slowclk, as shown in Figure 5, are neg-edge triggered. Scan-out data are captured into serializers though these half cycle paths during unload. This is done to avoid possible hold violation between the last scan flop in a scan chain on scan shift clock and the capturing flop in serializers on S/D_slowclk domains. If a scan chain ends with a neg-edge triggered flop, which violates Rule 2, the data on the flop will not be ready before the negative edge of S/D_slowclk, causing incorrect unload operations. Similarly, scan chains ending with unnecessary extraneous lockup latches without capturing flops can cause the same problem.

However, neg-edge triggered flops may not be eliminated from the design and they may need to be on scan chains for better coverage. To address these violations, the following schemes can be used by scan insertion tool:

- Scheme to fix Rule 1 violation: A pair of posedge/negedge dummy flops need to be added in front of the neg-edge flop starting a scan chain.

- Scheme to fix Rule 2 violation: A dummy pos-edge flop needs to be added in the scan chains consisting of all neg-edge flops.

We insert dummy flops to handle the shift timing issue. A pos-edge flop to a neg-edge flop order is normally considered as invalid chain order in DFT designs. The neg-edge flop always gets the same data as the pos-edge flop after load and hence it is treated as the slave of the pos-edge flop by ATPG tools. However, the data on the master stage of the pos-edge flop is lost during unload, causing pattern failing in simulation and on ATE. Normally, ATPG tools mask both flops to get the correct patterns behavior but losing coverage on them. The added pair of dummy flops are used to support correct shift operation. They can be non-scan flop or scan flops with D pin tied off. Any defects on them will cause shift failures on ATE and will be immediately noticed before capture patterns are applied, hence no capture coverage is lost. Adding a dummy pos-edge flop creates a neg-edge flop to a pos-edge flop order, which can be handled by ATPG tools properly.

5 Pattern generation and verification flow

Currently the ATPG tool we use does not fully support the S/D architecture. To overcome this limitation, a flow is created to perform pattern generation and verification. Design netlists need to be manipulated to remove the S/D modules for pattern generation. The internal nets connected to PSO/PSI ports on S/D modules are brought out as top level design ports. These ports are the scan-in/out ports defined in ATPG SPF (STIL Protocol File). The ATPG patterns are then generated on the manipulated netlists. In the ATPG environment, S/D_fastclk clocks are not defined as regular primary inputs and all test clocks (clocks directly from ATE clock ports to drive internal scan chains) are defined as clocks.

The generated STIL patterns cannot be simulated on the tape-out netlists or applied on silicon because the scan data contained in the pattern are associated with dummy PSI/PSO ports, not chip scan ports. These PSI/PSO ports need to be merged to chip scan-in/out ports by a STIL pattern merge tool. In merged STIL patterns, S/D_fastclk are the only clocks. Test clocks are not free running clocks but handled as scan in data with 0011 streams. By doing this, test clock pulses can be controlled precisely in the pattern to meet the shift requirements between S/D modules and internal scan chains.

The pattern merge tool also performs pattern padding to accommodate the scan data latency on S/D modules. Ot the deserializer side, it takes T cycles of S/D_fastclk for the data to be loaded into the S/D_fastclk shift registers. As the $(T + 1)^{th}$ clock pulse arrives, the S/D_slowclk is pulsed, capturing the data from the T-bit S/D_fastclk shift registers to the T-bit slow clock load registers. It then takes an additional 2 S/D_fastclk cycles for data to be available at the PSI's. Therefore, a minimum of $T + 3$ S/D_fastclk cycles are required before the first test clock can be pulsed to shift internal chains. Similarly on the serializer side, a minimum of $T + 1$ S/D_fastclk cycles are required before the scan out data are available for comparison on ATE. Details are ignored due to the lack of space. The number of scan data pipeline stages is also taken into account when padding is added.

6 Implementation and results

The S/D scan architecture has been implemented on NVIDIA's Fermi family GPUs and proven working properly on ATE. Since the mainstream ATEs generally support 200MHz on data channels, we set a target frequency of 50MHz as internal scan shift speed with $T = 4$. This T value is determined by our learning from a previous chip family and is design specific. Our scheme itself is scalable and can support other T values. Design data and silicon data of one GPU chip are shown in Table 2. Scan related data are from test compression mode. S/D is only inserted in test compression mode because all the ATPG production patterns use test compression, which is based on Synopsys Adaptive Scan.

The chip uses full scan design and has around 17.1M flops. It uses 195 scan-in and 395 scan-out pins. Note that in test compression mode the numbers of scan-in and scan-out pins are not equal. The longest scan chain with test compression has 256 flops. Without the S/D design, we would have needed 4 times as many scan pins. Such a large number of test channels exceeds our ATE's capacity. As an alternative, we can use long scan chains with at least 1000 flops, but that would increase our scan shift time by a factor of 4. Therefore, the use of S/D has significantly reduced test time without compromising test quality and chip area.

Since there are replicated modules inside the chip, basic scan data broadcast scheme is used on the instances of S/D modules. One scan-in pin sends identical data to all the replicated instances of the same module. As a result, the number of deserializers is more than that of the scan-in pins. Since each instance needs dedicated scan-out to observe its unique test response, the number of serializers is equal to the number of scan-out pins. The total overhead of deserializer/serializer logic is less than 0.05% of the whole chip.

The S/D_fastclk clocks can be easily timed to run at higher than 200MHz (50Mhz on internal scan chains) if one pipeline flop is added between S/D modules and scan pins. However in practice, this is limited by the power consumption. Significant power droop happens before this frequency can be reached. Hence we set the real ATE shift speed to 160Mhz (40Mhz on internal scan chains) to avoid power droop. Note that by properly manipulating patterns and chip config, we are able to alleviate the power droop so that 200Mhz can be achieved. Because power droop issue is beyond the scope of this paper, we will not discuss about it here. Without this scan architecture, we can only achieve a shift speed of 20Mhz.

Few neg-edge triggered flops are used in the design so only 8 pairs of dummy flops are needed to fix Rule 1 violations and only 1 flop is needed to fix Rule 2 violations. Since not using neg-edge triggered flops is a popular practice in industrial designs, the potential overhead for fixing design rules is ignorable.

7 Conclusions

In this paper, we described a comprehensive flow and detailed implementation of a time-division multiplexing scan architecture for test time reduction. We introduced the motivation and presented the overall architecture and specifications of the design. We illustrated details of the S/D logic. For practical usage, we also showed design rules associated with such scan architecture, as well as the flow for pattern generation and verification. Results from a Fermi family GPU design and silicon data showed that the area overhead of the S/D logic is ignorable, yet scan shift time can be reduced by a factor of 4 without compromising test quality.

Table 2. Design and ATE data on one GPU chip

Flop count	# of Scan In	# of Scan Out
17.1M	195	395
Max chain length	**Target shift speed**	**# of deserializers**
256	> 200/50Mhz	375
# of serializers	**# of S/D module flops**	**Real shift speed**
395	7740 (0.05% overhead)	160/40Mhz
# of Rule 1 fix flop	**# of Rule 2 fix flops**	**Non-S/D shift spee**
16	1	20Mhz

References

[1] A. Khoche. Test resource partitioning for scan architectures using bandwidth matching. , Digest of Workshop on Test Resource Partitioning, pp. 1.4.1C1.4.8, 2002.

[2] C. Krishna and N. A. Touba. Reducing test data volume using LFSR reseeding with seed compression. *Proc. Int. Test Conf.*, pp. 321-330, 2002.

[3] A. Pandey and J. Patel. Reconfiguration technique for reducing test time and test data volume in Illinois Scan Architecture based designs. *Proc. VLSI Test Symp.*, pp. 9-15, 2002.

[4] N. K. Jha and S. Gupta. Testing of Digital Systems. , Cambridge University Press, UK, 2003.

[5] J. H. Patel, S. S. Lumetta, and S. M. Reddy. Application of Saluja-Karpovsky compactors to test responses with many unknowns. *Proc. VLSI Test Symp.*, pp. 107-112, 2003.

[6] I. Pomeranz and S. M. Reddy. Test data compression based on input-output dependence. *IEEE Transactions on Computer-Aided Design of Integrated Circuits and Systems*, Vol. 22, pp. 1450-1455, 2003.

[7] J. Rajski, J. Tyszer, C. Wang, and S. M. Reddy. Convolutional compaction of test responses. *Proc. Int. Test Conf.*, pp. 745-754, 2003.

[8] E. H. Volkerink, A. Khoche, J. Rivoir and K. D. Hilliges. Modern Test Techniques: Tradeoffs, Synergies, and Scalable Benefits. *Journal of Electronic Testing: Theory and Applications*, Vol 19, 125-135, Apr. 2003.

[9] A. Sehgal, V. Iyengar and K. Chakrabarty. SOC test planning using virtual test access architectures. *IEEE Transactions on Very Large Scale Integration (VLSI) Systems*, Vol. 12, pp. 1263-1276, Dec. 2004.

[10] Q. Xu and N. Nicolici. Multi-frequency test access mechanism design for modular SOC testing. *Proc. Asian Test Symp.*, pp. 2-7, 2004.

[11] L-T Wang, et al. VirtualScan: A New Compressed Scan Technology for Test Cost Reduction. *Proc. Int. Test Conf.*, pp. 916-925, 2004.

[12] C. Liu, V. Iyengar, and D. K. Pradhan. Thermal-Aware Testing of Network-on-Chip Using Multiple Clocking. *Proc. VLSI Test Symp.*, pp. 46-51, 2006.

[13] F. Hussin, T. Yoneda and H. Fujiwara. NoC-Compatible Wrapper Design and Optimization under Channel-Bandwidth and Test-Time Constraints. *IEICE Transactions on Information and Systems.*, pp. 2008-2017, 2008.

Harmony Widget for X-Free Scan Testing

Dilip K. Bhavsar

Intel Corporation
Hudson, MA USA
Dilip.Bhavsar@Intel.com

Abstract—**This paper presents a simple innovative Design for Test (DFT) solution for removing one of the major sources of Xs (unknown outputs) during scan testing. The DFT called Harmony Widget is inserted at state elements that a scan test may load with illegal or unspecified state thus causing circuit to misbehave or produce indeterminate or unknown outputs (Xs). The DFT used is far simpler than the current practice in industry and blends with the prevailing scan design methodologies.**

Keywords-Scan, ATPG, Conention-Free ATPG, DFT, Testing.

I. INTRODUCTION

The presence of unknown outputs or Xs on circuit nodes impacts the effectiveness of scan test. First, Xs cause loss of coverage. Second, any X captured in Scan-flops must be masked during comparison or compression. When comparison or compression is performed on chip, the tester must load mask vectors for on-chip masking action. This can substantially increase the vector depth and test time and therefore test cost. Worst of all, the presence of Xs kills the opportunity of using Logic BIST (LBIST) and related test methodologies that use random stimuli for testing.

There are at least four known major sources of Xs in scan testing: Non-scan sequentials, Un-initialized arrays and their outputs, uncontrollable scan-test partition inputs, and Xs produced by scan pattern induced circuit misbehavior.

Non-scan sequential induced Xs are easily removed by flushing them out during ATPG and LBIST. Likewise simple DFT that initializes array outputs and then blocks array reads during scan testing can completely eliminate Xs from un-initialized arrays.

Interestingly, the third source of Xs has its origin in the DFT method. Large processor chips invariably divide the chip into several sub-chips and perform scan test on each as an independent scan test partition. Uncontrollable inputs to such partition become a major source of Xs. These too are easily eliminated by inserting DFT that simply blocks Xs or forces inputs to a known state or by using specially structured scan cells as in Scan Island partitioning discussed in [1].

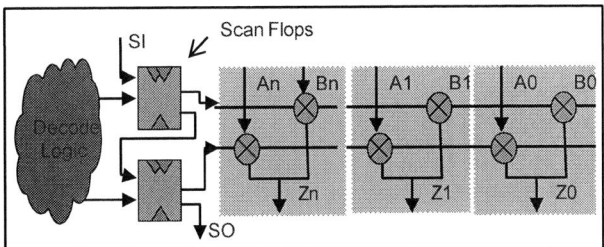

Figure 1: Contention Producing Pass Gate multiplexers (MUXes)

The fourth source of Xs and its elimination is the subject of this paper. These Xs arise when an ATPG or LBIST generated pattern loads an illegal state, that is, a state unreachable by normal operation of the circuit under test. One easily recognizable and frequently occurring example of such circuit in today's VLSI is the pass gate multiplexers (MUX) shown in Figure 1. When used in large data paths, their select control may be driven directly from flip-flops. The upstream decode logic guarantees that the select control satisfies the mutually exclusive and one-hot properties. That is, for the normal operation one and only one select line is active at any given time. ATPG or LBIST driven patterns in scanned flip-flops on the select lines may easily violate these properties. The resulting driver fights or floating MUX output nodes become Xs that downstream scan-flops may capture.

ATPG users may avoid such Xs by specifying constraints that prevent ATPG from generating patterns that violate the mutual exclusion and one-hot properties. Although some ATPG programs handle constraints more comfortably than others, most get bogged down and run inefficiently as the number of constraints increases. Experiments using popular commercial ATPG program on microprocessor class design have shown increase of vector count in the range of 4% to 40% when contention avoiding constraints were used [5]. This is the reason users sometimes run ATPG without constraints and pay the penalty in masking out X's while comparing responses on the tester. Unfortunately, working with constraints is not the option for users of LBIST which rely on random stimuli. Users of LBIST must resort to some DFT to eliminate contention and similar X producing circuit misbehavior.

The remainder of this paper discusses DFT for eliminating this class of X sources. Section 2 briefly reviews some DFT solutions reported in the industry. Section 3 is the main contribution of this paper. It reveals the novel solution called the Harmony Widget and discusses its distinctive characteristics relative to the current practice. Section 4 concludes the paper.

II. CURRENT INDUSTRY PRACTICE

Perhaps the simplest solution for preventing X-producing circuit behavior is the non-DFT method, namely, un-scanning the culprit flip-flops. This definitely eliminates Xs but the solution is not test-friendly. It increases the sequential depth of the circuit under test and impairs test coverage. The situation gets exponentially worse if the decode is distributed over many cycles.

There are also obvious solutions that force a safe value on one-hot selects. They work for test but are intrusive on design. They impair performance and worst defeat the very purpose for which designers drive pass gate MUX selects directly from flip-flops. Besides the obvious timing benefit, these flops help to reduce the transient contention current during normal switching of MUXes.

Solutions reported in [2]-[5] use elaborate DFT techniques that use additional scan-flops and some form of dedicated test decoders whose guaranteed one-hot outputs are worked into the mainline one-hot control.

In [2] Petaras et al use fairly complex encoder-decoder logic inserted in the serial scan path as shown in Figure 2. There is no impact on performance but the encoder-decoder circuit gets fairly complex as the number of select lines increases. Although the equivalent model abstracted for ATPG is very simple, it is not clear how it comprehends the difference in the model scan chain length and the physical scan chain length. This is perhaps the first reported solution in industry for eliminating this source of Xs.

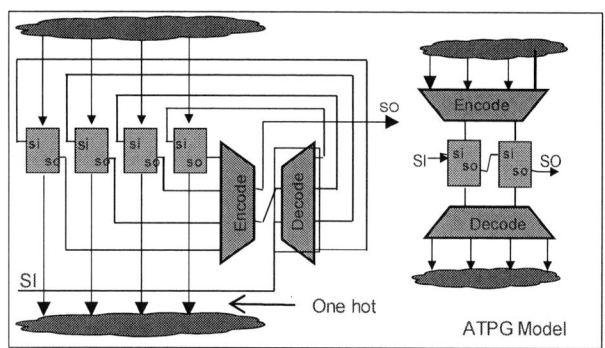

Figure 2: Pataras, et al Scheme (ITC'95)

Compared to the Pataras et al scheme, the scheme reported by Stanford researchers in [3] is more practical. In this scheme shown in Figure 3 the flops on the one-hot select lines are not scanned directly. Instead, the dedicated scan-flops in same number are added on the side. These shadow flops observe the one-hot select line outputs. Subsets of the shadow flop outputs (log(N), N = number of select lines) are

used to drive a dedicated test decoder whose outputs are then loaded into the mainline one-hot select flops. Besides the shadow scan-flops, the scheme requires two additional control signals.

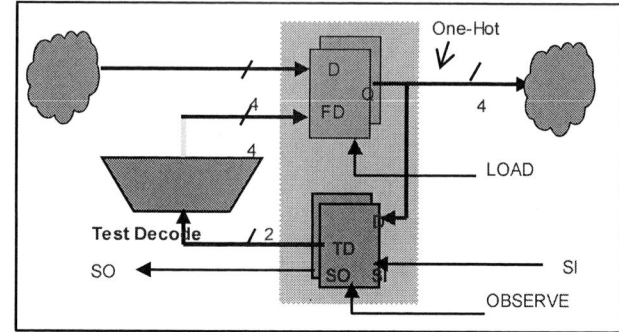

Figure 3: Stanford Scheme (ITC'97)

Another scheme deployed on some FreeScale ICs [4] is shown in Figure 4. This scheme scans the mainline one-hot flops but to kill contention it adds dedicated scan-flops and test decoder and fairly intrusive logic on select lines. This assures mutual exclusion property but sometimes patterns can violate the one-hot property and may float the MUX output. To solve this, the scheme uses a test pull down as shown in the figure. This scheme has a serious drawback because of the gates added on select control after the one-hot flops.

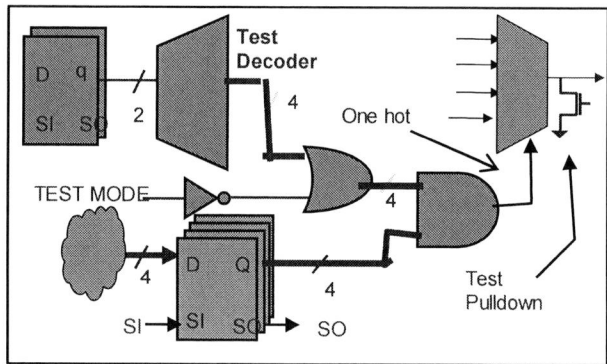

Figure 4: Pyron et al Scheme (ITC''99)

Finally, most recently, Giles, et al reported in [5] a scheme called Built-in Constraint Resolver (BICR) shown in Figure 5 that is similar to the Stanford scheme but takes only one additional control signal. The BICR is deployed on several recent microprocessors. The paper reports successful use of this DFT for running constraint-free ATPG and there have been further reports of its use for enabling LBIST.

978-1-61284-657-6/11 $26.00 © 2011 IEEE

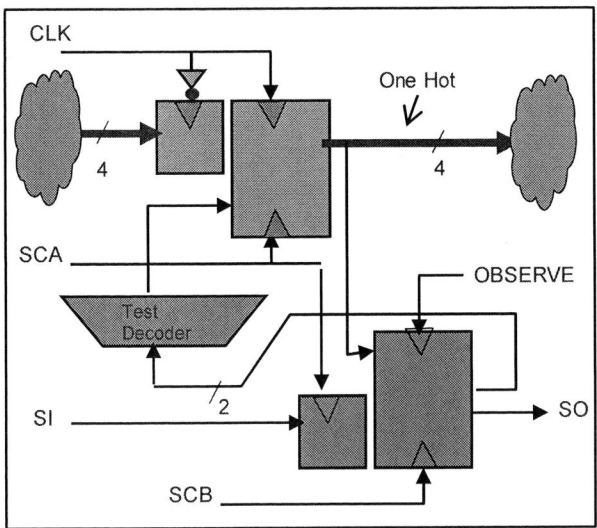

Figure 5: Built-in Constraint Resolver (ITC'05)

III. HARMONY WIDGET

Harmony Widget is simply the collection of enhanced scan-flops inserted at the one-hot MUX select control or any circuit situation subject to misbehavior during scan test. We will first focus on the LSSD style design which is the prevailing scan design style for high performance microprocessors. This enhanced scan-flop is nothing more than an ordinary scan-flop with a 2-to-1 multiplexer on the scan input port. For illustration, Figure 6 shows Harmony Widget on one-hot control for 4-to-1 MUX select lines.

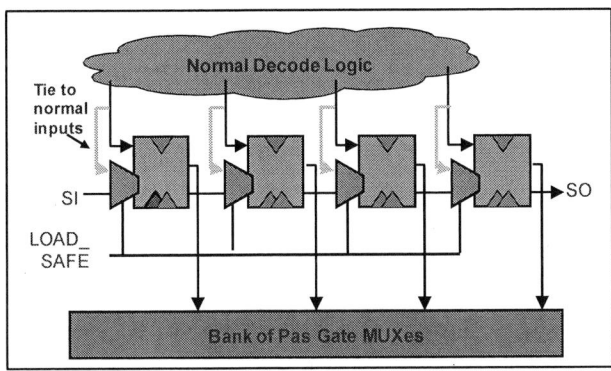

Figure 6: Harmony Widget

The 2-to-1 MUX selects between the usual shift data and the normal data from the upstream decode logic. LOAD_SAFE control, sourced from the chip scan system controller, decides what is loaded during shifting.

The scan chain incorporating the Harmony Widget operates just like a scan chain in an ordinary scan system with one difference; the scan controller asserts LOAD_SAFE signal during the last scan shift cycle. See Figure 7 for the timing diagram. This loads the normal scan vector normally

everywhere except in the Harmony Widgets. As seen in the diagram, during the last shift the Harmony Widget Scan-flops load from the normal upstream decode Logic with the guaranteed one-hot select values. Effectively, this action patches the random or ATPG generated bit patterns in-situ with a safe value already available in the upstream decoder logic. The downstream logic is thus driven with legal values and the subsequent scan-capture cycle captures an X-free response. Note that neither the normal flow of serial data load to the downstream scan-flops is perturbed, nor the scan chain length is affected.

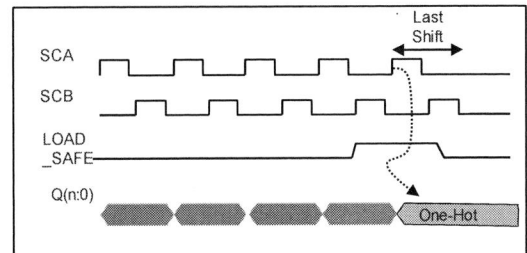

Figure 7: Harmony Widget Timing Diagram

By this time one obvious question in readers mind may be what if the upstream logic is a sequential decode happening over multiple cycles? The solution is to deploy Harmony Widget on all stages of the decode logic. There is no other special DFT or any other accommodation needed. The LOAD_SAFE action loads legal values in all stages of multi-cycle decode. This is shown in Figure 8. Notice that the path used for updating the safe value (shown in heavy line) is very likely a multi-cycle path. For the LSSD type scan system performing shift at slower rate than the normal clock rate this is not expected to be a problem.

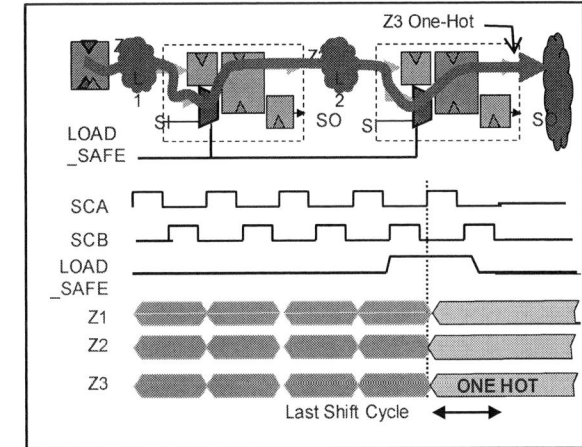

Figure 8: Harmony Widget on Multi-cycle Decoder

Note that once the safe value is loaded, the scan system is ready for capture cycles of arbitrary length burst. All subsequent capture cycles remain X-free for stuck-at as well as transition or path delay fault testing.

The design costs of Harmony Widget scheme are the special Scan-flop cells with the extra 2-to-1 MUX in the serial path and one LOAD_SAFE signal that is timed with respect to shift clock. The widget adds a small load to the normal select line but does not add any direct delay in the normal performance critical path. Compared to the current practice in industry, this is less hardware overhead and design complexity. One other cost is that the scan user must assert the LOAD_SAFE signal during the last shift or somehow indicate the start of the last shift cycle. This cost is similar in scope to the other methods that require assertion of special Load and Observe signals. For users of LBIST, it is fairly straightforward to integrate the LOAD_SAFE generation into the on-chip LBIST controller.

If ATPG wants to make good on the presence of the Harmony Widget and run ATPG without constraints, then it must be enhanced to recognize its presence and make use of it. This requirement is similar to the one discussed in prior work. In particular [5] discusses their accommodations in detail. It is important to note that if ATPG user so chooses, ATPG may be run normally with constraints. All user needs to do is tie-off the LOAD_SAFE signal to zero.

The Harmony Widget may be built as a library cell or constructed from normal library components as needed. That is, tool support may be used to identify sites and to automatically insert the Harmony Widget. This is also similar to the other reported methods.

The Harmony Widget does nothing for circuit misbehavior during shift operation. This misbehavior is not visible to the ATPG or the LBIST operation, but the associated circuit contention may affect long term reliability. If this becomes the concern, additional DFT, similar to the shadow flops used in Stanford or the BICR scheme must be added.

The Harmony Widget may also be applied to the MUX-D type scan system. It is interesting to note that the Harmony Widget may be simplified by dropping the MUX on the serial input pin. In other words, the Harmony Widget becomes the ordinary original scan-flop. The LAOD_SAFE signal, however, must be replaced by the special SCAN_ENABLE signal that de-asserts for the last shift cycle. That is, it de-asserts one cycle before the normal SCAN_ENABLE signal. Notice that in the MUX-D type scan system the path used for loading the safe data must meet the timing requirements. This may particularly become challenging if the shift operation is performed at full speed or if ATPG depends on using launch on shift coverage for transition fault testing.

IV. CONCLUDING REMARKS

This paper listed the major sources of Xs in scan testing and discussed DFT for eliminating them. It focused on one particular source, namely Xs resulting from the test pattern induced circuit under test misbehavior. This source consisting of circuits like pass gate MUX outputs, tri-state bus outputs, one hot CAM match-line outputs continues to challenge scan testing using ATPG and particularly LBIST. The paper reviewed the DFT currently in use in the industry for eliminating these Xs and then proposed its own novel solution called Harmony Widget. Harmony Widget offers some distinctive advantages over the current industry practice in making constraint-free ATPG and LBIST usable in designs with test pattern induced circuit misbehavior and Xs.

ACKNOWLEDGEMENT

The author is thankful to P. Pant, O. Mendoza, S. Sengupta, S. Patil and RJ Hayes for constructive suggestions including ideas for teaching ATPG about the Harmony Widget. Finally, the author wishes to thank T. Grutkowski for the incomplete thought-provoking conversation that led to the birth of the idea of the Harmony Road, Fort Collins, in an Airport Shuttle. (Therefore the name!)

REFERENCES

[1] Bhavsar, D. K. "Scan Islands: – A Scan Partitioning Architecture and its Implementation on the Alpha 21364 Processor," *VLSI Test Symposium*, May 2002.

[2] Pateras, S., M. S. Schmookler, "Avoiding unknown states when scanning mutually exclusive latches", *Int'l Test Conf.*, 1995.

[3] Mitra, S, L. Avra, E. J. McKluskey," Scan Synthesis for One-Hot Signals", *Int'l Test Conf.*, 1997.

[4] Pyron, C., M. Alexander, M., J. Golab; G. Joos; B. Long, R. Molyneaux, R. Raina, N. Tendolkar, "DFT Advances in Motorola's MPC7400, a PowerPC Microprocessor", *Int'l Test Conf.*, 2000.

[5] Giles, G., J. Irby, D. Toneva, K-H. Tai, "Built-In Constraint Resolution," *Int'l Test Conf.*, October 2005.

Localization of Damaged Resources in NoC Based Shared-Memory MP2SOC, using a Distributed Cooperative Configuration Infrastructure

Zhen Zhang*, Dimitri Refauvelet*, Alain Greiner*, Mounir Benabdenbi† and François Pecheux*
* *University Pierre et Marie Curie, LIP6-SoC laboratory, 4 place Jussieu, 75252 Paris, France*
{*zhen.zhang, dimitri.refauvelet, alain.greiner, francois.pecheux*}*@lip6.fr*
† *TIMA laboratory (Grenoble INP, CNRS, UJF), 46 avenue Felix Viallet, 38000 Grenoble, France*
mounir.benabdenbi@imag.fr

Abstract—In this paper, we present a software approach for localization of faulty components in a 2D-mesh Network-on-Chip, targeting fault tolerance in a shared memory MP2SoC architecture. We use a pre-existing and distributed hardware infrastructure supporting self-test and de-activation of the faulty components (routers and communication channels), that are transformed into "black hole". We detail the software method used to localize these "black holes", and centralize the information in a single point, where a modified global routing function can be defined. This embedded software makes an extensive use of a distributed fault-tolerant configuration firmware assisted by a Distributed Cooperative Configuration Infrastructure (DCCI), that is also presented. Finally, "black hole" detection and localization coverage is evaluated.

I. INTRODUCTION

According to the industrial forecast on high-performance computing issues [1], Network-on-Chip (NoC) based, shared memory, Massively Parallel Multi-Processor System-on-Chips (MP2SoCs) architecture, will soon be implemented in a single chip. However, with a high permanent failure rate [2] caused by poor manufacturing yields or aging defaults, fault-tolerance is a crucial issue that must be considered at a very early stage in the design.

As future MP2SoC architectures will contain a large number of replicated identical components, a simple fault-tolerant approach is to disable faulty components (such as a processor core, or an embedded memory bank), once their erratic behaviors have been detected, and to remap the embedded software application on the remaining operational hardware. Unfortunately, for the NoC itself, this approach is far from being sufficient. In order to save silicon area, and to minimize the network latency, most NoCs use dedicated routing algorithms, taking advantage of the regular micro-network topology. If a single component (a router or a communication channel) is faulty, the micro-network topology is modified and becomes irregular. If a packet is routed toward the faulty component, in the worst case, the whole NoC is blocked. Thus, the global routing function, and the NoC itself must be reconfigured to support the new topology.

In two previous works [3], [4], we presented a self-testable&cleanable, reconfigurable 2D-mesh NoC. The two key features are summarized below:

1) ***Self-testable&cleanable*** [4]: A fully distributed & de-centralized hardware built-in self-test (BIST) mechanism is integrated into the NoC. At power-on or system reboot, all NoC components are tested in isolation and in parallel. Each component that is found to be fault-free is enabled. All faulty components are disabled to prevent any fault propagation: A disabled component is configured to behave as a ***black hole***, that discards any incoming packet, and produces no outgoing packet.

2) ***Reconfigurable*** [3]: A reconfigurable, dead-lock free routing function (based on the X-First routing algorithm) has been defined. In each router, the reference X-First routing function can be modified through dedicated addressable reconfiguration registers (4 bits per router), to route the packets around the faulty components and bypass the black holes.

In conclusion, the distributed hardware BIST mechanism solves the problem of detection and de-activation of the faulty components in the NoC. Moreover, when the faulty components are localized, we have a general method to define a modified global routing function, and the NoC itself contains all the reconfiguration registers to implement the modified routing function.

But the hardware test and deactivation procedure is fully decentralized, and there is no centralization of the information about the localization of the faulty components. Therefore, we still have two problems to solve:

- ***Faulty components localization*** We need to centralize the information regarding the localization of the black holes (routers and channels), to be able to compute the global modified routing function.
- ***NoC reconfiguration*** We need a configuration bus to distribute the modified routing function in the reconfiguration registers of the fault-free routers.

In this paper we present a fully software approach for the localization problem. The only hypothesis is to have in each cluster (a cluster is a node in the 2D-mesh) a programmable processor and a ROM containing the fault tolerant configuration firmware. This configuration firmware is part of the Distributed Cooperative Configuration Infrastructure (DCCI), described in this paper.

After this introduction, section II presents the related work. Section III describes the generic 2D-mesh NoC based shared memory MP2SoC architecture, i.e our reference architecture. Section IV explains how the distributed configuration infrastructure is dynamically mapped onto this architecture on reset. Section V details the black hole localization procedure. And section VI presents experimental results for a mesh of dimension 4×4 that proves the effectiveness of the proposed approach.

II. RELATED WORK

The black hole model (proposed in [4]) is actually a functional fault model where the faulty components can be detected by means of a dedicated BIST approach, and de-activated prior any localization. Several papers present solutions for localization, [5], [6], [7], which rely on the use of ATE (Automatic Test Equipment) and TAM (Test Access Mechanism), to feed NoC inputs with external packets as the test vectors, and to analyze NoC outputs. Such approaches allow to test any deterministic end-to-end path (from an input to an output, defined by a deterministic routing algorithm such as X-First). The faulty components are identified by set intersection of faulty & fault-free paths, using an exclusive method.

In our 2D-mesh topology, any end-to-end path links a processor in a source cluster to a physical memory bank in a target cluster (a cluster is a node in the 2D-mesh). In a shared memory architecture, where any processor can address any memory location, such path can be tested by a simple software transaction, i.e a software task in the source cluster reads a data word mapped in the target cluster.

As we want to support "on the field" reconfiguration, we don't want to use an external ATE, and we propose an embedded and distributed software approach to detect the black holes.

It should be noted that our proposed strategy is different with the solution [8] that proposed a mechanism for discovering the faultless paths between an I/O port and the fault-free cores in a MP2SoC. This centralized discovering process is piloted by the smart I/O port, that is a critical resource. But, our proposed solution is fully distributed, it's achieved by an faultless embedded processor core, profiting the hardware redundancy of the MP2SoC architecture.

III. A NOC-BASED, SHARED MEMORY, MP2SOC ARCHITECTURE

As presented in Section I, in our previous works [4], [3], we designed and implemented a self-testable&cleanable, reconfigurable 2D-mesh micro-network DSPIN (Distributed Scalable Predictable Interconnect Network. The original DSPIN [9], [10] was designed by the LIP6 laboratory and was physically implemented by ST Microelectronics, to support MP2SoC architecture). In this paper, DSPIN is the self-testable&cleanable, reconfigurable version.

As shown in Fig.1.A, a DSPIN-based MP2SoC architecture is composed of a set of tiles called clusters.

As shown in Fig.1.B, a cluster may contain one or several processors (with their associated instruction and data caches), a local interconnect, an embedded RAM, an embedded ROM (for configuration firmware) and two routers (In order to avoid deadlocks in command/response traffic, each cluster contains

two independent routers implementing two separated sub-networks for commands and responses). In addition, some special clusters contain I/O ports controllers, used to access external mass storage devices. To each processor is associated a timeout mechanism: when it executes a memory load/store operation, the timeout mechanism is triggered. In the event the memory operation fails, the timeout generates an interrupt and the processor enters its exception mode.

Fig.1.C details the generic DSPIN router, that contains 5 ports (North, East, South, West & Local) interlinked as a full 5×5 crossbar. Each port contains two fifos, one for input and one for output. An input fifo in a given router, and the output fifo in the neighbor router define a point-to-point communication channel. In the following, an half-path (HP) is defined as the enumerated set of channels and routers involved in the carrying of a command packet (resp. response packet) from the initiator cluster (resp. target) to the target cluster (resp. initiator) through the NoC. A path (P) is the concatenation of two half-paths, one for the command (HPC) and one for the response (HPR).

After power-on or system reboot, and thanks to the hardware test & initialization mechanism, the fault-free components are enabled. The faulty ones are disabled and configured as black holes. The DSPIN NoC is not only cleaned from any evil failure propagation, but the fault-free routers are configured to implement the reference X-First routing function. Therefore, the NoC supports local communications, between a cluster and its neighbors clusters, as long as the corresponding communication channel is fault-free.

Finally, when there is more than one (fault-free) processor in a cluster at the end of the local BIST procedure, a local master is elected and can run the configuration firmware that is stored in the embedded ROM of each cluster.

IV. DISTRIBUTED COOPERATIVE CONFIGURATION INFRASTRUCTURE (DCCI)

During the classical (software) initialization of a system, the boot code that performs various tests and configuration tasks is generally located in a unique ROM, even in the case of a multi-cores architecture. Our configuration firmware (called CF in the following) is similar to the BIOS in a multi-cores PC, but in a possibly damaged MP2SoC, the hardware resources can not be trusted anymore, and chip initialization takes place in an uncertain world where a processor, a memory bank, a network interface controller, a router, or the boot ROM itself may be defective.

A. DCCI general principle

To remove this uncertainty, the key idea is to have one CF per cluster, to support a fully distributed approach, where a cluster is able to communicate and exchange information with its neighbor CFs, resulting in a Distributed Cooperative Configuration Infrastructure (DCCI).

During initialization, the role of the DCCI is to progressively build - only relying on local communications between neighbor clusters - a trusted tree of operational clusters, where

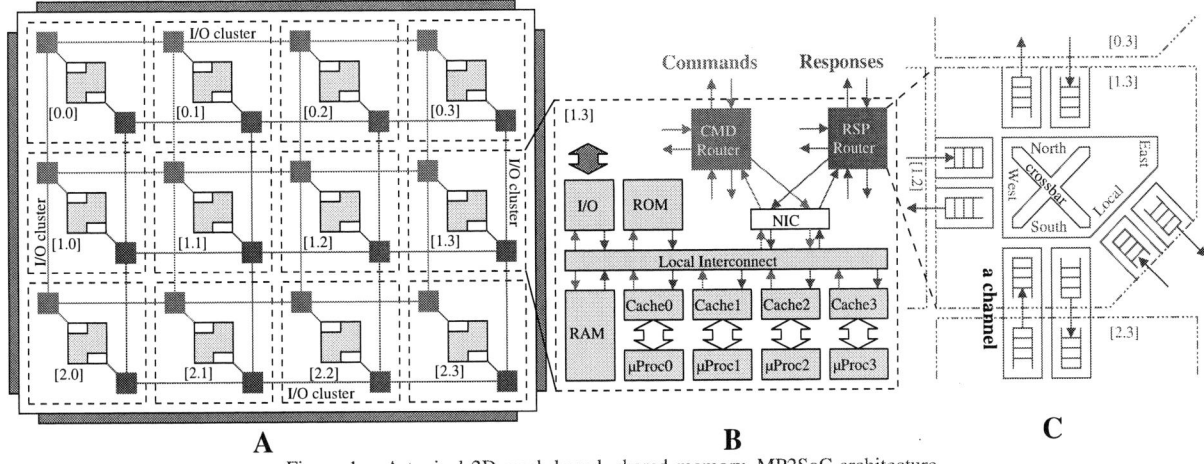

Figure 1. A typical 2D-mesh based, shared memory, MP2SoC architecture.

each operational cluster contains one operational processor running the CF. This tree is built in a bottom-up way, starting with operational clusters as leaves. This communication tree uses a limited part of the routing capabilities of the NoC (that has not yet been fully configured). It uses only the local communication channels between neighbor clusters for software based CF to CF communication through dedicated mailboxes.

The root of tree is a cluster determined by a distributed election process. The most important criterion in this election process is the capability of this root cluster to access the external mass storage where more exhaustive test programs and the final operating system itself are available, and can be loaded in the embedded RAM of the root cluster.

This software based communication tree can thus be seen as a slow and temporary communication infrastructure, dynamically constructed during the boot stage, using the NoC resources that have been identified as fault-free by the hardware BIST.

As soon as this tree is constructed, and the root is elected, this very unique master processor can access the external mass storage containing a virtually unlimited software stack outside the chip.

The master processor can use this communication tree to make any processor in the tree execute any specific software task (debug, fine-grain test), to propagate any configuration command to any child tree node, or read any status information. It can be used to complete the MP2SoC reconfiguration, and especially the configuration of the NoC itself.

B. Cluster self-test

After hardware reset, each cluster has to test itself by executing its local CF code, before it can try to participate in the distributed procedure to build the DCCI communication tree.

This local test is a 3 stages process:

1) *local intra-cluster test:* It is a first, coarse grain, functional software based test (such as presented in [11]) for all IPs belonging to the cluster, if a cluster is usable to

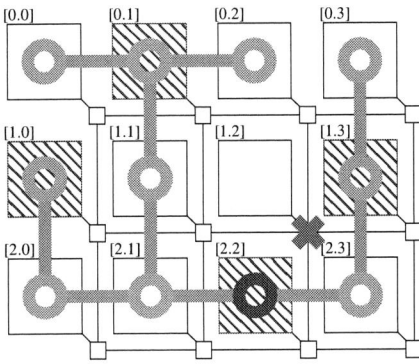

Figure 2. DCCI example in a damaged MP2SoC. At coordinates [1.2], at least one router (cmd or rsp) is faulty. A spanning DCCI communication tree is built as a result of each local CF task communicating only with its neighboring clusters. The tree node is presented as a circle, the tree root is at coordinates [2,2].

participate to the tree building procedure. For instance, a cluster which RAM does not pass its coarse march test is declared unusable.

2) *local leader election:* Second, an operational processor of the cluster is locally elected (the one with the smallest processor identifier). The elected processor represents the cluster with respect to the surrounding clusters. The other operational processors are put in idle state.

3) *access to the external memory:* The locally elected processor tries to establish a connection with the nearest external I/O controller, using the default configuration for the NoC infrastructure (standard X-first routing algorithm).

If a cluster pass successfully the two first steps, it is declared to be usable and the locally elected processor executes a specific CF code to discover its environment. Additionally, if the third test pass successfully, the cluster is a potential leader.

C. Spanning tree building

As stated before, the idea is to build a spanning tree (as shown in Fig.2) covering all connex clusters declared as usable. Each potential leader is a possible candidate to become the root, and the active processor in the elected cluster

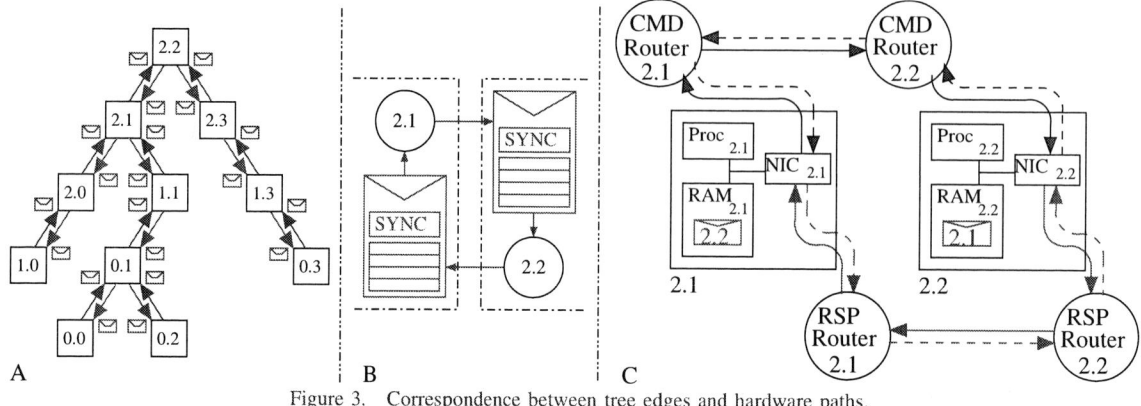

Figure 3. Correspondence between tree edges and hardware paths.

becomes the "chip leader". Fig.2 shows an example of a DCCI tree on a damaged MP2SoC (at least one router at coordinates [1.2] is faulty).

The spanning tree is built by concurrent aggregation of sub-trees. To manage concurrence between sub-trees, preference is given to clusters that are located nearby an access port to the external mass storage.

Fig.3.A presents a global representation of the final DCCI tree. To be connected, two neighbor clusters (such as clusters at coordinates [2.1] and [2.2]) must be able to communicate through a bidirectional link composed of two directed edges, for full duplex communication. Each edge corresponds to a software mailbox, as shown in Fig.3.B. In the followings, an edge from node/cluster A to node/cluster B is named e_{AB}.

A mailbox is a data structure in memory composed of a buffer for the message, and a synchronization flip-flop element. In the DCCI each mailbox is located in the local memory of the receiving node, and is shared by exactly one sender and one receiver. Consequently, two mailboxes are used for bidirectional communications.

Fig.3.C describes the whole set of hardware components involved in the bidirectional communication between two neighbor nodes using two mailboxes. As shown in the picture, each bidirectional communication channel uses 12 hardware channels and 4 routers. Each hardware element (channel or router) must be faultless. After the self-test, each cluster starts an handshake procedure to produce the list of neighbors clusters with which the mailbox communication is possible.

Considering two neighbors nodes A and B, this handshake algorithm performs the followings operations:

- Node A sends a PING message to node B using edge e_{AB}. This software transaction implies the use of HPC_{AB} (Half Path Command) and HPR_{AB} (Half Path Response).
- If node B receives the PING message it sends back an ALIVE message to node A using e_{BA}. This software transaction implies the use of HPC_{BA} and HPR_{BA}.
- If A receives a timeout for its PING message, and still receives the ALIVE response message, it means that HPR_{AB} is faulty. Thus, A sends a NOK message to B to inform it that HPR_{AB} is down.

The protocol is fully symmetric. If a timeout or a NOK message is received by a node, the two involved edges are declared has faulty. If the two paths between A and B are fault-free, the A-B connection can be used for tree construction and can potentially become active edges of the DCCI tree.

The characteristics of the resulting tree are the following:

- each node has been tested and is usable
- each edge of the tree has been tested and is usable
- there exists no edges between nodes that are not direct neighbors
- any tree path between any couple of clusters belonging to the tree is usable

D. Using the spanning tree

Like in [12], any message between 2 distant nodes N0 & N1 is propagated with the active cooperation of all intermediate nodes on the path between N0 & N1. Each intermediate node is actually acting as a software router. Therefore, and unlike [12], which uses probabilistic broadcast for fault tolerant communication, the approach presented herein is purely deterministic.

We previously indicated that a tree can be considered as a trusted launching pad for exploration and testing. By using end-to-end protocol or flooding protocol over the tree, we can send command, application test, or test results to one or to all nodes. For example, in the case of a more exhaustive NoC test application, we can dispatch the test application from the root to all tree nodes with flood message, execute application on every nodes, and retrieve results to the root.

V. THE "BLACK HOLE" LOCALIZATION PROCEDURE

A. General Principle

In a shared memory architecture, any path in the NoC links a processor (initiator cluster) to a physical memory bank (target cluster). The multi-threaded, distributed, software application for black hole localization is loaded by the master processor on all usable clusters, using the DCCI communication tree. All enabled NoC routers are initially configured to implement the reference X-First routing algorithm. The software application therefore tries to use the NoC routing infrastructure for inter-thread communications, but some packets may be lost in the black holes.

By collecting the informations stored in each cluster on successful and unsuccessful transactions (through the DCCI communication tree), the master processor (root of tree) is able to localize the black holes.

It should be noted that, in (DCCI) tree creation, some paths have been tested (the path between two neighboring clusters). The tree topology implicitly already contains informations of black hole locations, but these informations are very rough, so we must use a dedicated application of black hole localization to obtain most fine-grain informations.

B. Distributed Algorithm

A task (t), running in a cluster[y.x] (coordinates in the 2D-Mesh) makes a read transaction targeting a cluster [y'.x']. According to the X-First routing policy, a path P between two clusters define a unique set of routers and channels. If the transaction succeeds (both the command packet and the response packet), the task (t) receives the expected data, and registers the path (y.x/y'.x') as OK. If not, (t) receives a timeout interrupt, indicating a packet loss. Once all paths from cluster[y.x] to all other clusters have been tested, a list of FaultLess Paths (FLP) can be constructed. Therefore, two sets of FaultLess Routers FLR(y,x) and FaultLess Channels FLC(y,x) can be derived from the FaultLess Paths list (FLP).

$PathRegistration()$:

Require: $[Y.X]$ is the index of the source (the current cluster).
Require: FLP is the set of faultless paths seen by the source.
Require: FLC is the set of faultless channels seen by the source.
Require: FLR is the set of faultless routers seen by the source.

{In the following, $CHC/R_{YX\_N/S/E/W/L\_I/O}$ represents a channel. C/R means cmd or rsp subnetwork, YX means that the channel belongs to router[Y.X], N/S/E/W/L means the port that the channel belongs to, and I/O means input or output. The same, RTC/R_{YX} represents a router. C/R means cmd or rsp subnetwork, YX means the index of router. For example, $CHC_{YX\_L\_I}$ means the input channel of local port of cmd router [Y.X]. And RTC_{YX} means the cmd router [Y.X].}

1: Construction of FLP
2: $FLC \leftarrow NIL$
3: $FLR \leftarrow NIL$
4: **for** $i = 0$ to $|FLP|$ **do**
5:    $[y.x] \leftarrow FLP[i]$
   {Extract the target cluster index of a faultless path.}
6:    $FLC \leftarrow FLC \bigcup CHC_{YX\_L\_I} \bigcup CHC_{yx\_L\_O} \bigcup CHR_{yx\_L\_I} \bigcup CHR_{YX\_L\_O}$
7:    $FLR \leftarrow FLR \bigcup RTC_{YX} \bigcup RTC_{yx} \bigcup RTR_{yx} \bigcup RTR_{YX}$
8:    **if** $x > X$ **then**
9:       **for** $i = X$ to $x - 1$ **do**
10:          $FLC \leftarrow FLC \bigcup CHC_{Yi\_E\_O} \bigcup CHR_{yi\_E\_I}$
11:          $FLR \leftarrow FLR \bigcup RTC_{Y(i+1)} \bigcup RTR_{yi}$
12:       **end for**
13:    **else if** $x < X$ **then**
14:       **for** $i = X$ to $x + 1$ **do**
15:          $FLC \leftarrow FLC \bigcup CHC_{Yi\_W\_O} \bigcup CHR_{yi\_W\_I}$
16:          $FLR \leftarrow FLR \bigcup RTC_{Y(i-1)} \bigcup RTR_{yi}$
17:       **end for**
18:    **end if**
19:    **if** $y > Y$ **then**
20:       **for** $j = Y$ to $y - 1$ **do**
21:          $FLC \leftarrow FLC \bigcup CHC_{jx\_S\_O} \bigcup CHR_{jX\_S\_I}$
22:          $FLR \leftarrow FLR \bigcup RTC_{jx} \bigcup RTR_{(j+1)X}$
23:       **end for**
24:    **else if** $y < Y$ **then**
25:       **for** $j = Y$ to $y + 1$ **do**
26:          $FLC \leftarrow FLC \bigcup CHC_{jx\_N\_O} \bigcup CHR_{jX\_N\_I}$
27:          $FLR \leftarrow FLR \bigcup RTC_{jx} \bigcup RTR_{(j-1)X}$
28:       **end for**
29:    **end if**
30: **end for**
31: **return** FLC
32: **return** FLR

C. Localization and reconfiguration

The two sets FLR(y,x) and FLC(y,x) are distributed in each cluster[y.x]. These information data can be collected by the master processor, using the DCCI communication infrastructure, and merged in two global sets, GFLR and GFLC, as shown in Fig.4. Finally, according to this method, any router or communication channel that is not present in these GFLR and GFLC sets is a black hole. So, theoretically, the black hole Detection Coverage (DC) is 100%, for a defective NoC with any number of fault. And this is confirmed in the section VI by the experimental results.

It is worth noticing that the number of fault-free (usable) communication resources (routers or communication channels) identified by this procedure can be much larger than the resources used by the DCCI covering tree.

Once all the black holes are identified and localized, it is possible (if the number of faulty component is not too large) to compute a modified routing function, and to use the DCCI communication infrastructure to load this modified routing function into the addressable configuration registers distributed in the NoC.

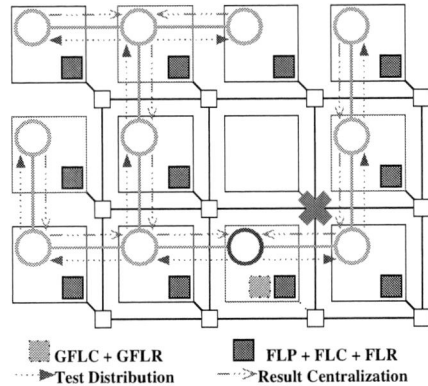

GFLC + GFLR   FLP + FLC + FLR
▶Test Distribution   ➤Result Centralization

Figure 4. Thanks to the DCCI communication tree, the root can load the test code from the external memory, and to distribute the code to each tree node, to do the black hole detection. Once all of detections have been achieved, the root can centralize the FLC and FLR from each node, and to merge these lists into two global sets, GFLC and GFLR.

VI. EXPERIMENTAL RESULTS

A. Detection Coverage

In this subsection, we present experimental results of Detection Coverage (DC) evaluation for the black hole localization procedure. We have simulated two types of fault in a 4×4 clusters MP2SoC, 1st, single fault injection (one faulty router or one faulty channel); 2nd, multi-faults injections. These fault injections were simulated on a dedicated C simulator.

1) Single Injection: As in a $M \times N$ DSPIN 2D-mesh, there are C cmd&rsp channels, and there are R cmd&rsp routers.
$$C = (M \times (N-1) \times 2 + N \times (M-1) \times 2 + M \times N \times 2) \times 2$$
$$R = M \times N \times 2$$

The Detection Coverage has been evaluated for all the situations where the NoC contains one single fault: one faulty router or one faulty channel defining a total of $(C+R)$ different faulty networks. In our example, with $M = 4, N = 4$, there are 160 channels and 32 routers.

In all cases, the black hole has been identified and located, resulting in a Detection Coverage of 100% for a single fault.

However, in some cases, some fault-free components are wrongly identified as black holes, which is explained in the following.

As shown in Fig.5 (a partial description of a cluster), there are some dependencies between channels. For example, channel 1 and channel 3 depends on each other, because any path between the local processor and a RAM of another cluster will contain this couple. Channel 1 is a black hole, channel 3 can not be tested with any faultless path, and it will be identified as a black hole. But this result is acceptable for the reconfigurable routing [3], since in this case the whole router is deactivated.

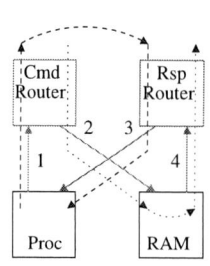

Figure 5. Dependency exists between channel 1 & channel 3 or between channel 2 & channel 4.

2) Multi-faults injection: The Detection Coverage of multi-faults injection has been evaluated for all the situations of

- 1 faulty router and 1 faulty channel
- 2 faulty routers
- 2 faulty channels
- 2 faulty routers and 1 faulty channel
- 1 faulty routers and 2 faulty channels
- 2 faulty routers and 2 faulty channels

All these simulations resulted in a 100% coverage.

B. Application Execution Time Evaluation

The execution time is an important issue for a software procedure that will be executed at each reboot of the system. For this experiment we simulated the complete procedure on a 4×4 2D-mesh MP2SoC architecture containing 16 processors, modeled with the cycle-accurate SoCLib virtual prototyping platform [13]. This architecture contained one single fault. The total time is 7.1×10^6 cycles (without hardware test process):

- Time for (DCCI) tree construction: 1.9×10^6 cycles
- Time for for test task distribution: 1.2×10^6 cycles
- Time for test execution: 3.5×10^6 cycles
- Time for test result centralization: 0.5×10^6 cycles

As this procedure is executed using the system clock, we obtain 0.014 second at 500Mhz. Which is fully acceptable.

C. Application Code Size

For a MIPS32 processor, the application code is split in: DCCI : 5 Kbytes; Test and localization procedure : 2.5 Kbytes. Embedding this application in a MP2SoC is thus affordable.

VII. CONCLUSION

In this paper, we presented a software approach to localize faulty hardware components in a 2D-mesh NoC used in a shared memory MP2SoC. The localization of faulty components is mandatory to implement an "on the field" reconfiguration mechanism supporting fault tolerance in the context of permanent failures. These faulty components can either be a

point-to-point communication channel, or a complete router. They can be transformed into black holes by a built-in self-test (BIST) mechanism that is the basis of our fault model. The localization algorithm relies on a Distributed Cooperative Configuration Infrastructure that dynamically builds a software based communication tree, covering all the nodes that have successfully passed the local BIST. This DCCI communication infrastructure is a distributed software mechanism that can be used as a configuration bus. It does not use the full routing capabilities of the NoC, but only the local communication channels between two neighbor nodes. The proposed black hole detection software algorithm has been evaluated in a 16 nodes MP2SoC architecture (4×4 mesh) modeled as a SystemC virtual prototype, in the framework of the cycle accurate SoCLib environment. It reaches a Detection Coverage of 100% in all tested cases. The same DCCI communication tree can be used to distribute the resulting modified routing functions to the fault-free routers.

It should be noted that, the method proposed in this paper, can be used in any shared memory multi-core architecture with a 2D-Mesh NoC.

REFERENCES

[1] International Technology Roadmap for Semiconductors. [Online]. Available: http://www.itrs.net

[2] S. Furber, "Living with Failure: Lessons from Nature?" in *Proc. of ETS'06, the 11th IEEE European Test Symposium*, 2006.

[3] Z. Zhang, A. Greiner, and S. Taktak, "A reconfigurable routing algorithm for a fault-tolerant 2D-Mesh Network-on-Chip," in *Proc. of DAC'08, the 45th Design Automation Conference*, 2008.

[4] Z. Zhang, A. Greiner, and M. Benabdenbi, "Fully Distributed Initialization Procedure for a 2D-Mesh NoC, Including Off Line BIST and Partial Deactivation of Faulty Components," in *Proc. of IOLTS'10, the 16th IEEE International On-Line Testing Symposium*, 2010.

[5] C. Grecu, P. Pande, B. Wang, A. Ivanov, and R. Saleh, "Methodologies and Algorithms for Testing Switch-Based NoC Interconnects," in *Proc. of DFT'05, the 20th IEEE International Symposium on Defect and Fault Tolerance in VLSI Systems*, 2005.

[6] K. Stewart and S. Tragoudas, "Interconnect Testing for Networks on Chips," in *Proc. of VTS'06, the 24th IEEE VLSI Test Symposium*, 2006.

[7] J. Raik, R. Ubar, and V. Govind, "Test Configurations for Diagnosing Faulty Links in NoC Switches," in *Proc. of ETS'07, the 12th IEEE European Test Symposium*, 2007.

[8] E. Kolonis, M. Nicolaidis, D. Gizopoulos, M. Psarakis, J. Collet, and P. Zajac, "Enhanced self-configurability and yield in multicore grids," in *Proc. of IOLTS'09, the 15th IEEE International On-Line Testing Symposium*, 2009.

[9] I. Panades, A. Greiner, and A. Sheibanyrad, "A Low Cost Network-on-Chip with Guaranteed Service Well Suited to the GALS Approach," in *Proc. of NanoNet'06. the 1st International Conference on Nano-Networks and Workshops.*, 2006.

[10] I. Miro-Panades, F. Clermidy, P. Vivet, and A. Greiner, "Physical implementation of the dspin network-on-chip in the faust architecture," in *Proc. of NoCS'08, the 2nd ACM/IEEE International Symposium on Networks-on-Chip*, 2008.

[11] D. Gizopoulos, A. Paschalis, and Y. Zorian, *Embedded processor-based self-test.* Kluwer Academic Pub, 2004.

[12] T. Dumitraş, S. Kerner, and R. Mǎrculescu, "Towards on-chip fault-tolerant communication," in *Proc. of ASPDAC'03, the 8th Asia and South Pacific Design Automation Conference*, 2003.

[13] LIP6 et al. SoClib. [Online]. Available: https://www.soclib.fr

978-1-61284-657-6/11 $26.00 © 2011 IEEE

Exponent Monitoring for Low-Cost Concurrent Error Detection in FPU Control Logic

Michail Maniatakos
EE Department
Yale University
michail.maniatakos@yale.edu

Yiorgos Makris
EE & CS Departments
Yale University
yiorgos.makris@yale.edu

Prabhakar Kudva
IBM T. J. Watson
Research Center
kudva@us.ibm.com

Bruce Fleischer
IBM T. J. Watson
Research Center
fleischr@us.ibm.com

Abstract—**We present a non-intrusive concurrent error detection (CED) method for protecting the control logic of a contemporary floating point unit (FPU). The proposed method is based on the observation that control logic errors lead to extensive datapath corruption and affect, with high probability, the exponent part of the IEEE 754 floating point representation. Thus, exponent monitoring can be utilized to detect errors in the control logic of the FPU. Predicting the exponent involves relatively simple operations, therefore our method incurs significantly lower overhead than the classical approach of duplicating the control logic of the FPU. Indeed, experimental results on the openSPARC T1 processor show that, as compared to control logic duplication, which incurs an area overhead of 17.9% of the FPU size, our method incurs an area overhead of only 5.8% yet still achieves detection of over 95% of transient errors in the FPU control logic. Moreover, the proposed method offers the ancillary benefit of also detecting 98.1% of datapath errors that affect the exponent, which cannot be detected via duplication of control logic. Finally, when combined with a classical residue code-based method for the fraction, our method leads to a complete CED solution for the entire FPU which provides a coverage of 94.4% of all errors at an area cost of 16.32% of the FPU size.**

I. INTRODUCTION

As aggressive scaling continues to push technology into smaller feature sizes, various design robustness concerns continue to arise. Among them, the frequent occurrence of transient errors [1] has resurfaced as a contemporary problem of interest. This problem is mainly attributed to strikes by neutrons or alpha particles and the corresponding single event upsets (SEUs) in memory bits, or single event transients (SETs) in combinational logic, which may potentially result in a soft error. However, several other factors such as design marginalities, Negative Bias Temperature Instability (NBTI), coupling, power supply noise, etc. [2], [3] also threaten the robustness of modern microprocessor units. The increasing severity of the above threats has spawned renewed efforts in developing cost-effective concurrent error detection (CED) methods for various key components of a circuit.

Floating point units (FPUs), in particular, are among the most crucial and hardest to protect [4], [5]. And while progress is being made on solutions using error detecting/correcting codes for the datapath portion of an FPU [6], [7], [8], little is known about its control logic, where either duplication [9] or Triple Modular Redundancy (TMR) [10] techniques are usually applied. Control logic errors might have catastrophic impact on the FPU output [11], [12], jeopardizing the application execution and providing the end-user with erroneous results. Furthermore, the size of control logic in modern FPUs is significant, often amounting up to 20% of the FPU size, thus rendering necessary the application of error detection methods.

In this study, we propose an alternative method to protect the control logic of an FPU by monitoring the exponent part of the floating point representation. Our method is based on the conjecture that a control logic error will incorrectly guide the datapath and, by extension, severely alter the expected outcome of the performed operation. As a result, it is highly likely that a control logic error will modify the value of the exponent portion of the floating point output. Given that it is relatively straightforward to calculate the correct exponent through simple operations, monitoring exponent correctness leads to an inexpensive yet very efficient CED method for the FPU control logic. Furthermore, it provides the ancillary benefit of detecting errors in the exponent part of the representation and, when combined with a residue code-based error detection method for the fraction, it results in a very low-cost CED solution for the entire FPU.

The rest of the paper is organized as follows: Section II briefly describes existing techniques for the protection of FPUs. Section III describes the proposed exponent monitoring-based CED method, followed by section IV where the development of the simulation-based experimental infrastructure and the actual CED implementation is presented. The merit figures of the proposed method, namely the attained coverage and incurred overhead, are assessed in section V, followed by conclusions in section VI.

II. ERROR DETECTION IN FPUs

Several error detection methods have been proposed for protecting FPUs. Most of them, however, target the datapath and have been ported from the corresponding techniques for integer arithmetic, while methods specifically designed to protect the FPU control logic have yet to be developed.

A. Datapath

The most popular technique for reliable arithmetic operations is residue codes [13], [14], [15]. Low-cost residue codes are single arithmetic error detecting codes with unidirectional error detecting capabilities. The efficiency of residue codes depends on the selection of the check base b. The higher the base the more errors the code will detect, yet also the more expensive the hardware overhead which will be incurred. Popular base selections are $b = 15$ (4 bits) and $b = 3$ (2 bits). In both cases the resulting *modulo* circuit is highly simplified and the theoretical error detection percentage is $1 - (1/2^4) = 93.4\%$ for $b = 15$ and $1 - (1/2^2) = 75\%$ for $b = 3$. Residue codes have been successfully applied to various designs [7], [16], [17].

978-1-61284-657-6/11 $26.00 © 2011 IEEE

Other techniques include Berger codes [18], [19] and two-rail checkers [20], [16]. Berger codes are optimal unidirectional error detecting codes. ALUs using Berger encoded operands have been shown to be strongly fault-secure [7], [11].

B. Control Logic

The simplest and most straightforward CED solution for random logic, such as the control logic of the FPU, is duplication [9]. The main advantage of duplication is simplicity and applicability to any given design. However, the $> 100\%$ area overhead (including the comparators) and the extra delay required for checking make duplication less appealing for modern FPUs. Furthermore, control in modern, pipelined FPUs is distributed across multiple components, necessitating manual and tedious effort to identify and replicate it.

Triple Modular Redundancy (TMR) [10] has similar properties to duplication, with the added advantage of error correction. However, the hardware overhead of $> 200\%$ (including the voter) makes it prohibitive for commercial designs.

III. Proposed CED Method

Our method is based on the conjecture that an error in the control logic will lead to extensive datapath corruption, which will propagate to the exponent part of the floating point representation. Numerous examples can be provided to show the impact of control errors on the exponent and justify our approach:

- *Special case control:* The control logic identifies whether the input operands are NaN (Not a Number), Infinity, 0, etc. Mishandling of special cases due to errors will result in a completely different output with an incorrect exponent. For example, any operation with NaN results in a NaN. If the control logic mistreats a normal operand as NaN, then the operation 3*5 will result in NaN (exponent = 255) instead of 15 (exponent = 130).
- *Algorithm stage control:* All floating point operations go through several stages before generating the final results. In case a stage is skipped or repeated (e.g. one more or one less division round is performed) due to a control error, the result will be incorrect and is likely to be reflected in the exponent.
- *Select lines:* Control logic is responsible for driving the correct operands to the datapath. In case an error occurs and the control drives a 0 instead of a 7, the operation 7*18 will result in a 0 (exponent = 0) instead of 126 (exponent = 133).
- *Operation control:* Along with the operands, the control logic also drives the signals for the operation selection. Therefore, if due to an error the operation changes, say from addition to subtraction, then the operation 2.0+1.9 will result in 0.1 (exponent of 123) instead of 3.9 (exponent of 128).

These are only a few examples of datapath corruption due to control logic errors, supporting our conjecture that errors in the FPU control logic can be detected by monitoring the datapath. Since the exponent part of the datapath is likely to be affected and fairly simple to calculate [21], we seek to develop a low-cost CED method for the control logic by predicting and verifying the exponent part of the floating point result.

A. Calculating the exponent

In this section, we discuss the exponent calculation for each of the three types of FPU functions, namely arithmetic operations, conversions and other operations. We note that the exponent is calculated independently of the fraction operation, hence the result is not exact since possible fraction normalization may affect the final value of the exponent.

1) Arithmetic operations: The first category is arithmetic operations, such as additions, subtractions, multiplications and divisions. We remind that the IEEE 754 representation of normalized floating point operands is $(-1)^s * 1.f * 2^e$, where s is the sign, f is the fraction and e the exponent. Thus, multiplication and division exponent calculation is simple, i.e.,

$$
\begin{aligned}
((-1)^{s_1} * 1.f_1 * 2^{e_1}) * ((-1)^{s_2} * 1.f_2 * 2^{e_2}) \\
= (-1)^{s_1+s_2} * 1.f_1 * 1.f_2 * 2^{e_1+e_2}
\end{aligned} \tag{1}
$$

for multiplication and

$$
\begin{aligned}
((-1)^{s_1} * 1.f_1 * 2^{e_1})/((-1)^{s_2} * 1.f_2 * 2^{e_2}) \\
= (-1)^{s_1+s_2} * 1.f_1/1.f_2 * 2^{e_1-e_2}
\end{aligned} \tag{2}
$$

for division. So we simply need to add (for multiplication) or subtract (for division) the exponents, operations which can be performed by the same hardware structure. In case the fraction overflows and needs to be normalized, the exponent needs to be adjusted accordingly. Hence, for arithmetic operations, we can only predict the exponent of normalized results with a ± 1 accuracy. Consequently, if an erroneous result differs from the correct result by 1, error masking will occur. However, our conjecture is that, in the presence of a control logic error, the datapath is corrupted extensively, hence the probability of such masking is very low. Indeed, the results presented in section V corroborate this conjecture.

For addition and subtraction, the exponent is the largest of the two operand exponents, therefore a simple comparator suffices to calculate it (similar to the multiplication/division cases, normalization may be required). This does not apply in the case of cancelation (i.e., when there is a subtraction of operands with equal exponents or exponents that differ by 1). In this case, the exponent can take a wide range of values and cannot be computed accurately without information from the fraction. In order to moderate cost, our CED method taps into the existing FPU hardware in order to obtain this information (rather than replicating it), hence error masking due to common mode failures may occur. Nevertheless, as we show in the results Section V, such masking is very small.

2) Conversions: Another common operation performed in FPUs is conversion from/to integer/floating point representations. The exponent of the results can be exactly calculated by appropriately offsetting the input operand. For floating point precision conversions (single to double and vice versa), the exponent needs to be offset by ± 896, since the actual exponent is $e_s - 127$ in single precision and $e_d - 1023$ in

TABLE I
OPENSPARC T1 FLOATING POINT INSTRUCTIONS

Operation	Result Exponent	Fraction Normalization
Addition	$max(e_1, e_2)$	Yes
Subtraction	$max(e_1, e_2)^1$	Yes
Multiplication	$e_1 + e_2$	Yes
Division	$e_1 - e_2$	Yes
Single to Double	$e_1 + 896$	No
Double to Single	$e_1 - 896$	No
Integer to Single	$MSB(i1) + 127$	No
Integer to Double	$MSB(i1) + 1023$	No
Negation	e_1	No
Absolute Value	e_1	No

[1] Equal or different-by-1 exponents may lead to cancelation.

double precision. Thus, for single to double conversion, the exponent is $e_d = (e_s - 127) + 1023$ and for double to single $e_s = (e_d - 1023) + 127$. For integer conversions, the exponent is a function of the most significant bit. Table I summarizes the exponent operation for different FPU functions.

3) Other operations: Modern FPUs usually implement more operations, such as absolute value, negation and comparison. In all these operations, the exponent is very simple to calculate. Negation/absolute value operations affect only the sign (i.e., the exponent is the same). Comparison operation results are implementation specific, as the output result is the comparison result and not a floating-point number. For example, SPARC ISA defines the exponent field of the output as 0, and the comparison result is stored in the flags field.

IV. EXPERIMENTAL SETUP

In order to assess the effectiveness of our method we built an extensive simulation infrastructure to perform error injection experiments. Since our target is transient errors, we need to perform a large number of injections; therefore, the infrastructure must support very fast simulations and error impact evaluation.

A. Test Vehicle

The test vehicle of our study is the register transfer-level (RTL) model of the openSPARC T1 microprocessor [22], the open source version of the UltraSPARC T1 microprocessor. The openSPARC T1 processor has eight SPARC processor cores which have full hardware support for four threads. Each SPARC CPU core can send a packet to the shared Floating Point Unit (FPU), using the cache-processor crossbar (CPX). Conversely, the FPU can send a packet to any one of the eight cores using the processor-cache crossbar (PCX). A floating point instruction is delivered from the cores to the FPU in either one- (single operand instructions) or two-packet transfer. One source operand is transferred in each cycle and the crossbar always provides a two-cycle transfer. In case of single operand instructions, an invalid transfer is produced in the second cycle [23].

Since the FPU is a single shared resource, each of the eight SPARC cores may have a maximum of one FPU instruction waiting to be executed. Thus, the FPU can hold up to 8 instructions at a given time. The FPU implements the SPARC V9 floating-point instruction set, and is fully compliant with

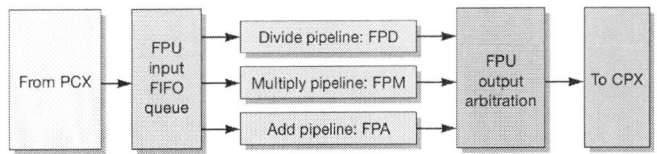

Fig. 1. T1 FPU Functional Block Diagram [23]

the IEEE 754 standard [24]. The floating point register file and floating point state register are in the SPARC core Floating point Front-end Unit (FFU), which is unique for every core (and not shared like the FPU).

The FPU includes three execution pipelines:

- Floating-point adder (FPA): Executes additions, subtractions, comparisons and conversions
- Floating-point multiplier (FPM): Executes multiplications
- Floating-point divider (FPD): Executes divisions

Incoming instructions are stored in a 16 entry x 155 bit FIFO queue (unless the FIFO is empty, in which case it is bypassed). In each cycle, one instruction may be issued from the FIFO and one instruction may complete and exit the FPU. Fig. 1 shows a block diagram with the three independent pipelines. Not-a-Number (NaN) source propagation is supported by steering the NaN source through the pipeline to the result.

B. CED implementation

The FPU is a shared resource with multiple floating-point instructions in flight. Moreover, the latency of some floating-point instructions (i.e. division) is variable and cannot be predicted a priori. Hence, it is not possible to predict which instruction should exit the FPU at each time. Instead, the exponents calculated for each incoming instructions are stored in an array which is indexed using the CPU ID of the outgoing instruction. The CPU ID is a unique identifier because each of the 8 cores can have a maximum of one outstanding FPU instruction. A thread with an outstanding FPU instruction switches out while waiting for the FPU result. This allows up to 8 instructions to be in the FPU. Therefore, storing the signatures requires a memory with 8 entries.

The block diagram of the CED implementation for the openSPARC T1 is presented in Fig. 2. This implementation applies to any pipelined FPU that executes multiple floating point instructions, such as the IBM PowerPC 405 FPU, Intel Pentium FPU and the SPARC T2. In case an FPU executes only one instruction at a time, the memory array is not needed and the output result can be checked directly.

The ± 1 exponent component presented in the diagram is needed for the arithmetic operations that may require fraction normalization, as explained in Section III-A1.

C. Experiment flow

Fig. 3 shows the data flow of our experiments. First, we use a python-based assembly generator which we developed to generate multi-threaded (MT) assembly utilizing floating point instructions. This synthetic workload is then simulated in the openSPARC T1 environment using sims, and Value Change Dump (VCD) traces are collected at the input of the FPU. These traces are then converted to a separate testbench

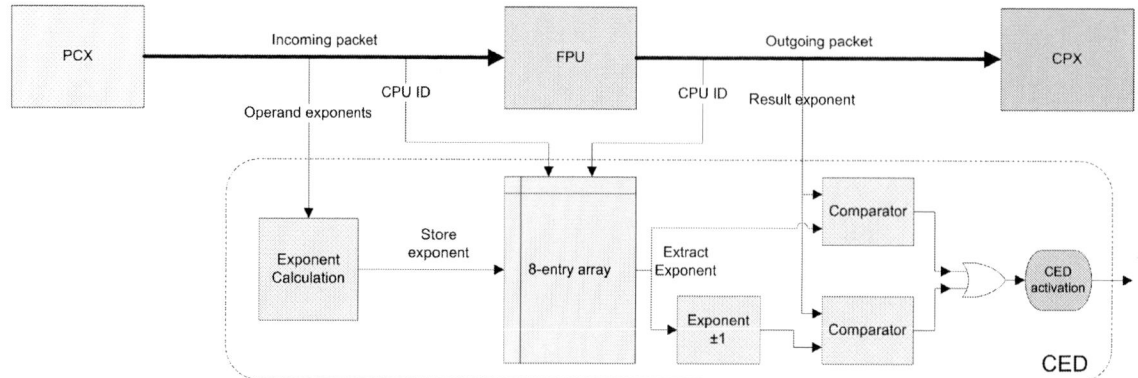

Fig. 2. Block Diagram of Proposed CED Structure

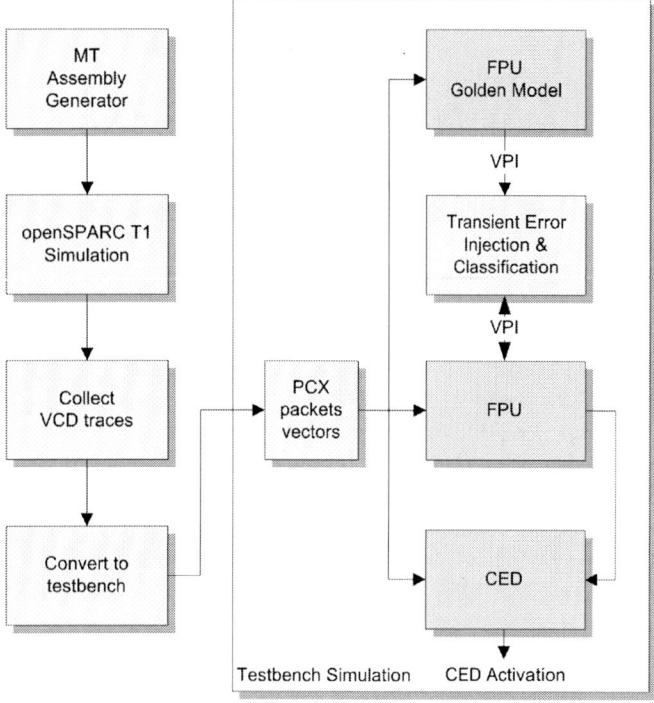

Fig. 3. Simulation Infrastructure

using Synopsys `vcat`. This testbench can be simulated using Synopsys `vcs` without the need to simulate the entire microprocessor model, leading to a 10x simulation speed-up.

The assembly generator can generate up to 32 different assembly files, one for each thread. The user can specify the percentage of the floating-point instructions in the file, as well as the desired number of instructions for each pipeline (FPA, FPM, FPD). Floating point registers are randomly selected for each instruction. 10 of the registers contain special values (NaNs, Inf, 0) to ensure operations with special numbers. Transient error injection is performed during simulation by mutating the microprocessor model for one clock cycle using the parallel saboteurs technique. An extensive description of the RTL error injection method can be found in [25].

The transient error injection is controlled by a python script through Verilog Procedural Interface (VPI) calls. The same script is used for error classification, with the help of a golden model that runs in parallel with the injected FPU model.

D. Hardware synthesis

In order to provide hardware overhead estimates, we synthesized the FPU model using Synopsys Design Compiler targeting a 90nm library. The timing constraints were set to a clock period of 1GHz, to match the running speed of the UltraSPARC T1. The total area of the synthesized FPU is $816,660\mu m^2$ ($587,947\mu m^2$ combinational, $181,110\mu m^2$ non-combinational and $47,601\mu m^2$ net interconnect area). Figure 4 shows the hierarchy of the FPU along with the area percentage of each main module. The `*_CTL` and `*_DP` blocks represent the pure control and the datapath portion of each module, respectively. The largest module is the 54x54 multiplier. The division pipeline is rather small (and, naturally, rather slow at the same time). The model also contains a few more very small modules, such as repeaters (to optimize timing) and scan-control modules, which are not shown in the figure. These modules along with the pipeline registers and the wiring add up to the remaining 23.3% of the FPU area. Overall, the control logic amounts to 16.1% of the FPU size.

V. EXPERIMENTAL RESULTS

This section describes the experimental results that support our conjectures. We simulated the openSPARC T1 microprocessor using two different types of synthetic workload: The first one (`FP-100`) consists of 100% floating point operations, to resemble applications with intense need for floating point calculations. The second (`FP-1`) consists of 1% floating point instructions, matching the profile of common applications that place very little demand on the FPU. On average, `FP-100` had 5 floating point instructions in the FPU (either executing or queued) and a maximum of 8 (one floating point instruction from each core). In contrast, `FP-1` had an average of 1 and a maximum of 4 floating point instructions in the FPU. For each of the two workloads, 5 million transient errors were injected, uniformly distributed over time and location across the FPU.

A. FPU Error Impact Analysis

The first set of results, shown in Table II, present statistics regarding the impact of injected errors on the FPU output. As expected in a transient error injection campaign, masking is very high; indeed, among the injected errors, only 2.13% for `FP-100` and 1.45% for `FP-1` reach the FPU output (i.e. non-masked errors). `FP-1` has fewer non-masked errors since

978-1-61284-657-6/11 $26.00 © 2011 IEEE

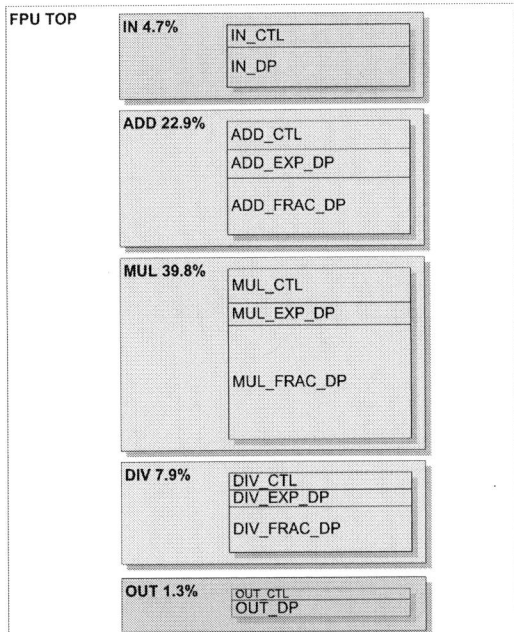

Fig. 4. FPU Hierarchy and Area Breakdown

TABLE II
ERROR CLASSIFICATION STATISTICS

| | FP-100 | | FP-1 | |
Module	# of Injected Errors	Non-Masked Errors	# of Injected Errors	Non-Masked Errors
fpu_top	495,585	2.78%	509,648	2.01%
in	86,501	3.97%	89,888	2.79%
in_ctl	33,117	12.13%	34,066	10.97%
in_dp	163,318	2.78%	167,595	1.87%
add	54,646	1.43%	55,874	1.28%
add_ctl	92,062	2.04%	94,832	1.76%
add_exp_dp	105,283	0.44%	108,368	0.57%
add_frac_dp	567,209	0.59%	582,034	0.43%
mul	75,006	1.98%	77,442	1.31%
mul_ctl	59,829	3.71%	61,907	2.81%
mul_exp_dp	54,652	2.29%	55,953	1.48%
mul_frac_dp	2,885,318	1.60%	2,961,476	0.89%
div	36,552	7.70%	37,228	5.51%
div_ctl	52,266	11.93%	53,915	9.53%
div_exp_dp	27,038	4.50%	27,421	3.34%
div_frac_dp	182,919	4.45%	189,131	3.83%
out	42,125	4.74%	43,234	3.49%
out_ctl	7,052	16.42%	7,229	15.83%
out_dp	86,758	4.37%	89,132	3.06%
Total	5,107,236	2.13%	5,246,373	1.45%

TABLE III
EXPONENT MONITORING VS. DUPLICATION

| FPU Control CED Method | Area Overhead | Coverage of | | |
		Control	Exponent	Fraction
Duplication	17.9%	100%	-	-
Monitor Exponent	5.8%	95.1%	98.1%	15%

fewer instructions flow through the FPU at the same time, thus increasing the chance for a transient error to strike on an inactive part. However, the distribution of non-masked errors is similar for the two different workloads.

The key observation from these results is that the average non-masked error rate of control modules (_CTL) for FP-100 is 9.2% (i.e., 1 out of 10 transients affect the FPU), a percentage that is is much higher than the average non-masked rate of datapath modules for the same workload, which is only 1.63%. In other words, errors in the control logic are six times more likely to affect the output than errors in the datapath. This supports our claim that, despite the control logic being smaller (i.e., 16.1% of the FPU size), protecting it is necessary for ensuring reliable FPU operation.

We also point out that the percentage of non-masked errors varies among the different control modules, with in_ctl, div_ctl and out_ctl having higher percentages. This is expected for the in_ctl and out_ctl modules, since all instructions have to go through them and they always contain critical information regarding instruction execution. As for div_ctl, division instructions have 10 times more latency than other instructions (up to 61 cycles), so the probability that the FPD pipeline will be occupied by a valid instruction during the workload execution is much higher.

B. Exponent Monitoring vs. Duplication for Control Logic

The second set of results, shown in Table III, compare the proposed exponent monitoring method to traditional duplication for performing CED in the FPU control logic.

In terms of area overhead, the cost of duplication is 17.9% of the FPU size, of which 16.1% is to replicate the control logic and 1.8% is to compare. In contrast, the proposed exponent monitoring incurs only one third of this cost, for a total of 5.8% of the FPU size. Both methods operate in parallel with the FPU and do not cause noticeable delay overhead.

In terms of effectiveness, exponent monitoring detects 95.1% of control logic errors, as opposed to the 100% coverage provided by duplication. The remaining 4.9% is attributed to control logic errors which only affect the fraction portion of the result and never propagate to the exponent, as we explained in section III-A1. However, exponent monitoring provides the ancillary benefit of also covering 98.1% of the errors that affect the exponent, which control logic duplication is unable to detect. Once again, the small error masking of 1.9% is because of the ±1 comparison and because of cancelation, as explained in section III-A1. The implication of this ancillary benefit is that no additional CED method is needed to cover the exponent portion. These results corroborate our conjecture that most errors in the control logic will result in an incorrect exponent and support the cost-effectiveness of our exponent monitoring-based CED method for FPU control logic.

C. Cost-Effective CED for Entire FPU

The last set of results examines the utility of exponent monitoring in developing a cost-effective CED method for the entire FPU. We note that, as shown in Table III, exponent monitoring also detects around 15% of the errors that only affect the fraction portion of the floating point representation, which the duplication method is unable to detect. Yet this percentage is very small, hence additional steps need to be taken in order to provide a complete FPU CED solution. To this end, we investigate how our method can be combined with a base-15 residue code for the fraction, in order to reduce the overall CED cost for the FPU. Specifically, we compare three alternative scenarios: (i) using base-15 residue codes for the

TABLE IV
COMPARISON OF CED SOLUTIONS FOR ENTIRE FPU

Detection Method			Coverage				Hardware Overhead
Control	Exponent	Fraction	Control	Exponent	Fraction	Total	
	Res-15	Res-15	0%	94.3%	94.3%	**59.2%**	**14.82%**
Duplication	Res-15	Res-15	100%	94.3%	94.3%	**96.2%**	**29.78%**
Exponent Monitoring		Res-15	95.1%	98.1%	94.3%	**94.4%**	**16.32%**

fraction and the exponent but leaving the control unprotected, (ii) adding control logic duplication to the above solution, and (iii) combining exponent monitoring with base-15 residue code only for the fraction (since the exponent is already covered).

The results reported in Table IV demonstrate two key points: First, if control is left unprotected, the overall fault coverage would be a mere 59.2%. This shows, once again, that protection of control logic is necessary in modern FPUs. Second, the proposed exponent monitoring method enables a complete FPU CED solution that provides almost equivalent coverage to the duplication-based solution (i.e., 94.4% vs. 96.2%), yet incurs almost half of the cost (i.e., 16.32% vs. 29.78%), thereby constituting a very appealing option.

VI. CONCLUSION

We presented a novel method for detecting transient errors in the control logic of a modern FPU. We demonstrated that errors in the control logic lead to an extensive corruption of the datapath and, by extension, have a high probability of affecting the exponent field of the operation output. Therefore, independently calculating and validating the exponent of the outgoing packet provides very high coverage to such errors. As demonstrated on the openSPARC T1 processor, the proposed exponent monitoring-based CED method costs less than one third of the cost of duplicating the control logic, while maintaining over 95% of its coverage. Moreover, in conjunction with a known residue code-based method for the fraction of the floating point representation, it facilitates a complete CED solution which offers over 94% coverage for the entire FPU at the cost of 16.32% of its size, which to our knowledge, constitutes the most cost-effective approach to date.

REFERENCES

[1] Y. Savaria, N.C. Rumin, V.K. Agarwal, and J.F. Hayes, "Soft-error filtering-A solution to the reliability problem of future VLSI digital circuits," in *IEEE Proceedings*, 1986, vol. 74, pp. 669–683.

[2] C. Metra, M. Favalli, and B. Ricco, "On-line detection of logic errors due to crosstalk, delay, and transient faults," in *International Test Conference*, 1998, pp. 524–533.

[3] T. Karnik, P. Hazucha, and J. Patel, "Characterization of soft errors caused by single event upsets in cmos processes," *IEEE Transactions on Dependable and Secure Computing*, vol. 1, no. 2, pp. 128–143, 2004.

[4] J. Gaisler, "Concurrent error-detection and modular fault-tolerance in a 32-bit processing core for embedded space flight applications," in *IEEE International Symposium on Fault-Tolerant Computing*, 1994, pp. 128–130.

[5] A. Naini, A. Dhablania, W. James, and D. Das Sarma, "1 GHz HAL SPARC64R Dual Floating Point Unit with RAS features," in *IEEE Symposium on Computer Arithmetic*, 2001, pp. 173–183.

[6] P. Eibl, A. Cook, and D. Sorin, "Reduced precision checking for a floating point adder," in *IEEE International Symposium on Defect and Fault Tolerance of VLSI Systems*, 2009, pp. 145–152.

[7] J.C. Lo, "Reliable floating-point arithmetic algorithms for error-coded operands," *IEEE Transactions on Computers*, pp. 400–412, 1994.

[8] S.M.H. Shekarian, A. Ejlali, and S.G. Miremadi, "A Low Power Error Detection Technique for Floating-Point Units in Embedded Applications," in *IEEE/IFIP International Conference on Embedded and Ubiquitous Computing, 2008. EUC'08*, 2008, vol. 1.

[9] TEMIC, "TSC692E Floating-point Unit User's Manual for Embedded Real Time 32 bit Computer (ERC32)," 1996.

[10] W.L. Gallagher and E.E. Swartzlander, "Fault-tolerant Newton-Raphson and Goldschmidt dividers using time shared TMR," *IEEE Transactions on Computers*, pp. 588–595, 2000.

[11] J.C. Lo, S. Thanawastien, T.R.N. Rao, and M. Nicolaidis, "An SFS Berger check prediction ALU and its application toself-checking processor designs," *IEEE Transactions on Computer-Aided Design of Integrated Circuits and Systems*, vol. 11, no. 4, pp. 525–540, 1992.

[12] G.G. Langdon and C.K. Tang, "Concurrent error detection for group look-ahead binary adders," *IBM Journal of Research and Development*, vol. 14, no. 5, pp. 563–573, 1970.

[13] A. Avizienis, "Arithmetic algorithms for error-coded operands," *IEEE Transactions on Computers*, vol. 22, no. 6, pp. 567–572, 1973.

[14] E. Kinoshita, H. Kosako, and Y. Kojima, "Floating-point arithmetic algorithms in the symmetric residue number system," *IEEE Transactions on Computers*, vol. 100, no. 23, pp. 9–20, 1974.

[15] A. Sasaki, "The Basis for Implementation of Ad idive Operations in the Residue Number System," *IEEE Transactions on Computers*, vol. 100, no. 17, pp. 1066–1073, 1968.

[16] D.A. Anderson and G. Metze, "Design of totally self-checking check circuits for m-out-of-n codes," *IEEE Transactions on Computers*, vol. 100, no. 22, pp. 263–269, 1973.

[17] A. Avizienis, G.C. Gilley, F.P. Mathur, D.A. Rennels, J.A. Rohr, and D.K. Rubin, "The STAR (self-testing and repairing) computer: An investigation of the theory and practice of fault-tolerant computer design," *IEEE Transactions on Computers*, vol. 100, no. 20, pp. 1312–1321, 1971.

[18] J.M. Berger, "A note on error detection codes for asymmetric channels," *Information and Control*, vol. 4, no. 1, pp. 68–73, 1961.

[19] M.A. Marouf and A.D. Friedman, "Design of self-checking checkers for Berger codes," in *IEEE International Symposium on Fault-Tolerant Computing*, 1978, vol. 8, pp. 179–184.

[20] M. Nicolaidis, "Self-exercising checkers for unified built-in self-test (UBIST)," *IEEE Transactions on Computer-Aided Design*, vol. 8, no. 3, pp. 203–218, 1989.

[21] D. Goldberg, "What every computer scientist should know about floating-point arithmetic," *ACM Computing Surveys (CSUR)*, vol. 23, no. 1, pp. 5–48, 1991.

[22] Sun Microsystems, "OpenSPARC T1 specifications," http://www.opensparc.net/opensparc-t1/index.html.

[23] Sun Microsystems, "OpenSPARC T1 Microarchitecture Specification," 2006.

[24] D. Stevenson et al., "IEEE standard for binary floating point arithmetic," *ACM SIGPLAN Notices*, vol. 22, no. 2, pp. 9–25, 1987.

[25] M. Maniatakos, N. Karimi, A. Jas, Tirumurti, and Y. Makris, "Instruction-level impact analysis of low-level faults in a modern microprocessor controller," *IEEE Transactions on Computers (TCOMP)*, 2010 (to appear).

978-1-61284-657-6/11 $26.00 © 2011 IEEE

Enhancing Online Error Detection through Area-Efficient Multi-Site Implications

N. Alves*, Y. Shi*, J. Dworak[††], R. I. Bahar*, K. Nepal[†]

*School of Engineering, Brown University, Providence, RI 02906

[††]Department of Computer Science and Engineering, Southern Methodist University, Dallas, TX 75205

[†]Electrical Engineering Department, Bucknell University, Lewisburg, PA 17837

Abstract—We present a new method to identify multi-site implications that can significantly increase the fault coverage of error-detecting hardware without increasing the area overhead. This method intelligently divides the input space about the functions of internal circuit sites and finds new valuable implications that can share gates in checker logic.

I. INTRODUCTION

Over the last several decades, aggressive scaling of integrated circuits has lead to higher packing densities, increased circuit performance, and lower production costs. Unfortunately, these reductions in feature size also mean that circuits and microprocessors are becoming more susceptible to errors arising from multiple sources, including excessive process variations, defects, wearout, and operational and environmental influences.

Online error detection aims to monitor circuit behavior at run time and detect deviations from its normal operating behavior while the device is in operation. The ideal online error detection scheme would detect all errors without requiring significant circuit modifications and would not negatively impact performance, power, or area overhead. Unfortunately, no scheme is ideal, and multiple tradeoffs need to be taken into consideration when choosing a particular implementation.

In the past, we have proposed a method for using the logic implications that occur naturally in circuits to provide varying degrees of coverage for online errors and reported its error-coverage/power/delay benefits when compared against other approaches, such as parity and logic duplication [1]. These logic implications can provide valuable reductions in the error rate while allowing the hardware overhead of the checker logic to be easily constrained within a desired hardware budget.

However, while the simple two-site implications we have previously studied can provide very good coverage of some errors, other errors may not be detectable by any *simple implication*. Expanding the implication set to include implications that involve additional circuit sites has the potential to cover some of these errors; however, the number of multi-site implications we could possibly consider is enormous, and this makes the identification and selection of such implications nontrivial. Intelligent methods are necessary to quickly identify new implications that have a reasonable chance of providing good error coverage.

In this paper, we propose a novel method for identifying such multi-site implications. In this method we divide the input space according to the functions realized by particular internal circuit sites and identify new *residual implications* that are valid in each portion of the space. The gating hardware that determines when a particular implication is valid can then be shared across multiple implications, significantly reducing the cost of the multi-site approach. We will show that these new *residual implications* can significantly increase error coverage for a constant hardware budget when combined with both the *simple implications* we described in [1] and a form of *checking functions* similar to that proposed in [2]. Furthermore, we will show that these new *residual implications* are especially useful when protecting circuitry that has been optimized for area and/or delay.

II. BACKGROUND

For over half a century, a wealth of techniques have been proposed and analyzed for the detection and correction of errors at run time. Some of these methods include coding techniques such as parity, Berger, and Bose Lin codes (e.g. [3]), executing portions of the code in redundant threads or in temporarily unused functional units (e.g. [4], [5]), and duplicating or tripling at least some of the circuit logic (e.g. [6]–[9]). Other online error detection schemes include Built-In Concurrent Self Test (BICST) [10] and Reduced Observation Width Replication (ROWR) [11] where prediction hardware is added to guess the appropriate response to a set of pre-computed test vectors. In addition, high-level functional assertions, such as those identified during functional verification, may also be hardcoded into the design to signal errors (e.g., [12]).

Other researchers have considered relationships that occur between flip-flops or between circuit sites at the gate level. For example, the authors of [13] investigated the protection of the control logic of a microprocessor though the use of relationships between functions of the flip-flops in a design. The authors of [2] proposed the use of checking functions, which identify circuit sites or functions of circuit sites that should always be equal to (or complements of) each other.

In the past, we have also investigated the use of relationships between sites at the gate level (i.e., logic implications) to identify internal circuit errors [1]. Logic implications arise naturally in digital circuits, and in their simplest form, they can be expressed in the following format:

$$siteA = x \rightarrow siteB = y, \text{ where } \{x, y\} \in \{0, 1\}$$

978-1-61284-657-6/11 $26.00 © 2011 IEEE

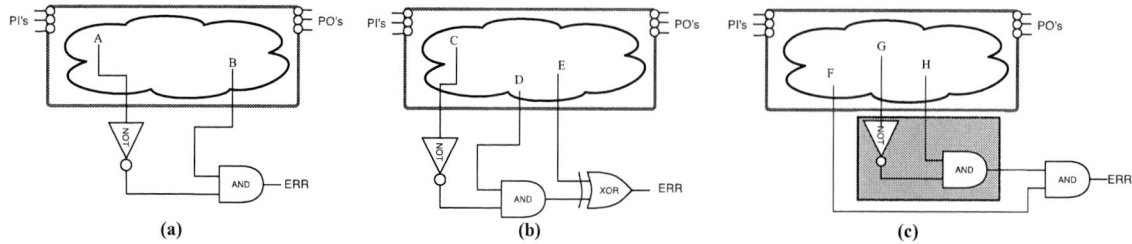

Fig. 1: Examples of error-detecting implications and checking functions implemented in parallel to the main logic circuitry. (a) A simple implication: $site(A) = 0 \rightarrow site(B) = 0$, (b) a checking function: $\overline{site(C)} \cdot site(D) \equiv site(E)$, and (c) a residual implication: $site(F) = 1 \rightarrow (site(G) = 0 \rightarrow site(H) = 0)$.

For example, in Figure 1(a), we show the simple implication $siteA = 0 \rightarrow siteB = 0$. In this case, checker logic can be inserted to flag an error should $siteA$ be set to logic zero and $siteB$ be set to a logic one.

However, while such implications have proven to be useful for online error detection, they also have several limitations. Specifically, each implication is generally only able to capture errors of a single polarity at a given site. Furthermore, it is often difficult to detect errors close to the primary outputs because we essentially "run out of logic" in which the implications can occur. Finally, it is also possible for no simple two-site implication to exist that can detect a particular error— especially when a circuit has been optimized to minimize area and/or delay.

Alternatively, the checking functions proposed in [2] may be used to detect errors with either polarity. However, they are dependent on the appearance of equalities between sites or functional combinations of sites naturally occurring in circuits. Unfortunately, many of these equality relationships are likely to be removed during logic minimization and thus checking functions may be less effective or prevalent in optimized circuits.

To address these limitations of previous methods, in this paper, we introduce a new method that efficiently identifies valuable multi-site logic implications. We will demonstrate that these new residual implications are capable of detecting many errors not covered by the approaches of [1], [2]. Furthermore, this new methodology is especially necessary when protecting circuits optimized for reduced area or delay.

III. METHODOLOGY

To detect errors not covered by simple two-site implications, additional relationships may be harnessed if more complex combinations of circuit sites are considered during the implication discovery process. In its most general form, a multi-site implication will encapsulate a relationship where some Boolean function of the values at a particular set of circuit sites *implies* the value of some Boolean function of other circuit sites. Thus, we have:

$$f(X_1, X_2, ... X_n) \rightarrow g(Y_1, Y_2, ... Y_m) \qquad (1)$$

where $X_1, ... X_n$ and $Y_1, ... Y_m$ are sites in the circuit and f and g Boolean logic functions.

Of course, the number of such implications grows much too quickly to be considered exhaustively even for very small

circuits. Furthermore, the need to realize the functions f and g with logic gates means that multi-site implications are inherently more expensive in terms of hardware overhead than two-site implications. As a result, f and g should be kept as small as possible. In this paper, we choose to replace the function g with the value of a single site. Thus, all multi-site implications considered are of the form:

$$f(X_1, X_2, ... X_n) \rightarrow Y_1 \qquad (2)$$

However, we still must determine an effective way of finding good choices for the function f that will increase coverage with low hardware overhead.

A. Residual implications

Intuitively, one of the problems we face when discovering logic implications is the fact that an implication must be valid for all possible input combinations. circuit, we cannot use it. However, faults are only detectable for a subset of all possible patterns. If we divide the input space, it is often possible to find additional implications that are valid in the subspace that were not valid in the space as a whole.

Shannon's Expansion Theorem provides a means of decomposing a logic function into two pieces about a particular input variable. Thus, the logic function $f(x_1, x_2, ... x_i, ... x_n)$ can be written in the form:

$$f = x_i' f_{x_i'} + x_i f_{x_i} \qquad (3)$$

Here $f_{x_i'}$ is the x_i' *residue* of f and is the function f when x_i is equal to 0. Similarly, f_{x_i} is the x_i residue and is the function f when x_i is equal to 1.

We apply this concept to the search for logic implications by choosing a site P in the circuit about which to divide the input space. (Note that in our implementation P does not need to be an input itself.) Specifically, we divide the input space into the part where P is equal to 0 and the part where P is equal to 1. Then, we find implications that are valid when P is equal to 0 as well as implications that are valid when P is equal to 1.

Because these implications are only valid in part of the input space, the error signal must be gated by the signal P in the checker logic. Specifically, an implication that is only valid when P is equal to 1 must be AND'ed with P and an implication that is only valid when P is equal to 0 must be AND'ed with P'. In the simplest form, this essentially produces a three-site implication consisting of the two sites that form a simple implication as well as the site P that

978-1-61284-657-6/11 $26.00 © 2011 IEEE   242

determines when the implication should be guaranteed to be valid.

For example, Figure 1(c) demonstrates how residual implications can be realized within the checker logic. The dashed box contains the simple implication $(site(G) = 0 \rightarrow site(H) = 0)$, which is only valid when the residual element $site(F) = 1$.

Generating multi-site implications in this way has several advantages. First, multiple implications that are all valid under the same conditions can share a single AND gate with P or P', thus minimizing the additional overhead required. Furthermore, if we can find appropriate choices of the site P, the amount of effort required to find and evaluate possible multi-site implications is much less than in the general case.

B. Our approach

Our approach to extend online error detection using logic implications can be summarized in the following steps:

1) Perform good circuit simulation on a gate level netlist of the circuit. Find all potential 2-site logic implications, validate them for all input combinations with a SAT solver, and remove all "weak" implications from further consideration via a compression algorithm. This step follows the flow presented in [1].
2) Find checking functions that are present in the circuit, using an approach similar to step 1.
3) Perform fault simulation on the two sets of logic relationships found in steps 1 and 2 and sort the elements into a single ordered list according to error coverage capability.
4) Determine optimal splitting points, calculate a set of residual implications, and calculate their error detection capability.
5) Select an appropriate subset of the identified implications and/or checking functions subject to the desired hardware budget.

While the execution time is heavily dependent on circuit-size and functionality, for the selected benchmarks, we were able to compute the outcome of each step in a few minutes. The remainder of this section describes these steps in detail.

1) Finding simple logic implications

Our implication discovery process follows the algorithm presented in [1] and begins with good circuit simulation of the netlist using a subset of input vectors. By performing pairwise compares between all circuit sites, simple two-site logic implications are extracted and then subsequently validated using a SAT solver. This list of valid implications is then compressed using the algorithms discussed in [1], and from this compressed list implications can be evaluated and ultimately selected to form part of the checker logic hardware.

2) Finding checking functions

In our implementation, checking functions like those described in [2], [14], are discovered by:

1) Comparing every pair of sites and determining if they always have equal (or complemented) logic values.
2) Determining, at each site g, all sites that are within a 10 site radius and finding all *equality functions* between every two internal sites, i and j in that radius. These equality functions are of the form $g = h_i^* \ OP \ h_j^*$, where $OP \in \{AND, OR, XOR\}$ and where h^* is the function realized at a site in either its complemented or uncomplemented state.

Figure 1(b) shows the checker logic implementation of a checking function where a simple relationship between two sites, C and D, equals a third site E, for all input vectors. We discover our possible checking functions from logic simulation values, and then proceed to validate these checking functions with a SAT solver.

3) Evaluating implications and checking functions

Once we have generated the two sets of logic relationships from steps 1) and 2), we need to create a single ordered list that ranks all elements within the sets by their ability to detect errors. In essence, we want an implication or checking function to be closer to the top of the ordered list when it is better at detecting errors than the implications or checking functions below it. Information regarding faults left uncovered by both simple implications and checking functions will subsequently be used to identify our new residual implications.

The procedure begins by including, one at a time, all implications and checker functions found in steps 1 and 2 in the checker logic for the circuit under test. Fault simulation using the stuck-at-fault model and 32k random input vectors is performed. Only stuck-at faults were used in our experiments as they are well-known, easy to model, and often used in related work. Fortunately, other error sources may cause behavior similar to stuck-at faults on the cycle when they are present. By detecting stuck-at faults effectively, we may be able to fortuitously detect these other types of errors. From this, we determine which checker logic elements are able to detect which faults for each vector, and how many times each fault causes an error at the circuit outputs. Once all faults are processed, we sort the checker logic elements in descending order according to number of fault detections with which they are associated and dynamically removed duplicated fault detections during the sorting. This procedure does not explicitly favor one fault over another; its sole purpose is to quickly process data and disregard redundant checker logic elements with lower fault coverage capabilities than other checker logic elements. These ideas can be implemented with the optimization algorithm described in [1].

4) Finding residual implications

The search for residual implications entails selecting a particular *splitting site* (P), fixing its logic state, and re-computing all implications using the same method as described in step 1). The newly generated implications that are not in the list of simple implications, are called residual implications.

ISCAS85					ITC99				
Circuit	PI	PO	Gates	Faults	Circuit	PI	PO	Gates	Faults
c432	36	7	160	864	b04C	77	74	698	3022
c499	41	32	202	998	b05C	35	70	855	4490
c880	60	26	383	1760	b06C	11	15	58	214
c1355	41	32	546	2710	b07C	50	57	429	1868
c1908	33	25	880	3816	b08C	30	25	178	768
c2670	233	140	1269	5340	b09C	29	29	174	702
c3540	50	22	1669	7080	b10C	28	23	191	878
c5315	178	123	2406	10630	b11C	38	37	653	3242
c6288	32	32	2406	12576	b12C	126	127	1084	4924
c7552	207	108	3512	15104	b13C	63	63	363	1426

TABLE I: Structural benchmark characteristics

Selecting a good set of splitting sites is of utmost importance, as they will dictate the quality of the subsequently generated residual implications. This selection is done with the information collected in step 3 after performing fault simulation with the simple 2-site implications. We select the splitting sites to be those sites whose associated faults are propagated the most and whose errors are not often detected by any simple implications. The vast majority of the splitting sites are located at the circuit inputs, or at sites very close to one.

5) Selecting and determining checker logic performance

To evaluate the quality of a particular checker logic set, we define a *probability of detection* metric. This probability quantifies how well a particular set of implications can detect online errors. It is calculated by initially performing stuck-at-fault simulation, with 32k random input vectors, and then counting how often an internal fault that is observable at at least one of the circuit's outputs is detected and flagged by the implication checker hardware. More formally, the probability of detection (P_{det}) is defined as,

$$P_{det} = \sum_{i=1}^{\#\text{faults}} \sum_{j=1}^{\#\text{vectors}} \left(\frac{E_{i,j}}{F_{i,j}} \right) \quad (4)$$

where, $F_{i,j}=1$ when fault i propagates to an output with vector j and 0 otherwise. Similarly, $E_{i,j}=1$ if the checker logic is violated when $F_{i,j}=1$ and 0 otherwise. This metric does not take into account faults occurring inside the checker logic.

IV. RESULTS

In this section we report the properties of the various checker logic implementations and their ability to detect faults. To determine the effectiveness of each implementation, we studied 20 circuits from two distinct benchmark suites, ISCAS85 [15] and ITC99 [16]. For the ITC99 circuits, we use the optimized gate-level combinational benchmarks available online [17]. Table I reports key structural properties for the selected benchmarks, such as the number of circuit inputs (PI), circuit outputs (PO), number of gates, and the number of stuck-at-faults.

A. Comparing checking functions with simple implications

Before assessing the effectiveness of our new residual implications, we first found the error coverage achievable using our implementations of both simple implications and checking functions. Through the procedures outlined in section III-B, we computed all simple implications and checking functions

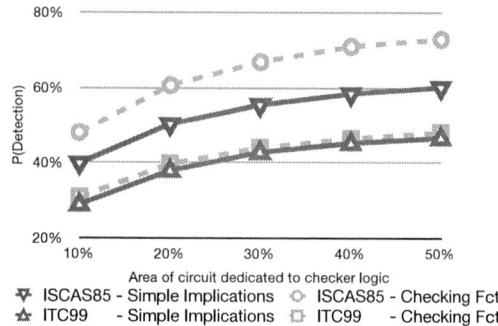

Fig. 2: Comparison in the probability of detecting an error when the checker logic is composed of either all simple implications or all checking functions.

for each benchmark. We then selected an appropriate subset of all simple implications or checking functions that would maximize coverage for a particular hardware budget. For these experiments, we approximated hardware overhead as the total number of extra standard cells added, compared to the original gate count of the circuit, where each simple implication is assumed to be implemented using a single standard cell and each checking function is implemented using two standard cells. In our experiments we allowed the checker logic to occupy an area equivalent to approximately 10%-50% of the circuit layout.

Once the checker logic elements were chosen, for each subset we used Eqn. 4 to calculate the probability that an error in a given clock cycle would be detected by the checker logic. This probability of detection is averaged across the ISCAS85 and ITC99 circuits, and the results are shown in Figure 2. Checking functions, represented with a dashed line, provide a higher probability of detection for the unoptimized circuits of the ISCAS85 suite. This is not an entirely surprising outcome because checking functions tend to exist in a circuit due to logic redundancies [2]. Since the combinational ITC99 benchmarks we selected have been optimized by their creators [17], it is likely that they contain fewer redundancies. This results in the checker logic created with a set of simple implications and that created with checking functions having roughly equivalent performance.

B. Residual implications and splitting nodes

We then began to investigate the effectiveness of residual implications. First, we investigated the impact of increasing the number of splitting nodes during the implication discovery process. Each additional splitting node provides a new way of dividing the input space into two pieces, and thus allows new residual implications to appear. Figure 3 shows the improvement in the probability of detection as we consider the inclusion of residual implications, generated by additional splitting sites, into the optimized set of simple implications. The plot, obtained after generating an optimized set of simple+residual implications for the ISCAS85 circuits, reveals an upwards trend in the probability of detection as the number of splitting sites increases. However, we found that increasing the number

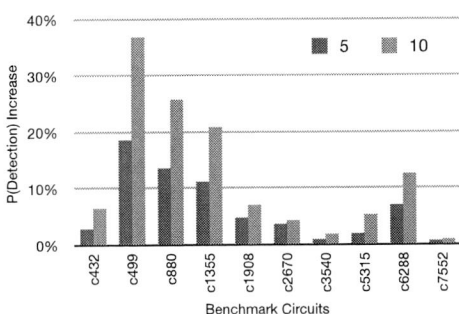

Fig. 3: Improvement in the probability of detecting an error as we allow 5 and 10 splitting sites. A 50% hardware budget for the checker logic was used for these experiments.

of splitting sites beyond 10 using our splitting site selection criteria did not necessarily increase the overall coverage in these circuits enough to justify the additional analysis cost. Thus, for the remaining experiments, we allowed 10 splitting sites to be the basis of our simple+residual implication set.

C. Combining all checker logic elements

We now consider four different scenarios for the checker logic implementation:

- simple two-site implications alone
- simple two-site implications and residual implications alone
- checking functions alone
- all three combined

In each case, a subset of the possible implications/checking functions are chosen to be included in the checker logic from the sorted list described in Section III-B. Figure 4 shows the propability of detection for the different types of checker hardware. The probability of detection values for each area overhead have been averaged across the ISCAS85 and ITC99 circuits separately. From this, we can observe that:

- While the optimized set of simple+residual implications is slightly inferior to the checking functions when applied to the ISCAS85 benchmarks, it is significantly better than either simple implications or checking functions when applied to the ITC99 benchmarks. This is encouraging because it indicates that residual implications can provide significant added value when applied to optimized circuitry with fewer redundancies.
- Because implications and checking functions have different fault-targeting properties, the combined checker logic is able to detect more errors than any individual checker logic implementation. By allowing only 10% of the hardware to be dedicated to the checker logic, we are able to detect errors on average almost 40% of the time for the ITC99 circuits and roughly 53% of the time for the ISCAS85 circuits.

Figure 5 shows the distribution for the type of implications in the optimized set of combined checker logic. Since checking functions were highly effective when applied to the ISCAS85

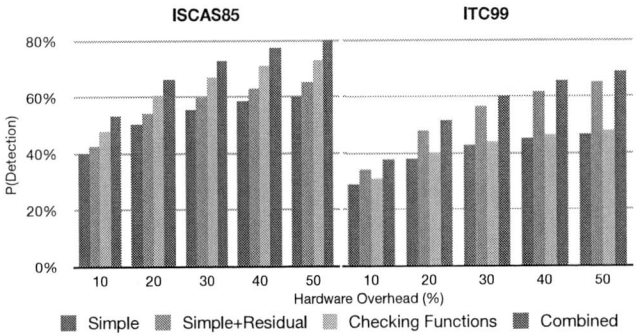

Fig. 4: Probability of detection for the various subsets of implications.

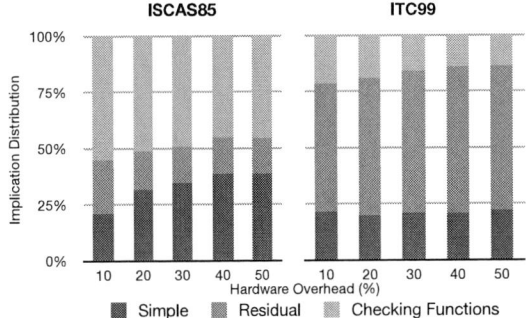

Fig. 5: Distribution of the types of relationships making up the checker logic when all three types are combined.

benchmarks, it is unsurprising that around half of the elements in the combined checker logic came from that group. The results are quite different for the circuits from the ITC99 benchmark suite, where residual implications became the major providers of error detection.

D. Fault coverage overlap for different implication sets

The combined checker logic showed us that we can get additional fault coverage by using checking functions together with implications. We also determined the fault coverage breakdown for different checker logic scenarios. We performed fault simulation on each ITC99 benchmark circuit, and for each propagated fault we determined which checker logic set would detect it. Using VennMaster [18], Figure 6 shows the relative number of faults covered by implications and checking functions. As expected, the faults covered by the simple+residual implications are not identical to the faults covered by checking functions. Figure 6 also highlights the fault coverage advantage of combining implications and checking functions in the checker logic, as it is able to cover most of the faults detected by the simple+residual implications and also most of the faults detected by the checking functions.

E. Residual implications on synthesized circuits

As reported above, we noticed that residual implications appear to be especially useful for circuits that have gone through some optimization procedure. To further investigate this hypothesis we re-synthesized ISCAS85 benchmarks, using

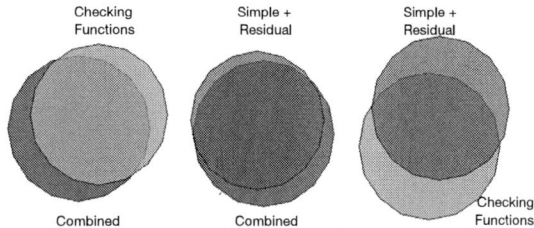

Fig. 6: Relative number of faults covered for all ITC99 benchmarks when 10% of hardware area is allocated to checker logic. Overlapping area indicates faults that are covered by more than one technique.

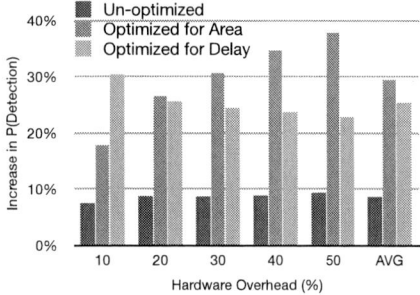

Fig. 7: Average improvement in probability of detection for ISCAS85 circuits using the simple+residual vs. the simple set of implications for different circuit optimizations.

Leonardo Spectrum, and optimized the circuits for area and delay. We then determined the relative impact on fault coverage that could be obtained when residual implications are added to the simple implication set for different overheads.

When we perform any logic re-optimization on a circuit, the internal structure will likely change, even though the logic functionality will remain the same. This implies that any previously calculated invariant relationship will not necessarily hold in a re-optimized circuit. For each re-optimized circuit, we re-computed all simple 2-site implications and residual implications, ensuring that the number of checker logic elements remain proportional to the new circuit area. Simulation results comparing optimized and unoptimized circuits are shown in Figure 7. In particular, we show the average improvement in probability of detection when we use the simple+residual implication set as opposed to the simple implications alone. As one can see, the residual implications for the area/delay optimized circuits become much more important contributors to the obtained error coverage when compared to non-optimized circuits.

V. CONCLUSIONS

In this paper we have introduced residual implications, a type of multi-site implication, that may be effectively used to detect faults not covered by simple implications or checking functions. We have presented an algorithm for finding and choosing a set of residual implications to be incorporated into the checker logic based on its ability to detect errors. While the exact amount of error detection provided from residual implications varies, our analysis has shown that significant

improvement in error detection is possible, especially in the case of the ITC99 benchmarks studied.

We also compared the behavior of residual implications against simple 2-site implications in circuits optimized for area or delay. When inserted into checker logic for these optimized circuits, the residual implications provided on average a more than 25% increase in the ability to detect errors when compared to using only simple 2-site implications alone.

Finally, we showed how residual implications can be combined with other concurrent online error detection methods, such as checking functions. With the implementation of hybrid checker logic we were able to significantly improve the online error detection capability. By allowing only a 10% area overhead for the hybrid checker logic, we were able to detect more than 50% of the errors in the unoptimized ISCAS85 benchmarks, and almost 40% of the errors in the ITC99 benchmark circuits. With a 40% area overhead, error detection was almost 80% and 65% respectively.

REFERENCES

[1] N. Alves, A. Buben, K. Nepal, J. Dworak, and R. I. Bahar, "A cost effective approach for online error detection using invariant relationships," *IEEE Trans. on Computer-Aided Design of Integrated Circuits and Systems*, vol. 29, no. 5, pp. 788–801, 2010.

[2] I. Pomeranz and S. M. Reddy, "Reducing fault latency in concurrent on-line testing by using checking functions over internal lines," in *DFT*, 2004, pp. 183–190.

[3] S. Mitra and E. McCluskey, "Which concurrent error detection scheme to choose?," in *ITC*, Oct. 2000, pp. 985–994.

[4] J. Ray, J. C. Hoe, and B. Falsafi, "Dual use of superscalar datapath for transient-fault detection and recovery," in *Intl. Symp. on Microarchitecture*, 2001, pp. 214–244.

[5] M. A. Gomaa and T. N. Vijaykumar, "Opportunistic transient-fault detection," in *ISCA*, June 2005.

[6] C. F. Webb and J. S. Liptay, "A high frequency custom CMOS S/390 microprocessor," *IBM Journal of Research and Development*, vol. 41, pp. 463–473, 1997.

[7] J. von Neumann, "Probabilistic logics and synthesis of reliable organisms from unreliable components," in *Automata Studies*. Princeton University Press, 1956, pp. 43–98.

[8] K. Mohanram and N. Touba, "Cost-effective approach for reducing soft error failure rate in logic circuits," in *ITC*, Oct. 2003, pp. 893–901.

[9] R. Sedmak and H. Liebergot, "Selective triple modular redundancy (STMR) based single-event upset (SEU) tolerant synthesis for FPGAs," *IEEE Trans. on Nuclear Science*, vol. 51, pp. 2957–2969, 2005.

[10] R. Sharma and K. Saluja, "An implementation and analysis of a concurrent built-in self-test technique," in *FTCS*, June 1988, pp. 164–169.

[11] P. Drineas and Y. Makris, "Concurrent fault detection in random combinational logic," in *ISQED*, Mar. 2003, pp. 425–430.

[12] Y. Abarbanel, I. Beer, L. Glushovsky, S. Keidar, and Y. Wolfsthal, "FoCs: automatic generation of simulation checkers from formal specifications," in *CAV*, 2000, pp. 538–542.

[13] R. Vemu, A. Jas, J. Abraham, S. Patil, and R. Galivanche, "A low-cost concurrent error detection technique for processor control logic," in *DATE*, March 2008, pp. 897–902.

[14] I. Pomeranz and S. M. Reddy, "On finding functionally identical and functionally opposite lines in combinational logic circuits." *VLSI Design Conf*, pp. 254–259, 1996.

[15] F. Brglez and H. Fujiwara, "A neutral netlist of 10 combinational benchmark circuits and a target translator in fortran," in *ISCAS*, 1985.

[16] F. Corno, M. S. Reorda, and G. Squillero, "Rt-level ITC'99 benchmarks and first ATPG results," *IEEE Des. Test*, vol. 17, no. 3, pp. 44–53, 2000.

[17] Reorda. (2010, September) ITC99 benchmarks. [Online]. Available: http://www.cad.polito.it/downloads/tools/itc99.html

[18] S. Chow and P. Rodgers, "Constructing area-proportional venn and euler diagrams with three circles," in *Euler Diagrams Workshop 2005*, August 2005. [Online]. Available: http://www.cs.kent.ac.uk/pubs/2005/2354

978-1-61284-657-6/11 $26.00 © 2011 IEEE

Session 11

978-1-61284-657-6/11 $26.00 © 2011 IEEE

Dynamic Scan Clock Control for Test Time Reduction Maintaining Peak Power Limit

Priyadharshini Shanmugasundaram and Vishwani D. Agrawal
Auburn University, Auburn, AL 36849
pzs0012@auburn.edu, vagrawal@eng.auburn.edu

Abstract—We dynamically monitor per cycle scan activity to speed up the scan clock for low activity cycles without exceeding the specified peak power budget. The activity monitor is implemented either as on-chip hardware or through presimulated and stored test data. In either case a handshake protocol controls the rate of test data flow between the automatic test equipment (ATE) and device under test (DUT). The test time reduction accomplished depends upon an average activity factor α. For low α, about 50% test time reduction is analytically shown. With moderate activity, $\alpha = 0.5$, simulated test data gives about 25% test time reduction for ITC02 benchmarks. For full scan s38584, the dynamic scan clock control reduced the test time by 19% when fully specified ATPG vectors were used and by 43% for vectors with don't cares. BIST with dynamic clock showed about 19% test time reduction for the largest ISCAS89 circuits in which the hardware activity monitor and scan clock control required about 2-3% hardware overhead.

Index Terms—Scan test, test time reduction, test power, on-chip activity monitor, adaptive test clock

I. INTRODUCTION

Full scan [1], a commonly used method for testing digital VLSI circuits, spends a large fraction of the test time for loading (scan-in) and unloading (scan-out) test data in flip-flops that are chained as shift registers. During this process, random combinational logic activity can produce large unintentional power consumption resulting in power supply noise and heating. If this consumption is higher than that of the normal functional operation for which the circuit is designed [2] the test can cause yield loss [3]. Therefore, scan testing is carried out at a slower speed than the normal operation. The scan clock frequency is determined based on the maximum power consumption the circuit under test can withstand. The power P dissipated at a node is given by [4]:

$$P = \frac{1}{2}CV^2\alpha f \qquad (1)$$

where C is the capacitance of the node, V is supply voltage, f is clock frequency and α is a node activity factor.

$$\alpha = Number\ of\ transitions\ per\ clock\ cycle \qquad (2)$$

The activity factor α for a clock signal is 2 because there are two (rising and falling) transitions per cycle. For a combinational node, α ranges between 0 (no transition) and 1 (a toggle every clock cycle). In the worst case, scan clock frequency f_{test} can be based on the maximum activity, i.e., $\alpha = 1$, so that the test power can never exceed the power limit. Therefore,

$$P_{budget} = \frac{1}{2}CV^2 f_{test} \qquad (3)$$

where P_{budget} is the maximum power dissipation the circuit can withstand without malfunctioning. Thus,

$$f_{test} = \frac{2P_{budget}}{CV^2} \qquad (4)$$

This research is supported in part by the National Science Foundation Grant CNS-0708962.

P. Shanmugasundaram is presently with NVIDIA, Santa Clara, CA 95050.

In general, the worst case assumption of $\alpha = 1.0$ can be relaxed. Although all vector bits are scanned in and scanned out at this frequency, many may not cause the maximum activity. It is possible to scan those in at higher clock frequencies without exceeding the power budget. When the number of transitions in the circuit reduces to a fraction $\frac{1}{i}$ of the maximum,

$$P = \frac{1}{2}CV^2\frac{1}{i}f_{test} \qquad (5)$$

From Eq. (3) and Eq. (5),

$$\frac{P}{P_{budget}} = \frac{1}{i} \qquad (6)$$

The capacitance and the voltage are constant for a node and so the power is proportional to the product of activity and frequency. Since the circuit can withstand a power P_{budget}, the frequency can be multiplied by i, and the power dissipated in every cycle can still be kept within the allowed limit. Girard [3] defines peak power as the highest energy consumed during one clock period divided by the clock period and the average power as the total energy consumed during test divided by the test time. Since the power must never exceed P_{budget} in any clock cycle, both peak power and average power will be below P_{budget} in spite of the increased shift frequency. Also, instantaneous peak power [3] is consumed right after the application of the clock edge. This power depends on the vectors scanned in and is unaffected by changes in the scan clock frequency. Hence, it can be reduced only by changing the test vectors. In this work we assume that the vectors conform to the instantaneous peak power requirement.

During scan tests, gates are either driven by outputs of the scan flip-flops or by primary inputs. Primary inputs do not change during scan in and scan out. Thus, scan chain activity is a direct measure of the test power and by *monitoring and controlling* this activity, we can speed up the test as well as limit the test power. That is the idea presented in this paper.

Section II discusses previous work on test time optimization. Section III discusses implementations of the proposed technique. Section IV gives a mathematical analysis of the scheme. Section V explains the experimental results obtained. Section VI discusses the conclusion of this work.

II. PREVIOUS WORK

Many test time reduction methods for scan circuits use compression. In a simple compression technique, the number of scan chains is increased reducing the number of flip-flops per chain. This reduces the time for shifting the input vector bits through scan flip-flops resulting in an overall reduction in test time. However, compression techniques require alterations in the design and may also suffer from linear dependencies.

One compression technique keeps the functionality of the ATE intact by moving the decompression task to the circuit under test [5]. Another technique [6] uses a dynamically

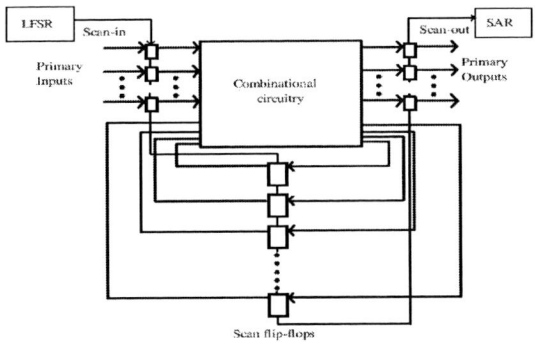

Fig. 1.   Test-per-scan built-in self-test (BIST).

reconfigurable scan tree that applies a part of the test sequence in scan tree mode and the other part in single scan mode. Reference [7] describes decompression hardware for test pattern compression. References [7] and [8] use compression algorithms with concurrent application of compaction and compression. Reference [9] gives a compression technique with embedded deterministic test logic on chip to provide vectors for the internal scan chains. Reference [10] employs alternating run-length codes [11] for test data compression.

Reference [12] employs a two phase testing strategy where the first phase is a scan-less phase for easy-to-detect faults and the second phase is a scan phase for hard to detect faults. Scan is performed only until all effective test bits are shifted to the right position and until all fault-affected response bits are shifted out. Reference [13] uses genetic algorithms to obtain compact test sets, which limit the scan operations. References [14] and [15] reduce test application time by generating a test for a sequential circuit using combinational test generation and sequential test generation adaptively. Reference [16] proposes a strategy to identify flip-flops to be removed from scan chains to increase the observability of the circuit so that faults activated during scan cycles can be observed at a primary output. The original technique of this paper whose details and some implementations are reported in recent documents [17], [18] can be additionally applied to any scan circuit that may include other methods mentioned above.

III. IMPLEMENTATION

A. BIST circuit with a single scan chain

We add flip-flops at primary inputs and outputs and as shown in Figure 1 connect all flip-flops into a single scan chain. A linear feedback shift register (LFSR), a signature analysis register (SAR) and a BIST controller are added to the circuit to implement the test per scan BIST architecture [16]. BIST vectors are scanned in and combinational outputs are captured through scan flip-flops. Application of a vector includes scanning in LFSR bits into flip-flops, normal mode capture and scan out (overlapped with next scan in) into SAR.

The proposed dynamic frequency control is shown in Figure 2. As test vectors are scanned in, the activity (or inactivity) in the scan chain is monitored at the first flip-flop of the chain. The entering transitions ripple through other flip-flops in subsequent cycles. Inversions along the chain do not change this activity. When a transition passes through an inverting flip-flop, a rising transition becomes a falling transition and vice-versa, leaving the number of transitions unchanged.

An XNOR gate between the input and output of the first flip-flop monitors the activity. The output of the XNOR gate is 0 when a transition enters the scan chain and is 1 when

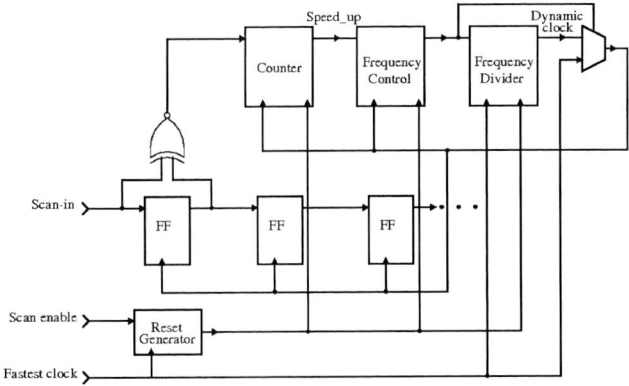

Fig. 2.   Schematic of proposed dynamic frequency control.

a no transition enters. The XNOR output is fed to a counter, which counts up for each 1, i.e., a non-transition. The counter is set to 0 at the start of every scan in sequence. According to Eq. (6), the scan frequency can be raised as the number of non-transitions entering the scan chain increases. This is accomplished through frequency control and frequency divider blocks in Figure 2. We assume that the response captured from the combinational circuit for the previous vector has a transition density of 1, i.e., the scan chain is filled with alternating 1s and 0s before scan-in begins. This pessimistic worst-case assumption guarantees that the power budget shall not be exceeded. Correspondingly, the scan in of each vector begins with the slowest frequency, f_{test}, permitted by the power budget for $\alpha = 1$. The f_{test} clock is the lowest frequency generated by the frequency divider that divides the frequency of an externally supplied fast tester clock. The frequency control circuit monitors the state of the counter. As the count goes up it lowers the frequency division ratio of the clock divider in several steps.

The reset generator in Figure 2 applies a reset signal to the counter, frequency control block and frequency divider at the positive edge of the scan enable signal, i.e., at the start of scan-in for every combinational vector. Since the frequency divider cannot generate a f/1 (divide by 1) clock, a multiplexer selects either the frequency divider output or the fastest clock.

Let us consider a circuit with 1000 flip-flops. If the slowest scan clock period based on the power budget is 80ns and we raise the frequency in 8 steps, then a modulo 125 (1000/8) counter will be implemented. *Assuming the worst-case activity by the captured states*, every scan-in is started with the 80ns clock and counter set to 0. The count goes up by 1 at every clock in which a non-transition enters the scan chain. When the count reaches 125, the counter is reset and the frequency divider generates a 70ns clock to scan-in the subsequent bits. The counter may again count up to 125 and the clock period would be reduced to 60ns. This process repeats until all 1000 bits are scanned in. Thus, if the input were a series of 1000 1s, the first 125 bits are scanned in at a clock of period 80ns, the second 125 bits at 70ns, until the last 125 bits are scanned in using a clock period 10ns. If the scan-in bits were a series of alternating 0s and 1s, the counter would never count up since there are no non-transitions entering the scan chain and hence the entire scan-in will use the 80ns clock. Notice that due to the *worst-case assumption* we start each scan-in with slowest clock and so the activity monitor only raises the clock rate without ever having to lower it during the same scan-in.

Clearly, a bit stream with fewer transitions will be scanned in faster than one with many transitions. Don't cares in deter-

978-1-61284-657-6/11 $26.00 © 2011 IEEE   249

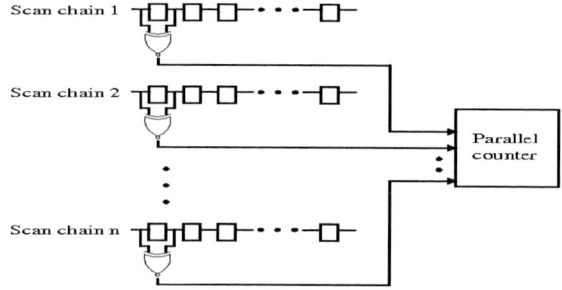

Fig. 3. Dynamic frequency control for multiple scan chains.

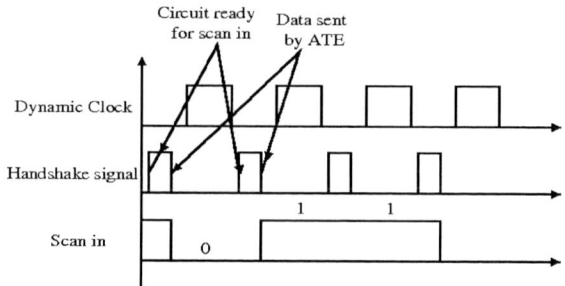

Fig. 4. Handshake protocol between ATE and circuit under test (CUT).

ministic ATPG patterns can be filled in such that the number of transitions is minimum [19]. Also, techniques to generate BIST patterns with low transition densities [20] may be useful. This technique would perform well for such patterns.

B. BIST circuit with multiple scan chains

When the circuit has multiple scan chains, the activity of all chains must be monitored. As shown in Figure 3, XNOR gates are added across the input and output of the first flip-flop in every scan chain. Outputs of XNOR gates are supplied to a parallel counter [21] that counts up by the number of 1s at its input. The rest of the circuitry remains unaltered and still resembles Figure 2. When the count reaches a certain threshold value, the frequency is stepped up and the counter is reset. Except for the use of the parallel counter the control scheme is similar to that in Figure 2.

C. Circuit tested with external ATE

Suppose we perform power analysis through simulation for scan sequences and use that information to scan in vectors at appropriate speeds. Ideally, we may scan in every bit with a customized minimum clock period such that the power dissipated in each clock cycle is pushed up to the maximum limit. Down sides are long simulation runs that must be repeated each time tests are modified and a large amount of per clock information to be stored in the ATE. Since tests are constrained by ATE memory capacity we explore alternatives.

A dynamic control of scan frequency for circuits tested by ATE can be similar to that used in BIST. However, BIST patterns are generated and applied under the control of the same on-chip dynamic clock. When patterns are supplied by an ATE and the activity monitor and dynamic clock control are implemented on the circuit under test (CUT), the CUT must transmit the clock information back to the ATE so that it can send the test data at a continuously changing rate. This problem is similar to that of communication between two systems operating at different clock frequencies. It can be solved by using a handshake protocol [22] that facilitates communication between asynchronous digital systems.

A simple handshake protocol is illustrated in Figure 4. When the circuit is ready to scan in data, a synchronizer, either residing on the chip or on the tester head, toggles a handshake signal. The ATE acknowledges this by scanning in the next bit into the scan-in pin, scanning out the next bit from the scan-out pin and toggling the handshake signal. The synchronizer recognizes the toggle on the handshake signal and accepts the new scan-in bit. This is shown in Figure 5.

We can reduces the hardware overhead required for the dynamic scan clock control if the data on activity is pre-generated by simulation. This data can be used by the test program. Thus, the chip would have an additional input

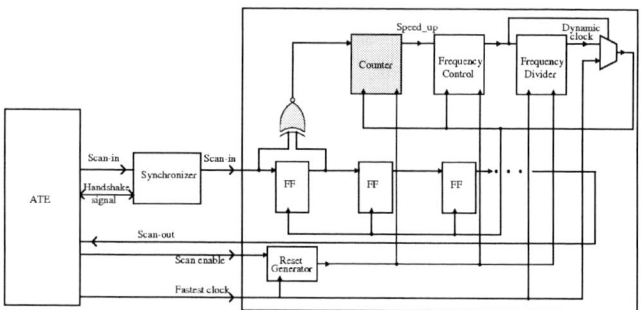

Fig. 5. Dynamic scan clock control for external ATE.

TABLE I
DETERMINATION OF CLOCK CYCLE RANGE FOR DIFFERENT FREQUENCIES.

S. No.	Clock period	Number of non-transitions		Clock cycles	
		Lower limit	Upper limit	Lower limit	Upper limit
1	vT	0	$\lceil \frac{N}{v} \rceil$	0	$\lceil \frac{N}{Av} \rceil$
2	$(v-1)T$	$\lceil \frac{N}{v} \rceil$	$\lceil \frac{2N}{v} \rceil$	$\lceil \frac{N}{Av} \rceil$	$\lceil \frac{2N}{Av} \rceil$
.	.	.	.	.	.
i	$(v-i+1)T$	$\lceil \frac{(i-1)N}{v} \rceil$	$\lceil \frac{iN}{v} \rceil$	$\lceil \frac{(i-1)N}{Av} \rceil$	$\lceil \frac{iN}{Av} \rceil$
.	.	.	.	.	.
v	T	$\lceil \frac{(v-1)N}{v} \rceil$	$\lceil \frac{vN}{v} \rceil$	$\lceil \frac{(v-1)N}{Av} \rceil$	$\lceil \frac{vN}{Av} \rceil$

pin (Speed_up) controlling the frequency control block. The frequency control block then receives its input directly from the ATE instead of from the counter. The ATE signals the frequency control block when the activity in the scan chains is low enough to speed up scan shift. This eliminates the XNOR gates at the front end of every scan chain and the counter that monitors the activity.

For circuits with multiple scan chains, an XNOR gate was added at the front end of every scan chain as shown in Figure 3 and a parallel counter monitored the activity and triggered the frequency control block once the activity threshold was reached. If simulation results are used to determine activity, the implementation does not change and the frequency control block that will then be driven directly by the ATE. In both cases, the use of compression does not affect the implementation since activity is monitored in every scan chain.

IV. ANALYSIS

From this point forward, α refers to the activity factor in the scan chain. Let N be the number of flip-flops, A be the non-transition density in the scan chain, $A = 1 - \alpha$, v be the number of frequencies and T be the time period corresponding to the fastest clock. The period of the fastest scan clock is v times shorter than the slowest clock. Therefore, the period of the slowest clock is given by vT. If the vectors were scanned in at the slowest clock, the total scan-in time per vector would

TABLE II
SCAN-IN TIME REDUCTION VS. NUMBER OF SCAN CLOCK SPEEDS FOR
ACTIVITY FACTOR $\alpha = 0.5$.

Number of scan	Test time reduction (%)		
clock speeds	Simulation	Eq. (9)	Eq. (11)
1	0.00	0.00	0.00
2	0.34	0.00	0.00
4	12.64	12.50	12.50
8	18.78	18.75	18.75
16	22.03	21.90	21.88
32	23.56	23.48	23.44
64	25.17	24.26	24.22
128	27.41	24.66	24.61

TABLE III
SCAN-IN TIME REDUCTION VS. ACTIVITY FACTOR α FOR 8 SCAN-IN
CLOCK SPEEDS.

Activity	Test time reduction (%)		
factor, α	Simulation	Eq. (9)	Eq. (11)
0	43.75	43.75	43.75
0.1	38.63	38.85	38.75
0.2	34.00	33.95	33.75
0.3	28.97	28.99	28.75
0.4	23.51	23.94	23.75
0.5	18.78	18.75	18.75
0.6	14.92	14.04	13.75
0.7	9.60	9.36	8.75
0.8	4.79	4.68	3.75
0.9	0.00	0.00	0.00
1	0.00	0.00	0.00

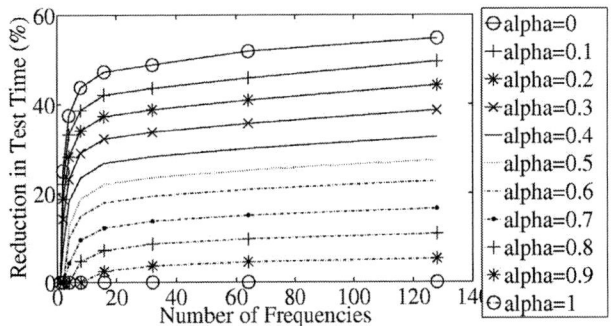

Fig. 6. Test time reduction vs. number of scan clock frequencies.

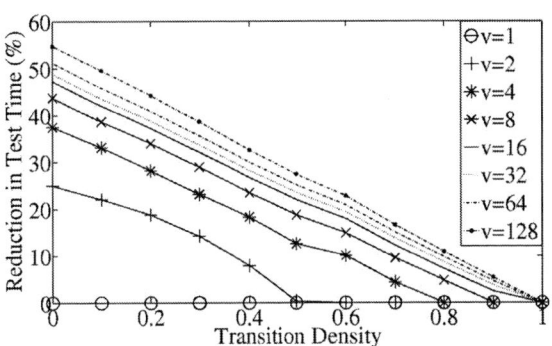

Fig. 7. Test time reduction vs. activity factor for various number of clock frequencies.

be NvT. The number of non-transitions in the scan-in bits equals AN. Thus, AN non-transitions occur in N cycles and a non-transition occurs every $\frac{1}{A}$ cycles.

The scan frequency is increased only after the counter counts up to $\frac{N}{v}$. If x is the maximum number of speeds the scan clock will reach within any scan-in sequence, then

$$\frac{N}{v}x = AN \tag{7}$$

$$x = Av \tag{8}$$

The total scan-in time per combinational vector is the sum of all clock periods used. The test time at each frequency is given by the product of the number of cycles run at that frequency and clock period. These values are given in Table I. Total time per vector is given by

$$\sum_{i=1}^{Av}\{\{\lceil \frac{iN}{Av}\rceil - \lceil \frac{(i-1)N}{Av}\rceil\}(v-i+1)T\} \tag{9}$$

where v is usually chosen as a power of 2 because we can design a divide by 2^n frequency divider with n flip-flops. If N was also chosen as a power of 2, the formula reduces to

$$\text{Total time per vector} = \sum_{i=1}^{Av}\{(\frac{N}{Av})(v-i+1)T\}$$

$$= (\frac{N}{Av})(v.Av - \frac{Av(Av+1)}{2} + Av)T \tag{10}$$

Time per vector if a single speed is used is NvT, and

$$\text{Reduction in test time} = \frac{\{NTv - NT(v - \frac{Av+1}{2} + 1)\}}{NTv}$$

$$= \frac{A}{2} - \frac{1}{2v} = \frac{(1-\alpha)}{2} - \frac{1}{2v} \tag{11}$$

A C program was written to generate random vectors for a circuit with 1000 flip-flops. The test time reduction for these

vectors was estimated, and compared with the values obtained from the formula. Table II shows the test time reduction versus number of frequencies for an activity factor of 0.5. Table III shows the variation of test time reduction with activity factor when the number of frequencies is 8. Both tables compare the test times estimated for random vectors (column 2), with those obtained from the accurate formula Eq. (9) (column 3) and from the approximate formula Eq. (11) (column 4). Figure 6 shows the test time reduction as a function of the number of frequencies for different values of activity factor. Figure 7 gives the test time reduction as a function of α for different numbers of frequencies.

Figures 6 and 7 show that for a chosen number of frequencies, vectors with lower activity achieve higher reduction in test time. The test time reduction increases when the number of frequencies increases. The test time initially reduces rapidly for 8 frequencies and after that the reduction is gradual.

V. EXPERIMENTAL RESULTS

In verilog netlists of the ISCAS89 benchmark circuits flip-flops were added at all primary inputs and primary outputs. All flip-flops were converted to scan types and chained together. Thus, the number of flip-flops in the circuit would be the sum of the number of primary inputs, number of primary outputs and number of D-type flip-flops. A 23-bit linear feedback shift register (LFSR), a 23-bit signature analysis register (SAR), and a test-per-scan BIST controller were implemented [23], [24]. A single bit output of the LFSR supplied the scan input and the scan output was fed into the SAR. A suitable number for random patterns to achieve sufficient fault coverage for each circuit [25] was incorporated into the BIST controller. The sequential circuit along with the BIST circuitry was treated as

TABLE IV
REDUCTION IN TEST TIME FOR ISCAS89 CIRCUITS - TEST PER SCAN
BIST WITH SINGLE SCAN CHAIN.

Circuit	Number of scan flip-flops	Number of frequencies	Test time reduction (%)	Increase in area (%)
s27	8	2	7.49	14.72
s298	23	4	14.57	16.25
s344	35	4	13.48	15.06
s349	35	4	13.81	13.38
s382	30	4	13.20	12.24
s386	20	4	15.25	15.29
s400	30	4	13.18	11.36
s420	35	4	13.81	13.02
s444	30	4	13.18	11.07
s510	32	4	14.30	7.14
s526	30	4	13.18	11.12
s526n	30	4	13.15	11.34
s641	78	4	13.15	11.81
s713	77	4	12.88	11.86
s820	42	4	13.20	10.69
s832	42	4	13.23	11.10
s838	67	4	13.51	11.73
s953	68	4	13.83	10.60
s1196	46	4	13.24	10.65
s1238	46	4	13.24	10.64
s1423	96	4	13.60	8.77
s1488	33	4	12.61	10.25
s1494	33	4	12.56	10.34
s5378	263	4	13.03	6.65
s9234	286	4	14.01	5.82
s13207	852	8	19.00	3.98
s15850	761	8	18.97	3.23
s35932	2083	8	18.74	2.55
s38417	1770	8	18.83	3.14
s38584	1768	8	18.91	2.13

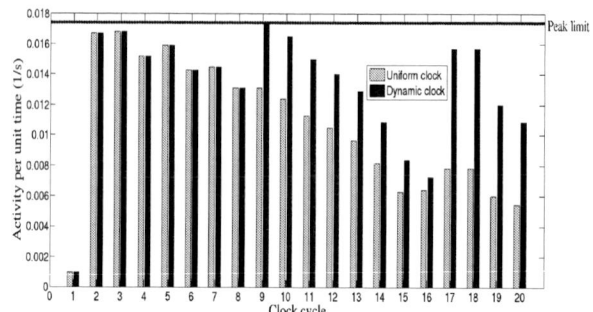

Fig. 8. Activity vs. number of clock cycle for s386 test.

the core circuit for test time and area analysis. The counter, frequency control circuitry, and frequency divider circuitry for dynamic frequency control were implemented as shown in Figure 2. The number of frequencies for each circuit was chosen according to the size of the circuit or the number of scan flip-flops.

ModelSim from MentorGraphics was used to simulate the circuits with and without the dynamic frequency control circuitry. The time required for test application was recorded in each case. DesignCompiler, a synthesis tool from Synopsys, was used to analyze the area of the circuits with and without the dynamic frequency control circuitry.

Since the LFSR generates pseudo random patterns, the activity factor is about 0.5. From (8), $x = 0.5v$, and hence, the number of frequencies the circuit will run at, is half the chosen number of frequencies. This corresponds to a clock period of $(0.5v + 1)T$ from Table I. However, during power analysis, the next higher frequency is taken into consideration, in order to obtain pessimistic data. Thus, power analysis is done for a clock period of $0.5vT$, i.e., for a clock having twice the lowest frequency. Therefore, the power dissipated by the circuit for an activity factor 0.5 at every node and operating at twice the lowest frequency, was estimated for every circuit. The dynamic frequency control circuitry was included in this analysis.

Table IV shows the results. The number of frequencies chosen for each circuit is shown in column 3. The percentage reduction in test time with respect to the test time for the core circuit is shown in column 4 and the percentage increase in area with respect to the area of the core circuit is shown in column 5. At any node, the capacitance and the voltage are constant. From (1), the power dissipated at any node is proportional to the product of activity and frequency. Thus, the activity per unit time is a direct measure of power dissipated in the circuit. Therefore, an analysis to find activity per unit time was performed on the s386 benchmark circuit. The Synopsys power analysis tool, PrimeTime PX, was used. The activity per unit time in every cycle was found for the circuit for

a scan vector with an activity factor of 1. The peak among these values was set as the limit for activity per unit time. The values of activity per unit time of the circuit in every cycle were found for a vector with an activity factor of 0.25 using uniform clock and dynamic clock methods. The results are shown in Figure 8. Notably, the activity per unit time in every cycle is closer to the peak limit when dynamic clock method is used. Also, the peak limit is never exceeded in both methods. A reduction of 11.25% was observed when the dynamic clock method was used.

The results for multiple scan chain implementation would be very similar to that obtained for single scan chain. The test time will not vary much since the activity of the circuit will be very similar in both single and multiple chain implementations. However, there would be a marginal increase in area due to the additional XNOR gates at the first flip-flop of every scan chain and also due to the use of a parallel counter as opposed to the simple counter used for the single scan chain.

These results for reduction in test time conform to the theoretical results given in Figures 6 and 7. Two trends are clearly observed in Table IV. As circuit size increases, the area overhead drops and test time reduction improves. These circuits are not very large from today's standard and we can expect better results as predicted by the analysis.

To estimate the test time reduction for larger circuits, an accurate mathematical analysis was applied to ITC02 circuits. Test time reduction was computed for best ($\alpha \approx 0$), moderate ($\alpha = 0.5$) and worst ($\alpha \approx 1$) cases of scan chain activity factors. The test-per-scan BIST was assumed. Table V shows the results. The number of scan flip-flops in column 2 is the sum of number of inputs, number of outputs and number of flip-flops. The number of frequencies for circuits are shown in column 3. The test time reductions achieved for best, moderate and worst case activity factors are shown in Columns 4, 5 and 6, respectively. Evidently, more test time reduction can be achieved in larger circuits. The reduction in test time varies from 0% for patterns causing very high activity to 50% for patterns with almost no activity.

When external tests are used and an ATPG tool generates them, the vectors may have very few care bits. The don't care bits can be filled in using heuristics [26] to minimize scan transitions. Then, a dynamic control of scan clock will provide a large reduction in test time. This is illustrated using the ISCAS89 benchmark s38584. The Synopsys ATPG tool TetraMAX was used to generate two sets of vectors, a set of 961 vectors with no don't care bits and another set of 14,196 vectors with don't care bits. Figure 9 shows the activity vs. number of vectors distributions for the these sets. The vector set without don't cares has an activity factor around 0.5 and the vector set with don't care bits has an activity factor around

TABLE V
REDUCTION IN TEST TIME FOR ITC02 CIRCUITS.

Circuit	Number of scan flip-flops	Number of frequencies	Test time reduction (%)		
			$\alpha \approx 0$	$\alpha = 0.5$	$\alpha \approx 1$
u226	1416	8	46.68	18.75	0
d281	3813	16	46.74	21.81	0
d695	8229	32	48.28	23.36	0
h953	5586	32	48.32	23.38	0
g1023	5253	32	48.19	23.32	0
f2126	15593	64	49.15	24.18	0
q12710	26158	128	49.45	24.53	0
p22810	29006	128	49.52	24.57	0
p34392	23005	128	49.53	24.57	0
p93791	96916	512	49.72	24.81	0
t512505	76714	512	49.85	24.87	0
a586710	41411	256	49.73	24.77	0

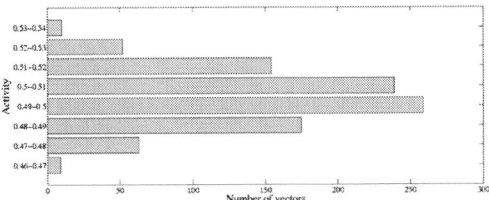

(a) 961 vectors without don't care bits.

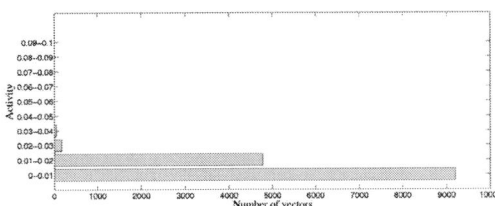

(b) 14,196 vectors with don't care bits.

Fig. 9.   Distribution of s38584 vectors according to their activity factor.

TABLE VI
REDUCTION IN TEST TIME IN s38584 CIRCUIT

Without don't care bits		With don't care bits	
Number of patterns	Reduction in time (%)	Number of patterns	Reduction in time (%)
961	18.8	14196	43.14

0.01. The don't care bits in the second set were filled using a minimum transition heuristic [26]. Reductions in test time achieved for both test vector sets are shown in Table VI.

In another typical scenario, a test set may initially contain few (say, 10%) high activity ($\alpha = 0.5$) vectors. These resemble fully-specified random vectors and achieve about 70-75% fault coverage. The latter 90% vectors then detect about 20-25% hard-to-detect faults and contain many don't cares, which may be filled in for reduced ($\alpha \leq 0.05$) activity. The adoptive test will be potentially beneficial in such cases.

VI. CONCLUSION

Reduction of test application time in power-constrained testing by adoptively adjusting the scan frequency to the circuit activity is demonstrated. On-chip hardware, whose overhead reduces as the circuit becomes large, provides the adoptive control. Self-test as well as external ATE test can be made adoptive. The technique is particularly beneficial when the peak circuit activity during test is very high but the average activity is quite low.

REFERENCES

[1] M. L. Bushnell and V. D. Agrawal, *Essentials of Electronic Testing for Digital, Memory and Mixed-Signal VLSI Circuits*. Springer, 2000.

[2] K. M. Butler, J. Saxena, A. Jain, T. Fryars, J. Lewis, and G. Hetherington, "Minimizing Power Consumption in Scan Testing: Pattern Generation and DFT Techniques," in *Proc. Int. Test Conf.*, pp. 355–364, 2004.

[3] P. Girard, "Survey of Low-Power Testing of VLSI Circuits," *IEEE Design & Test of Computers*, vol. 19, pp. 80–90, May–June 2002.

[4] V. Dabholkar, S. Chakravarty, I. Pomeranz, and S. M. Reddy, "Techniques for Minimizing Power Dissipation in Scan and Combinational Circuits During Test Application," *IEEE Trans. CAD*, vol. 17, pp. 1325–1333, Dec. 1998.

[5] I. Bayraktaroglu and A. Orailoglu, "Test Volume and Application Time Reduction through Scan Chain Concealment," in *Proc. Des. Automation Conf.*, pp. 151–155, 2001.

[6] Y. Bonhomme, T. Yoneda, H. Fujiwara, and P. Girard, "An Efficient Scan Tree Design for Test Time Reduction," in *Proc. IEEE European Test Symp.*, pp. 174–179, 2004.

[7] I. Bayraktaroglu and A. Orailoglu, "Decompression Hardware Determination for Test Volume and Time Reduction through Unified Test Pattern Compaction and Compression," in *Proc. 21st IEEE VLSI Test Symp.*, pp. 113–118, 2003.

[8] I. Bayraktaroglu and A. Orailoglu, "Concurrent Application of Compaction and Compression for Test Time and Data Volume Reduction in Scan Designs," *IEEE Trans. Computers*, vol. 52, pp. 1480–1489, Nov. 2000.

[9] J. Rajski, J. Tyszer, M. Kassab, N. Mukherjee, R. Thompson, H. Tsai, A. Hertwig, N. Tamarapalli, G. Mrugalski, G. Eide, and J. Qian, "Embedded Deterministic Test for Low Cost Manufacturing Test," in *Proc. Int. Test Conf.*, pp. 301–310, 2002.

[10] A. Chandra and K. Chakrabarty, "Reduction of SoC Test Data Volume, Scan Power and Testing Time Using Alternating Run-Length Codes," in *Proc. Int. Conf. Computer Aided Design*, pp. 673–678, 2002.

[11] A. Chandra and K. Chakrabarty, "Frequency-Directed Run-Length (FDR) Codes With Applicatin to System-on-A-Chip Test Data Compression," in *Proc. 19th IEEE VLSI Test Symposium*, pp. 42–47, 2001.

[12] W. J. Lai, C. P. Kung, and C. S. Lin, "Test Time Reduction in Scan Designed Circuits," in *Proc. European Des. Automation Conf.*, pp. 489–493, 1993.

[13] E. M. Rudnick and J. H. Patel, "A Genetic Approach to Test Application Time Reduction for Full Scan and Partial Scan Circuits," in *Proc. Int. Conf. VLSI Design*, pp. 288–293, Jan. 1995.

[14] S. Y. Lee and K. K. Saluja, "Test Application Time Reduction for Sequential Circuits with Scan," *IEEE Trans. CAD*, vol. 14, pp. 1128–1140, Sept. 1995.

[15] S. Y. Lee and K. K. Saluja, "An Algorithm to Reduce Test Application Time in Full Scan Designs," in *Proc. Int. Conf. CAD*, pp. 17–20, 1992.

[16] H. C. Tsai, S. Bhawmik, and K.-T. Cheng, "An Almost Fullscan BIST Solution - Higher Fault Coverage and Shorter Test Application Time," in *Proc. Int. Test Conf.*, pp. 1065–1073, Oct. 1998.

[17] P. Shanmugasundaram, "Test Time Optimization in Scan Circuits," Master's thesis, Auburn University, Dec. 2010.

[18] P. Shanmugasundaram and V. D. Agrawal, "Dynamic Scan Clock Control in BIST Circuits," in *Proc. 43rd Southeastern Symp. System Theory*, Mar. 2011.

[19] R. Sankaralingam, R. R. Oruganti, and N. A. Touba, "Static Compaction Techniques to Control Scan Vector Power Dissipation," in *Proc. 18th IEEE VLSI Test Symp.*, pp. 35–40, Apr. 2000.

[20] S. Wang and S. K. Gupta, "LT-RTPG: A New Test-Per-Scan BIST TPG for Low Heat Dissipation," in *Proc. Int. Test Conf.*, pp. 85–94, Sept. 1999.

[21] E. E. Swartzlander, Jr., "A Review of Large Parallel Counter Designs," in *Proc. IEEE Computer Society Annual Symposium on VLSI*, pp. 89–98, Feb. 2004.

[22] W. J. Dally and J. W. Poulton, *Digital Systems Engineering*. Cambridge University Press, 1998.

[23] V. D. Agrawal, C. R. Kime, and K. K. Saluja, "A Tutorial on Built-In Self-Test, Part 1: Principles," *IEEE Design & Test of Computers*, vol. 10, pp. 73–82, Mar. 1993.

[24] C. Stroud, *A Designer's Guide to Built-In Self-Test*. Springer, 2002.

[25] F. Brglez, D. Bryan, and K. Kozminski, "Combinational Profiles of Sequential Benchmark Circuits," in *Proc. Int. Symp. Circuits and Systems*, pp. 1929–1934, May 1989.

[26] N. Badereddine, P. Girard, S. Pravossoudovitch, C. Landrault, and A. Virazel, "Minimizing Peak Power Consumption during Scan Testing: Test Pattern Modification with X Filling Heuristics," in *Proc. Int. Conf. on Design and Test of Integrated Systems in Nanoscale Technology*, pp. 359–364, Sept. 2006.

Structural Tests of Slave Clock Gating in Low-power Flip-flop

Baosheng Wang, Jayalakshmi Rajaraman, Kanwaldeep Sobti, Derrick Losli, and Jeff Rearick

Advanced Micro Devices, Inc., 1 AMD Place, Sunnyvale CA 94085, USA

{FirstName.LastName}@amd.com

Abstract

A novel slave clock-gating technique in [5] is designed to save power when the master and slave latches of a low-power flip-flop reach certain correlated states (e.g., both latches are at logic 0 or 1). Testing this clock-gating circuit is essential for power-sensitive applications, but is also very challenging. This is because power consumption increase is its only defective behavior, and it involves cell internal states, both of which are unfriendly to general automatic test-pattern generation (ATPG). This paper proposes an innovative method to test the slave clock-gating circuitry structurally with slight modification of the flop cell. The implementation on a two-latch version of a level-sensitive scan design (LSSD) flip-flop and its capability of extending to other types of flip-flop cells are presented.

Keywords: *Slave Clock Gating, Low-power Flip-flop, Structural Test, Diagnosis*

1. Introduction

The sequential circuits in a system-on-a-chip (SoC) contribute significantly to the chip's power dissipation, mainly because the clocks of those sequential circuits toggle frequently during application time. Recent studies indicate that the clock signals in digital systems consume up to 45% of the system power [1-3]. Thus, the circuit power can be reduced dramatically by decreasing clock-switching activities.

Clock gating [4] is an effective and well-known technique for minimizing dynamic power in sequential circuits (e.g., flip-flops). The flip-flop clock-gate controls can be from its internal nodes [5-6] or implemented with external circuits (e.g., finite state machines (FSMs) [7]). In [5], the slave-latch clock is turned off when the flop master and slave latches stay at the same state values. According to the experimental results in [5], after applying the slave clock-gating technique, the power dissipation of a flip-flop can be reduced by 60% when its input data signal switching activity is low. Compared with results in [6], in which both data and clock are gated, the slave clock-gate technique in [5] is more practical in terms of area and timing penalties. This gating technique is applicable to all general scannable flip-flops (i.e., those based on MUX-D [8], traditional level-sensitive scan design (LSSD) [9], and two-latch version of LSSD [10]).

Although the low-power flip-flop in [5] is scannable (i.e., it has full controllability and observability on the data path), it lacks the controllability on the clock path and its power-reduction feature is not structurally testable. For example, if the clock-control circuit is stuck-on (i.e., the clock gater is always on), it is impossible to differentiate the logic behavior of a faulty flip-flop from that of a good one. Their only electrical difference is that a faulty flop dissipates more power than a good one when the master and slave latches stay at their correlated states. However, when they are deeply embedded in a million-gate circuit, it is very expensive and, hence, economically infeasible to measure each low-power flop current, even when utilizing function patterns with high switching activities. It may be argued that traditional automatic test-pattern generation (ATPG) tools might be powerful enough to figure out an appropriate pattern to isolate the clock gater stuck-on fault without circuit modifications. Unfortunately, those tools usually assume flops in unknown states during scan tracing, and fail even before starting pattern generation.

To the best of our knowledge, the current open literature offers no circuits and/or methods to test the power-reduction feature structurally for this type of flip-flop. For the first time, this proposal presents a low-cost structural-test solution to detect slave clock-gating circuit stuck-on conditions in such a low-power flip-flop. With dedicated input patterns and slight modification of the low-power flop cell, the proposed method can catch such stuck-on defects and diagnose defective low-power flip-flops. Specially, we implement our proposal on the two-latch version of an LSSD cell due to its best trade-off factor among speed, power, and area. However, our proposal can be extended to the other two types of flip-flop cells (see Section 5).

In the rest of this paper, Section 2 explains the low-power flip-flop in [5], and motivations of structural test on its slave clock-gating circuitry. The proposed method, circuit implementation, and input pattern generation are presented in detail in Section 3. In Section 4, experimental results demonstrate its effectiveness and evaluations on the new cell's impacts. Section 5 extends the proposal to the traditional LSSD and MUX-based flip-flop cells, and Section 6 concludes the paper.

2. The Low-power Flip-flop and Motivations of Structural Tests

978-1-61284-657-6/11 $26.00 © 2011 IEEE

Figure 1 depicts the block diagram for a two-latch version of the low-power LSSD flip-flop in [5]. The slave clock-gating circuit is a comparator that compares the states of master and slave latches to determine whether the slave clock should be on or off. Depending on applications, this comparator can be an OR, XOR, or a NAND in the actual implementation. Obviously, with a stuck-at-1 (SA1) fault on Node A, this low-power LSSD flop will behave normally in functionality, but consume more power than a good flop.

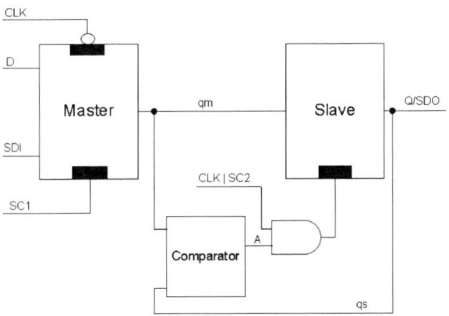

Figure 1. Block Diagram of a Low-power Flip-flop in [5]

Detecting this SA1 fault is essential for manufacturing test, especially for battery-driven semiconductor chips and other applications with military-level quality requirements. This is because Node A has high likelihood of shorting with power (see the circuit and layout diagram of an OR comparator in Figure 2), and there would be noticeable test escapes if such a test is not performed.

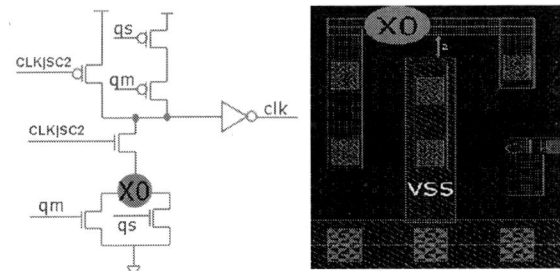

Figure 2. Circuit and Layout Diagram of an OR Comparator

For example, assuming a typical gate has four faults, a SoC with 50 million logic gates (excluding memory arrays) can have 200 million faults. If 200,000 low-power flops are utilized in this SoC, 200,000 SA1 faults could escape the regular manufacturing test. In other words, under an ideal environment in which we obtain perfect coverage elsewhere, the fault coverage on logic gates in this case would be 1 – 200k/200m = 99.9%. According to the Seth-Agrawal yield model [11], the values of defect per million (DPM) in this case would be 24.3, 15.1, and 7.1 respectively when the logic yield is 85%, 90%, and 95%. For applications with military-level DPM requirements, those escapes are unacceptable.

Unfortunately, functional patterns and logic differentiation cannot detect SA1 faults of Node A in the clock-gating

circuitry. As technology continues to scale down and leakage current becomes dominant in the sub-100-nm era, IDDQ test is also ineffective to detect such faults. Some design-for-manufacture (DFM) methods combined with a sufficient burn-in process might alleviate this problem, but they cannot measure their test quality. Furthermore, this fault is untestable using existing ATPG methods because internal clock-gating is currently not recognized by commercial ATPG tools in which scan tracing often fails:

- The slave-latch clock is determined by the input and output values of the slave latch, and ATPG tools have problems resolving these values during scan. With the current capability of ATPG tools, the internal clock-gating circuits go unmodeled and, hence, untested.

- Even if future ATPG tools are able to resolve scan tracing problems, there is no controllable signal present in the circuit that entails any test coverage.

In summary, the low-power flip-flop with slave-clock gating in [5] has poor testability on its power-reduction feature. A new structural test solution is warranted.

3. The Proposal

Our proposal is based on two conditions that are required to invoke an SA1 fault at Node A in Figure 1:

1. The comparator logic output is constrained to 0 and the clock is pulsed, and

2. Node qm is set to the un-correlated value of Node qs.

In such a situation, a good machine will not clock the slave latch and update its data, but a bad machine will. To achieve this, a controlling signal is required on Nodes qm and qs such that it disables the output of the comparator logic when pulsing clock. Fortunately, this control signal can be found internally either within the cell or through appropriate spare gates. Without losing generality, the two-latch version of an LSSD cell and OR implementation as the comparator demonstrates the proposed method.

3.1 The New Low-power Flip-flop Cell

As explained in [10], the two-latch version of LSSD consists of a two-port latch L1 for scan and system operation, respectively, and a one-port latch L2 shared by both scan and system operations. The latch L2 clock is a multiplexed version of system and scan clocks that are toggled depending on the mode of operations (i.e., functional or scan). To test the SA1 fault at Node A, we make use of an inverted SC1 signal to disable the comparison logic so the comparator output Node A is 0 during test. Meanwhile, we design a dedicated sequence and patterns to work with this modified flop cell. The block diagram of the new flip-flop cell with DFT augmentation is shown in Figure 3(a), while the intended waveform is shown in Figure 3(b).

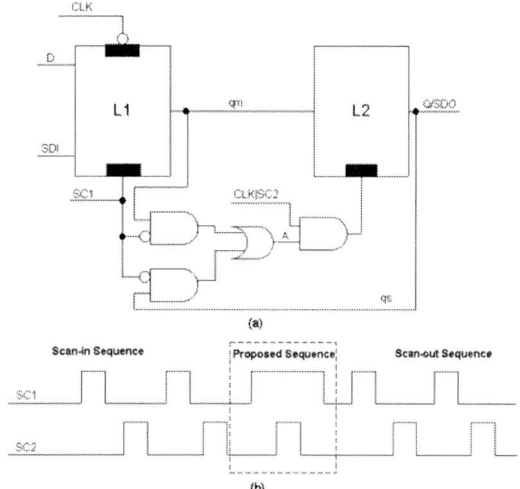

Figure 3. Two-latch version of LSSD with DFT Augmentation: (a) circuit (b) waveform

The new cell with DFT augmentation works as follows:

1. Shift in all zeros through the entire scan chain to initialize all scan cells. This assures every scan cell value is deterministic.

2. Assert SC1 while applying a 1 at SDI so Nodes qm and qs have an opposite value (i.e., qm = 1 and qs = 0).

3. Assert SC2 while holding SC1. For a good machine, due to the SC1 inversion signal disabling the comparison, L2 latch will not be updated. On the contrary, a bad machine with an SA1 fault at Node A will have L2 latch updated (i.e., from 0 to 1). It is preferable to enable SC1 first, before asserting SC2, for stable results.

4. De-assert SC2 while continuing to hold SC1 for some time. The comparison will be re-enabled, and Node A's SA1 fault will escape the test if SC1 and SC2 are de-asserted at the same time, or if SC1 is de-asserted earlier than SC2.

5. De-assert both SC1 and SC2, then shift out the cell under test for comparison. As explained in Step 3, in this case, if the shift out value is 1, this cell has an SA1 at Node A; otherwise, it is fault-free.

Since SC1 is always 0 during function mode, the modified flip-flop cell works as usual (i.e., the comparison logic is not affected).

3.2 The Patterns

Although modeling techniques could utilize ATPG tools to generate patterns to detect the slave clock-gating faults by manipulating the control signal, we propose generating patterns manually to minimize the pattern counts. For the same purposes in Section 3.1, the proposed flip-flop cell

shown in Figure 3 explains the proposed pattern generation method.

Depending on detection requirements (i.e., test or diagnosis), the pattern generation process is different. For manufacturing test, a checkerboard-like pattern (010101...) would be sufficient. If the scan cells in a scan chain happen to be all low-power flip-flops, the regular checkerboard pattern will be applied. The detection process in this situation is illustrated in Figure 4.

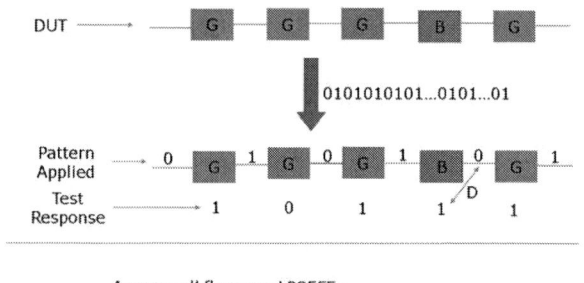

Figure 4. Detection Process on a Scan Chain with All Low-power Flip-flops

After initializing all scan cells with 0s, simply applying the checkerboard pattern can detect if this scan chain has defective low-power flop cells. The proposed SC2 pulse should be long enough to guarantee all cells in that scan chain can be flushed at one time.

In reality, a scan chain can have regular flops, low-power flops, and even scannable A-phase latches [10]. In this case, the checkerboard-like pattern will vary, depending on the relative locations of low-power flops and other types of scan cells. Generally, since the proposed sequence is a flush operation on scan cells other than low-power flops, we apply the same values between the scan input and output for a scan cell that is not a low-power flop. In other words, as long as we can create an opposite value between the input and the output for a low-power flop, our proposal will be effective. Figure 5 demonstrates a scenario of such a proposed pattern:

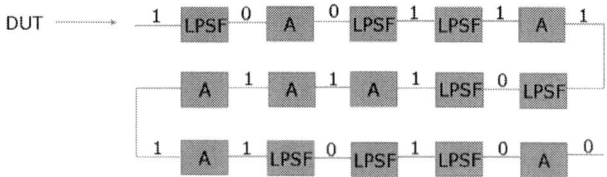

Figure 5. Proposed Pattern on a Scan Chain with Mixed Types of Scan Cells

In Figure 5, A represents a scan cell other than a low-power flop, while LPSF represents a low-power flop. Generally, after a SoC is taped out, all scan-chain scenarios should be known, so all appropriate patterns can be generated.

To diagnose all defective low-power flops in a scan chain, the pattern-generation process will be more complex if multiple defective low-power flops are in a row for that scan chain. Without loss of generality, we assume a case in which a scan chain consists of M consecutive defective low-power flip-flop cells. The first pattern applied would be the default checkerboard-like pattern: 010101.... To identify the odd-positioned defective cells, the second pattern would be 00110011.... If we need N total number of patterns, the nth pattern would repeat the following sequence across all low-power flops (apply the same values on the regular flop and latch-based scan cells):

$$\{0\}*2^{(n-1)}\{1\}*2^{(n-1)}, n = 1, ..., N$$

The total pattern count N can be derived from the data in Table 1 (also verified through real examples):

Table 1. Pattern-count Calculation

# (defective low-power cells in a row)	# Patterns
1	1
2	2
3	2
4	3
5	3
...	...
M	$N = Log_2(M + 1)$

Based on the data in Table 1, the maximum number of patterns required for a thorough diagnosis would be log_2(the maximum number of low-power cells in a row for all scan chains).

In Figure 6, we diagnose a scan chain with four (M = 4) defective low-power flop cells through three (N = 3) patterns:

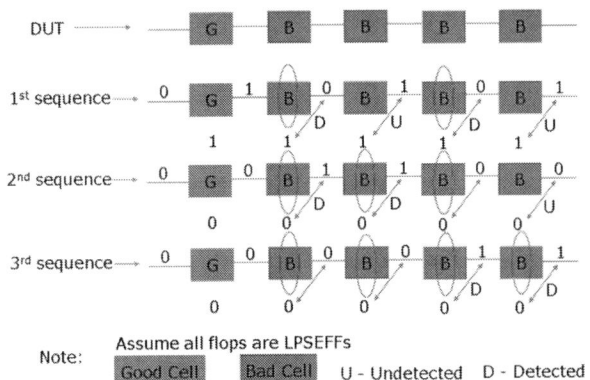

Figure 6. Diagnosis Process on a Scan Chain with Four Consecutive Defective Low-power Flip-flops

3.3 The Circuits

Although the block diagram in Figure 3 shows two extra AND gates required, there are actually two additional transistors required after merging the proposed gates with the existing comparator. Figure 7 shows the transistor-level AND gate implementation in which the transistors MP_SC1b and MN_SC1b are the only extras. The evaluation analysis results (see the following section 4) show that MP_SC1b can be a weak PMOS transistor to provide a weak 1. As a result, it can be overcome by the defect in the NMOS stack (SA1 node in Figure 1 transformed to SA0 in the NMOS stack). In a good machine, the weak 1 at Node A will shut off the slave-latch clock.

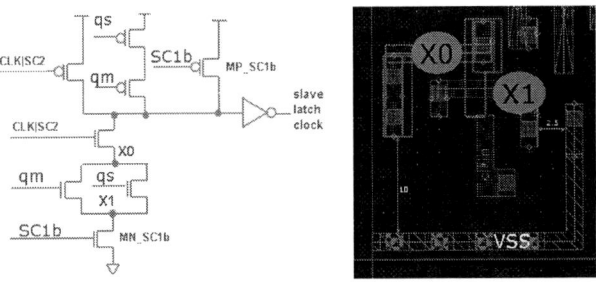

Figure 7. Proposed Transistor-level Implementation and Layout

One may argue the second AND gate to control Node qm in Figure 3 is redundant because Node qs is initialized as 0 during proposed test sequence. However, removing that gate does not change the physical area because our proposal only requires two extra transistors for implementing the disable mechanism.

4. Experimental Results

Our validations include two steps: RTL pattern verification and transistor-level fault simulation.

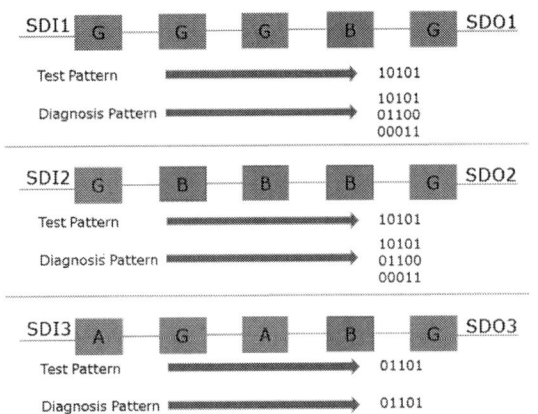

Figure 8. Test Cases and Proposed Test/Diagnosis Patterns

To verify the patterns at RTL, we built a test case with three scan chains. Both Chains 1 and 2 contain five consecutive low-power flops, while Chain 3 includes two scan cells that are not low-power flops, as shown in Figure 8.

We inject faulty cells by replacing the low-power flops with the regular ones. Based on the pattern-generation theory in Section 3.2, our test bench -- which embeds the test/diagnosis patterns shown in Figure 8 -- successfully reports the number of faulty cells and their locations.

The transistor-level fault analysis is performed on a single cell and mainly ensures the extra transistors would not introduce any new untestable faults. Fault injection is done through adding large resistance on the node under test. The test case set-up and the fault analysis results are shown in Figure 9 and Table 2, respectively.

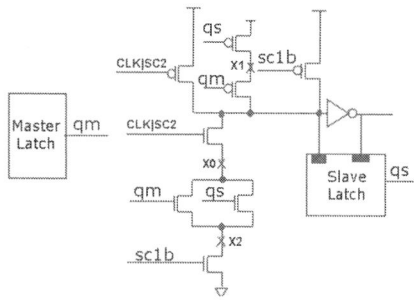

Figure 9. Fault Analysis Setup

Table 2. Fault Analysis Results

Fault	Old Cell	New Cell
X0 SA1	testable	testable
X0 SA0	**untestable**	**testable**
X0 slow to fall	testable	testable
X0 slow to rise	N/A	N/A
X1 SA1	possibly testable	possibly testable
X1 SA0	possibly testable	possibly testable
X1 slow to fall	N/A	N/A
X1 slow to rise	untestable	untestable
X2 SA1	N/A	**testable**
X2 SA0	N/A	**testable**
X2 slow to fall	N/A	**testable**
X2 slow to rise	N/A	**N/A**

From Table 2, the key fault X0 SA0 becomes testable after cell DFT augmentation, and there are no new untestable faults.

To further evaluate the new cell, we laid out the new cell, estimated possible yield impact, calculated possible power increase, and analyzed its robustness from a DFM perspective on AMD's next-generation microprocessor core in [10]. The summary of our results:

- Total area increase due to the new cell: 0.27%

- Yield loss due to additional area: 0.025%

- Insignificant change in power profiles

- Proposed design is more robust in terms of printability and critical area distribution

5. Extension of the Proposal to Other Types of Low-power Flip-flops

5.1 Traditional LSSD Cell

For conventional LSSD three-latch scan architecture, system slave L2 is the same as scan master. Both system master and scan slave are separate latches. Such a generic architecture is shown in Figure 10(a). Our testing methodology can be applied to this architecture as well.

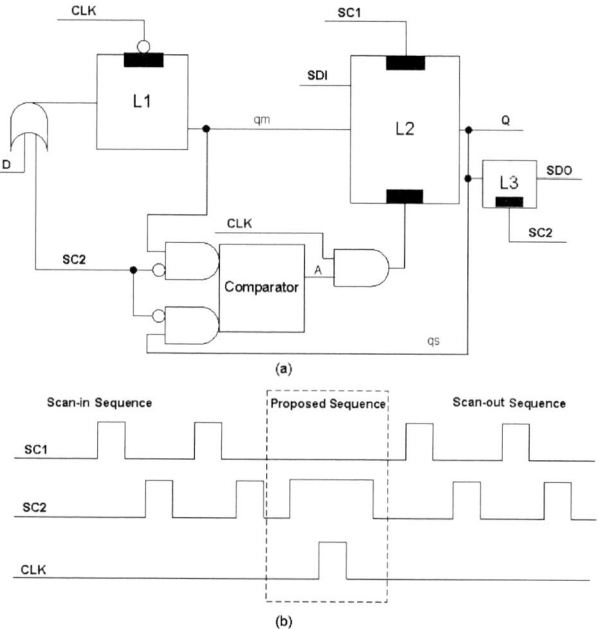

Figure 10. Method Extension to Traditional LSSD Flip-flop Cell: (a) circuit (b) waveform

The signal to execute the disable mechanism can be a logical inversion of SC2. In our testing methodology, SC2 was held high while CLK was pulsed as shown in Figure 10(b). The SC1 was held low to disable the scan master. A good machine will not clock the system slave, while a bad machine with such a fault will. Hence, a good machine would not update the data at the output of system slave, while a bad machine would.

To facilitate the test-pattern generation, an OR gate controlled by SC2 is added in the function data path. Since SC2 is always low during functional operation, there would be no functional impacts, but there may be a slight timing impact on the data path. During our proposed test mode, SC2 was held high so the function slave latch L2 had its data input forced to 1. Since L2 was initialized to 0 during scan, we can apply the proposed methodology to detect the slave clock-gating stuck-on conditions. The detailed pattern generation process would be the same as that for the two-latch version of LSSD (see Section 3.2).

5.2 MUX-D Scan Cell

A MUX-D type flip-flop works almost the same as the two-latch version of LSSD, except there are no separate scan clocks; functional and scan operations share the same clock. For this type of flop, an external pin low-power test (LPT) has to be introduced for the flop cell. As already described, this cell pin will be routed globally for all low-power flip-flops with slave clock-gating circuitry and controlled through direct input port, JTAG register bit, etc. Fortunately, this global signal is not timing-critical and no special routing requirements need to be enforced.

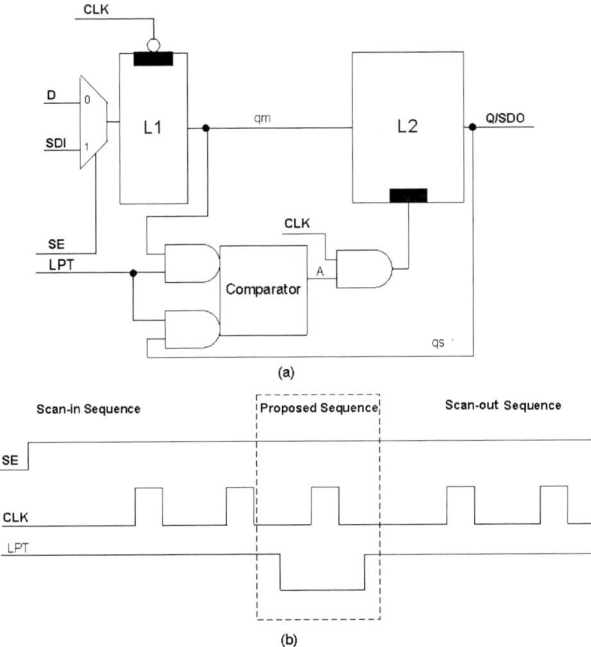

Figure 11. Method Extension to MUX-D Type Flip-flop Cell: (a) circuit (b) waveform

6. Conclusion

Slave-clock gating in [5] is a novel technique to reduce clock power of flip-flops significantly. Due to lack of controllability on its clock paths, the gater stuck-on conditions cannot be tested structurally. This proposal provides a low-cost solution to test internal-clock gating stuck-on faults structurally. It can also diagnose all possible faulty cells with a manageable count of patterns. Evaluation results show the proposed new cell has negligible penalties on area, timing, routing, power, and yield without introducing new untestable faults or reducing DFM robustness.

7. Acknowledgements

We thank anonymous reviewers at Advanced Micro Devices, Inc. for their insightful comments and suggestions on this work. We also thank AMD layout and

DFM teams for helping evaluate the physical impacts of the new cells.

8. References

[1] G. Palumbo, F. Pappalardo, and S. Sannella, "Evaluation on power reduction applying gated clock approaches," *IEEE International Symposium on Circuits and Systems (ISCAS)*, Vol. 4, pp. 85-88, 2002.

[2] R.Y. Chen, N. Vijaykrishnan, and M.J. Irwin, "Clock power issues in a System-on-a-Chip Designs," *Proceedings of IEEE Computer Society Workshop on VLSI*, pp. 48-53, 1999.

[3] S.D. Naffziger, "Power Optimizations for a 45nm Processor Core," an invited talk at University of Berkley, Nov. 12, 2007, http://bwrc.eecs.berkeley.edu/Classes/ICDesign/icseminar/icseminar_f07.

[4] Q. Wu, M. Pedram, and X. Wu, "Clock-gating and its application to low power design of sequential circuits," *IEEE Transactions on Circuits and System I: Fundamental Theory and Applications*, Vol. 47, Issue 3, pp. 415-410, March 2000.

[5] S.D. Naffziger, *Low Power Flip Flop Through Partially Gated Slave Clock*, US Patent #7772906, 2010.

[6] A.G.M. Strollo and D. De Caro, "Low power flip-flop with clock gating on master and slave latches," *IEEE Electronics Letters*, Vol. 36, Issue 4, pp. 294-295, Feb. 2000.

[7] R. Bhurada and Y. Manoli, "Complex clock gating with integrated clock gating logic cell," *International Conference on Design & Technology of Integrated System in Nanoscale Era (DTIS)*, pp. 164-169, Sept. 2007.

[8] M.L. Bushnell and V.D. Agrawal, *Essentials of Electronic Testing for Digital, Memory & Mixed-Signal VLSI Circuits*, Kluwer Academic Publishers, ISBN: 0-7923-7991-8, 2000.

[9] T. Wood, "Test and debug features of the AMD-K7TM microprocessor," *Proceedings of IEEE International Test Conference*, 1999, pp. 130-136.

[10] M. Yilmaz, B. Wang, J. Rajaraman, T. Olsen, K. Sobti, D. Elvey, J. Fitzgerald, G. Giles, and W. Chen, "The Scan-DFT Features of AMD's Next-Generation Microprocessor Core," accepted for *International Test Conference (ITC)*, November 2010.

[11] V.D. Agrawal, S.C. Seth, and P. Agrawal, "Fault Coverage Requirement in Production Testing of LSI Circuits," in *IEEE Journal of Solid-Sate Circuits*, Vol. SC-17, No. 1, pp. 57-61, Feb. 1982.

Revival of Partial Scan: Test Cube Analysis Driven Conversion of Flip-Flops

Nader Alawadhi
Computer Science Department
Kuwait University

Ozgur Sinanoglu*
Computer Engineering Department
New York University - Abu Dhabi

Abstract

Increasing complexity of integrated circuits has forced the industry to abandon partial scan, which necessitates a computationally demanding and unaffordable sequential ATPG, and to rather adopt full scan despite its costs. In this paper, we propose a partial scan scheme driven by a computationally efficient test cube analysis. We tackle the challenges associated with the identification of the conditions to restore the controllability and observability compromised due to partial scan, and with the formulation of these conditions in terms of test cube operations. Upon the identification of a maximal-sized set of scan flip-flops that are converted to non-scan, a simple post-processing of the test cubes helps compute the values to be loaded into the scan flip-flops, eliminating the need to re-run ATPG while at the same time ensuring the quality of full scan. The proposed scheme combines the simplicity of the conventional ATPG flow with the area, performance, test time, and test power reduction benefits of partial scan. The proposed test cube analysis driven partial scan scheme is orthogonal and thus fully compatible with other test cost reduction techniques, such as test data compression and test power reduction, which can be applied in conjunction.

1 Introduction

Increasing complexity of integrated circuits has forced the industry to abandon partial scan, which removes scan multiplexers corresponding to some of the flip-flops, disconnecting them from the scan path. As a result, controllability and observability of these flip-flops are compromised, necessitating sequential ATPG, where these flip-flops are controlled and observed through functional paths. The computational cost of sequential ATPG cannot be afforded, given the complexity of integrated circuits today. Consequently, the industry has given up on partial scan, and adopted full scan despite its costs.

Full scan incurs area, performance and test costs. Insertion of as many multiplexers as the number of flip-flops in the design obviously imposes a considerable area cost. Furthermore, these multiplexers are inserted on functional paths, resulting in critical path prolongation by a multiplexer delay, and hence degrading the performance of the design timing-wise. Full scan also incurs significant test costs. Every test pattern consists of as many bits as the number of flip-flops in

the design, translating into high test time and test data volume. Another problem that full scan imposes is the excessive switching activity during test, as all the flip-flops are active during shift operations. Elevated levels of power dissipation [1] occur during testing, which, if overlooked, can cause reliability issues.

A computationally efficient partial scan can be a remedy of the problems of full scan. Removal of scan multiplexers, and thus taking some of the flip-flops off the scan path [2]:

- Reduces area cost.

- Potentially improves the critical paths of the design, and thus, enhances functional performance.

- Reduces the scan path length, and thus, decreases test time and test data volume. It is a form of test data compression.

- Reduces switching activity during testing and leashes power dissipation and IR drop, as only the scan flip-flops toggle during shift operations while the non-scan flip-flops are inactive during shift.

An extensive amount of research has been conducted in partial scan design. The previously proposed techniques in this field can be classified mainly into three categories: structure-based techniques that typically involves breaking the cycles and/or reducing scan depth [3, 4, 5, 6, 7, 8, 9, 10, 11, 12, 13], testability-based techniques that select scan flip-flops based on testability improvements [7, 8, 14, 15, 16, 17, 18, 19, 20, 21, 22, 23], and test generation-based techniques which intertwine test generation and scan flip-flop selection [24, 25, 26, 27, 28, 29]. Other partial scan techniques include those driven by layout constraints [7], timing constraints [30], re-timing [4, 31], and toggling rate of flip-flops and entropy measures [32]. These techniques typically necessitate the utilization of sequential ATPG to generate test patterns on the partially scanned design or combinational ATPG with time frame expansion (if all cycles are broken), not only failing to comply with the existing design/test flow that industry utilizes today but also incapable of ensuring the quality of full scan.

In this paper, we propose a computationally efficient and design flow compliant partial scan scheme that can deliver all the aforementioned area, test cost reduction, and performance benefits while ensuring the quality of full scan. The proposed scheme is driven by an analysis of test cubes, which have been generated by a combinational ATPG tool, and identifies the

---

*Email: ozgursin@nyu.edu, Phone: +1 (347) 309 8079

978-1-61284-657-6/11 $26.00 © 2011 IEEE

flip-flops that can be converted to non-scan while retaining the test quality intact. Upon the conversion of scan flip-flops to non-scan, a simple post-processing of the test cubes helps compute the values to be loaded into the remaining scan flip-flops, eliminating the need to re-run ATPG. This way, the proposed partial scan scheme combines the simplicity of the conventional ATPG flow, and the area, performance and test cost reduction benefits of partial scan. *Ensurance of the quality of full scan and the elimination of the need to re-run ATPG on the partially scanned design render the proposed scheme fully compliant with the design and test flow, and, to the best of our knowledge, uniquely differentiable from the previously proposed partial scan techniques*; as these features exist in full scan, and not in the other partial scan approaches proposed earlier, we confine our comparisons against full scan only.

The challenges engendered in inferring, from only a given set of test cubes, a maximal-sized set of flip-flops that can be converted to non-scan consist of the identification of the conditions to restore the controllability and observability compromised due to partial scan, and of the formulation of these conditions in terms of test cube operations. By tackling these challenges, we propose a partial scan scheme that offers yet another benefit; as the structural details about the design are not required in this analysis, and instead, the proposed tool operates only on a set of test cubes, partial scan implementation can be out-sourced with no intellectual property considerations. It is also noteworthy that the proposed test cube analysis driven partial scan scheme is orthogonal and thus fully compatible with other test cost reduction techniques, such as test data compression [33] and test power reduction [1], which need to be applied subsequently on the test cubes processed by the proposed partial scan technique.

The remainder of the paper is organized as follows. In Section 2, we describe the proposed partial scan scheme, with particular focus on the clocking scheme. We follow this discussion up with the proposed test cube analysis in Section 3. The experimental results, future research work to enhance the proposed scheme further, and concluding remarks are presented in Sections 4, 5, and 6, respectively.

2 Proposed Partial Scan Scheme

Conversion of a scan flip-flop to a non-scan flip-flop can be accomplished by removing the associated scan multiplexer and re-routing the scan chain around the flip-flop, bypassing it. The end-result is area cost reduction and potentially performance enhancement due to the removal of the multiplexer, in addition to the test time, data volume and power dissipation reductions due to the shortened scan chain; yet, controllability and observability of the converted flip-flop are compromised with the removal of the multiplexer.

In order to preserve test quality, the effect of the scan to non-scan conversion should be nullified by restoring the compromised controllability and observability. The latter is easier gain back via a simple tap off of the output of the flip-flop as

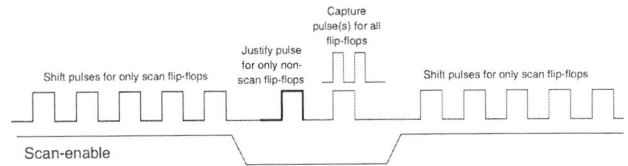

Figure 1. Clocking in the proposed partial scan scheme

an observation point. Observing the content of the flip-flop through the observation point subsequent to each capture operation suffices to restore the observability compromised due to scan to non-scan conversion. Furthermore, the observation points corresponding to multiple non-scan flip-flops can be compacted together via a logic cone analysis [34] in order to reduce the associated area cost while retaining error detection level intact; error masking can be prevented by compacting the outputs of the flip-flops that have disjoint input cones. The compacted observation points can be multiplexed onto the primary outputs, or alternatively feed an existing or a dedicated compactor/MISR along with the scan chain(s).

The compromised controllability is more challenging to restore though. With the removal of the scan multiplexer, the non-scan flip-flop should be justified, through the functional path driving the flip-flop, to the value required by a test pattern. In order to render a simple test cube analysis sufficient for the identification of whether and how this justification can be accomplished, we constrain any such justification to span only a single time frame; in the proposed scheme, a single clock pulse received only by the non-scan flip-flops justify them to the required value. As the associated functional paths are driven by the scan flip-flops, the *justify pulse* is applied after the shift pulses (upon the completion of the shift-in operations, and thus upon the load of the scan flip-flops) and before the capture pulse(s) (so that the non-scan flip-flops are also loaded through the functional paths prior to capture). The clocking of the flip-flops in the proposed partial scan scheme is provided in Figure 1; aside from the newly inserted justify pulse, this clocking scheme is identical to that of the traditional scan-based scheme, and can be implemented via simple clock gating rather than separate dedicated clocks for the two groups of flip-flops. The shift pulses drive only the scan flip-flops, the justify pulse drives only the non-scan flip-flops, and the capture pulse(s) are received by both the scan and non-scan flip-flops. Compared to full scan testing, the same pattern is applied to the circuit under test prior to the capture pulse(s), and with a careful feed of the observation points to the outputs/compactor/MISR, the same response is observed. Both static (stuck-at) and dynamic (launch-off-capture at-speed) types of testing are supported as can be seen in Figure 1, while launch-off-shift type of testing requires drastic changes in the proposed clocking scheme.

In this scheme, a non-scan flip-flop may possibly receive an unintended value upon the justify pulse due to a defect in the functional path driving this flip-flop. While such a de-

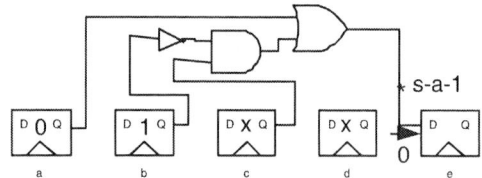

Figure 2. Justification of a flip-flop through functional paths

fect can be implicitly detected in the captured response subsequently, its detection is ensured by observing the observation points twice: once after the justify pulse, and once after the capture pulse.

3 Proposed Test Cube Analysis

The simplicity of the proposed partial scan scheme in justifying a non-scan flip-flop enables a test cube analysis driven identification of flip-flops that can be converted to non-scan. In this test cube analysis, we adhere to the constraint that fault coverage must remain intact. In other words, the test cube analysis identifies a subset of flip-flops to be converted to non-scan by ensuring that **all test cubes** can still be applied intact. Thus, the combinational ATPG process conducted to generate the test cubes need **not** be repeated. In this section, we present the proposed test cube analysis, which can be applied as a post-ATPG process.

We first introduce the following terminology. The test cubes of the design, which represent the values to be loaded into the flip-flops of the circuit for detecting the faults of a particular type, are denoted by $TC[i][j]$, where $0 \leq i < Num\_cubes$ and $0 \leq j < Num\_inputs$; $TC[i][j]$ denotes the binary value of jth input (Primary Input (PI) or Pseudo-Primary Input (PPI)) in the ith test cube, Num_cubes denotes the number of test cubes, and Num_inputs denotes the total number of PIs and PPIs.[1] We next define a flip-flop *justification cube*, $JC_v[j]$, which denotes the bit sequence for justifying a flip-flop j to a value v (0 or 1) through the functional paths, where $0 \leq j < Num\_inputs$.

A design fragment consisting of a single logic cone is provided in Figure 2 where the functional logic driving the rightmost flip-flop e is shown. This flip-flop can be justified to 1 by setting the leftmost flip-flop a to 1, or bc to 01. Similarly, $ab = 01$ or $ac = 00$ sets flip-flop e to 0. *The condition for justifying a flip-flop to a value is identical to the detection condition for the fault on the D-input of the flip-flop that is stuck-at the complementary value*; the activation of the fault necessitates the flip-flop to be justified to the value complementary to the stuck-at value, while there is no propagation requirement for the fault as the flip-flop is an observable point. In other words, the test cube for the s-a-0 fault on the D-input of e is $1xxxx$ or $x01xx$, denoting the condition for justifying e to 1.

---

[1]PPIs correspond to the output of the flip-flops that drive the combinational logic.

Similarly, the test cube for the s-a-1 fault on the D-input of e is $01xxx$ or $0x0xx$, denoting the condition for justifying e to 0, as illustrated in Figure 2. In this example, $JC_0[e] = 01xxx$ or $0x0xx$, and $JC_1[e] = 1xxxx$ or $x01xx$. The justification information for each flip-flop is actually embedded within the set of test cubes.

Single flip-flop conversion: There are two conditions to be satisfied in order to convert a flip-flop f to non-scan:

- $JC_v[f]$ can be merged[2] with $TC[i]$ for all i such that $TC[i][f] = v$. In other words, there must be no 0-1 conflicts between a test cube that requires f to be at v, and the condition for justifying f to v.

- $JC_v[f][f] = x$ for $v = 0$ and $v = 1$. In other words, the justification condition for f should not specify itself to a value, creating a circular dependency; if f is converted to non-scan, it can only be justified by controlling *other* scan flip-flops.

For the example above, flip-flop e can be converted to non-scan if the first condition is met, as the second condition is satisfied; neither $JC_0[e]$ nor $JC_1[e]$ require e to be specified. If all the test cubes that require e to be at 0 merge with $JC_0[e]$, and all the test cubes that require e to be at 1 merge with $JC_1[e]$, then e can be converted to non-scan. For instance, a test cube $0x010$, which specifies e as 0, is compatible $JC_0[e] = 01xxx$. Therefore, if 0101 is loaded into the other flip-flops a, b, c and d in four shift cycles, a subsequent justify pulse received by only e would load 0 into e, delivering the desired bits of the test cube into all the flip-flops.

Pair conversion: It is possible that two flip-flops that can be converted individually cannot be converted together due to conflicting justification conditions. Next, we define the conditions for converting two flip-flops f_1 and f_2 simultaneously:

- Single flip-flop conversion conditions are met for both f_1 and f_2.

- $JC_{v_1}[f_1]$ can be merged with $JC_{v_2}[f_2]$, if $\exists_i$ such that $TC[i][f_1] = v_1$ and $TC[i][f_2] = v_2$. In other words, if the two bits corresponding to f_1 and f_2 are both specified by a test cube i, then the associated justification cubes of f_1 and f_2 should be non-conflicting.

- $JC_v[f_1][f_2] = x$ and $JC_v[f_2][f_1] = x$ for $v = 0$ and $v = 1$. In other words, the justification cube for either flip-flop should not specify the other flip-flop, creating a circular dependency; if both flip-flops are converted non-scan, they can only be justified by controlling *other* scan flip-flops.

The next question that can be raised is how can we maximize the number of flip-flops converted to non-scan, as commensurate benefits in area cost, test time, test data volume, and test power dissipation will be reaped. The single flip-flop

---

[2]Two cubes can be merged together if and only if the two cubes never have complementary values in the same bit position.

978-1-61284-657-6/11 $26.00 © 2011 IEEE

conversion conditions can be used to identify all the candidate flip-flops that can potentially be converted, while pair conversion condition introduces a notion of compatibility between two flip-flops. This compatibility notion can be extended to a group of flip-flops as follows.

Group conversion: A group of flip-flops f_k can all be converted to non-scan if the following conditions hold:

- Single flip-flop conversion conditions are met for each flip-flop in the group.

- For each of the test cube $TC[i]$ that specifies some of the bits in the group, the justification cubes corresponding to the specified flip-flops should all be non-conflicting, and thus, mergeable.

- The justification cube for none of these flip-flops should specify any other flip-flops in the group.

It can be observed that group conversion is a direct extension of pair conversion. If pair conversion conditions are met for every pair of flip-flops within a group, then the group conversion conditions automatically hold. The underlying reason is the natural extension of pairwise to group compatibility of cube merge operations; for instance, if cubes c_1 and c_2, c_1 and c_3, and c_2 and c_3 can merge, then c_1, c_2 and c_3 can all merge together.

The problem of identifying a maximal-sized group of flip-flops that can all be converted to non-scan can thus be mapped to the maximum independent set problem [35]. A conflict graph can be formed, wherein the nodes correspond to the flip-flops that satisfy the single flip-flop conversion conditions. An edge denoting a conflict is inserted between two nodes that fail the pair conversion conditions. A maximal-sized group of independent nodes[3] represents all pairwise compatible flip-flops, namely, a group of flip-flops that can all be converted to non-scan. Since the independent set problem is known to be NP-Complete, efficient heuristics can be utilized to identify near-optimal solutions.

The test cube analysis to create the conflict graph, on which the maximum independent set algorithm is executed, is illustrated on an example with 18 test cubes and seven flip-flops in Figure 3. Out of the seven flip-flops, only two of them, b and e, cannot meet the single flip-flop conversion conditions; $JC_0[b] = 01xx1x0$ requires b to be specified, and $JC_0[e] = 0xx1xx1$ cannot be merged with the fourth test cube $x0x10x0$, which specifies e to be 0. The conflict graph is thus formed with five nodes corresponding to the remaining flip-flops. In this graph, nodes a and g are conflicting, as $JC_0[a] = x1xxxx1$ specifies g. Also, a and f are conflicting, as the test cube $11x101x$ specifies both a and f as 1's, and $JC_1[a] = xxx1xx0$ and $JC_1[f] = x10x0x1$ cannot merge. The only pair of flip-flops that are compatible are a and c, as no test cube specifies both of them at the same time, and as their justification cubes do not specify each other. As a result,

---
[3]An independent group of nodes denote a group of nodes with no edge connecting any node to any other node in the group.

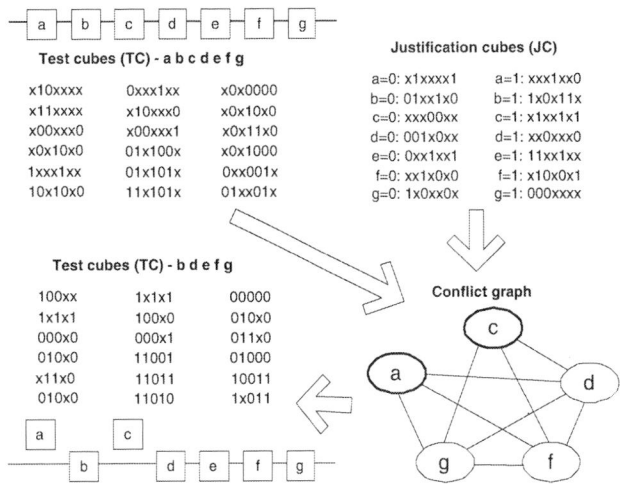

Figure 3. Test cube analysis mapped to the independent set problem

the maximum independent set is a and c, both of which can be converted to non-scan by removing the two scan multiplexers.

The same figure also shows the bits to be loaded into the scan flip-flops b, d, e, f, and g; these new cubes are obtained by merging the original test cubes with the justification cube of the non-scan flip-flop specified by the test cube, and by removing the bits of a and c. Each of the new test cubes requires five shift cycles, as opposed to seven, and a subsequent justify pulse received only by a and c load the required values into these non-scan flip-flops. During shift cycles, only five flip-flops (and their clock lines) potentially toggle, while the other two flip-flops preserve their values throughout the shift cycles as they not clocked during this period of time.

4 Experimental Results

We have implemented the proposed test cube analysis tool and applied it on a variety (ISCAS89 and ITC99) of academic benchmark circuits, and in this section, we present the results, which mainly consist of the number of flip-flops that can be converted to non-scan **without losing any fault coverage**. We have executed our tool with the test cubes of stuck-at faults, while we note that this analysis can be applied with any underlying fault model.

Table 1 provides the results of the proposed partial scan scheme. The first two columns provide the name of the benchmark circuit and the number of flip-flops, while column 3 presents the number of flip-flops that satisfy the single flip-flop conversion conditions and can thus be converted to non-scan individually; this number denotes the number of nodes in the conflict graph of the proposed test cube analysis. Column 4 presents the number of flip-flops that can be converted to non-scan altogether, while column 5 provides the same number in percentage with respect to the number of flip-flops, and column 6 provides the run-time of the analysis. The number

| | | Proposed scan to non-scan conversion | | | |
Circuit	Flip-flops	Single	Multiple	(%)	Run-time (s)
s713	19	6	6	**31.6**	<1
s953	29	23	23	**79.3**	<1
s1423	74	2	2	**2.7**	<1
s3271	116	6	3	**2.6**	<1
s3330	132	93	52	**39.4**	<1
s3384	183	111	46	**25.1**	<1
s4863	104	102	48	**46.2**	<1
s5378	179	130	72	**40.2**	<1
s6669	239	193	86	**36.0**	<1
s9234	228	22	17	**7.5**	<1
s13207	669	283	202	**30.2**	5
s15850	597	72	50	**8.4**	2
s35932	1728	42	41	**2.4**	7
s38417	1636	514	312	**19.1**	215
s38584	1452	65	43	**3.0**	13
b20	490	352	111	**22.7**	203
b21	490	359	160	**32.7**	208
b22	735	461	202	**27.5**	579

Table 1. Single and multiple flip-flop conversion.

given in column 4 denotes the size of the maximally-sized independent set in the conflict graph. For s5378, for instance, the proposed test cube analysis shows that 130 out of 179 flip-flops satisfy the single flip-flop conversion conditions, and can be converted to non-scan; 72 of these 130 flip-flops can be simultaneously converted to non-scan, as this group of 72 flip-flops (40.2%) satisfies the group conversion conditions.

It is important to note that the percentage flip-flop conversion ratio, provided in column 5, also denotes expected reductions in **test time, test data volume, and test power with respect to full-scan**, while retaining fault coverage intact.[4] It is difficult, however, to quantify the exact area cost savings, as the cost of the observation points depends on the scan configuration (number of POs, chains, and the compactor/MISR, if any); we expect the savings due the scan multiplexers removed by the proposed scheme to outweigh the cost of observation points, leading to some overall area savings.

The results show that the proposed test cube analysis approach is capable of converting 30-40% of flip-flops to non-scan for seven circuits, while the conversion percentage is poor (2-3%) in four circuits, from which two are small and two are among the largest, deducing no direct conclusions regarding the effectiveness versus size. For one circuit, 23 out of 29 flip-flops are converted, resulting in almost 80% conversion ratio. For the remaining six circuits, the proposed tool attains around 8% conversion for two of the circuits, and 19-28% for the other four. The effectiveness of the proposed test cube analysis approach directly depends on the care bit distribution in test cubes and justification cubes, which reflects the cone structure and input-output connectivity of the design.

---

[4]Proposed scheme may increase the care bit ratio upon cube merge operations, and may degrade the effectiveness of x-fill techniques in reducing scan-in power.

5 Future Research Work

The results presented in the previous section can be potentially improved further. One degree of freedom that can be exploited is the multiplicity of different conditions to justify a flip-flop to a value; the stuck-at fault at the input of the flip-flop can have multiple cubes detecting it. Various approaches can be taken to benefit from such a flexibility. The "best" cube can be selected for each justification condition in an effort to include more nodes (more flip-flops satisfying the single flip-flop conversion conditions) in the conflict graph or to have fewer edges (more pairs of flip-flops satisfying the pair conversion conditions) in the conflict graph. Alternatively, multiple justification cubes can be utilized for each flip-flop to increase the chances of satisfying the single flip-flop conversion conditions; as long as a compatible justification cube can be identified for each test cube, the single flip-flop conversion conditions are satisfied. In such a case, the pair and group conversion conditions need to be revised properly. Furthermore, techniques incorporated into ATPG so as to properly manipulate decision order may help produce test cubes that increase the effectiveness of the proposed partial scan scheme.

In this paper, we strictly adhere to the constraint that all test cubes can still be applied even after the conversions, perfectly preserving the fault coverage. This constraint can be relaxed to tolerate only a minor coverage loss, but in return to eliminate many conflicts in the graph, thereby increasing the number of scan to non-scan conversions.

Another direction is the re-formulation of the problem to prioritize performance savings. By removing the scan multiplexers from the critical paths, the functional performance of the design can be enhanced timing-wise. For this purpose, the proposed framework can be extended to incorporate a timing analysis, prioritizing the removal of all the multiplexers on critical paths, in addition to maximizing the number of other flip-flops that can be converted together.

6 Conclusions

Partial scan has been abandoned by the industry, as it necessitates sequential ATPG to recover the controllability and observability loss. Full scan has been adopted instead, despite the area, performance, and test costs it incurs. In this paper, we propose a test cube analysis driven partial scan scheme. The proposed technique operates only on a set of test cubes generated by a combinational ATPG tool, and identifies a maximum number of flip-flops that can be converted to non-scan while delivering the quality of full scan.

By identifying the conditions to recover the controllability and observability compromised due to partial scan, and by formulating these conditions via test cube operations, we enable a computationally efficient partial scan scheme that is compatible with the conventional ATPG flow. Upon the identification of the flip-flops that can be converted to non-scan, the test cubes are post-processed to ensure the delivery of the original set intact into all the flip-flops. This simple post-processing step eliminates the need for an ATPG re-run.

978-1-61284-657-6/11 $26.00 © 2011 IEEE

The proposed partial scan scheme combines the simplicity of the conventional (full scan-based) ATPG flow, and the area, performance, test time, and test power reduction benefits of partial scan. The removal of scan multiplexers delivers area as well as performance savings, while the shortening of the scan path translates into test time and data volume reductions. Furthermore, as the non-scan flip-flops are inactive during shift operations, power dissipation in the scan path, in the combinational logic, and in the clock tree are all reduced. The proposed partial scan scheme can be applied in conjunction with test compression and test power reduction techniques to drive the test costs down even further.

References

[1] P. Girard, "Survey of low-power testing of VLSI circuits," *IEEE Design Test of Computers*, vol. 19, no. 3, pp. 80–90, May. 2002.

[2] J. Rearick, "The case for partial scan," *International Test Conference*, p. 1032, nov. 1997.

[3] P. Ashar and S. Malik, "Implicit computation of minimum-cost feedback-vertex sets for partial scan and other applications," *Design Automation Conference*, pp. 77–80, Jun. 1994.

[4] S.T. Chakradhar, A. Balakrishnan, and V.D. Agrawal, "An exact algorithm for selecting partial scan flip-flops," *Design Automation Conference*, pp. 81–86, Jun. 1994.

[5] K.-T. Cheng and V.D. Agrawal, "A partial scan method for sequential circuits with feedback," *IEEE Transactions on Computers*, vol. 39, no. 4, pp. 544–548, Apr. 1990.

[6] K.-T. Cheng, "Single clock partial scan," *IEEE Design Test of Computers*, vol. 12, no. 2, pp. 24–31, 1995.

[7] V. Chickermane and J.H. Patel, "An optimization based approach to the partial scan design problem," *International Test Conference*, pp. 377–386, Sep. 1990.

[8] V. Chickermane and J.H. Patel, "A fault oriented partial scan design approach," *International Conference on Computer-Aided Design*, pp. 400–403, Nov. 1991.

[9] R. Gupta and M.A. Breuer, "The ballast methodology for structured partial scan design," *IEEE Transactions on Computers*, vol. 39, no. 4, pp. 538–544, Apr. 1990.

[10] A. Kunzmann and H. J. Wunderlich, "An analytical approach to the partial scan design problem," *Journal of Electronic Testing: Theory and Applications*, vol. 1, pp. 163–174, 1990.

[11] D.H. Lee and S.M. Reddy, "On determining scan flip-flops in partial-scan designs," *International Conference on Computer-Aided Design*, pp. 322–325, Nov. 1990.

[12] J. Park, S. Shin, and S. Park, "A partial scan design by unifying structural analysis and testabilities," *International Symposium on Circuits and Systems*, vol. 1, pp. 88–91, 2000.

[13] S.-E. Tai and D. Bhattacharya, "A three-stage partial scan design method using the sequential circuit flow graph," *International Conference on VLSI Design*, pp. 101–106, Jan. 1994.

[14] M. Abramovici, J.J. Kulikowski, and R.K. Roy, "The best flip-flops to scan," *International Test Conference*, p. 166, Oct. 1991.

[15] V. Boppana and W.K. Fuchs, "Partial scan design based on state transition modeling," *International Test Conference*, pp. 538–547, Oct. 1996.

[16] P. Kalla and M. Ciesielski, "A comprehensive approach to the partial scan problem using implicit state enumeration," *IEEE Transactions on Computer-Aided Design of Integrated Circuits and Systems*, vol. 21, no. 7, pp. 810–826, Jul. 2002.

[17] K.S. Kim and C.R. Kime, "Partial scan by use of empirical testability," *International Conference on Computer-Aided Design*, pp. 314–317, Nov. 1990.

[18] P. S. Parihk and M. Abramovici, "Testability-based partial scan analysis," *Journal of Electronic Testing: Theory and Applications*, vol. 7, pp. 47–60, Aug. 1995.

[19] G.S. Saund, M.S. Hsiao, and J.H. Patel, "Partial scan beyond cycle cutting," *International Symposium on Fault-Tolerant Computing*, pp. 320–328, Jun. 1997.

[20] E. Trischler, "Incomplete scan path with an automatic test generation methodology," *International Test Conference*, pp. 153–162, 1980.

[21] D. Xiang, S. Venkataraman, W.K. Fuchs, and J.H. Patel, "Partial scan design based on circuit state information," *Design Automation Conference*, pp. 807–812, Jun. 1996.

[22] D. Xiang and J.H. Patel, "A global algorithm for the partial scan design problem using circuit state information," *International Test Conference*, pp. 548–557, oct. 1996.

[23] D. Xiang and J.H. Patel, "Partial scan design based on circuit state information and functional analysis," *IEEE Transactions on Computers*, vol. 53, no. 3, pp. 276–287, Mar. 2004.

[24] V.D. Agrawal, K.-T. Cheng, D.D. Johnson, and T.S. Lin, "Designing circuits with partial scan," *IEEE Design Test of Computers*, vol. 5, no. 2, pp. 8–15, Apr. 1988.

[25] M.S. Hsiao, G.S. Saund, E.M. Rudnick, and J.H. Patel, "Partial scan selection based on dynamic reachability and observability information," *International Conference on VLSI Design*, pp. 174–180, Jan. 1998.

[26] H.-C. Liang and C. L. Lee, "An effective methodology for mixed scan and reset design based on test generation and structure of sequential circuits," *Asian Test Symposium*, pp. 173–178, 1999.

[27] X. Lin, I. Pomeranz, and S.M. Reddy, "Full scan fault coverage with partial scan," *Design, Automation and Test in Europe*, pp. 468–472, 1999.

[28] I. Park, D. S. Ha, and G. Sim, "A new method for partial scan design based on propagation and justification requirements of faults," *International Test Conference*, pp. 413–422, Oct. 1995.

[29] S. Sharma and M.S. Hsiao, "Combination of structural and state analysis for partial scan," *International Conference on VLSI Design*, pp. 134–139, 2001.

[30] J.-Y. Jou and K.-T. Cheng, "Timing-driven partial scan," *International Conference on Computer-Aided Design*, pp. 404–407, Nov. 1991.

[31] D. Kagaris and S. Tragoudas, "Retiming-based partial scan," *IEEE Transactions on Computers*, vol. 45, no. 1, pp. 74–87, Jan. 1996.

[32] O. Khan, M. L. Bushnell, S. K. Devanathan, and V. D. Agrawal, "Spartan: A spectral and information theoretic approach to partial scan," *International Test Conference*, p. Paper 21.1, 2007.

[33] N.A. Touba, "Survey of test vector compression techniques," *IEEE Design Test of Computers*, vol. 23, no. 4, pp. 294–303, Apr. 2006.

[34] Z. You, J.H., M. Inoue, J. Kuang, and H. Fujiwara, "A response compactor for extended compatibility scan tree construction," *International Conference on ASIC*, pp. 609–612, Oct. 2009.

[35] R. E. Tarjan and A. E. Trojanowski, "Finding a maximum independent set," *SIAM Journal of Computing*, vol. 3, pp. 537–546, 1977.

Memory-Based Embedded Digital ATE

[1]Dongsoo Lee, [1]Sang Phill Park, [2]Ashish Goel, and [1]Kaushik Roy

[1]Electrical and Computer Engineering, Purdue University, West Lafayette, IN 47907, USA

[2currently at] Broadcom Corporation, USA

{dslee, sppark, ashishg, kaushik}@purdue.edu

Abstract — This paper presents *memory-based embedded digital ATE* (Automatic Test Equipment) – a new logic BIST methodology that can deliver deterministic test stimuli and stores output responses on a chip. The proposed scheme consists of test data compression logic and a new on-chip SRAM structure, which is operated as a ROM when the logic BIST mode is on. The new BIST-oriented RAM (BRAM) implements ROM features in the BIST mode and incurs no performance penalty in the normal SRAM mode of operation. BRAM can be designed by inserting an additional word line in a row to a conventional SRAM bit-cell (no increase in bit-cell area). BRAM stores the compressed test vectors that can be transmitted to on-chip decompressors during test mode. BRAM also accepts compacted output responses. Experimental results show that BRAM performs stable and high-performance ROM operations in the BIST mode. Run-length coding can be incorporated into the proposed test data compression to reduce test data volume further. Test data volume and fault coverage on ISCAS89 benchmark show that the proposed test methodology can be used as a stand-alone BIST scheme while providing test quality of deterministic testss.

I. INTRODUCTION

Scaling of transistor dimensions and higher levels of integration has led to higher performance and reduced cost per transistor. However, the test and verification cost increased steadily over the past several years. To ease testing and to reduce test cost, Design For Testability (DFT) methodologies have been extensively studied. Among them, scan design and automatic test pattern generation (ATPG) have been widely adopted as reliable and efficient test techniques. Despite providing high fault coverage, large memory requirement and limited interface for an external automatic test equipment (ATE) to deliver test stimuli and store responses, demand alternative techniques such as test data compression [1-3] and built-in self-test (BIST) [4], [5].

Conventional logic BIST architectures assume a pseudo-random pattern generator (PRPG) and output response analyzer (ORA) [4]. Researchers have proposed linear feedback shift register (LFSR) [2] and ring generators [3] as efficient PRPG architectures with low area overhead. Ideal BIST architectures require only one start signal at an external interface, resulting in large reduction of external test cost. However, due to random pattern resistance faults, PRPG cannot achieve high enough fault coverage [4]. To increase test quality, test point insertion [4] and hybrid BIST techniques [5] have been suggested. The problem with test point insertion is that it is intrusive to the circuit under test (CUT), leading to negative impacts on area and delay. Hybrid BIST reduces the amount of test data and increases fault coverage using deterministic information from an external ATE [5]. Note that it is not a stand-alone methodology and assumes special control signals and interface.

Another major problem with previous BIST approaches is that any unknown (X) values fed into an ORA may corrupt the signature produced by the ORA. Hence, there is a need for bounding the number of X's, X-masking logic before ORAs, or X-tolerant ORAs [3], [4]. Indeed, industrial designs using BIST show that they need significant amount of time and manpower to make a circuit BIST-ready, while still demanding considerable assistance of ATPG for test stimuli [4].

Test data compression techniques have been suggested as a practical solution to reduce test cost [1-3]. The rationale behind this approach is that test vectors generated by ATPG contain a large number of unspecified bits which can be assigned at random. Decompressors receive compressed test stimuli as inputs and provide all-specified test vectors to the scan chains. Compression ratio is determined by the portion of unspecified bits and the capability of decompressors to generate various output combinations. However, a stand-alone BIST methodology is still needed for numerous applications which require high reliability and a diagnosis function at the system level.

The key innovation of this paper is the concept of a new logic BIST methodology called memory-based embedded digital ATE. The proposed scheme stores deterministic test data into a new on-chip SRAM structure, which plays the role of a ROM in the BIST mode operation. This new SRAM architecture, called the BIST-oriented RAM (BRAM), implements ROM features by adding an additional word line in a row and deciding proper via positions based on the test data to be stored. Note that a single mask layer of via layout is able to control test data in BRAM. Moreover, BRAM does not incur any performance loss in the normal SRAM mode operation. An array of BRAM can be used as standard SRAM memory array once the BIST operation is complete. The array of BRAM receives compacted test responses from the CUT while the next BRAM array continues to deliver compressed test stimuli to on-chip decompressors. This feature enables the proposed scheme to provide not only high fault coverage but also better diagnosability compared to previous logic BIST architectures. Note that BRAM is more than a simple ATE since it allows at-speed operations and complex synchronization due to the predictable latency between on-chip memory and test logic. In addition, rapid increase of on-chip SRAM memory space in scaled technology generations and advanced test data compression techniques make the proposed test scheme very effective in future generation processors. The proposed scheme also can be very efficient in multi-core architectures since the same set of test vectors in BRAMs can be used multiple times.

It is worth noting that few approaches to combining RAM and ROM features into a design are available in the literature [6-9]. Stacked bitlines scheme, based on multiple interconnect

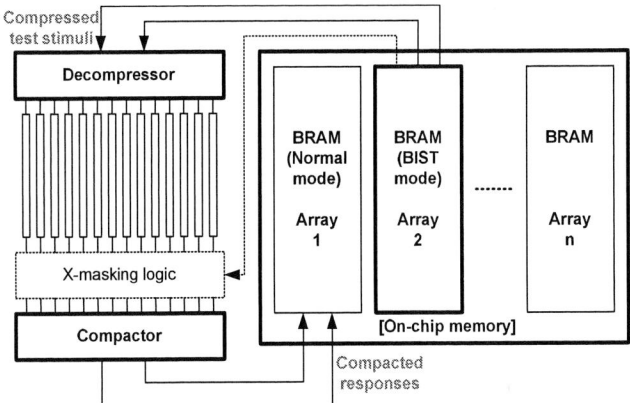

Figure 1. Memory-based embedded digital ATE architecture.

layers, allows the intermixing of SRAM and ROM in a memory array [6]. In [7], multiple supply power rails are selectively connected to SRAM bit-cells to determine ROM data. Or additional transistors are added to bit-cells for the same purpose in [8] and [9]. All of these works call for large area and/or performance penalty in a significant way. On the other hand, BRAM is able to maintain the same bit-cell area and performance of conventional SRAMs.

Stand-alone self-test at the board level without an external ATE has been suggested by researchers. An on-line self-test scheme using non-volatile memory as test data storage and architectural support has been proposed in [10]. Dedicated programmable chips for testing a target chip have also been proposed [11]. However, these methods are restricted to board level only with the assumption of having special external interface. Note that our proposed test scheme is not only valid on-chip but also suitable for board-level design/test.

The rest of this paper is organized as follows. Section 2 presents the basic idea and the overview of the proposed test scheme. The proposed memory structure, BRAM, is introduced in Section 3. Design considerations of BRAM are discussed in Section 4. Section 5 describes a linear test data compression architecture/algorithm which is combined with run-length coding to reduce test data to be recorded in BRAM. We conclude this paper in Section 6.

II. BASIC IDEA

The proposed logic BIST architecture consists of a decompressor, an output response compactor, and arrays of BRAM, as shown in Fig. 1. If we need to block unknown values produced by the CUT, an X-masking logic can be inserted between the CUT and the compactor. A decompressor accepts a small number of inputs from BRAM and produces much larger number of outputs to be fed into the internal scan chains. Note, the volume of test data to be stored is highly dependent on the test data compression algorithm. If the size of BRAM is larger than or equal to the volume of compressed test data, the fault coverage of the proposed logic BIST methodology is equal to that of test data compression techniques that use an external ATE.

BRAM stores the compressed test data in the BIST mode of operation. Note that BRAM also performs SRAM functions in the normal mode. It also follows the layout topology and design considerations of conventional SRAMs. The schematic

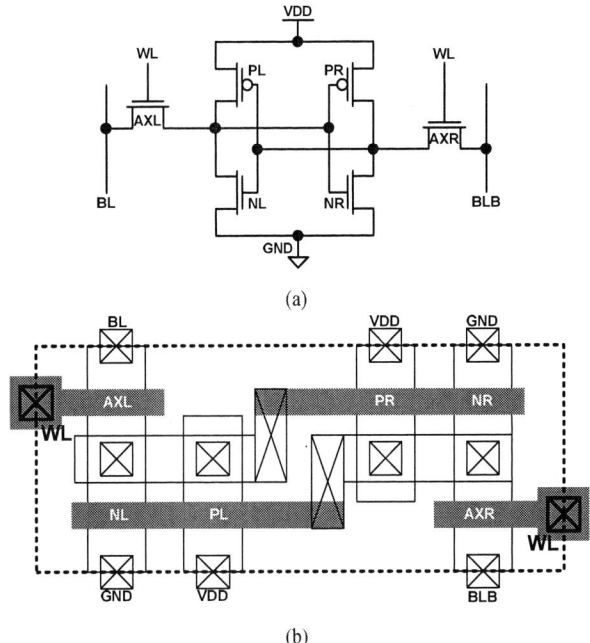

Figure 2. The schematic and thin-cell layout of conventional 6T-SRAM.

and thin-cell layout of a conventional 6T SRAM bit-cell is shown in Fig. 2 [19]. Note that the contact for wordline (WL) signal is shared by two neighboring SRAM bit-cells. For each row, there is one metal line of WL. Hence, gate signals of all the access transistors in the same row are turned on and off simultaneously. The layout topology including sizing of each transistor of BRAM bit-cell remains the same as that of the conventional 6T SRAM bit-cell.

The major difference from the conventional SRAM is that BRAM contains two WLs for each row. A gate signal of each access transistor is connected to one of two WLs. Note that two-poly thin-cell layout topology of SRAM allows two metal lines (metal-3 in general) of WLs to be embedded without any area penalty on the bit-cell. Depending on the values to be stored as test data, the corresponding via (between metal-2 and metal-3) positions are determined. In the normal SRAM operation, the two WLs are always turned on and off simultaneously. On the other hand, only one WL is turned on during the write procedure in the BIST mode so as to selectively write values to the cells where the access transistor is connected to the corresponding WL. Note that one mask of via layer determines test data, and hence, the design complexity is similar to standard 6T memory.

Once the BRAM array applies the test stimuli to the CUT, it returns to normal SRAM mode of operation, and records output responses compacted by a space compactor, such as an X-compactor [12]. The output response from the BRAM can be observed through a simple scan chain. It is also possible to embed a multiple input signature register (MISR) in order to minimize the volume of output responses to be observed.

III. BIST-ORIENTED MEMORY STRUCTURE (BRAM)

As noted earlier, the BRAM is used for both normal and test mode of operations. In this section we consider the different operation mode of BRAM.

978-1-61284-657-6/11 $26.00 © 2011 IEEE

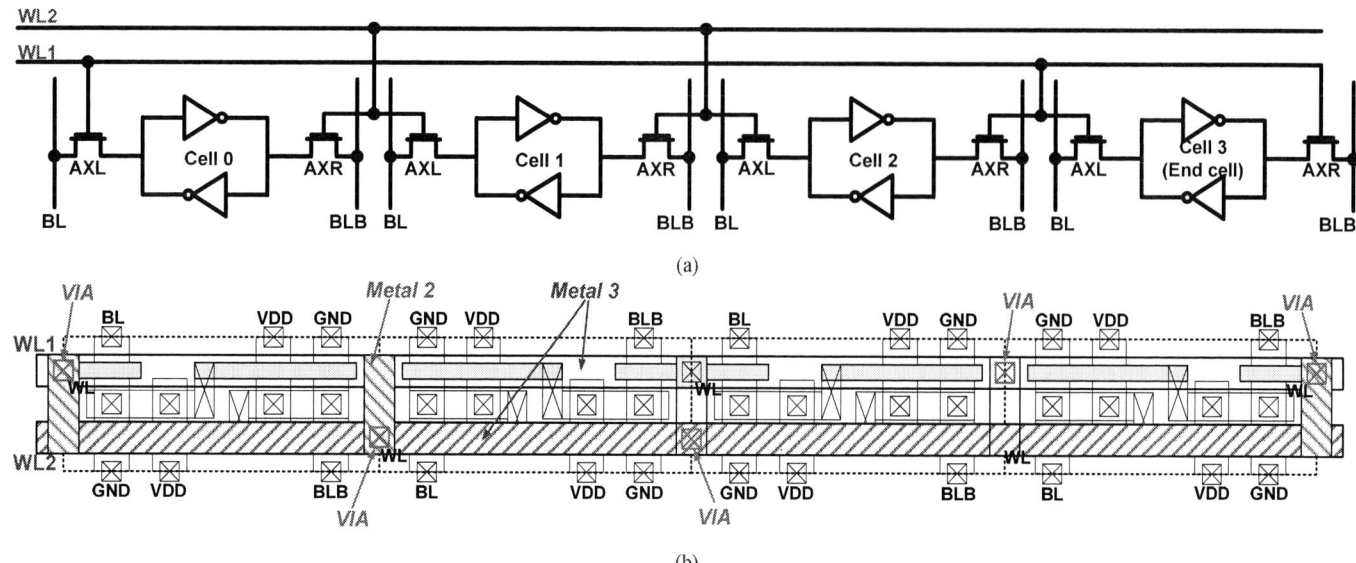

(a)

(b)

Figure 3. Schematic and layout of BRAM bit-cells storing '1001'.

A. Normal SRAM mode of operation

Conventional 6T thin-cell layout has enough routing space for two separate wordlines (WL1 and WL2) since the height of memory cell is determined by the pitch of two poly-silicon lines [19]. As a result, we can easily split the conventional wordline into two without changing the sizing of transistors in the 6T bit-cell.

In the normal operations of SRAM array, the two wordlines (WL1 and WL2) are switched simultaneously for both read and write operations. This leads to the same functions as the conventional 6T SRAM array.

B. BIST mode of operation

After writing test stimuli into BRAM, the read operation of BRAM in the BIST mode is performed in the same way as that of conventional 6T SRAM array. Writing test stimuli using the two wordlines is done in two steps:

Step 1: Write 1's to all the bit-cells while WL1 and WL2 are turned on. (BL=1, BLB=0, WL1=WL2=1)

Step 2: Write 0's to all the bit-cells with WL1 off and WL2 on. (BL=0, BLB=1, WL1=0, WL2=1)

An example schematic and layout of BRAM to store test stimulus of '1001' is shown in Fig. 3. After step 1, the row stores '1111'. Following step 2, the row of Fig. 3 stores 1001

since only the left access transistors (AXLs) of cell 1 and 2 are turned on.

As seen from the figure, the sizing of the cells need not be modified and can be the same as conventional 6T memory array. Only the gates of access transistors are selectively connected to WL1 or WL2 depending on the test stimuli to be read in the BIST mode. In the case of a bit-cell storing a '1'('0'), AXL is driven by WL1(WL2). The right access transistor (AXR) is driven by the wordline connected to AXL in the right-side neighboring cell. The last cell, which is at the end of each row of the array, AXL and AXR are connected together.

IV. DESIGN CONSIDERATIONS: BRAM

Compared to conventional SRAM design, the proposed BRAM has the following issues:

- If two consecutive bit-cells have different test data to be stored, step 2 of BRAM write procedure performs write operation as a 5T SRAM cell, since only one access transistor is turned on. Such writing can lead to a "write stability" problem in the bit-cells.
- Redundancy technique, which is widely used to correct defective SRAM cells in arrays, is not applicable to BRAM in the BIST mode of operation.
- BRAM requires two wordline drivers to drive two separate wordlines. This may lead to area/power overhead.

In this section, we analyze above issues and present solutions.

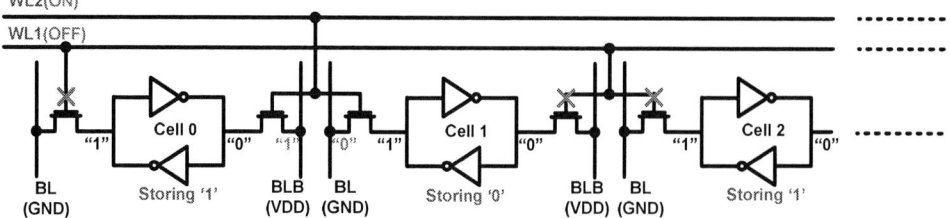

Figure 4. An example for step 2 in Section 3 when the test data to be stored is '101'.

978-1-61284-657-6/11 $26.00 © 2011 IEEE   268

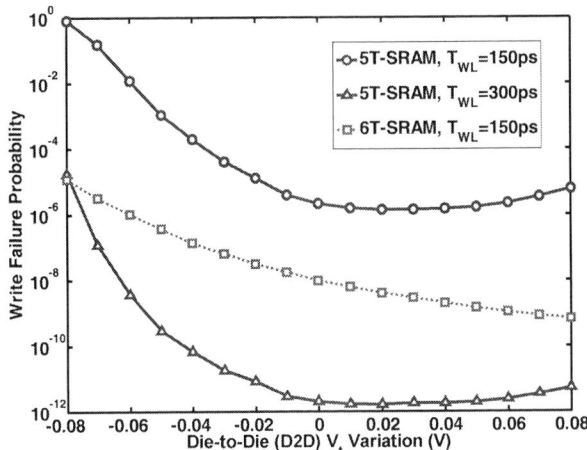

Figure 5. Write-'0' failure probability with die-to-die (D2D) V_t shift.

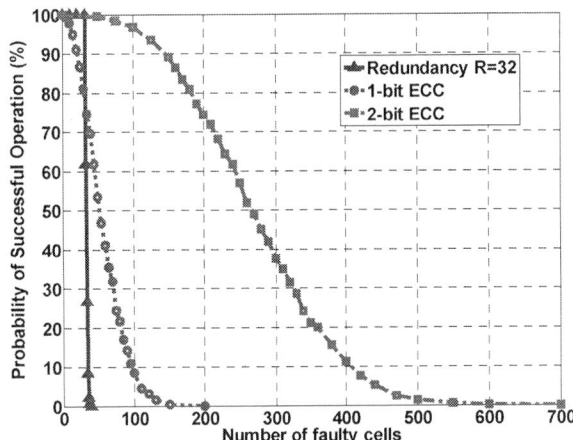

Figure 6. Successful operation probability with various number of faulty cells.

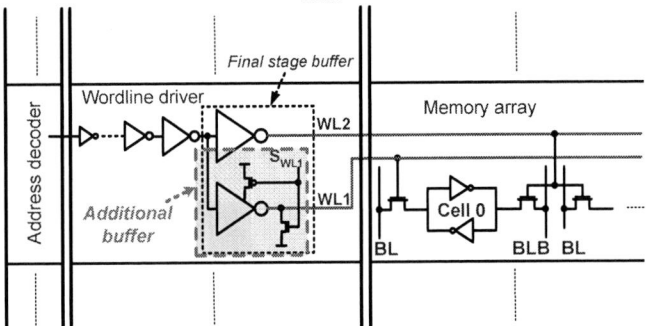

Figure 7. Wordline driver design for BRAM with additional final-stage buffer.

A. Selective flipping data in BRAM bit-cells – Write stability

Lets us consider the write stability problems using an example. Fig. 4 shows an example of write operation (step 2 of section 3) when three consecutive bit-cells store '101' as test stimulus. First, all the bit-cells are written to '111' during step 1 (conventional 6T write operation). Then, only cell 1, which is mapped to '0' in the test stimulus, is re-written through AXL in step 2 (equivalent 5T write operation for "0": BL=0, AXL=on, AXR=off). Note, cell 0 has to resist being re-written by BLB (5T write-"1": BLB=1, AXL=off, AXR=on) since we do not want to flip cell 0 during step 2. In other words, writing a '0' should occur successfully and writing a '1' should **not** occur. Difficulty in write-"1" through a NMOS access transistor is a well-known bottleneck for 5T SRAMs [13]. Interestingly, the bottleneck is a desired requirement in our BRAM.

Due to the random variations in process parameters in the nanometer regime, SRAMs suffer from mismatches in the strength between the neighboring transistors. Such mismatches can lead to access failure (the cell cannot be read within the access time), read failure (the cell data is flipped during read operation – destructive read), and write failure (the cell data cannot be written successfully) [14]. Note, BRAM operates under the same conditions as 6T SRAM except for the write operation in the step 2. Hence, the failure probabilities of write-'0' and write-'1' cases in the step 2 have to be analyzed to ensure correct BIST mode of operation.

To investigate the write-'0' failure probability in 5T SRAM, we follow the modeling of failure probabilities given in [14] and [15]. The write failure probability (P_{WF}) is expressed as

$$P_{WF} = P\left(T_{WRITE} > T_{WL}\right) = 1 - \Phi_{T_{WRITE}}\left(T_{WL}\right)$$

where T_{WRITE} is the required time for successful write, T_{WL} is the duration when WL is high, and Φ_x is the cumulative distribution function of x. We conducted circuit simulations using HSPICE and PTM 32nm models [16] assuming 0.9V of supply voltage and 40mV of standard deviation of transistor threshold voltage (V_t). We assume that the transistor widths are (Fig. 2) 50nm for PL and PR, 100nm for NL and NR, and 75nm for AXL and AXR, respectively.

Fig. 5 shows the write-'0' failure probability with various die-to-die V_t variations for 5T SRAM and 6T SRAM. As can be observed, write-'0' failure probability of 5T SRAM is

higher than that of 6T SRAM if T_{WL} is the same. However, successful write-'0' through single access transistor can be ensured by simply extending T_{WL} as shown in Fig. 5. The two write steps in the BIST mode can be done with a slower clock speed – note that the test compression logic is inactive during the write.

We also obtained the write-'1' failures under different process corners – simulation results show that none of the cases lead to successful write-'1' operation. As mentioned earlier, unsuccessful write-'1' in 5T SRAM is a requirement for correct operation of BRAM in step 2.

B. ECC technique to improve yield

Redundancy techniques, that are used to improve yield of SRAMs, can also be used in BRAM in the normal mode of operation. Noting that BRAM functions as a ROM in the BIST mode, redundancy scheme cannot be applied to BRAM under BIST operation. It is because re-mapping of column or row addresses in a ROM is not possible. Since error-correction coding (ECC) is an effective way to improve yield of a ROM [16], ECC can be used in BRAM in the BIST mode of operation. Fig. 6 shows the probability of successful operation of a memory array versus three different fault tolerant architectures – (a) 32 redundant rows, (b) 1-bit ECC, and (c) 2-bit ECC. We assumed a 64Kbytes memory array which has 512 rows, 256-bit block size, and 4 blocks for each row. To apply 1-bit ECC to each block, additional 10 parity bits are added to each block. This means block size is shrunk to 246 bits (ROM size becomes 61.5Kbytes). In the case of 2-bit

978-1-61284-657-6/11 $26.00 © 2011 IEEE   269

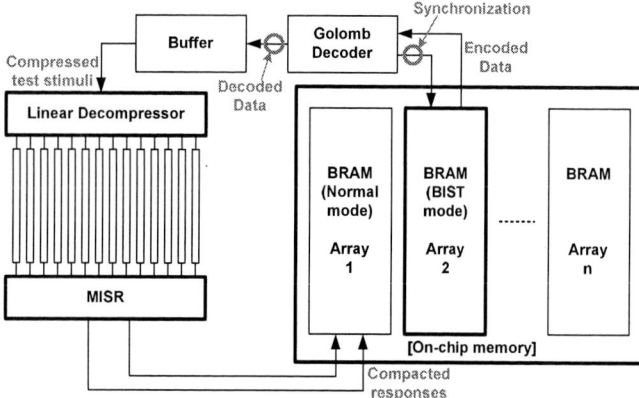

Figure 8. Proposed logic BIST architecture to combine linear decompressor and Golomb coding using BRAM.

ECC, additional 18 bits are required for each block, leading to 238-bit block size (ROM size becomes 59.5Kbytes). As can be seen from Fig. 6, ECC can improve yield of memory significantly. Note, ECC logic can be implemented in pipelined memory read path in the proposed BIST architecture, ensuring the same throughput during test.

C. Wordline driver design

The proposed BRAM requires two wordline drivers to supply two wordlines with the same drive strength of conventional SRAMs. However, additional wordline driver may lead to large area/power overhead. Instead, the same drive strength for each wordline is achieved by addition of a final-stage buffer in a single wordline driver as shown in Fig. 7. The drive strength of the final buffer is determined by the worst case test stimuli patterns such as all zeros or ones (all access transistors in a row are connected to either WL1 or WL2, respectively). Considering the worst case, the additional buffer in BRAM has the same size of the final-stage buffer as in conventional arrays. S_{WL1} in Fig. 7 controls the power supply gate of the final buffer to turn off WL1 only during step 2 in the BIST mode. SRAM array layout from a commercial memory compiler shows wordline driver taking less than 10% of the total SRAM area. The area of final stage buffers is approximately 20% of a wordline driver. Hence, the area increase due to the additional buffers in our BRAM can be less than 2% in comparison to the same size of SRAM array.

V. LOGIC BIST ARCHITECTURE BASED ON BRAM, LINEAR DECOMPRESSOR AND RUN-LENGTH CODING

Test cubes generated by automatic test pattern generator (ATPG) usually have only small portion of specified bits. The unspecified bits can be assigned at random [1]. Depending on how the random bits are assigned 0/1 values, test compression architectures can be broadly classified into three types: code-based, linear-decompression-based, and broadcast-scan-based schemes [1]. The linear-decompression-based techniques show high encoding efficiency with simple on-chip decompressors (linear decompressors consist of XOR gates and Flip-Flops) [3]. Compressed test data can be obtained by solving linear equations [2]. Whenever test cubes are merged by compaction algorithms, merged test cube is checked to determine whether it can be compressed by linear decompressors (otherwise, the compaction of corresponding merged test cube is abandoned). If there is no more remaining test cube to be merged, linear decompressors generate an all-specified test vector. Then, we perform a fault simulation and drop all detected faults from the fault list. Therefore, the next merging procedure is done using reduced number of undetected faults. As a result, the number of detected faults and specified bits decreases in test cubes which are merged later.

Based on this observation, the compressed test vector is selected to involve as many zeros as possible. Many zeros give us an opportunity to adopt run-length coding, which maps variable-length runs of zeros in a compressed test vector to variable-length codewords. In our proposed logic BIST architecture, run-length coding performs an additional compression to further reduce the test volume before linear decompression. Golomb coding is a well-known run-length coding which provides good compression ratio with low hardware resources [18]. If a certain bit of a test vector compressed by a linear decompressor is zero, the counter for the number of zeros counts up. When the next bit is one, the counter is encoded by Golomb coding and the corresponding codeword is stored into the BRAM. After storing the codeword, the counter is reset to zero and we repeat the procedure for all bits in a test vector. The length of codeword of high run-length of zeros is shortened by Golomb coding.

Note that code-based test compression algorithms based on invariable codeword, including Golomb coding, demand a synchronization signal for communication with an external

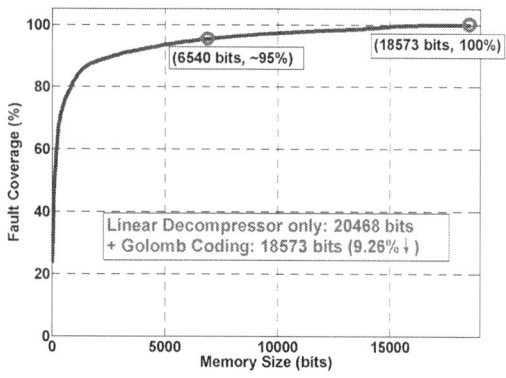

Figure 9. Fault coverage of s38584 using proposed architecture.

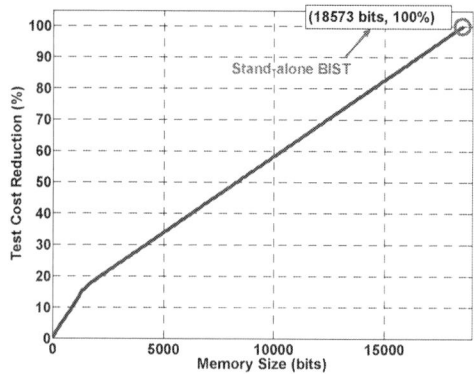

Figure 10. External test cost reduction of s38584 using proposed architecture.

TABLE I. EXPERIMENTAL RESULTS ON ISCAS89 BENCHMARK CIRCUITS.

Circuit	Scan cells	N_{out}	vec	vol1	vol2	m
s5378	214	10	173	4844	4386	4
s9234	247	6	241	11568	10409	4
s13207	700	21	311	12440	11779	4
s15850	611	13	215	11395	10880	4
s35932	1763	32	43	2709	2524	4
s38417	1664	15	302	35334	32520	4
s38584	1464	24	301	20468	18573	4

ATE. However, the latency of the synchronization signal from Golomb decoder to an ATE can be unpredictable since it is difficult to anticipate timing of I/O pads and signals outside the chip. On the other hand, the synchronization signal can be easily employed in the proposed test methodology since the latency between on-chip BRAM and decompressors can be estimated during design time. The entire logic BIST architecture based on linear decompressor and Golomb coding is shown in Fig. 8. Golomb decoder is implemented using a counter and a finite-state macine (FSM) [18]. If the size of a word in BRAM is larger than the number of inputs of decompressor, a buffer is inserted in between Golomb decoder and a linear decompressor.

We implemented the proposed test methodology (linear-decompression-based architecture and Golomb coding) in C programming language. Atalanta [20] ATPG tool was used to generate test cubes for the ISCAS89 benchmark circuits. Bit-stripping technique [21] was used to erase unnecessarily specified bits. Hope [22] was used as a fault simulator to remove detected faults from the fault list. For all benchmark circuits, we detected all detectable faults. Linear decompression algorithm tries to find a solution which has high run-length of zeros in compressed test vectors. However, Golomb coding may suffer from long size of encoded data due to low run-length of zeros. Note that the number of possible compressed test vectors increases if merged test cubes are more sparsely specified. Therefore, we applied Golomb coding to last 20% of compressed test vectors which have small number of specified bits. Fig. 9 describes fault coverage of s38584 using the proposed architecture when the group size for Golomb coding is 4. If we assume that BRAM size is larger than or equal to 18573 bits, we achieve 100% stuck-at fault coverage, resulting in stand-alone BIST. Golomb coding provides additional reduction on test volume by 9.26% compared to linear decompression only. Compared to the case when we utilize linear decompressor and external ATE, the test cost reduction is presented in Fig. 10. For all circuits, Golomb coding reduces test volume as shown in Table I, where N_{out} is the ratio of the number of outputs to the number of inputs in a linear decompressor, vec is the number of test vectors, vol1 is the test volume when we use a linear decompressor, vol2 is the test volume when we use a linear decompressor and Golomb coding, and m is the group size for Golomb coding.

VI. CONCLUSION

We present a new logic BIST methodology that consists of test data compression logic and a new on-chip SRAM, which can store and deliver the test data without external support of

ATEs. The new SRAM, called BRAM, does not incur area penalty on a bit-cell.

As the size of on-chip memory increases and more cores are incorporated into a chip, the proposed test methodology provides higher fault coverage, reducing the external tester cost. Experimental results on ISCAS89 benchmark circuits show that the proposed test methodology can provide fault coverage similar to that of deterministic tests, if memory size is large enough.

Acknowledgement: The research was funded in part by Semiconductor Research Corporation and by National Science Foundation under Grant No. CCF-1018205.

REFERENCES

[1] N. A. Touba, "Survey of test vector compression techniques," *IEEE Design & Test of Computers*, vol. 23, pp. 294-303, April 2006.

[2] B. Koenemann, "LFSR-coded test patterns for scan designs," *Proc. European Test Conference*, pp. 237-242, 1991.

[3] J. Rajski, J. Tyszer, M. Kassab, and N. Mukherjee, "Embedded deterministic test," *IEEE Trans. CAD*, vol. 23, pp. 776-792, May 2004.

[4] C. Hetherington *et al.*, "Logic BIST for large industrial designs: Real issues and case studies," *Proc. ITC*, pp. 358-367, 1999.

[5] A. Jas, C. V. Krishna, and N. A. Touba, "Weighted pseudorandom hybrid BIST," *IEEE Trans. VLSI Systems*, vol. 12, pp. 1277-1283, Dec. 2004.

[6] T. L. Brandon, D. G. Elliott, and B. F. Cockbum, "Using stacked bitlines and hybrid ROM cells to form ROM and SRAM-ROM with increased storage density," *IEEE Trans. Circuits Syst. I, Reg. Papers*, vol 53, pp. 2595-2605, Dec. 2006.

[7] T. Matsumura and M. Yoshimoto, "Semiconductor memory device usable as static type memory and read-only memory and operating method therefor," U.S Patent 5 365 475, Nov. 15, 1994.

[8] G. M. Ansel *et al.*, "Read only/random access memory architecture and methods for operating the same," U.S. Patent 5 5880 999, Mar. 9, 1999.

[9] S. M. Gold and M. Lamere, "Combining RAM and ROM into a single memory array," U.S. Patent 6 438 024, Aug. 20, 2002.

[10] Y. Li, S. Makar, and S. Mitra, "CASP: concurrent autonomous chip self-test using stored test patterns," *Proc. DATE*, 2008, pp. 885-890.

[11] A. W. Hakmi, H.-J. Wunderich, V. Gherman, M. Garbers, and J. Schloffel, "Implementing a scheme for external deterministic self-test," *Proc. VLSI Test Symp.*, pp. 101-106, 2005.

[12] S. Mitra and K. S. Kim, "X-Compact: an efficient response compaction technique," *IEEE Trans. CAD*, vol. 23, pp. 421-432, Mar. 2004.

[13] S. Nalam and B. H. Calhoun, "Asymmetric sizing in a 45nm 5T SRAM to improve read stability over 6T," *Proc. CICC*, 2009, pp. 709-712.

[14] S. Mukhopadhyay, H. Mahmoodi, K. Roy, "Modeling of failure probability and statistical design of SRAM array for yield enhancement in nanoscaled CMOS," *IEEE Trans. CAD*, vol. 24, pp. 1859-1880, Dec. 2005.

[15] P. Ndai, A. Goel, and K. Roy, "A scalable circuit-architecture co-design to improve memory yield for high-performance processors," *IEEE Trans. VLSI*, vol. 18, pp. 1209-1219, Aug. 2010.

[16] T. Shinoda, Y. Ohnishi, and H. Kawamoto, "A 1Mb ROM with on-chip ECC for yield enhancement," *ISSCC Dig. Tech. Papers*, 1983, pp. 158-159.

[17] Predictive Technology Model (PTM), Nanoscale Integration and Mudeling (NIMO) Group, ASU, 2007 [Online]. Available: www.eas.asu.edu/~ptm

[18] A. Chandra and K. Chakrabarty, "System-on-a-chip test-data compression and decompression architectures based on Golomb codes," *IEEE Trans. CAD*, vol. 20, pp. 355-368, Mar. 2001.

[19] M. Khare *et al.*, "A high performance 90nm SOI technology with 0.992μm^2 6T-SRAM cell," *IEEE IEDM Dig. Tech. Papers*, 2002, pp. 8-11.

[20] H. K. Lee and D. S. Ha, "On the generation of test patterns for combinational circuits," Technical Report 12_93, Depart. Of Electrical Eng., Virginia Polytechnic Institute and State University.

[21] R. Sankaralingam and N. A. Touba, "Controlling peak power during scan testing," *Proc. VLSI Test Symp.*, pp. 153-159, 2002.

[22] H. K. Lee and D. S. Ha, "HOPE: An efficient parallel fault simulator," *Proc. DAC*, pp. 336-340, 1992.

A Unified Test Architecture for On-Line and Off-Line Delay Fault Detections

Songwei Pei [1,2], Huawei Li [1*], and Xiaowei Li [1]

[1] Key Laboratory of Computer System and Architecture, Institute of Computing Technology,
Chinese Academy of Sciences, Beijing 100190, China
[2] Graduate University of Chinese Academy of Sciences, Beijing, China
{peisongwei, lihuawei, lxw}@ict.ac.cn

Abstract—**This paper proposes a unified delay test architecture, in which the design resources for on-line delay fault detection can be reused to support off-line delay testing. A stability checker, which has low hardware overhead, is presented to monitor the stability violation from each critical combinational output. A global error generator, which is shared among stability checkers, can produce a *global error* signal from individual stability checkers to indicate whether a delay fault appears. A local scan enable generator is incorporated into the scan chain to support scan-based off-line delay testing. Experimental results are presented to validate the effectiveness of the proposed approach.**

Keywords- delay fault detection; stabilty checker;on-line testing

I.   INTRODUCTION

Rapidly shrinking of transistor dimensions has resulted in increased count of components being crammed onto modern integrated circuits (IC). Meanwhile, with the increasing prominent deep submicron (DSM) effects, timing-related defects are more prone to occur in ICs. Due to the enhanced controllability and observability for circuit internal signals, structural scan-based delay testing established itself as a cost-effective way for delay fault detection. Typically, a pair of test vectors <*V1, V2*> is required to detect a delay fault. The first vector *V1*, referred to as initialization vector, can be used to initialize the circuit to a predefined state via the scan chains. The second vector *V2*, referred to as launch vector, can be used to launch a transition at the target circuit line and propagate the corresponding fault effect to the observable outputs. The circuit response to the launch vector can then be at-speed captured to compare with the golden response to verify the correctness of circuit timing requirements [3].

Depending on the way of applying the second vector, two different approaches are widely practiced for standard scan-based designs [3]. In the first approach, referred to as launch-on-capture (LOC), the launch vector is obtained by capturing the circuit response to the initialization vector. In the second approach, referred to as launch-on-shift (LOS), the launch vector is obtained by one bit scan-in of the initialization vector. Figure 1(a) and Figure 1(b) illustrate the signal waveforms of the LOC and LOS approaches respectively. It should be noticed from Figure 1(a) that the LOC approach can provide enough time for the scan enable signal (*SEN*) to settle low after the last scan-in cycle. Therefore, the LOC approach has a

relaxed constraint on the *SEN* signal and hence has a low implementation cost. However, the delay fault coverage achieved by the LOC approach is modest due to the stringent functional dependency between the initialization and launch vector. It is important to note from Figure 1(b) that the *SEN* signal for the LOS approach should switch from high to low at functional frequency after launching the *V2* vector. Hence, the requirement of a timing critical *SEN* signal is apparently posing the great obstacle to adopt the LOS approach. However, the LOS approach has the advantage of providing better fault coverage as compared to the LOC approach. Moreover, it can typically achieve the same fault coverage to that of the LOC approach with significantly fewer test pattern counts [4].

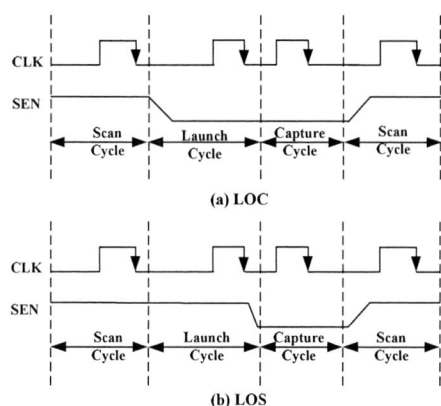

Figure 1. Scan-Based Delay Testing

A simple way to provide a timing critical *SEN* signal for supporting LOS delay testing is to design it as a clock tree. In [5], multi-stage scan enable signals are pipelined and the respective scan enable signal with satisfied timing constraint is then used to drive partial scan cells in local regions. In [6], a last transition generator (LTG) is designed and incorporated into the scan chains to generate local scan enable signals that can switch at functional frequency. In [7], a delay test scan flip-flop with a clock alignment logic added into the standard muxed-D scan cell is proposed to support LOS delay testing. However, all of the above techniques suffer from considerable extra hardware overheads and design efforts.

On the other hand, it is well known that in deep sub-micron process technologies, the manufactured chips are more vulnerable to environmental influences, such as crosstalk, power supply noise, soft error etc. [2, 8]. However, noise induced faults typically depend on the activation conditions

---

* To whom correspondence should be addressed.

978-1-61284-657-6/11 $26.00 © 2011 IEEE

and hence can hardly be detected by off-line delay testing. Furthermore, aging effects can also degrade circuit performance continuously. Therefore, on-line delay testing plays an especially important role in detecting circuit timing failure during normal functional applications [1]. This is especially important for the critical applications.

Several techniques have been proposed to detect circuit timing failure caused by delay faults by constantly monitoring the stability violation of each circuit combinational output. In [10, 11], a two-transistor checker is designed to monitor the late transitions of a combinational output during a given checking period to detect timing failure. However, this technique suffers from a low noise margin. In [12], a sensing circuit is proposed to detect delay faults. Customized rationed capacitors are required to be implemented in this technique. Moreover, the skew happened between the two control clock phases may induce the fail detection of an existing delay fault. In [13], a concurrent checker is proposed, which enables for on-line timing error detection. By inserting the concurrent checker with an incorporated exclusive OR (XOR) gate at each combinational output, the delayed signal transitions after the sampling time can then be detected. However, the requirement of an extra XOR gate for each concurrent checker further burdens the hardware overhead of this approach. Moreover, an extra logic network with significant amount of area overhead is also required to analyze the error signals generated from each combinational output to produce a global error signal. In [14], the sensing checkers based on respective two-phase and signal-phase schemes are presented for on-line detection of delay faults. Similarly, the stability checking technique is adopted in [15, 16] to monitor the combinational output signal transitions during the guard band interval to predict circuit timing failure.

To summarize, there is an urgent need to develop effective on-line and off-line delay fault detection techniques to satisfy the growing demand for reliability of modern chips. However, traditionally the challenges of providing robust on-line delay testing and supporting effective LOS approach are treated separately. Therefore, only a cumbersome combination of the respective previous approaches can handle the two challenges, thereby resulting in a large area overhead and design complexity. Rather than exploiting separate solutions, there is a considerable interest in developing a unified delay test architecture with low hardware overhead to achieve the two goals.

In this paper, a unified delay test architecture is presented, which consists of stability checkers, a global error generator, and a local scan enable (*LSEN*) generator being embedded into the scan chain. It enables not only on-line delay fault detection during normal functional applications, but also off-line delay testing (either LOS or LOC) without truly implementing a timing critical *SEN* signal. Compared to the previous techniques, the proposed delay fault detection methods have much lower hardware overhead and design complexity.

The remainder of the paper is organized as follows. In Section II, the signal behavior for stability checking is analyzed. Section III describes the proposed test architecture and illustrates its applications to on-line and off-line delay testing. The experimental results are presented in Section IV. Section V concludes the paper.

II. SIGNAL BEHAVIOR ANALYSIS FOR STABILITY CHECKING

In this paper, we consider the designs with falling-edge-triggered flip-flops. However, the proposed technique can also be applicable to rising-edge-triggered flip-flop based designs with a little modification. As we know, a fully synchronous circuit is synchronized by a system clock signal, which can be denoted as *CLK*. A combinational logic output signal feeds to the input of a flip-flop.

Theoretically, for a fault free combinational logic output signal S, there exists at least one stable period, denoted as $T_S = (t_1, t_2)$, during which no transition would be occurred. The t_1 and t_2 represent the start and end times of the stable period respectively. The existence of a stable period is due to the fact that the combinational logic output signals should keep stable at least for a flip-flop setup time before the falling edge of system clock to capture valid data in the fault free case. Further, all combinational outputs will remain stable for the following flip-flop's clock-to-q time plus the minimum circuit propagation delay after the sampling instant [10].

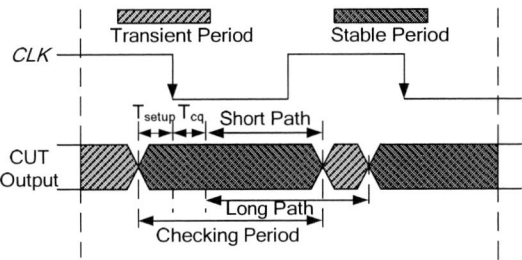

Figure 2. Combinational Output Signal Waveform

Mathematically, the stable period T_S can be expressed as:

$$T_S = ((T_c - T_{setup}), (T_c + T_{cq} + T_{commin})) \quad (1)$$

Where T_{setup} is the setup time of flip-flop, T_{commin} is the propagation delay of the shortest path in the circuit, T_{cq} is the flip-flop's clock-to-q time, and T_c is the instant when the falling edge of system clock arrives.

The timing waveform and the corresponding stable period for a fault free combinational output signal are shown in Figure 2. If the circuit under test operates correctly, all the combinational outputs should remain stable during the stable period. Otherwise, at least one combinational output would transits during this period. Hence, the fundamental principle of the stability checking for delay fault detection is to observe the stability violations of combinational outputs during the stable period. In practice, a signal which is used to indicate the stable period is indispensable. If an extra signal is customized for this purpose, significant hardware overhead would be incurred. To address this problem, the negative half cycle of system clock itself can be used as a cost-effective way to indicate the checking period for signal stability checking [11]. However, it is important to note from (1) that the propagation delay of the circuit shortest path should satisfy the following constraint to avoid false positive induced by the short path effects [11, 16] :

$$T_{commin} > T_{clk}/2 \quad (2)$$

Where T_{clk} represents the clock period of *CLK*. This requirement is also consistent with various timing optimization

978-1-61284-657-6/11 $26.00 © 2011 IEEE   273

methods to design high performance chips [9, 16, 17]. In practice, this requirement can be alleviated by adjusting the duty cycle of *CLK*.

III. THE PROPOSED DELAY TEST ARCHITECTURE

The key principle of the proposed method is to detect delay faults by constantly monitoring the stability violations of combinational outputs during the checking period. First, we propose the unified delay test architecture and implementations of its components. Then, the applications of the proposed test architecture to on-line and off-line delay testing are discussed.

A. General view and implementations

Figure 3 shows the proposed unified delay test architecture, which consists of two major parts:

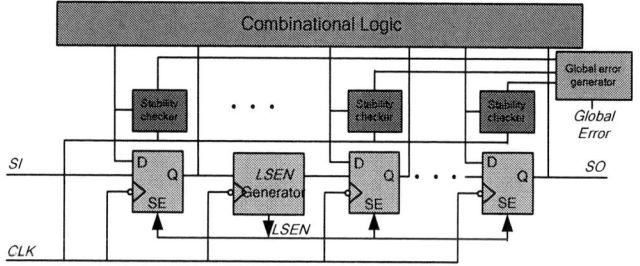

Figure 3. The Proposed Unified Delay Test Architecture

- **(1) Stability Checker and Global Error Generator.**

Figure 4(a) is the transistor level implementation of the stability checker, which can be inserted at each critical circuit combinational output to detect the corresponding signal's stability violation during the checking period. Figure 4(b) is the transistor level implementation of the global error generator, which can be used to generate a *global error* signal to indicate the circuit timing failure when any stability violation is committed.

The circuit structure of the stability checker is similar to that of the split-output latch, where the transistor gated by the clock signal is inserted into the output of the inverter [18]. A similar structure is also previously employed by the concurrent checker in [13] to detect delay faults, in which the inconsistence of states between the pair of circuit nodes isolated by a clock gated transistor is used to indicate the occurrence of a delay fault. The same technique is also employed to design the novel stability checker in this work. However, there are several advantages as compared to the concurrent checker [13] in terms of hardware overhead and design effort. The following presents the operational principles and the new features of the designed stability checker and the global error generator.

When the *CLK* signal has a logic high value, the transistor of *M1* with a low threshold voltage would be conducted. Hence *S1* and *S2* would have the same logic values, 1s or 0s, equaling the negative of the combinational output CO_1. When *CLK* switches to a logic low value, either *S1* or *S2* would become floating and keep its previous value (the negative of CO_1), while the other node has a logic 1 or 0 by connecting to either *VDD* or *GND* depending on the logic value of CO_1. Meanwhile, if a transition occurs at the combinational output

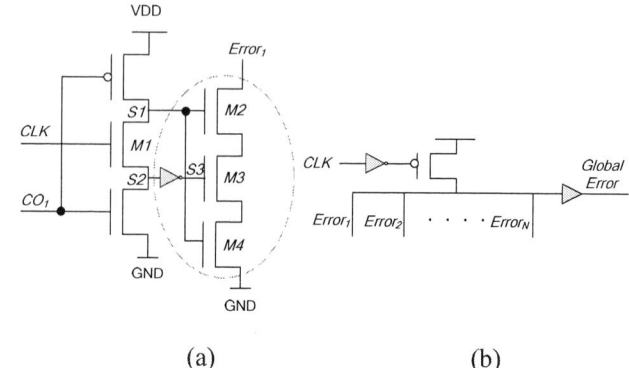

(a)   (b)

Figure 4. (a) Stability Checker   (b) Global Error Generator

CO_1 when *CLK* is 0, the value of the floating node would be changed by connecting to either *VDD* or *GND*, while the other node would become floating.

Clearly, when the *CLK* signal has a logic high value, a high impedance path would be created between $Error_1$ and *GND* due to the different logic values of *S1* and *S3* (inversion of *S2*). The $Error_1$ node would then be charged to logic high value by using the global error generator as shown in Figure 4(b). If transitions are occurred at CO_1 during the checking period (negative half cycle of *CLK*), both *S1* and *S3* would then have logic high values, and hence the $Error_1$ would be discharged to logic low value due to the created short path between *GND* and $Error_1$. Otherwise, the logic value of $Error_1$ would keep the previously logic high value because of floating.

Consequently, the logic low value of $Error_1$ can be used to indicate the stability violation of the combinational logic output CO_1. For each critical combinational output CO_i ($1 \le i \le N$), a stability checker can be inserted and the $Error_i$ can be used to drive the *global error* signal. The de-assertion of the *global error* signal can then indicate that at least one of combinational outputs suffers from delay faults.

The following presents two new features of the stability checker and global error generator:

1. Based on the key observation of that if transitions are occurred at a combinational output during the checking period of *CLK*, both the states of *S1* and *S3* in the corresponding stability checker, which are initialized with different states when *CLK* is 1, would change to the same and become high. Hence, $Error_i$ signal, which drives the *global error* signal, can be realized by creating the pull-down network using the transistors of *M1*, *M2*, and *M3* as shown in Figure 4(a). Therefore, unlike the concurrent checker, an additional XOR gate which typically consumes 12 transistors [19] can be avoided in the stability checker to indicate a stability violation for each combinational output. Such implementation brings significant reduction of hardware overhead

2. It traditionally requires high hardware overhead to design a logic network for producing a global timing error signal from the large numbers of local error signals generated by individual combinational outputs. However, it is important to note that either short path or high impedance would be created between *GND* and $Error_i$ in the designed stability checker depending on whether the signal arrived late or not. Hence the $Error_1$, $Error_2$... and $Error_N$ can be directly connected to the global error generator to generate a *global*

978-1-61284-657-6/11 $26.00 © 2011 IEEE   274

error signal with attractive hardware overhead and design complexity.

- **(2) Local Scan Enable (*LSEN*) Generator.**

It is usually preferred to use a slow speed scan clock provided by an external ATE to load the test stimuli and unload the test response and use a high speed functional clock generated by the on-chip PLL to perform the at-speed delay testing. Typically, the *SEN* signal conducts itself as the selection signal to choose the proper test clock. When the *SEN* signal is 0, the high speed functional clock is selected to send on the system clock tree. Otherwise, the low speed scan clock is sent on the system clock tree [20]. Thereby, traditionally it can hardly guarantee that the time interval between launch and capture edges of test clock to satisfy the at-speed frequency property for LOS delay testing due to the different selected clock sources.

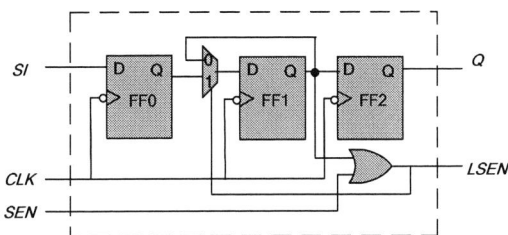

Figure 5. Schematic of *LSEN* Generator

Figure 5 shows the *LSEN* generator, which adds a *FF0* from the LTG [6] to avoid the state restriction posed by FF1 to the previous regular scan cell of the LTG in the scan chain. As shown in Figure 3, rather than using the *SEN* signal, the *LSEN* signal generated from the *LSEN* generator can be used to control the scan cells in the proposed test architecture.

The operational principle of the *LSEN* generator can be referred from [6]. The following presents the fundamental differences for adopting the *LTG* cell in [6] and the *LSEN* generator in this work.

The purpose of incorporating the *LTG* cell into the scan chain in [6] is to generate multiple local *LSEN* signals that can switch at functional speed to support LOS delay testing. Hence, design iterations are indispensable to add enough *LTG* cells to be incorporated into the scan chains to satisfy the tight timing constraint. Therefore, the design process would become very complex. Moreover, the large numbers of extra *LTG* cells would suffer from significant hardware overhead. In this work, however, the LOS delay testing can be conducted by observing the stability violations of combinational outputs rather than to capture the test response triggered by launch vector with a timing critical scan enable signal. Therefore, actually, lots of extra *LSEN* generators can be abandoned from the proposed unified architecture based LOS delay testing. The reason of introducing only one *LSEN* generator incorporated into the scan chain is to facilitate selecting high speed functional clock with *SEN* signal to guarantee the at-speed frequency property between launch and capture edges for LOS delay testing. The generated *LSEN* signal can be used to control scan cells to shift one bit from V_1 to obtain the launch vector V_2. Clearly, the hardware overhead that only one *LSEN* generator being incorporated into the proposed architecture can be ignored.

Detailed description of the proposed unified delay test architecture will be given in the following subsection.

B. *The Proposed Unified Architecture Based Delay Testing Methods*

As mentioned in the previous sections, the proposed unified architecture targets for effective on-line and off-line delay fault detection. The following analyzes the proposed unified architecture based delay testing methods.

- **On-line delay testing:** During the normal functional application, the *SEN* signal is set to 0. Hence, the high speed functional clock provided by the on-chip PLL is sent on the clock tree and the combinational outputs are captured into the data inputs of scan cells. Thereby, if a stability violation is occurred at a combinational output during the checking period caused by various reasons, such as crosstalk noise, power supply noise, and soft error, etc., the corresponding stability checker and the global error generator will be triggered. Hence, the *global error* signal will be discharged to 0 to indicate the occurrence of the circuit timing failure.

- **Off-line delay testing:** As mentioned before, the requirement of a timing critical scan enable signal poses the great obstacle to adopt the traditional LOS delay testing. The following describes the alternate way to implement LOS delay testing.

Figure 6 shows the timing diagram of LOS delay testing based on the proposed architecture. When the *SEN* signal is 1, the low speed scan clock is employed to shift the test vector V1 into the scan chain. The FF1 of *LSEN* generator gets 1 and the *SEN* signal switches from 1 to 0 after the last shift-in cycle. Hence, the *LSEN* signal keeps a logic high value during the

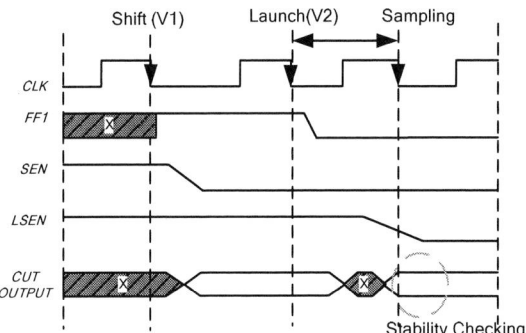

Figure 6. Timing Diagram of LOS Delay Testing

launch cycle and the test vector *V2* can be obtained by shifting 1 bit from *V1* by using the launch edge selected from the high speed functional clock. Although the *LSEN*, which is generated by *FF1+SEN*, is not timing critical and cannot switch at functional speed between the launch and sampling cycles, the existence of delay fault can also be indicated by the *global error* signal after observing any stability violation from combinational outputs during the checking period. Further, unlike the traditional LOS delay testing, it does not require to compare the test response with golden counterpart by using the proposed approach due to the fact that the existence of delay faults can be directly indicated by the *global error* signal. The scenario for LOC delay testing based on the proposed architecture is similar but by shifting 0 to FF1 of *LSEN* generator in the last shift-in cycle. It should be noted that the traditional LOC delay testing and stuck-at fault testing can also be conducted in the traditional way by ignoring the proposed test architecture.

IV. EXPERIMENTAL RESULTS

The proposed unified architecture, which enables both on-line and off-line delay testing, was implemented using SMIC 90nm CMOS technology [19] in this work. The experimental results consist of three main parts: 1) simulated waveforms for on-line delay testing; 2) simulated waveforms for off-line delay testing; 3) the hardware overhead of the proposed architecture.

Figure 7 shows the signal waveforms of the proposed architecture for on-line delay testing, which are obtained by HSPICE simulation.

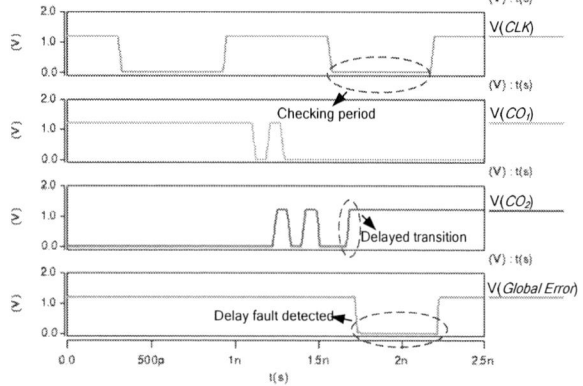

Figure 7. Simulated Waveform for On-Line Delay Testing

The propagation delay of the shortest circuit path is assumed larger than a half cycle of functional clock. For the sake of graphic clarity, only two combinational output signals, denoted by CO_1 and CO_2, are shown in Figure 7. It can be noted from Figure 7, during the circuit normal functional application, only occur delayed transitions at the combinational outputs during the checking period, the *global error* signal will be de-asserted to denote the occurrence of a timing failure.

Figure 8 shows the signal waveforms of the proposed architecture for LOS delay testing. Clearly, when the *SEN* signal is 1, the slow speed scan clock *SCLK* is sent on the system clock *CLK* for scanning-in the test vector *V1*. When the *SEN* signal switches to 0, the high speed functional clock *FCLK* is sent on the *CLK* for launching test vector *V2* and indicating the checking period. The launch to sampling edge of the *CLK* is therefore ensured to equal a functional clock period. During the last shift-in and launch cycles, 1 and 0 are scanned into to the FF1 of *LSEN* generator respectively. Thereby the *LSEN* signal switches from 1 to 0 only after the arrival of launch edge of *CLK*. The *V2* is 1 bit shifted from *V1* because the *LSEN* signal has logic 1 value before the arrival of launch edge. Although the *LSEN* signal switches from 1 to 0 very slowly after that, the delay fault is also detected by observing the delayed transition occurred at CO_2 during the checking period.

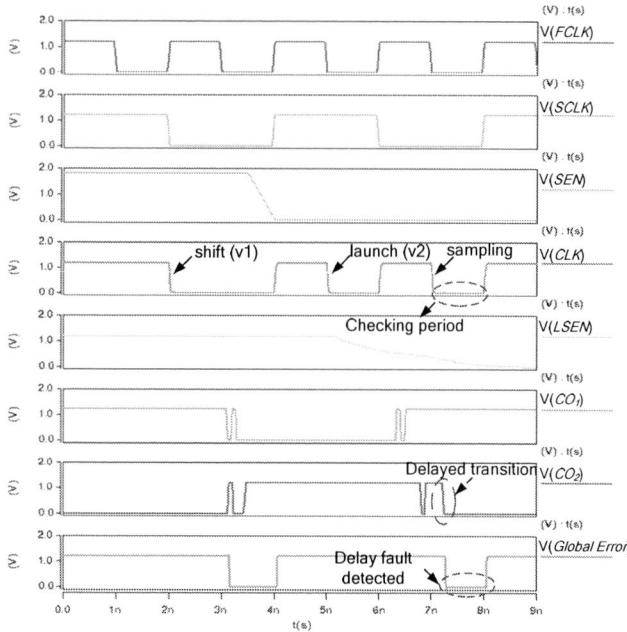

Figure 8. Simulated Waveform for Off-Line Delay Testing

The existence of a delay fault is indicated by the de-assertion of *global error* signal. The HSPICE waveform for LOC delay testing is omitted in the paper due to the similar scenario to that of the LOS approach.

In order to evaluate the hardware overhead, the proposed test architecture was incorporated into several IWLS 2005 full-scan based benchmark circuits respectively. Table I presents the benchmark circuit profiles and the experimental results for the proposed test architecture. The circuit names of the sample benchmark circuits are shown in column 1. The numbers of flip-flops and primary outputs are given in column 2 and column 3 respectively. The benchmark circuits are synthesized by using the Synopsys Design Compiler synthesis tool targeting the SMIC typical 90nm CMOS technology. The circuit clock names are listed in columns "clock domain 1" and "clock domain2", while the corresponding clock periods are listed in columns "P1" and "P2" respectively. The slack of the longest circuit path in each clock domain is equal to 10% of the clock period. The column under "#CE" represents the total number of critical outputs (primary output or pseudo primary output) for the benchmark circuits in all clock domains. In this experiment, if there exists a circuit path, which has a slack lesser than 20% of the corresponding clock period, ending at a circuit output, the circuit output is then considered as a critical output. The numbers of circuit critical outputs are obtained by using the Synopsys PrimeTime tool. For each critical output of the circuit under test, a stability checker is inserted. The last three columns give the area overhead of the proposed test architecture for each benchmark circuit, the total area of each benchmark after incorporation of the proposed test architecture, and the hardware overhead percentage of the proposed test architecture.

Figure 9 compares the hardware overhead of the proposed stability checker with that of concurrent checker (CC) [13], aging resistant stability checker (ARSC) [15], and stability violation based fault detection (SVFD) [16] in terms of

978-1-61284-657-6/11 $26.00 © 2011 IEEE

TABLE I. EXPERIMENTAL RESULT FOR THE PROPOSED TEST ARCHITECTURE

Circuit	#flip-flops	#PO	Circuit Clocks				#CE	Area Overhead (μm^2)	Total Area (μm^2)	Area Overhead (%)
			Clock domain1	P1 (ns)	Clock domain 2	P2 (ns)				
pci_bridge32	3313	208	wb_clk_i	1.2	pci_clk_i	2.3	1236	4070	99713	4.08%
usb_funct	1743	122	clk_i	1.1	phy_clk_pad_i	1.85	135	444	61427	0.72%
ac97_ctrl	2229	50	bit_clk_pad_i	1	clk_i	1.5	503	1656	62650	2.64%
mem_ctrl	1126	153	mc_clk_i	1	clk_i	1.6	334	1100	42629	2.58%
des_perf	8808	65	clk	1.1	--	--	1356	5058	405056	1.25%
aes_core	530	130	clk	1.6	--	--	50	165	69207	0.24%
wb_conmax	578	1416	clk_i	1.6	--	--	280	922	70139	1.31%
systemcaes	670	129	clk	1.75	--	--	183	603	35816	1.68%

transistor counts. Even though the numbers of transistors for delay element incorporated in the ARSC and the XOR protector adopted in the SVFD are ignored, the ARSC and SVFD also require at least 14 and 20 transistors respectively. In this comparison, the hardware overhead occupied by the logic network for producing a global error signal from individual combinational outputs is not taking into account. Or else, the proposed global error generator, which has very low hardware overhead and can be shared among the stability checkers, will make the hardware overhead of the proposed test architecture more attractive.

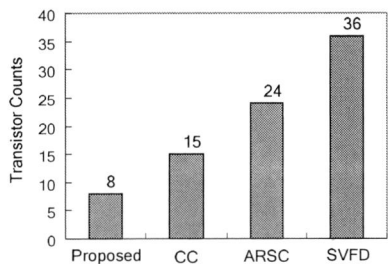

Figure 9. Comparisons

V. CONCLUSIONS

We described a unified test architecture for both on-line and off-line delay fault detections. The designed stability checker is inserted at each critical combinational output, which is used to detect the delayed transitions of combinational output signal during the checking period. The *error* signal generated by each stability checker can be directly connected to the global error generator to produce a *global error* signal for indicating the occurrence of delay faults. By shifting a logic ONE to the FF1 of the *LSEN* generator in the last shift-in cycle, LOS delay testing can be supported by the proposed architecture without the need of a timing critical scan enable signal. HSPICE simulation verified that the proposed design can support on-line and off-line delay testing effectively. Experimental results also show the proposed delay test architecture has a very low hardware overhead.

ACKNOWLEDGMENT

This work was supported in part by National Natural Science Foundation of China (NSFC) under grant No. (60776031, 60633060, 60921002), and in part by National Basic Research Program of China (973) under grant No. (2011CB302501, 2011CB302503).

REFERENCES

[1] C.Metra, M.Favalli, B.Ricco, "Self-checking detection and diagnosis of transient, delay, and crosstalk faults affecting bus lines ," IEEE Transactions on Computers,vol.49.no.6,pp.560-574,2000.

[2] H. Li, P.Shen, and X.Li, "Robust Test Generation for Precise Crosstalk-induced Path Delay Faults," Proceedings of VTS ,2006,pp. 300-305.

[3] A.Krstic and K.T.Cheng, "Delay Fault Testing for VLSI Circuits," Boston, MA:Kluwer,1998.

[4] S.Wang, X.Liu, and S.T.Chakradhar, "Hybrid delay scan: a low hardware overhead scan-based delay test technique for high fault coverage and compact test sets,"Proceedings of DATE, 2004, pp. 1296-1301.

[5] Synopsys Application Note, "Tutorial on Pipelining Scan Enables".

[6] N.Ahmed, C.P.Ravikumar, M.Tehranipoor, and J.Plusquellic, "At-Speed Transition Fault Testing With Low Speed Scan Enable," Proceedings of VTS, 2005, pp.1-6.

[7] G.Xu, A.D.Singh, "Low Cost Launch-on-shift Delay Test with Slow Scan Enable" Proceedings of ETS, 2006, pp. 9-14

[8] S.Murali, L.Benini, et al., "Analysis of error recovery schemes for networks on chips", IEEE Design & Test of Computers, Vol.22,No.5,pp.434-442,2005.

[9] N.V.Shenoy, R.K.Brayton, et al., "Minimum padding to satisfy short path constraints", Proceedings of ICCAD,pp.156-161,1993.

[10] P.Franco and E.J.McCluskey, "Delay testing of digital circuits by output waveform analysis", Proceedings of ITC, 1991, pp.798-807.

[11] P.Franco and E.J.McCluskey, "On-Line Delay Testing of Digital Circuits", Proceedings of VTS, 1994, pp.167-173.

[12] M.Favalli, P.Olivo, M.Damiani, and B.Ricco, "Novel design for testability schemes for CMOS IC's," IEEE J.Solid-State Circuits, vol.25,pp. 1239-1245, 1990.

[13] K.Raahemifar and M.Ahmadi, "Design-for-Testability techniques for detecting delay faults in CMOS/BiCMOS logic families", IEEE Transactions on circuits and systems-□Analog and digital signal processing, Vol.47, No.11, pp.1279-1290, 2000.

[14] M.Favalli and C.Metra, "Sensing circuit for on-line detection of delay faults," IEEE Transactions on VLSI,vol.4,no.1,pp.130-133,1996

[15] M.Agarwal, B.C.Paul,M.Zhang,and S.Mitra, "Circuit Failure Prediction and Its Application to Transistor Aging," Proceedings of VTS, 2007, pp. 277-286

[16] G.Yan, Y. Han, and X. Li, "A Unified Online Fault Detection Scheme via Checking of Stability Violation", Proceedings of DATE, 2009, pp. 496-501

[17] B.Taskin and I.S.Kourtev, "Delay insertion method in clock skew sheduling,"IEEE Transactions on CAD,vol.25,no.4,pp.651-663,2006.

[18] J.Yuan and C.Svensson, "High-Speed CMOS circuit Technique," IEEE JSSC, Vol.24,No.1, pp.62-70,1989.

[19] "SMIC 90nm LOGIC 90LL RVT 1.2V Advantage ™ v1.0 Standard Cell Library Databook,"January 2008, Revision 1.0.

[20] S.Pei, H. Li, and X.Li, "An On-Chip Clock Generation Scheme for Faster than-at-Speed Delay Testing ," Proceedings of DATE, 2010,pp.1353-1356.

978-1-61284-657-6/11 $26.00 © 2011 IEEE

Design For Bit Error Rate Estimation of High Speed Serial Links

Ujjwal Guin
Dept. of ECE
Temple University
Philadelphia, PA 19122
ujjwal.guin@temple.edu

Chen-Huan Chiang
Supply Chain Engineering
Alcatel-Lucent
Murray Hill, NJ 07941
chen-huan.chiang@alcatel-lucent.com

Abstract

High speed serial links, consisting of SerDes devices, require the Bit Error Rate (BER) to be at the level of 10^{-12} or lower. The excessive test time for comparing each captured bit for error detection in the traditional BER measurement and the costly instrumentation are major drawbacks for high volume production test of SerDes devices. In this paper, we propose a design for BER estimation methodology which includes a new BER estimation method, a simple BER test system which incorporates a novel design of time-to-digital converter (TDC).

Unlike the previous BER estimation methods [12, 13, 15], our proposed BER estimation method incorporates the total jitter (TJ) spectral information of the serial data with the TJ spectral information of the recovered clock, not with the jitter transfer characteristics of the CDR circuit. Hence, its accuracy does not depend on the deterministic jitter (DJ) in the serial data stream of the SerDes. Meanwhile, it still benefits from using the TJ spectral information for efficient BER estimation without excessive test time just as the previous BER estimation methods.

A novel TDC design is proposed for the implementation of the BER test system, where the TJ spectral information of a SerDes under test can be accurately estimated from the known TJ distribution of a golden SerDes. The TDC design measures the delay between the golden and test SerDes devices and converts it into the digital format to be used in the proposed BER estimation method. The experimental results demonstrate that the test time of the proposed BER test system is in the order of seconds, which translates into the test time savings of more than hundred times compared to the traditional BER measurement.

1 Introduction

The data transfer rate in the modern systems has increased significantly since the adoption of high speed serial communication links. For example, 40 Gbps (giga bits per sec) and 100 Gbps systems are available for the Gigabit Ethernet, whose parallel bus predecessor, Fast Ethernet, was at 100 Mbps in 1995. I/O buses are available up to 8 GT/s (gigatransfer per sec) bit rate for PCI express that was 133 Mbps for its parallel bus counterpart, PCI, in the early 90's. The backbone of the serial link architecture consists of the serializer and de-serializer (SerDes) devices, where the clock is embedded in the data and transmitted asynchronously

(self-synchronous links).

The bit error rate (BER) is a widely accepted quality measurement of serial links. The BER specification of a high quality SerDes is extremely small and is usually in the order of 10^{-12} or smaller. The test time to acquire such a small BER is very high and usually in terms of hours. It is because we need to compare each captured bit for error detection for as long as the time taken to capture at least 10 errors, i.e., $10 * 10^{12}$ bits of data.

Most of the BER Testers (BERTs) measure the total jitter (TJ) at a given BER using bathtub curves by integrating the extrapolated TJ probability density function(pdf) [15, 16]. Typical test times are approximately 20 minutes at 10 Gbps, and a little more than one hour at 2.5 Gbps, for a TJ measurement that was done with a confidence level of better than 90% at the 10^{-12} BER level [16]. It is a significant improvement for the BER estimation, compared to the traditional bit-by-bit comparison for calculating BER. However, it is still far from a solution to high volume testing of the SerDes.

To expedite the BER testing, a BER estimation technique for high-speed serial interfaces that incorporates a linear clock and data recovery (CDR) circuit was presented [13]. It utilizes the jitter spectral information extracted from the transmitted data and some key characteristics of the CDR circuit in the receiver. This technique, however, is not applicable to bang-bang (BB) CDR circuits [1, 12, 14], which are widely used in today's high speed serial link architectures because of their advantages in high speed implementations. It is because the BB phase detector in a CDR circuit behaves non-linearly with respect to the input jitter and its jitter transfer function varies significantly with respect to the jitter magnitude. To consider the non-linear BB CDR, another similar BER estimation technique was proposed in [12] based on the spectral information of jitter and the jitter transfer characteristics of the BB CDR circuit.

In this paper, we first propose a new BER estimation method incorporating the TJ spectral information of the serial data and the TJ spectral information of *the recovered clock*. Hence, its accuracy does not depend on the deterministic jitter (DJ) present in the serial data stream of the SerDes. In addition, because of using the TJ spectral information of the recovered clock, the proposed BER technique is independent of whether a linear or non-linear CDR phase detector is used. Secondly, we propose a simple test system consisting of a golden SerDes and a SerDes under test (which will be referred as *test SerDes*, where the TJ spectral information of the test SerDes can be accurately estimated from the known TJ distribution of a golden SerDes.

The primary challenge to implement such a proposed test system is the delay measurement between the golden SerDes and the test SerDes. In Gbps range, the unit interval (UI) is in several hundred picoseconds. Due to the jitter, the UI varies by a small amount and is in the order of picoseconds. A novel time-to-digital converter (TDC) is proposed to meet the challenge to measure the delay difference in the order of picoseconds between the golden SerDes and the test SerDes.

The design and implementation of the TDC are the most challenging issue in this research. The idea of using the principle of time-to-digital conversion for delay measurement has been introduced in many prior schemes. Time-to-digital conversion based on CMOS tapped delay lines was used in [5, 6, 9, 10, 17]. It employs a technique where the delay is converted into a pulse, which is used to charge a capacitor. The final voltage of the capacitor indicates the delay between the two signals. The main drawback is that the leakage of the charge in the capacitor can cause errors during the test. Again, low charging time of the capacitor makes the design susceptible to noise in the high frequency range for the jitter measurement.

A on-chip delay measurement based on timing characterization technique is presented in [18]. The propagation delay of the paths are converted into a pulse using a XOR gate and then sampled with a high-speed clock. A counter is incremented at the active clock edge when the output of the XOR gate is logic-1. The value of the counter gives the delay. To achieve a good resolution, say $\frac{1}{100}$, the synchronous clock must run at least 100 times higher than the system clock. This is practically impossible when the serial link is running in Gbps range.

In [2, 3], a single ended tapped delay line (SEDL) based on CMOS technology is presented. A differential or balanced tapped delay lines or vernier delay line (VDL) along with SEDL are presented in [7, 8]. A VDL circuit is made of two delay chains. The intermediate tap points are fed to the inputs (which could be the clock and data input of a D flip-flop) of a chain of storage elements. The main advantage of the VDL over the SEDL is the resolution of the delay measurement. The resolution the of SEDL is limited by the minimum achievable delay in given technology whereas the resolution of a VDL is determined by the delay difference between the delay elements in the two delay chains. However, the VDL circuit suffers from its size because of the large number of storage elements present in it. Furthermore, the size of the circuit will become larger if higher resolution is required.

We have developed two types of TDC designs [11]: one is binary SEDL and the other is binary VDL. Both are binary versions of the prior SEDL and VDL designs, meaning the delay at each stage is no longer a unit delay but a delay of 2^i buffers for the i-th stage. The delay measurement resolution of the former is equal to the buffer delay; while the resolution of the latter could be adjusted within the TDC design. The TDC design using binary SEDL and the detail description of how to consider the delay of each building block (such as buffer, MUX, and D flip-flop) for both types of TDC designs can be found in [11]. We will intro-

duce a 3-stage TDC using the binary VDL in this paper.

2 Proposed BER Estimation Method

For BER estimation using the widely accepted dual-Dirac model [15], the spectral information, i.e., the variance (σ^2) and mean (μ), of the TJ distribution are the parameters that must be estimated accurately. As a standard industry-wide assumption, the TJ is universally accepted to be modeled as a random variable. If we add any two independent random variables, their density functions will be convolved and the resultant distribution will be spread. In this research, we have selected *the subtraction of the TJ random variable of the golden SerDes and that of the test SerDes* as the resultant TJ distribution to model the difference of the golden and test SerDes devices. In our proposed test methodology, there are three major assumptions:

- *The TJ distribution of the golden SerDes and test SerDes are independent due to the process variation.*

- *The spectral information of the TJ distribution of the golden SerDes is measured previously with the help of a conventional testing method.*

- *TJ is a stationary phenomenon, i.e., a measurement of the spectral information on a given system, taken over an appropriate interval, will give the same result regardless of what time interval is initiated.*

The notations for the variance and mean of the of the TJ distribution of the golden SerDes, test SerDes and the resultant are described in the Table 1. As the TJ distributions of the golden SerDes and test SerDes are independent, the mean and variance of the above three distributions are related as:

$$\sigma_{d,r}^2 = \sigma_{d,g}^2 + \sigma_{d,t}^2 \tag{1}$$

$$\mu_{Rd,r} = \mu_{Rd,g} + \mu_{Rd,t} \tag{2}$$

$$\mu_{Ld,r} = \mu_{Ld,g} + \mu_{Ld,t} \tag{3}$$

$$\sigma_{c,r}^2 = \sigma_{c,g}^2 + \sigma_{c,t}^2 \tag{4}$$

$$\mu_{Rc,r} = \mu_{Rc,g} + \mu_{Rc,t} \tag{5}$$

$$\mu_{Lc,r} = \mu_{Lc,g} + \mu_{Lc,t} \tag{6}$$

Unlike the previous BER estimation methods [12, 13, 15], our proposed BER estimation method incorporates the TJ spectral information of the serial data with the TJ spectral information of *the recovered clock*, not with the jitter transfer characteristics of the CDR circuit. In [13], as the clock varies from its ideal sampling position due to the jitter transfer characteristics of the CDR and random jitter generated in the CDR circuit, the formulas for the estimation of BER are different in the different regions of the jitter transfer function of the CDR circuit. To develop a common formula for the estimation of BER in different regions of the CDR circuit, the TJ present in the recovered clock is incorporated

Parameter	:	Description (RJ: Random Jitter)
$\sigma^2_{d,g}$	:	Variance of RJ in the serializer output of Golden SerDes
$\sigma^2_{c,g}$	:	Variance of RJ in the recovered clock of Golden SerDes
$\mu_{Rd,g} - \mu_{Ld,g}$	:	DJ in the serializer output of Golden SerDes
$\mu_{Rc,g} - \mu_{Lc,g}$	:	DJ in the recovered clock output of Golden SerDes
$\sigma^2_{d,t}$	:	Variance of RJ in the serializer output of Test SerDes
$\sigma^2_{c,t}$	:	Variance of RJ in the recovered clock of Test SerDes
$\mu_{Rd,t} - \mu_{Ld,t}$	:	DJ in the serializer output of Test SerDes
$\mu_{Rc,t} - \mu_{Lc,t}$	:	DJ in the recovered clock output of Test SerDes
$\sigma^2_{d,r}$	:	Variance of RJ in the resultant data jitter
$\sigma^2_{c,r}$	:	Variance of RJ in the resultant clock jitter
$\mu_{Rd,r} - \mu_{Ld,r}$	:	DJ in the resultant data jitter
$\mu_{Rc,r} - \mu_{Lc,r}$	:	DJ in the resultant clock jitter

Table 1: Notations.

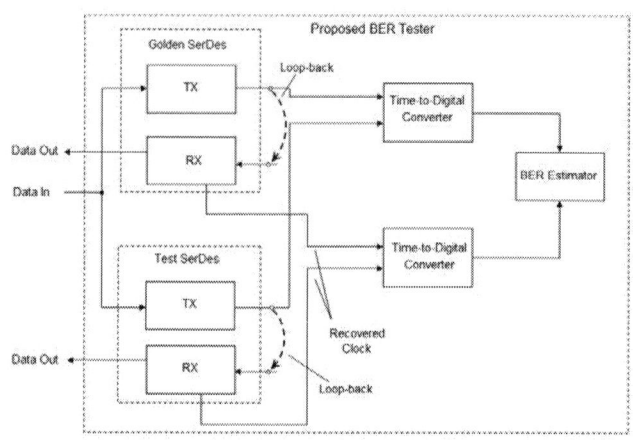

Figure 1: The proposed BER test system.

in our BER estimation. As a result, the BER would be independent of the regions of the jitter transfer characteristic of the CDR circuit and can simply be estimated by the following equation,

$$BER = 2 * \rho_T * Q\left(\frac{T/2 - A_{eff}}{\sigma_{eff}}\right) \qquad (7)$$

where,

ρ_T is the transition density and it is equal to 1 for clock like pattern (i.e., the periodic "1010" data stream) [15],

$$Q(x) = \frac{1}{\sqrt{2\pi}} \int_x^\infty \exp^{-\frac{x^2}{2}} \, \mathrm{d}x,$$

$$\sigma_{eff} = \sqrt{\sigma^2_{d,t} + \sigma^2_{c,t}},$$

and

$$A_{eff} = \mu_{Ld,t} - \mu_{Lc,t}$$

3 Proposed BER Test System

The proposed BER test system is illustrated in Figure 1. It comprises of a golden SerDes and a test SerDes. The spectral information, variance (σ^2_d, σ^2_c) and mean (μ_L, μ_R) of the golden SerDes are previously measured by the conventional method using oscilloscope or BERT. It is assumed that the spectral information would be constant during the testing. This SerDes is permanently placed in the test system. The main idea of the proposed test methodology is to estimate the spectral information of the test SerDes based on the known spectral information of the golden one. The common test data is fed to both the SerDeses. The length of the paths from the data to the SerDeses must be equal to eliminate the delay difference. The serial output of the two transmitters are fed to a TDC. The length of the paths must be of equal as well. To obtain the recovered clocks, the serial output of the transmitter is looped back to the receiver of the same SerDes. The recovered clock of each receiver is fed to a second TDC. (We assume that the recovered clock is an available pin of

the SerDes device.) The output of the TDC is the phase differences of the two input signals in binary format. The histogram of the output of the TDC will be resultant TJ distribution.

Finally, the BER estimator, currently a software tool, collects the data from the TDC, makes the histogram and calculates the BER of test SerDes. Based on the histogram, the BER estimator calculates all the spectral information such as σ^2, μ_L and μ_R using the Expectation Maximization (EM) algorithm [4, 19] in MATLAB. Hence, the BER can be estimated using the complete set of spectral information of the test SerDes via the Equation 7.

3.1 Proposed Binary Vernier Delay Line as Time-to-Digital Converter

The basic concept for the proposed TDC is based on "successive approximation". The digitized output of the converter is set to "1" initially. It remains the same in a particular stage when a successful comparison is made; otherwise, it will be reset to "0". The comparison is carried out bit by bit from the most significant bit (MSB) to least significant bit (LSB).

For the simplicity to describe how the design works, we assume all the circuit components, except the buffers, have no delay. The detail description of how the delay of each building components (such as buffer, MUX, and D flip-flop) affects the operation of the TDC design can be found in [11].

An example of a three-stage binary VDL TDC is shown in Figure 2. It consists of two delays chains. The upper chain has the three buffers, denoted by BUF4L, BUF2L and BUF1L, each with binary delay of $4t_{buf\_low}$, $2t_{buf\_low}$ and t_{buf_low}, respectively. Similarly, the lower chain also has three buffers, denoted by BUF4H, BUF2H and BUF1H, with binary delay of $4t_{buf\_high}$, $2t_{buf\_high}$ and t_{buf_high}, respectively. The D and Clk input of the DFF is coming from the output of BUFiL and BUFiH respectively, where i denotes the stage number. All the transitions in X and Y must be at the positive clock edges. Before the delay measurement (or time-to-digital conversion), all the flip-flops must be reset and the MUXes of all the stages will select the D1 input. The resolution of this binary VDL TDC is $t_{res} = t_{buf\_high} - t_{buf\_low}$.

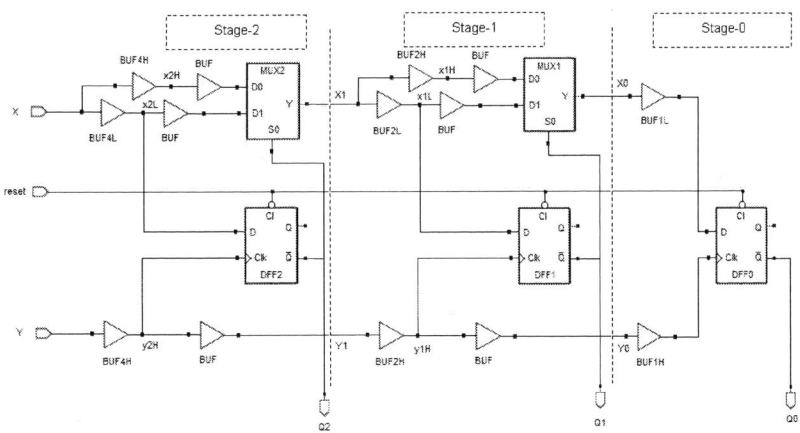

Figure 2: A 3-stage binary VDL TDC.

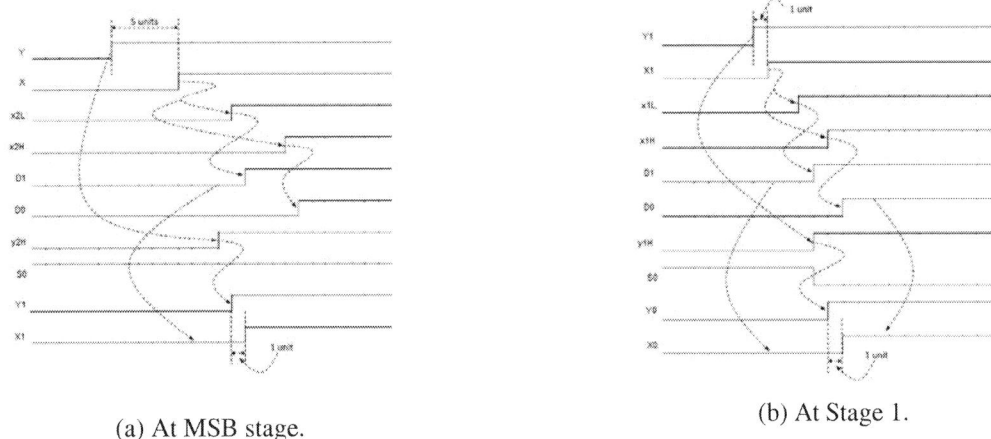

(a) At MSB stage.

(b) At Stage 1.

Figure 3: Timing waveforms of the example 3-stage binary VDL TDC.

Now, let's examine how the delay between the input X and Y is to be measured. Assuming that the minimum buffer delay of the lower chain, i.e., t_{buf_high}, is 2 units and that of the upper chain, i.e., t_{buf_low}, is 1 unit. Thus the resolution is $2 - 1 = 1$ unit. Assuming that the *phase difference* between Y and X (i.e., the time of the transition of Y ahead of the transition of X) is T. To illustrate how the TDC works, we consider the value of

Figure 3 (a) shows the timings of the MSB stage (or Stage 2) of the TDC. At the inputs of TDC, the rising edges of Y and X come at t_0 and $t_0 + 5$ respectively. The output of the BUF4H and BUF4L, i.e., $x2H$ and $x2L$, in the upper chain would be shifted version of X and the amount of shifts are 8 units and 4 units respectively. Again, the BUF4H output, i.e., y2H, in the lower chain will be shifted Y by 8 units. The D0 and D1 input of the MUX2 will be the shifted version of $x2H$ and $x2L$ and the amount of shifting is the inertial delay of the MUX which is assumed to be of 1 unit for simplicity. The rising of the clock will come before the rising edge of D that will lead the DFF2 to remain its previous state, i.e., reset. As a result, the MUX will select D1 input through out this stage as selection input, S0, of MUX2 does not change. As a result, the delay difference in the next stage will be the difference of the input delay and delay difference between BUF4H and BUF4L, i.e., 1 unit. In this stage,

a true comparison has been made and the output of this stage i.e., $Q2$ will be 1.

Figure 3 (b) shows the stage 1 of the conversion process with the input delay of 1 unit. The output of the BUF2H and BUF2L, i.e., $x1H$ and $x1L$, in the upper chain would be shifted version of $X1$ and the amount of shifts are 4 units and 2 units respectively. Again, the BUF2H output, i.e., y1H, in the lower chain will be shifted $Y1$ by 4 units. As a result, the rising edge of the clock comes after the rising edge of data that leads the flip-flop (DFF1) changes its state i.e., $Q = 1$ and $\overline{Q} = 0$ to set from reset. Thus, MUX2 selects $D0$ input which leads to the output delay difference of this stage remains same as the input delay difference i.e., 1 unit. In this stage a false comparison has been made and the output, $Q1$ of this stage will be 0. In the LSB stage (or stage 0), the input delay difference (1 units) is equal to the delay difference of BUF1H and BUF1L. As a result, DFF0 will remain at the same state i.e., $Q = 0$ and $\overline{Q} = 1$ that leads to the output of this stage $Q0$ will be 1. The final output corresponds to the delay of 5 units between Y and X becomes $Q2Q1Q0 = 101$ which is the binary equivalent of 5.

978-1-61284-657-6/11 $26.00 © 2011 IEEE

4 Experimental Results

The proposed binary VDL TDC is implemented in Verilog for behavioral simulation and SPICE for performance analysis [11]. A behavioral model of the proposed BER test system (including TDC and EM algorithm) is also developed in MATLAB to validate our proposed BER estimation method. The TJ in the output data of the golden SerDes is modeled as a Gaussian distribution in MATLAB. The spectral information of the TJ present in the data of the golden SerDes is taken from [15]. The spectral information, i.e., μ_L, μ_R and σ, of the golden SerDes are -13 ps, 13 ps and 5.7 ps respectively. Similarly, the TJ of the test SerDes is also modeled as another Gaussian distribution in MATLAB. We assumed that the spectral information of the test SerDes are -14 ps, 14 ps and 5.5 ps for μ_L, μ_R and σ respectively.

The experiment procedure is to apply the spectral information of both golden and test SerDeses into the behavioral model of the proposed BER system. Because the proposed BER test system focuses on the difference between the golden and test SerDeses, any data can be applied to the proposed BER test system. Here, the input data applied to the test system is clean and free from jitter.

We will then estimate the spectral information of the resultant TJ distribution in the proposed BER test system using MATLAB simulation. The spectral information of the test SerDes can then be calculated using Equations 1, 2 and 3. In the process, we also validate our proposed BER estimation method by comparing the obtained spectral information of the test SerDes with their assumed values.

The TJ distribution of the golden SerDes and test SerDes and the resultant TJ distribution are generated in MATLAB and shown in the Figure 4.

We have run the MATLAB simulation model for each set of data samples for more than 20-50 times to estimate μ and σ using our proposed BER estimation methodology. Although the TJ distributions of both golden and test SerDeses are modeled as random variables, the spectral information of the resultant TJ seems repeating after very six times. Table 2, 3 and 4 summarize the results. We ran the proposed model to capture 10^4 samples, $2*10^4$ samples, $5*10^4$ samples and 10^5 samples to estimate the TJ spectral information. Table 2 represents the estimated μ_R values with different samples and time intervals. Based on the six measurements, the maximum variation of μ_R with respect to the actual value is 2.8% when the captured samples are 10^4. However, the estimation accuracy increases significantly when larger number of samples are captured. For example, the maximum variation is reduced to 0.86% when the captured samples are 10^5.

Table 3 and 4 summarize the estimation of μ_L and σ of the TJ distribution of the test SerDes. The maximum variation of μ_L with respect to the actual value is 2.7% and 0.79% when the captured samples are 10^4 and 10^5 respectively and the maximum variation of σ with respect to the actual value is 5.65% and 0.73% when the captured samples are 10^4 and 10^5 respectively.

Figure 5 shows the scattered plot of the data from the Table 2, 3 and 4. The x-axis represents the number of transmitted bits and y-axis represents the spectral information. It clearly shows

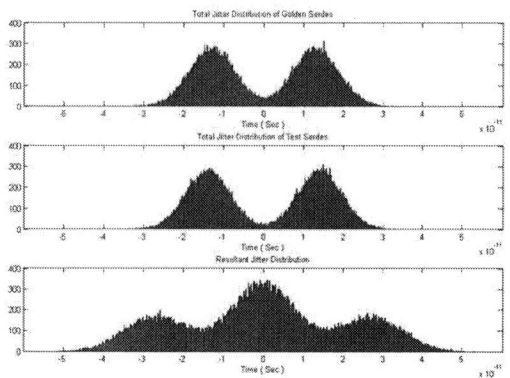

Figure 4: a) The TJ distribution of the Golden SerDes; b) The TJ distribution of the Test SerDes; c) The resultant TJ distribution

	10^4	% Err	$2*10^4$	% Err	$5*10^4$	% Err	$10*10^4$	% Err
1	14.30 ps	2.14	14.29 ps	2.07	13.90 ps	0.71	14.05 ps	0.36
2	14.11 ps	0.79	13.96 ps	0.29	13.81 ps	1.36	13.88 ps	0.86
3	13.61 ps	2.79	13.94 ps	0.43	14.01 ps	0.07	14.03 ps	0.21
4	14.30 ps	2.14	14.29 ps	2.07	13.90 ps	0.71	13.93 ps	0.50
5	14.28 ps	2.00	13.69 ps	2.21	14.05 ps	0.36	14.01 ps	0.07
6	14.06 ps	0.43	14.00 ps	0.00	14.01 ps	0.07	13.92 ps	0.57

Table 2: Estimation of μ_R

	10^4	% Err	$2*10^4$	% Err	$5*10^4$	% Err	$10*10^4$	% Err
1	-13.88ps	0.86	-14.12ps	0.86	-14.03ps	0.21	-14.11ps	0.79
2	-14.32ps	2.29	-14.21ps	1.50	-14.00ps	0.00	-13.94ps	0.43
3	-14.29ps	2.07	-13.95ps	0.36	-13.88ps	0.86	-14.02ps	0.14
4	-13.88ps	0.86	-14.12ps	0.86	-14.00ps	0.00	-13.90ps	0.71
5	-14.38ps	2.71	-13.95ps	0.36	-14.1ps	0.71	-14.01ps	0.07
6	-13.81ps	1.36	-14.10ps	0.71	-14.15ps	1.07	-13.90ps	0.71

Table 3: Estimation of μ_L

	10^4	% Err	$2*10^4$	% Err	$5*10^4$	% Err	$10*10^4$	% Err
1	5.22ps	4.93	5.48ps	0.36	5.49ps	0.18	5.47ps	0.49
2	5.18ps	5.65	5.25ps	4.47	5.41ps	1.49	5.54ps	0.73
3	5.27ps	4.07	5.59ps	1.76	5.52ps	0.36	5.53ps	0.55
4	5.22ps	4.92	5.48ps	0.32	5.50ps	0.13	5.51ps	0.19
5	5.68ps	3.42	5.58ps	1.63	5.44ps	0.99	5.51ps	0.21
6	5.24ps	4.65	5.72ps	4.04	5.45ps	0.80	5.50ps	0.03

Table 4: Estimation of σ

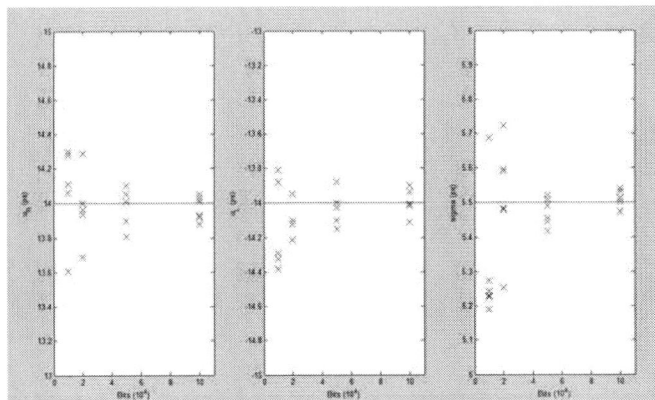

Figure 5: Estimation of mean and variance

that variations of estimated TJ spectral information are reduced significantly when larger number samples are captured. The convergence signifies the higher estimation accuracy.

The testing time corresponding to capture 10^5 samples would be $10^5/(2.5 * 10^{+9}) = 0.4 * 10^{-4}$ seconds when the link runs at 2.5 Gbps (such as PCIe Generation I). The estimation time for the spectral information using EM algorithm in MATLAB takes 25-30 seconds on a PC with Intel Core 2 Duo at 2 GHz with 3 GB RAM. However, this time could be much lesser if the EM algorithm is implemented in hardware. As a result, we can conclude that the total estimation time is the order of seconds which makes high volume test possible.

In the above, we have validated the proposed BER test system by showing that it accurately estimates the spectral information of the test SerDes. However, we could not estimate the BER because the spectral information of TJ present in the recovered clock is not available in the literature. We are currently working on an opportunity to collect such information. Once the spectral information of TJ in the recovered clock is obtained, with the spectral information of the test SerDes estimated by our proposed BER estimation, BER of the test SerDes can be easily derived from Equation 7.

5 Conclusion

In this paper, we have proposed a novel BER estimation methodology, consisting of a new BER estimation method and a BER test system. To implement that proposed BER test system, we have designed a novel binary TDC.

The proposed BER estimation method makes it possible to measure BER accurately without the prior knowledge of jitter present in the serial data stream. This is the significant improvement over the dual-Dirac model.

The novel binary VDL TDC directly converts the delay between the two signals in the digital format. Two significant improvements over the prior TDCs are the improved resolution and the compact design. It can measure delay in the range of picoseconds.

The experimental results not only validate the accuracy of the

estimated spectral information, but also show more than hundreds times of improvement in test time saving when comparing with the traditional BER measurement. We are currently working on the validation of the BER estimation once the spectral information of the received clock can be obtained.

References

[1] J. D. H. Alexander. Clock Recovery From Random Binary Signals. *Electronics Letters*, 11:541–542, October 1975.

[2] M. Bazes. A novel precision MOS synchronous delay line. *IEEE Journal of Solid State Circuits*, 20:1265–1271, December 1985.

[3] M. Bazes and R. Ashuri. A Novel CMOS Digital Clock and Data Decoder. *IEEE Journal of Solid State Circuits*, 27:1934–1940, December 1992.

[4] S. Borman. The Expectation Minimization Algorithm. A short tutorial, July 2004. Available online (9 pages) http://www.seanborman.com/publications/EM_algorithm.pdf.

[5] A. M. C. Ljuslin, J. Christiansen and O. Klingsheim. An Integrated 16-channel CMOS Time to Digital Converter. *IEEE Transactions on Nuclear Science*, 41:1104–1108, August 1994.

[6] J. Christiansen. An Integrated CMOS 0.15ns Digital Timing Generator for TDC's and Clock Distribution Systems. *IEEE Transactions on Nuclear Science*, 42:753–757, August 1995.

[7] J.-F. Genat. High resolution time-to-digital converters. Nuclear Instruments and Methods in Physics Research Section A: Accelerators, Spectrometers, Detectors and Associated Equipment, May 1992.

[8] J.-F. Genat and F. Rossel. Ultra high-speed time-to-digital converter. United States Patent no. 4719608, January 1988.

[9] M. S. Gorbics, J. Kelly, K. M. Roberts, and R. L. Sumner. A high resolution multihit time to digital converter integrated circuit. *IEEE Transactions on Nuclear Science*, 44:379–384, June 1997.

[10] C. T. Gray, W. Liu, W. A. M. V. Noije, T. A. Hughes, and R. K. Cavin. A Sampling Technique and its CMOS Implementation with 1 Gb/s Bandwidth and 25 ps Resolution. *IEEE Journal of Solid State Circuits*, 29:340–349, March 1994.

[11] U. Guin. Design For Bit Error Rate Estimation of High Speed Serial Links. Master's thesis, Department of Electrical and Computer Engineering, Temple University, Philadelphia, 2010.

[12] D. Hong and K.-T. T. Cheng. Bit-Error Rate Estimation for Bang-Bang Clock and Data Recovery Circuit in High-Speed Serial Links. In *Proceedings IEEE VLSI Test Symposium*, 2008.

[13] D. Hong, C.-K. Ong, and K.-T. T. Cheng. Bit Error Rate Estimation for High Speed Serial Links. *IEEE Transaction on Circuits and Systems*, 53(12):2616–2627, December 2006.

[14] K. Kundert. Verification of Bit-Error Rate in Bang-Bang Clock and Data Recovery Circuits. White Paper, 2010. Available online (22 pages) http://www.designers-guide.org/Analysis/bang-bang.pdf.

[15] M. Mueller, R. Stephens, and R. McHugh. Jitter Analysis: The dual-Dirac Model, RJ/DJ, and Q-scale. White Paper, Agilent Technologies, 2004. Available online (16 pages) http://cp.literature.agilent.com/litweb/pdf/5989-3206EN.pdf.

[16] M. Mueller, R. Stephens, and R. McHugh. Total Jitter Measurement at Low Probability Levels, Using Optimized BERT Scan Method. White Paper, Agilent Technologies, 2005. Available online (18 pages) http://cp.literature.agilent.com/litweb/pdf/5989-2933EN.pdf.

[17] T. Rahkonen and J. T. Kostamovaara. The use of stabilized CMOS delay lines for the digitization of short time intervals. *IEEE Journal of Solid State Circuits*, 28:887–894, August 1993.

[18] C. Su, Y.-T. Chen, M.-J. Huang, G.-N. Chen, and C.-L. Lee. All Digital Built-In Delay and Crosstalk Measurement for On-chip Buses. In *Proceedings of the conference on Design, automation and test in Europe*, March 2000.

[19] C. Tomasi. Estimating Gaussian Mixture Densities with EM - A Tutorial. Available online (8 pages) http://www.cs.duke.edu/courses/spring04/cps196.1/handouts/EM/tomasiEM.pdf.

978-1-61284-657-6/11 $26.00 © 2011 IEEE

Session 12

978-1-61284-657-6/11 $26.00 © 2011 IEEE

An Efficient Test Data Reduction Technique Through Dynamic Pattern Mixing Across Multiple Fault Models

S. Alampally[1], R. T. Venkatesh[2], P. Shanmugasundaram[2], R. A. Parekhji[1] and V. D. Agrawal[2]
[1]Texas Instruments, Bangalore (India) and [2]Auburn University (USA).

Abstract: ATPG tool generated patterns are a major component of test data for large SOCs. With increasing sizes of chips, higher integration involving IP cores and the need for patterns targeting multiple fault models for better defect coverage in newer technologies, the issues of adequate coverage and reasonable test data volume and application time dominate the economics of test. We address the problem of generating compact set of test patterns across multiple fault models. Traditional approaches use separate ATPG for each fault models and minimize patterns either during pattern generation through static or dynamic compaction, or after pattern generation by simulating all patterns over all fault models for static compaction. We propose a novel ATPG technique where all fault models of interest are concurrently targeted in a single ATPG run. Patterns are generated in small intervals, each consisting of 16, 32 or 64 patterns. In each interval fault model specific ATPG setups generate separate pattern sets for their respective fault model. An effectiveness criterion then selects exactly one of those pattern sets. The selected set covers untargeted faults that would have required the most additional patterns. Pattern generation intervals are repeated until required coverage for faults of all models of interest is achieved. The sum total of all selected interval pattern sets is the overall test set for the DUT. Experiments on industrial circuits show pattern count reductions of 21% to 68%. The technique is independent of any special ATPG tool or scan compression technique and requires no change or additional support in an existing ATPG system.

Keywords: ATPG optimizations, pattern merge, test data volume reduction, composite fault models.

1. INTRODUCTION

Manufacturing test contributes a significant portion to the overall cost of an IC and the tester time required for test application is a critical entity for cost minimization. Due to the larger designs and process variations seen in the advanced technology nodes, the number of fault models to be considered for minimizing test escapes has gone up. Consequently there is an increase in test data volume (TDV) and test application time (TAT). An account of the costs of testing on an ATE is given in [1] and a cost model proposed in [2] gives a good explanation of the cost metrics. TDV and TAT are popular test cost metrics and are proportional to the number of patterns required to get the desired coverage, the maximum scan chain length and the number of inputs and outputs. An insight on TDV for SOCs is provided in [3].

Test compression techniques listed by [4] have been used to contain TDV and TAT by reducing the scan chain lengths and increasing the number of chains by providing extra hardware for test data distribution among these chains. They basically exploit the fact that only 2-5% of the bits in the pattern are 'care bits' and the rest can be compressed [4]. The work in [5] explores the solution space for maximizing compression on some large industrial designs. However, there are limits as explained in [6]. Procedures such as test point insertion, mixing combinational and sequential compression approaches and more complex test compression logic can potentially improve the compression, but at the expense of additional hardware complexity. In order to achieve further reduction in test data volume, we have to look beyond the on-chip compression techniques. Reducing the number of patterns through 'pattern reuse' can be a good solution.

In this work, we target test pattern reduction with the help of a simple ATPG flow improvement across multiple fault models. Effective and efficient pattern sets are chosen based on the proposed metric. We intend to exploit both 'care' and 'don't care' bits in a pattern to efficiently combine and compact the pattern sets of different models into one single set. The technique has been evaluated across complex models such as path delay, dynamic bridging and small delay defect driven faults in addition to the stuck-at and transition delay fault models. The results show a comparison with existing pattern optimization approaches. The rest of the paper is organized into six sections. Prior work in the area of multiple fault model pattern optimizations is reviewed in Section 2. The motivation for the work in this paper is explained in Section 3. A detailed description of the methodology is provided in Section 4. Sections 5 and 6 provide the results on large SOC designs and observations are made based on these results. Section 7 concludes the paper.

2. TEST PATTERN REUSABILITY

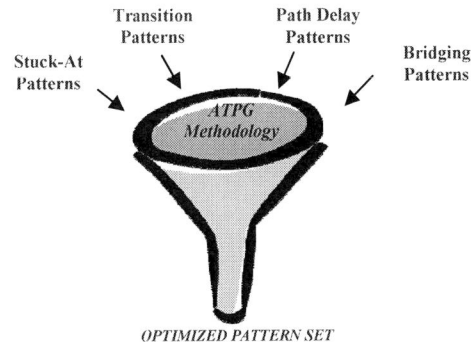

Figure 1: Pattern optimization across multiple fault models.

Test pattern reuse is a general term associated with the ability of patterns of one fault model to detect faults belonging to another fault model, thereby enabling reusability. This characteristic can be exploited to reduce the final pattern count when multiple fault models are considered. As fault models under consideration increase in number there is more room for pattern reuse as pattern-sets associated with different models have varied care-bit densities. Figure 1 illustrates a general pattern optimization approach across fault models by exploiting pattern reusability. One fine example is that of transition fault patterns for the 'Launch off Shift' (LOS) scan approach (as against 'Launch off Capture' (LOC)) and stuck-at patterns. Experimental results in [7] show the effectiveness of stuck-at ATPG patterns for transition fault detection.

A pattern selection approach for the combination of stuck-at, transition and I_{DDQ} models is discussed in [8]. Hybrid LP-ILP technique is used to optimize patterns after their generation. The solution to the LP formulation gives a pattern set that ensures the desired coverage across all the models under consideration. Though this technique optimizes the pattern count, the exponential complexity of linear programming is not suitable for large designs. Another technique proposed in [9] considers path delay, transition and stuck-at models on large SoC designs. The algorithm starts with ATPG on the path delay faults and later uses the generated patterns to detect transition faults. The remaining transition faults are targeted through ATPG and the process is continued till stuck-at faults are also targeted. Though both the works provide good pattern optimization, the optimization is limited due to the non-dynamic nature of fault model selection. The methodology proposed in this paper makes this selection variable during the ATPG process and enables further pattern count reduction for a reasonable run time overhead.

3. MOTIVATION

In a deterministic ATPG process, patterns are generated to target a set of faults that are specific to a fault model. At the end of the process, we get pattern sets and the detected faults for each fault model. It has also been found that reuse of test patterns is possible amongst the fault models. The proposed concurrent ATPG algorithm exploits these facts to minimize the final test set:

- Since patterns have a large number of unspecified or 'don't care' bits in them, opportunities exist for augmenting the random pattern detectability across fault models.
- Transition test patterns have been found to provide good coverage for stuck-at faults. The converse is true as well. The usage of the stuck-at patterns for detecting delay faults depends on whether a 'Launch off Shift'

(LOS) or a 'Launch off Capture' (LOC) scheme is being used [10]. In the former case, this is straightforward due to similar clocking used. In a concurrent ATPG flow, patterns can be generated for all other models and then the resulting optimized pattern set can be used to fault simulate on the stuck-at model. The remaining stuck-at faults can be detected with patterns.

- Since path-delay patterns target cumulative delay faults on an entire path, transition faults on nodes on the path under test are detected by the same pattern. If the number of detectable faults with the path-delay fault pattern set is high, the transition fault coverage with those patterns will also be proportionately high. This can also significantly reduce the transition fault pattern-set. On the other hand, transition patterns typically do not give a good coverage on path delay faults since they typically propagate the transition through the shortest (easiest) paths. If small delay defect (SDD) model is considered instead of transition fault model, the probability of getting path delay coverage improves since the timing slack information forces SDD ATPG to propagate transition through longer paths.

- Dynamic bridging fault model requires a transition on the victim node. Hence there exists a high probability of detecting transition faults while targeting the dynamic bridging fault model.

These observations indicate how we can reuse patterns targeted for one fault model for detecting faults in another fault model. Significant pattern count reduction can be achieved if ATPG is performed simultaneously on all fault models. Since current ATPG tools target only one fault model at a time, we create a flow to support such a *concurrent ATPG* process in our work. The ATPG run is split into small intervals. In each interval, patterns are generated targeting all fault models under consideration separately. And the pattern set for each fault model is cross fault simulated on the other fault models. The pattern-set that has highest effectiveness across all fault models is selected for that interval. This process is repeated till final coverage levels are achieved for all fault models. In the next section, we will describe the effectiveness algorithm in detail. The size of the final pattern-set P_{total} can be denoted as:

$$P_{total} \le \sum_i P_i \text{, where } i \in F$$

where F is the set of fault models considered and P is the set of patterns for each fault model.

4. ALGORITHM FOR CONCURRENT ATPG

The proposed flow is shown in Figure 2. The following abbreviations have been used for convenience:

N:– Number of fault models.

FC_cum_n:– Cumulative fault coverage for model 'n'.

FC_curr_n:– Fault coverage of model 'n' in the current interval.

$FC\_curr_{n/m}$:– Fault coverage of model 'm' after fault simulation with model 'n' patterns.

P_n:– Pattern set of model 'n'.

F_n:– Fault set for model 'n' at the start of an interval.

F_n^*:– Fault set for model 'n' after ATPG.

$F_{n/m}$:– Fault set after fault simulation of model 'n' patterns on model 'm'.

I:– Number of patterns in an interval.

FFC_n:– Final fault coverage for model 'n'.

$SP_{n/m}$:– Saved patterns for fault model 'm' after fault simulation of model 'n' patterns on model 'm'.

SP_n:– Total saved patterns across all models if model 'n' patterns are chosen in an interval.

Information for concurrent ATPG like the 'list of fault models to be considered for optimization', 'number of patterns per interval' and 'fault coverage limit' for each model is provided to start with. With this initial data for all the fault models in hand, we begin an interval by generating the specified number of patterns (size of interval) for each fault model and then use these pattern sets to fault simulate on each of the other fault models under consideration. The effectiveness of each of the pattern sets is then evaluated using a metric. Pattern sets chosen to be optimal (using this metric) at the end of an interval are saved along with corresponding detected fault sets (for all the models). Undetected faults for all fault models are targeted at the start of the next interval. This process of pattern generation is continued till the desired coverage for all the fault models is obtained.

The metric used for determining the most effective pattern set at the end of an interval is related to the amount of pattern savings that is realized across all the fault models. The number of saved patterns denoted by SP_n for a fault model 'n' is defined as the total number of the patterns saved across all the other fault models when patterns of fault model 'n' are chosen at the end of an interval. In other words, we need not generate SP_n number of patterns if we use fault model 'n' patterns since they detect faults across other fault models, which otherwise requires SP_n additional patterns with targeted ATPG. Depending upon the speed and accuracy requirements, we can adopt either an approximate or an accurate metric for determining the SP_n.

a. Accurate metric

To calculate the accurate metric of savings with patterns P_n, fault simulation is performed on the undetected faults of fault model 'm' with P_n. ATPG is then run for fault model 'm' on a fault set containing only the faults detected by P_n. The number of patterns required to get equivalent detection on model 'm' through ATPG is termed as $SP_{n/m}$. The $SP_{n/m}$ denotes number of patterns saved (need not be generated) by selecting patterns of set P_n for the final pattern set in the interval. This process is repeated on all other fault models with pattern set P_n. The sum is denoted as SP_n.

b. Approximate metric

To calculate the approximate metric of pattern set P_n, fault simulation is performed on fault model 'm' with P_n on undetected faults in the current interval to get $FC\_curr_{n/m}$. To get the effectiveness of the set P_n, a simple ratio-metric calculation is performed. If I patterns from set P_m are required to get FC_curr_m coverage, the number of model 'm' patterns that would be needed to get $FC\_curr_{n/m}$ is calculated as:

$$SP_{n/m} = (FC\_curr_{n/m} / FC\_curr_m) * I$$

This number represents saving achieved by selecting P_n over P_m in the interval. The individual savings on each fault model with pattern set P_n are summed up to get SP_n.

This metric is calculated for all fault models and the pattern set having highest SP is selected for the current interval. The approach using the accurate metric consumes more CPU time compared to that using the approximate metric since it involves an extra ATPG step for effectiveness calculation. It was also observed that the basic algorithm (as described in Figure 2) can be speeded up by parallelizing certain independent parts of the flow. Though fault simulation has to always follow pattern generation for each model in the flow without any room for concurrency, individual model runs can be parallelized as all information is available at the start of the interval. This parallelization has also been incorporated into the algorithm to reduce run times.

Table 1: Sample iteration of the flow with approximate metric for an interval limit of 32 patterns.

	Stuck-at (1)	Transition (2)
Step 1	$FC\_curr_1 = 0$, F_1	$FC\_curr_2 = 0$, F_2
Step 2	$FC\_curr_1 = 58$, P_1	$FC\_curr_2 = 48$, P_2
Step 3	$FC\_curr_{2/1} = 30$	$FC\_curr_{1/2} = 30$
Step 4 (approx)	$SP_{2/1} = SP_2 = 16.55$ (32 * $FC\_curr_{2/1}/ FC\_curr_1$)	$SP_{1/2} = SP_1 = 20$ (32* $FC\_curr_{1/2}/ FC\_curr_2$)
Conclusion: After Step 4, $SP_1 > SP_2$, Stuck-at fault pattern set is more effective than transition fault pattern set		

Table 1 provides a snapshot of the pattern generation and fault simulation flow when stuck-at and transition models are used. Step 1 initializes the fault coverage numbers for fault models 1 and 2 to zero. In Step 2, pattern generation is performed on both fault models with pattern count limited to the interval size. In Step 3, fault simulation on the other fault models is performed (e.g. stuck-at fault patterns for current interval are simulated against transition faults and vice-versa). In Step 4, the metric to evaluate the best pattern set for the current interval is calculated. The metric can either be an approximate one or an accurate one. The pattern set that achieves the highest pattern savings is chosen in that interval.

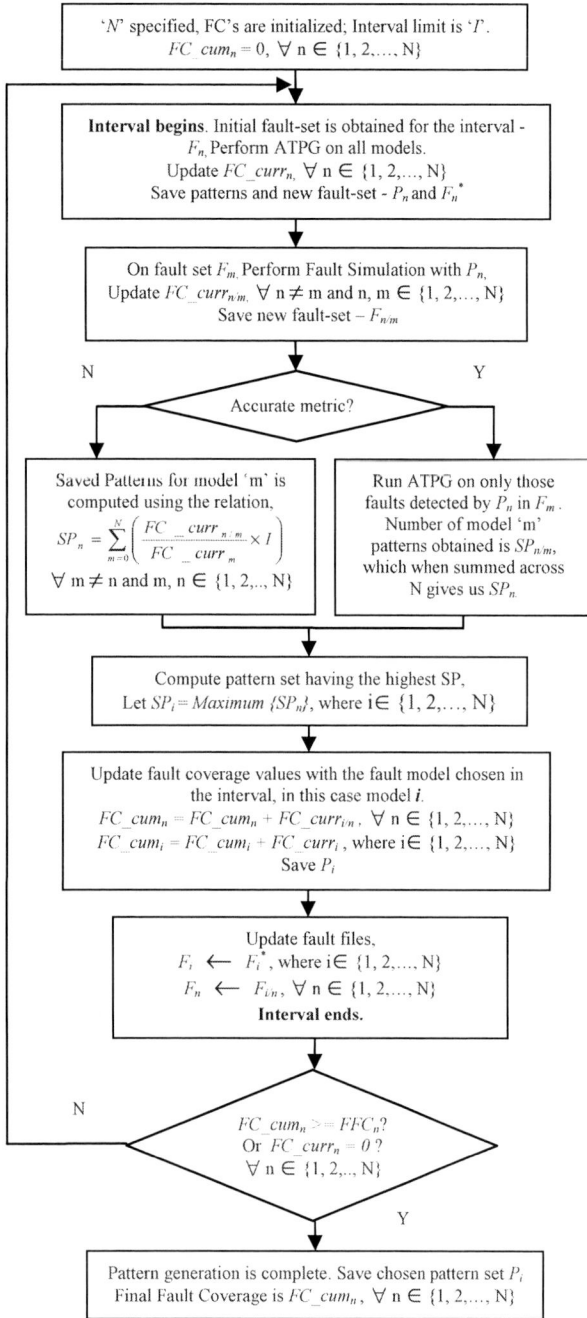

Figure 2: Basic concurrent ATPG flow.

5. EXPERIMENTAL RESULTS

Experiments were carried out on some SoCs in Texas Instruments to evaluate the practical benefits explained in the previous sections. The methodology was tried on several different fault model combinations. In addition to uncompressed ATPG, designs with combinational (Synopsys' DFTMAX) [11] and sequential (Synopsys' DBIST) [12] compression techniques were used. LOC and LOS at-speed approaches were used, but since LOS tends to provide more benefits over LOC in terms of pattern count and coverage levels, the former was chosen whenever we had an option to choose between the two approaches. The fault models considered were stuck-at, transition / small delay defect (SDD), bridging and path delay. TetraMAX [13] was used for ATPG and fault simulation. Table 2 lists the size of the designs used in the experiments. Percentage of pattern count reduction was taken as the metric for evaluation of the results. In order to reduce the run times the approximate metric was used.

Table 2: Statistics of designs used in the experiments.

Design	Flip-flop count	Gate count (in millions)	ATPG technique
A	219574	4	LOC
B	240000	2.5	LOS
C	33792	0.4	LOS

The results shown in Table 3 were obtained for Designs A and B by bypassing the compression features available for them. The unoptimized results are nothing but a sum of the pattern counts of each model. The results are compared against both the unoptimized runs and the existing optimization technique that is used by a number of teams at Texas Instruments [9]. Pattern count reductions of up to 45% were obtained when compared to the approach of combining patterns from standalone ATPG runs. A comparison with [9] showed improvements of up to 26% on an average. From the table, it can be observed that optimal benefit is achieved when stuck-at and transition faults are considered. This is mainly due to the fact that the two models have large fault-sets and pattern counts as compared to either bridging or path delay fault models. The scope for reuse increases with increase in number of patterns due to matching care bits and increased probability of random fault detection. Design A could not be simulated for the combination of stuck-at and transition models because it uses the LOC approach for at-speed testing. It can be seen that the model which has the largest standalone pattern count forces a limit on the compaction possible. For Design B, it can also be observed that dynamic pattern mixing results in lesser pattern count than the transition fault model. We attribute this anomaly to the new ATPG run that gets fired for every interval in dynamic pattern mixing mode. ATPG tool starts with a new random seed in each interval, increasing the chances of detecting more faults than with that of using single seed for entire pattern generation.

Table 4 provides the results for all the three designs in scan compression mode. Designs A and C uses a combinational compression technique (DFTMAX) whereas B uses a sequential compression technique (DBIST). Design B was evaluated using pattern intervals as opposed to pattern numbers where each pattern interval was made up of 32 patterns. The overall benefits over [9] were reduced when compared to the same design set in the uncompressed mode, but a look at the table shows a 30% average reduction against the unoptimized set. Stuck-at and transition models again seem to dominate for the same reasons mentioned before.

6. OBSERVATIONS

The designs used for experimentation were varied in size and so were the ATPG techniques used on each of them. This variety has helped to arrive at the following observations:

a) As shown in Table 3 and Table 4, the advantage with concurrent ATPG is less in the presence of scan compression. This can be attributed to: (i) The care bit availability with scan compression is lesser, leaving less scope of pattern-reuse. (ii) The don't care bits are also less random, due to increased correlation. The results with combinational and sequential compression differ since don't care bits in the former are more correlated.

b) Figure 3, 4 and 5 give a good account of the pattern mixing across fault models that occurs during the course of a concurrent ATPG run for one of the combinations with Design A. As observed from the graphs, the fault model chosen by the algorithm varies frequently during the ATPG process indicating that it is beneficial to employ this technique against the existing optimization [9].

c) Path delay faults require relatively higher percentage of specified bits compared to other models. Reuse of path delay patterns is not very effective as the fault simulation coverage with these patterns on other models doesn't yield high benefits. It can be clearly observed from Figure 5 that the path delay model gets de-prioritized due to the combined dominance of the transition and dynamic bridging models and requires its own patterns. Situations like this force an approach that combines the method in [9] with concurrent ATPG. In this case, the path delay patterns can be generated with a standalone run and can then be used for fault simulation with transition and dynamic bridging fault models.

d) Run times were large mainly due to the frequent ATPG and fault simulation processes at each interval. The ATPG run times can be reduced by increasing the interval size. In most fault model combinations with concurrent ATPG, there would be only one model left for coverage improvement towards the end of the process. Once this state is reached the pattern limitation can be taken off to allow the ATPG to run till the desired coverage is reached for that model. This will reduce run times.

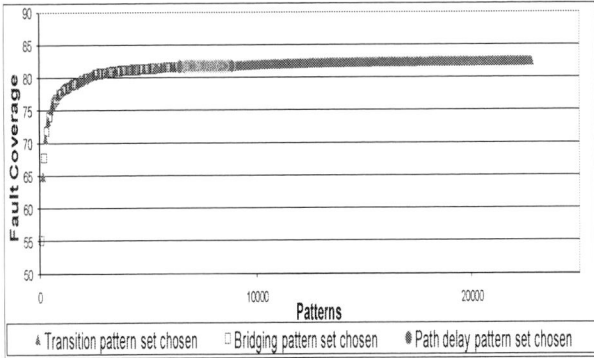

Figure 3: Transition fault coverage when run along with dynamic bridging and path delay fault models.

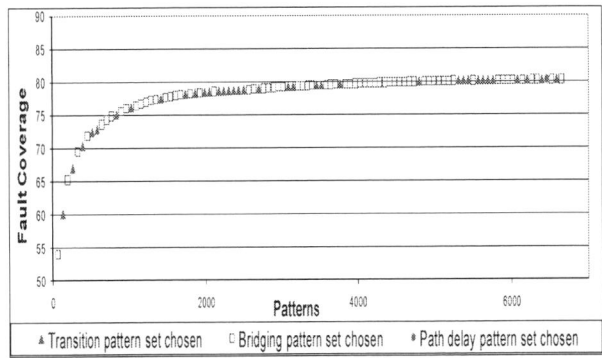

Figure 4: Bridging fault coverage when run along with transition and path delay fault models.

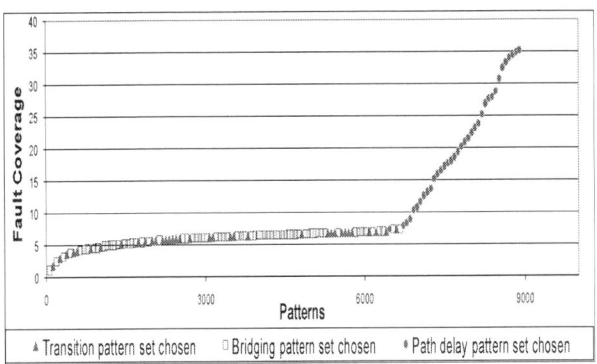

Figure 5: Path delay fault coverage when run along with transition and bridging fault models.

7. CONCLUSION

For large SOCs, structural test patterns obtained using ATPG tools continue to dominate the test time. As the number of fault models being targeted increases, the number of such patterns increases too. This paper presents a methodology for reducing the pattern count across multiple fault models. ATPG is performed in steps and various fault models are concurrently targeted. Pattern sets for a given fault model are incrementally generated and simulated across the other fault models. Stuck-at, transition, path delay and bridging fault patterns have been considered for experiments, with and without scan compression. This approach has been applied to several industrial designs in Texas Instruments and benefits from 21% to 68% are seen with and without compression, as compared to the conventional approach of just adding the patterns across all fault models. When compared to existing optimization technique, benefits with scan compression are in the range of 3% to 14%, whereas they are 17% to 36% without scan compression. This method does not require any modification to the ATPG flow. Other ideas for further optimizations have also been identified in the paper and will be explored in future work.

References

[1] J. Bedsole, R. Raina, A. Crouch and M. S. Abadir, "Very Low Cost Testers: Opportunities and Challenges," *IEEE Design and Test of Computers*, vol. 18, no. 5, pp. 60-69, 2001.

[2] S. Wei, P. K. Nag, R. D. Blanton, A. Gattiker and W. Maly, "To DFT or not to DFT?" *Proc. Intl. Test Conf.*, pp. 557-566, 1997.

[3] O. Sinanoglu, E. J. Marinissen, A. Sehgal, J. Fitzgerald and J. Rearick, "Test Data Volume Comparison: Monolithic vs. Modular

[4] N. A. Touba, "Survey of Test Vector Compression Techniques," *IEEE Design & Test of Computers*, vol. 23, no. 4, pp. 294-303, 2006.

[5] S. Alampally, J. Abraham, R. A. Parekhji, R. Kapur and T. W. Williams, "Evaluation of Entropy Driven Compression Bounds on Industrial Designs," *Proc. Asian Test Symp.*, pp. 13-18, 2008.

[6] T. W. Williams, "The Limits of Compression," *Proc. Intl. Test Conf.*, pp. 1-2, 2008.

[7] W. Kawamura and T. Onodera, "Experimental Results of Transition Fault Simulation with DC Scan Tests," *Proc. Asian Test Symp.*, pp. 212, 2007.

[8] N. Yogi and V. D. Agrawal, "N-Model Tests for VLSI Circuits," *Proc. Southeastern Symposium on System Theory*, pp. 242-246, 2008.

[9] S. Goel and R. A. Parekhji, "Choosing the Right Mix of At-Speed Structural Test Patterns: Comparisons in Pattern Volume Reduction and Fault Detection Efficiency," *Proc. Asian Test Symp.*, pp. 330-336, Dec. 2005.

[10] I. Park and E. J. McCluskey, "Launch-on-Shift-Capture Transition Tests," *Proc. Intl. Test Conf.*, pp. 1-9, 2008.

[11] Synopsys, Inc., *DFTMAX Compression User Guide*, Version E-2010.12.

[12] Synopsys, Inc., *SoCBIST Deterministic Logic BIST User Guide*, version 2005.09.

[13] Synopsys, Inc., *TetraMAX User Guide*, Version V-2010.03, 2010.

Table 3: Pattern statistics for all the designs in non compression mode.

| Design | Fault model combinations | Test coverage % | Pattern Count | | | % Reduction w.r.t | |
			Unoptimized	Optimized using [9]	Concurrent ATPG	Unoptimized	[9]
A	Transition	96.91	14590	14059	22784	20.91	17.24
	Dynamic Bridging	90.79	12592	11666			
	Path delay	37.45	1806	1806			
	Final Pattern Count		28808	27531			
A	Transition	96.7	13919	13482	18752	28.78	27.58
	Dynamic Bridging	90.79	12412	12412			
	Final Pattern Count		26331	25894			
A	Small Delay	96.03	12896	2784	8768	67.69	46.06
	Dynamic Bridging	90.79	12412	11666			
	Path delay	37.45	1806	1806			
	Final Pattern Count		27144	16256			
B	Stuck-at	96.41	1535	1535	4448	45.12	36.23
	Transition	91.97	6570	5441			
	Final Pattern Count		8105	6976			

Table 4: Pattern statistics for all the designs in compression mode.

| Design | Fault model combinations | Test coverage % | Pattern Count (DFTMAX) / Intervals (DBIST) | | | % Reduction w.r.t | |
			Unoptimized	Optimized using [9]	Concurrent ATPG	Unoptimized	[9]
A (DFTMAX)	Transition	96.91	25056	14048	33250	26.72	3.25
	Dynamic Bridging	89.56	20320	20320			
	Final Pattern Count		45376	34368			
B (DBIST)	Stuck-At	96.49	326	35	1084	31.21	13.9
	Transition	91.60	1214	1214			
	Static Bridging	70.36	17	3			
	Dynamic Bridging	61.84	19	7			
	Final Pattern Count		1576	1259			
C (DFTMAX)	Stuck-At	99.03	4706	498	11392	30.91	7.24
	Transition	95.29	11784	11784			
	Final Pattern Count		16490	12282			

Low Coverage Analysis using Dynamic Un-testability debug in ATPG

Kameshwar Chandrasekar, Surendra Bommu and Sanjay Sengupta

{kameshwar.chandrasekar, surendra.k.bommu, sanjay.sengupta}@intel.com

Intel Corporation, Santa Clara, CA 95054

Abstract

In this paper, we propose an automated technique to identify the reasons for un-testable faults and, an interactive Low Coverage Analysis flow to expedite the coverage analysis step, in scan ATPG. We seamlessly use an implication graph to keep track of the reasons that are responsible for each conflict encountered during ATPG. As ATPG progresses, for each fault, all the reasons arising from ATPG constraints are logged systematically. Then, we use a low coverage analysis flow to cumulatively analyze the faults and reasons / ATPG constraints. We integrated the proposed technique into the production scan ATPG flow at Intel. The proposed technique resolved up to 15% coverage gap on real micro-processor designs in a few hours. Potentially, this would have, otherwise, taken a few days of manual effort with considerable design knowledge.

1 Introduction

Over the past decade, we have seen a significant increase in the complexity of hardware designs. Designers are encouraged to do modular design that facilitates plug and play functional blocks in the hardware. The environment for the modular blocks and also partially implemented designs are encoded as design constraints to facilitate co-development efforts in the industry.

Specifically, Automatic Test Pattern Generation (ATPG) is performed on functional blocks, in the design, to alleviate the test generation complexity. Automatic Test Pattern Generation (ATPG) is used to generate scan patterns that can potentially detect the manufacturing defects on the chip. The clocks, interface signals, partially implemented blocks and sequential values are typically specified as ATPG constraints. These ATPG constraints are based on design assumptions and can potentially affect the test coverage of scan patterns. The test coverage target for scan patterns is usually very high to cover the scanned logic in the design. For designs with low coverage, the DFT engineer performs coverage analysis to understand the reasons for low coverage and resolve the coverage issues, as appropriate.

In ATPG: Given a circuit, a fault-list and design constraints, the tool attempts to generate a test suite that detects the faults in the circuit. At the end of ATPG, the tool generally reports the test coverage and the classification of each fault. At a high level, ATPG classifies each fault as either (i) detected – generates a test vector, (ii) un-testable – no test vector exists or (iii) aborted – unresolved due to resource constraints. The detected faults directly contribute to the actual fault coverage. On the other hand, the un-testable and aborted faults are responsible for the coverage gap between the target and actual test coverage.

The aborted faults can be potentially resolved by increasing the resource parameters for ATPG. The un-testable faults could be further classified into circuit un-testable and constraint un-testable faults. The circuit un-testable faults, also called as redundant faults, are un-testable because of the nature of the circuit and do not have any significant scope to improve the test coverage. The constraint un-testable faults are un-testable because of the ATPG constraints. In the early stages of the design, the ATPG constraints are usually conservative, if not incorrect, since the implementation will not be mature enough. The DFT engineer may have the liberty to relax the constraints / add additional DFT features, at a later point of time, to improve the test coverage for the design. The reasons for the un-testable faults would equip the DFT engineer to make informed decisions while looking for avenues to improve the test coverage.

In this work, we propose an automated technique to identify the reasons for un-testable faults and use them in an interactive low coverage analysis flow to address the coverage gap. The contributions of this work are:

1. We use an Implication Graph reasoning technique to identify the ATPG constraints that are responsible for making a fault un-testable.

2. We specifically identify the gates at which propagation paths for an un-testable fault are blocked for each fault.

3. We integrate the proposed technique into a low coverage analysis flow.

The rest of the paper is organized as follows: In the next Section, we review the previous work and the basics of our ATPG tool. In Section 3, we explain the proposed technique to identify the un-testability constraints and propagation blocking gates for each fault. In Section 4, we integrate the proposed technique into an interactive Low Coverage Analysis flow to perform a cumulative and systematic coverage debug. In Section 5, we present the first industrial results on real micro-processor designs. Finally, we conclude the paper in Section 6.

2 Previous Work

Various algorithms have been proposed, over the years, to tackle the basic ATPG problem. The first few deterministic algorithms were: Roth's 5-valued D-algorithm, 9-valued algorithm, PODEM and FAN. Please refer to [1] for a detailed study on the basics of these algorithms and the ATPG guidance heuristics. Cheng [2] proposed the split circuit model for ATPG, where the good machine and faulty machine values for all gates in the circuit are treated separately.

In [3], [4], the ATPG algorithm was accelerated by learning more necessary value assignments and unique sensitization points. Further, using Recursive learning [5], we could do an in-depth analysis to learn more implications. Additionally, explicit search space pruning methods were proposed in [6] and [7] to learn from equivalent search spaces and backtrack from redundant search spaces during ATPG search. In [8], AND-OR reasoning graphs were used to identify necessary and non-conflicting assignments during ATPG. The learning is done specifically for each set of justification points to identify extra local implications. Similar reasoning is also done in Recursive Learning [5], to identify more implications during ATPG. In [9], conflict driven heuristics and implication learning were proposed for D-algorithm which uses a cache to store the implications.

On a parallel front, Larrabee [10] introduced Boolean Satisfiability (SAT) based ATPG that models the circuit as a Conjunctive Normal Form (CNF) formula and uses SAT algorithms to solve the justification and propagation problems. The non-chronological backtracking and conflict-clause learning in [11] fuelled a break-through in SAT. The SAT-based techniques were borrowed in [12] to improve ATPG. In [13], the non-chronological backtracking and conflict learning was performed with an AND-OR reasoning graph and proved to be a very effective approach for ATPG on industrial desgins.

All these afore-mentioned techniques attempt to improve the test generation capability of the ATPG engine, that basically target to resolve a fault as soon as possible. On the other hand, after generating patterns, the test coverage report is typically analyzed. If any coverage difference exists between the actual and target coverage, we need to address the coverage gap and fix it iteratively until the target coverage is achieved. This is mostly done in the industry based on manual debug and design experience, by post-processing the fault statistics and the output fault list [14] and [15].

In this work, we have integrated an un-testability root-causing technique into our base-line ATPG tool to expedite the low coverage analysis step in scan ATPG flow. Our work is built upon an industrial ATPG engine with in-built conflict learning techniques and supports partial scan designs [13].

2.1 Base line ATPG tool

Our sequential ATPG tool implements D-algorithm style ATPG in conjunction with single-path sensitization on a split-circuit model. First, the fault effect is propagated to an observable gate (across time frames, if required) through D-frontiers. During this propagation phase, the resulting J-Frontiers are collected. Then, the J-Frontiers are justified, one time frame at a time, to find the test vector that detects the fault.

We use the circuit segment in Figure 1, fault: I_4 s-a-0 and ATPG constraint: $I_5=1$, to explain the concepts in this paper. First, during propagation phase, the D-frontiers are identified for the propagation path, G_8-G_{10}-G_{15}-G_{16}-G_{17}-O_1. The alternative D-Frontiers are G_{13} (to G_{15}) and G_{18}, G_{19} (to O_1). The corresponding J-frontiers to justify the off-path inputs, $G_4=1$, $G_7=0$, $G_{12}=1$, $G_{14}=Z$, $G_{11}=1$ are added to a J-Frontier list. Then, in the justification phase, we justify the J-frontiers one-by-one. Suppose, we first pick the J-frontier $\underline{G_{12}=1}$ to be justified. We can justify $G_{12}=1$ with $G_9=1$ or $G_5=1$. If we choose G_9, it results in implications $G_6=1$, $G_1=1$, $G_2=1$, $G_3=1$. The new J-frontier list is: $G_4=1$, $G_7=0$, $\underline{G_1=1, G_2=1, G_3=1}$, $G_{14}=Z$, $G_{11}=1$. In this way, we try to justify the gate values all the way to primary inputs or control points.

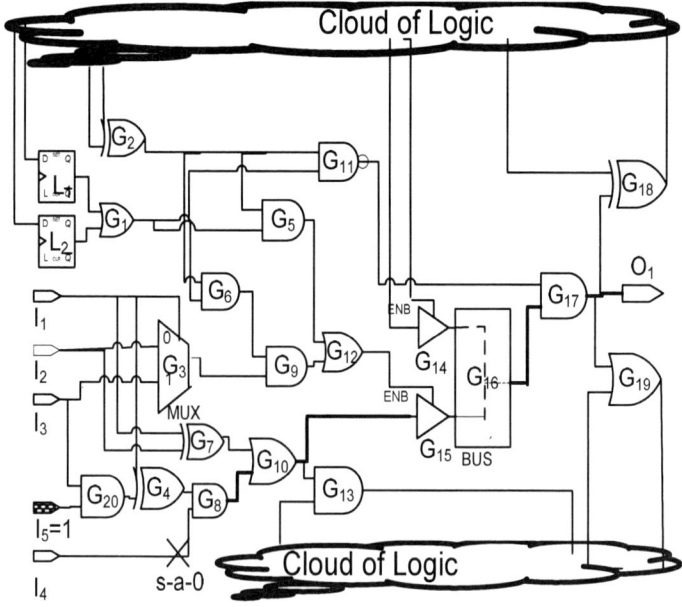

Figure 1: Running Example to explain un-testability-root causing

2.2 AND/OR reasoning framework

In Figure 2, we demonstrate the AND/OR reasoning graph for the example in Figure 1. We use the following conventions in the graph: (i) the rectangles represent D-frontier nodes, hexagons represent J-frontier nodes and ovals represent other implication nodes. (ii) we use solid lines to show AND relations, dotted lines to show OR relations and arrows to show implications in the graph (iii) a node with thicker boundary represents the chosen decision among other OR choices. Each decision is assigned a decision level (prefixed by @), which represents the chronological order of decision making. Each implication node gets the maximum decision level among all its implicants i.e. fanin nodes.

For the fault in Figure 1, the propagation path is represented by the nodes N_1-N_3-N_5-N_7-N_9-N_{11} in the propagation phase – exhibit an AND relationship. The D-frontier choices have an OR relationship. The off-path J-Frontiers have an AND relation to the D-frontier.

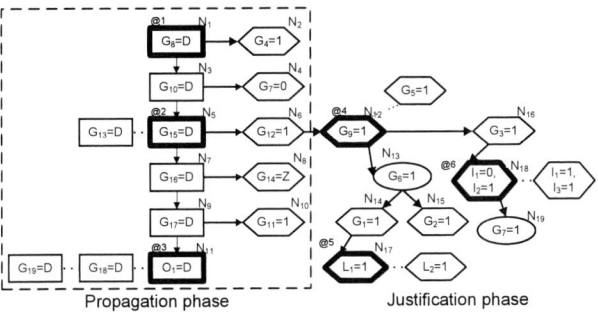

Figure 2: AND/OR Framework for ATPG

In justification phase, consider justification of $G_{12}=1$ at node N_6. The OR choices $G_9=1$ OR $G_5=1$ are linked by dotted lines. We choose $G_9=1$ in the example (thick boundary line) and keep $G_5=1$ as an alternative. Next, consider the justification of $G_9=1$ at node N_{12}. The resulting implications, $G_3=1$, $G_6=1$, $G_1=1$ and $G_2=1$, are linked by an AND relationship as shown by solid lines. The AND/OR-implication graph is constructed only for the active portion of the circuit during ATPG. Hence, the size of the implication graph is generally small as compared to the size of the circuit.

2.3 Conflict-diagnosis

When we reach a conflict scenario in ATPG, starting from the conflicting nodes, we can traverse backwards in the implication graph to identify the set of implicants that are actually responsible for the conflict. This step is called the *conflict diagnosis* step. Figure 2 actually shows a conflict scenario. The gate values in nodes N_{19} and N_4 conflict for gate value at G_7 and represent the conflicting nodes. During backward traversal for conflict

diagnosis, we start from the conflicting nodes and stop at a node if it is (i) a decision node or, (ii) an implication node at a lower decision level than the highest decision level of the conflicting nodes. In the above example, we identify that N_{18} and N_4 are responsible for the conflict. The node N_4 is the actual *reason* for the conflict arising from the decision made at N_{18}. We build the implication graph and do conflict diagnosis to perform non-chronological backtracking and conflict clause learning in our ATPG tool. We refer the reader to [13] for more details on these advanced learning techniques.

3 Un-testability constraints identification

The basic idea is as follows: While attempting to generate a test vector for a fault, the ATPG tool encounters different conflict scenarios (similar to the one shown above). The reasons for all the conflict scenarios, for a given un-testable fault, will explain the un-testability of the fault. The external ATPG constraints that contribute to these conflicts would provide an opportunity to make the fault testable. We embed the identification of these external design constraints into the ATPG algorithm.

The different types of ATPG constraints are: internal cut points, black box instances, latch initialization properties, functional constraints, scan-cell constraints, etc. In order to account for these constraints in ATPG, we create a graph node for each constraint and associate it to the corresponding gate in the circuit. We assign the least decision level – say 0 - to these design constraints. During implication graph construction, these graph nodes become part of the implication graph. In conflict scenarios, the constraints that cause the conflict will be part of the implicants to the conflicting nodes. In the conflict diagnosis step, when we encounter graph nodes with decision level 0 or their implications, they automatically trickle up to the root of the implication graph. At the end of ATPG, when the fault is declared un-testable, all the implicants at decision level 0 can be collected as the reasons for un-testability. This technique is independent of the type of design constraints and can be extended, as necessary, to other types of design constraints as well. It should be noted that the reasons are collected with no significant implementation over-head to the base line ATPG algorithm. The additional performance/memory overhead is dependent on the number of ATPG constraints and the mechanism used to log the un-testability reasons identified for every fault.

Consider the scenario in Figure 2, where we are currently at decision level 6, justifying $G_3=1$ (at node N_{16}). We have two choices [($I_1=0$, $I_2=1$), ($I_1=1$, $I_3=1$)] to justify $G_3=1$. The current decision is C_1: ($I_1=0$, $I_2=1$). It is obvious that N_{19} and N_4 represent conflict values at gate G_7 and we have to backtrack. After we do a conflict

978-1-61284-657-6/11 $26.00 © 2011 IEEE

diagnosis step, as explained earlier, we identify that N_{18}: ($I_1=0$, $I_2=1$) and N_4: ($G_7=0$) are the implicants for the conflict. Therefore, ($I_1=0$, $I_2=1$) cannot be justified because of the reason-set: $R_1=\{N_4 @1\}$.

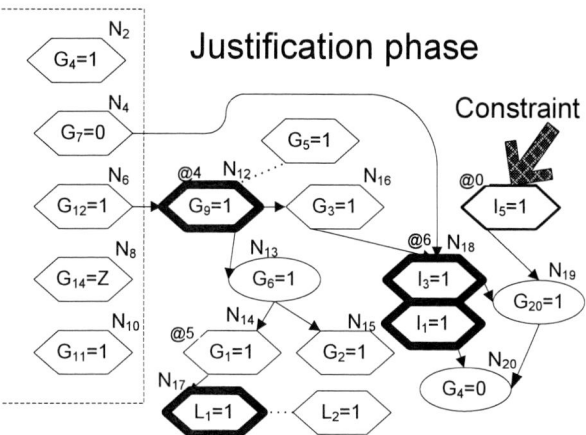

Figure 3: Untestability Constraint Identification

Now, we make the next choice, C_2: ($I_1=1$, $I_3=1$) as shown in Figure 3 (only the required and changed nodes in the justification phase are shown). The reason for the previous conflict, N_4, implies this alternative choice and is added as a fan-in to N_{18}. It can be seen from the implication graph in Figure 3 and the circuit in Figure 1 that: (i) the decision ($I_3=1$) and the circuit constraint ($I_5=1$) imply ($G_{20}=1$) and (ii) ($G_{20}=1$) and ($I_1=1$) imply ($G_4=0$). This again leads to a conflict at gate G_4 and the corresponding conflict reasons are: $\{I_5 @0, N_2 @1, N_{18} @6\}$. This completes all the choices at decision level 6.

Then, the reasons identified for the conflict scenario at the J-Frontier G_3 at N_{16} are: $\{N_{12} @4, N_4 @1, N_2 @1, I_5@0\}$. Now, we can non-chronologically backtrack to the decision node at the maximum decision level $N_{12} @4$ and make the alternate choice $G_5=1$. The other reasons (in lesser decision levels) imply the alternate choice and are added as fan-ins. In this process, the constraint graph nodes will also be added as implicants. These implicants will be carried over to other conflict scenarios in lesser decision levels if they contribute to the conflicts.

For an un-testable fault, as we continue this conflict diagnosis step for lesser decision levels, the relevant constraint graph nodes or their implications will trickle up to the root of the AND-OR graph. At the end of ATPG, all the constraints that contribute to the un-testability of the fault can be collected as fan-ins to the root of the implication graph. It should be noted that we skip the constraints contributing to redundant conflicts. When backtracking non-chronologically, the redundant conflict search spaces are pruned without conflict diagnosis step (for example: Node N_{17} in Figure 3).

The identification of un-testability constraints is independent of the ATPG tool and can be extended to any ATPG algorithm by learning from the conflict scenarios. However, different set of conflict reasons can explain the un-testability of a given fault. These different sets of reasons are sensitive to the ATPG algorithm and also the guidance heuristics – analogous to the different test patterns for a given testable fault. We may get an over-approximation of the un-testability constraints if we perform conflict diagnosis on redundant conflict scenarios during the search. Albeit, in practice, low coverage debug is performed cumulatively on all un-testable faults and the common reasons serve effectively to address the coverage gap.

3.1 Blocking gate identification

During ATPG, we implicitly try to propagate the fault through every possible fan-out path in the circuit. We stop at the D-frontier to justify the off-path inputs and propagate the fault past the D-frontier. Effectively, every D-frontier gate serves as a light-house to propagate the fault effect to an observe point.

In case of an un-testable fault, the fault would have been blocked at a specific D-frontier in every propagation path or a set of propagation paths. These D-frontier gates are marked as blocking gates for the fault, and the reasons from the conflict scenario of all off path J-Frontiers are stored as the reasons for the block. This information is useful for coverage analysis to analyze the root-cause for low coverage at a specific D frontier.

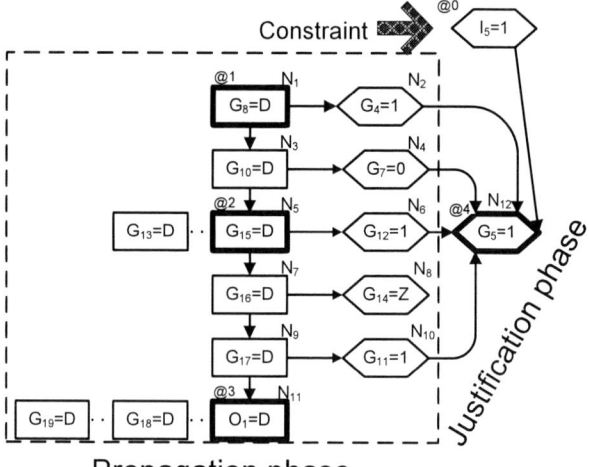

Figure 4: Blocking gate identification

Consider the scenario in Figure 4. We have exercised both the choices at N_{12} (i.e $G_9=1$ and $G_5=1$). After doing a conflict diagnosis on node N_{12}, it is identified that we can backtrack to the D-frontier at node $N_5 @2$. At this point, all the constraints in the reason set for the conflict scenario are responsible for blocking the fault effect at

978-1-61284-657-6/11 $26.00 © 2011 IEEE

the D-frontier. We store the information that the external constraint $I_5=1$ @ 0 is responsible for the conflict at $G_{15}=D$. In this way, we can log the debug information for all blocking gates and use it for the coverage analysis during low coverage debug. It should be noted that we have also pruned the search space for the D-frontier choices at node N_{11} @3 and continued without affecting non-chronological backtracking.

4 Un-testability Debug Analysis flow

In Low Coverage Analysis [14, 15], typically, we pick a fault based on design knowledge or concentration of un-testable faults in instances. After analyzing the fault in ATPG, there is a post-processing step to confirm the reasons with the designers and integrate the learning back to ATPG. Therefore, it is critical that we identify the right reasons earlier to improve the through put time for Low Coverage Analysis. The techniques in the previous section enable us to identify the right reasons for each fault. In order to do a cumulative and systematic analysis of all un-testable faults, we enabled a sorting and ranking mechanism in the tool to identify the faults/reasons with high coverage impact.

As part of the ATPG suite, we have developed a set of interactive commands that (i) analyze the un-testability reasons and the un-testable faults, as post-atpg step, (ii) sorts and ranks them based on user defined criteria and, (iii) helps to expedite the low coverage debug. We can do a cumulative and systematic analysis on a set of un-testable faults based on the fault status or reason type or fault location. Then, we can address the critical reasons reported by these interactive commands in order to improve the fault coverage. Specifically, the interactive commands enable to:

- report the top reasons that are responsible for most of the coverage loss

- display the faults corresponding to specific set of reasons / instance hierarchies

- select and debug an un-testable fault that explains a majority of the coverage gap.

In Figure 5, we show the flow to effectively perform Un-testability Debug Analysis. In Figure 5, the grey boxes require interactions with designers and the white boxes are based on the new commands in the ATPG suite. First, we have to understand the coverage gap between the target and the current coverage. If the coverage gap is high, we can use the interactive commands to analyze the reasons and determine the faults / reasons with high coverage impact. These can lead to three fan-outs: (i) the faults are not scan-targeted faults – example, faults on the scan path, faults targeted by other DFT techniques, (ii) the given ATPG constraints are incorrectly specified

and need a fix – example, incorrectly initialized sequentials and, (iii) the constraints identified by ATPG are valid and we need to live with the coverage gap – example VSS, GND. Based on the decision, made in the previous step, we can re-estimate coverage and continue iteratively until the coverage gap is resolved.

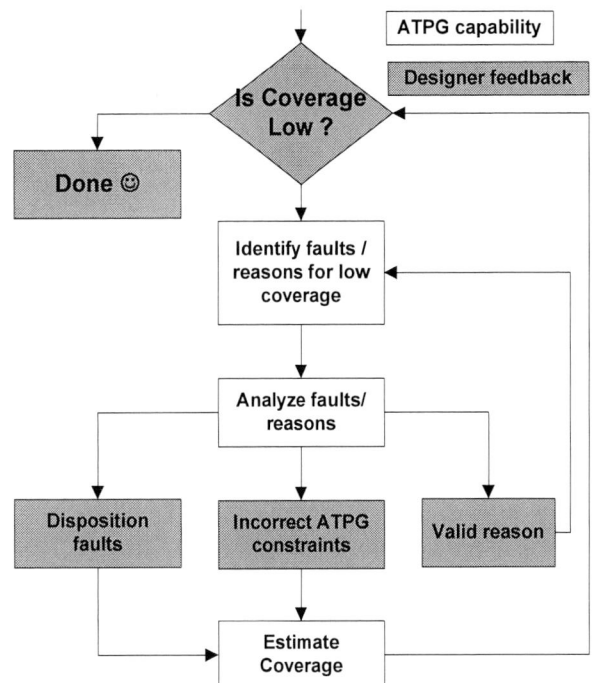

Figure 5: Un-testability debug analysis

6 Experimental Results

The proposed techniques were implemented in Intel's in-house sequential ATPG tool and used in the scan ATPG flow for industrial micro-processor designs. We replaced the traditional approach of manual analysis of the fault lists and ATPG reports with our semi-automated method. In previous generation products, the throughput time for low coverage analysis was dependent on the experience of the DFT engineer and the opportunistic reasons that were identified earlier. With our method, we have enabled a systematic and cumulative approach to low coverage analysis.

In Table 1, we report the impact of our techniques on real micro-processor designs from the production scan ATPG flow. Column 1 represents the design, Column 2 is design size, Column 3 lists coverage gap explained by our technique, Column 4 lists the major type of ATPG constraint reported and Column 5 reports the explanation from a design perspective. We enabled our technique, as a debug ATPG run, on designs with low coverage issues and instance specific fault lists. We were able to resolve 7%-15% coverage gap on real life micro-processor designs. The cumulative analysis on all faults with our

978-1-61284-657-6/11 $26.00 © 2011 IEEE

ranking and reasoning mechanism was practically useful to narrow down the most effective reasons.

Design	Size	Gap	Constraint	Explanation
D1	6.3M	7%	Sequential latch initialization	Disabled functional block
D2	596K	11%	Clock signals	Un-targeted clock domain
D3	21K	15%	Internal cut-point	In-correct assumption
D4	20K	15%	In-sufficient test cycles	In-adequate scan
D5	20K	15%	Insufficient test cycles	In-correct clocking

Table 1: Low Coverage analysis on industrial designs

We performed controlled experiments with same ATPG parameters to demonstrate the over-head data in Table 2 for high-coverage to low-coverage designs. There is 52% memory and 3.5X performance over-head for D3 with highest 35% un-testables. It is seen that the memory and performance over head is sensitive to the % un-testables and size of the design. Nevertheless, it should be noted that Low Coverage Analysis is enabled as part of the debug flow and does not affect the performance and capacity of regular flow (without low coverage analysis).

D	Unt	Size	Run time (s)			Memory (M)		
		#gates	Regular	Debug	Over-head	Regular	Debug	Over-head
D1	13%	6.3M	2616	3883	48%	4.9G	5.96G	21%
D2	35%	603K	614	2800	356%	613M	932M	52%
D3	3%	453K	398	483	21%	419M	560M	33%
D4	3%	120K	1771	1800	1.6%	259M	287.9	11%
D5	20%	21K	88	108	22%	101M	111M	9.9%

Table 2: Over head of logging un-testability reasons

7 Conclusion

We proposed a novel technique to do un-testability root-causing in ATPG and use the identified reasons to do low coverage analysis. The technique is embedded along with the ATPG algorithm and has minimal implementation overhead. The technique moves the onus on the tool to perform low coverage debug with less dependency on manual intervention. On real life microprocessor designs, this technique has resolved up to 15% coverage gap during low coverage debug step in scan ATPG flow at Intel Corporation.

8 Acknowledgements

We would like to acknowledge the ATPG team at Intel Corporation for their support throughout this work.

9 References

[1] M. Abramovici, M.A. Breuer and A.D. Friedman, "Digital Systems Testing & Testable Design", *IEEE Press*, 1990

[2] Cheng, "Split circuit model for test generation", *Proceedings of IEEE DAC*, 1988, pp. 96-101

[3] M.H. Schulz, E. Trischler and T.M. Sarfert, "SOCRATES: A Highly Efficient Automatic Test Pattern Generation System", *IEEE Transactions on CAD*, Vol. 7, No. 1, 1988, pp. 126-137.

[4] I. Hamzaglou and J. Patel, "New Techniques for Deterministic Test Pattern Generation", *Proceedings IEEE VLSI Test Symposium,* 1998, pp. 446-452

[5] W. Kunz and D.K. Pradhan, "Recursive Learning: A new implication technique for efficient solutions to CAD problems – Test, Verification and Optimization", *IEEE Transactions of Computer Aided Design of Integrated Circuits and Systems,* 1994, pp. 1143-1158

[6] J. Giraldi and M. Bushnell, "EST: The new Frontier in Automatic Test pattern Generation", *Proceedings of IEEE DAC,* 1990, pp. 667-673

[7] J.P. Marques Silva and K.A. Sakallah "Dynamic Search-Space Pruning Techniques in Path Sensitization", *Proceedings of IEEE DAC,* 1994

[8] H. Cox and J. Rajski, "On necessary and non-conflicting assignments in Algorithmic Test Pattern Generation", *IEEE Transactions of Computer Aided Design of Integrated Circuits and Systems,* Vol 13, No. 4, April, 1994, pp. 515-530

[9] Chen Wang et al., "Conflict driven techniques for improving Deterministic Test Pattern Generation", *Proceedings of IEEE ICCAD,* 2002, pp. 87-93

[10] T. Larrabee, "Test pattern generation using Boolean Satisfiability", *IEEE Transactions of Computer Aided Design,* Vol. 11, Jan 1992, pp. 4-15

[11] J.P. Marques Silva and K.A. Sakallah, "GRASP – A new search algorithm for satisfiability", *Proceedings of IEEE ICCAD,* 1996, pp. 220-227

[12] Lu et al., "An Efficient Sequential SAT solver with improved search strategies", *Proccedings of IEEE Conference on DATE,* 2005, pp. 1102-1107

[13] S. Bommu, K. Chandrasekar, R. Kundu and S. Sengupta, "CONCAT: Conflict Driven Learning Techniques in ATPG for Industrial designs", Proceedings of IEEE ITC, 2008, pp. 1-10

[14] L. Chang, "Systematic Methodology with DFT rules reduces fault coverage analysis", www.eetimes.com, August 2001

[15] R. Fisette, "Debugging Low Test Coverage situations", www.electronicdesign.com, Nov. 2009

Prediction of Compression Bound and Optimization of Compression Architecture for Linear Decompression-based Schemes

Jia Li
School of Software,
Tsinghua University
Beijing, China
Email: jiali@mail.tsinghua.edu.cn

Yu Huang
Mentor Graphics Corporation
300 Nickerson Rd., Marlboro,
MA 01752, USA
Email: yu_huang@mentor.com

Dong Xiang
School of Software,
Tsinghua University
Beijing, China
Email: dxiang@tsinghua.edu.cn

Abstract—On-chip linear decompression-based schemes have been widely adopted by industrial circuits nowadays to effectively reduce the ever increasing test data volume and test time. Though they can easily achieve relatively high compression ratio, there is a bound of effective compression ratio for these compression schemes. Prior work tried to address this problem by trying different compression architectures to identify this compression bound. However, they can not predict this compression bound efficiently. In this paper, we will first analyze the correlation between the effective compression ratio and the compression architecture, thus to predict that compression bound efficiently. In addition, this paper will also propose how to design the compression architecture for target effective compression ratio with one-pass calculation, which was usually done by a time-consuming try-and-error process as well in the current DFT flow. Experimental results show the accuracy of the prediction and the effectiveness of the compression architecture design.

Keywords-test compression; linear decompression-based; test compression optimization; compression bound prediction;

I. INTRODUCTION

Large test data volume has become a more and more serious concern for the semiconductor testing with the increasing levels of integration and new deep submicron fault models of current SoCs [1]. To address this problem, various test compression techniques have been proposed in the literature, by exploiting the "don't-care" bits (i.e., X-bits) in the test patterns generated by ATPG tool [2].

According to [2], existing test compression techniques fall broadly into three categories [3–5]: 1) Code-based schemes; 2) Linear Decompression-based schemes and 3) Broadcast-scan-based schemes. Generally, Linear Decompression-based schemes, such as (Linear Feedback Shift Register) LFSR [6] and Embedded Deterministic Test (EDT) [7] have been widely adopted by industry designs since they can use a generic decompressor architecture across multiple designs with very little control logic, and can exploit the X-bits more efficiently for test compression.

Researches found that the maximum compression ratio of the above test compression schemes will not exceed the entropy bound decided by the percentage of care bits of a given

This work was supported in part by the National Science Foundation of China under grants No. 60425203, 60910003 and 61006017, in part by the National 863 Project under grant No. 2009AA01Z129, in part by the Key Laboratory of Computer System and Architecture, ICT, CAS(ICTARCH200902), and in part by China Postdoctoral Science Foundation under grants No. 20100470014.

test cube [8], especially for Linear Decompression-based schemes [9]. The compression bounds of industrial designs with embedded test architecture has been exploited in [10] by using elaborated experimental data, which discussed the various parameters that may result in test tradeoffs.

However, there is still no theoretical analysis which can determine the maximum compression obtainable for given test cubes of the circuit under test (CUT). The correlation between the effective test compression ratio and the compression architecture for these schemes are still not well revealed, and there is still no useful guidance for compression architecture design for achieving required effective test compression ratio.

To address the above problems, this paper will analyze the correlation between effective test compression ratio and the compression architecture, thus to prediction the achievable effective compression ratio for the given test cubes. Moreover, we can guide the design of compression architecture for target effective compression ratio based on the prediction.

This paper is organized into six sections. Section II reviews the compression architecture and gives the calculation of effective compression ratio (ECR) of Linear Decompression-based schemes, analyzes the basic parameters that may affect ECR, and demonstrates the motivation of this work. The prediction of ECR with given compression architecture is discussed in Section III. Based on the prediction, Section IV shows how to guide the design of compression architecture for achieving target ECR. The experimental results are presented in Section V, and Section VI concludes the paper.

II. BACKGROUND

A. Linear Decompression Architecture

Fig. 1 shows the basic architecture of Linear Decompression-based schemes [7] containing an on-chip **linear decompressor** and an on-chip **selective compactor**. Various combinational or sequential elements are utilized for the decompressor and compactor logic [10]. The scan cells in the design are partitioned into multiple short scan chains to reduce the test application time. Compressed test cubes are delivered to the decompressor through external channels, and the compacted test responses are also transferred out through the external channels to the ATE. The test data volume reduction is achieved by the reduction of test data on the external channels compared to the test data on the

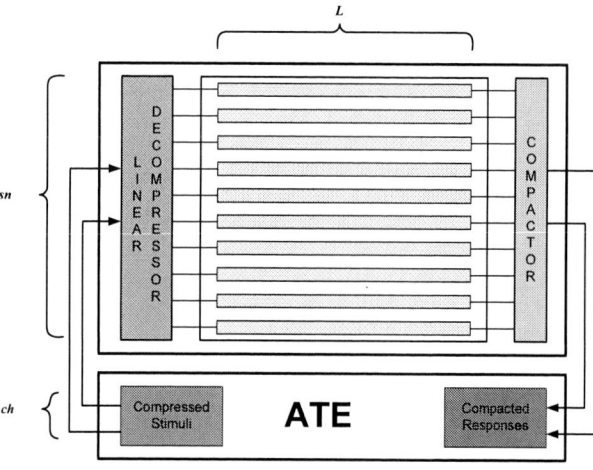

Figure 1. On-chip linear decompression-based scheme.

internal scan chains, e.g. in Fig. 1, there are two external channels and ten internal scan chains, which means only two compressed test bits are transferred from the ATE to provide the test stimuli for ten scan cells and the test responses of ten scan cells are compacted to only two bits in each shift cycle.

From Fig. 1 we can find that the key parameter of the on-chip compression architecture is the ratio between the number of internal scan chains sn and the number of the external channel ch: $R = sn/ch$. However, it is not always better to select higher R for achieving higher compression ratio. There exists a limit for the selection of R decided by the care-bit density of the given test cubes of the CUT. If it is higher than the limit, the test pattern count of the new test set will increase compared to the original test set and the test coverage of the new test pattern set may decrease dramatically, which makes the compression architecture "breakdown" [10].

To show the total test data volume reduction with Linear Decompression-based schemes considering both the test data reduction per test pattern and the test pattern count increment, the calculation of effective compression ratio (ECR) should be conducted to show the total test data volume reduction by the compression techniques.

B. Calculation of Effective Compression Ratio (ECR)

For example in Fig. 1, suppose the maximum length of the internal scan chains is L, the number of internal scan chains is sn, the number of external channels is ch, the pattern count of the original test set without compression and the new test set after compression are op and np, respectively. Its ECR should be:

$$ECR = \frac{sn * L * op}{ch * L * np} = \frac{sn * op}{ch * np}$$
$$= R * \frac{op}{np} \qquad (1)$$

From the above equation we can see that increasing the number of internal scan chains and reducing the external channel count can help increasing the first factor of the

formula: $R = sn/ch$. However, if R is too high, as analyzed in previous subsection, there might be more test patterns, thus reducing the second factor of the formula: op/np, and leading to lower ECR. Besides np, the other variables are all known with given compression architecture. Therefore, it is important to reveal the correlation between np and the given compression architecture for effective prediction of ECR. According to analysis of the previous subsection, np is correlated to both the density of care-bits in the given test cubes and R, which will be analyzed in Section III.

C. Motivation

Prior work show that there exists a maximum compression bound for scan-based compression techniques [8], and [10] has shown the existing evidence of this compression bound by varying the parameters of compression architecture designs. However, there is still no theoretical analysis that could be used to determine the maximum ECR obtainable, and the compression architecture is usually determined by a time-consuming try-and-error process in the current DFT flow [10]. Therefore, in this work, we will give a theoretical analysis on prediction of ECR, and reveal the correlation between the compression architecture design and ECR. The contribution of this work includes:

1) **Estimate ECR for given test cubes with given compression architecture.** By analyzing the correlation between R and np, we can obtain the estimation formula for ECR considering both the density of care-bits of the test cubes and R.
2) **Predict the maximum ECR bound for given test cubes.** The maximum compression bound is theoretically analyzed and deduced from the prediction of ECR, which is proven to be only correlated to the 'entropy' – care-bit density of given test cubes.
3) **Optimize the compression architecture for target ECR.** With the estimation formula of ECR, R can be directly computed for the target ECR, which can be used to guide the compression architecture design without the time-consuming try-and-error process.

III. PREDICTION OF EFFECTIVE COMPRESSION RATIO WITH GIVEN COMPRESSION ARCHITECTURE

As mentioned earlier, the key point of ECR prediction is to predict the value of np. In this section, the prediction of np will first be analyzed, and then be integrated for ECR prediction. In the rest of this paper, we use ring generator as an example of linear decompressor. Without losing generality, the same flow can be applied to other linear decompressors as well.

A. Prediction of np with Given Compression Architecture

Since the decompressor can be regarded as a linear system represented by a $M \times N$ matrix [7], and its rank should be lower than $min(M, N)$. As the matrix after the Gaussian-Jordan elimination shown in Fig. 2, $M = sn*L$ equals to the total test bits of one test pattern; $N = D + ch*L$, where D is the size of the ring generator and $\{S_1, S_2, ..., S_D\}$ is the initialized state of the ring generator. Since $(D + ch*L) \ll$

978-1-61284-657-6/11 $26.00 © 2011 IEEE

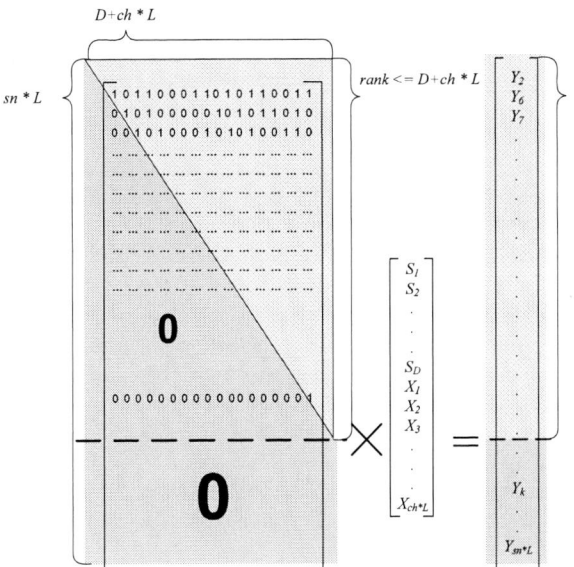

Figure 2. Linear system of the decompressor.

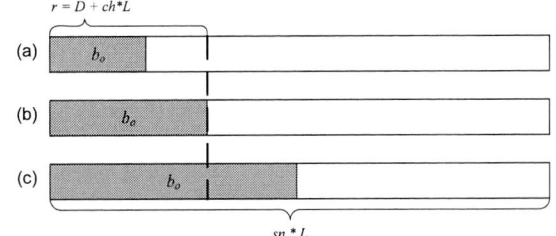

Figure 3. Different care-bits densities and the rank of the linear system.

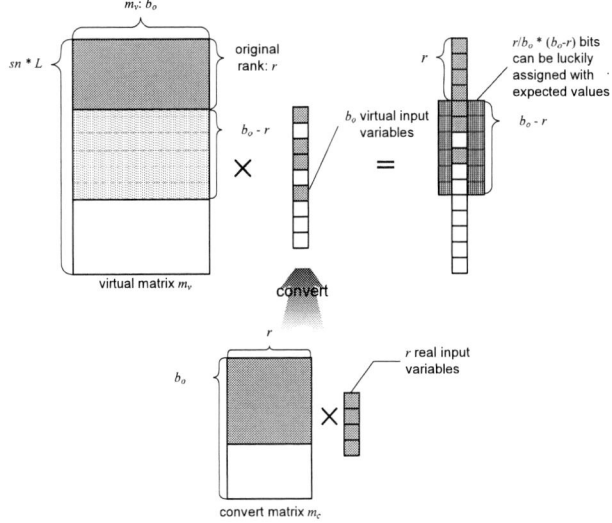

Figure 4. Virtual Matrix conversion.

$(sn * L)$ in normal compression settings, the rank of this system should be lower than $(D + ch * L)$, which means in $\{Y_1, Y_2, ..., Y_{sn*L}\}$, the number of independent variables is no greater than $(D+ch*L)$, while the other variables will be decided by the value of these independent variables. In real test compression cases, if there are fewer than $(D + ch * L)$ care-bits in one test pattern, there is a very high probability that they can be compressed. However, if the care-bit count of one test pattern is higher than $(D + ch * L)$, there must be dependent relationship between some bits.

Suppose the number of total care-bits of the original test set is $care$, to achieve the same test coverage with the new test pattern set after compression, the number of total care-bits in the new test set should also be $care$. If the average care-bits per test pattern of the original test set and the new test set are b_o and b_n, respectively, the new test pattern count should be:

$$np = \frac{care}{b_n} = \frac{b_o * op}{b_n} \qquad (2)$$

To estimate b_n, we need to consider different situations of care-bits density of the original test set as shown in Fig. 3: (a) $b_o < r$; (b) $b_o = r$ (c) $b_o > r$, where r is the rank of the decompressor. For situations (a) and (b), since b_o care-bits can all be decompressed by selecting them as members of the maximal independent system, $b_n = b_o$. But for (c), only r in b_o care-bits can be directly decompressed as their expected logic values, for the rest $(b_o - r)$ care-bits, there are possibilities for them to be luckily assigned with the expect logic values:

Lemma 1: The possibility of one care-bit in $(b_o - r)$ care-bits to be luckily assigned with its expected logic value is r/b_o.

Proof: As illustrated in Fig. 4, suppose there is a virtual matrix m_v containing $sn * L$ lines and b_o columns. The

rank of m_v is approximately b_o, because $b_o \ll sn * L$. It means b_o care-bits can be decompressed with b_o input free variables. However, since there are only $r = D + ch * L$ real input variables, these b_o virtual input variables can be converted from r real input variables by a convert matrix m_c containing b_o lines and r columns. In this matrix, besides the r lines that can be directly solved from r care-bits in the test pattern, the rest lines are decided by these r lines. In general, for each line of m_v represented by b_o virtual input variables, among these virtual input variables there are only r real input variables. Therefore in the original matrix, the possibility for one care-bits which is not in the maximal independent system to be luckily assigned with its expected logic value is r/b_o. ∎

So the number of effective test bits per pattern with compression can be estimated as:

$$b_n = r + \frac{r(b_o - r)}{b_o} = r(2 - \frac{r}{b_o}) \qquad (3)$$

Now we can predict the new pattern count np with Eq.(2) when $b_o > r$:

$$np = \frac{b_o * op}{r(2 - r/b_o)} = \frac{b_o^2 * op}{r(2b_o - r)} \qquad (4)$$

Note that, when $b_o \leq r$, np should be approximately equals to op.

B. ECR Prediction Formula

As analyzed in the previous section, the number of test patterns may only increase when $b_o > r$, so the prediction of ECR should also be conducted under the different conditions below:

1) If $b_o \leq r$: When np should be approximately equals to op, so ECR can be simply obtained by: $ECR = sn/ch = R$;

2) If $b_o > r$: With the prediction of np when $b_o > r$, it is easy to obtain the prediction of ECR now by Eq.(1):

$$ECR = \frac{sn}{ch} * \frac{op}{np} = \frac{sn * r(2b_o - r)}{ch * b_o^2}$$
$$= \frac{sn(D + ch * L)(2b_o - D - ch * L)}{ch * b_o^2} \quad (5)$$

To understand the above equation more clearly, it can be simplified as below: Since D is usually much smaller than $ch * L$, it can be omitted from this equation. Suppose the average number of care-bits in each shift cycle of the original test cubes is c, b_o can be represented as $c*L$, and the density of care-bits in the original test cubes can be defined as: $\rho = b_o/(sn*L) = c*L/(sn*L) = c/sn$. Now the prediction of ECR becomes:

$$ECR = \frac{sn * ch * L(2c * L - ch * L)}{ch * (c * L)^2}$$
$$= \frac{sn * (2c - ch)}{c^2} = \frac{2}{\rho} - \frac{1}{R * \rho^2} \quad (6)$$

Since ρ and b_o are both constants, the value of ECR is only correlated to R: it grows with the increment of R, however, its limit (maximum value) should be:

$$ECR_{max} = \frac{2}{\rho} \quad (7)$$

IV. GUIDANCE FOR COMPRESSION ARCHITECTURE DESIGN

It is usually difficult and time-consuming for existing Linear Decompression-based schemes to decide the number of internal scan chains or the number of external channels for achieving target ECR. Based on the prediction formula of ECR proposed in this paper, $R = sn/ch$ can be easily calculated now to guide the compression architecture design including selection of sn and ch for achieving the target ECR value. There are basical guidelines for designing the compression architecture:

1) R should not be too small.

From the previous section we can find that, when $b_o \leq r$, the value of ECR is only correlated to the ratio between sn and ch. Therefore, if ch is fixed, it is better to have higher sn, or if sn is fixed, reducing ch can produce higher ECR. If $b_o = ch * N/sn = N/R$, where N is the total scan cell count of the CUT, which infers $R = N/b_o = 1/\rho$. Therefore, if ch is fixed, the minimum internal scan chain count for compression without pattern increment is $sn_{min} = ch/\rho$, which should be the minimum internal scan chain count; and if sn is fixed, the maximal ch without pattern increment is

$ch_{max} = sn * \rho$, which infers the maximal external channel count. In another word, when the internal scan chain count is higher than sn_{min} or the external channel count is lower than ch_{max}, ECR will grow linearly with R.

2) R should not be too high.

If the ratio between sn and ch is too high, when the new care-bit count after the decompression is lower than the total stimulus bit count for one fault, this fault will become untestable by the new test set, thus cause test coverage loss [10].

3) How to decide R for target ECR?

For most common cases, R should be decided in the region between the lower and upper bound determined by 1) and 2), respectively, to achieve higher effective compression ratio without test coverage loss. From Eq.(7) we know that the maximum ECR can be achieved is $2/\rho$, and from 1) we know that as long as $R < 1/\rho$, $ECR = R$, therefore for the target ECR varies in region of $[1/\rho, 2/\rho]$, R can be decided by Eq.(6) as below:

$$ECR = \frac{2}{\rho} - \frac{1}{\rho^2 * R}$$
$$\implies R = \frac{1}{\rho(2 - ECR * \rho)} \quad (8)$$

Since b_o, ρ, N and ECR are all given constants, it is easy to get the value of R, and then ch and sn can be decided with given constraints.

V. EXPERIMENTAL RESULTS

To evaluate accuracy of the proposed ECR prediction formula, we have conducted several sets of experiments on larger ITC'99 benchmark circuits. The statistics of the experimental circuits and their test cubes without compression targeting at stuck-at faults generated by a commercial tool are listed in Table. I, the number of scan cells in these circuits are shown in the column under "N", the average number of care-bits for each test pattern of each circuit is given in the column under "b_o", the original test pattern counts of these circuits are listed in the column under "op", and the densities of care-bits in these test cubes: $\rho\% = b_o/N \times 100\%$ are calculated as shown in the column under "$\rho\%$".

The key point for validating the accuracy of the proposed ECR prediction is to validate the accuracy of prediction of np. As long as np is well predicted, since R and op are both known constants, it is easy to obtain ECR by Eq.(1). So first of all, with given care-bit density of the test cubes and different settings of R, we will compare the predicted np and the real np to see if the prediction is accurate enough. After validating the accuracy of the prediction of ECR, it is possible now to guide the design of the compression architecture with R calculated by Eq.(8) for target ECR. Therefore, in the second set of experiments, for different target ECR values, we will guide the compression architecture design according to the corresponding R, and compare the real ECR with the target ECR to see if they match.

Table I
STATISTICS OF THE EXPERIMENTAL CIRCUITS.

circuits	N	b_o	op	$\rho\%$
b20	430	156	415	36.3%
b21	430	151	430	35.1%
b22	613	217	416	35.4%
b17	1319	272	635	20.6%
b18	3023	560	675	18.5%
b19	6056	1083	690	17.9%

A. Experiments on prediction of np

To validate the accuracy of ECR prediction for given test cubes, two sets of experiments are conducted: In the first set, the circuits are designed with fixed external channel count ch and variable internal scan chain count sn, while in the second set, the compression architecture of the circuits are designed with fixed sn and variable ch.

In Table II, ch is fixed as 4, and sn for each scan chain are tried with different settings. Since when $sn < ch/\rho$, np should be approximately equal to op according to the analysis in previous section, in this set of experiment, sn are all selected higher than ch/ρ to show the increment of np as R grows. In this table, the number of internal scan chains are shown in the column under "sn", the predicted np calculated by Eq.(4) are listed in the column under "np_{pred}" and the real np obtained by ATPG for the circuit with the given compression architecture done by a commercial ATPG tool are listed in the column under "np_{real}". The difference between the prediction and the real case are given in the column under "$|\triangle\%| = |(np_{pred} - np_{real})|/np_{real}$", and the fault coverage achievable by ATPG of each circuit with the compression architecture is listed in the column under "$fc\%$".

From this table we can see that in most cases, the difference between np_{pred} and np_{real} is very small, however, there exist some cases that the predicted np is about 30-40% higher than the real np ($sn = 40$ for circuit b20 and b21, $sn = 60$ for circuit b22). From the observation of $fc\%$ of these cases, we can find that, the fault coverage of these cases are lower than other sn settings, which means selecting these sn values for these circuit will cause fault coverage drop, in another word, some faults can not be detected with these compression architecture designs. Without detecting these faults, the test pattern count will also drop compared with the predicted test pattern count with which all the faults will assume to be detected.

Similarly, Table III gives the experimental results for the circuits designed with fixed $sn = 25$ and variable ch settings. From this table we can also find that besides the settings with fault coverage drop, the predicted np values are very close to the real np values.

The correlation between ECR and R for given test cubes can be seen clearly from Fig. 5. For circuit b22, 9 different R values are selected for experiments: R is in the range of [2-10]. The maximum compression bound shown as "bound" is estimated by $2/\rho$. ECR_p shows the predicted ECR value with given R and ρ, and ECR_r gives the real ECR value

Table II
PREDICTION OF np WITH FIXED ch AND VARIABLE sn.

$ch=4$	sn	np_{pred}	np_{real}	$|\triangle\%|$	$fc\%$
b20	12	418	492	15.1%	98.81%
	20	520	538	3.4%	98.81%
	30	691	677	2.2%	98.65%
	40	873	658	32.7%	98.04%
b21	12	431	492	12.4%	98.64%
	20	528	537	1.7%	98.63%
	30	699	660	5.9%	98.49%
	40	880	625	41.9%	97.64%
b22	12	417	493	15.3%	98.39%
	20	513	598	14.2%	98.39%
	40	857	845	1.5%	98.27%
	60	1219	904	34.9%	97.70%
b17	20	636	636	0%	97.18%
	40	864	798	8.3%	97.18%
	60	1171	1150	1.9%	97.18%
	80	1490	1441	3.4%	97.18%
b18	25	688	686	0.3%	97.38%
	50	997	1025	2.8%	97.37%
	100	1752	1929	9.2%	97.38%
	150	2526	2713	6.9%	97.36%
b19	25	698	709	1.6%	97.07%
	50	993	1047	5.1%	97.07%
	100	1737	1969	11.8%	97.08%
	200	3268	3696	11.6%	97.06%

Table III
PREDICTION OF np WITH FIXED sn AND VARIABLE ch.

$sn=25$	ch	np_{pred}	np_{real}	$|\triangle\%|$	$fc\%$
b20	2	1058	683	54.8%	97.56%
	4	604	618	2.3%	98.75%
	6	469	460	1.9%	98.82%
	8	421	439	4.1%	98.82%
b21	2	1065	617	72.6%	97.15%
	4	611	617	1.0%	98.55%
	6	478	467	2.3%	98.64%
	8	433	441	1.7%	98.64%
b22	2	1038	883	17.5%	98.05%
	4	595	666	10.7%	98.38%
	6	464	487	4.7%	98.39%
	8	420	454	7.5%	98.39%
b17	2	1015	944	7.6%	97.18%
	4	669	650	2.9%	97.18%
	6	635	629	1.0%	97.18%
	8	635	629	1.0%	97.18%
b18	2	997	1019	2.2%	97.38%
	4	688	686	0.3%	97.38%
	6	675	656	2.9%	97.37%
	8	675	657	2.7%	97.37%
b19	2	993	1058	6.1%	97.07%
	4	698	709	1.8%	97.07%
	6	690	678	1.6%	97.08%
	8	690	679	1.6%	97.08%

978-1-61284-657-6/11 $26.00 © 2011 IEEE

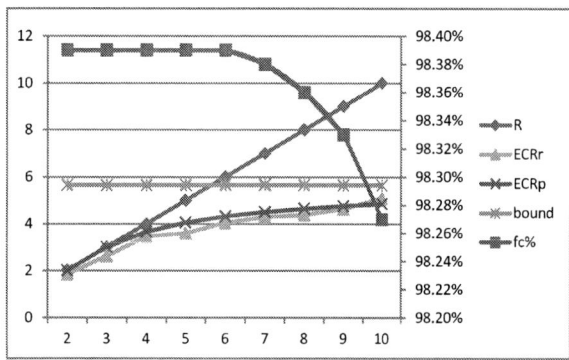

Figure 5. Correlation between R, ECR and the compression bound of b22.

Table IV
COMPARISON BETWEEN TARGET ECR AND REAL ECR.

	ECR_t	R_{cal}	ECR_r	$|\triangle\%|$	$fc\%$
b20	3.5	3.8	3.38	3.4%	98.82%
	4.5	7.5	4.60	2.2%	98.65%
b21	3.5	3.7	3.54	1.1%	98.64%
	4.5	6.8	4.65	3.3%	98.48%
b22	3.5	3.7	3.31	5.4%	98.39%
	4.5	6.9	4.15	7.8%	98.38%
b17	7	8.7	7.50	7.1%	97.18%
	9	33.7	10.42	15.7%	97.08%
b18	7	7.7	7.03	0.4%	97.38%
	9	16.2	8.28	8.0%	97.38%
b19	7	7.5	6.83	2.4%	97.07%
	9	14.3	8.35	7.2%	97.07%

by experimental results. From this figure we can see that when R is in the region $[1, 1/\rho]$, ECR is approximately equal to R, and when R is higher than $1/\rho$, it will grow slower and will not exceed the "bound". As R grows to a limit, the fault coverage drops rapidly which will make the compression infeasible. We can also verify from this figure that the predict ECR is very close to the real ECR value.

B. Experiments on compression architecture design

With the guidance of ECR prediction, we can decide the design of the compression architecture for the CUT now. From our analysis, to achieve effective compression, the target ECR should be chosen between $1/\rho$ and $2/\rho$. Therefore, for the experimental circuits, we've arbitrarily chosen some target ECR values in this region and check the real ECR obtainable with R calculated by Eq. (8).

From Table IV we can see that in most cases, the difference between the target ECR (ECR_t) and the real ECR (ECR_r) obtained with the compression architecture design guided by R_{cal} calculated by Eq. (8): $|\triangle\%| = (ECR_r - ECR_t)/ECR_t$ is below 10% when the fault coverage is not dropped. However, for compression architecture with relatively higher R which will cause fault coverage drop, i.e. $ECR_t = 9$ for circuit b17, the real ECR will be higher than the predict ECR because of the test pattern reduction caused by the faults that can not be detected by the new test pattern set. From the experimental results we can also

conclude that the real maximum compression bound should be decided by the impact of R on fault coverage, since we do not want to lose fault coverage with on-chip compression schemes. When the fault coverage stays similar with that without compression, the proposed ECR estimation can always work well. Therefore a more realistic application is to combine the prediction of ECR proposed in this paper and the prediction of fault coverage proposed in the prior work [8].

In the current DFT flow, the ratio between sn and ch is usually decided by several iterations of try-and-error process. For example, if ch is fixed, sn will be tried to vary in a wide range to find the configuration for achieving the best ECR without fault coverage loss, which is very time-consuming. With the proposed guidance for linear compression architecture design, we can first decide the target ECR in the range of $[1/\rho, 2/\rho]$, and then the corresponding R for achieving this target ECR can be decided by a one-pass calculation.

VI. CONCLUSION

To explore the limitation of Linear Decompression-based test compression schemes and to provide guidance for design the compression architecture effectively, this paper proposed the prediction of Effective Compression Ratio (ECR) with given test cubes and the ratio between the number of internal scan chains and external channels. By analyzing the prediction formula, this work has also concluded the maximum compression bound obtainable for given test cubes. Furthermore, we can also obtain the approximate ratio R between sn and ch for achieving target ECR, which can be used to guide the design of the compression architecture efficiently, while in the current DFT flow, R is usually decided by a time-consuming try-and-error process. Experimental results proved the accuracy of the prediction and the effectiveness of the guidance. Future work will also take the impact of other factors on ECR into account to consider the compression bound for more real cases.

REFERENCES

[1] International SEMATECH. *The International Technology Roadmap for Semiconductors (ITRS): 2009 Edition.* http://www.itrs.net/links/2009ITRS/Home2009.htm, 2009.

[2] N. A. Touba. Survey of Test Vector Compression Techniques. *IEEE Design & Test of Computers*, 23(4):294–303, Jul.-Aug. 2006.

[3] D. Das and N.A. Touba. Reducing Test Data Volume Using External/LBIST Hybrid Test Patterns. In *Proceedings IEEE International Test Conference (ITC)*, pages 115–122, 2000.

[4] K.J. Balakrishnan and N. A. Touba. Relating Entropy Theory to Test Data Compression. In *Proceedings IEEE European Test Symposium (ETS)*, pages 94–99, 2004.

[5] A. Chandra and K. Chakrabarty. Frequency-Directed Run-Length (FDR) Codes with Application to System-on-a-Chip Test Data Compression. In *Proceedings IEEE VLSI Test Symposium (VTS)*, pages 114–121, 2001.

[6] B. Koenemann. LFSR-Coded Test Patterns for Scan Designs. In *Proceedings IEEE European Test Conference (ETC)*, pages 237–242, 1991.

[7] M. Kassab J. Rajski, J. Tyszer and N. Mukherjee. Embedded deterministic test. *IEEE Transactions on Computer-Aided Design*, 23(5):776–792, May 2004.

[8] T. W. Williams. The Limits of Compression. In *Proceedings IEEE International Test Conference (ITC)*, Panel 3.5, 2008.

[9] D. Czysz, G. Mrugalski, J. Rajski, and J. Tyszer. Low Power Embedded Deterministic Test. In *Proceedings IEEE VLSI Test Symposium (VTS)*, 2007.

[10] S. Alampally, J. Abraham, R. A. Parekhji, R. Kapur, and T. W. Williams. Evaluation of Entropy Driven Compression Bounds on Industrial Design. In *Proceedings IEEE Asian Test Symposium (ATS)*, pages 13–18, 2008.

Multi Domain Test:
Novel Test Strategy to reduce the Cost of Test

Yasuhiro Takahashi
S&S Professional Service
Verigy Japan K.K.
9-1, Takakura-cho, Hachioji, Tokyo, 192-0023 Japan
yasuhiro.takahashi@verigy.com
TEL: +81-42-631-8200

Akinori Maeda
Center Of Experts
Verigy Japan K.K.
9-1, Takakura-cho, Hachioji, Tokyo, 192-0023 Japan
akinori.maeda@verigy.com
TEL: +81-42-631-8326

Abstract—The Multi-Domain-Test is the new test strategy to resolve problems and limitations of the Multi-Site-Test and the Concurrent-Test. By this novel test strategy, test time can be reduced down to 50% of the Single-Site-Test with almost the same amount of tester resources. Cost Of Test (COT) can be lower than the Multi-Site-Test that is well used at productions.

Keywordst; Multi Domain Test, Concurrent Test Multi Site Test, Costs Of Test (COT), Test System Configuration

I. INTRODUCTION

All of the dies in wafers must be tested to reduce the assembly cost, and all of the shipped devices must pass the final production test to keep the quality of devices high. The cost of these tests can not be ignored, and lowering the Costs Of Test (COT) is one of the endless challenges of the semiconductor production. The major factors of the COT are the test time of the device and the cost of the ATE.

The Multi-Site-Test is the test strategy to test the plural same devices with one ATE at the same time. The test time of the Multi-Site-Test is almost same as the Single-Site-Test, therefore, the test time per device of the Multi-Site-Test becomes '1/N' of the Single-Site-Test. This strategy has been used from old days for the memory device tests, but nowadays, this strategy is used for almost any semiconductor devices to reduce the COT.

The Multi-Site-Test requires 'N' times of the tester resources. If the ATE has less tester resources than the required, some tests must be performed serially and the test time per device is increased. This requirement of 'N' times of tester resources increases the cost of the ATE. The Multi-Site-Test of a large SOC device requires huge amount of tester resources, and the cost of this ATE must be very expensive.

The test time of the Multi-Site-Test is not exactly same as the Single-Site-Test. The Multi-Site-Efficiency (MSE) is widely used to show how the test time of the Multi-Site-Test is same as the Single-Site-Test. MSE is calculated as shown in (1).

$$\frac{N - \dfrac{Tm}{Ts}}{N-1} \times 100 \qquad (1)$$

Where 'N' is the number of devices tested at the same time, 'Tm' is the test time at the Multi-Site-Test and 'Ts' is the test time at the Single-Site-Test. In general, the MSE of DC and Digital tests are more than 95% but the MSE of analog mixed signal tests is about 60% to 85%.

In the last few years, the Concurrent-Test has been implemented [3], [4], [5], [6]. This is another test strategy to reduce the COT. In general, each test item is performed one by one serially in time. The Concurrent-Test executes multiple test items at the same time. For example, the IP core A and the IP core B in a device are tested at the same time. It is obvious that the Concurrent-Test reduces the test time.

But the Concurrent-Test has several limitations. One is independency of the IP cores. If two IP cores are tested at the same time and both cores are connected together internally, the state change of one IP core by the test causes unexpected state changes to another IP core, and it affects the test results. Next limitation is degradation of performance by the interference. Even if two IP cores are independent, the noise generated by the test for one IP core affects the parametric performance of another IP core through the GND or power supplies. The last limitation is the common block. If tested IP cores share the same circuit block, states of this common circuit block must be same for those IP core tests. If state is different, those IP cores can not be tested at the same time.

II. CONCEPT OF MULTI DOMAIN TEST

To further reduction of the COT, we enhanced the Concurrent-Test. The Figure 1 shows the simplified concept of the proposed Multi-Domain-Test.

There are two new defined words for the Multi-Domain-Test. They are "Domain" and "Pseudo Single Device".

A. Domain

The Domain is one group of IP cores or circuit blocks in the device to be tested. Two or more domains (Multi-Domain) cover entire circuit blocks in the device. Each domain is tested at one of the multiple locations (sockets/insertion) on the load board.

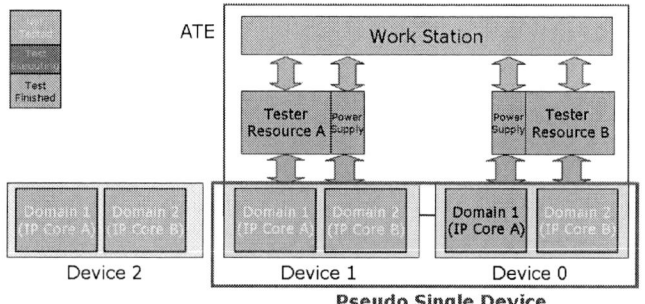

Figure 1.   Multi Domain Test

The Figure 2 shows the two domains example. The Domain 1 contains the analog circuits and the interface circuits. The Domain 2 includes the digital logic circuits. On the load board, there are two sockets (locations). At the left hand side socket, the tests for the Domain 1 are executed, and at the right hand side socket, the tests for the Domain 2 are executed.

B. Pseudo Single Device

The Pseudo Single Device is one group of devices on the load board. The ATE treats this group as one single device. In case of the above example, two devices form the Pseudo Single Device.

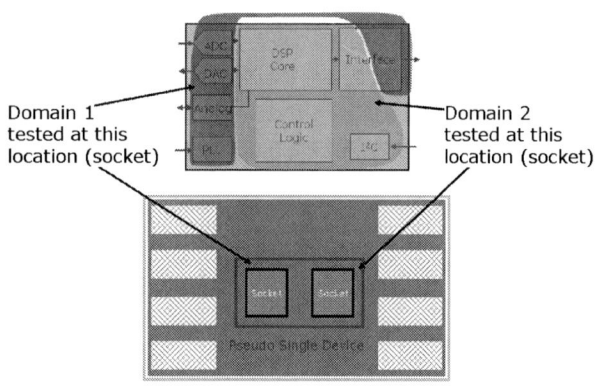

Figure 2.   Two Domains Example

C. Detail of Multi Domain Test

To explain the details of the Multi-Domain-Test, we assume the two domains example. The device has two IP cores, the IP core A and the IP core B. The IP core A is assigned to the Domain 1 and the IP core B is assigned to the Domain 2. On the load board, there are two sockets; the Domain 1 is tested at the left hand side socket and the Domain 2 is tested at the right hand side socket. The ATE has the Tester Resource A, the Test Resource B, the power supplies and the work station.

At the beginning, the Device 0 and the Device 1 form the Pseudo Single Device (shown in the Figure 1), and the ATE executes the Concurrent-Test for the IP core A and the IP core B. The Domain 1 (IP core A) in the Device 1 is tested by the Tester Resource A, and at the same time, the Domain 2 (IP core B) in the Device 0 is tested by the Tester Resource B.

Therefore, from the device point of view, just the regular test is executed, and no Concurrent-Test is executed. This means that the Multi-Domain-Test has no limitations caused by the Concurrent-Test and it can be applied to any devices like non-concurrent testable devices.

As shown in the Figure 4, after all of the tests are done, the Device 0 is binned and the Device 1 is moved to the right hand side socket. Now, the new device, Device 2 is inserted to the left hand side socket. The Device 1 and the Device 2 form the Pseudo Single Device as shown in the Figure 5, and this situation is same as the situation shown in the Figure 1. Then, the ATE executes the Concurrent-Test. These sequences are repeated until all of the devices are tested.

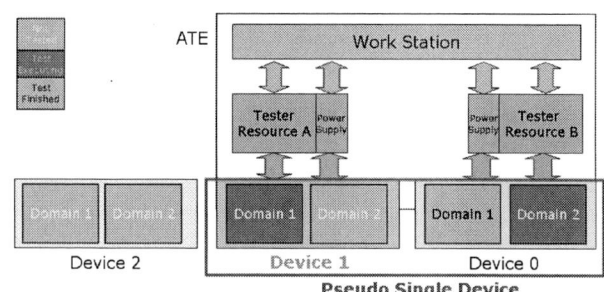

Figure 3.   Multi Domain Test Details -1

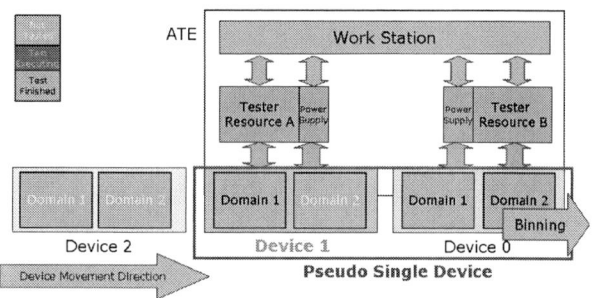

Figure 4.   Multi Domain Test Details -2

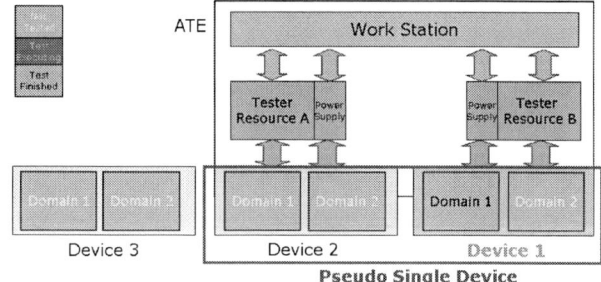

Figure 5.   Multi Domain Test Details -3

D. ATE Configuration of Multi Domain Test

The Figure 6 shows the ATE configurations of the Single-Site-Test, the Concurrent-Test, the Multi-Site-Test and the Multi-Domain-Test. It is obvious that the relation of the costs of these ATE configurations is:

978-1-61284-657-6/11 $26.00 © 2011 IEEE

(Single Site Test) = (Concurrent Test)

< (Multi-Domain Test)

<< (Multi-Site Test)   (2)

The delta of costs between the Single-Site-Test (Concurrent-Test) and the Multi-Domain-Test is not so large. Only difference is the cost of the additional power supplies. But the differences between the Single-Site-Test and the Multi-Site-Test are large because it needs two times of the test resources. This means that the ATE cost of the Multi-Domain-Test is quite lower than the Multi-Site-Test if the device is a large SOC and it requires many test resources.

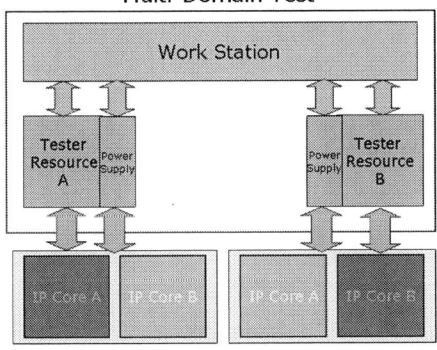

Figure 6.   **ATE Configuration**

E. Test Time of Multi Domain Test

The test time of the Multi-Domain-Test is the longest test time among the test time of each Domain. At the example shown in the Figure 1, we assume that the test time of the

Domain 1 is 'Ta' and the Domain 2 is 'Tb'. Also, we assume 'Ta' is longer than 'Tb'. To calculate the 'per device test time', we assume that the five devices are tested by the Multi-Domain-Test. It needs six test steps to complete the tests of all five devices as shown in the Table 1.

At the first test step (Step 0), the Device 0 is placed into the left test location (location 1). The right test location (location 2) has no device. Next, the Device 0 is moved to the test location 2, and the Device 1 is inserted into the test location 1 (Step 1). These steps are repeated four times (Step 1 to Step 4). At the last test step (Step 5), the Device 4 is located at the test location 2 and no device is at the test location 1. Totally, the time of testing five devices is '6*Ta'. The 'per device test time' is '1.2*Ta'.

Table 1.   **Test Steps of 5 Devices by Multi Domain Test**

Location	1 (Left)	2 (Right)	Test Time
Tested Domain at this location	Domain 0	Domain 1	
Step 0	Device 0	--	Ta
Step 1	Device 1	Device 0	Ta
Step 2	Device 2	Device 1	Ta
Step 3	Device 3	Device 2	Ta
Step 4	Device 4	Device 3	Ta
Step 5	--	Device 4	Ta

Ta: Test Time of Domain 1

If number of devices is N, the 'per device test time' of the Multi-Domain-Test, 'Tpd' is:

$$Tpd = (1 + 1/N) * Ta \qquad (3)$$

Because the test time of the Single-Site-Test is 'Ta+Tb', the 'per device test time' of the Multi-Site-Test, 'Tpm' is:

$$Tpm = (Ta + Tb) * (1 - 0.5 * Em/100) \qquad (4)$$

Where 'Em' is the MSE. The test time of the Concurrent-Test, 'Tpc' is described as:

$$Tpc = (Ta + Tb) / (1 + Ec/100) \qquad (5)$$

Where 'Ec' is the Concurrent-Efficiency and 'Ec' is defined as:

$$Ec = (Ts - Tpc) / Tpc * 100 \qquad (6)$$

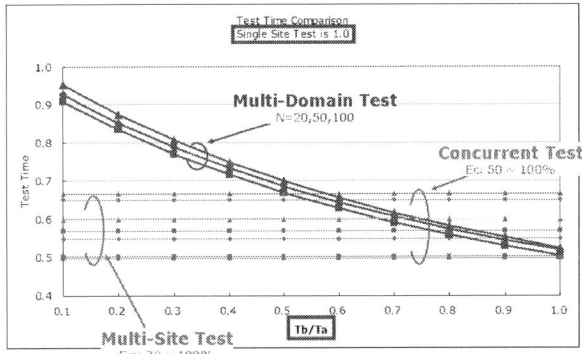

Figure 7.   **Test Time Comparison**

The Figure 7 shows the 'per device test time' of the Multi-Site-Test, the Concurrent-Test and the Multi-Domain-Test normalized by the test time of the Single-Site-Test. The X-axis is the ratio of 'Ta' and 'Tb'. As shown in this graph, the test time of the Multi-Domain-Test is largely affected by the ratio of 'Ta' and 'Tb'. The Multi-Domain-Test can achieve the same level of the test time reduction ratio as the Multi-Site-Test and the Concurrent-Test when the ratio of 'Ta' and 'Tb' is closed to 1.

III. COT ANALYSIS

The COT of the Single-Site-Test, the Dual-Site-Test and the Dual-Domain-Test were calculated for two example SOC devices: the SOC A and the SOC B. The Table 2 shows the details of the test condition of the SOC A and the SOC B. For the Dual-Domain-Test, the Domain 1 is assigned to the digital blocks and the Domain 2 is assigned to the analog blocks, and the test time ratio of the digital tests and the analog tests, 'Tdigital/Tanalog', is 1.45.

Table 2. Details of Example SOCs

Device Information

	SOC A	SOC B
Number of pins	496+	248+
Power Supply	12	6
HSIO < 3.6Gbps	16	8
HSIO < 1.8Gbps	16	6
Digital < 800Mbps	64	32
Digital < 400Mbps	384	192
Analog Input < 500MHz	1	1
Analog Input < 200MHz	1	1
Analog Output < 100MHz	2	2

Test Time (sec)

	SOC A and SOC B
Single Site	54.0
Digital	32.0
Analog	22.0
Index	0.7
Dual Site	61.4
Index	0.9
Multi Site Efficiency	86.3%
Digital	94.0%
Analog	75.0%
Dual Domain	32.0
Index	1.0
Tdigital / Tanalog	1.45

We assume that the test time of both of the SOC A and the SOC B at the Single-Site-Test are same, and the overall MSE is 86.3 % at the Dual-Site-Test. For the COT calculation, costs of the ATE and the handler, depreciation span, utilization, operational cost, uptime, expected yield and profit margin are estimated.

The Figure 8 shows the COT calculation result. In this graph, the COT is normalized as the Single-Site-Test is 1.0, and it clearly shows that the Dual-Domain-Test is lower COT than the Dual-Site-Test.

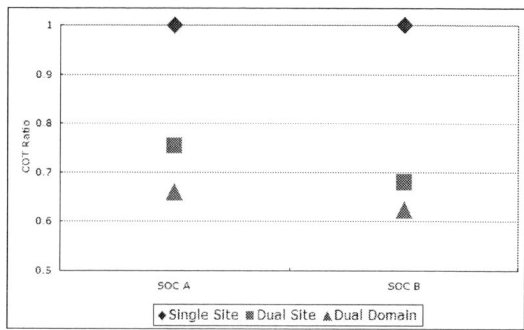

Figure 8. COT comparison

Now, the COT of the SOC A are calculated varying the test time and the MSE while keeping the 'Tdigital/Tanalog' is 1.45. The Figure 9 shows points where the COT of the Dual-Site-Test and the Dual-Domain-Test are same. The below area from the line is that the Dual-Domain-Test is lower COT than the Dual-Site-Test. And the upper area from the line is that the Dual-Site-Test is lower COT than the Dual-Domain-Test. In this condition, the Dual-Domain-Test is lower COT when the single site test time is more than 5 sec and the MSE is lower than 85%.

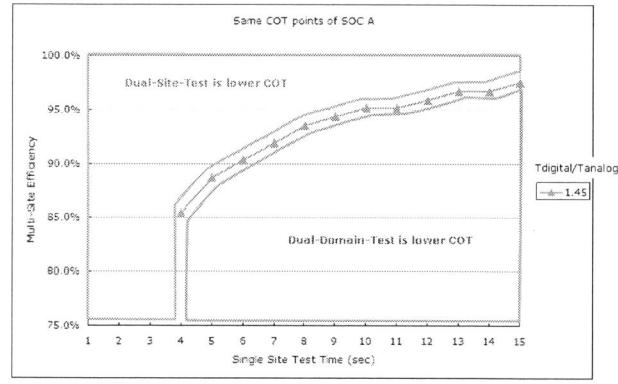

Figure 9. Same COT

As shown in the Figure 7, the test time of the Multi-Domain-Test is varied by 'Ta/Tb', and in this COT analysis, 'Ta/Tb' is 'Tdigital/Tanalog'. The Figure 9 shows same COT points when 'Tdigital/Tanalog' is 1.45. Same COT points while varing 'Tdigital/Tanalog' are calculated for the SOC A and the SOC B. The Figure 10 and the Figure 11 shows results. In these graphs, 'Tdigital/Tanalog' is varied from 0.5 to 1.45 in 0.125 steps.

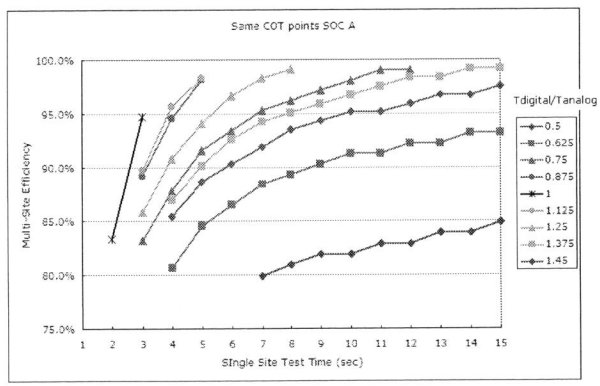

Figure 10. **Same COT points of SOC A**

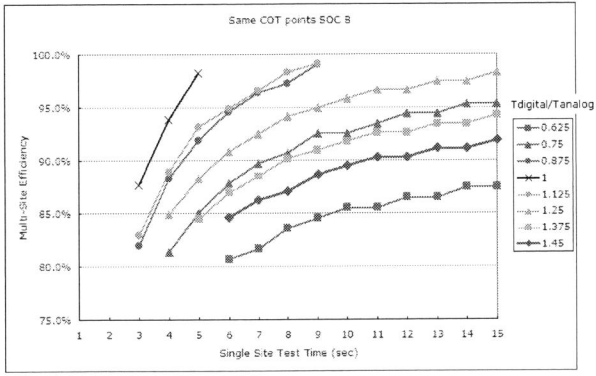

Figure 11. **Same COT points of SOC B**

These COT analysis shows that the Multi-Domain-Test is suitable for the high-end SOC device that tend to be longer test time and lower MSE. For the low-end SOC device, the COT of the Multi-Domain-Test is not low enough because of the short test time. In case of the Multi-Site-Test, one handler index is for several devices, but one device needs one handler index for the Multi-Domain-Test.

IV. DISCUSSION

The biggest challenge to implement this Multi-Domain-Test is the device movement complexity by the handler. The Multi-Domain-Test is one of the multi-insertion solutions and the handler needs to move devices from one socket to another socket on the DUT board. It is possible for the high end handlers to move devices as required by the Multi-Domain-Test. But it needs 2 to 3 seconds to move devices. The recent high end handlers have two sets of movements, and one movement prepares the devices while another movement pushes the devices to the sockets while the test is executing. Therefore, if the test time is more than 2 to 3 seconds, device positioning is prepared by another movement while the test is executing. The details of the device movement are shown in the Figure 13.

As described in previous section, the test time reduction ratio of this Multi-Domain-Test is varied by the ratio of 'Ta' and 'Tb'. The shortest test time is achieved when 'Ta' equals to 'Tb'. In general, several test items are executed for the Domain

1 and also the Domain 2. And, 'Ta' and 'Tb' are the total sum of the test time of each test item of the Domain 1 and the Domain 2 respectively. If the test items are freely assigned to the Domain 1 and the Domain 2, there should be the assignment of the test items that makes 'Ta' is very close to 'Tb'. In this best assignment, the circuit blocks of the Domain 1 and the Domain 2 may be overwrapped; this means that same circuit blocks are assigned to both of the Domain 1 and the Domain 2 as shown in the Figure 12. In this Figure, the Control Logic block is assigned to both of the Domain 1 and the Domain 2.

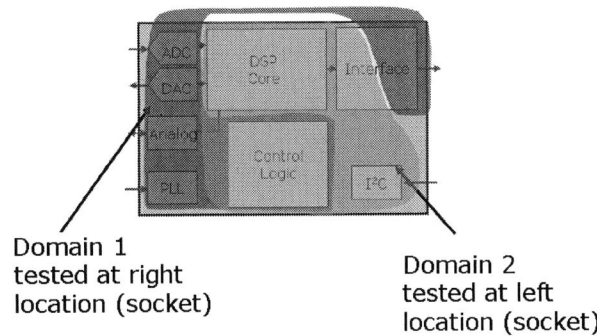

Figure 12. Domain Assignments with Shared Block

This Domain assignment affects the system configuration. If some circuit blocks are overwrapped in the Domain 1 and the Domain 2, it may require more test resources and this increases the cost of the ATE. The same situation happens if the device has many common pins for both of the Domain 1 and the Domain 2. But, even though the cost of the ATE is higher, the test time can be the shortest and the low COT can be achieved.

As described here, the Multi-Domain-Test has the two axis flexibility, the test resources and the test item assignments (Ta/Tb). Therefore, the COT is optimized flexibly according to the production requirements by selecting the test resources and the assignment of the test times to the Domains.

V. CONCLUSION

Currently, the Multi-Site-Test is the main stream of the test strategy to reduce the COT of the device, and the Concurrent-Test is actually implemented at the production recently. The newly proposed test strategy, the Multi-Domain-Test is the third choice of the test strategy for lower COT, and this strategy is the uniquely enhanced multi-insertion Concurrent-Test. In this new strategy, the multiple IP cores in different devices are tested at the same time. Therefore, no interference and no interaction between tested IP cores are existed, and there is no common block issue. Because this new test strategy does not have any limitations that original Concurrent-Test has, it can be implemented any kind of devices even if they are no DFT designed for the Concurrent-Test [1], [2], [4].

The full test items are tested within multiple insertions, reduced number of test items are tested at each insertion. This shortens the test time and reduces the required test resources. This reduces the COT of the device more than 10% compared to the Multi-Site-Test. And this can reduces the size of circuits on the load board and they can be optimized for performance.

978-1-61284-657-6/11 $26.00 © 2011 IEEE

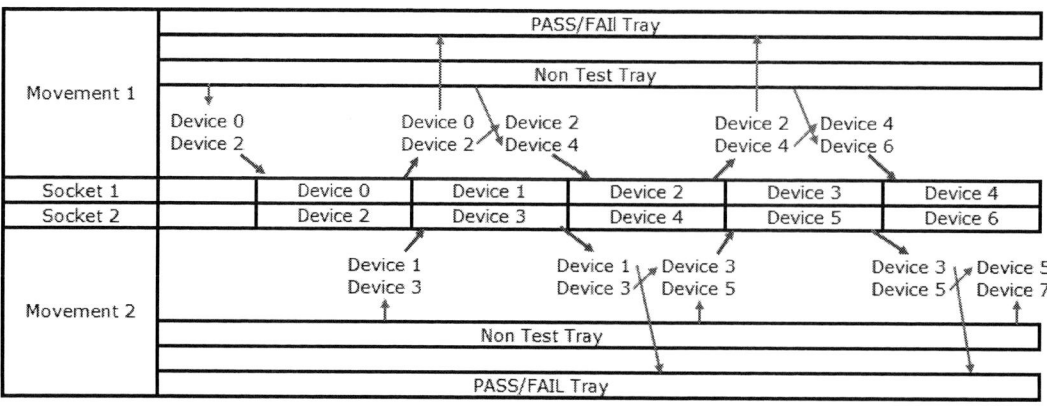

Device 0 and Device 1 are dummy devices

Figure 13.   Details of device movement

Further more, this new strategy opens the way to reduce the COT of the very large scale SOC to that the Multi-Site-Test can not be applied due to the limitation of the maximum number of the test resources.

As described above, the Multi-Domain-Test is one kind of the Concurrent-Test. It is not practicable to implement the Multi-Domain-Test with more than two domains. The Dual-Domain can be easily implemented, but the Triple-Domain or the Quad-Domain is very hard. Further reduction of COT than the Dual-Domain-Test, the Multi-Site Multi-Domain-Test is possible. The Two-Sites Dual-Domain-Test and the Four-Sites Dual-Domain-Test are feasible and four and eight devices are located on the load board respectively. Because the required test resources are less than the Quad-Site-Test and the Octal-Site-Test, the Two-Sites Dual-Domain-Test and the Four-Sites Dual-Domain-Test have less chance to meet the limitation of the maximum number of the test resources.

ACKNOWLEDGEMENTS

Author thanks Hideo Okawara of Verigy for providing technical hints and suggestions, and Masaru Ajiro and Mitsuhiro Ogura for managing this project.

REFERENCES

[1] K. Nikila and R. Parekhji, "DFT for Test Optimisations in a Complex Mixed-Signal SOC – Case Study on TI's TNETD7300 ADSL Modem Device" *Proceedings IEEE Int. Test Conf.*, 2004, Paper 27.3.

[2] F. Perez, and M. Vogt, "Concurrent Test Implementation on a 2G/3G Baseband Device", Verigy *VOICE 2008*, 2008, crf-19

[3] R. Myers, and A. Salagianis, "Implementation of a Concurrent Test Plan for a System in Package Device", Verigy *VOICE 2007*, 2007, TTM

[4] S. Molavi and T. McPheeters, "Concurrent Test Implementations" *16th IEEE Asian Test Symposium, 2007*

[5] R. Dorsch, R. Rivera, H. Wunderlich and M. Fischer, "Adapting an SoC to ATE Concurrent Test Capabilities" *Proceedings IEEE Int. Test Conf.*, 2002, Paper 41.2.

[6] E. Volkerink, A. Khoche, L. Kamas, J. Rivoir and H. Kerkhoff "Tackling Test Trade-offs from Design, Manufacturing to Market using Economic" *Proceedings IEEE Int. Test Conf.*, 2001, Paper 40.1.

2011 29th IEEE VLSI Test Symposium

Low-Cost Diagnostic Pattern Generation and Evaluation Procedures for Noise-Related Failures

Junxia Ma[1], Nisar Ahmed[2] and Mohammad Tehranipoor[1]
[1]ECE Department, University of Connecticut, Storrs, CT, 06269
[1]{junxia, tehrani}@engr.uconn.edu
[2]Texas Instruments, Dallas, TX, 75243, n-ahmed@ti.com

ABSTRACT

As technology feature geometries shrink, failures caused by signal integrity issues have become prominent during test. To avoid the time consuming silicon inspection and reduce the engineering cost and effort for failure analysis, a fast and cost-effective diagnostic flow is proposed in this paper. The flow targets delay faults and can be used to (1) identify noise-related failures with a quiet pattern and (2) evaluate the failed pattern in terms of its noise-induced delay to help identify the root cause of failure. A novel procedure is developed to generate a quiet pattern to help differentiate sources of the failure. The quiet pattern targets the same physical defects as the failed pattern but offers much lower noises level. A pattern evaluation procedure is used to evaluate the noise-induced delay. The proposed procedures are implemented on ITC'99 b19 benchmark. Simulation results demonstrate the effectiveness of the proposed procedure in identifying the failure mechanism. The noise-induced path delay for both failed patterns and diagnostic quiet patterns are thoroughly evaluated.

Keywords: *Diagnosis, Delay Test, Crosstalk, Power Supply Noise, Quiet Pattern Generation.*

I. INTRODUCTION

As technology advances, it is vitally important for the semiconductor industry to shorten the time-to-market and deal with yield, yield loss and escape more effectively. Fast and effective fault diagnosis techniques are essential to help improve the yield and product quality [1]. Industry surveys show that more than 70% of all IC designs need one or more respins [2], in spite of the large amount of resources devoted to design, validation and verification at every step. Fast, accurate and low-cost diagnosis is always in need to find the root cause of the failures and provide feedback to the designers.

Timing failures are often the result of a combination of weak points in a design and silicon abnormalities [2], which reduce the noise immunity of the design and expose it to signal integrity (SI) issues. Design weak points are due to imperfect design, i.e., poor power planning, inadequate power vias, and long parallel interconnects; while silicon abnormalities are caused by manufacture errors, i.e., missing vias, via voids, resistive opens or vias for power/ground lines and signal interconnects. A poor power planning, resistive open power lines/vias, or missing power vias can incur on-chip power droop for some test vectors. The power droop can impact a gate(s) on a critical path and may cause timing failure. Crosstalk noises introduced by large parasitic coupling capacitances between long parallel interconnects can also impact the path delay.

Such failures are switching dependent and may only be excited with certain test vectors as inputs. Mostly small-delay defects (SDDs) manifest such problems and the accumulative SDDs cause timing

failures. Currently the commercial diagnosis tools are noise unaware. If a test part fails due to excessive power supply noise and/or crosstalk noise, with sufficient fail/pass test patterns counts, the tool can report a list of suspect pins for corresponding failure model (i.e., slow-to-rise/fall faults). However, it has no information of the failure reason for the suspects. Although for pure IR-drop failures, changing test supply voltage can help find the failure reason; it does not help for crosstalk noise related failures, because crosstalk noise has no direct connection with supply voltage. Since no physical defects can be observed under microscope for noise related failures, it is vain to check the suspects under microscope. Besides, laser-based timing analysis [3] can only be performed on device's active diffusion regions; while interconnects usually go through multiple metal layers, so it is very time-consuming and most of the time impossible for failure analysis engineers to inspect the silicon to identify the root cause.

Diagnosis for physical defects such as stuck-at, bridge, short and delay faults have been extensively investigated in the past decade. There are many sophisticated tools and procedures that can effectively point to the location of physical defects based on the collected failure log from tester [4] [5]. There has been less work devoted to developing effective procedures to address noise-related failures. In [6], Killpack et al. discussed the causes of at-speed failures in microprocessors. The relative importance of IR-drop and crosstalk compared with defect issues in observed speed-path failures was addressed. Saxena et al. presented a case study for an IR-drop induced failure in scan-based at-speed test [7]. Mehta et al. proposed a methodology to diagnose delay defects in presence of crosstalk [8]. A test pattern generation method to identify IR-drop failures during launch-off-shift (LOS) test was proposed in [9]. It minimizes launch and capture mode transitions in LOS test to reduce test mode IR-drop; however, it does not take into account crosstalk-induced delay. In practice, Shmoo plots developed based on sweeping the voltage and frequency during test pattern application are used for failure root cause analysis. However, changing frequency and voltage will only change the voltage drop characteristics and other sources of noise is neglected. For instance, as technology scales, crosstalk-induced delay could be as much as IR-drop induced delay in the circuit.

Motivated by the reasons presented above, we propose fast, low-cost diagnostic pattern generation and evaluation procedures that are applicable to both transition delay faults (TDF) and path delay faults (PDF), and can be used in both launch-off-shift (LOS) and launch-off-capture (LOC) test schemes. It avoids the time consuming silicon inspection and reduces the engineering cost and effort in failure analysis. The proposed diagnostic procedures can be used to: (1) Identify the noise caused failures; (2) Evaluate the noise strength for the failed patterns; (3) Grade patterns during path delay fault testing to ensure selection of high quality patterns to improve the quality of manufacturing test and minimize escape.

The rest of the paper is organized as follows. In Section II, an example is used to discuss the problem and application targeted by

---

* This work is supported by National Science Foundation grant CCF-0811632.

978-1-61284-657-6/11 $26.00 © 2011 IEEE

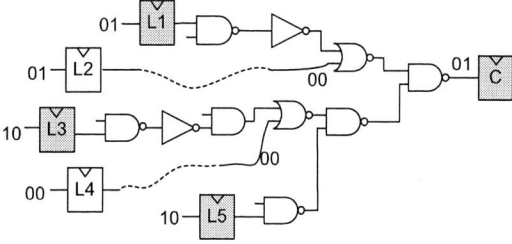

Fig. 1. An example for path delay fault detection.

this work. Section III presents the overall view of the fast, low-cost noise-related failure diagnosis flow. The diagnostic pattern generation is presented in Section IV. Pattern evaluation is covered in Section V. Simulation results to demonstrate the effectiveness and efficiency of the proposed procedures are shown in Section VI. Finally, in Section VII, we present the concluding remarks.

II. PROBLEM STATEMENT

An example of delay fault detection is shown in Figure 1. For the purpose of illustration, assume that when a test pattern is applied, transitions are launched at *L1*, *L2*, *L3* and *L5*. Among them, transitions starting from *L1*, *L3* and *L5* can be propagated to the capture scan flip-flop (SFF) *C* while transition from *L2* is blocked by the combinational logic between *L2* and *C*. Thus, paths *L1 -> C*, *L3 -> C*, and *L5 -> C* are sensitized by the test pattern applied. The following observations are made from the above example:

- Any physical defect and/or excessive noise-induced faults along these paths, which are long enough to cause SFF *C* to capture a wrong value, can be detected by the test pattern applied.
- If a wrong value is captured by SFF *C*, it must be caused by either physical defects, excessive noise on the detectable paths, or both (e.g. a small-delay defect combined with noise).

Assume that SFF *C* is the flip-flop where error is observed. Physical defects on any of the testable paths ending at *C* and associated with the failed test pattern on tester, including both robust and non-robust testable paths, could be the failing reason. The cumulative noise-induced delay along the long path could cause the failure too. There are two scenarios for the *noise-related failure*:

(I) failure is caused by *excessive* noise introduced by bad noisy test patterns.

(II) failure is caused by combined effect of design weak points, manufacture errors and pattern induced noise; and noise is within *acceptable* range, which means it is comparable with (sometimes higher considering design margins) functional mode noise. It is the chip under test that has low noise immunity.

In this paper, the diagnostic procedure we propose can be used to:

- Distinguish noise-related failures from that caused by physical defects.
- Differentiate the two scenarios discussed above if the failure mechanism is identified as noise-related reason. This is done by performing the proposed pattern evaluation procedure and quantify the noises' impact on path delay.

A novel pattern generation procedure is developed to generate a diagnostic pattern (i.e., quiet pattern) to help differentiate the failures caused by noise from those caused by physical defects. The quiet pattern targets the same physical defects as the original failed pattern but introduces much lower noise levels. Once the noise induced failures are identified, a pattern evaluation procedure is used to check each noise level (e.g., power supply noise and

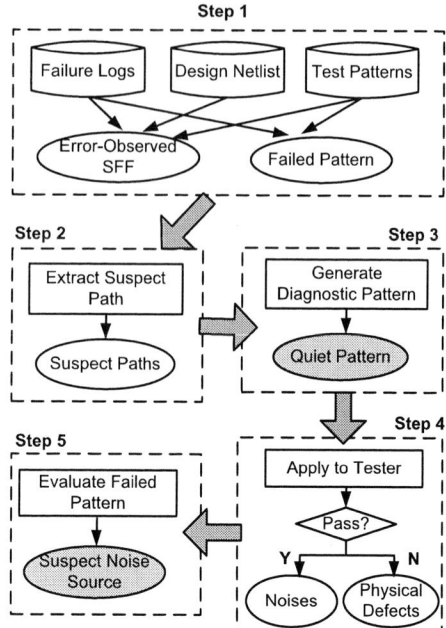

Fig. 2. Proposed flow for noise-related failure diagnosis.

crosstalk) of the failed pattern. If the noise level of the failed pattern is within acceptable range, the failure information can be used as feedback to design engineers to help better perform SI analysis, power-distribution network design, and timing closure. If failures are caused by excessive noises, the failed test pattern must be replaced with one or multiple low-noise test patterns.

III. DIAGNOSTIC FLOW

Figure 2 shows the overall noise-related failure diagnosis flow consisting of five major steps, namely (1) Error-observed SFF and failed pattern identification; (2) Suspect paths extraction; (3) Diagnostic pattern generation; (4) Physical/non-physical defects identification; and (5) Failed pattern evaluation. Each step is briefly described in the following.

Step 1: Error-observed SFF and failed pattern identification: The SFF observing error is the capture flip-flop of suspect paths. The input to this step is the failure log collected on tester, the corresponding test pattern set and design netlist. By parsing the failure log, we can easily obtain (i) the failed pattern ID, (ii) observation node, and (ii) observation cycle ID. The pattern ID identifies the failed pattern from the test pattern set; the observation node is the output pin name where the error is observed on the tester. The error-observed SFF can be identified by either custom script or using commercial diagnosis tools.

Step 2: Suspect paths extraction: There are two methods that can be used to extract the suspect paths: (a) By backtracking the simulation pin data from the error-observed SFF, the sensitized paths can be identified as the suspect paths. (b) With the knowledge about error-observed SFF and its corresponding test pattern, we can then extract all the paths ending at the error-observed SFF (also called capture SFF) that are within specified slack threshold. By running fault simulation those paths that are activated during test by the failed pattern can be identified. These paths are taken as initial suspect paths. This is the method we used in this work.

Step 3: Diagnostic (quiet) pattern generation: Diagnostic patterns are generated to differentiate the failure mechanism for the suspect paths. In this paper, the diagnostic pattern is also called "quiet" pattern because of its low noise property. It can detect the same delay faults on the suspect paths as original failed pattern. That is, the newly generated quiet pattern activates all the suspect physical fault sites in the same way as the failed pattern does. The pattern generation procedure is discussed in details in Section IV.

Step 4: Physical/non-physical defects identification: The quiet pattern, which has noise level much lower than the failed pattern as well as the functional mode, is applied to the tester again under the same test environment (i.e., supply voltage, temperature and frequency). If the quiet test pattern passes on the tester, it proves that the original pattern failed due to noise-related reason, because the only difference between the quiet pattern and the original failed pattern is their noise levels. A pattern evaluation procedure will be performed to quantify the noise induced delay, and further differentiate two noise-related failure scenarios (I) and (II) as discussed in Section II. If the quiet test pattern also fails, then the chip must have failed the original test due to physical defects since the low-noise pattern excludes the noise reason.

Step 5: Failed pattern evaluation: If the quiet pattern passes the test and noise-related reason is identified as the failure mechanism, the failed pattern (here, also called "noisy" pattern) is then examined with our pattern evaluation procedure. The contribution on path-delay increase from noises, such as power supply noise and crosstalk, will be quickly evaluated and represented with pattern quality metrics: Q_{Xtalk} and Q_{PSN}. By comparing the failed pattern's Q values with the mean of the functional patterns, we can conclude whether the failure is caused by bad test patterns with excessive noise (scenario I) or combined effect of design weak points and/or manufacture errors, and pattern induced noise on the chip (scenario II).

IV. DIAGNOSTIC QUIET PATTERN GENERATION PROCEDURE

As discussed above, a quiet pattern is needed to distinguish the failure mechanism between physical defects and noises related problems. The quiet pattern should fulfill the following requirements:

(1) It needs to detect all the physical defects in the same way as the failed "noisy" pattern detects on the suspect paths;

(2) It must have a much lower noise level compared with the failed pattern.

If we add path delay faults on the suspect paths and run automatic test pattern generation (ATPG) tool, the tool cannot guarantee to generate a new pattern that detects all the faults. This is because some faults are detected by the don't-care bits, filled randomly, in the original failed pattern. Simulation results show that even if the tool can generate a pattern targeting all the suspect paths sometimes, it cannot fulfill requirement (1) in most cases. The new pattern cannot reproduce the same detection conditions for these physical defects. To solve this problem, *test relaxation* is used to identify the don't-care bits from the failed pattern and re-fill them by a low switching activity filling method.

Test relaxation techniques for different applications have been discussed in [10]-[12]. However, in this work, since the constraint for test relaxation is simpler, which is to detect path delay faults on the suspect paths, don't-care bit identification is easier. We develop a new test relaxation algorithm which runs quite fast on large designs. The proposed algorithm is implemented in a *TCL* script so that it is executable in a commercial ATPG tool.

As can be seen from Figure 3, design netlist and test rule files are read as inputs and path delay faults on suspect paths are taken as the fault candidates. We use the same design rules as that used during original test pattern generation to make sure the test environment for the quiet pattern is the same as the original pattern that failed on the tester. The faults are then analyzed by the ATPG tool. This is the

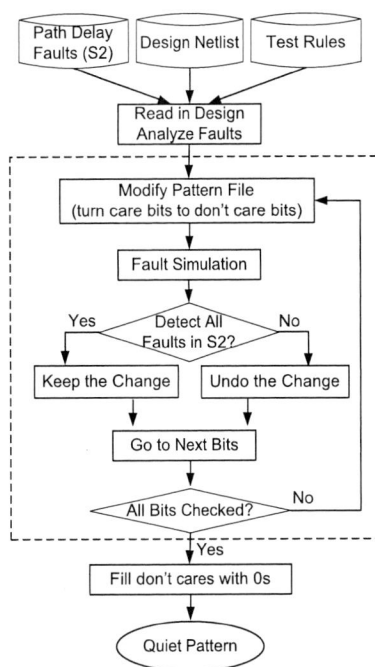

Fig. 3. Diagnostic quiet pattern generation procedure.

most time consuming part for large designs. However, in this flow, it only needs to be done once. By developing such a *TCL* script and running test relaxation procedure within the ATPG tool, we eliminate the needs to run design compiling, rule checking, and fault analysis for each fault simulation run. This brings a major advantage on the proposed pattern generation method – *it can be performed on large designs very fast*.

During the test relaxation, we tentatively replace care bits with don't-care bits, then based on the fault detection results of fault simulation, we decide whether to keep the changes to the bit value or not. In this way, all the bits for scan and the primary inputs in the pattern file are checked and don't-care bits are identified. A quiet pattern is then generated by filling the don't-care bits with low switching filling methods, fill-0/1/adjacent methods.

The scan chains are processed one by one. As there is always a large number of don't-care bits, for each scan, we replace all the bits with don't-care bits in the first iteration for the selected scan chain. Fault simulation is performed with the modified pattern file. If all the path delay faults are detected, the replacement made in the previous iteration will be accepted; otherwise, the changes will be discarded. In the latter case, the length of bits to be replaced for next iteration will be reduced by half. So a shorter new replacement is made and the process is repeated until all the don't-care bits in the pattern file are identified and replaced. The complexity of this algorithm is $O(Nlogn)$, where N is the number of total scan chains, and n is the average scan chain length. The entire test relaxation process can finish in a short time for even large designs. The CPU run time results will be presented in Section VI.

V. PATTERN EVALUATION

Once the failing reason is identified as noise-related by the quiet pattern, it is beneficial to know (1) how much impact each noise source has on path delay and (2) the noise level of each effect. This information is important for identifying the root cause of the failure, i.e., whether it fails due to bad design (weak points and/or manufacture errors) or bad test pattern.

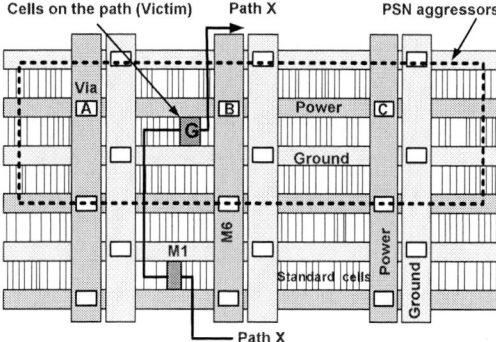

Fig. 4. Aggressor cells identification for power supply noise.

In this work, power supply noise and crosstalk are considered as the two major noise sources that negatively impact signal integrity and can potentially cause silicon failure and escape. A pattern evaluation procedure is used to grade the failed pattern and report the impact of these two noises on path delay. In this phase, the failed patterns refer to those that caused the failure due to noise.

The failed patterns are graded against the suspect paths. Pattern quality metrics Q_{PSN} and Q_{Xtalk} [13] are used to quantify the power supply noise and crosstalk, respectively. The pattern quality metric indicates the amount of noise-induced path delay increase. For each suspect path, we calculate its Q_{PSN} and Q_{Xtalk} values. When comparing these values with that of the functional patterns, we can determine whether the failure is caused by excessive pattern-induced noise or low noise immunity of the chip under test. If the failure is caused by low noise immunity of the design, the pattern evaluation results can help locate the weak point/area in the layout.

A. Aggressor Identification for Power Supply Noise

Simulations have been performed to analyze localized IR-drop effects caused by switching cells that are in close proximity to one another in [14], which shows the closer the neighboring cell is to the switching cell, the larger the voltage drop created by the switching cell and experienced by the neighboring cell. As shown in Figure 4, cell G is a cell on path under test. We refer to cell G as "victim" cell and the neighboring cells as "aggressors" because their switching activity can impact the voltage drop and performance of the victim cell. If there are m gates on a critical path, we consider all of them to be victim cells; some may be in the same row and others in different rows. As shown in the dashed box in Figure 4, cells directly adjacent to G and those that extend beyond the left and right of *Via A* and *Via C*, respectively, are classified as aggressor cells. The block with dashed lines in Figure 4 shows all the aggressor cells for victim cell G.

B. Aggressor Identification for Crosstalk Noise

To evaluate the impact of crosstalk effects from the aggressors, we need to identify the aggressor nets for the critical paths, and this requires knowledge of the physical design. We use the extracted coupling capacitance of each of the nets to identify those that will have a significant effect on the victim net [15]. A distributed *RC* model is used as the interconnect model during parasitic extraction. We use a minimum coupling threshold during 3D extraction of the layout to prune aggressors with coupling capacitance smaller than the threshold. Using the coupling threshold will reduce the complexity of our analysis by filtering some of the neighboring nets that have almost no effect on the victim path. This will eliminate nets that may be near each other but are routed perpendicularly.

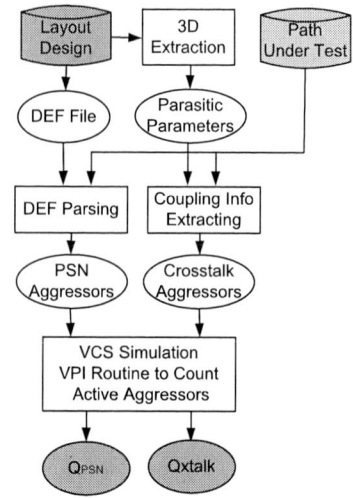

Fig. 5. Pattern evaluation procedure.

C. Pattern Quality Value Calculation

Figure 5 depicts the flow of our pattern evaluation procedure. The inputs are layout design and the suspect paths. For each path under consideration, we can identify its power supply noise aggressors by parsing the Design Exchange File (DEF) file. DEF file is generated from the layout design and contains the physical placement and routing information of all the elements in the circuit. We developed an in-house tool DEF file parser in C program. Since the timing analysis tool stores the parasitic coupling information, custom scripts have been developed to report this for each net of the critical path to assist in identifying the crosstalk aggressor nets. A verilog programming language interface (called VPI) routine is used in gate level simulation to monitor the transition of all the aggressors and calculate Q's.

The pattern quality value (Q) for power supply noise is the sum of the weighted switching activity (WSA) [16] value of all N_{aggr_cell} aggressor cells, which can be expressed by

$$Q_{PSN} = \sum_{k=0}^{N_{aggr\_cell}} s_k \cdot WSA_k, \text{ where}$$

$$WSA_k = \tau_k + \phi_k f_k \qquad (1)$$

$$s_k = \begin{cases} 1, & \text{Transition on gate output} \\ 0, & \text{No transition} \end{cases}$$

The *WSA* of switching gate k depends on its gate weight τ_k, the number of fan-out, f_k, and the fan-out load weight, ϕ_k.

In a similar way, we calculate Q for crosstalk noise by

$$Q_{Xtalk} = \sum_{i=0}^{N_{aggr\_net}} d_i \cdot C_i \cdot f(\Delta t), \text{ where}$$

$$d_i = \begin{cases} 1, & \text{Opposite transition} \\ 0, & \text{No transition} \\ -1, & \text{Same transition} \end{cases} \qquad (2)$$

$$f(\Delta t) = \begin{cases} 1, & t_1 < \Delta t < t_2 \text{ and } \Delta t = t_a - t_v \\ 0, & \text{otherwise} \end{cases}$$

For all N_{aggr_net} aggressor nets, the equation considers the direction of the transition with respect to the associated nets of the targeted path, d, the amount of coupling between the two nets, C_i, and a timing window $f(\Delta t)$ to take into account the arrival time difference (Δt) of the transitions on aggressors (t_a) and victims (t_v). A positive value of Q_{Xtalk} indicates that the targeted path will experience slow-down due to induced crosstalk effects from the switching aggressors.

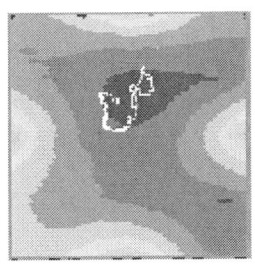
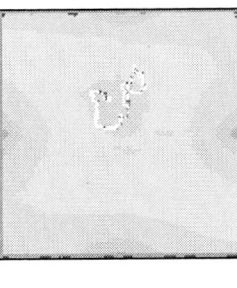

(a)   (b)

Fig. 6. IR drop plots for (a) failed (noisy) pattern and (b) quiet pattern with the longest suspect path highlighted.

TABLE I

COMPARISON RESULTS FOR NOISY AND QUIET PATTERNS FOR
THE LONGEST SUSPECT PATHS IN *b19* BENCHMARK.

Metric	Noisy Pattern	Quiet Pattern	Reduction
Path Delay	7.38 *ns*	6.89 *ns*	6.6% (35.0%[1])
Q_{PSN}	2203	1342	39.1%
Q_{Xtalk}	7.422	1.521	79.5%
IR_{ave}	104.1 *mV*	59.2 *mV*	43.1%
Worst IRdrop	107.8 *mV*	61.0 *mV*	42.3%
Switching Power	67.2*mW*	27.3 *mW*	59.4%
Total Switches	549300	225787	58.9%

[1] This percentage (35%) stands for the reduction on noise-induced path delay.

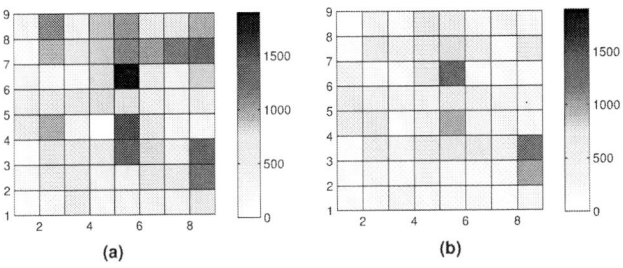

(a)   (b)

Fig. 7. *WSA* plots for (a) failed (noisy) pattern and (b) quiet pattern. Layout is partitioned into 9x9 regions according to the power distribution network structure.

Similarly, a negative Q_{Xtalk} indicates that the targeted path will experience speed-up.

VI. RESULTS

We implemented the proposed diagnostic pattern generation and pattern evaluation procedures on benchmark *b19*, the largest circuit among ITC'99 benchmarks [17], containing about 200K gates and more than 6K flip-flops. The physical layout was designed using 180nm Cadence Generic Standard Cell Library [18] with 1.8 V as the typical supply voltage.

Because of the lack of test chips and testers, to validate and analyze the effectiveness and complexity of the proposed method, we use simulation to "diagnose" the failures observed at the scan output pins. The effectiveness of the quiet pattern will be demonstrated by comparing its path delay with that of the failed pattern.

For a bit mismatch occurs on a scan output pin on b19, we locate the error-observed SFF using the failed pattern ID, cycle ID and observed pin name, as discussed in Step 1 in Section III. By running a custom developed script with timing analysis tool, we find there are a total of 3843 paths connecting to the capture SFF. We consider both slow-to-rise and slow-to-fall path delay faults for all these paths, and run fault simulation against the failed pattern using ATPG tool. 98 paths out of the 3843 paths are reported as detected; these paths are taken as the suspect paths. We run our test relaxation procedure using the flow presented in Section IV. Out of the 6206 bits in both scan chain inputs and primary inputs, 276 are identified as care bits and the remaining are don't-cares. That is, 95.5% of the input bits can be used for making the pattern a low switching pattern using filling methods. Here, we use 0-fill technique to fill in don't-care bits; other low switching X-filling methods can be easily incorporated in this flow since a large portion of the pattern contains X's. In this way, a quiet pattern is generated such that it ensures detection of the same transition and path delay faults as the failed pattern.

Figure 6 presents the IR-drop plots on power pins for the failed noisy pattern and the diagnostic quiet pattern. In this layout design, four pairs of power and ground pads are placed in the middle of

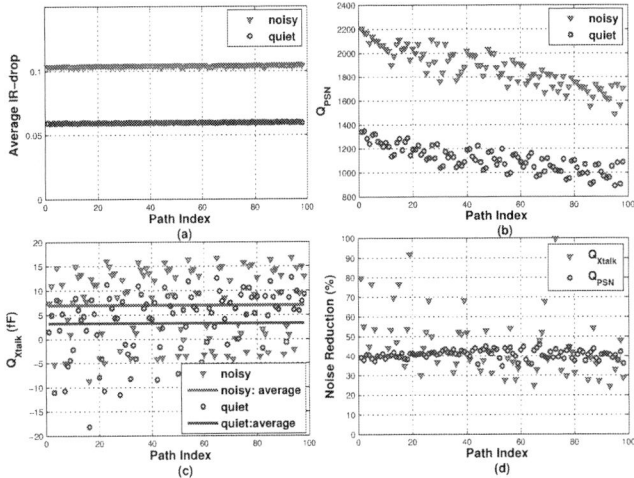

Fig. 8. Comparison between noisy pattern and quiet pattern on 98 suspect paths for (a) average IR-drop; (b) Q_{PSN}; (c)Q_{Xtalk} and (d) Percentage of noise reduction.

each side. The longest suspect path is highlighted on the plots to show its physical location on the layout. The darkest color in Figure 6(a) represents IR-drop values over 90 *mV* on power pin, which is 5% of the power supply voltage. When comparing the two plots, we can see that the quiet pattern incurs much lower IR-drop in the design and around the suspect path.

IR-drop plots depend both on current distribution and the resistance of the power distribution network. The layout-aware weighted switching activity, i.e., WSA, represents the switching current distribution. Figure 7 illustrates the dynamic WSA plots for both the failed noisy pattern and the diagnostic quiet pattern on b19 layout. We partition the design into 9x9 regions according to the location of power straps and power rings on the layout [16]. We develop a VPI routine and run it with gate-level simulation to calculate the sum of switching gates' WSA for each region. From the two WSA plots it is observed that the quiet pattern considerably reduces the switching current.

Since we do not have a tester to apply the patterns and measure the paths delay, we run fast SPICE simulation to simulate and compare the path delay of the failed and quiet patterns. Table I compares the path delay and other metrics of the longest suspect paths between noisy and quiet patterns. The reduction percentages are listed in Column 4. For path delay, 6.6% reduction is calculated when compared with noisy pattern's path delay; and 35% is the reduction on noise-induced path delay. Noise-induced path delay is calculated by comparing path delays of each case with that of the ideal case 5.98 *ns*, i.e., path delay without noises.

Figure 8 compares average IR-drop IR_{ave}, pattern quality values Q_{PSN} and Q_{Xtalk} of noisy pattern and quiet pattern on 98 suspects paths. The x-axis presents the path index; the paths are sorted based

TABLE II

NOISY AND QUIET PATTERNS COMPARISON FOR LONGEST SUSPECT PATH OF 5 FAILURES IN *b19* BENCHMARK.

| Path | Noisy Pattern | | | | | Quiet Pattern | | | | | X bits |
	Delay (ns)	IR_{ave} (mV)	Q_{PSN}	Q_{Xtalk} (fF)	Switch Power (mW)	Delay (ns)	IR_{ave} (mV)	Q_{PSN}	Q_{Xtalk} (fF)	Switch Power (mW)	(%)
1	8.334	106.9	1194	13.28	135.3	7.815	49.9 (53%)	266 (77%)	3.962 (70%)	83.8 (38%)	98.7
2	8.742	78.6	517	18.88	122.9	8.440	47.0 (40%)	342 (34%)	0.005 (99%)	84.8 (31%)	98.8
3	8.009	105.4	995	5.365	122.1	7.604	50.0 (52%)	265 (73%)	-3.962 (174%)	83.9 (31%)	98.8
4	7.952	102.8	262	-6.230	127.6	7.572	51.0 (50%)	244 (68%)	-13.47 (116%)	81.7 (36%)	98.9
5	5.146	81.0	531	23.20	138.3	4.944	38.7 (52%)	50 (90%)	22.2 (4%)	77.2 (44%)	98.5

on their slack. Path *1* is the longest path with the least slack. From the results shown in Figure 8(a), it can be seen that the quiet pattern reduces the IR_{ave} by more than 40% on all the 98 suspect paths. Since these suspect paths are located close to each other on the layout, in this case, their IR_{ave} values are very close. The pattern evaluation procedure is performed for all the 98 suspect paths. Figure 8(b) shows the power supply noise pattern quality values Q_{PSN} of the suspect paths. Similarly, the crosstalk pattern quality values Q_{Xtalk} are shown in Figure 8(c). The percentage of noise reduction when comparing quiet pattern with noisy pattern for each suspect path is presented in Figure 8(d).

From these figures we can see that, for Q_{PSN}, longer paths reported by static timing analysis tool, that have smaller path index numbers, have higher Q_{PSN} values when compared with short paths shown with large index numbers in Figure 8. Comparing with power supply noise, the Q_{Xtalk} values for the suspect paths are more random. In some cases, short paths may have same amount of crosstalk noise as long paths. This makes sense because crosstalk depends on the routing, the direction of transitions on aggressors as well as the arrival time of the aggressors. The quiet pattern reduces the switching activities around suspect paths, which effectively reduces the aggressor nets and gates' switchings and thus reduces both noises. Figure 8(d) shows significant noise reduction on quiet pattern compared with noisy pattern. The high percentage of noise reduction validates the diagnostic function of the quiet pattern.

We run the proposed pattern generation and evaluation procedures for five other error-observed SFFs. A failure log is generated for five randomly selected SFFs. A quiet pattern is generated for each case. The paths delay of the original failed patterns and the diagnostic quiet patterns are compared. Table II summarizes the comparative performance of the noisy patterns and quiet patterns. The IR_{ave}, Q_{PSN} and Q_{Xtalk} are calculated over the longest suspect paths. The percentages of reduction for each metric are also given in paired parentheses. The large percent of don't-care (X) bits (listed in last column) enables the quiet pattern to reduce the noise effectively.

The simulations were performed on an x86 server architecture, running a Linux OS, 8 CPU cores clocked at 2.826 GHz, and 32GB of RAM. The average CPU run time for quiet pattern generation is 3m22s. The the time to run pattern evaluation procedure over one path is 15.92s; while the time to run it over 98 paths is 16m33s.

VII. CONCLUSIONS

With the increase of power supply noise and crosstalk noise in nanometer designs, diagnosis for noise-related failures during both first silicon and manufacturing test has become an important and challenging task. A fast and low-cost diagnostic flow is proposed in this paper to address this challenging problem. A diagnostic quiet pattern is generated to differentiate the noise-related failures from physical defects. The quiet pattern can detect the same physical defects as the original failed pattern for those that can be captured by the error-observed SFF. A pattern evaluation procedure is also presented in this work to evaluate the failed pattern to quantify the

noisy impact on path delay. The information obtained from pattern evaluation can be used to speed up the failure analysis process. The results have demonstrated the effectiveness of the proposed flow in reducing switching noise with the diagnostic quiet pattern and analyzing the patterns that fail due to noise.

REFERENCES

[1] J. A. Waicukauski and E. Lindbloom, "Failure Diagnosis of Structured VLSI," in *IEEE Trans. on Design and Test of Computers*, vol.6, no.4, pp.49-60, August 1989.

[2] R. Aitken and E. J. Marinissen, "Guest Editors' Introduction: Addressing the Challenges of Debug and Diagnosis," in *IEEE Design & Test of Computers*, vol. 25, no. 3, pp. 206-207, May 2008.

[3] J. A. Rowlette and T.M. Eiles, "Critical Timing Analysis in Microprocessors Using Near-IR Laser Assisted Device Alteration (LADA)," in *Proc. of International Test Conference (ITC'03)*, pp. 264-273, 2003.

[4] M. Sharma, W. Cheng, T. Tai, Y. S. Cheng, W. Hsu, C. Liu, S. M. Reddy and A. Mann, "Faster Defect Localization in Nanometer Technology based on Defective Cell Diagnosis," in *Proc. of International Test Conference (ITC'07)*, pp. 1-10, 2007.

[5] V. J. Mehta and K. H. Tsai, "Timing-Aware Multiple-Delay-Fault Diagnosis," in *IEEE Trans. on Computer-Aided Design of Integrated Circuits and Systems*, vol. 28 , no. 2, pp. 245-258, 2009.

[6] K. Killpack, S. Natarajan, A. Krishnamachary, and P. Bastani, "Case Study on Speed Failure Causes in a Microprocessor," in *IEEE Design & Test of Computers*, pp. 224-230, May/June 2008.

[7] J. Saxena, K. M. Butler, V. B. Jayaram, S. Kundu, N. V. Arvind, P. Sreeprakash, and M. Hachinger, "A Case Study of IR-Drop in Structured At-Speed Testing," in *Proc. of Intl. Test Conf.*, pp. 1098-1104, 2003.

[8] V. J. Mehta, M. Marek-Sadowska, K. H. Tsai and J. Rajski, "Timing Defect Diagnosis in Presence of Crosstalk for Nanometer Technology," in *Proc. of International Test Conference (ITC'06)*, pp. 1-10, 2006.

[9] M. Chen and A. Orailoglu, "Cost-effective IR-drop Failure Identification and Yield Recovery through a Failure-adaptive Test Scheme," in *Proc. Design, Automation, and Test in Europe (DATE'10)*, pp. 63-68, 2010.

[10] S. Kajihara and K. Miyase, "On identifying dont care inputs of test patterns for combinational circuits," in *Proc. Int. Conf. Computer-Aided Design*, pp. 364-369, 2001.

[11] A. El-Maleh and A. Al-Suwaiyan, "An efficient test relaxation technique for combinational & full-scan sequential circuits," in *Proc. 20th IEEE VLSI Test Symp.*, pp. 53-59, 2002.

[12] H. Furukawa , X. Wen, K. Miyase, Y. Yamato, S. Kajihara, P. Girard, L.-T. Wang, and M. Tehranipoor, "CTX: A Clock-Gating-Based Test Relaxation and X-Filling Scheme for Reducing Yield Loss Risk in At-Speed Scan Test," in *Proc. 17th Asian Test Symposium*, pp. 397-402, 2008.

[13] J. Ma, J. Lee, N. Ahmed, P. Girard, and M. Tehranipoor, "Pattern Grading for Testing Critical Paths Considering Power Supply Noise and Crosstalk Using a Layout-Aware Quality Metric," in *Proc. of GLSVLSI'10*, pp. 127-130, 2010.

[14] J. Ma, J. Lee, and M. Tehranipoor, "Layout-Aware Pattern Generation for Maximizing Supply Noise Effects on Critical Paths, in *Proc. IEEE VLSI Test Symposium (VTS)*, pp. 221-226, 2009.

[15] J. Lee and M. Tehranipoor, "A Novel Test Pattern Generation Framework for Inducing Maximum Crosstalk Effects on Delay-Sensitive Paths," in *Proc. IEEE International Test Conference (ITC)*, pp. 1-10, 2008.

[16] J. Lee, S. Narayan, and M. Tehranipoor, "Layout-Aware, IR-drop Tolerant Transition Fault Pattern Generation," in Proc. *Design, Automation, and Test in Europe (DATE'08)*, pp. 1172-1177, 2008.

[17] http://www.cerc.utexas.edu/itc99-benchmarks/bench.html.

[18] http://crete.cadence.com, 0.18um standard cell GSCLib library version 2.0, Cadence Design System, Inc., 2005.

978-1-61284-657-6/11 $26.00 © 2011 IEEE

Sigma-Delta Modulation Based Wafer-Level Testing for TFT-LCD Source Driver ICs

W.-A. Lin[1], C.-C. Lee[3], and J.-L. Huang[1,2]
[1]Graduate Institute of Electronics Engineering
[2]Department of Electrical Engineering
National Taiwan University, Taipei 106, Taiwan
[3]Himax Inc., Taiwan
email: jlhuang@ntu.edu.tw

Abstract—Output variation testing of TFT-LCD source driver ICs is very expensive and time-consuming due to the large amount of analog output channels and levels to measure. This paper presents a low-cost on-scribe-line BIST technique for wafer-level source driver IC testing. Based on the BIST structure and the sigma-delta modulation principle, we propose a two-stage test flow and construct a generalized test cost function to find the optimal test setup parameters.

Keywords-TFT-LCD, source driver, sigma-delta modulation, mixed-signal testing, built-in self-testing.

I. INTRODUCTION

In recent years, thin-film transistor liquid crystal displays (TFT-LCDs) have become the mainstream in the consumer electronics market. A typical TFT-LCD consists of the light source, polarizers, the liquid crystal, the color filter, and the circuit plate. Among them, the circuit plate receives and stores the image information. The circuit plate behaves much like the 1-T DRAM; it utilizes gate drivers to activate one row at a time and source drivers to write image information to the activated row.

This work relates to the manufacturing testing of the source driver ICs which is basically a large DAC array, in which each DAC drives one column of the TFT-LCD display. Specifically, the focus is the output voltage variation specifications. The output variation specification requires that all the output levels that each channel produces be within the respective acceptable range. While simple, validating this specification is costly due to the following factors.

1) The need of high pin-count probe cards to access all the output channels (as well as other input and control pins).
2) The large amount of analog voltages to measure. For example, without considering the output polarity and other image quality enhancement features, a 720-channel 8-bit source driver requires more than 18K analog voltage measurement.
3) The wide spread dynamic range and high measurement resolution. Consider the output driving range of 8 to 13.5 V. A 2 mV variation tolerance specification requires typically 13 to 14-bit test resolution.

These factors will get even worse as the demands for higher image resolution, higher pixel depths, and larger area display continue.

Built-In Self-Test (BIST) is one promising solution to these challenges. DAC BIST methods have been widely researched, for example, in [6], [1], [7], [2], [4], [5], [8]. However, applying these techniques to source driver IC testing is non-trivial because (1) the driver IC's possess a non-linear I/O transfer curves, and (2) access to the driver IC output channels leads to unacceptable area overhead. In [3], an on-scribe-line BIST architecture for wafer-level source driver IC testing is proposed. It utilizes the scribe line area to implement both test access mechanism and an analog-to-digital converter (ADC). The advantages include (1) nearly no performance and area overhead to the device, and (2) substantial reduction of the probe count. The main limitation is the elongated test time.

This work improves the on-scribe-line BIST architecture from [3] in the following aspects.

1) This work adopts the more flexible sigma-delta modulation (SDM) analog-to-digital conversion principle; this empowers one to explore the tradeoff between test accuracy and time. We also modify the SDM ADC input end to reduce its input dynamic range (from a few volts down to a few tens of millivolts); this improves the overall BIST robustness and reduce the design complexity.
2) Based on the SDM-based BIST architecture, we propose a two-stage testing strategy which significantly reduces the test time.
3) We build a cost function model for the two-stage test strategy; this allows one to identify the optimal test parameters with respect to the manufacturing test cost models.

The rest of the paper is organized as follows. In II, we describe the necessary background and related works. Sec. III and Sec. IV present the proposed SDM-based BIST architecture and the two-stage test strategy, respectively. Then, Sec. V presents the cost function model for one to identify the optimal test parameters. Finally, Sec. VI shows the simulation results and Sec. VII concludes this work.

978-1-61284-657-6/11 $26.00 © 2011 IEEE

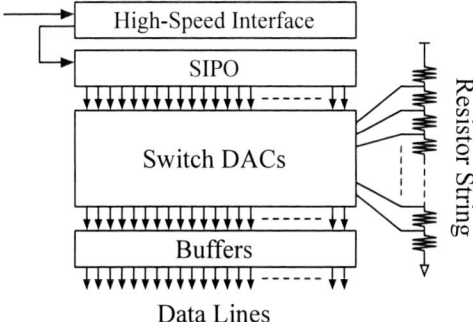

Fig. 1. The source driver IC architecture.

II. PRELIMINARIES

Fig. 1 illustrates one possible source driver IC implementation. The DAC array occupies most of the chip area and perimeter. The DAC outputs are properly buffered to drive the data lines of the TFT pixel array. The high-speed interface and the SIPO (serial-in-parallel-out) units receives the control and image information.

A. Source Driver Test Specifications

Among the source driver IC specifications, we focus on the DAC output variations test as it is the most time consuming.

Consider a K-channel B-bit source driver IC. Let $v_k(c)$ denote the output voltage of channel k with respect to input code $c \in \{0, 1, \ldots, 2^B - 1\}$. The average output of code c is then defined as

$$A(c) = \frac{\sum_{i=1}^{K} V_k(c)}{K} \quad (1)$$

The two specifications are as follows.

$$|A(c) - V_{ideal}(c)| < A_{TH} \quad (2)$$
$$|v_k(c) - V_{ideal}(c)| < V_{TH} \quad (3)$$

In (2), A_{TH} is the average variation tolerance; in (3), V_{TH} is the per-channel variation tolerance. The following analysis only considers the V_{TH} specification—if a device passes this test, we have the necessary information to perform the A_{TH} test.

B. On-Scribe-Line Wafer-Level BIST [3]

Fig. 2 illustrates the concept of on-scribe-line wafer-level BIST; the idea is to realize the BIST circuitry on the scribe line so as to avoid area and performance overhead. The BIST circuitry consists of the channel selection logic (CSL), the low-cost digitizer, and the controller. The CSL connects each channel output pad to the analog test bus via a digitally controlled analog switch; this reduces the required analog probe count from K to just a few reference voltages (for the BIST circuitry). In [3], the digitizer is a single-slope integrating ADC tailored to meet the source driver test accuracy requirement.

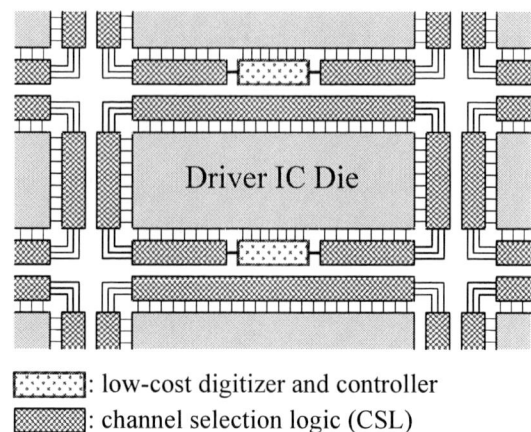

▨▨▨ : low-cost digitizer and controller
▨▨▨ : channel selection logic (CSL)

Fig. 2. The built-on-scribe-line BIST floorplan.

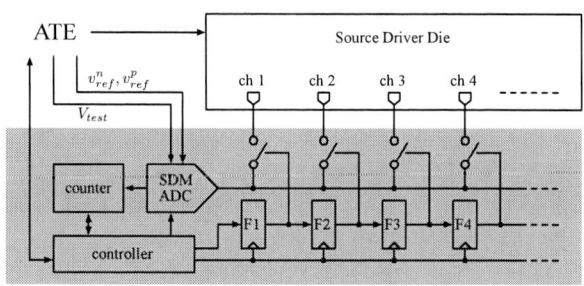

Fig. 3. The SDM-based BIST architecture.

III. THE SDM-BASED BIST ARCHITECTURE

From hardware point of view, this work utilizes the same wafer-level on-scribe-line BIST architecture as [3]; the main difference is the SDM-based digitizer. Because of its over-sampling nature, SDM ADC empowers one to adjust the measurement accuracy at will; this allows us to develop more sophisticated and efficient test strategy, for example, the two-stage approach in this work.

A. The Basic Test Flow

The overall SDM-based BIST architecture is illustrated in Fig. 3. In this BIST architecture, the SDM ADC digitizes $v_k(c) - V_{ideal}(c)$ instead of $v_k(c)$ to reduce the input dynamic range (from a few volts to a few tens of millivolts); this leads to less design efforts and more robustness. Measuring $v_k(c) - V_{ideal}(c)$ consists of three steps.

1) ATE sets V_{test} to $V_{ideal}(c)$ and DAC inputs to c.
2) The controller connects the k-th channel output to the analog bus by setting flip-flop F_k to 1 and the other flip-flops to 0.
3) The SDM ADC starts digitization.

The basic test flow measures the output variation for every channel and for every input code. ATE controls the DAC inputs, and CSL is responsible for channel selection. To go though all the output channels, the controller first sets F_1 to 1 and resets all the other flip-flops to 0; this connects channel 1

978-1-61284-657-6/11 $26.00 © 2011 IEEE

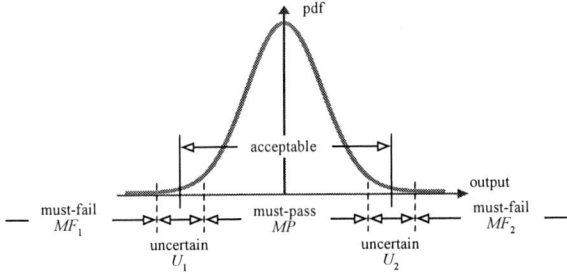

Fig. 4. Instance classification with low resolution measurement.

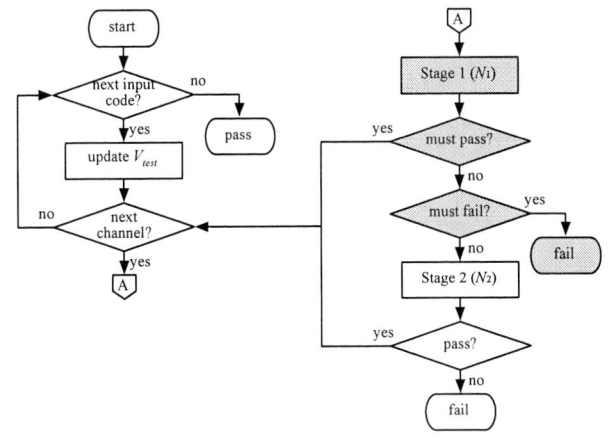

Fig. 5. The two-stage test flow.

to the analog bus. Then, the controller ticks the flip-flop clock once to advance the 1 token in the shift register and selects the next channel.

B. The SDM ADC

To fit into the scribe line area, we opt for the 1-bit first-order SDM ADC for its simplicity and robustness. Because we utilize the SDM ADC to perform DC measurement, the decimation filter is simply a counter that counts the number of ones in the SDM ADC output bit stream. Let N be the number of sampling cycles and $M_k(c)$ the measurement result, i.e., the counter value, for channel k and input code c. One can derive $\Delta v_k(c) = v_k(c) - V_{ideal}(c)$ as follows.

$$\Delta v_k(c) = \frac{M_k(c)}{N} \cdot FSR + v_{ref}^n - v_{ref}^p \qquad (4)$$

In (4), $FSR = 2\left(v_{ref}^p - v_{ref}^n\right)$ is the SDM ADC full-scale range; the resulting test resolution is $R = \frac{FSR}{N}$.

In practice, one can pre-compute the acceptable ranges of $M_k(c)$ and $\sum_{k=1}^{K} M_k(c)$ to determine whether the source driver IC meets the specifications in (2) and (3), respectively.

Using the SDM ADC, one can easily enhance the test resolution by increasing N. While the single-slope integrating ADC (used in [3]) also allows tradeoff between test time and accuracy, this requires that the analog circuitry operate in a broader range and incurs more design efforts.

C. Discussions

The test resolution of the above setup is $\frac{FSR}{N}$; this indicates that we can enhance test accuracy by increasing N or decreasing FSR.

Increasing N is straightforward; the cost is longer test time. On the other hand, FSR must cover the input dynamic range and cannot be arbitrarily reduced. In the proposed BIST, we measure the output voltage deviation instead of the output voltage. Thus, we can reduce FSR from the source driver output range to the deviation tolerance, which is just a few tens of millivolts.

IV. THE TWO-STAGE TEST STRATEGY

The proposed two-stage test strategy explores the tradeoff between test resolution and test time (by adjusting the number of sampling cycles) to reduce the overall test cost.

A. Motivation

From the past manufacturing testing data, it is observed that, after a few rounds of process and design tuning, the source driver IC output variations exhibit a Gaussian distribution with the mean close to zero as shown in Fig. 4. If the standard deviation of the output variations is small, one can identify most of the good ICs even with low-resolution measurement, i.e., the instances within the "must-pass" region in Fig. 4. Similarly, one can recognize the instances in the "must-fail" region with low-resolution measurement. However, for instances in the "uncertain" region due to low measurement resolution, one has to resolve to higher-resolution measurement to make more confident pass/fail decision.

Based on the observations, a two-stage flow with low-resolution stage 1 and high-resolution stage 2, will reduce the overall test cost if (1) the low-resolution test cost is substantially lower than the high-resolution test cost (note that the two-stage flow also introduces test cost overhead), and (2) the number of "must-pass" instances is sufficiently large. The 1-bit first-order SDM ADC is an ideal candidate for the two-stage flow because it is simple and its test resolution $R = \frac{FSR}{N}$ is inversely proportional to the test time.

B. The Two-Stage Test Flow

As shown in Fig. 4, we divide the possible output variation into five regions, MF_1, U_1, MP, U_2, and MF_2. U_1/U_2 has a width of R and centers at the high/low acceptable boundary.

The two-stage test flow is depicted in Fig. 5; it differs from the basic test flow in that (1) it includes the low-resolution stage 1 to quickly identify the "must-pass" and "must-fail" instances, and (2) the test process is terminated once an out-of-specification output voltage is detected.

1) Stage 1: The SDM ADC samples the output variation for N_1 cycles; this corresponds to a measurement resolution of $R_1 = \frac{FSR}{N_1}$. Note that FSR extends the test specification by two times the maximum stage 1 test error. If the measurement results is in MP, this output voltage meets the specification; if in MF_1 or MF_2, it fails the specification and the test process is terminated.

978-1-61284-657-6/11 $26.00 © 2011 IEEE

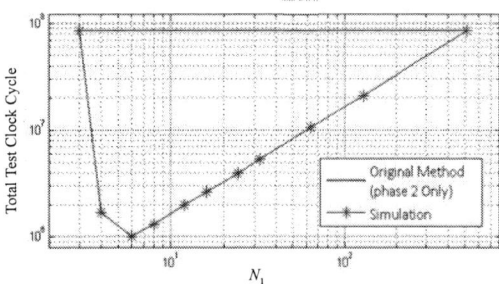

Fig. 6. The two-stage flow speedup.

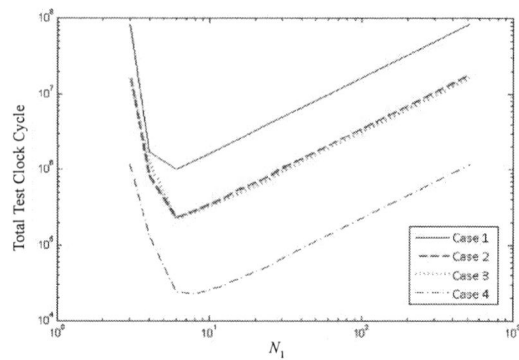

Fig. 7. Test cycle count under different output variation conditions.

TABLE I
SIMULATION SETUP

device under test	642-channel 8-bit source driver
output variation	Gaussian distribution with variance σ^2
specification	the variation be within $[-5\sigma, 5\sigma]$
N_2	512

2) Stage 2: The SDM ADC samples the output variation for N_2 cycles ($N_2 > N_1$); this corresponds to a measurement resolution of $R_2 = \frac{FSR}{N_2}$. If the measurement result is in the acceptable range, this output voltage meets the specification; otherwise, it fails and the test process is terminated.

In the test flow, stage 1 contributes to the overall test cost reduction. However, the achievable reduction depends on the actual output voltage distribution. With a small standard deviation, most pass/fail decisions are made in stage 1 with low test time; this reduces the overall test cost. However, this may not be the case for a new process line or product. Using the SDM ADC as the digitizer, one can adjust N_1 according to previous manufacturing test results or even completely skip stage 1.

Intuitively, one may determine N_2 according to the *original* test accuracy requirement. In this two-stage flow, one can utilize a higher test accuracy, i.e., larger N_2 to improve the test yield without incurring too much test time overhead because only a small portion of instances enters stage 2. (Note that there are also uncertain regions in stage 2.)

Fig. 6 compares the test clock cycles of the two-stage flow and the original one, i.e., without stage 1. (The simulation setup is summarized in Table I.) From the plot, the total test clock cycle count is lowest when $N_1 = 6$ and the corresponding speedup is around 80. Fig. 7 further shows that the required test clock cycles under the following four conditions.

Case 1   The condition in Table I.
Case 2   The standard deviation is increase to $\sigma' = 1.2\sigma$.
Case 3   The mean of output variation is σ instead of 0.
Case 4   The combination of case 2 and case 3.

As one can see, the optimal N_1 are the same except in case 4. In case 4, if we use $N_1 = 6$ as the other three cases, the resulting test clock cycle count is very close to the optimal value. This shows the robustness of the two-stage approach.

Now that Fig. 6 and Fig. 7 validate the feasibility of the two stage flow, we will proceed to construct the cost function based on which one can find the optimal value of N_1.

V. THE COST FUNCTION

In our model, the total test cost $\mathcal{C}_{total}$ consists of the test process cost $\mathcal{C}_{process}$, the overkill cost $\mathcal{C}_{overkill}$, and the test escape cost $\mathcal{C}_{escape}$.

$$\mathcal{C}_{total} = \mathcal{C}_{process} + \mathcal{C}_{overkill} + \mathcal{C}_{escape} \qquad (5)$$

When deriving the cost function, we make the following assumptions.

- Each output voltage is a Gaussian random variable whose mean is the ideal value.
- All the output voltages are independent and have the same standard deviation.
- The quantization noise is a uniform distribution random variable.
- The performed test is wafer-level test. Dies that pass the test will be packaged and tested again.

A. Test Process Cost

The test process cost can be further decomposed as follows.

$$\mathcal{C}_{process} = \mathcal{C}_{setup} + N_{chip} \cdot (T_{ov} + T_{other}) \cdot s_{runtime} \qquad (6)$$

In (6), $\mathcal{C}_{setup}$ is the test setup cost and is in general fixed, N_{chip} is the number of chips under test, T_{ov} is the average output variation test time, T_{other} is the test time associated with other tests, and $s_{runtime}$ is the per unit time test cost. Note that we assume that the other tests consume a fixed test time.

With the two-stage test strategy, the average output variation test time is as follows.

$$T_{ov} \approx \sum_{c=0}^{2^B-1} \sum_{k=1}^{K} P_k(c)(N_1 + P_{phase2} \cdot N_2) \qquad (7)$$

Because the test process ends if any $\Delta v_k(c)$ measurement fails the specification, we use $P_k(c)$ to approximate the probability that $\Delta v_k(c)$ is measured in the two-stage test flow, i.e., the source driver IC has passed all previous tests.

$$P_k(c) = \left(\frac{2}{\sqrt{2\pi\sigma^2}} \int_0^{V_{TH}} e^{-\frac{x^2}{2\sigma^2}} \, dx \right)^{c \cdot K + k - 1} \tag{8}$$

P_{phase2} is the probability that the test flow has to enter stage 2 to make the pass/fail decision.

$$P_{phase2} = \frac{2}{\sqrt{2\pi\sigma^2}} \cdot \int_{V_{TH}-\frac{R_1}{2}}^{V_{TH}+\frac{R_1}{2}} e^{-\frac{x^2}{2\sigma^2}} \, dx \tag{9}$$

B. Overkill and Test Escape Cost

The average overkill and test escape costs are

$$\mathcal{C}_{overkill} = N_{chip} \cdot P_{overkill} \cdot s_{overkill} \tag{10}$$
$$\mathcal{C}_{escape} = N_{chip} \cdot P_{escape} \cdot s_{escape} \tag{11}$$

where $P_{overkill}$ and P_{escape} are the overkill and test escape probabilities, respectively, and $s_{overkill}$ and s_{escape} denote the incurred cost due to overkill and test escape, respectively.

In this cost model, we assume that overkill and test escape are due to the quantization noise in stage 2. Furthermore, we assume that the quantization noise is uniformly distributed. Thus, we have

$$P_{overkill} = P_{goodchip} \cdot \sum_{i=1}^{2^B \cdot K} P_{p|g}^{i-1} \cdot P_{f|g} \tag{12}$$

$$= P_{goodchip} \frac{P_{f|g} \cdot \left(1 - P_{p|g}^{2^B \cdot K}\right)}{1 - P_{p|g}} \tag{13}$$

where $P_{goodchip}$ is the probability that the chip under test is good, $P_{p|g}$ is the probability that an output passes the test given that it is a good output, and $P_{f|g}$ is the probability that an output fails the test given that it is a good output. Let P_g be the probability that an output meets the test specification, i.e,

$$P_g = \frac{2}{\sqrt{2\pi\sigma^2}} \int_0^{V_{TH}} e^{-\frac{x^2}{2\sigma^2}} \, dx \tag{14}$$

Then, $P_{goodchip}$, $P_{p|g}$, and $P_{f|g}$ are as follows.

$$P_{goodchip} = P_g^{2^B \cdot K} \tag{15}$$

$$P_{p|g} = \frac{P_{gp}}{P_g} \tag{16}$$

$$P_{f|g} = \frac{P_{gf}}{P_g} \tag{17}$$

Here, P_{gp} and P_{gf} are the probabilities that a good output passes and fails the test, respectively. Assuming that the quantization noise in stage 2 is uniformly distributed, we have

$$P_{gp} = \frac{2}{\sqrt{2\pi\sigma^2}} \int_0^{V_{TH}-\frac{R_2}{2}} e^{-\frac{x^2}{2\sigma^2}} \, dx$$
$$+ \frac{2}{\sqrt{2\pi\sigma^2}} \int_{V_{TH}-\frac{R_2}{2}}^{V_{TH}} e^{-\frac{x^2}{2\sigma^2}} \cdot \frac{\frac{R_2}{2}+V_{TH}-x}{R_2} \, dx \tag{18}$$

$$P_{gf} = \frac{2}{\sqrt{2\pi\sigma^2}} \int_{V_{TH}-\frac{R_2}{2}}^{V_{TH}} e^{-\frac{x^2}{2\sigma^2}} \cdot \frac{\frac{R_2}{2}-V_{TH}+x}{R_2} \, dx \tag{19}$$

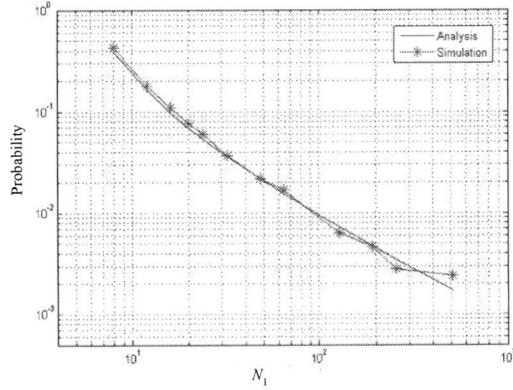

Fig. 8. Overkill probability.

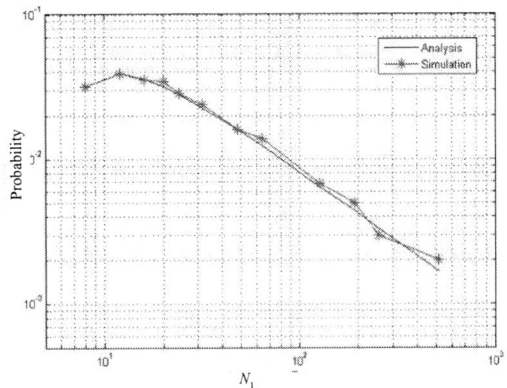

Fig. 9. Test escape probability.

One can derive the test escape probability P_{escape} in a similar manner.

$$P_{escape} = \left(P_g \cdot P_{p|g} + P_b \cdot P_{p|b}\right)^{2^B \cdot K} - \left(P_g \cdot P_{p|g}\right)^{2^B \cdot K} \tag{20}$$

Here $P_b = 1 - P_g$ is the probability that a chip is bad, P_{bp} is the probability that a bad output passes the test.

$$P_{bp} = \frac{2}{\sqrt{2\pi\sigma^2}} \int_{V_{TH}}^{V_{TH}+\frac{R_2}{2}} e^{-\frac{x^2}{2\sigma^2}} \cdot \frac{\frac{R_2}{2}+V_{TH}-x}{R_2} \, dx \tag{21}$$

VI. SIMULATION RESULTS

We first validate the derived $P_{overkill}$ and P_{escape}. Then, we perform numerical simulations to obtain the best combination of N_1 and N_2.

A. Validating $P_{overkill}$ and P_{escape}

Fig. 8 and Fig. 9 compares the overkill and test escape probabilities obtained from the derived function and from simulation. The experimental setup is the same as those in Table I except that we enlarge the standard deviation to $\sigma' = 1.35\sigma$ to increase $P_{overkill}$ and P_{escape}. Both plots show good agreement.

TABLE II
ADDITIONAL TEST COST PARAMETERS

manufacturing cost	1.0 unit per die
test time cost	10^{-11} unit per clock cycle
$s_{overkill}$	1.0 unit
s_{escape}	3.0 unit

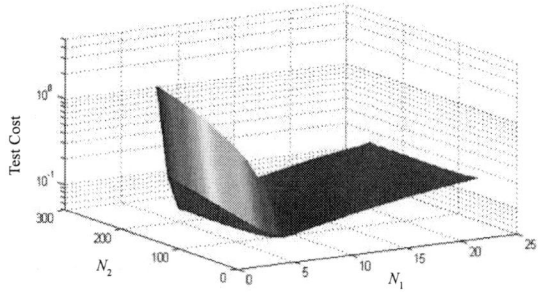

Fig. 10. The optimal N_1-N_2 combination.

B. N_1 Optimization

Table II lists the test cost parameters (in addition to Table I) needed to optimize N_1 for test cost reduction. For convenience, we use the per die manufacturing cost as the unit cost. The per clock cycle test time cost corresponds to the condition that 5,000 clock cycles at 20 MHz costs the price of a die. $s_{overkill}$ equals the manufacturing cost; s_{escape} is 3 times the die price to include the package cost. (Note that we assume that the post-packaging testing will identify all the bad chips.)

We employ numerical methods to find the optimal combination of N_1 and N_2—analytical approaches are infeasible because the cost function does not have a closed form expression and N's are integers. The results are shown in Fig. 10. In this example, the total test cost is more sensitive to N_1 than N_2. The low impact of N_2 is due to the low output variation compared to the test specification ($[-5\sigma, 5\sigma]$); as a result, very few instances enter stage 2.

VII. CONCLUSION

This paper presents a wafer-level BIST architecture for TFT-LCD source driver output variation testing. The architecture utilizes the 1-bit first-order SDM ADC as the digitization engine; this empowers us to develop the two-stage test strategy which takes advantage of the fact that most instances in a mature process are very close to the ideal value. We construct a cost function for the proposed two-stage test flow. Simulation results show that it is robust under different output variation distributions. In the future, we will further optimize the test flow and performs silicon validation.

REFERENCES

[1] K. Arabi, K. Kaminska, and M. Sawan. On chip testing data converters using static parameters. *IEEE Transactions on VLSI Systems*, 6(3):409–419, 1998.

[2] S. Chang, C. Lee, and J. Chen. BIST scheme for DAC testing. *IEEE Electronics Letters*, 38(15):776–777, 2002.

[3] J.-J. Huang, C.-C. Li, and J.-L. Huang. Testing LCD source driver IC with Built-on-Scribe-Line test circuitry. In *Asian Test Symposium*, pages 117–122, 2008.

[4] J.-L. Huang, C.-K. Ong, and K.-T. Cheng. A BIST scheme for on-chip ADC and DAC testing. In *Design, Automation & Test in Europe*, pages 216–220, 2000.

[5] V. Kerzerho, P. Cauvet, S. Bernard, F. Azais, M. Comte, and M. Renovell. A novel DFT technique for testing complete sets of ADCs and DACs in complex SiPs. *IEEE Design & Test of Computers*, 23(3):234–243, 2006.

[6] H. Son, J. Jang, Y. Kim, K. Kim, I. Kim, and S. Kang. A BIST architecture for multiple DACs in an LTPS TFT-LCD source driver IC. In *International SoC Design Conference*, pages 120–123, 2009.

[7] Y. Wen and K. Lee. BIST structure for DAC testing. *IEEE Electronics Letters*, 34(12):1173–1174, 1998.

[8] H. Xing, D. Chen, and R. Geiger. On-chip at-speed linearity testing of high-resolution high-speed DACs using DDEM ADCs with dithering. In *Electro/Information Technology*, pages 117–122, 2008.

Session 13

978-1-61284-657-6/11 $26.00 © 2011 IEEE

Practical Signal Processing at Mixed Signal Test Venues
– Trend Removal, Noise Reduction, Wideband Signal Capturing –

Hideo Okawara
Verigy Japan, K.K.   Tokyo, Japan

This is an embedded tutorial. It consists of two sections. The first section reviews several basic background topics for understanding the practical application examples presented in the second section. Most mixed signal tests employ the FFT-based processing. Firstly the relationship of the sampled waveform and the frequency spectrum is discussed by reviewing the DFT and IDFT equations. The point in it is that a real number waveform has the complex conjugate spectrum. The coherent condition is crucial in the FFT-based testing, which is described as $Ft/Fs=M/N$. Ft and Fs denote the test signal and the sampling frequencies. M and N denote the numbers of signal cycles and sampling points in the unit test period. The point is that M and N are mutually prime integers. The coherent waveform reconstruction is a useful tool in various aspects. You can utilize it not only for verification of test plan but also the original waveform reconstruction from the scrambled waveform, which is often utilized in under-sampling. If the measurement condition violates the Nyquist sampling theorem, it is called under-sampling. Waveform samplers perform under-sampling to capture very high frequency signals and analyze the characteristic parameters of the test signal.

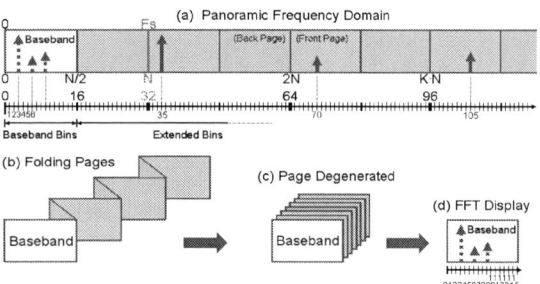

Figure 1   Page Concept

The page concept (Figure 1) is introduced to understand what under-sampling is and how a sampler works to capture high frequency signals. The panoramic frequency domain can be considered as a series of $N/2$-size pages concatenated. When the under-sampling is performed, the frequency domain is folded page-wise and degenerated in the baseband page. FFT provides the spectrum in the baseband page.

$$\frac{Ft}{Fs} = \frac{M}{N} = K + \frac{Mx}{N} \qquad (1)$$

The coherent condition can be extended as Equation (1) for under-sampling. Mx should be an odd integer within $-N/2$

and $N/2$ for $N=2^n$. $|Mx|$ is the aliased bin number. When under-sampling with the condition of $Mx=1$, you can capture a single cycle waveform directly in the baseband. So it is good for waveform analysis. When $Mx>1$, you can capture the fundamental tone at the bin Mx in the spectrum. If the signal spectrum is spread wider than the page-size, you should carefully make your test plan. The practical guideline is $N>2L$ and $\Delta F=Nx \cdot Fs/N$. L denotes the number of total spectrum lines to be captured. ΔF denotes the frequency resolution of lines. Nx should be an odd integer.

In the second section, there are three practical application examples presented. The first example (Figure 2) is the trend removal. The DC offset drift in AC-coupled measurement is removed with using the FFT&IFFT method. The second example (Figure 3) reduces noise on small signals measured by a passive resistive probe in the TDR experiments. It is a frequency domain filtering by the FFT&IFFT method. The third example (Figure 4) demonstrates the planning procedure of wideband signal under-sampling. High-speed serial bit stream is measured by a sampler. The guideline introduced in the first section is verified step by step.

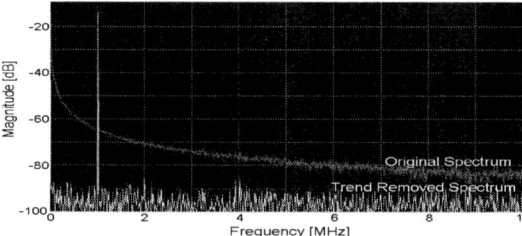

Figure 2   Trend Removed Signal Spectrum

Figure 3   Noise Reduced Signal

Figure 4   Reconstructed PRBS Waveform

Special Session: Hot Topic: Smart Silicon

**Organizers: LeRoy Winemberg, Freescale Semiconductor
Mohammad Tehranipoor, ECE Department, University of Connecticut**

As process geometries drop below 65nm, the differences between design models and manufactured silicon becomes unacceptably large. Up to this point, the industry-wide solution has been to use excessive guardbanding to compensate for this difference. However, the use of excessive guardbanding is expensive as it tends to leave a lot of performance on the table, makes timing closure more difficult, is typically inaccurate, and in the end usually leads to lost revenue. Is the use of on-chip "sensors" or adaptive circuits the solution?

Better pre-silicon characterization techniques and data collection, STA, SSTA, and other methods have all been tried to help close this gap between model and silicon reality but with increasing random variability and increasing impact of aging effects (NBTI/PBTI/HCI) on design reliability a gap exists that is expensive in terms of time and money. Another approach that has been proposed is to use a variety of more "sophisticated" on-chip sensors/monitors (i.e. embedded circuits that are more than just simple ring oscillators) to collect data from the manufactured silicon itself. The benefit of this approach is that the data collected on-chip by these circuits/sensors would still be used to tune the design models for subsequent designs as proposed above. The idea can be taken a step further in that these sensors could also be used by the design to adapt itself to aging/reliability effects that can vary both temporally and spatially but not always in a uniform fashion, as well as process variations, such that the design would continue to operate at an optimum point. Also, there is the possibility of self-correction which could possibly be used in manufacturing test to increase yield, in the field to increase quality, or both.

One downside of this approach is the area overhead of these small circuits and their support infrastructure. Also, if these self-tuning circuits are not designed properly (i.e. with a proper understanding of the impact they can have on the design of the "resilient" circuit itself) or not functioning properly (giving inaccurate measurements), the financial benefit from the sensor-based self-tuning/adaptation can be lost. For example, what "knobs" should be available to the sensor/monitor self-adaption/tuning? Supply voltage? Frequency? Adaptive body bias? Control of power islands? Others? The wrong balance/mix could result in the loss of performance, power and reliability. In addition, such embedded circuits must be carefully tested and accessed for configuration and reading out the collected data.

The goal of this Hot Topic Session is to discuss this cutting-edge topic that is being researched by several teams in both academia and industry, and debate which is the best approach for sub-65nm silicon designs. The point of debate will be what embedded circuits make the most sense (aging, enablement of more aggressive design, characterization, diagnosis, debug, etc.). The talks will answer major questions such as:
1. Are these embedded circuits needed and why?
2. How are they going to be accessed, tested and calibrated?
3. What are the impacts (benefit and cost) on silicon, design and test engineers?
4. Is standardization needed?
5. How will these circuits communicate with each other (again, should there be a standard – e.g. iJTAG)

1. [Invited] Yin and Yang of Embedded Sensors for Post-Scaling-Era
Anne Gattiker, IBM Research Austin

As gains in integrated circuit power and performance achieved through scaling of feature sizes slows, system and circuit design must carry more of the burden for improvement. Aggressive design decreases guardbands, increasing the risk of circuit failure. This paper discusses the increasing need for embedded on-chip sensors to enable aggressive design and also to help counter the attendant increased risks.

2. Accelerating ASIC Debug, Diagnostics and Quality
Matthias Kamm, Cisco

Time to market pressures and modern SOC designs are creating challenges during device debug and bring-up. Early access to technology is desirable, but this can lead to increased uncertainty and risk. Test chips can mitigate this risk, but IP content is highly variable across a technology node. This makes it difficult to predict all worst case combinations and conditions which can occur during a product ramp. Simple guardbanding methodology risks impacting yield, delivery and ability to ramp. These challenges drive new requirements for device IP structures to monitor, screen and diagnose. Simple go/nogo test is no longer sufficient; new methods and test IP must quickly isolate and diagnose problems. As the gap between structural and functional test widens, this becomes critical to meet aggressive schedule requirements. This talk will discuss on-chip IP, diagnostics tools and methods, and measurement capability. ATE as well as board level approaches will be reviewed.

3. Why is Access to Embedded Instrumentation so Critical Today
Al Crouch, Asset-Intertech

As Moore's Law continues forward with unstoppable inertia, more and more logic is integrated onto single die. And now it is suspected that the more economical way to keep Moore's graph on track is to start stacking bare die on top of each other using Through-Silicon-Vias (TSV's). This brings the test problem to the forefront with standard access to embedded features being one of the most critical requirements. Why isn't the testing of bare die on the wafer to render Known-Good-Die sufficient for modern chips? Because the impact of nanometer process nodes (DFM and more parametric fault behavior), complexity of the die (which may require embedded test and debug instruments), and the stacking of die (stressful operations and new fault/defect opportunities) result in a complex environment that may need to be tested and evaluated at each step of the assembly process to meet the yield, volume, and economic goals of the overall product. To address this continually growing complex environment, new standards such as P1687 and P1838 are being added to existing standards such as 1149.1/6/7 and 1500 to create a complete test, debug-diagnosis, characterization environment that supports not only the IC-centric issues of bare-die through package test, but may also allow board and system access to enable economically feasible solutions to the No-Trouble-Found and tuning of High-Speed Signal Lanes on boards. This presentation will describe the possible access mechanisms and some of the drivers, tradeoffs and efficiencies that may result from different choices and configurations.

2011 29th IEEE VLSI Test Symposium

Invited Paper: Yin and Yang of Embedded Sensors for Post-Scaling-Era

Anne Gattiker

IBM Research Austin

gattiker@us.ibm.com

Abstract

As gains in integrated circuit power and performance achieved through scaling of feature sizes slows, system and circuit design must carry more of the burden for improvement. Aggressive design decreases guardbands, increasing the risk of circuit failure. This paper discusses the increasing need for embedded on-chip sensors to enable aggressive design and also to help counter the attendant increased risks.

1 Introduction

As CMOS scaling slows, more and more we look to circuit and system design to continue the trends of ever-increasing power-performance improvement. However, new aggressive design practices decrease the cushion traditionally provided by guardbanding. Embedded sensors are used both to enable and guide adaptive circuit operation and, perhaps ironically, to counter the risks associated with the attendant decreases in guardbands.

Figure 1 shows data from a microprocessor chip that provides illustration and motivation for allowing adaptive choice of operating conditions. The x-axis represents chip delay as 1/FMAX, where FMAX is the measured maximum frequency at which the chip operates without error [1]. The y-axis plots the change in chip delay (1/FMAX) as power supply voltage, VDD, is changed by approximately 15%. Points with lower y-axis values change less.

Focusing first on just the x-axis, the graph clearly shows there is a distribution of chips that run at different speeds. A system that operates this chip at a pre-defined speed would have to run at at least 6, but likely would be specified to run even slower to accommodate lot-to-lot variation. Such variation could easily produce chips 15% slower. Running at those high values of delay is inefficient in two ways. First, the fastest chips could run much faster, so system ship frequency may be unnecessarily limited. Second, the faster chips will have high static leakage, incommensurate with their speed.

For microprocessors used in computers, including many laptops and servers, the system can take advantage of faster chip operating speeds. To improve efficiency over running all chips in the system at a specified speed, one option is to run each chip at its tested FMAX. Standard practice would be to add a guardband, but each increment

of guardband sacrifices performance, so chip and system makers typically strive to make the guardband as small as possible. Under the pre-defined-speed scenario, a large fraction of the chips operate in the system with significant headroom. That headroom protects the system. For example, if chips have small defects, their effect may be within the margin between the speed at which the chip is capable of running and the speed at which it actually runs. In that case the chip could operate error-free in the system despite the presence of a defect. Similarly, the headroom protects the system from test incompleteness. E.g., if the path delay coverage of the speed-test is only 90%, for chip failure in the system, a problem needs to occur on one of the uncovered paths *and* to be big enough not to be protected by the margin between potential and actual speed of operation. The likelihood of that event is decreased as the margin increases. Similarly, the headroom can protect chips from failing due to reliability-related wear-out mechanisms, such as hot-carrier injection (HCI) and negative bias temperature instability (NBTI). The headroom can also protect defect-free chips from failing due to noise problems, such as power supply droop. Running each chip in the system according to its tested FMAX decreases that headroom, and increases the risk of failure.

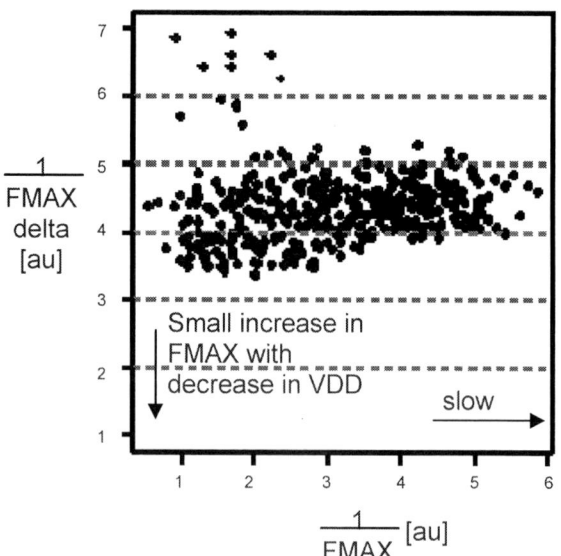

Figure 1 1/FMAX (maximum operating frequency) dependence on VDD vs. 1/FMAX for microprocessor [1].

Note that to save power, system makers frequently want to run the chips not just a their fastest rated speed, but also at the lowest VDD at which they can achieve that speed.

978-1-61284-657-6/11 $26.00 © 2011 IEEE

Focusing now on the y-axis as well as the x-axis, it is clear that there is also a distribution in terms of how much the chips slow down with a decrease in VDD. Assuming the test is carried out at only the higher of the two VDDs, the chips' VDD-sensitivities will be unknown. In the system, all chips will run at a similarly decreased VDD which would presumably have been determined based on characterization hardware that undergoes extra testing. The result in practice of lowering all chip VDDs similarly is that chips with larger deltas in FMAX with VDD will suffer disproportionately large decreases in headroom. That decrease makes those chips more susceptible to power supply noise and reliability-related degradation. It also makes them more susceptible to defects, both because of the decrease headroom directly, and also because the effects of many types of defects are enhanced when the chip is operating at low-VDD [2-5]. In fact the data of Figure 1 shows outlier points for which a decrease in VDD causes a larger increase in FMAX than the bulk of the population. Those points may represent chips with defects, and those chips are more likely to fail due to the defects when the chip is operating at reduced VDD.

There are significant power and performance benefits to tailoring operating conditions according to chip capabilities. The benefits extend to tailoring chip operation according to need based on workload. The benefits come at the expense, however, of increased risk of failure. Embedded sensors are key enablers of guiding the tailoring of operating conditions. At the same time, they are finding very productive additional use as measures to mitigate the risks posed by the tailoring. The following sections of this paper describe each of these two uses of embedded sensors.

2 Sensors and Actions to Guide Adaptive Operation

This sections discusses embedded sensor use for guiding choice of operating conditions. We focus our attention on the POWER7™ microprocessor to illustrate emerging capabilities. The POWER7 is an eight-core chip with embedded DRAM [6].

2.1 Sensors
One of the key parameters determining chip operation efficiency is temperature. The chip contains 44 on-chip digital thermal sensors [DTSs], with five in each core. The digital thermal sensors are implemented using a bandgap diode voltage comparator and are calibrated during manufacturing test carried out at a known temperature. They are used for enabling the chip to self-protect in emergency situations and thermal throttling [7-9].

The chip also includes distributed critical path monitors [CPMs] [10-11]. The CPMs provide operating frequency measurements for a sample of path types meant to mock the composition of critical paths. The path types in the CPM implementation described in [10] are nand4, nor3, adder, wire-dominated and passgate. The paths can be monitored individually or combined to mock a hybrid path. Together they provide up to 14 path combinations. They are placed near DTSs, which are placed in potential thermal hot spots and areas of high current draw. Those areas in turn are potentially susceptible to power supply droop and transistor slow-down due to localized heating and as a result may contain the speed-limiting critical paths. The CPMs provide real-time feedback on the frequency-voltage relationship given the workload that is actually running and in the real operating environment. They determine circuit timing margin and can be used to assist in choosing the chip's optimal frequency and voltage settings [7-11].

The chip also contains activity and event counters. The counters monitor processor core, memory hierarchy and main memory access activity. They provide performance, utilization and activity measurements and are used to direct power/performance trade-off decisions and techniques [7-9].

2.2 Actions
Actions related to the data provided by embedded sensors start with initial sorting of chips according to their power/performance capabilities as determined during chip manufacturing test. Other actions relate to real-time adaptation during system operation. The POWER7 system uses a dedicated off-chip microcontroller whose sole purpose is to manage system energy. Actions that can be taken include descent into idle states and frequency-voltage scaling [7-9]. When the system is in a state such that there is no work for a particular processor to do, the chip can enter one of three idle states. The nap state is optimized for wake-up time. Sleep mode provides more savings at a higher latency. "Heavy" sleep mode puts all cores in sleep mode and reduces the voltage of all cores to the minimum required for retention.

When the processor does have work to do, the chip can actuate per-core frequency scaling, guided by the embedded sensors described above. [9] reports a 30% benefit when changing from eight cores running at FMAX to one core running at FMAX with the other seven running at FMIN. The highest frequency core determines the voltage. This per-core frequency scaling allows tuning when there is a different workload on each core. It allows less-utilized partitions to run at lower frequencies, while heavily-utilized partitions maintain peak performance. It also allows each partition to run under a different energy-saving policy.

978-1-61284-657-6/11 $26.00 © 2011 IEEE

Additional exploratory mechanisms available in POWER7 hardware include low activity detection, a power proxy and autonomic circuit timing margin feedback control [8,9]. The low activity detection mechanism implements a hardware frequency-control loop that measures active cycles/instruction throughput over an interval as short as a few microseconds. It very quickly reduces processor frequency in response to a drop in instruction throughput that may occur for example while servicing memory bound workloads that do not need full processor compute frequency while waiting on data. Next, the power proxy allows coarse, e.g., chip-level power measurements, to be used to infer finer, e.g., core-level power consumption. The inferences are based on expected power consumption of monitored architectural events and can be used to shift power to cores or other components that need it the most, especially when operating under a power cap. Weights for calculating event power consumption are adjusted for leakage, temperature and voltage. Finally, CPMs can be used to support dynamic guardbanding which uses the CPMs to optimize frequency or voltage.

3 Using Embedded Sensors to Mitigate Aggressive-Design Risk

3.1 Hardware characterization to guide adaptive operation

One of the challenges in determining optimal operating conditions is that manufactured chips can vary widely in their attributes. As described in Section 1, Figure 1 provides an example, as chips have different sensitivities in terms of delay to changes in VDD. Figure 2 provides another example. Here we look at hardware data from two different manufacturing lots [12]. Lot 1 (circles) is more sensitive in terms of delay to changes in VDD than Lot 2 (x's). The difference indicates hardware in Lot 1 has a higher threshold voltage. Note that consistent behavior is seen in Figure 2b which shows lower IDDQ versus delay for Lot 1. The composition of the delay for a given chip has a strong influence on its power and performance reactions to changes in operating conditions such as voltage and temperature. Embedded sensors can give excellent insight into that composition [1,12-14]. Ring oscillators, in particular, are easy to measure in terms of delay. While it is difficult to measure IDDQ of an embedded ring oscillator, Figure 2 illustrates that change in delay with VDD can be used as a proxy to understand threshold voltage and in turn leakage characteristics. Since decreasing leakage is one of the key goals of adaptive circuit operation, such insight is very useful.

3.2 High-fidelity test

As mentioned in Section 1, circuits may become more susceptible to small defects as their operation is tuned according to chip capability and/or operating conditions. Embedded sensors can be used to help ensure high-fidelity test that counters the enhanced defect-sensitivity risk. [6]

describes how embedded temperature sensors are used to optimize test yields and device operation. These sensors enable real-time measurements of device temperature during testing. Other on-chip monitors, such as the skitter circuit can be used in a similar way to ensure a firm understanding of the power supply voltage present during test [15]. In "oscilloscope mode" the skitter circuit can provide an actual waveform of the power supply voltage seen at sample points on the circuit. Insight into VDD behavior at test can be used to guard against overkill and underkill during operational tests, especially LBIST [15].

3.3 Reliability monitoring

Section 1 discusses the fact that chips become more susceptible to failure due to reliability-related wear-out mechanisms as guardbands are decreased through aggressive design. The CPM described in Section 2 can be used to monitor delay degradation over the life of a processor. Other NBTI monitors are described in [16-20]. A challenge for NBTI monitors is ensuring representative aging based on activity seen by the actual circuitry.

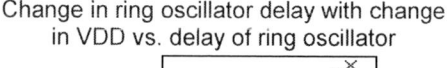

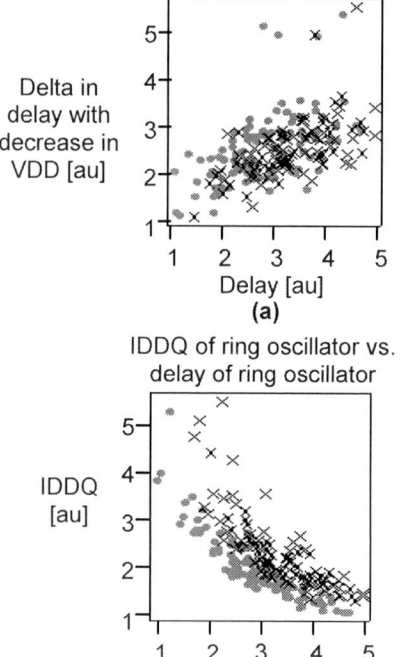

Figure 2 Characterization of ring oscillator behavior for two manufacturing lots.

3.4 Process characterization for manufacturing feedback

Figure 1 and its description in Section 1 highlighted the variation in chip performances and their reactions to changed operating conditions due to manufacturing variation. Figure 2 shows a similar phenomenon. Because schemes for guiding adaptive behavior can be complex, and they take time to implement, the ideal situation would be if all chips were manufactured identically. Clearly today's processes do not deliver that ideal, but a key role of embedded sensors in the post-scaling era is to present an accurate characterization of variability seen by manufactured chips and to present that characterization in a way that can be used to guide corrective process actions. Figure 3 shows an example where a wafer map of ring oscillator delay has been processed to take out systematic across reticle variation and an extracted linear plane to highlight remaining sources of across wafer variation [21]. A wafer map like this can be used to understand the sources of across-chip variation and rectify them. Mitigating manufacturing variability, in turn, enables simpler and more accurate adaptive operation tuning.

Processed wafer map of ring oscillator delay

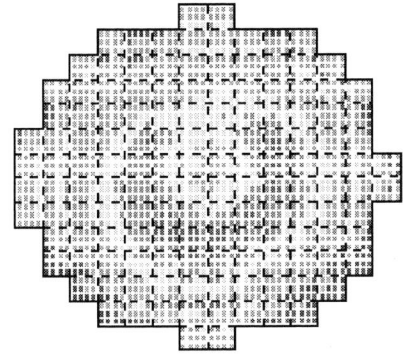

Fas ▪▪▪▪ ▪▪▪▪ ▫▫▪▪ Slow

Figure 3 Delay wafer map with systematic across-reticle and variation modeled by linear plane removed [21].

Conclusions

Today's chips are helping maintain traditional improvements in power-performance in part by tailoring operating conditions to manufactured hardware characteristics and workload. Embedded sensors support that adaptive behavior. They also help mitigate the risks posed by the attendant decreased guardbands by providing high-fidelity hardware characterization, aiding robust test, monitoring wear-out and providing feedback to decrease manufacturing variability.

Acknowledgements

The author gratefully acknowledges Karthick Rajamani, Juan Rubio and Rahul Rao for insightful discussion.

References

[1] A. Gattiker, M. Bhushan and M. Ketchen, "Data Analysis Techniques for CMOS Technology Characterization and Product Impact Assessment," Intl. Test Conference, 2006.

[2] H. Hao and E. McCluskey, "Very-Low Voltage Testing for Weak CMOS Logic ICS", Intl. Test Conference, 1993.

[3] S. Ma, P. Franco and E. McCluskey, "An Experimental Chip to Evaluate Test Techniques: Experimental Results." Intl. Test Conference e, 1995.

[4] J. Chang, and E. McCluskey, "Quantitative. Analysis of Very-Low-Voltage Testing," VLSI Test Symposium, 1996.

[5] S. Ma, P. Franco and e. McCluskey, "An Experimental Test Chip to Evaluate Test Techniques Experimental Results," International Test Conference, 1995.

[6] J. Crafts, D. Bogdan, D. Conti, D. Forlenza, O. Forlenza, W. Huott, M. Kusko, E. Seymour, T. Taylor and B. Walsh, "Testing the IBM POWER7TM 4 GHz Eight Core Microprocessor," International Test Conference, 2010.

[7] M. Floyd, S. Ghiasi, T. Keller, K. Rajamani, F. Rawson, J. Rubio and M. Ware, "System Power Management Support in the IBM POWER6 Microprocessor[TM]," IBM Journal of Research & Development, Volume 61, No. 6, Nov. 2007.

[8] M. Ware, K. Rajamani, M. Floyd, B. Brock, J. Rubio, F. Rawson and J. Carter, "Architecting for Power Management: The IBM Power7 Approach," IEEE Intl. Symposium on High-Performance Computer Architecture, January 2010

[9] M. Floyd, B. Brock, M. Ware, K. Rajamani, A. Drake, C. Lefurgy and L. Pesantez, "Adaptive Energy Management Features of the POWER7[TM] Processor," Hot Chips 22, August 23, 2010.

[10] A. Drake, R. Senger, H. Doegun, G. Carpenter, S. Ghiasi, T. Nguyen, N. James, M. Floyd and V. Pokala, "A Distributed Critical-Path Timing Monitor for a 65nm High-Performance Microprocessor", Intl Solid-State Circuits Conference, 2007.

[11] A. Drake, R. Senger, H. Singh, G. Carpenter and N. James, "Dynamic Measurement of Critical-path Timing," IEEE Intl. Conf. on Integrated Circuit Design and Technology, 2008.

[12] M. Bhushan, A. Gattiker, M. Ketchen and K. Das, "Ring Oscillators for CMOS Process Tuning and Variability Control," IEEE Trans. On Semiconductor Manufacturing, Vol. 19, No. 1, 2006.

[13] M. Ketchen, M. Bhushan and D. Pearson, "High Speed Test Structures for In-Line Process Monitoring and Model Calibration," Intl. Conf. on Microelectronic Test Structures, 2005.

[14] M. Ketchen and M. Bhushan, "Product Representative 'At-Speed' Test Structures for CMOS Characterization," IBM Journal of Res. & Dev., Vol. 4/5, 2006.

[15] R. Franch, P. Restle, N. James, W. Huott, J. Friedrich, R. Dixon, S. Weitzel, K. Van Goor and G. Salem, On-Chip Timing Uncertainty Measurements on IBM Microprocessors," International Test Conference, 2007.

[16] J. Keane, T. Kim, C. Kim, "An on-chip *NBTI* sensor for measuring. PMOS threshold voltage degradation", IEEE Intl. Symposium on Low Power Electronics and Design, 2007.

[17] M. B. Ketchen, M. Bhushan, and R. Bolam, "Ring oscillator based test structure for NBTI analysis," in Proc. IEEE Int. Conf. Microelectronic Test Structures, 2007.

[18] K. Stawiasz, K. Jenkins and P-F. Lu , "On-Chip Circuit for Monitoring Degradation Due to NBTI", IEEE North Atlantic Test Workshop, 2008.

[19] X. Wang and M. Tehranipoor, "Low-Cost On-Chip Structures for Measuring NBTI Effects, Variations, Path Delay, and Noise," SRC TECHCON, Poster Presentation, 2010.

[20] A. Ghosh , R. Rao , R. Brown , Ching-te Chuang, "On-chip Negative Bias Temperature Instability Sensor using Slew Rate Monitoring Circuitry," ACM/IEEE Intl. Symposium on Low Power Electronics and Design, 2009.

[21] A. Gattiker, "Unraveling Process Variability for Process/Product Improvement," Intl. Test Conference, 2008.

Special Session: Hot Topic
Design and Test of 3D and Emerging Memories

Organizer: Cheng-Wen Wu, National Tsing Hua University / ITRI

In this hot topic session, we are including three talks that cover design of reliable, emerging memories, with emphasis on 3D memories, DRAM and non-volatile memories. We will also introduce a new class of memory, the Storage Class Memory (SCM), and discuss its reliability issues.

Talk I: 3D Memory Design

Dave Dunning, Intel

The evolution and advance of computer systems over the last few decades has been impressive; often citing Moore's Law which predicts the increase over time in the number of transistors that can be built into integrated circuits. Over the past many years, these improvements in integrated circuit process have allowed for roughly proportional increases in CPU performance as well as DRAM bit density increases. This talk examines these trends and discusses which are anticipated to continue, which are not.

For most market segments, cost per bit has been the most important metric used to designing the memory subsystem for CPU-based systems. To date, the DRAM vendors have done a great job in using process improvements to continually drive down the cost per bit while meeting the performance and capacity requirements. Many optimizations have made associated with single chip DRAM devices. Those optimizations center around the two dimensional construction of the arrays of storage capacitors and single transistor switches accessed through row and column strobes; this maximizes the array efficiency and also minimizes the number of pins/bumps required to read and write to the devices. The DRAM devices are usually packaged together into modules. Standards have been specified and adopted which have also helped reduce the cost per bit for DRAM-based solutions.

In the last few years, we have started seeing new metrics gaining in importance associated with DRAM. Those metrics include a form of energy (or power) as well as form factor associated with memory subsystem. This talk will discuss what aspects of DRAM design will be affected by the increased focus on reducing the read/write energy as well as the changes required to reduce the memory subsystem volume while still meeting the performance and capacity needs of future systems. This talk will also discuss the benefits and challenges of transitioning from two dimensional (2D) devices to three dimensional (3D) modules using through silicon via (TSV) technology. TSVs certainly present many technological advantages; they arguably present an equal number of test, manufacturing and therefore cost challenges. Lastly, the role of future standards will also be discussed.

"This research was, in part, funded by the U.S. Government. The views and conclusions contained in this document are those of the authors and should not be interpreted as representing the official policies, either expressed or implied, of the U.S. Government."

Talk II: New Design Considerations for 3D-Memories

Frederick Chen, ITRI

As line dimensions no longer exceed the electron mean free path, scaling of electronic devices below 40 nm is a difficult challenge. Memory applications are particularly constrained due to their reliance on scaling increase the number of devices in a given area. Use of the vertical dimension is therefore gaining more popularity as an option for increasing memory density. A review of currently investigated three-dimensional (3D) memory systems will be presented. Today most work is focused on the following areas: 3D NAND Flash memory (vertical channel bit string), 3D cross-point BEOL non-volatile memory, vertical stacking of memory transistor planes on the same wafer and vertical stacking of memory arrays on different wafers. While the use of the third dimension in principle allows no limit to increasing density within a given area, consideration must also be given to the added area dedicated to decoding, algorithm logic, error correction, memory control and other peripheral functions. Thus, solutions for reducing the footprint of these non-memory components are essential. We will consider scaling vs. stacking in terms of complexity and cost. Besides footprint, thermal considerations are also important. Any 3D system will have different thermal boundaries from a conventional 2D system, where the heat sink is always nearby on one side. In a 3D system, the layers above and below any active device layer can act as heat sources, which will limit the necessary transfer of heat. Practical solutions for thermal isolation between transistor planes will be considered. In addition, thermal issues for different alternative non-volatile memories fabricated in the BEOL will be considered. These memories will also be constrained by the impact of the 3D architecture on disturbance of adjacent cells and capacitive effects. Some general conclusions will be given as directions for future 3D memory development.

Talk III: Making Storage-Class Memory Practical in Future Memory Hierarchy

Hsien-Hsin Lee, Georgia Institute of Technology

A new class of memory called Storage Class Memory (SCM) is emerging to be considered an integral part of the main memory hierarchy. These memories are non-volatile, scalable down to nanometer range, consuming lower power with multi-level cell capability. Despite these prominent properties, their adoption is hindered by their operational reliability, primarily due to their low write endurance. The situation is exacerbated under the scenarios of deliberately designed malicious attacks. In this talk, I will discuss our recent research findings for improving SCM's reliability via two enabling hardware techniques: taper-proof wear-out management and self-healing capability. The first approach aims at randomizing and obfuscating the write patterns of these memories with one additional level of address remapping within the memory module. The second method presents a novel multi-bit error recovery mechanism that can self-repair multi-bit stuck-at faults and continue to use these memories even if some cells are worn out and become unusable. With these techniques, we will be able to extend the lifetime of these SCM to approach the theoretical limit and even continue to operate them when faulty cells are present.

Author Index

Magdy S. Abadir.................................2
Jacob A. Abraham........................….90
Mohamed Abu-Rahma.....................46
Vishwani D. Agrawal...........64, 248,285
Nisar Ahmed..................................309
S. Alampally................................…285
Nader Alawadh..............................260
N. Alves.......................................241
Karim Arabi...................................46
Yuuki Araga..............................…70
Fabrice Auzanneau.........................140
R. I. Bahar...................................241
Aritra Banerjee..............................58
Fang Bao......................................78
Kanad Basu..................................14
Mounir Benabdenbi.........................229
Dilip K. Bhavsar............................225
R. D. (Shawn) Blanton....................172
Surendra Bommu...........................291
Yannick Bonhomme........................140
Sreejit Chakravarty.........................160
Victor Champac......................184, 203
Kameshwar Chandrasekar............…291
Abhijit Chatterjee...........................58
Ji-Jan Chen..................................20
Ching-Yi Chen...............................146
Chen-Huan Chiang.........................278
Kyoung Youn Cho..........................213
Yung-Fa Chou...............................20
Shu-Hsuan Chou............................116
B. Courtois..........................99, 209
Rudrajit Datta...............................134
Shyam Kumar Devarakond................58
Yan Dong...................................…116
J. Dworak....................................241
K. Enokimoto................................166
Samuel Evain................................140
Bruce Fleischer..............................235
Sreenivas Gangadhar......................197
Anne Gattiker..........................90,324
Valentin Gherman..........................140
P. Girard.....................................166
Alain Giulieri................................191
Ashish Goel.................................266
Olivier Goncalves......................…191
Alain Greiner................................229

Julien Guilhemsang.........................191
Ujjwal Guin..................................278
Takushi Hashida............................70
Yu-Jen Huang...............................20
Yu Huang...................................…297
J.-L. Huang..................................315
Olivier Héron................................191
Eun Jung Jang...............................90
Zhongwei Jiang.............................52
Vinayaka Jyothi.............................105
S. Kajihara..................................166
B. Kaminska.........................…99,209
Ramesh Karri........................100,105
M. A. Kochte................................166
Prabhakar Kudva...........................235
Ding-Ming Kwai............................20
Dongsoo Lee.................................266
C.-C. Lee....................................315
Jin-Fu Li.....................................20
Huawei Li....................................272
Xiaowei Li...................................272
Jia Li...297
Sung Kyu Lim...............................26
Chih-He Lin..................................146
W.-A. Lin....................................315
Chunsheng Liu..............................219
Derrick Losli.................................254
Junxia Ma....................................309
Akinori Maeda...............................303
Amitava Majumdar..........................116
Yiorgos Makris..............................235
Michail Maniatakos.........................235
C. V. Martins................................203
Anne Meixner................................178
Prabhat Mishra..............................14
Rajesh Mittal................................154
K. Miyase....................................166
Jesus Moreno................................184
Makoto Nagata...............................70
Prakash Narayanan..........................154
Sani Nassif...................................90
Karthikeyan Natarajan......................219
M. H Neishaburi.............................8
K. Nepal.....................................241
H. Okawara..................................322
Sule Ozev....................................178

Shreepad Panth..............................26
R. A. Parekhji.............................285
Sang Phill Park...........................266
Nehal Patel...............................116
Franc͜ois Pecheux..........................229
Songwei Pei...............................272
Ke Peng....................................78
Sumanth Poddutur..........................154
Irith Pomeranz........................84, 128
Jayalakshmi Rajaraman.....................254
Jeyavijayan Rajendran.....................105
Jeff Rearick..............................254
Dimitri Refauvelet........................229
Michel Renovell...........................184
Kurt Rosenfeld............................100
Kaushik Roy...............................266
Puneet Sabbarwal..........................154
Samah Mohamed Saeed........................40
K. S. KIM – Samsung.......................113
Amit Sanghani.............................219
M. Santos.................................203
Y. Sato....................................96
J. Semião.................................203
Shreyas Sen................................58
Sanjay Sengupta...........................291
Rajamani Sethuram..........................46
Ramamurthy Setty..........................116
Priyadharshini Shanmugasundaram...........248
P. Shanmugasundaram.......................285
Shyh-Shyuan Sheu..........................146
Y. Shi....................................241
Hsiu-Chuan Shih...........................146
Geoff Shofner..............................78
Ozgur Sinanoglu...................40,105,260
Suraj Sindia...............................64
Eshan Singh................................32
Virendra Singh.............................64
Vivek Singhal.............................154
Arani Sinha...............................116
Kanwaldeep Sobti..........................254
Rajagopalan Srinivasan....................213
Nik Sumikawa...............................2
Yasuhiro Takahashi........................303
Wing Chiu Tam.............................172
Mohammad Tehranipoor.....78,160,166,309,323
I. C. Teixeira............................203
J. P. Teixeira............................203
Nur A. Touba..............................134
Spyros Tragoudas..........................197
S. Vasudevan..............................211
J. C. Vazquez.............................203

R. T. Venkatesh...........................285
Nicolas Ventroux..........................191
D. M. H. Walker............................52
Li-C. Wang..................................2
Zheng Wang.................................52
Jing Wang..................................52
Seongmoon Wang............................122
Baosheng Wang.............................254
X. Wen................................114, 166
Dragoljub (Gagi) Drmanac....................2
LeRoy Winemberg.....................2,78,323
Cheng-Wen Wu.......................20,146,328
Dong Xiang................................297
Y. Yamato.................................166
Bo Yang...................................219
Ender Yilmaz..............................178
Zhen Zhang................................229
Wei Zhao..................................160
Zeljko Zilic................................8

9781612846576